# Essentials of Human Nutrition

# Essentials of Human Nutrition

Sixth Edition

Edited by

**Jim Mann**
**A. Stewart Truswell**
**Leanne Hodson**

Great Clarendon Street, Oxford, OX2 6DP,
United Kingdom

Oxford University Press is a department of the University of Oxford. It furthers the University's objective of excellence in research, scholarship, and education by publishing worldwide. Oxford is a registered trade mark of Oxford University Press in the UK and in certain other countries



Third Edition 2007
Fourth Edition 2012
Fifth Edition 2017



Published in the United States of America by Oxford University Press
198 Madison Avenue, New York, NY 10016, United States of America

British Library Cataloguing in Publication Data
Data available

Library of Congress Control Number: 2023932436

ISBN 978-0-19-886667-1

Printed in the UK by
Bell & Bain Ltd., Glasgow

# Preface

When *Essentials of Human Nutrition* was first published by Oxford University Press (OUP) in 1997, there was no certainty that the book would find a place amongst the several excellent textbooks of human nutrition. Now, a quarter of a century later, with the publication of the sixth edition, we consider that our original vision for the book has been justified. Despite the proliferation of sources of nutrition information, there is strong evidence that *Essentials* is acknowledged as a leading source of up-to-date nutrition knowledge. The medium-sized textbook is available in three languages and sales suggest that it continues to have widespread popularity in hard copy and electronic format. The fifth edition was highly commended in the Medicine category of the 2018 British Medical Association Book Awards and described as *'the most comprehensive and student friendly introduction to the subject, built on carefully edited contributions from an international team of experts'*.

We believed then, as we do now, that there is a need for a book in which internationally acknowledged experts in their fields describe what they regard as those aspects of their topics that are essential to the understanding and practice of human nutrition. Authors were asked not only to provide the essential information but to do so in a way that would be appropriate to the needs of the disparate groups requiring such knowledge. In addition to students embarking on a university course in human nutrition and those in training in the health and food science professions, there are many others who are increasingly appreciating the importance of nutrition as a key determinant of individual and public health. Many of our clinical colleagues in medicine, dentistry, pharmacy, nursing, and physiotherapy, and school teachers, provided strong encouragement for this project, since they too required a simple reference text, having received little formal training in nutrition. Health professionals and food scientists need to be able to disentangle scientifically established nutrition principles from the morass of misinformation now widely available in the public domain. We also believed that the book would be of value to those in the fitness industry and, last but not least, individual members of the public who have sufficient knowledge of biology and chemistry and who wish to be informed of the essentials of human nutrition.

When writing and reviewing the chapters for this edition, we were once again reminded of the need for regular updating of texts like *Essentials*. Since the publication of the fifth edition there have been several instances in which promising findings have not been confirmed, where new research has substantiated or quantified earlier observations, or where misunderstandings have been clarified. For example, in the past five years a number of randomized controlled trials have not confirmed several of the multiple health benefits that had been claimed for supplemental doses of vitamin D. On the other hand, convincing evidence now exists for remission of type 2 diabetes following appreciable weight loss. The argument regarding the optimal proportion of total energy that should be derived from carbohydrate has been resolved by the demonstration that it is quality rather than quantity of carbohydrate that determines health outcomes. These are but a few of the many important issues that have been resolved by research that has been reported over the past few years, and are included in the new edition. The book is not intended to be a detailed reference volume but each chapter includes further reading for those wishing to extend the information provided in the text.

We are immensely grateful to the 37 continuing and 14 first-time contributors to the sixth edition. The professional and personal lives of all have, at least to some extent, been affected by the Covid pandemic. Despite the additional workload incurred as a consequence, almost all those invited to update or rewrite chapters agreed to do so. In the very few instances where this was not possible, the editors undertook this role. Jim Mann and Stewart Truswell were delighted that Leanne Hodson agreed to join them as an editor of the sixth edition.

We hope that all readers of our book will find it useful in their study of human nutrition or a useful reference when seeking the *'Essentials'* of the discipline that, as both an art and a science, is central to the maintenance of good health of individuals and populations, as well as to the treatment of people suffering from one or more diet-related diseases.

**Professor Jim Mann**
**Professor A. Stewart Truswell**
**Professor Leanne Hodson**

## New to this edition

- Rewritten or extended chapters: Introduction; Carbohydrates; Lipids; Protein; Water, Electrolytes, and Acid–base Balance; Other Biologically Active Substances in Plant Foods: Phytochemicals; Nutrition, Genetics, and Personalized Nutrition; Overweight and Obesity; The Eating Disorders: Anorexia Nervosa, Bulimia Nervosa, Binge-Eating Disorder, and OSFED; Food Groups; Nutritional Recommendations for the General Population; Infant Feeding and Eating Well for Toddlers; Nutrition and Ageing; Dietary Assessment; Assessment of Nutritional Status; Biomarkers; Nutritional Consequences of Poverty and Food Insecurity in Developed Countries.
- Several new case studies.

# Acknowledgement

The editors acknowledge the pivotal role that Lesley Day has played in the production of this book. She has been the editorial assistant for this edition as well as the three previous editions. She has assisted with the revision of many of the chapters following the editorial process and reformatted all the chapters to ensure compliance with the standardized format required by the publisher. She has been the point of contact for all the contributors and the publisher. It is unlikely that there would have been a sixth edition of *Essentials* without her organizational skills and support in every step of its production.

# Resources for lecturers

The online resources for lecturers are available at: www.oup.com/he/mann-truswell-hodson6e.

Careful editing of contributions from an international team of experts makes *Essentials of Human Nutrition* the most complete and student-friendly introduction to the field of human nutrition.

Adopting lecturers can access the following **online resources**:

- figures and tables from the text, arranged by chapter, for use in lectures to support efficient, effective teaching preparation.

# Contents

## Part 4 Nutrition-related Disorders



## Part 5 Foods



## Part 6 Changing Food Habits

# Abbreviations

| | |
|---|---|
| AA | arachidonic acid |
| AAS | atomic absorption spectrophotometry |
| ACE | angiotensin-converting enzyme |
| ACP | acyl carrier protein |
| ADH | alcohol dehydrogenase |
| ADH | antidiuretic hormone |
| ADI | acceptable daily intake |
| AGP | α1-acid glycoprotein |
| AGRP | agouti-related proteins |
| AI | adequate intake |
| AIDS | acquired immunodeficiency syndrome |
| ALA | α-linolenic |
| ALDH | aldehyde dehydrogenases |
| AML | acute myeloid leukaemia |
| AMP | adenosine monophosphate |
| AN | anorexia nervosa |
| AOAC | Association of Official Analytical Chemists |
| APOE | apolipoprotein E |
| APOSS | Aberdeen Prospective Osteoporosis Screening Study |
| AQP | aquaporins |
| ATP | adenosine triphosphate |
| BED | binge-eating disorder |
| BM | body mass |
| BMC | bone mineral content |
| BMD | bone mineral density |
| BMI | body mass index |
| BMP | bone morphogenic protein |
| BMR | basal metabolic rate |
| BMS | breast milk substitute |
| BN | bulimia nervosa |
| BP | blood pressure |
| BSE | bovine spongiform encephalopathy |
| BST | bovine somatotrophin |
| C | cholesterol |
| CAP | Common Agricultural Policy |
| CAR | Central African Republic |
| CBT | cognitive behavioural therapy |
| CCK | cholecystokinin |
| CE | cholesterol ester |
| CF | cystic fibrosis |
| CHD | coronary heart disease |
| CMAM | community-based management of acute malnutrition |
| CMP | cows' milk protein |
| CNS | central nervous system |
| CO | cyclo-oxygenase pathway |
| CoA | co-enzyme A |
| CPAP | continuous positive airway pressure |
| CRB | cellular binding protein |
| CRP | C-reactive protein |
| CSA | childhood sexual abuse |
| CT | computerized tomography |
| CVD | cardiovascular disease |
| DALY | disability-adjusted life years |
| DASH | dietary approaches to stop hypertension |
| DDT | dichlorodiphenyltrichloroethane |
| DE | digestible energy |
| DEXA | dual energy X-ray absorptiometry |
| DFEs | dietary folate equivalents |
| DGA | Dietary Guidelines for Americans |
| DGLA | dihomo-γ-linolenic acid |
| DHA | docosahexaenoic acid |
| D-IT | di-iodotyrosine |
| DIT | diet-induced thermogenesis |
| DLW | doubly labelled water |
| DP | degree of polymerization |
| DRI | dietary reference intakes |
| DRV | dietary reference values |
| DV | daily value |
| EABV | 'effective' arterial blood volume |
| EAR | estimated average requirement |
| EASD | European Association for the Study of Diabetes |
| EBM | evidence-based medicine |
| ECF | extracellular fluid |
| ECG | electrocardiogram |
| EDNOS | eating disorder not otherwise specified |
| EFA | essential fatty acid |
| EFSA | European Food Safety Authority |
| EGRA | erythrocyte glutathione reductase activity |
| EPA | eicosapentaenoic acid |
| ERFE | erythroferrone |
| F&V | fruit and vegetables |

| | |
|---|---|
| FAD | flavin adenine dinucleotide |
| FAO | Food and Agriculture Organization |
| FASD | foetal alcohol spectrum disorders |
| FDA | Food and Drug Administration |
| FEP | free erythrocyte protoporphyrin |
| FFA | free fatty acid |
| FFM | fat-free mass |
| FFPI | FAO Food Price Index |
| FFQ | food frequency questionnaires |
| FIVE | familial isolated vitamin E deficiency |
| FMN | flavin mononucleotide |
| FSA | Food Standards Agency |
| FSANZ | Food Standards of Australia and New Zealand |
| GABA | γ-aminobutyric acid |
| GAG | glycoproteins and glycosaminoglycans |
| GTC | green tea catechins |
| GE | gross energy |
| GFR | glomerular filtration rate |
| GGT | γ-glutamyl transferase |
| GHG | greenhouse gas |
| GI | glycaemic index |
| GIP | gastric inhibitory peptide |
| GL | glycaemic load |
| GLC | gas–liquid chromatography |
| GLP | glucagon-like peptide |
| GLUT | glucose transporter |
| GM | genetically modified |
| GMP | guanosine monophosphate |
| GPx | glutathione peroxidase |
| GRAS | generally regarded as safe |
| GST | glutathione S-transferase |
| GTP | guanosine triphosphate |
| GWAS | genome-wide association studies |
| HACCP | hazard analysis of critical control points |
| HDL | high-density lipoprotein |
| HERP | Human Exposure/Rodent Potency Index |
| HFE | haemochromatosis protein |
| HG | hyperemesis gravidarum |
| HH | hereditary haemochromatosis |
| HIV | human immunodeficiency virus |
| HJV | haemojuvalin |
| HPLC | high-performance liquid chromatography |
| IBS | irritable bowel syndrome |
| ICCIDD | International Council for the Control of Iodine Deficiency Disorders |
| ICF | intracellular fluid |
| IDA | iron deficiency anaemia |
| IDD | iodine-deficiency disorders |
| IDDM | insulin-dependent diabetes |
| IDL | intermediate-density lipoprotein |
| IF | intrinsic factor |
| IFG | impaired fasting glucose |
| IGF | insulin-like growth factor |
| IGT | impaired glucose tolerance |
| IHD | ischaemic heart disease |
| IL | interleukin |
| IOM | Institute of Medicine |
| IOTF | International Obesity Taskforce |
| IPT | interpersonal psychotherapy |
| IRP | iron-responsive protein |
| IU | international units |
| IYCF | infant and young child feeding |
| JECFA | Joint Expert Committee on Food Additives |
| KKP | key kitchen person |
| LBM | lean body mass |
| LBW | low birth weight |
| LCHF | low-carbohydrate, high-fat |
| LCP | long-chain ω3 polyunsaturated |
| LDL | low-density lipoprotein |
| LL | lipoprotein lipase |
| LMIC | low- and middle-income countries |
| LO | lipoxygenase |
| LOH | loop of Henle |
| LPL | lysophospholipid |
| LRAT | lecithin:retinol acyltransferase |
| LT | leukotrienes |
| MCHC | mean cell haemoglobin concentration |
| ME | metabolizable energy |
| MEOS | microsomal ethanol oxidizing system |
| MG | monoglyceride |
| MGP | matrix Gla protein |
| MIRA | multi-cluster/sector initial rapid assessment |
| MIT | mono-iodotyrosine |
| MPOD | macular pigment optical density |
| MSG | monosodium glutamate |
| MTHFR | methylene tetrahydrofolate reductase |
| MUAC | mid-upper arm circumference |
| MUFA | monounsaturated fatty acids |
| MZ | meso-zeaxanthin |

| | |
|---|---|
| NAD | nicotinamide-adenine-dinucleotide |
| NADP | nicotinamide-adenine-dinucleotide phosphate |
| NAFLD | non-alcoholic fatty liver disease |
| NASH | non-alcoholic steatohepatitis |
| NCD | non-communicable disease |
| NDO | non-digestible oligosaccharides |
| NE | niacin equivalents |
| NEAT | non-exercise activity thermogenesis |
| NGO | non-governmental organization |
| NIDDM | non-insulin-dependent diabetes |
| NMES | non-milk extrinsic sugar |
| NOEL | no observed effect level |
| NPRQ | non-protein respiratory quotient |
| NPY | neuropeptide Y |
| NRV | nutrient reference values |
| NSP | non-starch polysaccharide |
| NTBI | non-transferrin-bound iron |
| NTD | neural tube defects |
| OCHA | Office for the Coordination of Humanitarian Affairs |
| OR | odds ratio |
| OSFED | other specified feeding or eating disorder |
| PAF | platelet-activating factor |
| PAH | phenylalanine hydroxylase |
| PAL | physical activity level |
| PAPS | 3′-phosphoadenosine-5′-phosphosulphate |
| PArH | polycyclic aromatic hydrocarbons |
| PBM | peak bone mass |
| PCB | polychlorinated biphenyls |
| PEM | protein-energy malnutrition |
| PG | prostaglandins |
| PGI | prostacyclin I |
| PICC | peripherally inserted central catheter |
| PIVKA | protein induced by vitamin K absence |
| PKU | phenylketonuria |
| PL | phospholipid |
| $PLA_2$ | phospholipase $A_2$ |
| PLP | pyridoxal 5′-phosphate |
| PLps | pancreatic lipase |
| PPAR | peroxisome proliferator-activated receptor |
| p.p.m. | parts per million |
| PTH | parathyroid hormone |
| PUFA | polyunsaturated fatty acid |
| PVC | polyvinyl chloride |
| PYY | peptide YY |
| RAAS | renin–angiotensin–aldosterone system |
| RAR | retinoic acid receptors |
| RBP | retinol binding protein |
| RCT | randomized controlled trial |
| RDA | recommended dietary allowance |
| RDI | recommended dietary intake |
| RDS | rapidly digestible starch |
| RE | retinol equivalent |
| RES | reticuloendothelial system |
| RNI | recommended nutrient intake |
| RS | resistant starch |
| RUTF | ready-to-use therapeutic food |
| RXR | retinoid X receptors |
| SAM | severe acute malnutrition |
| SCFA | short-chain fatty acids |
| SDG | sustainable development goals |
| SDS | slowly digestible starch |
| SELECT | Selenium and Vitamin E Cancer Prevention Trial |
| SEPP | selenoprotein P |
| SGLT1 | sodium glucose cotransporter 1 |
| SHBG | sex hormone-binding globulin |
| SLR | systematic literature reviews |
| SMAD | signalling molecules |
| SNP | single-nucleotide polymorphisms |
| SSRI | selective serotonin reuptake inhibitor |
| TAG | triacylglycerol |
| TBW | total body water |
| TE | total energy |
| TEE | total energy expenditure |
| TFA | *trans*-unsaturated fatty acids |
| TfR | transferrin receptor |
| Tg | thyroglobulin |
| THF | tetrahydrofolatereductase |
| TNF | tumour necrosis factor |
| TPP | thiamin pyrophosphate |
| TSH | thyroid-stimulating hormone |
| TTR | transthyretin |
| $TxA_2$ | tromboxane $A_2$ |
| UDP | uridine diphosphate |
| UIL | upper intake level |
| UNAIDS | United Nations Programme on HIV/AIDS |
| USI | universal salt iodization |

| | |
|---|---|
| UV | ultraviolet |
| vCJD | variant Creutzfeldt–Jakob disease |
| VDR | vitamin D receptor |
| VLCD | very-low-calorie diet |
| VLDL | very-low-density lipoprotein |
| VLED | very low-energy diet |
| WHA | World Health Assembly |
| WHI | Women's Health Initiative |
| WHO | World Health Organization |
| WKS | Wernicke–Korsakoff syndrome |

# Contributors

**Professor Margaret Allman-Farinelli**
Professor of Dietetics
University of Sydney, Australia

**Professor Annie S. Anderson**
Professor Emerita of Public Health Nutrition
University of Dundee, UK

**Professor Philip J. Atherton**
Professor of Clinical, Metabolic & Molecular Physiology
University of Nottingham, UK

**Associate Professor Kim Bell-Anderson**
Nutritional Physiologist
University of Sydney, Australia

**Dr Kathryn E. Bradbury**
Senior Research Fellow, School of Population Health
University of Auckland, New Zealand

**Dr Paul Brent**
Director, Global Food & Chemical Risk Assessment & Risk Management Solutions, Gold Coast Queensland, Australia
Adjunct Professor
University of Laval, Canada

**Professor Louise M. Burke**
Chair of Sports Nutrition
Australian Catholic University, Canberra, Australia

**Professor Clare Corish**
Professor of Clinical Nutrition & Dietetics
University College Dublin, Republic of Ireland

**Dr Helen Crawley**
Public health nutrition policy expert with senior advisory/policy/research roles for national and international organisations, UK

**Dr Colleen S. Deane**
Lecturer in Muscle Cell Biology
University of Southampton, UK

**Dr Gethin H. Evans**
Deputy Head, Department of Life Sciences Reader
Manchester Metropolitan University, UK

**Dr Suzie Ferrie**
Critical Care Dietitian, Department of Nutrition and Dietetics
Royal Prince Alfred Hospital, Sydney, Australia

**Professor Nita G. Forouhi**
Programme Leader and MRC investigator
Director of Organisational Affairs, School of Clinical Medicine
University of Cambridge, UK

**Dr Meika Foster**
Director, Edible Research, NZ
Honorary Senior Research Fellow, Department of Human Nutrition
University of Otago, New Zealand

**Assistant Professor Ohood Hakim**
Assistant Professor in Clinical Nutrition
King Abdulaziz University, Jeddah, Saudi Arabia

**Professor Anne-Louise Heath**
Professor in Human Nutrition
University of Otago, New Zealand

**Professor Leanne Hodson**
Professor of Metabolic Physiology & BHF Senior Research Fellow in Basic Science
University of Oxford, UK

**Dr Lewis J. James**
Senior lecturer in Nutrition
Loughborough University, UK

**Professor Susan A. Jebb**
Professor of Diet and Population Health
University of Oxford, UK

**Professor Martijn B. Katan**
Emeritus Professor of Nutrition
Vrije Universiteit Amsterdam, The Netherlands

**Professor Timothy J. Key**
Professor of Epidemiology & Deputy Director of the Cancer Epidemiology Unit
University of Oxford, UK

**Dr Albert Koulman**
Head of the Nutritional Biomarker Laboratory
University of Cambridge, UK

**Dr Helen Lambert**
Lecturer in Public Health Nutrition
Department of Nutrition, Food and Exercise Science
University of Surrey, UK

**Professor Tim Lang**
Emeritus Professor of Food Policy
City, University of London, UK

**Professor Susan A. Lanham-New**
Head of the Department of Nutritional Sciences
University of Surrey, UK

**Professor Helen Leach**
Emeritus Professor in Anthropology
University of Otago, New Zealand

**Professor Mike Lean**
Chair, Human Nutrition
Glasgow Royal Infirmary, University of Glasgow, UK
Adjunct Professor, University of Otago, New Zealand
Visiting Professor, University of Sydney, Australia

**Associate Professor Philippa Lyons-Wall**
Associate Professor, School of Medical and Health Sciences
Edith Cowan University, Perth, Australia

**Professor Patrick MacPhail**
Professor Emeritus in Medicine
University of the Witwatersrand, South Africa

**Professor Jim Mann KNZM**
Professor in Medicine & Human Nutrition
University of Otago, New Zealand

**Dr Pamela Mason**
Independent public health nutritionist, UK

**Associate Professor Rachael McLean**
Associate Professor of Public Health & Epidemiology
University of Otago, New Zealand

**Professor Anne-Marie Minihane**
Professor in Nutrigenetics
University of East Anglia, UK

**Dr Silke Morrison**
NZ Registered Dietitian
University of Otago, New Zealand

**Associate Professor Winsome R. Parnell**
Associate Professor of Human Nutrition
University of Otago, New Zealand

**Professor Robert Peveler**
Emeritus Professor in Liaison Psychiatry
University of Southampton, UK

**Professor Andrew M. Prentice**
Head of Nutrition & Planetary Health Theme, MRC Unit
The Gambia, West Africa and London School of Hygiene &
Tropical Medicine

**Associate Professor Anna Rangan**
Associate Professor of Nutrition and Dietetics
University of Sydney, Australia

**Dr Andrew Reynolds**
Senior Research Fellow, Department of Medicine
University of Otago, New Zealand

**Professor Sian Robinson**
Professor, AGE Research Group
Newcastle University, UK

**Professor Samir Samman**
Honorary Professor in Nutrition & Metabolism
University of Sydney, Australia

**Professor C. Murray Skeaff**
Emeritus Professor in Human Nutrition
University of Otago, New Zealand

**Professor Sheila Skeaff**
Professor in Human Nutrition
University of Otago, New Zealand

**Dr Claire Smith**
Senior lecturer in Human Nutrition
University of Otago, New Zealand

**Professor Ross C. Smith**
Emeritus Professor in Surgery
University of Sydney, Australia

**Professor Boyd Swinburn**
Professor of Population and Global Health
University of Auckland, New Zealand

**Professor Rachael Taylor**
Research Professor and Karitane Fellow in early childhood obesity
University of Otago, New Zealand

**Professor Christine D. Thomson**
Emeritus Professor in Human Nutrition
University of Otago, New Zealand

**Professor David I. Thurnham**
Emeritus Professor in Human Nutrition
Ulster University, UK

**Professor A. Stewart Truswell AO**
Boden Professor Emeritus in Human Nutrition
University of Sydney, Australia

**Dr Hannah Turner**
Consultant Clinical Psychologist
Southern Health NHS Foundation Trust, Southampton, UK

**Dr Wilma Waterlander**
Assistant Professor, Department of Public and Occupational Health
Amsterdam, UMC, The Netherlands

**Professor Dr Bernhard Watzl**
Director of Institute for Physiology and Nutritional Biochemistry
Max Rubner Institute, Germany

**Assistant Professor Daniel J. Wilkinson**
Assistant Professor of Analytical Mass Spectrometry and Informatics
University of Nottingham, UK

**Professor Peter Williams (deceased; 1950–2022)**
Consultant Nutritionist and Dietitian
Adjunct Professor, Nutrition and Dietetics
University of Canberra, Australia

# Part 1

# Introducing Human Nutrition

# Introduction

A. Stewart Truswell, Jim Mann, and Leanne Hodson



## 1.1 Food for human nutrition

This book is about what we consider to be the essentials of human nutrition. The science of nutrition deals with all the effects on people of any component found in food. This starts with the physiological and biochemical processes involved in nourishment—how substances in food provide energy or are converted into body tissues (e.g. adipose tissue) and the diseases that result from insufficiency or excess of essential nutrients (malnutrition). The role of food components in the development of chronic degenerative diseases, notably coronary heart disease, cancer, diabetes, and obesity, are major targets of research activity nowadays. The scope of nutrition extends to any effect of food on human function: foetal health and development, resistance to infection, mental function, and athletic performance. There is growing interaction between nutritional science and molecular biology, which may help to explain the action of food components at the cellular level and the diversity of human biochemical responses.

Nutrition is also about why people choose to eat the foods they do, even if they have been advised that doing so may be unhealthy. The study of food habits thus overlaps with the social sciences of psychology, anthropology, sociology, and economics. Dietetics and community nutrition are the application of nutritional knowledge to promote health and well-being. Dietitians advise people how to modify what they eat in order to maintain or restore optimum health and to help in the treatment of disease. People expect to enjoy eating the foods that promote these things; and the production, preparation, and distribution of foods provides many people with employment.

A healthy diet means different things to different people. Those concerned with children's nutrition—parents, teachers, and paediatricians—aim to promote healthy growth and development. For adults in affluent communities, nutrition research has become focused on attaining optimal health and 'preventing' (which mainly means 'delaying') chronic degenerative diseases of complex causation, especially obesity (Chapter 17), cardiovascular diseases (Chapter 19), cancer (Chapter 20), and diabetes (Chapter 21). These chronic diseases have also become major causes of ill health and premature death in many developing countries (Chapter 18), where they may coexist with malnutrition (Chapter 18) and even periods of famine.

Apart from behavioural and sociological aspects of eating, there are two broad groups of questions in human nutrition, with appropriate methods for answering them.

*Firstly*, what are the essential nutrients, the substances that are needed in the diet for normal function of the human body? How do they work in the body and from which foods can we obtain each of them? Many of the answers to these questions have been established.

*Secondly*, can we delay or even prevent the chronic degenerative diseases by modifying what we usually eat? These diseases, like coronary heart disease, have multiple causes, so nutrition can only be expected to make a contribution—causative or protective. While some questions remain there is now sufficient evidence to justify nutritional recommendations which have the potential to reduce the risk of several chronic diseases.

## 1.2 Essential nutrients

Essential nutrients have been defined as chemical substances found in food that cannot be synthesized at all or in sufficient amounts in the body, and are necessary for life, growth, and tissue repair. Water is the most important nutrient for survival. By the end of the nineteenth century, the essential amino acids in proteins had been mostly identified, as well as the major inorganic nutrients, such as calcium, potassium, iodine, and iron.

The period 1890–1940 saw the discovery of 13 vitamins, organic compounds that are essential in small amounts. Each discovery was quite different; several involve fascinating stories. The research methods have comprised observations in poorly nourished humans, animal experiments, chemical fractionation of foods, biochemical research with tissues in the laboratory, and human trials.

Animal experiments played major roles in discovering which fraction of a curative diet was the missing essential food factor and how this fraction functions biochemically inside the body. The white laboratory rat is widely used, but it is not suitable for experimental deficiency of all nutrients; the right animal model has to be found. Lind demonstrated as early as 1747, in a controlled trial on board *HMS Salisbury*, that scurvy could be cured by a few oranges and lemons, but progress towards identifying *vitamin C* had to wait until 1907, when the guinea pig was found to be susceptible to an illness like scurvy. Rats and other laboratory animals do not become ill on a diet lacking fruit and vegetables; they make their own vitamin C in the liver from glucose.

For *thiamin* (vitamin $B_1$) deficiency, birds provide good experimental models. The first step in the discovery of this vitamin was the chance observation in 1890 by Eijkman in Java, while looking for what was expected to be a bacterial cause of beriberi, that chickens became ill with polyneuritis on a diet of cooked polished rice but stayed well if they were fed cheap unhusked rice. Human trials in Java, Malaysia, and the Philippines showed that beriberi could be prevented or cured with rice bran. A bird that is unusually sensitive to thiamin deficiency, a type of rice bird, was used by Dutch workers in Java to test the different fractions in rice bran. The anti-beriberi vitamin was first isolated in crystalline form in 1926. It took another 10 years of work before two teams of chemists in the USA and Germany were able to synthesize vitamin $B_1$, which was given the chemical name thiamin.

To find the cause of *pellagra*, which was endemic among the rural poor in the south-eastern states of the USA at the beginning of this century, Goldberger gave restricted maize diets to healthy volunteers, some of whom developed early signs of the disease. However, the missing substance, niacin, could not be identified until there was an animal model—in this case, 'black tongue' in dogs.

In the 1920s, linoleic and linolenic acids were identified as essential fatty acids, followed by the development of analytical techniques for determining micro-amounts of trace elements in foods and tissues. In this way, the other group of essential micronutrients emerged—the trace elements copper, zinc, manganese, selenium, molybdenum, fluoride, and chromium.

There is an additional group of food components, such as dietary fibre, that are not considered to be essential, but which are important for maintenance of health and also for reducing the risk of chronic disease.

## 1.3 Relation of diet to chronic diseases

The realization that environmental factors, including dietary factors, are of importance in many of the chronic degenerative diseases that are major causes of ill health and death in affluent societies has been a relatively recent discovery. The nutritional component of these is more difficult to study than with classical nutritional deficiency diseases because chronic diseases have multiple causes and take years to develop. It may be a 'risk factor', rather than a direct cause, but for some of these diseases there is sufficient evidence to show that dietary change can appreciably reduce the risk of developing the condition. The scientific methods for investigating chronic diseases, their causes, and treatment and prevention differ from those used for studying adequacy of nutrient intakes.

Often the first clue to the association between a food or nutrient and a disease comes from observing striking differences in incidence of that disease between countries (or groups within a country) and these differences are then found to correlate with differences in intake of dietary components. Sometimes dietary changes over time in a single country have been found to coincide with changes in disease rates. Such ecological observations give rise to hypotheses (theories) about possible diet–disease links, rather than proof of causation, because many potential causative factors may change in parallel with dietary change and it is very difficult to disentangle the separate effects.

Animal experiments, because they are usually short term, and diets are often not compatible with typical human intakes, are not as useful for investigating diet and chronic diseases and can be misleading. More information has come, and continues to come, from well-designed (human) *epidemiological* studies that investigate subjects either after diagnosis of the disease (retrospective studies) or before diagnosis (prospective studies).

*Retrospective* or *case–control studies* are quicker and less expensive to carry out, but are less reliable than prospective studies. In retrospective studies a series of people who have been diagnosed with cancer of the large bowel, for example, will be asked what they usually eat, or what they ate before they became ill. These are the 'cases'. They are compared with at least an equal number of 'controls', who are people without bowel cancer, but of the same age, gender, and, if possible, social conditions. Weaknesses of the method include the possibility that:

- the disease may affect food habits
- the cases cannot recall their diet accurately before the cancer really started
- the controls may have some other disease (known or latent) that affects their dietary habits
- food intakes are recorded by cases and controls in a different way (bias).

*Prospective* or *cohort studies* avoid the biases involved in asking people to recall past eating habits. Information about food intake and other characteristics is collected well before the onset of disease. Large numbers of people must be interviewed and examined; they must be of an age at which bowel cancer (for example) starts to be fairly common (middle aged) and in a population that has a fairly high rate of this disease. The healthy cohort thus examined and recorded is then followed up for 5 years or more. Eventually, a proportion will be diagnosed with bowel cancer and the original dietary details of those who develop cancer can be compared with the diets of the majority who have not developed the disease. Usually a number of dietary and other environmental factors are found to be more (or less) frequent in those who develop the disease. These, then, are apparent risk factors, or protective factors, but they are not necessarily the operative factors. For example, initial dietary fibre intake may be lower in those who have subsequently developed bowel cancer. However, before concluding that low intakes are a risk factor (or that high intakes are protective) it is necessary to carry out further analyses of the data to ensure that the observation is not a result of 'confounding'; that is, explained by a relationship between dietary fibre and some other factor that is more likely to be causal.

Prospective studies usually provide stronger evidence of a diet–disease association than case-control studies, and where several prospective studies produce similar findings from different parts of the world, this is impressive evidence of association (positive or negative), but it is still not final proof of causation. If an association is deemed not due to bias or confounding, is qualitatively strong, biologically credible, follows a plausible time sequence, and especially if there is evidence of a dose–response relationship, it is likely that the association is causal. However, there are some disadvantages with regard to cohort studies. The prospective follow-up of large numbers of people (usually thousands or tens of thousands) is a complicated and costly exercise. Furthermore, assessing dietary intake at one point in time may not provide a true reflection of usual intake. It is also conceivable that a dietary factor operating before the study has started, perhaps even in childhood, may be responsible for promoting a disease.

Definitive proof that a dietary characteristic is a direct causative or protective factor may come from *randomized controlled trials* (RCTs). Those in the experimental group are typically prescribed a dietary regimen that involves altering the distribution of macronutrients or intakes of other essential nutrients from food sources or supplements, while the controls continue to follow their usual diet with, when appropriate, a placebo capsule or tablet. Disease (and death) outcomes in the two groups are compared. Such trials have the advantage of being able to prove causality, as well as the potential cost-benefit of the dietary change. However, they are costly to carry out because, as with prospective studies, it is usually necessary to study large numbers of people over a prolonged period of time.

Quite often, a single trial or single prospective study does not, in itself, produce a definitive answer, but by combining the results of all completed investigations in a *meta-analysis*, more meaningful answers are obtained. For example, a much clearer picture has emerged regarding the role of dietary factors in the aetiology of coronary heart disease from meta-analyses of prospective studies or clinical trials.

*Mendelian randomization analysis* is an approach to investigate, in observational studies, associations between disease outcomes and modifiable or endogenous exposures that are strongly associated, without the potential biases and confounding that are often associated with observational studies. Such analyses may permit confident conclusions regarding causality without the need for RCTs or in situations where randomized trials are not feasible. Mendelian randomization analyses are based on the principle that genotypes are assigned randomly when passed from parents to offspring during meiosis and the assumption that choice of partner is not based on genotype. Thus, population genotype distribution should be unrelated to the confounders that prevent confident conclusions from observational epidemiological studies. Mendelian randomization has been described as a 'natural' RCT. The causal relationship between alcohol and oesophageal cancer, as well as the causal effect of alcohol intake on high blood pressure and risk of hypertension, have been confirmed using genetic data on the ALDHH2 (aldehyde dehydrogenase 2 family) gene. The lack of an association between the C-reactive protein genotype and disease risk has shown that C-reactive protein, a non-specific marker of systemic inflammation, is not a causal factor for coronary heart disease.

In addition to epidemiological studies and trials, much research involving the role of diet in chronic degenerative disease has centred around the effects of diet on modifying risk factors, rather than the disease itself. For many chronic diseases, there are biochemical markers of risk. High plasma cholesterol, for example, is an important risk factor for coronary heart disease. Innumerable studies have examined the role of different nutrients and foods on plasma cholesterol or other risk factors. Such studies are generally cheaper and easier to undertake than epidemiological studies and RCTs with disease outcome, because fewer people can be studied over a relatively short period of time. They have helped to find which foods lower cholesterol and so should help protect against coronary heart disease. It is this information that has formed the basis of the public health messages dating back to the late 1960s that have contributed to the decline in the incidence of coronary disease in most affluent societies over the past 40 years (see Chapter 19).

# 1.4 Tools of the trade

The epidemiological methods described above are of course not unique to nutritional science. However, as with any other science or profession, nutrition has its own specialized techniques and technical terms. Those that are frequently used in research and professional work are introduced here. They are described in more detail later in the book.

## 1.4.1 Measuring food and drink intake

Which foods (and drinks) does a person or a group of people usually eat (and drink), and how many grams of each per day? Unless the subject is confined within a special research facility under constant observation, the answer can never be 100% accurate. Information about food intake is subjective and depends on memory; people do not always notice or know the exact description of the foods they are given to eat (especially in mixed dishes). When asked to record what they eat, they may alter their diet. People do not eat the same every day, so it is difficult to obtain a profile of their usual diet.

The different techniques used are described in Chapter 37. One set of methods estimates the amounts of food produced and sold in a whole country, and divides these by the estimated population. This is 'food disappearance' or 'food moving into consumption'. Obviously, some of this food is wasted, and some is eaten by tourists and pets. The main value of the data is for following national trends and seeing if people appear to be eating too little or too much of some foods.

Other methods capture the particular foods and amounts of them that individuals say they actually eat. These methods rely either on the subjects' memories or on asking them to write down everything they eat or drink for (usually) several days.

## 1.4.2 Food composition tables

Food tables that have comprehensive food composition data are used to calculate people's nutrient intakes from their food intake estimates. Ideally, food tables would contain all the usual foods eaten in a country and give average numbers for the calories (food energy), the major essential nutrients, and other important food components (e.g. dietary fibre) of each food, measured by chemical analysis. In many smaller, less affluent countries there is no complete set of food composition data, so 'borrowed' data are used from one of the Western countries (e.g. the UK, USA, or Germany) for which reliable data are available. Most food tables are also available as computer software, which greatly speeds up the computations (see Chapter 36).

## 1.4.3 Dietary reference values and guidelines

Computer software packages for dietary analyses generate printouts that show what a subject or groups of individuals have eaten in terms of nutrients. This does not mean anything unless comparison is made with normative dietary reference values. Two sets are used, but they differ somewhat from country to country. For essential nutrients (protein, vitamins, minerals) the reference is in a table of *recommended nutrient intakes* (*dietary reference intakes* in the USA). For some nutrients that are not essential, but which may be related to disease risk (e.g. saturated fat, which is related to risk of coronary heart disease), advice relates to an acceptable upper limit or range of intakes expressed as a percentage of total energy. *Dietary guidelines* translate such advice into information about recommended food choices or dietary patterns (see Chapter 27).

### 1.4.4 Biomarkers: Biochemical tests

If an individual (or group) is found to eat less than the recommended intake of nutrient 'A', then they may not necessarily have any features of 'A' deficiency. The food intake may have been under-reported or may have been only temporarily less than usual, and there are large body stores for some nutrients. Some people may have lower requirements than average. On the other hand, individuals can suffer deficiency of nutrient 'A' despite an acceptable intake if they have an unusually high requirement, perhaps because of increased losses from disease. For many nutrients, biochemical tests using blood or urine are available to help estimate the amount of the nutrient functioning inside the body (see Chapters 38 and 39). Furthermore, *biomarkers* provide a more objective, and often more economical, method for estimating intake of some food components (e.g. urinary sodium for salt) than food intake measurement, and they can also be used to check the reliability of subject histories or records of food intake.

### 1.4.5 Human studies and trials

Most of the detailed knowledge in human nutrition comes from a range of different types of human experiments. They may last from hours to years and may include just a few subjects or thousands of subjects. Controlled trials (mentioned in section 1.3 above), which are often undertaken to further pursue observations made in epidemiological studies, are typically used to define more clearly the relationships between nutrients and disease risk factors. Some studies and trials that are used to further understand relevant physiology and metabolism are unique to nutritional science.

The following are examples:

- *Absorption studies*: Some nutrients are poorly absorbed, and absorption of nutrients is better from some foods than from others. Many studies have been done to measure *bioavailability*—that is, the percentage of the nutrient intake that is available to be used inside the body. After a test meal, the increase in some nutrients can be measured in blood or urine. Isotopes may be needed to label the nutrient to help determine its bioavailability.
- *Metabolic studies*: Typically only one component of the diet is changed at a time, and the result is measured in a change in blood or excreta (urine and/or faeces). One type is the *balance experiment*. This may measure, for example, the intake of calcium, and its excretion in urine and faeces. Because of the minor variability of urine production and the major variability of defecation, these measurements have to be made for a metabolic period of several days every time any dietary change is made.

## 1.5 Food and the environment

Concern that human food production might not deliver sufficient food for the world's population was first expressed in the 18th century. The extent to which food production and delivery can adversely affect the environment has been known for several decades. However, more recent evidence of the contribution to climate change, notably from increased global demands for meat and dairy products, which is associated with an increase in greenhouse gas production, has resulted in greatly increased awareness of the relevance. The issue was first acknowledged in a fairly brief chapter in the 4th edition of our book, expanded in the 5th edition, and is now considered in more depth in two important chapters in this edition. Important examples of dietary changes in affluent societies include a reduced intake of red meat (especially beef and lamb) and dairy, because of the contribution by ruminants to methane emission, and increased reliance on plant-based foods, whenever possible locally produced and from crops that do not have excessive water requirements. Such changes have the potential to appreciably improve human health and reduce environmental impact.

## 1.6 Interpreting and translating the results of nutrition research

Interpreting and translating the results of nutrition research into advice for populations and individuals is a complex matter. Genes may affect the function of nutrients, and gene expression may in turn be influenced by nutrient intake. Thus, there is individual variation in the way people absorb, metabolize, and respond to different nutrients. There is also the possibility that changes observed over a short time period may not persist indefinitely because humans may adapt to dietary change. Furthermore, when one component is added or removed from the diet, there are usually consequential changes to the rest of the diet, as some other food is put in its place. An apparent effect of removing one food may, at least in part, be due to the action of its replacement or to the energy deficit that results if it is not replaced. Before recommending dietary change, it is imperative that nutritionists consider not only the role of individual nutrients as determinants of health and disease but also diet as a whole, the complex family and societal dynamics of dietary change, and the environmental impact, in order to ensure that overall benefit will accrue from any changes that are made. The approaches that are used to make nutrition recommendations for individuals and populations are to be found in the introductory section to Part 4, *Nutrition-related Disorders*.

# Part 2

# Energy and Macronutrients

# Carbohydrates

## Andrew Reynolds and Jim Mann



Carbohydrates are a diverse group of compounds with the general formula $(CH_2O)n$. In plants they are synthesized from water and carbon dioxide using the sun's energy during photosynthesis. Animals have limited capacity to synthesize carbohydrates but make lactose and oligosaccharides for milk. Following digestion of disaccharides and most starches, simple sugars such as glucose ($C_6H_{12}O_6$) are absorbed in the small intestine. They are then transported to the tissues to be oxidized back to water and carbon dioxide, by which process the host gains energy for cellular processes. Simple sugars not immediately required as a source of energy may be converted to glycogen, which is stored in liver and muscle. When dietary carbohydrate is not available as an energy source and stored glycogen has been depleted, glucose can be made from lactate, glycerol, and some amino acids.

Carbohydrates are the most important source of food energy in the world. The major sources of dietary carbohydrate worldwide are cereal grains (primarily rice, wheat, and maize), with refined sugar, root crops (potatoes, cassava, yams, sweet potatoes, and taro), pulses, vegetables, fruit, and milk products contributing less to overall energy intake (see Fig. 2.1). These data for grain intake are lower than crop production statistics, as grain (especially maize and barley) is also grown for animal feed, seed, or for brewing and distilling. Carbohydrate-containing foods, with the exception of sugar, contribute important amounts of protein, vitamins, minerals, phytochemicals, sterols, and antioxidants to the diet.

World production of grain, sugar cane, vegetables, and fruits has increased over the last 20–30 years (see Chapter 31). Production of root crops, pulses, and sugar beet has changed little on a worldwide basis, while pulse production has decreased in some Asian countries and root crop production has fallen in Europe. The proportional reduction of root crops and pulses may have nutrient intake consequences in these areas.

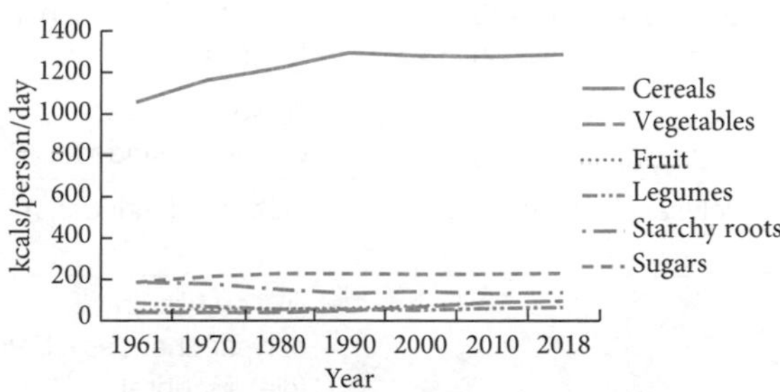


Fig. 2.1 Principal carbohydrate sources of food energy (kcal/person/day) from major carbohydrate sources (1961–2018) for the whole world.

*Source:* FAOSTAT Food Balance Sheets. Available at: faostat3/fao/org/download/FB/FBS/E (Accessed 12 April 2022).

At first glance, these data suggest that overall food production is keeping pace with population growth, which has continued in most parts of the world. Increased production, however, is due largely to altered agricultural practices, principally the increased use of fertilizers, a practice that cannot continue indefinitely. Indeed, in Africa, cereal production is already inadequate and it seems conceivable, indeed likely, that there are major problems ahead, particularly with the global rise in temperature leading to the frequent failure of arable crops.

Population surveys permit a broad overview of carbohydrate consumption in different parts of the world. As a percentage of total energy intake, the contribution from all carbohydrates ranges from about 40% to over 80%, representing about 250–400 g carbohydrate/day. More affluent countries, such as those in North America, Europe, and Australasia, are at the lower end of the range, and less affluent countries at the higher end. Starch accounts for 20–50% and sugars 9–27% of total energy. In Western countries, 20–25% of total energy is derived from sugars, with about one-third from naturally occurring sources (vegetables, milk, fruits, and juices) and two-thirds from added sugars, especially sucrose. In North America, a relatively high proportion of free sugars comes from corn syrup solids or high-fructose corn syrups, so that sucrose intake—but not total sugar intake—is lower than in most European countries.

## 2.1 Classification and terminology of dietary carbohydrates

Carbohydrates are a diverse group of compounds that have different physiological properties, which determine their health effects. They may be classified on the basis of their chemical structure (Table 2.1). However, their effects on health also depend upon the structure of the food and the site and rate of

**Table 2.1 Classification of dietary carbohydrates by chemical structure**

| Class (DP)[a] | Subgroup | Principal components | Absorption site |
|---|---|---|---|
| Sugars (1–2) | (i) Monosaccharides | Glucose, fructose, galactose | Small intestine |
| | (ii) Disaccharides | Sucrose, lactose, maltose | Small intestine |
| | (iii) Polyols (sugar alcohols) | Mannitol, xylitol, isomalt | Small and large intestines |
| Oligosaccharides (3–9) (short-chain carbohydrates)[b] | (i) Malto-oligosaccharides (α-glucans) | Maltodextrins | Small intestine |
| | (ii) Non-α-glucan-oligosaccharides | fructo- and galacto-oligosaccharides, inulin | Large intestine |
| Polysaccharides (≥10) | (i) Starch (α-glucans) | Amylose, amylopectin, modified starches | Small and large intestines |
| | (ii) Non-starch polysaccharides (NSPs) (dietary fibre) | Cellulose, pectin, β-glucans, glucomannans | Large intestine or not absorbed |

[a]Degree of polymerization or number of monomeric (single-sugar) units

[b]See IUB-IUPAC definition and Cummings et al. (1997)

*Adapted from:* Macmillan Publishers Ltd: Nishida, C., Nocito, F.M., and Mann, J. (2007) Joint FAO/WHO Scientific Update on Carbohydrates in Human Nutrition. *Eur J Clin Nutr*, **61**(Suppl. 1), S1–137.

digestion and absorption (Table 2.1). This has led to alternative classifications and a variety of terms used to describe carbohydrates. Sugars, starch, and dietary fibre provide most dietary carbohydrate, and have identifiable chemical structures.

Oligosaccharides are included in the chemical classification of carbohydrates and have a degree of polymerization (DP) of 3–9. The division between oligosaccharides and polysaccharides at DP 10 is somewhat arbitrary, as there is a continuum of molecular size from sugars to complex polymers of DP 100,000 or more. In reality, the division is made analytically by defining oligosaccharides as carbohydrates other than mono- and disaccharides that remain in solution in 80% (w/v) ethanol. Important non-α-glucan-oligosaccharides are inulin and fructo-oligosaccharides, which are the storage carbohydrates in artichokes with small amounts of the lower-molecular-weight varieties found in wheat, rye, asparagus, and members of the onion, leek, and garlic families. They may also be produced industrially. Terms which have been suggested to describe physiological and other properties are discussed in the context of these chemical classes.

### 2.1.1 Sugars

Sugars are the simplest carbohydrates in structure, and comprise monosaccharides, disaccharides, and sugar alcohols. The three principal monosaccharides are glucose, fructose, and galactose (see Fig. 2.2). These monosaccharides are the building blocks of naturally occurring di-, oligo-, and

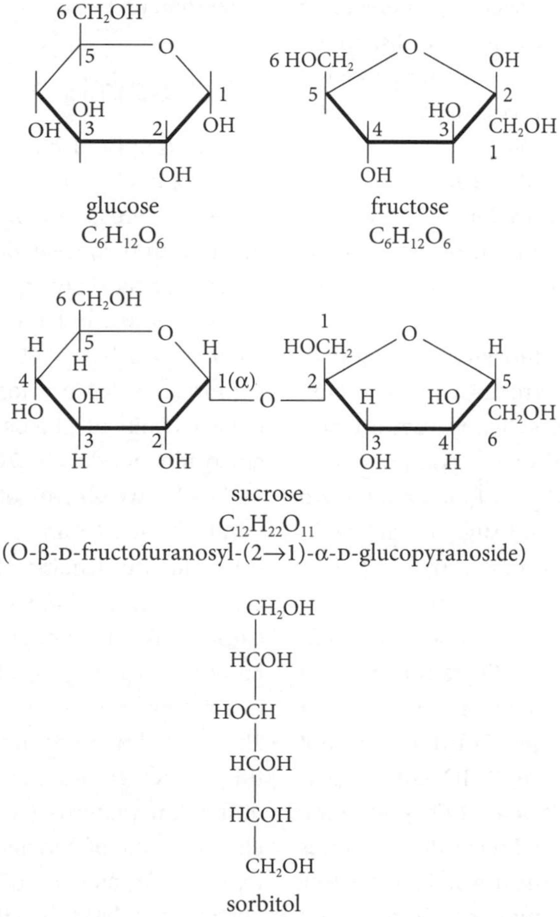

glucose
$C_6H_{12}O_6$

fructose
$C_6H_{12}O_6$

sucrose
$C_{12}H_{22}O_{11}$
(O-β-D-fructofuranosyl-(2→1)-α-D-glucopyranoside)

sorbitol

Fig. 2.2 The structure of mono- and disaccharides.

polysaccharides. Free glucose and fructose occur in honey and cooked or dried fruit (invert sugar), and in small amounts in raw fruit, berries, and vegetables such as carrots, onions, swedes, turnips, and tomatoes. Corn syrup (a glucose syrup produced by the hydrolysis of corn starch) and high-fructose corn syrup (which contains glucose and fructose) are used by the food industry, particularly in the production of carbonated drinks.

The principal disaccharides are sucrose (αGlc(1→2)β-Fru) and lactose (β-Gal(1→4)βGlc) (see Fig. 2.2). Sucrose is widely found in fruit, berries, and vegetables, and can be extracted from sugar cane or sugar beet. Lactose is the main sugar in milk. Maltose, a disaccharide derived from starch, occurs in sprouted wheat and barley. Trehalose (α-Glc(1→4)α-Glc) is found in yeast, in fungi (mushrooms), and in small amounts in bread and honey. It is used by the food industry as a replacement for sucrose, where a less sweet taste is desired. Most mono- and disaccharides are able to provide glucose as an energy source for the tissues following the process of digestion and absorption in the small intestine.

The polyols, such as sorbitol, are alcohols of glucose and other sugars. They are found naturally in some fruits and made commercially by using aldose reductase to convert the aldehyde group of the glucose molecule to the alcohol.

Sugars can also be characterized as free, added, or intrinsic. Intrinsic sugars are those incorporated within the cell walls of plants; that is, they are naturally occurring and are always accompanied by other nutrients. Sugars added to foods are known as either free sugars or added sugars. Lactose in milk does not fall readily into the intrinsic or free category, but milk has important nutritional properties, so the term non-milk extrinsic sugar (NMES) was introduced in the UK to indicate the group of sugars other than intrinsic and milk sugars that should be identified in the diet. This term has not gained widespread use. The WHO/FAO Expert Consultation on *Diet, nutrition and the prevention of chronic diseases* (WHO Technical Report Series 916) in 2003 recommended the use of the term 'free sugars', which refers to all 'monosaccharides and disaccharides added to foods by the manufacturer, cook and consumer, plus sugars naturally present in honey, syrups and fruit juices'. This has now been recommended for use in the UK and is broadly comparable with the concept of 'added' sugars, a term used in the United States and some other countries.

Analytically, it is not readily possible to distinguish in a processed food which sugars might have been added and which are naturally present in a constituent (e.g. fruit) of the processed food. While absorption rates differ between food sources of sugars, there is probably little difference in the way they are handled physiologically. Of course, the nutritional profile of these two foods will be very different. For labelling purposes, therefore, a further category 'total sugars' has been suggested, which includes all sugars from whatever source in a food and is defined as all monosaccharides and disaccharides other than polyols.

## 2.1.2 Starch

Starch comprises most of the carbohydrates consumed globally. Starch is the storage carbohydrate found in cereals, potatoes, cassava, legumes, and bananas and consists only of glucose molecules. It occurs in a partially crystalline form in granules and comprises two polymers: amylose (DP ~ $10^3$) and amylopectin (DP ~ $10^4$–$10^5$). Most cereal starches comprise 15–30% amylose, which is a non-branching helical chain of glucose residues linked by α-1,4 glucosidic bonds, and 70–85% amylopectin, a high-molecular-weight, highly branched polymer containing both α-1,4 and α-1,6 linkages (Fig. 2.3). Some waxy starches (maize, rice, sorghum, barley) comprise mostly amylopectin. In animals and humans, carbohydrate is stored in liver and muscle in limited amounts as glycogen, which has a structure similar to amylopectin, but is even more highly branched. The crystalline form of the amylose and amylopectin in starch granules confers on them distinct X-ray diffraction patterns (A, B, and C). The A type is characteristic of cereals (rice, wheat, maize), the B type is characteristic of potato, banana, and high-amylose starches, while the C type is intermediate between A and B, and found in legumes. In their native

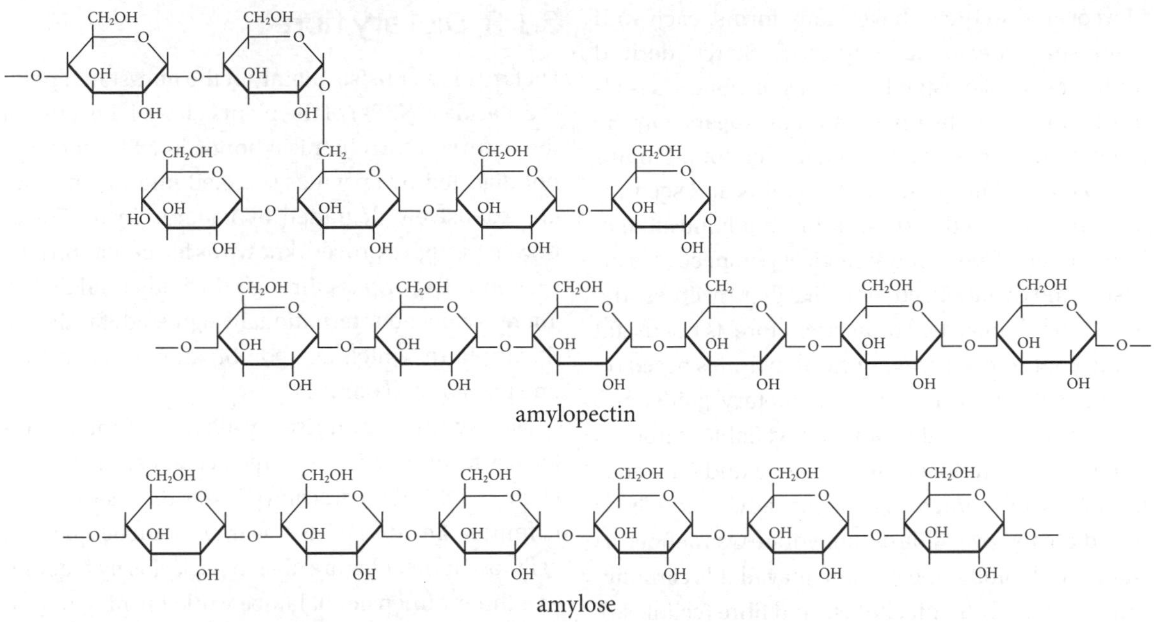

amylopectin

amylose

Fig. 2.3 The structure of amylose and amylopectin.

(raw) form, the B starches are resistant to digestion by pancreatic amylase, but the crystalline structure is lost when starch is heated in water (gelatinization), thus permitting more rapid digestion to take place. Recrystallization (retrogradation) takes place to a variable extent after cooking and cooling, and is in the B form. Starch that is incompletely digested in the small intestine is known as 'resistant starch.'

Because many starches do not have the functional properties needed to impart or maintain desired qualities in food products, some have been modified, by either chemical processing or plant breeding techniques, to obtain these properties. Various processes are used to modify starch, the two most important being substitution and cross-linking. Substitution involves etherification or esterification of a relatively small number of hydroxyl groups on the glucose units of amylose and amylopectin. This reduces retrogradation, which is part of the process of bread going stale, for example. Substitution also lowers gelatinization temperature, provides freeze-thaw stability, and increases viscosity. Cross-linking involves the introduction of a limited number of linkages between the chains of amylose and amylopectin. The process reinforces the hydrogen bonding that occurs within the granule. Cross-linking increases gelatinization temperature, improves acid and heat stabilities, inhibits gel formation, and controls viscosity during processing. Techniques of plant breeding, including genetic modification, can also be used to alter the proportions of amylose and amylopectin in starchy foods. For example, high-amylose corn starch requires higher temperatures for gelatinization and is more prone to retrogradation.

Other terms related to dietary starches include 'complex carbohydrate,' which was first introduced in 1977 in *Dietary goals for the United States.* Complex carbohydrate consumption, such as whole grains, vegetables, and fruit, was encouraged as part of a healthy diet. Subsequently, the term came to be used to distinguish between sugars and longer carbohydrate chains such as starch and fibres. Complex carbohydrate has, however, never been formally defined and eventually it became equated with starch. It has not proved to be a useful term to identify 'healthy' food because many fruits and vegetables that are important sources of micronutrients are low in polysaccharides and starch.

Moreover, starch can have many forms, each with contrasting metabolic properties. Starch derived from most cooked starchy cereals or potatoes is almost as rapidly absorbed as many sugars. On the other hand, some starch, such as that found in unripe bananas and partly milled grains and seeds, is fairly resistant to digestion in the small intestine of humans, and from a physiological perspective is digested and metabolized more like dietary fibres. The term 'complex carbohydrate', therefore, is not useful in the context of classifying carbohydrates based on their chemical structure or in dietary guidelines. Other terms, 'available and unavailable carbohydrate', were introduced by McCance and Lawrence in 1921, while preparing dietary advice for those with diabetes. Available carbohydrate was defined as 'starch and soluble sugars' and unavailable carbohydrate as 'mainly hemicellulose and fibre (cellulose)'. The concepts glycaemic index and glycaemic load (see section 2.4) are more widely applied and follow on from McCance and Lawrence's original ideas.

### 2.1.3 Dietary fibre

Dietary fibre consists mainly of the non-starch polysaccharides (NSPs) of the plant cell wall. Because of the nature of the chemical bonds in NSPs, they are not digested by enzymes secreted into the gut, but are extensively degraded by bacteria in the lower bowel through a process known as fermentation (see section 2.3.4) or pass through the body undigested. There is now an internationally agreed definition of dietary fibre, which is a combination of chemistry and physiology (Box 2.1).

Dietary fibre comprises a mixture of many molecular forms of NSPs, of which cellulose, a straight-chain β-1,4-linked glucan (DP $10^3$–$10^6$), is the most common. Because of its linear, unbranched nature, cellulose molecules are able to pack closely together in a three-dimensional lattice work, forming microfibrils. These form the basis of cellulose fibres, which are woven into the plant cell wall and give it structure. Cellulose comprises 10–30% of the NSPs in foods.

#### Box 2.1 Definition of dietary fibre agreed at Codex Alimentarius in 2009

Dietary fibre means carbohydrate polymers with ten or more monomeric units, which are not hydrolysed by the endogenous enzymes in the small intestine of humans and belong to the following categories:

- edible carbohydrate polymers naturally occurring in food as consumed
- carbohydrate polymers that have been obtained from food raw material by physical, enzymatic, or chemical means, and which have been shown to have a physiological effect of benefit to health as demonstrated by generally accepted scientific evidence to competent authorities
- synthetic carbohydrate polymers that have been shown to have a physiological effect of benefit to health as demonstrated by generally accepted scientific evidence to competent authorities.

There are two footnotes to the definition, one of which allows countries to include oligosaccharides (DP 3–10) as fibre, the position taken by the European Commission, and the other that allows 'fractions of lignin and/or other compounds when associated with polysaccharides in the plant cell walls and if these compounds are quantified by the AOAC gravimetric methods'.

Given that the main body of epidemiological and experimental data regarding the health benefits of fibre derives from studies of diets based on fruit, vegetables, and whole grains, it is important to emphasize the benefits of the polysaccharides of the plant cell wall, which are characteristic of such diets.

There is experimental evidence that some carbohydrate polymers extracted from food and synthetic polymers may have physiological effects, such as cholesterol lowering. However, there is a lack of long-term epidemiological evidence of health benefit, so bullet points two and three in Box 2.1 emphasize the need for 'generally accepted scientific evidence' to be accepted by 'competent authorities' before such polymers can be labelled as fibre.

*Source:* Cummings, J.H., Mann, J.I., Nishida, C., and Vorster, H.H. (2009) Dietary fibre: an agreed definition. *Lancet*, **373**, 365–6.

By contrast, the hemicelluloses are a diverse group of polysaccharide polymers that contain a mixture of hexose (6C) and pentose (5C) sugars, often in highly branched chains. Mostly they comprise a backbone of xylose sugars with branches of arabinose, mannose, galactose, and glucose, and have a DP of 150–200. Typical of the hemicelluloses are the arabinoxylans found in cereals.

Pectins are forms of dietary fibre common to all cell walls. They are primarily β-1,4-D-galacturonic acid polymers, although they usually contain 10–25% other sugars, such as rhamnose, galactose, and arabinose, as side chains. Some 3–11% of the uronic acids have methyl substitutions, which improve the gel-forming properties of pectin, as used in jam making.

Chemically related to the cell wall NSPs—but not strictly cell wall components—are the plant gums and mucilages. Plant gums are sticky exudates that form at the sites of injuries to plants. They are mostly highly branched, complex uronic acid-containing polymers. Gum Arabic, named after the Arabian port from which it was originally exported to Europe, comes from the acacia tree and is one of the better-known plant gums. It is sold commercially as an adhesive, and used in the food industry as a thickener and to retard sugar crystallization.

Plant mucilages are botanically very different in that they are usually mixed with the endosperm of the storage carbohydrates of seeds. Their role is to retain water and prevent desiccation. They are polysaccharides similar to hemicelluloses, of which guar gum, from the cluster bean, and carob gum are similar β-1,4-D-galactomannans with α-1,6-galactose single-unit side chains. Again, they are widely used in the pharmaceutical and food industries as thickeners and stabilizers in salad creams, soups, and toothpastes.

The algal polysaccharides, which include carageenan, agar, and alginate, are all NSPs extracted from seaweeds or algae. They replace cellulose in the cell wall and have gel-forming properties. Carageenan and agar are highly sulphated, and the ability of carageenan to react with milk protein has led to its use in dairy products and chocolate. Maltodextrins and the non-α-glucan-oligosaccharides, two groups of food oligosaccharides, meet the definitions of a dietary fibre.

The terms 'soluble' and 'insoluble' fibre developed out of the early chemistry of NSPs. However, solubility is a spectrum and this separation is not chemically distinct; it can be changed by simply altering the pH of the extraction, and the physiological roles of soluble and insoluble NSPs in foods are not clear. Much of the early work on soluble fibre, which suggested that it had good cholesterol-lowering properties, was from studies using pure polysaccharides such as pectin, guar, psyllium, and milling fractions like oat bran. All plant cell wall/NSP complexes, as they exist in foods, contain a 'soluble' fraction. While it may be useful to encourage the use of some foods with a relatively high 'soluble' fibre fraction (e.g. cooked dried beans, chickpeas) by those with diabetes and raised cholesterol levels, ascribing specific physiological properties to, or making dietary recommendations for, soluble and insoluble fibre is not justified at present.

## 2.2 Measurement of dietary carbohydrates

In North America, total carbohydrate in the diet has traditionally been estimated by measuring moisture, fat, protein, and ash, and subtracting the sum of these from the total dry weight of the food. The remaining dry weight is deemed to be carbohydrate 'by difference'. In Europe and Australasia, direct measurement of carbohydrates is done. The 1997 and 2006 consultations by the Food and Agriculture Organization (FAO) and the World Health Organization (WHO) use the latter approach because the figure arrived at by difference includes non-carbohydrate components (lignin, tannins, waxes, and some organic acids) and compounds the analytical errors of all the other determinations. Moreover, by difference does not allow a detailed characterization of dietary carbohydrate, which is necessary for understanding the relationship between carbohydrates and health.

A single comprehensive system to measure all the carbohydrate fractions in the diet does not exist. Thus, the amount of information given in food tables worldwide varies considerably. Comprehensive analytical values are listed in the UK McCance and Widdowson Food tables (Public Health England, 2021) and include total carbohydrate measured as the sum of all 'available carbohydrates,' which includes free sugars and complex carbohydrates but not fibres such as NSPs. Few published data are available on the resistant starch content of food because such values are highly dependent on food handling prior to analysis and on physiological variables (Englyst et al., 1992).

In brief, the methodology for assessing total carbohydrate is as follows. After homogenization of a diet or food, and lipid extraction, if necessary, free sugars (mono- and disaccharides) can be extracted into aqueous solutions and measured by gas-liquid chromatography (GLC) or high-performance liquid chromatography (HPLC). Enzymic methods also exist for individual sugars. Oligosaccharides or short-chain carbohydrates are more difficult to determine, and are measured as those carbohydrates other than free sugars that are soluble in 80% ethanol. Starch and maltodextrins are hydrolysed and NSPs precipitated with ethanol. Fructans are hydrolysed enzymatically and the monosaccharide constituents reduced to acid-stable alditol derivatives, while the remaining oligosaccharides are hydrolysed with sulphuric acid and measured as alditol acetates by GLC.

Starch is solubilized and then hydrolysed with a combination of amylolytic enzymes, and the increase in glucose is measured either enzymatically, colorimetrically, or by GLC. Starch can be subdivided analytically into rapidly digestible (RDS), slowly digestible (SDS), and resistant starch (RS). RDS and SDS relate to the rate of release of glucose from starch and are relevant to postprandial glycaemia (see section 2.4).

A universally agreed single method for the measurement of dietary fibre/NSP has evaded the best intentions of Codex Alimentarius, the FAO supported Food Code and standards, and at present a list of 17 methods is approved (see CODEX Alimentarius Meetings, 2011, CCMAS32 Report).

## 2.3 Digestion and absorption of carbohydrates

### 2.3.1 Sugars and starch

The mono- and disaccharides, maltodextrins, and most starches are digested and absorbed from the upper part of the small intestine. They are hydrolysed to their constituent monosaccharides before transport across the mucosa (see Fig. 2.4). Several classes of carbohydrate resist digestion in the small intestine and pass into the large bowel, where they are fermented by the resident microbiome (see section 2.3.4). These are principally some of the sugar alcohols, lactose in many populations, most oligosaccharides except maltodextrins, some starches, and all NSPs. There are thus two distinct pathways for carbohydrate digestion, with differing effects on insulin secretion and end-products for absorption.

Carbohydrate digestion starts in the mouth, where salivary α-amylase is secreted. However, its activity is substantially inhibited by low pH when ingested food enters the stomach and so is relatively unimportant compared with that resulting from pancreatic amylase in the small intestine. Amylase is the only active carbohydrate-digesting enzyme produced by the pancreas. α-Amylase hydrolyses the α-1,4 bonds, but only those that are not at the ends of a molecule or next to α-1,6 branch points, so α-amylase produces a mixture of glucose, maltose, maltotriose, and α-limit dextrins, which are presented to the small intestine mucosa. The surface area of the small intestine is about 200 $m^2$ because it is covered with microvilli. These microvilli extend into the unstirred water-layer phase of the intestinal lumen. The microvillus layer is known as the brush border and in

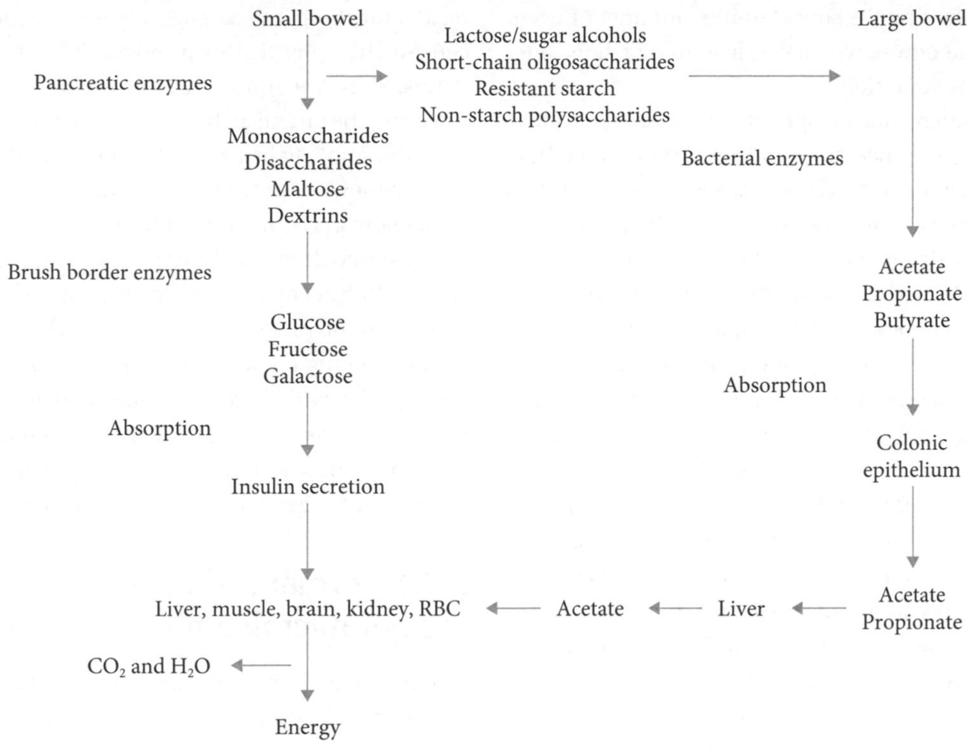


Fig. 2.4 The two principal pathways of carbohydrate digestion and absorption. RBC, red blood cells.

it are the three principal enzymes that complete digestion of carbohydrates to monosaccharides: these are glucoamylase (α-glucosidase) and sucrose-isomaltase. These enzymes are able to reduce the products of starch digestion to glucose monomers and sucrose to glucose and fructose, and lactase (β-galactosidase), which hydrolyses lactose to glucose and galactose. The expression of β-galactosidase is retained in the brush border after weaning in only a minority of world populations, mainly those in northern climates. In close association with these enzymes are transporters that move the monosaccharides into the portal blood. Glucose and galactose are both absorbed into the enterocyte by a process of active transport facilitated by sodium glucose cotransporters (SGLT1). Sodium is pumped from the cell to create a sodium gradient between the intestinal lumen and the interior of the cell. The resultant sodium gradient drives the cotransporter so that one molecule of sodium and one molecule of glucose or galactose are transported into the cytoplasm of the enterocyte against a concentration gradient. Glucose is pumped out of the enterocyte and into the intracellular space by glucose transporter 2 (GLUT 2). This is one of a large family of more than 12 facilitative GLUTs that are found in tissues throughout the body, which transport d-glucose down its concentration gradient, a process described as facilitated diffusion. Fructose is taken up from the gut lumen by a similar process of facilitated transport by glucose transporter 5 (GLUT 5), which may also be the means by which it exits the enterocyte. Glucose uptake by SGLT1 triggers glucagon-like peptide-1 (GLP1) and gastric inhibitory peptide (GIP) secretion, which ensures that the magnitude of the hormone stimulus to pancreatic b-cells coincides with the appearance of glucose in the bloodstream.

Sugar alcohols such as sorbitol, mannitol, xylitol, and erythritol have no specific transport mechanism and are absorbed by simple diffusion. At low amounts this works well, but as the amount ingested increases, so the transport capacity of the small intestine is overwhelmed and they partly pass into the large bowel. Because of their relatively low molecular weight, they retain considerable amounts of water within the bowel, which can lead to diarrhoea after excess consumption.

Any starch that escapes the normal small intestine digestive processes is called resistant starch (RS). Starch may resist digestion because it is enclosed in whole grains or is otherwise physically inaccessible (RS1), or because it is present in the B-type crystalline form (RS2), e.g. banana starch, or because it has retrograded after cooking and cooling (RS3). Some modified starch (e.g. hydroxypropyl or acetylated starch) will also resist digestion (RS4). The amount of RS in the diet is not accurately known because measuring it is technically difficult. This is because almost any handling of starchy foods from diet collections (i.e. mixing with water, homogenization, freezing, or cooling) will affect RS content. Present estimates for countries with Westernized diets are in the range of 3–10 g/day. Clearly, this is very diet dependent. A couple of relatively unripe bananas will readily provide 20 g RS. A biscuit made with potato flour would give 10 g. The amount of RS that escapes digestion also varies among people, partly dependent on transit time through the small intestine.

### 2.3.2 Oligosaccharides

Oligosaccharides are carbohydrates that have a degree of polymerization (DP) of 3–9. The nature of their chemical bonds means they are not susceptible to either pancreatic or brush border hydrolysis and so they pass entirely into the large intestine. The lower molecular weight species (DP 3–5) have the potential to provide an osmotic gradient in the small intestine, and if taken in large quantities (15–30 g) can cause disturbances in gut function. They are better known for their propensity to produce gas in the colon as a result of their rapid fermentation. Some of these oligosaccharides have been shown to have prebiotic properties (see section 2.3.5).

### 2.3.3 Dietary fibre (non-starch polysaccharides)

NSPs escape digestion in the small intestine and pass into the large bowel, where they either are fermented or go on to pass through the body undigested. The reason for their resistance to digestion is partly because of their physical form and partly their chemical bonds. They are almost entirely found in plant cell walls and the chemical bonds in these molecules are not susceptible to brush border or pancreatic digestive enzymes. For example, the bonds in cellulose are principally β-1,4 in contrast to α-1,4 in starch. This minor stereochemical difference is sufficient to prevent hydrolysis by pancreatic amylase. The amount of NSPs in the diet varies (10–36 g/day), with higher amounts being characteristic of vegetarian diets or populations with high vegetable and fruit intake. High NSP intake is not characteristic of diets in developing countries because rice and maize are low in NSPs and vegetables and fruit are in short supply.

### 2.3.4 Fermentation and the gut microbiome

Virtually all carbohydrate that enters the large bowel is fermented by the commensal bacteria of the colon, which are present at densities of up to $10^{12}$ cells/g contents. The advent of technology that allows the sequencing of 16S ribosomal RNA (16SrRNA) genes from gut contents has greatly expanded our knowledge of the gut microbiome in recent years. Many thousands of species are now known to exist in a community called the gut microbiome. Two major divisions dominate, the Bacteroidetes and Firmicutes. Uniquely in the body, their metabolism is primarily anaerobic, although the microbiome is very active metabolically, and study of its products has given rise to the science of metabolomics and study of the gut metabolome. The gut microbiota play a number of important roles in human health, including protecting against pathogens, development of the gut immune system, vitamin synthesis, and metabolism of xenobiotics, and they may be involved in complex gut–brain communication. Claims have also been made for certain characteristics of the microbiome to be important in the determination of energy metabolism and obesity.

However, the principal function of the gut microbiome is to complete the digestion of carbohydrate that escapes breakdown in the small intestine and it is the availability of these substrates that dominates the metabolism of the microbiome. Recovery of oligosaccharides in faeces is effectively nil, indicating complete breakdown in the large bowel, while 80–90% of RS and NSP are also fermented. Only very resistant retrograded starches survive partly, along with highly lignified cell wall material, such as that found in cereal bran. The extent of fermentation also depends to some extent on host factors such as colonic transit time and the composition of the microbiome. Microcrystalline cellulose may resist fermentation because of its highly condensed structure, an observation that led to the belief that cellulose was not digested in the human gut. This is because microcrystalline cellulose was used in many early experiments of cellulose digestion. However, cellulose naturally present in the cell walls of food is completely fermented unless it is in a highly lignified structure. Other polysaccharides of the plant cell wall are also readily fermented, even when given in purified forms such as pectin or guar gum. This process is facilitated by the ability of these latter substances to form gels readily accessible to the microbiota.

Microbial fermentation is an anaerobic process and produces quite different end-products from aerobic metabolism. As Fig. 2.5 shows, these include principally the short-chain fatty acids (SCFAs) acetate, propionate, and butyrate, which are the two-, three-, and four-carbon fatty acids of the same chemical series that includes C12–C22 fatty acids, which are the major lipids of the diet. The SCFAs are, however, much more water soluble and are rapidly absorbed. Butyrate is the major energy source for the colonic epithelial cell, in contrast to glutamine for the small intestine and glucose for most other tissues. Butyrate also has differentiating properties in the cell, arresting cell division through its ability to regulate gene expression. This property provides a credible link between the dietary intake of fermented carbohydrates, such as NSPs, and protection against colorectal cancer.

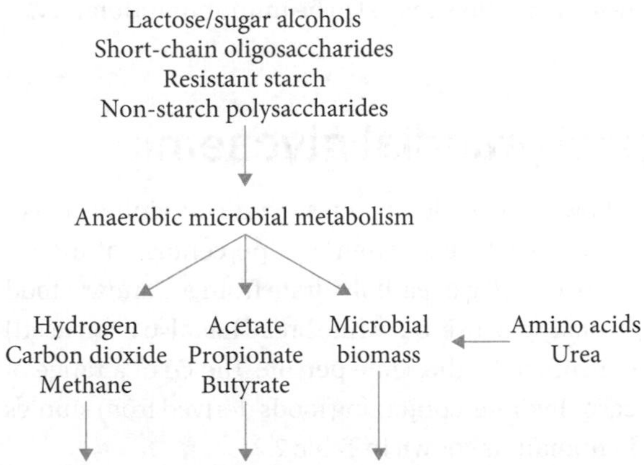

Fig. 2.5 Fermentation of carbohydrate in the large bowel.

Propionate is absorbed and passes to the liver, where it is taken up and metabolized aerobically. This molecule is not seen as having significant regulatory properties in humans, although it may moderate hepatic lipid metabolism. However, in ruminant animals, propionate is crucial to life because it is used to synthesize glucose in the liver.

Acetate is the major SCFA produced in all types of gut fermentation and the molar ratio of acetate to propionate to butyrate is around 60:20:20. Acetate is rapidly absorbed, stimulating sodium absorption, and passes to the liver and then into the blood, from where it is available as an energy source. Fasting blood acetate levels are about 50 μ/L, rising 8–12 hours later to 100–300 μmol/L after meals containing fermentable carbohydrate. Acetate is rapidly cleared from the blood with a half-life of only a few minutes and is metabolized principally by skeletal and cardiac muscle, and the brain. Acetate spares free fatty acid oxidation in humans and its absorption does not stimulate insulin release. Another precursor of blood acetate is alcohol.

Thus, fermentation is an integral part of digestion and provides energy, which is up to 70% of the available energy in equivalent monosaccharides. By convention, however, the energy value of oligosaccharides has been set as 2 kcal/g, and the same value could usefully be applied to RS and NSPs. The total energy provided from fermentation in the human is probably only 5% of total energy requirements, although it could be more, depending on diet.

Fermentation of amino acids, derived from protein, also occurs in the large bowel and, in addition to SCFAs, yields ammonia, amines, phenols, and sulphur compounds. Fermentation also gives rise to the gases hydrogen and carbon dioxide. Much of the hydrogen is converted to methane by bacteria, and both hydrogen and methane are excreted in breath and flatus. Gas production, especially if rapid, is one of the principal complaints of people unused to eating foods containing significant amounts of fermentable carbohydrate. Another product of fermentation is microbial biomass or microbial growth. These bacteria are excreted in faeces and this is one of the principal mechanisms of laxation by NSPs especially. Also produced from fermentation is lactate. This occurs usually during rapid fermentation of soluble carbohydrates such as oligosaccharides. Both D- and L-lactate are produced, and both are absorbed. Ethyl alcohol is also produced during fermentation, although it is more characteristic of fermentation by yeasts, as in brewing and wine making.

### 2.3.5 Prebiotics

Prebiotics are novel foods largely comprised of dietary fibres that are digested in the large intestine. Prebiotics were defined by an FAO Technical Meeting in 2007 as 'a non-viable food component that confers a health benefit on the host associated with modulation of the microbiota.' Prebiotics alter the balance of the gut microbiome to one with more bifidobacteria and lactobacilli. Bacteria from these groups are important in maintaining the gut barrier to infection, are almost entirely non-pathogenic, synthesize B vitamins, and are predominantly saccharolytic (i.e. break down carbohydrate). Apart from altering the balance of the flora, prebiotics are also fermented in the large bowel, producing short-chain fatty acids.

The best-established prebiotics are fructo-oligosaccharides, inulin, and galacto-oligosaccharides. Prebiotic oligosaccharides have been shown to reduce the risk of traveller's diarrhoea, and animal studies have demonstrated clear anti-inflammatory properties in the gut. Perhaps more surprisingly, prebiotic carbohydrates increase calcium absorption and bone mineral density in adolescents. It has been suggested that they are beneficial in the prevention of cardiovascular disease and cancer, and are immunomodulatory, but further research is needed in these areas.

The oligosaccharides of breast milk have long been credited with being the principal growth factor for bifidobacteria in the infant gut and thus primarily responsible for these bacteria dominating the microbiota of the gut of breastfed babies. In this context, milk oligosaccharides are acting as prebiotics. Bifidobacteria can grow on milk oligosaccharides as their sole carbon source, while lactobacilli may not be able to do so. The similarities between milk oligosaccharide structure and epithelial cell surface carbohydrates in the gut suggest that milk oligosaccharides may act as soluble receptors for gut pathogens and thus form an essential part of colonization resistance. They may also be immunomodulatory.

## 2.4 Carbohydrate foods and postprandial glycaemia

Plasma glucose levels rise 5–45 minutes after any meal that contains sugars or digestible starch, and usually return to fasting levels 2–3 hours later. This rise in blood glucose is known as the postprandial glycaemic response, and depends upon the rate and extent of digestion, absorption, and clearance from the plasma. The glycaemic index (GI) of a carbohydrate-containing food is a useful standardized method to measure postprandial glycaemia in controlled conditions. The GI is defined as the incremental area under the blood-glucose curve following a portion of a test food containing 50 g of carbohydrate, expressed as a percentage of the response to 50 g of carbohydrate from a standard food (usually glucose, but white bread has also been used) consumed by the same person. The GI of a range of carbohydrate-containing foods derived from studies in humans is shown in Table 2.2.

Postprandial glycaemia is influenced by a number of attributes of foods (see Table 2.3). As a general rule, foods that are rich in glucose-containing sugars and rapidly digested starches,

**Table 2.2 The average glycaemic index (GI) of 62 common carbohydrate-containing foods**

| | GI | | GI | | GI |
|---|---|---|---|---|---|
| **High-carbohydrate foods** | | **Breakfast cereals** | | **Vegetables** | |
| White breads | 70–95 | Cornflakes | 81 | Potato, boiled | 78 |
| Whole wheat/ wholemeal bread | 58–74 | Wheat flake biscuits | 69 | Potato, instant mash | 87 |
| Rye breads | 41–86 | Instant oat porridge | 79 | Carrots, boiled | 39 |
| Specialty grain bread | 40–76 | Porridge, rolled oats | 55 | Potato, French fries | 63 |
| Wheat roti | 62 | Rice porridge/congee | 78 | Sweet potato, boiled | 63 |
| Chapatti | 52 | Millet porridge | 67 | Pumpkin, boiled | 64 |
| Corn tortilla | 46 | Muesli | 57 | Plantain/green banana | 55 |
| White rice, boiled | 73 | | | Taro, boiled | 53 |
| Brown rice, boiled | 68 | **Fruit and fruit products** | | Vegetable soup | 48 |
| Barley | 28 | Apple, raw | 36 | | |
| Sweetcorn | 52 | Orange, raw | 43 | **Dairy products and alternatives** | |
| Spaghetti, white | 49 | Banana, raw | 51 | Milk, full fat | 39 |
| Spaghetti, wholemeal | 48 | Pineapple, raw | 59 | Milk, skim | 37 |
| Rice noodles | 53 | Mango, raw | 51 | Ice cream | 51 |
| Udon noodles | 55 | Watermelon, raw | 76 | Yoghurt, fruit | 41 |
| Couscous | 65 | Dates, raw | 42 | Soy milk | 34 |
| | | Peaches, canned | 43 | Rice milk | 86 |
| **Snack products** | | Strawberry jam/jelly | 49 | | |
| Chocolate | 40 | Apple juice | 41 | **Legumes** | |
| Popcorn | 65 | Orange juice | 50 | Chickpeas | 28 |
| Potato crisps | 56 | | | Kidney beans | 24 |
| Soft drink/soda | 59 | **Sugars** | | Lentils | 32 |
| Rice crackers/crisps | 87 | Fructose | 15 | Soya beans | 16 |
| | | Sucrose | 65 | | |
| | | Glucose | 103 | | |
| | | Honey | 61 | | |

*Adapted from:* Atkinson, F.S., Foster-Powell, K., and Brand-Miller, J.C. (2008) International tables of glycemic index and glycemic load values. *Diabetes Care*, **31**, 2281–3.

such as white bread or cornflakes, will have a relatively large glycaemic response because they are rapidly digested and absorbed. On the other hand, foods such as pulses that are rich in slowly digested and resistant starch, non-starch polysaccharides, and oligosaccharides will have a low GI. Foods containing fructose also have a relatively low GI because fructose is taken up by the liver and contributes very little to blood glucose. Carbohydrate-containing foods that are also high in fat and protein have a low GI because these other components delay gastric emptying.

GI is useful when comparing foods belonging to the same group, for example, different types of breads for which GI might range from relatively low (e.g. some rye breads) to around 100 in the case of

Table 2.3 **Food factors influencing glycaemic response**

| **Amount of carbohydrate** |
|---|
| *Nature of the monosaccharide components* |
| Glucose |
| Fructose |
| Galactose |
| *Nature of the starch* |
| Amylose |
| Amylopectin |
| Crystalline form A, B, or C |
| Starch–nutrient interaction |
| Resistant starch |
| *Physical form/cooking/food processing* |
| Degree of starch gelatinization/retrogradation |
| Particle size |
| Food form |
| Cellular structure |
| Fibre (affects physical form and viscosity) |
| *Other food components* |
| Fat and protein |

some white breads. Indices at the high or low extremes of the range are probably a true reflection of the glycaemic response and may therefore be helpful when making food choices. However, small differences in published values are probably of little relevance, since most GI values are based on the study of relatively few healthy individuals where there is considerable inter- and intra-individual variation in glycaemic response. Furthermore, some foods advertised as having a low GI may have a high content of sucrose, high fructose corn syrup, and/or fat, including saturated fat.

The postprandial glycaemic response to a food or meal is determined by the amount of carbohydrate consumed, the GI of each food, and the nature of the other components of the meal. A product or food may have a high GI, but if only a small quantity is consumed, the total amount of glucose available as an immediate energy source is limited. The concept of glycaemic load (GL) was introduced to quantify the overall glycaemic effect of a portion of food. GL is the product of the amount of available carbohydrate in a typical serving and the GI of the food divided by 100. The ranking of foods according to GL is in most instances similar to that for GI.

Along with the blood glucose response to carbohydrate-containing meals, there is an insulin response, which is important in the control of metabolism. It often reflects the glycaemic response, but can, for some foods, be more exaggerated, so the insulin response after apple juice is much greater than that after whole apples, while the glycaemic responses are similar. The presence of fat and protein in a meal increases insulin responses.

Dietary GI or GL is a metric that attempts to characterize an individual's diet in terms of a GI or GL value. This is quite different from the calculation of the postprandial glycaemic response following the intake of a pre-specified weight of carbohydrate-containing food in a controlled environment. Calculations are usually based on self-reported dietary intakes from food diaries or Food Frequency Questionnaires. Dietary GI and GL do not appear to be useful as health metrics, as higher or lower values have not been confirmed as determinants of health outcomes in the general population.

## 2.5 Carbohydrate metabolism

The concentration of glucose in the blood of adults is generally controlled within the range 4.0–5.5 mmol/L. However, when a carbohydrate-containing meal is ingested, the level may temporarily rise to as high as 7.5 mmol/L and fall to as low as 3.0–3.5 mmol/L during fasting in normoglycaemic individuals. When levels are about 10 mmol/L or greater, such as occurs in poorly controlled diabetes, or even at lower levels in some people, glucose is present in the urine (glycosuria). Close regulation

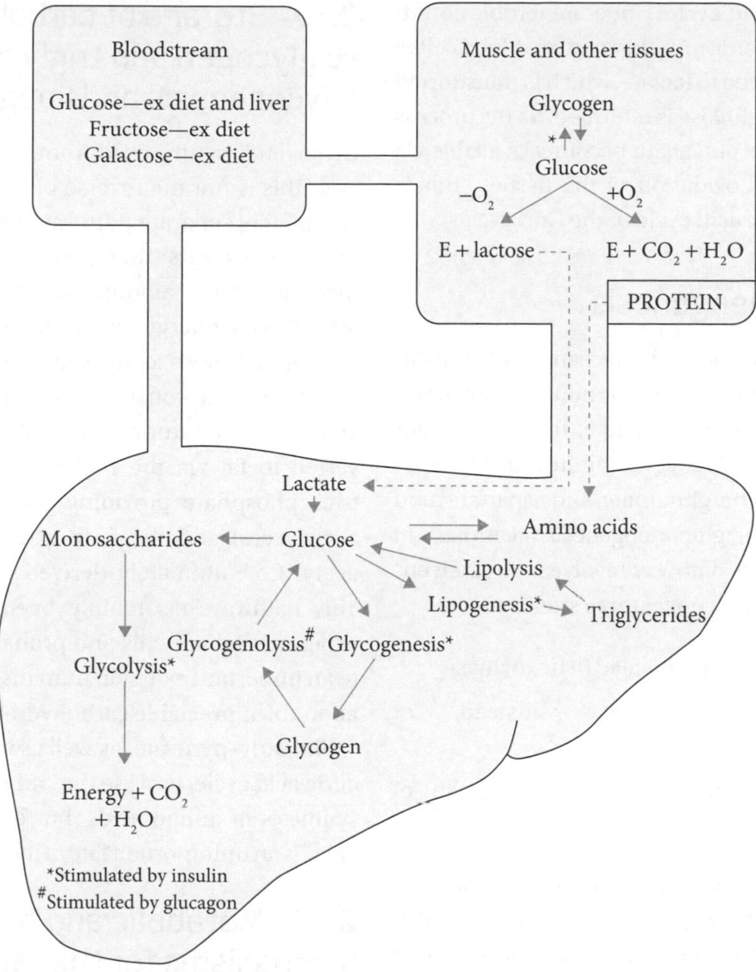


Fig. 2.6 Carbohydrate metabolism and its control.

of blood glucose is necessary, because the brain requires a continuous supply, although it can adapt to lower levels or use ketone bodies from fat breakdown if adaptation occurs slowly. Erythrocytes rely almost totally on glucose. Several metabolic pathways are involved in the utilization of glucose as an energy source and the maintenance of glucose homeostasis. The pathways are described in detail in textbooks of biochemistry and are summarized briefly in Fig. 2.6.

## 2.5.1 Glycolysis

Following absorption, glucose, fructose, and galactose are transported to the liver via the hepatic portal vein. Fructose and galactose are very rapidly converted to glucose. Fructose may also enter the glycolytic pathway directly. Given the rapid conversion, very low concentrations of both these sugars are detected in the blood immediately after ingestion.

Glucose is taken up from blood and its breakdown, glycolysis, occurs in the cytosol of all cells and may occur in the presence (aerobic) or absence (anaerobic) of oxygen. The pathway involves the metabolism of hexoses to pyruvate. Many reactions catalysed by different enzymes are involved, but glycolysis is regulated by three enzymes that catalyse non-equilibrium reactions—namely, hexokinase (glycokinase), phosphofructokinase, and pyruvate kinase. Under aerobic conditions, pyruvate is transported into the mitochondrion and is then oxidatively decarboxylated to acetyl co-enzyme A (CoA), which enters the citric acid cycle. Under anaerobic conditions (e.g. in exercising muscles or in red blood cells), pyruvate is converted to lactate, which is transported to the liver, where glucose is reformed via the process of gluconeogenesis and again becomes available via the circulation for oxidation in the tissues. This is known as the lactic acid cycle or the Cori cycle.

### 2.5.2 Gluconeogenesis

Gluconeogenesis includes all mechanisms and pathways responsible for converting non-carbohydrates to glucose. In addition to lactate, the other major substrates are the glucogenic amino acids (especially glycine, alanine, glutamate, and aspartate) and glycerol. In the liver, gluconeogenesis takes place in the cytosol, and this pathway involves the same enzymes as glycolysis except at three sites:

- glucose 6-phosphatase instead of hexokinase
- fructose 1,6-biphosphatase instead of phosphofructokinase
- phosphoenolyruvate carboxykinase instead of pyruvate kinase.

Triglycerides of adipose tissue are continuously undergoing hydrolysis to form free glycerol, which cannot be utilized by adipose tissue, and therefore diffuses out into the blood and reaches the liver. In the liver it is first converted to fructose 1,6-biphosphate, before being converted to glucose.

### 2.5.3 Glycogenolysis

Glycogen represents the principal storage form of carbohydrate in animals and is present mainly in the liver and muscle. In the liver, its major function is to service other tissues via formation of glucose, when dietary sugars are not immediately available as an energy source between meals or while fasting. In muscle it serves only the needs of that organ by providing an immediate source of metabolic fuel. Glycogenolysis is the pathway by which the glycogen stores are converted via glucose 1-phosphate and glucose 6-phosphate to glucose. After 12–18 hours of fasting, the liver becomes totally depleted of glycogen, whereas muscle glycogen is depleted only after physical activity.

### 2.5.4 Storage of carbohydrates as glycogen and triglyceride (glycogenesis and lipogenesis)

Glycogen is synthesized from glucose by glycogenesis. This is not the reverse of glycogenolysis, but a completely separate pathway that usually operates for several hours after a carbohydrate-containing meal, when the amount of ingested carbohydrate far exceeds energy requirements for the tissues. Glycogen stores become saturated at around 1000 g. If carbohydrate consumption continues after glycogen stores are saturated, carbohydrate may be converted to fat via the pathway of lipogenesis with triosephosphate providing the glycerol moiety of acylglycerol and the fatty acids synthesized from acetyl CoA ultimately derived from carbohydrate. This pathway has mainly been demonstrated in experimental animals and probably does not occur to an important extent in humans, other than in situations of appreciable carbohydrate overfeeding.

Similarly, pyruvate, as well as intermediates of the citric acid cycle, provide the carbon skeletons for the synthesis of amino acids, but conversion to amino acids is an unimportant fate of ingested carbohydrate.

### 2.5.5 Metabolic and hormonal mechanisms for the regulation of blood glucose levels

The maintenance of stable levels of glucose in the blood is one of the most carefully regulated homeostatic mechanisms in the body (see Fig. 2.6). Insulin, secreted by the β-cells of the islets of Langerhans in the pancreas, plays a central role in regulating blood glucose. About 40–50 units (15–20% of the total amount stored) are produced daily. Insulin secretion is stimulated by rising blood glucose levels, as well as by amino acids, free fatty acids, ketone bodies, glucagon, and secretin. Insulin lowers blood glucose by facilitating its uptake into insulin-sensitive tissues and the liver by enhancing the activity of glucose transporters. Insulin also stimulates the storage of glucose as glycogen (glycogenesis) and enhances the metabolism of glucose via the glycolytic pathway. The action of glucagon, secreted by the α-cells of the islets of Langerhans, opposes that of insulin.

Secretion occurs in response to hypoglycaemia (low blood glucose levels). Glucagon stimulates glycogenolysis by activating the enzyme phosphorylase and enhances gluconeogenesis. Hypoglycaemia also stimulates the secretion of adrenaline by the chromaffin cells of the adrenal medulla and acts by stimulating the phosphorylase and, hence, glycogenolysis. Thyroid hormones, glucocorticoids, and growth hormone have a lesser effect than insulin has on blood glucose, glucagon, and adrenaline in healthy individuals. Resistance to the effects of insulin occurs in diabetes and obesity.

# 2.6 Carbohydrates and gut disorders

## 2.6.1 Lactose malabsorption

The universal presence of lactose in milk means that all newborn mammalian species have the appropriate enzyme lactase (β-1,4-galactosidase) in the brush border to deal with this sugar. However, after weaning, lactase activity declines rapidly in all species, except some human populations. These are the traditional milk-drinking people who have their ancestors in the Middle East and North India. In practice this means the majority of Northern Europeans and populations deriving from them. Individuals who do not express lactase can tolerate small amounts of milk in their diet, but large amounts lead to unabsorbed lactose, which exerts an osmotic effect in the small intestine with fluid and sugar entering the large bowel. Here, partial fermentation occurs and there is often rapid gas production, causing abdominal pain and an osmotic diarrhoea. The use of yoghurts and other fermented milk products, as well as the use of preparations of the enzyme lactase, may improve lactose tolerance.

## 2.6.2 Other carbohydrate intolerances and inborn errors of metabolism

There are other less frequent clinical disorders in which digestion or absorption of sugars is disturbed, resulting in symptoms similar to those of lactase deficiency. Most frequently they are seen in children secondary to underlying gastrointestinal disease, especially severe gastrointestinal infections in undernourished children. They may also be congenital and, although rare, such conditions may be life-threatening. Three examples of these disorders are: sucrase–isomaltase deficiency, which is associated with watery diarrhoea following consumption of sucrose-containing foods; alactasia, or the total absence of lactase in infancy, which is accompanied by diarrhoea associated with milk (note the difference between this condition and lactose intolerance); and glucose–galactose malabsorption (diarrhoea after eating glucose, galactose, or lactose). The diagnosis is usually suspected on the basis of clinical observations, and is confirmed by sugar tolerance tests and measurement of breath hydrogen. Several different inborn errors of galactose and fructose metabolism are associated with faltering growth and a range of serious clinical outcomes in infants if the conditions are not managed. Such infants are likely to thrive if the relevant sugar (lactose, fructose) is withdrawn from the diet.

## 2.6.3 Large bowel function and its disorders

Bowel habit is determined largely by the amount of carbohydrate that enters the colon and by transit time. Fibre, essentially the NSPs, is the major controller of faecal bulk, with other carbohydrates, such as RS and oligosaccharides, mildly laxative. NSPs from different sources have been shown to have contrasting effects on stool weight, with NSPs from wheat (particularly wheat bran) being the most potent laxative, increasing stool output by around 5 or 6 g/day per gram of NSPs consumed from this source. Following close on the heels of wheat sources are NSPs from vegetables and fruit (4 g/g of NSPs fed), after which come the gums, oats, and legumes, and probably the least laxative of the NSP sources, pectin. The effect of NSPs on bowel habit is modified principally by gut transit time, which is an innate control of large bowel function.

The mechanism whereby fermentable carbohydrates affect bowel habit is now well established.

The notion that fibre acts like an inert sponge in the colon has now been superseded because we know that NSPs are extensively metabolized by colonic bacteria. Faecal weight, therefore, increases through a variety of mechanisms that include increased bacterial biomass (Fig. 2.5), increased hydration of stool mass associated with more rapid transit time, and the presence of unmetabolized cell wall material, particularly lignified forms such as that present in bran.

It follows, because NSPs are such effective laxatives, that they have been used very widely in the management of constipation. There are many causes of constipation, but the commonest one is low NSP-containing diets, often associated with a sedentary lifestyle, travel, and, sometimes, therapeutic diets, such as those used for weight reduction. Other physiological causes of low stool weight are pregnancy, some phases of the menstrual cycle, and old age, where a combination of low food intake and lack of physical activity are probably most important. Irritable bowel syndrome can include constipation and many drugs have constipation as a side effect. Diets high in NSPs are well established in the management of constipation. There is no universal prescription, but the amount required is one that will produce a satisfactory bowel habit. The aim should be to increase the patient's NSP intake to 18–24 g/day. Those who fail to respond to this sort of dietary change should be carefully screened for more serious causes of the problem. Easy ways of increasing NSP intake include:

- increasing bread intake to 200 g/day and changing to 100% whole grain
- eating a whole grain breakfast cereal
- increasing vegetable and fruit intake to 400 g/day
- eating more legumes, such as beans and peas, although this may produce problems with gas due to oligosaccharide fermentation
- using bulk laxatives such as ispagula and sterculia.

Irritable bowel syndrome (IBS) is one of the most common disorders seen in the gastroenterology clinic and has two main presenting features, namely abdominal pain and altered bowel habit. It is, however, a very diverse syndrome with no clear aetiology. While the cause is unknown, NSPs help in the management of constipation-predominant IBS. However, wheat bran is not universally beneficial in this condition, possibly because it is thought that a significant number of IBS patients are wheat-intolerant without having the diagnostic features of coeliac disease. Furthermore, changing people onto significantly increased NSP intakes leads to excess gas production and IBS patients may have a gut that is unusually sensitive to gas.

Colonic diverticular disease is another condition that benefits from carbohydrate in the diet, particularly NSPs. A diverticulum is a pouch that protrudes outwards from the wall of the bowel and is associated with hypertrophy of the muscle layers of the large intestine, particularly the sigmoid colon. Diverticular disease is very common in industrialized societies, the prevalence rising with age to about 30% of people over the age of 65. Many people with diverticula do not have symptoms, but those that do complain of lower abdominal pain and changes in bowel habit. High-NSP-containing diets were introduced in the 1960s and their use revolutionized the management of this condition. Wheat bran is thought to be more effective than other sources of NSPs or bulk laxatives, although bran is not a panacea and may aggravate gas production, feelings of abdominal distension, and incomplete emptying of the rectum.

## 2.7 Carbohydrate intake and non-communicable disease

The effects of carbohydrate in relation to human health are often considered in relation to the total amount consumed or the proportion of total energy provided by carbohydrate. This has led to risks and benefits being attributed to high and low intake. Epidemiological evidence from prospective

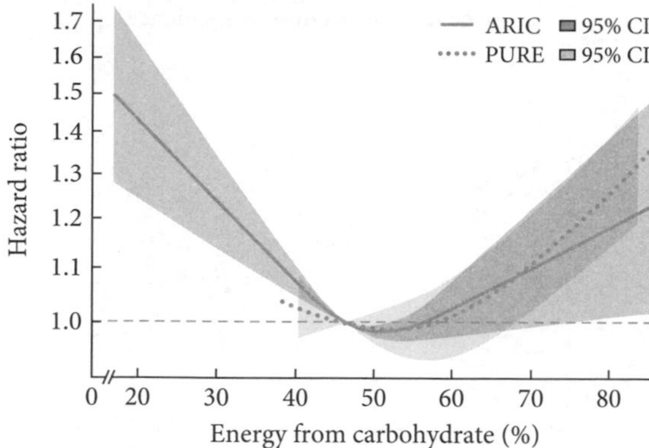


Fig. 2.7 Carbohydrate intake and mortality.

*Source:* Seidelmann, S.B., Claggett, B., Cheng, S., et al. (2018) Carbohydrate intake and mortality: a prospective cohort study and meta-analysis. *Lancet Public Health*, **3**: e419–28.

observational studies indicates both low and high carbohydrate intakes are associated with higher mortality, as shown in Fig. 2.7. The least risk was observed with intakes between 40% and 70% total energy. Considering only total carbohydrate does not take into account the very different health effects of sugars, starch, and dietary fibre. Meta-analyses of fibre intakes have shown appreciable health benefits associated with higher intakes, regardless of the other carbohydrate components of the diet. Higher intakes of fibre are associated with a 15% (9–21%) reduction in premature mortality, with a dose-response relationship evident. Dose-response relationships between dietary fibre and important health outcomes are shown in Fig. 2.8. Similar risk reductions have been observed with high-fibre foods, such as wholegrains, vegetables, fruits, and legumes. Sugars, on the other hand, have been convincingly linked to dental caries and obesity and as a consequence the comorbidities associated with obesity. In addition there may be an independent effect on cardiometabolic risk. Rapidly digested starches may have effects similar to those of sugars.

## 2.7.1 Obesity

The avoidance of obesity depends upon maintaining energy balance (see Chapters 5 and 17). However, the nature of dietary carbohydrate appears to influence energy balance. Diets rich in dietary fibre tend to promote satiety, and reduce the risks of excessive weight gain. Furthermore, randomized controlled trials (RCTs) have shown the potential of such diets or those high in whole grains to promote weight loss in those who have already become overweight or obese. Conversely, a high intake of energy-dense foods, whether they be high in fats or free sugars or both, increases the risk of excessive fat accumulation. There is particular concern around the potential of a high intake of free sugars in beverages and fruit juices to contribute to the increased risk of obesity in childhood. The physiological effects of energy intake on satiation and satiety appear to differ between solid foods and fluids. High-energy beverages (invariably high in free sugars) may be less 'sensed' than comparable energy intakes from solid foods, perhaps because of reduced gastric distension and more rapid transit times. As a result, there may be a failure to adjust food intake to take into account the energy derived from beverages. The potential for free sugars to promote overweight and obesity has been a major justification for the recent more restrictive recommendations regarding intake. Intake of sugar-sweetened beverages is especially discouraged. The history of sugar and its potential effects on human health is summarized in the case study (Box 2.3). There has also been recent interest in the potential of diets low or very low in total carbohydrate to facilitate weight loss to a greater extent than can be achieved on moderate to high carbohydrate diets. However, RCTs that have continued for a year or longer suggest that, in the medium to long term, distribution of macronutrients has little influence on the magnitude of weight reduction. Furthermore, many of the trials that have suggested benefits of appreciable carbohydrate restriction have not taken into account the relative proportions of total carbohydrate provided by sugars, starch, and dietary fibre.

## 2.7.2 Lipids, the metabolic syndrome, and cardiovascular disease

For many adults, increasing the consumption of appropriate carbohydrate-containing foods such as whole grains, vegetables, fruit, and legumes will

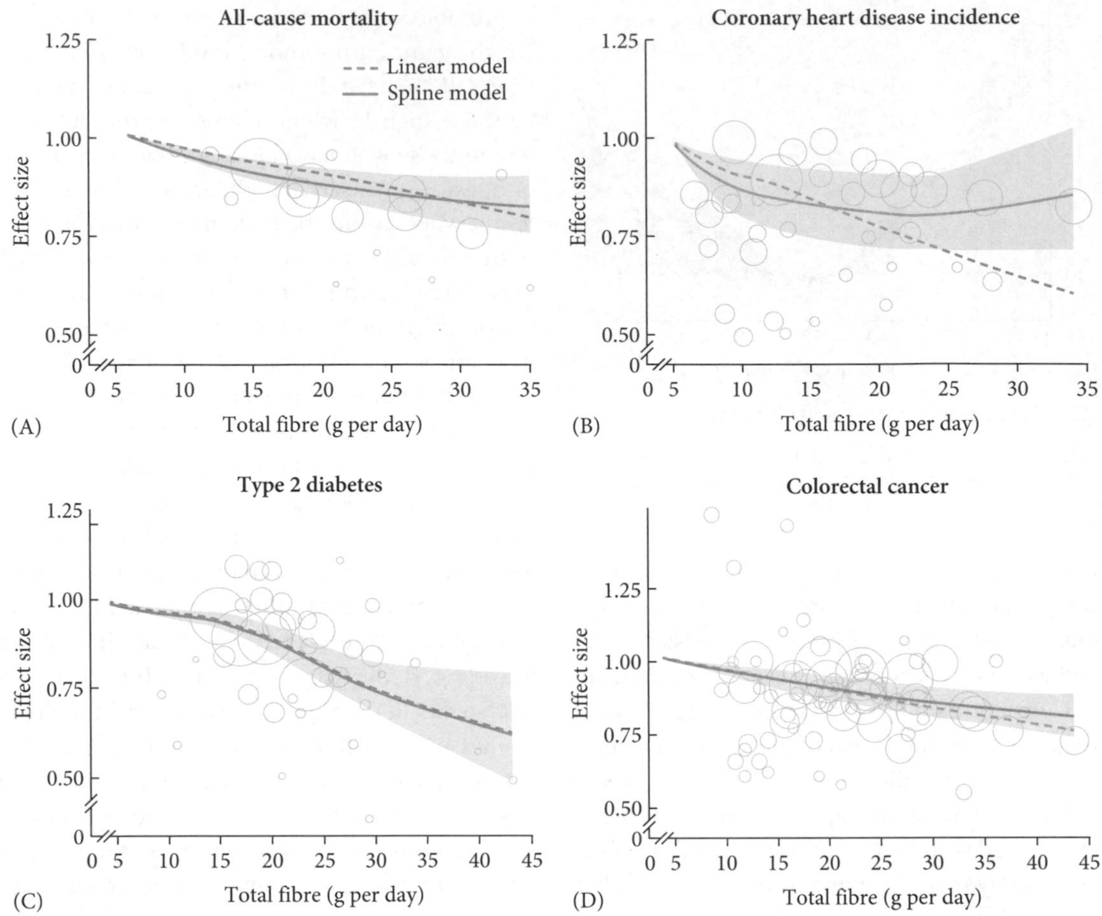


**Fig. 2.8** Dose-response relationships between dietary fibre and health outcomes. (A) Total fibre and all-cause mortality. 68,183 deaths over 11.3 million person-years. Assuming linearity, a risk ratio of 0.93 (95% CI 0.90–0.95) was observed for every 8 g more fibre consumed per day. (B) Total fibre and incidence of coronary heart disease. 6449 deaths over 2.5 million person-years. Assuming linearity, a risk ratio of 0.81 (0.73–0.90) was observed for every 8 g more fibre consumed per day. (C) Total fibre and incidence of type 2 diabetes. 22,450 cases over 3.2 million person-years. Assuming linearity, a risk ratio of 0.85 (0.82–0.89) was observed for every 8 g more fibre consumed per day. (D) Total fibre and incidence of colorectal cancer. 20,009 cases over 20.9 million person-years. Assuming linearity, a risk ratio of 0.92 (0.89–0.95) was observed for every 8 g more fibre consumed per day.

*Source:* Reynolds, A., Mann, J., Cummings, J., et al. (2019) Carbohydrate quality and human health: a series of systematic reviews and meta analyses. *The Lancet*, **393**(10170), 434–445.

facilitate the reduction in saturated fatty acids that are causally linked to cardiovascular disease and increase the intake of antioxidants and other cardioprotective nutrients (see Chapter 19). High intakes of whole grains and dietary fibre (see Fig. 2.8) have been shown in prospective studies to be associated with reduced cardiovascular risk. However, high intakes of free sugars and rapidly digested starches may be associated with high insulin levels, high triglycerides and very-low-density lipoproteins, and reduced high-density lipoproteins. Large amounts of fructose or sucrose have especially been implicated as a cause of hypertriglyceridaemia and other features of the metabolic syndrome, including hyperuricaemia and gout. However, a high carbohydrate intake when derived from vegetables, intact fruits, and whole grains does not appear to have long-term adverse effects on the overall lipoprotein profile or other indicators of cardiometabolic risk.

### 2.7.3 Blood glucose and diabetes

Foods high in fibre, whole grains, and those with a low glycaemic response reduce the risk of developing type 2 diabetes. Diets high in sucrose may be associated with increased insulin resistance and, by contributing to the energy density of the diet, lead to the development of being overweight and obese, the principal risk determinant of type 2 diabetes. Foods with a high content of dietary fibre are associated with improvements in glycaemic control in both type 1 and type 2 diabetes, as well as improvement in several other cardiovascular risk factors. The majority of patients with type 2 diabetes are overweight and restriction of free sugars, along with other energy-dense foods, will facilitate weight loss. This is a principal goal in the treatment of this condition, since it is almost invariably associated with improved control of blood glucose, and other clinical and metabolic abnormalities associated with diabetes. There is no convincing long-term evidence for the benefit of diets low or very low in carbohydrate in the management of people who have already developed diabetes (see Chapter 21).

### 2.7.4 Cancer

Carbohydrates are not generally implicated in either the cause or prevention of the major cancers, with the exception of colorectal cancer. In 1969, Burkitt pointed out that those countries where colorectal cancer risk was low, principally African countries, had high intakes of fibre and large stool bulk. He suggested that lack of fibre was the cause of large bowel cancer because it allowed slow transit time through the colon and permitted cancer-forming chemicals to accumulate in the large bowel. We now know that there are other important contributors to colorectal cancer risk, in particular lack of physical activity, obesity, and high intake of red and processed meat, but nevertheless dietary fibre remains an important protective element of the diet. Meta-analyses of epidemiological studies have shown a protective effect of fibre on colorectal cancer incidence, both in terms of higher versus lower intakes and in dose-response analyses (Fig. 2.8). This is further discussed in Chapter 20.

### 2.7.5 Dental caries

Dental diseases are a costly burden to healthcare services, accounting for between 5% and 10% of total healthcare expenditure worldwide. Dental caries rates appear to be declining in some relatively affluent countries, but prevalence is increasing in many developing countries. Dental caries has a complex aetiology. The condition occurs because of demineralization of enamel and dentine by organic acids formed by bacteria in dental plaque through the anaerobic metabolism of sugars derived from the diet.

Free sugars and readily digestible starches are involved, although lactose, sugar alcohols, oligosaccharides, and NSPs are less acidogenic than other carbohydrates. Dental caries occurs infrequently when the consumption of free sugars is below 15–20 kg per person per year, equivalent to a daily intake of 40–55 g per person or 6–8% of total energy intake. The role of free sugars in dental caries is a major justification for the recommendation to radically reduce intake. Frequency of consumption is particularly relevant, hence the recommendation that foods and drinks containing free sugars should be limited to a maximum of four times per day. Fluoride protects against dental caries, reducing rates by 20–40%, but it does not eliminate the condition.

## 2.8 Energy values and recommended intakes of carbohydrates

### 2.8.1 Energy value of carbohydrates

Dietary carbohydrate has traditionally been assigned an energy value of 4 kcal/g (17 kJ/g), though when carbohydrates are expressed as monosaccharides, the value of 3.75 kcal/g (15.7 kJ/g) is usually used. However, because NSP, RS, and some oligosaccharides are not digested in the small intestine and the process of fermentation by the gut microbiome diverts some carbohydrate energy into microbial growth, these carbohydrates provide the body with less energy. This energy comes from SCFAs, the main product of fermentation, which are rapidly absorbed from the colon. The energy value of all carbohydrate requires reassessment, but until this has been carried out, the FAO/WHO consultation recommends that the energy value of carbohydrates that reach the colon be set at 2 kcal/g (8 kJ/g).

### 2.8.2 Recommended intakes of carbohydrates

The minimum amount of carbohydrate required to avoid ketosis is considered to be about 50 g/day. Glucose is an essential energy source for the brain, red blood cells, and the renal medulla, the daily requirement being about 180 g/day. Approximately 130 g/day can be produced in the body from non-carbohydrate sources by gluconeogenesis, hence, the amount of 50 g/day recommended intake. If this is not provided, these organs can adapt by utilizing ketones derived from fatty acid oxidation as a source of energy. A state of ketosis is undesirable because cognitive function may be impaired and, in pregnant women, the foetus may be adversely affected. During pregnancy and lactation, the minimum carbohydrate required should probably be about 100 g/day.

Most people consume appreciably more than 100 g carbohydrate daily, with intakes ranging from 200 to 400 g/day. Dietary reference values do not always include recommendations for total carbohydrate intake, but assume that carbohydrate provides the energy deficit after specifying intake ranges for protein and fat. However, the FAO/WHO Expert Consultation on carbohydrates (1998) suggested that at least 55% of total energy should be derived from carbohydrate obtained from a variety of food sources. A wide range, up to 75% of total energy, was regarded as acceptable. A significant adverse effect on health is possible with higher levels of intake because adequate amounts of protein, fat, and other essential nutrients may be excluded. A comparable range was suggested by the WHO/FAO Expert Consultation on *Diet, nutrition and the prevention of chronic diseases* (2003). The FAO/WHO Scientific Update on Carbohydrates (Nishida et al., 2007) recommended a reduction of the lower limit of the acceptable range to 50%. The most recent international dietary reference values from the European Food Safety Authority (EFSA, 2019) and the 2018 Nordic Nutrition Recommendations suggest that for both adults and children, intake of total carbohydrate should range from 45% to 60% total energy. The UK Scientific Advisory Committee on Nutrition (SACN) recommends a population average of approximately 50% total energy. The range derives from the recommended ranges for other macronutrients and compatibility with dietary practices associated with favourable health outcomes. These more recent recommendations are more in keeping with current dietary practices, since in most Western countries, intake of total carbohydrate is typically below 50% of food energy. Some recommendations have suggested 'lower', 'low', or 'very low' intakes of total carbohydrate to facilitate the energy deficit required for weight loss or maintenance of weight loss. Such advice is largely based on benefits observed in short-term studies, many of which have not taken type of dietary carbohydrate into account. There is also no evidence relating to long-term safety of very low carbohydrate ketogenic diets. Intakes of total carbohydrate below 45% total energy are acceptable provided recommendations for dietary fibre and other macronutrients are met. Definitive guidelines regarding appreciable reductions of total carbohydrate require further evidence.

In 2015, the WHO strongly endorsed the earlier recommendation to reduce individual intake of free sugars to below 10% total energy and made a conditional recommendation suggesting that additional benefit is likely to accrue from further reduction to 5% or less total energy. SACN has recommended a population intake of free sugars that should not exceed 5% total energy. Draft recommendations issued by the European Food Safety Authority in 2021 suggest that intakes of free or added sugars should be as low as possible to reduce risk of dental caries and non-communicable diseases. The suggestion is based on the linear relationship between free or added sugars and risk of disease with no clear cut-off. Recommendations from different countries are inconsistent and use several approaches to defining sugars. Food-based dietary guidelines, such as those developed in South Africa (Box 2.2), may be more useful. The role of sugar in human health during the past 50–60 years is summarized in the case study (see Box 2.3).

**Box 2.2** Example of dietary guidelines for sugars intakes in South Africa (and some other countries)

**Dietary guidelines for sugars intakes**

1 Use food and drinks containing sugar sparingly and not between meals.
2 Use little or no sugar.
3 Enjoy foods that contain sugar as a treat on special occasions.
4 Try not to use sugar and sugary drinks more than four times a day, and only during mealtimes.
5 Brush teeth twice a day with fluoride toothpaste.
6 Rinse the mouth after eating or drinking sweet things.

**Box 2.3** Case study

Sugar cane was first cultivated in Papua New Guinea some 10,000 years ago and sugar beet has been grown in cooler climates for the past 250 years. Production has gradually increased so that, in many countries, sugar is a major energy source, providing as much, if not more, energy than starchy vegetables and fruit. A desire for sweet foods, and their generally widespread availability, account for this trend. The term 'sugar' is mainly used to refer to sucrose extracted from sugar cane or beet, but it is found in vegetables, fruit, and berries, along with small amounts of free glucose and fructose, the constituent monosaccharides of sucrose. These naturally occurring sugars, which are incorporated within the cell walls of plants (sometimes referred to as intrinsic sugars) and milk sugar (lactose) are not of concern when considering potentially adverse health-related issues. However, corn syrup and high-fructose corn syrup are increasingly used in confectionary and manufactured foods in some countries, most notably the USA. They may contribute a substantial proportion of total dietary sugar and, therefore, warrant consideration when examining health-related issues.

Sugar has long been regarded as a potential cause of diabetes, not surprisingly, given the body's inability to adequately handle ingested sugar and the loss of sugar in the urine; indeed, the condition was frequently referred to in lay terms as 'sugar diabetes'. In the 1950s, fairly convincing evidence emerged from Sweden suggesting a strong association with dental caries that was later confirmed to be related particularly to frequency of consumption. In the 1960s and early 1970s, many research publications (including a number in major international journals) appeared suggesting that sugar not only was a strong determinant of diabetes risk but also was related to several cardiovascular risk factors (notably cholesterol, triglycerides, and blood pressure) and also clinical cardiovascular disease. The claims were based upon cross-sectional epidemiological data, trends over

*(Continued)*

**Box 2.3 Case study (*Continued*)**

time, and experimental studies in which the effect of feeding very large quantities of sucrose on risk factors were examined.

In contrast to these observations, more carefully controlled studies in the 1970s suggested that the epidemiological data had been misinterpreted. Studies involving isocaloric comparisons of more physiological intakes of sucrose with starches did not generally have an adverse effect on risk factors. Further reassurance regarding potential adverse health consequences of sucrose followed, with a series of studies reporting no deterioration in glycaemic (blood glucose) control or lipid levels in people with diabetes when modest amounts of sucrose were incorporated into their diets. It is noteworthy, although not always appreciated, that the sucrose in the experimental diets replaced relatively rapidly digested starches, that the amounts were modest, and that in all other respects the diets complied with the dietary guidelines for diabetes at the time (i.e. they were high in fibre-rich vegetables, fruits, and whole grains, and low in fat, especially saturated fat). Although, in reality, the observations are only reassuring in the context in which the research was undertaken, this series of observations tended to allay concerns regarding adverse consequences of sugar, except with regard to the link between amount and frequency of consumption and dental caries. Two exceptions to the otherwise reassuring findings did not appear to cause undue concern. Some individuals with hypertriglyceridaemia appeared to be sensitive to even fairly modest changes in sucrose intakes. They were considered to have 'sucrose-induced hypertriglyceridaemia', but the condition was believed to be relatively rare. Furthermore, free-living individuals who were asked to replace their usual dietary sucrose intake with carbohydrate foods rich in starch tended to lose weight and their triglyceride levels fell as they were unable to fully replace calories from this energy-dense source.

During the past decade the tide has once again started to turn. Obesity has reached epidemic proportions in many countries and the metabolic syndrome (of which hypertriglyceridaemia is frequently a feature) has been reported in around one-quarter of the adult population in some societies. While the precise role of sugars as a cause of obesity and the metabolic syndrome, and in their management, remains to be established, there is no doubt that excessive consumption of energy-dense foods is associated with an increased risk of obesity and that sugar, corn syrup, and high-fructose corn syrup contribute to the energy density of confectionary products, snacks, and convenience foods. There is consistent evidence that a high intake of sugar-sweetened beverages increases the risk of excessive weight gain in children, and there is emerging evidence that sugar may be associated with weight gain and some adverse metabolic effects in those with the metabolic syndrome, especially the potential to increase triglycerides and uric acid. In the light of these observations and the increased risk of dental caries, especially associated with frequent consumption of sucrose, many sets of dietary guidelines include advice to limit the intake of sugar and corn syrups. The WHO has firmly recommended that intake of free sugars by individuals should be no greater than 10% total energy and that additional benefit may accrue from further reduction to no more than 5% total energy. The UK Scientific Advisory Committee on Nutrition has recommended a population average intake of no more than 5% total energy.

The role of dietary sucrose has once again become a focus of intense interest and research.

Recommendations regarding dietary fibre also tend to vary because of lack of agreement as to what should be included as dietary fibre and how to measure it. The WHO/FAO Expert Consultation (2003) and EFSA dietary reference values recommend at least 25 g/day dietary fibre (20 g NSP). SACN (2015) has recommended an average population intake of 30 g/day dietary fibre. The assumption is that the dietary fibre will be derived from vegetables, fruits, and whole grains, rather than supplements.

### 2.8.3 Carbohydrate through the life cycle

During the first 6 months of life, exclusive breastfeeding is recommended by the WHO, with lactose being the major source of carbohydrate and accounting for about 40% of milk energy. The concentration of lactose is tailored to the needs of the maturing neonatal and infant gut, especially while colonic microbiome and pancreatic amylase production are developing. For infants fed on formulas, the carbohydrate content should be as similar as possible to breast milk. Significant spill-over of carbohydrate into the neonatal colon occurs where, in breastfed infants, a largely saccharolytic flora dominated by bifidobacteria are responsible for fermentation and salvage of energy through SCFAs and lactate absorption.

Preschool children can be fed the same foods as adults, but because of their rapid growth, they require a more energy-dense diet. Food should, therefore, provide more than 30% of energy from fat and less than 50% from carbohydrate. Thus, the change from breast milk, which provides more than 50% of the energy from fat, to an adult-type diet should be gradual. The preschool years are a particularly important time to teach the avoidance of sugary drinks and puddings, because of the known association with the onset of obesity.

During pregnancy and lactation, the minimum requirement for carbohydrate is doubled to around 100 g/day, but most women achieve adequate intakes by consuming a wide variety of carbohydrate-containing foods. Particular attention to carbohydrate intake may also be appropriate for some groups of older adults, as with children, to ensure adequate intake of total energy and some essential nutrients.

### 2.8.4 Carbohydrates and physical performance

Adults and children undertaking regular physical activity will have greater energy requirements than those who are sedentary. A high-carbohydrate diet in the day preceding endurance-type physical activity ('carbohydrate loading') enhances physical performance, as may a high carbohydrate pre-event meal and carbohydrate supplementation in the form of beverages containing free sugars (see Chapter 40). This is presumably achieved as a result of glycogenesis and accumulation of maximum glycogen stores. Similarly, carbohydrate intake after an event can aid recovery by replenishing depleted glycogen stores. Low-intensity recreational physical activity or bouts of activity less than 90 minutes' duration does not require carbohydrate supplementation. Some people may consume unnecessary extra energy in this way.

## 2.9 Future directions

While there is now *greater* clarity relating to the different effects on human health of the three major groups of dietary carbohydrate (sugars, starch, and dietary fibre), debate continues regarding the optimal range of intakes of total carbohydrate. Future research that attempts to resolve this issue must take into account the nature of dietary carbohydrate, potential adverse effects of very low carbohydrate ketogenic diets, as well as the potential adverse environmental and health consequences of radical reductions of carbohydrate, which would inevitably be associated with increases in protein and fat.

Further research is needed on interactions between the microbiome and the host, and the potential for dietary change to moderate these interactions. Methodologies for the measurement of various dietary carbohydrates are likely to be refined and, given the advances in the field of molecular biology, it seems highly likely that there will be further insights into the understanding of the molecular basis of many of the issues discussed in this chapter.

### Further reading

1. **Åberg, Mann, Neumann, et al.** (2020) Whole-grain processing and glycemic control in type 2 diabetes: a randomized crossover trial. *Diabetes Care*, **43**(8), 1717–3.
2. **Atkinson, F.S., Foster-Powell, K., and Brand-Miller, J.C.** (2008) International tables of glycemic index and glycemic load values. *Diabetes Care*, **31**, 2281–3.

3. **Churuangsuk, Hall, Reynolds, et al.** (2022) Diets for weight management in adults with type 2 diabetes: an umbrella review of published meta-analyses and systematic review of trials of diets for diabetes remission. *Diabetologia*, **65**, 14–36.

4. **CODEX Alimentarius Meetings** (2011) CCMAS32 Report. Available at: https://www.fao.org/fao-who-codexalimentarius/meetings/archives/en/?y=2011&mf=07 (accessed 10 March 2023).

5. **Cummings, J.H., Beatty, E.R., Kingman, S.M., Bingham, S.A., and Englyst, H.N.** (1996) Digestion and physiological properties of resistant starch in the human large bowel. *Br J Nutr*, **75**, 733–47.

6. **Englyst, H.N., Kingman, S.M., and Cummings, J.H.** (1992) Classification and measurement of nutritionally important starch fractions. *Eur J Clin Nutr*, **46**, S33–50.

7. **FAO/WHO** (1998) *Carbohydrates in human nutrition.* Report of a Joint FAO/WHO Expert Consultation, Paper 66. Rome: Food and Agricultural Organization, World Health Organization.

8. **Gilbert, J.A., Blaser, M.J., Caporaso, J.G., et al.** (2018) Current understanding of the human microbiome. *Nature Medicine*, **24**, 392–400.

9. **Mann, J.I., and Cummings, J.H.** (2009) Possible implications for health of the different definitions of dietary fibre. *Nutr Metab Cardiovasc Dis*, **19**, 226–9.

10. **Nishida, C., Nocito, F.M., and Mann, J.** (2007) Joint FAO/WHO Scientific Update on Carbohydrates in Human Nutrition. *Eur J Clin Nutr*, **61**(Suppl. 1), S1–137.

11. **Public Health England** (2021) *Composition of foods integrated dataset (CoFID).* Available at: https://www.gov.uk/government/publications/composition-of-foods-integrated-dataset-cofid (accessed 10 March 2023).

12. **Reynolds, Akerman, Kumar, et al.** (2022) Dietary fibre in hypertension and cardiovascular disease management: systematic review and meta-analyses. *BMC Medicine*, **20**, 139.

13. **Reynolds, Akerman, Mann** (2020) Dietary fibre and whole grains in diabetes management: systematic review and meta-analyses. *PLoS Medicine*, https://doi.org/10.1371/journal.pmed.1003053

14. **Reynolds, A., Mann, J., Cummings, J., et al** (2019) Carbohydrate quality and human health: a series of systematic reviews and meta analyses. *The Lancet*. **393**(10,170), 434–45.

15. **Seidelmann, S.B., Claggett, B., Cheng, S., et al.** (2018) Carbohydrate intake and mortality: a prospective cohort study and meta-analysis. *Lancet Public Health*, **3**, e419–28.

16. **Te Morenga, L., Mallard, S., and Mann, J.** (2013) Dietary sugars and body weight: Systematic review and meta-analyses of randomised controlled trials and cohort studies. *BMJ*, **346**, e7492. doi: 10.1136/bmj.e7492

17. **The Diabetes and Nutrition Study Group of the European Association for the Study of Diabetes (DNSG)** (2023) Evidence-based European recommendations for the dietary management of diabetes. *Diabetologia*, **66**, 965–85.

18. **WHO/FAO** (2003) *Diet, nutrition and the prevention of chronic diseases.* Report of a Joint WHO/FAO Expert Consultation. Geneva: World Health Organization.

19. **WHO** (2015) *Sugars intake for adults and children.* Geneva: World Health Organization.

# Lipids

C. Murray Skeaff, Jim Mann, and Leanne Hodson



Lipids are a group of compounds that dissolve in organic solvents such as petrol or chloroform but are usually insoluble in water. The most obvious lipids in food and nutrition are edible oils, which are liquid at room temperature, and fats, which are solid at room temperature. Many people in high-income countries regard fats and oils as foods that should be avoided as far as possible because of their perceived role in the development of obesity, coronary heart disease, and diabetes. However, in addition to enhancing the flavour and palatability of food, lipids make an important contribution to adequate nutrition—some of them important to reduce the risk of heart disease. They are major sources of energy; some are essential nutrients because they cannot be synthesized in the body, yet they are required for a range of metabolic and physiological processes and to maintain the structural and functional integrity of all cell membranes. Lipids are also the only form in which the body can store energy for a prolonged period of time. These stored lipids in adipose tissue also serve to provide insulation, help to control body temperature, and afford some physical protection to internal organs. Lipids include the fat-soluble vitamins. Triacylglycerols, also referred to as triglycerides, make up the bulk of dietary (and stored) lipid, with phospholipids and sterols making up nearly all the remainder.

## 3.1 Naturally occurring dietary lipids

Naturally occurring dietary lipids are derived from a wide variety of animal and plant sources, including animal adipose tissue (the visible fat on meat, lard, and suet); milk and products derived from milk fat (cream, butter, cheese, and yoghurt); vegetable seeds, nuts, oils, and products derived from them (e.g. margarines); eggs; plant leaves; and fish oil. Many sources of dietary lipid are visible and obvious, while others are less so, for example those found in the muscle of lean meat, avocado, nuts, and seeds, as well as those in processed or home-prepared foods such as pies, cakes, biscuits, and chocolates. In most Western countries, dietary lipid provides 30–40% of total dietary energy. In Asian countries and throughout the economically developing world, the proportion of energy derived from dietary lipids is usually much lower.

### 3.1.1 Glycerides and fatty acids

Triacylglycerols make up about 95% of dietary lipids. A triacylglycerol molecule is formed from a molecule of glycerol (a three-carbon alcohol) with three fatty acids attached (Fig. 3.1). Fatty acids consist of an even-numbered chain of carbon atoms with hydrogens attached, a methyl group at one end, and a carboxylic acid group at the other (Fig. 3.2). The carbon atoms are classically numbered from the carboxyl carbon (carbon number 1). The methyl end carbon is known as the n minus (n–) or omega (ω) carbon atom. The physical and biological properties of triacylglycerols are determined by the nature of the constituent fatty acids.

Saturated fatty acids are those in which carbon–carbon bonds are fully saturated with hydrogen atoms (i.e. four hydrogens per carbon–carbon bond). When two hydrogens are absent, the carbons form double bonds with each other and monounsaturated (a single double bond) or polyunsaturated (two or more double bonds) fatty acids result. Double bonds in polyunsaturated fatty acids are always separated by one $CH_2$ (methylene) group. Fatty acids can be described by their common name, their chemical name, their full or simplified chemical structure, or a shorthand notation in which the first number indicates the number of carbon atoms and the second the number of double bonds (Fig. 3.2). For monounsaturated and polyunsaturated fatty acids, a third descriptor indicates the position of the

$$
\begin{array}{l}
\text{H} \qquad \text{HO–C(=O)–}(CH_2)_{16}CH_3 \\
\text{H–C–OH} \quad + \\
\text{H–C–OH} + \text{HO–C(=O)–}(CH_2)_7CH{=}CHCH_2CH{=}CH(CH_2)_4CH_3 \\
\text{H–C–OH} \quad + \\
\text{H} \qquad \text{HO–C(=O)–}(CH_2)_7CH{=}CH(CH_2)_7CH_3
\end{array}
$$

glycerol

free fatty acids

$$
\begin{array}{l}
\text{H} \\
\text{H–C–O–C(=O)–}(CH_2)_{16}CH_3 \\
\text{H–C–O–C(=O)–}(CH_2)_7CH{=}CHCH_2CH{=}CH(CH_2)_4CH_3 + 3H_2O \\
\text{H–C–O–C(=O)–}(CH_2)_7CH{=}CH(CH_2)_7CH_3 \\
\text{H}
\end{array}
$$

triacylglycerol (triglyceride)

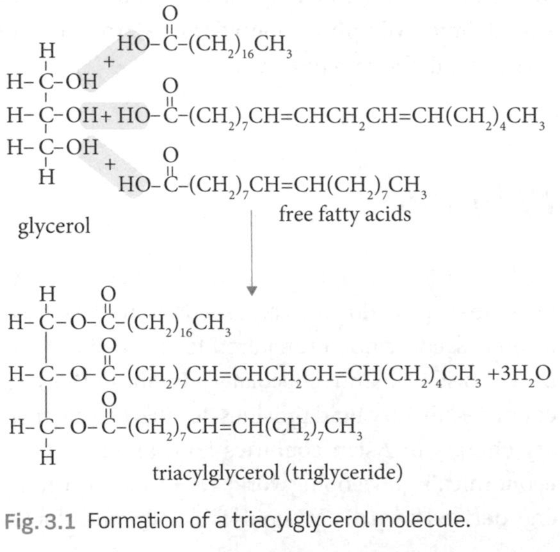

Fig. 3.1 Formation of a triacylglycerol molecule.

first double bond relative to and including the methyl end. Inserting a double bond in a saturated fatty acid reduces its melting point. For this reason, fats (e.g. butter) containing a predominance of saturated fatty acids are usually solid at room temperature while oils (e.g. soybean oil) containing a predominance of polyunsaturated fatty acids are liquid at room temperature. The position of the unsaturated bonds in mono- and polyunsaturated fatty acids has a profound influence on their nutritional properties and health effects. The position of the first double bond relative to the methyl end indicates the 'family' to which the unsaturated fatty acid belongs. Polyunsaturated fatty acids in which the first double bond is three carbon atoms from the methyl end of the carbon chain are called n-3 or ω3 fatty acids, and those in which the first double bond is next to the sixth carbon atom are n-6 or ω6 fatty acids. The third important family is the n-9 or ω9 group, in which the first double bond is next to the ninth carbon atom from the methyl end. In the body, fatty acids of one 'family' cannot be converted into those of another 'family'. Fatty acids are sometimes classified by carbon chain-length and referred to as short-chain (i.e. fewer than eight carbons), medium-chain (8–12 carbons), long-chain (14 or more carbons), or very long-chain (22 or more) fatty acids. Almost all fatty acids have chain lengths with an even number of carbon atoms; however, there are a couple of odd-chain length fatty acids, which are present in very small amounts in fats from ruminant animals. A list of the fatty acids of nutritional interest is given in Table 3.1.

A single triacylglycerol molecule may contain either three identical fatty acids or, more frequently, a combination of different fatty acids. It is important to appreciate that while one fatty acid, or class of fatty acids (e.g. saturated), might predominate in a particular food, most foods contain a wide range of fatty acids. Occasionally in naturally occurring glycerides, only one or two fatty acids are attached to a glycerol molecule. These are called monoacylglycerols (monoglycerides) and diacylglycerols (diglycerides). The major food sources of fatty acids and sources of triacylglycerols that contain them are shown in Table 3.2, and the detailed fatty acid compositions of some fats and oils are given in Table 3.3.

Common name: stearic acid  Chemical name: octadecanoic acid
Fatty acid notation: 18:0

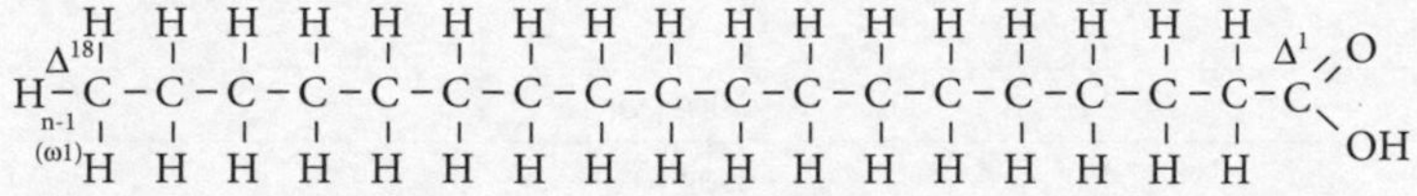

$CH_3(CH_2)_{16}COOH$

Common name: oleic acid  Chemical name: $\Delta^9$-octadecenoic acid
Fatty acid notation: 18:1n-9, or 18:1ω9, or 18:1$\Delta^9$

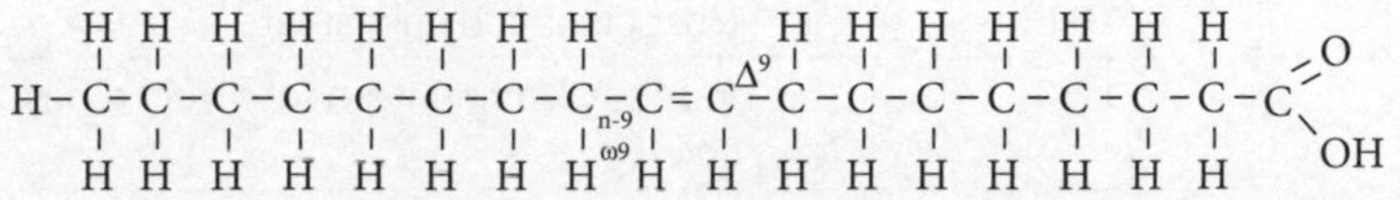

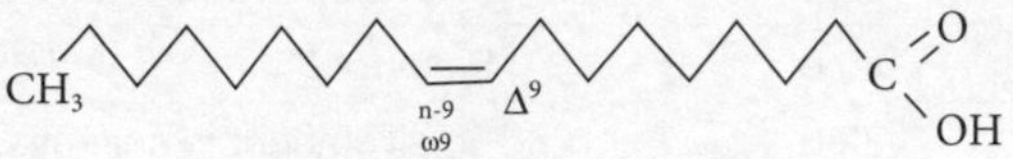

$CH_3(CH_2)_7CH{=}CH(CH_2)_7COOH$

Common name: linoleic acid  Chemical name: $\Delta^{9,12}$-octadecadienoic acid
Fatty acid notation: 18:2n-6, or 18:2ω6, or 18:2$\Delta^{9,12}$

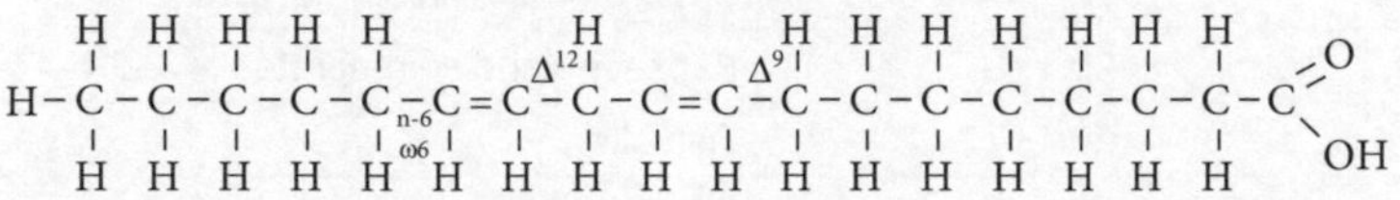

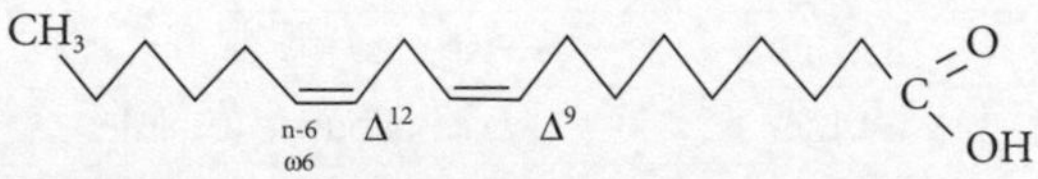

$CH_3(CH_2)_4CH{=}CHCH_2CH{=}CH(CH_2)_7COOH$

Common name: linolenic acid  Chemical name: $\Delta^{9,12,15}$-octadecatrienoic acid
Fatty acid notation: 18:3n-3, or 18:3ω3, or 18:3$\Delta^{9,12,15}$

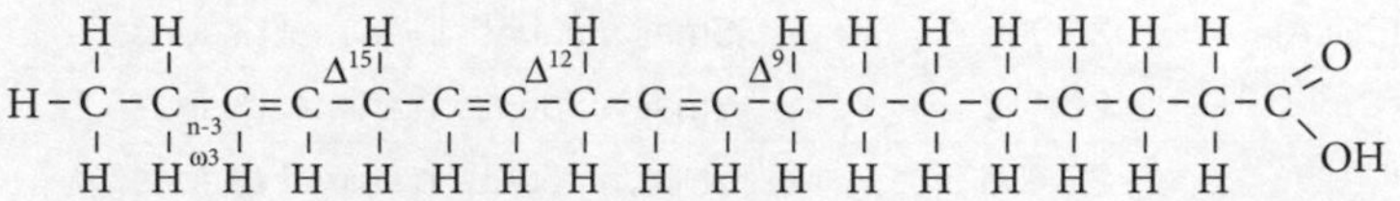

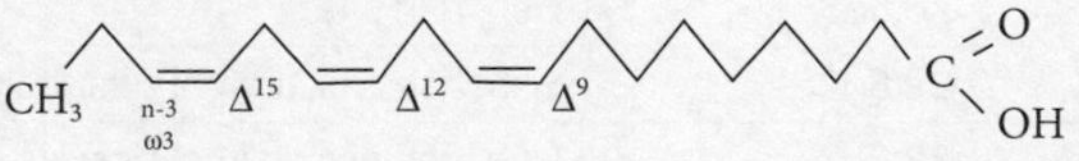

$CH_3CH_2CH{=}CHCH_2CH{=}CHCH_2CH{=}CH(CH_2)_7COOH$

Fig. 3.2 Names and structures of some common fatty acids.

Table 3.1 **Fatty acid names and occurrence**

| Common name | Nomenclature | Occurrence |
|---|---|---|
| **Saturated** | | |
| Acetic | 2:0 | Vinegar |
| Butyric | 4:0 | Dairy fat |
| Caproic | 6:0 | Dairy fat |
| Caprylic | 8:0 | Palm kernel oil |
| Capric | 10:0 | Dairy fat, coconut oil |
| Lauric | 12:0 | Coconut oil |
| Myristic | 14:0 | Dairy fat, coconut oil |
| Pentadecanoic | 15:0 | Small amounts in fats from ruminant animals (e.g. cow) |
| Palmitic | 16:0 | Most plant and animal fats |
| Margaric | 17:0 | Very small amounts in fats from ruminant animals (e.g. cow) |
| Stearic | 18:0 | Most plant and animal fats |
| Arachidic | 20:0 | Peanuts |
| Behenic | 22:0 | Small amount in animal fats |
| Lignoceric | 24:0 | Plant cutin |
| **Monounsaturated** | | |
| Palmitoleic | 16:1ω7 | Fish and animal fats |
| Oleic | 18:1ω9 | All plant and animal fats |
| cis-Vaccenic | 18:1ω7 | Small amounts in animal fats |
| Eicosenoic | 20:1ω9 | Rapeseed and animal fats |
| Gadoleic | 20:1ω11 | Fish oils |
| Erucic | 22:1ω9 | Rapeseed, animal tissue |
| Cetoleic | 22:1ω13 | Fish oils |
| Nervonic | 24:1ω9 | Animal tissue (brain) |
| Hexacosenoic | 26:1ω9 | Minute amounts in animal tissues |
| **Polyunsaturated** | | |
| Linoleic (LO) | 18:2ω6 | Plant oils: cottonseed, sesame, soybean, corn, safflower |
| α-Linolenic (LN) | 18:3ω3 | Plant oils: soybean, mustard, walnut, linseed, rapeseed |
| γ-Linolenic (GLA) | 18:3ω6 | Plant oils: evening primrose, borage, blackcurrant |
| Dihomo-γ-linolenic (DGLA) | 20:3ω6 | Small amounts in animal tissues |
| Arachidonic (AA) | 20:4ω6 | Small amounts in animal tissues |
| Adrenic | 22:4ω6 | Small amounts in animal tissues |
| Eicosapentaenoic acid (EPA) | 20:5ω3 | Fish, fish oils |
| Docosapentaenoic (DPA) | 22:5ω3 | Fish, fish oils, animal tissues (brain) |
| Docosahexaenoic (DHA) | 22:6ω3 | Fish, fish oils, animal tissues (brain) |

Table 3.2 **Fat and cholesterol content of some common foods**[a]

| Food item | Common serving size | Total fat (g) | SFA (g) | MUFA (g) | PUFA (g) | Chol (mg) |
|---|---|---|---|---|---|---|
| | | per 100 g edible portion | | | | |
| Skimmed milk | 1 cup (260 g) | 0.4 | 0.3 | 0.1 | 0.0 | 4 |
| Yoghurt | 1 pot (150 g) | 2.4 | 1.5 | 0.6 | 0.1 | 8 |
| Cottage cheese | 1/2 cup (120 g) | 3.5 | 2.2 | 0.9 | 0.1 | 9 |
| Whole milk | 1 cup (260 g) | 4.0 | 2.4 | 1.1 | 0.1 | 12 |
| Ice cream | 1 cup (143 g) | 10.8 | 6.5 | 2.3 | 0.3 | 30 |
| Cheddar cheese | 1 × 2 cm cube (22 g) | 35.2 | 22.3 | 8.4 | 0.8 | 107 |
| Cream | 1 tbsp (15 g) | 40.0 | 24.9 | 10.1 | 1.3 | 104 |
| Wholemeal bread | 1 slice (22 g) | 1.7 | 0.4 | 0.4 | 0.6 | 1 |
| Toasted muesli | 1 cup (110 g) | 16.6 | 7.7 | 5.0 | 2.9 | 0 |
| Egg | 1 medium (32 g) | 11.6 | 3.4 | 4.6 | 1.2 | 412 |
| Baked potato | 1 potato (90 g) | 0.2 | 0.0 | 0.0 | 0.1 | 0 |
| Potato crisps | 1 packet (50 g) | 33.4 | 14.3 | 13.8 | 3.8 | 1 |
| Cauliflower | 1 stem + flower (90 g) | 0.2 | 0.0 | 0.0 | 0.1 | 0 |
| Lentils | 1/2 cup (100 g) | 0.5 | 0.1 | 0.1 | 0.2 | 0 |
| Peanuts | 1/3 cup (50 g) | 49.0 | 9.2 | 23.4 | 13.9 | 0 |
| Cashew nuts | 18 cashews (28 g) | 51.0 | 8.3 | 25.4 | 15.1 | 0 |
| Sole | 1 fillet (51 g) | 1.2 | 0.3 | 0.4 | 0.3 | 53 |
| Mackerel | 1 fillet (89 g) | 2.9 | 0.8 | 0.8 | 0.9 | 53 |
| Salmon (tinned) | 1/2 cup (120 g) | 8.2 | 2.0 | 3.1 | 2.1 | 90 |
| Sausage | 1 serving (79 g) | 25.2 | 11.3 | 10.8 | 1.2 | 48 |
| Beef blade steak (lean) | 1 steak (216 g) | 5.0 | 2.2 | 1.9 | 0.2 | 60 |
| Beef mince | 1/2 cup (130 g) | 13.8 | 5.7 | 5.4 | 0.5 | 68 |
| Chicken breast (lean, no skin) | 1 breast (192 g) | 5.5 | 1.7 | 2.5 | 0.6 | 66 |
| Fried chicken | 1 wing (37 g) | 28.4 | 8.7 | 13.4 | 2.7 | 116 |
| Pork loin steak (lean) | 1 fillet (98 g) | 2.3 | 0.9 | 0.9 | 0.2 | 68 |
| Lamb midloin chop | 1 chop (50 g) | 5.7 | 2.5 | 2.0 | 0.2 | 66 |
| Pizza | 1 slice (57 g) | 10.5 | 4.5 | 3.3 | 1.8 | 13 |
| Hamburger | 1 burger (204 g) | 15.6 | 5.7 | 5.4 | 2.4 | 22 |
| Muesli bar | 1 bar (32 g) | 19.4 | 9.1 | 7.2 | 1.9 | 1 |
| Biscuit | 1 biscuit (12 g) | 30.0 | 19.2 | 6.2 | 1.2 | 98 |
| Salad dressing | 1 tbsp (16 g) | 48.3 | 7.0 | 11.1 | 28.1 | 0 |
| Palm oil | 1 tbsp (14 g) | 98.7 | 44.7 | 41.1 | 8.2 | 0 |
| Olive oil | 1 tbsp (14 g) | 99.6 | 16.6 | 65.3 | 11.8 | 0 |
| Sunflower seed oil | 1 tbsp (14 g) | 99.7 | 11.7 | 21.1 | 61.9 | 0 |

[a]The fatty acid composition of manufactured foods is likely to be brand dependent and show considerable variation. Chol, cholesterol; MUFA, monounsaturated fatty acids; PUFA, polyunsaturated fatty acids; SFA, saturated fatty acids.

**Table 3.3 Fatty acid composition of plant and animal fats**

| Food item | 4:0 | 6:0 | 8:0 | 10:0 | 12:0 | 14:0 | 16:0 | 18:0 | 16:1ω7 | 18:1ω9 | 18:2ω6 | 18:3ω3 | 20:4ω6 | 20:5ω3 | 22:6ω3 | 20:1ω11 | 22:1ω13 | 22:5ω3 |
|---|---|---|---|---|---|---|---|---|---|---|---|---|---|---|---|---|---|---|
| | Percentage of total acids | | | | | | | | | | | | | | | | | |
| **Plant fats** | | | | | | | | | | | | | | | | | | |
| Olive | – | – | – | – | – | – | 12 | 2 | 1 | 72 | 11 | 1 | – | – | – | – | – | – |
| Palm | – | – | – | – | 0 | 1 | 42 | 4 | 0 | 43 | 8 | 0 | – | – | – | – | – | – |
| Canola | – | – | – | – | – | – | 5 | 1 | 2 | 56 | 24 | 10 | – | – | – | – | – | – |
| Safflower | – | – | – | – | – | – | 8 | 3 | 0 | 13 | 76 | 0 | – | – | – | – | – | – |
| Sunflower | – | – | – | – | – | 0 | 6 | 6 | 0 | 33 | 53 | 0 | – | – | – | – | – | – |
| Avocado | – | – | – | – | – | – | 12 | – | 3 | 75 | 9 | 0 | 0 | – | – | – | – | – |
| Soybean | – | – | – | – | 0 | 0 | 10 | 4 | 0 | 25 | 52 | 7 | – | – | – | – | – | – |
| Coconut | – | – | 8 | 7 | 48 | 16 | 9 | 2 | – | 7 | 2 | – | – | – | – | – | – | – |
| **Animal fats** | | | | | | | | | | | | | | | | | | |
| Butter | 4 | 2 | 1 | 3 | 3 | 11 | 28 | 16 | 1 | 26 | 1 | 2 | – | – | – | – | – | – |
| Beef | – | – | – | – | – | 3 | 28 | 13 | 7 | 43 | 2 | 1 | 1 | – | – | – | – | – |
| Chicken | – | – | – | – | – | 1 | 27 | 7 | 7 | 41 | 14 | 1 | 1 | – | – | – | – | – |
| Lamb | – | – | – | – | – | 6 | 25 | 22 | 1 | 40 | 3 | 3 | – | – | – | – | – | – |
| Pork | – | – | – | – | – | 2 | 27 | 13 | 4 | 41 | 8 | 1 | 1 | – | – | – | – | – |
| Salmon | – | – | – | – | – | 5 | 19 | 4 | 6 | 23 | 1 | 1 | 1 | 8 | 11 | 8 | 5 | 3 |
| Trout | – | – | – | – | – | 2 | 37 | 13 | 5 | 17 | 1 | – | – | 3 | 11 | 2 | 2 | 2 |

## 3.1.2 Phospholipids

Phospholipids comprise a relatively small proportion of total dietary lipid. The four major phospholipids comprise a diglyceride in which the third position of the glycerol molecule is occupied by a phosphoric acid residue to which one of four different base groups is attached (choline, inositol, serine, or ethanolamine). Along with sphingomyelin, these four phospholipids make up more than 95% of the phospholipids found in the body and in foods. The structure of the most abundant phospholipid in nature, phosphatidylcholine (also known as lecithin), is shown in Fig. 3.3.

Phospholipids occur in virtually all animal and vegetable foods: liver, eggs, peanuts, soya beans, and wheatgerm are very rich sources. The base group endows the phospholipid with a polar region soluble in water, while the fatty acids constitute a non-polar region, insoluble in water. This amphipathic nature—having both polar and non-polar characteristics—of the phospholipid enables it to act at the interface between aqueous and lipid media, so they make excellent emulsifying agents. The structural integrity of all cell membranes and lipoproteins is dependent, among other factors, on

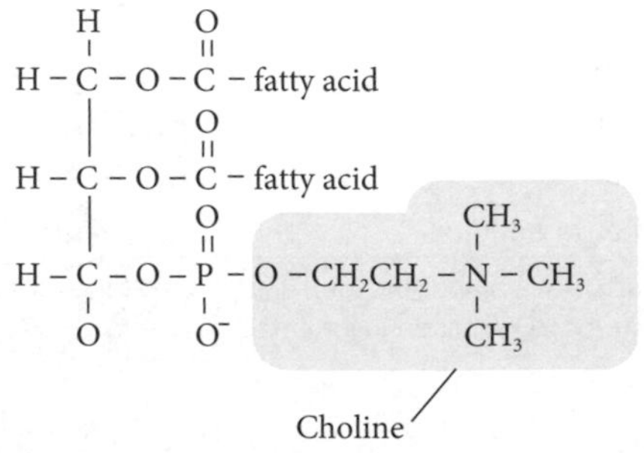

Fig. 3.3 Structure of phosphatidylcholine.

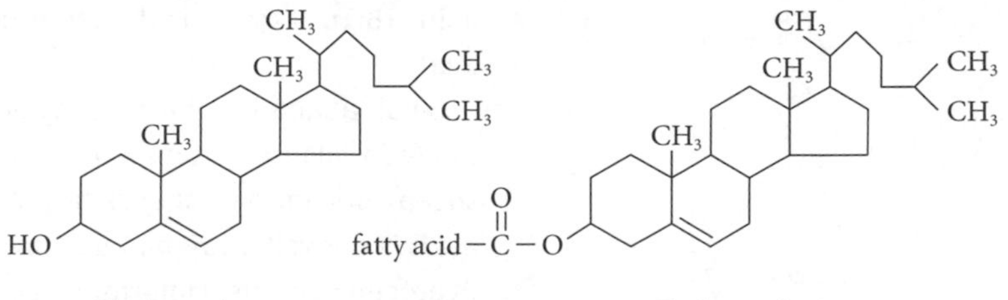

Fig. 3.4 Structure of cholesterol and cholesterol ester.

the amphipathic nature of the constituent phospholipids. Phospholipids are also an important source of essential fatty acids.

### 3.1.3 Sterols

Sterols are also built up from carbon, hydrogen, and oxygen, but in these lipid compounds (unlike triacylglycerols and phospholipids), the carbon, hydrogen, and oxygen atoms are arranged in a series of four rings with a range of side chains. Cholesterol is the principal sterol of animal tissues and is found only in animal foods, especially eggs, meat, dairy products, fish, and poultry. Cholesterol in food often has a fatty acid attached to it, thus forming a cholesterol ester (Fig. 3.4). Approximate quantities of cholesterol in some common foods are given in Table 3.2. The major sterols of plants (group name phytosterols) are β-sitosterol, campesterol, and stigmasterol. Cholesterol plays an important structural role in membranes and lipoproteins, and functions as a precursor of bile acids, steroid hormones, and vitamin D.

### 3.1.4 Other constituents of dietary fat

Dietary fats may also contain small quantities of other lipids, including fatty alcohols, gangliosides, sulphatides, and cerebrosides, as well as vitamin A, vitamin E (tocopherols, tocotrienols), carotenoids (α- and β-carotene, lycopene, and xanthophylls), and vitamin D (see Chapters 11, 13, and 14, respectively).

## 3.2 Dietary fats altered during food processing

The food industry incorporates fats and oils into margarines, biscuits, cakes, chocolates, pies, sauces, and other manufactured food products. In addition to using naturally occurring lipids, food manufacturers use fats and oils that have been altered by the process of hydrogenation, adding hydrogen atoms to the double bonds in mono- or polyunsaturated fatty acids in order to increase the degree of saturation of the fatty acids in the oil (i.e. reduce the number of double bonds) and consequently increase the melting point of the fat. Through this process, a polyunsaturated oil that is liquid can be converted into a fat that is solid at room temperature. Partial hydrogenation of oils is used by manufacturers to produce a fat consistency appropriate to the texture of the desired food and to prolong the stability and shelf-life of the food product. Until relatively recently, margarines for home use normally contained partially hydrogenated fats; in recent years, margarines are now manufactured without such fats using a process called interesterification and blending. Partial hydrogenation also changes the configuration of some of the remaining double bonds from the natural *cis* configuration to a *trans* configuration. *cis*-Mono- and polyunsaturated fatty acids have the two hydrogen atoms attached to the carbons on the same side of the double bond and the molecule bends at the double bond. In *trans*-fatty acids, the hydrogens are placed on opposite sides of the double bond and the molecule stays straight at the double bond (see Fig. 3.5). *Trans*-unsaturated fatty acids

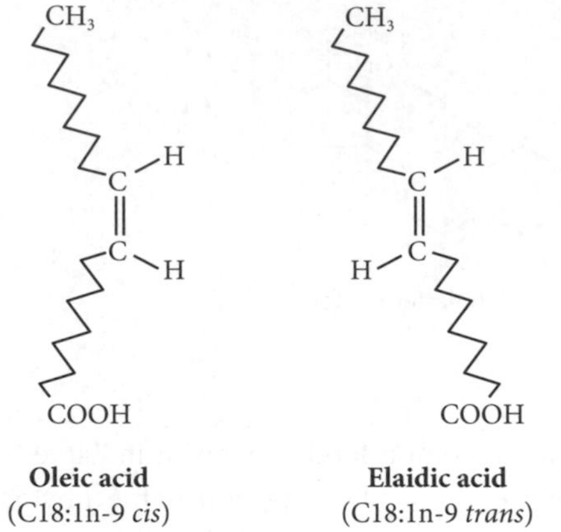

Oleic acid
($C18{:}1n\text{-}9$ *cis*)

Elaidic acid
($C18{:}1n\text{-}9$ *trans*)

Fig. 3.5 Structure of a *cis* and a *trans* monounsaturated fatty acid.

behave biologically like saturated rather than like *cis*-unsaturated fatty acids. The bulk of *trans*-fatty acids in hydrogenated fats are monounsaturated (elaidic acid, 18:1n-9 *trans*, is the *trans* equivalent of oleic acid).

Small quantities of *trans*-fatty acids are found naturally in fats from ruminant animals (e.g. cows and sheep) but most of the dietary intake of *trans*-fatty acids is derived from manufactured foods containing hydrogenated fats. Unfortunately, in most countries information about the relative proportions of *cis*- and *trans*-fatty acids is not available for many foods, especially manufactured products, so it is not possible at present to quantify the total amount of *trans*-unsaturated fatty acids in the diet. *Trans*-fatty acids have in the past made up 5–10% of fatty acids in soft margarines but most now contain less than 1–2%. Hard margarines contain up to 40–50% of fatty acids in the *trans*-form and their use by food manufacturers in some countries continues to be common in processed foods such as cakes, pastry, pies, biscuits, and crackers (Box 3.1).

# 3.3 Digestion, absorption, and transport

## 3.3.1 Digestion

Triacylglycerols must be hydrolysed to fatty acids and monoacylglycerols before they can be absorbed. In children and adults, the process starts in the stomach, where the churning action helps to create an emulsion. Fat entering the intestine is mixed with bile and further emulsified so that lipids are reduced to small bile acid-coated droplets that disperse in aqueous solutions and provide a sufficiently large surface area for the digestive enzymes to act. Bile acids facilitate the process of emulsification because they are amphipathic. Lipase enzymes secreted by the pancreas split by hydrolysis each triacylglycerol molecule, removing the two outer fatty acids, which can be absorbed

### Box 3.1 *Trans*-fatty acids

- Clinical trials have shown that TFAs behave like saturated fat rather than like *cis*-unsaturated fatty acids in that they are associated with increases in serum total and low-density lipoprotein (LDL) cholesterol. In addition, they have the ability to reduce high-density lipoprotein (HDL) cholesterol.
- Epidemiological studies show convincingly that TFAs are associated with increased risk of coronary heart disease.
- In 2003, legislation introduced in Denmark made it illegal for any food to contain more than 2% of its fat content as industrially produced TFA. Doing so can result in a maximum penalty of 2 years' imprisonment.
- In 2006, labelling of TFA in food containing 0.5 g or more per serving became mandatory in the USA.
- In 2008, New York City '*banned*' the use of TFA in restaurant food; the TFA content must be less than 0.5 g per serving.
- In 2018, the World Health Organization (WHO) launched a 'REPLACE' action plan to eliminate TFA from the global food industry by 2023.
- These measures, with potentially important health benefits, have largely resulted from nutrition 'activists' publicizing the results of experimental and epidemiological research.

with the remaining monoacylglycerol. Some monoacylglycerols (about 20%) are rearranged so that the lipase enzymes remove the third fatty acid. Phospholipids are hydrolysed by a phospholipase and cholesterol ester by cholesterol ester hydrolase. In the newborn, the pancreatic secretion of lipases is low, and fat digestion is augmented by lingual lipase secreted from the glands of the tongue and by a lipase present in human milk. The products of lipid digestion, along with other minor dietary lipids, such as fat-soluble vitamins, coalesce with bile acids into microscopic aggregates known as mixed micelles.

## 3.3.2 Absorption

Glycerol and fatty acids with a chain length of less than 12 carbon atoms can enter the portal vein system directly by diffusing across the enterocytes (cells lining the wall of the small intestine). On the other hand, monoglycerides, fatty acids, cholesterol, lysophospholipids, and other dietary lipids diffuse from the mixed micelles into the enterocytes of the small intestine, where they are resynthesized into triacylglycerols, phospholipids, and cholesterol esters in preparation for their incorporation into chylomicrons (Fig. 3.6). In general, absorption is efficient,

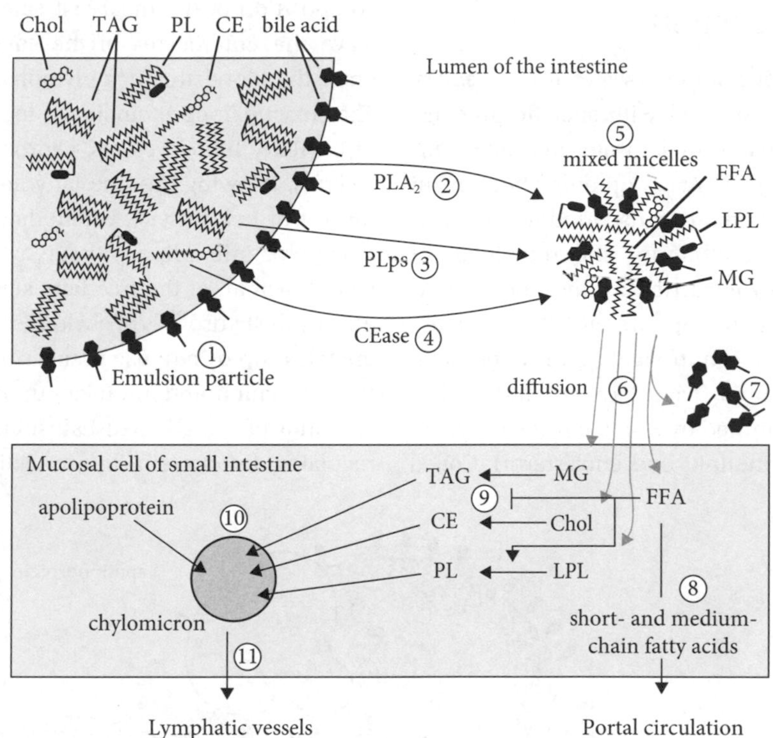


Fig. 3.6 Digestion and absorption of dietary lipids: (1) Dietary lipid leaves the stomach and enters the upper region of the small intestine where bile acids, released from the gallbladder, surround and coat droplets of fat to form emulsion particles. The emulsion particles provide the surface area for the pancreatic enzymes to degrade the dietary lipids. (2) Phospholipase $A_2$ ($PLA_2$) breaks down each phospholipid (PL) into a free fatty acid (FFA) and a lysophospholipid (LPL). (3) Pancreatic lipase (PLps) converts triacylglycerol (TAG) into a monoglyceride (MG) and two free fatty acids. (4) Cholesterol esterase (CEase) splits cholesterol ester (CE) into free cholesterol (Chol) and a free fatty acid. (5) The products of lipid digestion coalesce with bile acids into mixed micelles. (6) The mixed micelles move close to the mucosal cell surface where the lipids diffuse down a concentration gradient into the mucosal cells. (7) Bile acids are not absorbed. (8) Short- and medium-chain fatty acids move immediately into the portal circulation where they are transported in the blood bound to albumin. (9) To maintain the concentration gradient necessary for lipid diffusion, the breakdown products of lipid digestion are resynthesized into their parental lipids. (10) The lipids are combined with apolipoproteins, synthesized in the mucosal cells, to form chylomicrons. (11) Chylomicrons leave the mucosal cell via the lymphatic vessels.

with greater than 95% of dietary lipid absorbed (triacylglycerols, phospholipids, and fat-soluble vitamins). Cholesterol, other sterols, and β-carotene are only partially absorbed (less than 30%).

Diseases that impair the secretion of bile (e.g. obstruction of the bile duct), that reduce secretion of lipase enzymes from the pancreas (e.g. pancreatitis or cystic fibrosis), or that damage the cell lining of the small intestine (e.g. coeliac disease) can lead to severe malabsorption of fat. Under such circumstances, medium-chain triacylglycerols can be better tolerated and are often used as part of the dietary treatment.

### 3.3.3 Lipid transport

Since lipids are not soluble in water, it is necessary for them to be associated with specific proteins, the apolipoproteins, to make water-miscible complexes. Free fatty acids make up only about 2% of total plasma lipid and are transported in the blood as complexes with albumin. The remainder of lipid in the plasma is carried as lipoprotein complexes (lipid + protein = lipoprotein). The structure of a lipoprotein is given in Fig. 3.7. They consist of a core of neutral lipid (triacylglycerol and cholesterol esters) surrounded by a single surface layer of polar lipid (phospholipid and cholesterol). Coiled chains of apolipoproteins extend over the surface. Lipoproteins, which are identified according to their density (Table 3.4), contain apolipoproteins (apo A, apo B, apo C, and apo E), which play important roles in determining the functions of the lipoproteins. Each has a distinct physiological role (Table 3.5) and when present in inappropriate amounts (too high or too low) has different adverse health consequences.

*Chylomicrons* transport lipids of dietary origin, so they consist predominantly of triacylglycerols. Chylomicrons are abundant in the blood after eating food, particularly fatty food, but are scarce in fasting blood. The fatty acid composition of the lipids in chylomicrons is largely determined by the composition of the meal just eaten. Chylomicrons leave the enterocytes of the small intestine and enter the bloodstream via lymph vessels (Fig. 3.8). The enzyme lipoprotein lipase, located on the walls of capillary blood vessels, hydrolyses the triacylglycerols, allowing the free fatty acids to move into muscle or heart tissue, where they can be used for energy, or into adipose tissue, where they can be stored. Not all of the free fatty acids that are liberated by the hydrolysis of chylomicron triacylglycerol are taken up by adipose tissue; some spill over into the circulation and are taken up by the liver. Consumption of a meal increases insulin levels, which upregulates lipoprotein lipase activity in adipose

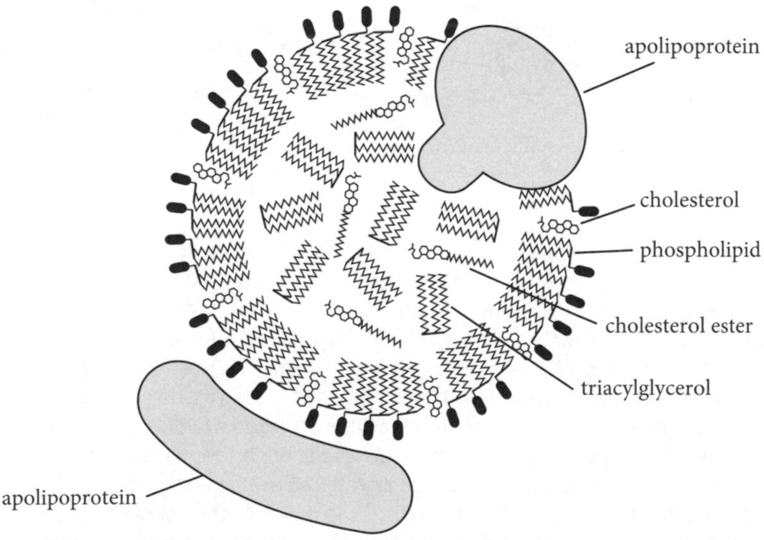


Fig. 3.7 Structure of a plasma lipoprotein.

**Table 3.4 Composition of human plasma lipoproteins**

| Class | Density | Composition (weight %) | | Percentage of total lipid (weight %) | | | | | Major apoproteins |
|---|---|---|---|---|---|---|---|---|---|
| | | Protein | Lipid | TAG | PL | CE | Chol | FFA | |
| Chylomicrons[a] | <0.95 | 2 | 98 | 88 | 8 | 3 | 1 | – | AI, AIV, B-48, Cs, and E from HDL |
| Very-low-density lipoprotein (VLDL)[b] | 0.95–1.006 | 10 | 90 | 56 | 20 | 15 | 8 | 1 | B-100, Cs, and E |
| Intermediate-density lipoprotein (IDL)[c] | 1.006–1.019 | 11 | 89 | 29 | 26 | 34 | 6 | 1 | B-100, E |
| Low-density lipoprotein (LDL)[c] | 1.019–1.063 | 21 | 79 | 13 | 28 | 48 | 10 | 1 | B-100 |
| Lipoprotein(a) (Lp(a))[b] | 1.05–1.12 | 31 | 69 | 11 | 29 | 48 | 11 | 1 | B-100, apo(a) |
| High-density lipoprotein $HDL_2$[a] | 1.063–1.125 | 33 | 68 | 16 | 43 | 31 | 10 | – | As, Cs, E |
| High-density lipoprotein $HDL_3$[b] | 1.125–1.210 | 57 | 43 | 13 | 46 | 29 | 9 | 6 | As, Cs, E |
| Albumin[b] | >1.281 | 99 | 1 | – | – | – | – | 100 | Albumin |

[a]Origin: intestine.
[b]Origin: liver.
[c]Origin: VLDL.
CE, cholesterol ester; chol, cholesterol; FFA, free fatty acids; PL, phospholipid; TAG, triacylglycerol (triglyceride).

**Table 3.5 Functions of human plasma lipoproteins**

| Class | Function |
|---|---|
| Chylomicrons[a] | Transport dietary lipids from intestine to peripheral tissues and liver |
| Very-low-density lipoprotein (VLDL)[b] | Transports lipids from liver to peripheral tissues |
| Intermediate-density lipoprotein (IDL)[c] | Precursor of LDL |
| Low-density lipoprotein (LDL)[c] | Transports cholesterol to peripheral tissues and liver |
| Lipoprotein(a) (Lp(a))[b] | Uncertain. Associated with coronary heart disease risk |
| High-density lipoprotein $HDL_2$[a] | Removes cholesterol from tissues and transfers it to the liver or other lipoproteins |
| High-density lipoprotein $HDL_3$[b] | |
| Albumin[b] | Transports free fatty acids from adipose tissue to peripheral tissues |

[a]Origin: intestine.
[b]Origin: liver.
[c]Origin: VLDL.

tissue. In contrast, heart and muscle lipoprotein lipase activity is up-regulated in the fasted state. Exercise also up-regulates lipoprotein lipase in muscle. During a short life in the circulation (15–30 minutes) more than 90% of the triacylglycerol in the chylomicron is removed. The resulting chylomicron remnant is cleared from the circulation by the liver. The fat-soluble vitamins (A, D, E, and K) are delivered to the liver as part of the chylomicron remnant.

*Very-low-density lipoproteins (VLDLs)* are large, triacylglycerol-rich particles synthesized in the liver from fatty acids derived from adipose tissue,

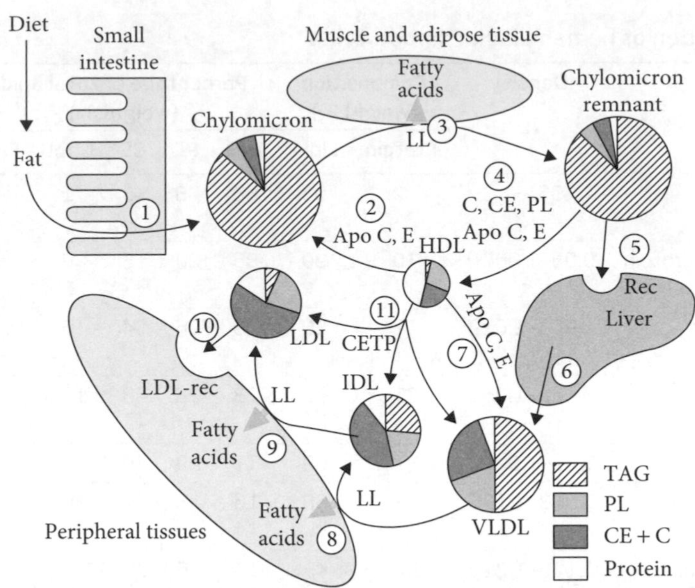


Fig. 3.8 Lipid transport and lipoprotein metabolism: (1) Chylomicrons transport recently ingested fats into the blood. (2) Upon entering the blood, chylomicrons pick up apolipoproteins C and E (apo C, E) from high-density lipoprotein (HDL). (3) Apolipoprotein C activates lipoprotein lipase (LL) on the walls of the capillaries, causing triacylglycerol to be broken down to glycerol and three fatty acids. The fatty acids are taken up primarily by adipose and muscle tissue. (4) During breakdown of triacylglycerol (TAG), some cholesterol (C), cholesterol ester (CE), and phospholipids (PL), along with apo C and E, pinch off to form HDL. (5) Following degradation of 70–80% of the chylomicron's TAG, the resulting chylomicron remnant binds to receptors (rec) on the liver cells and is removed from the circulation. (6) Lipids synthesized in the liver and those delivered to the liver by chylomicron remnants are packaged into very-low-density lipoproteins (VLDL) and secreted into the blood. (7) VLDL picks up apo C and E from HDL. (8) LL, activated by apo C, breaks down VLDL TAG and the fatty acids are transferred to peripheral tissue (mainly muscle and adipose), resulting in the formation of intermediate-density lipoprotein (IDL). (9) Nearly all of the TAG is removed from IDL, producing a cholesterol-rich LDL. (10) Cholesterol is delivered to the cells when LDL binds to LDL-receptors (LDL-rec) and is taken up into the tissues. (11) Cholesterol ester transfer protein (CETP) redistributes cholesterol esters from HDL to VLDL, IDL, and LDL.

dietary fat, and those synthesized within the liver from non-lipid precursors and triacylglycerol stores within the liver. The liver synthesizes triacylglycerol-rich VLDL in a spectrum of sizes, which is influenced by an individual's genotype and phenotype (i.e. sex, age, and adiposity) along with dietary macronutrient composition; the size and composition of the particle influences its metabolic fate. VLDLs function as a vehicle for delivery of fatty acids to the heart, muscles, and adipose tissue, lipoprotein lipase again being needed for their liberation. Following removal of much of the triacylglycerol from VLDL, the remaining remnant particles are intermediate-density lipoproteins (IDL), which are the precursors of low-density lipoprotein.

*Low-density lipoprotein (LDL)* is the end-product of VLDL metabolism and its lipid consists largely of cholesterol ester and cholesterol. Its surface has only one type of apolipoprotein, apo B100. LDL carries about 70% of all cholesterol in the plasma. LDL is taken up by the liver and other tissues by LDL receptors.

*High-density lipoprotein (HDL)* is synthesized and secreted both by the liver and intestine. A major function of HDL is to transfer apolipoproteins C and E to chylomicrons so that lipoprotein lipase can break down the triacylglycerols in the lipoproteins. HDL also plays a key role in the reverse transport of cholesterol, i.e. the transfer of cholesterol back from the tissues to the liver. HDL can be divided into two subfractions of different densities: $HDL_2$ and $HDL_3$.

*Lipoprotein(a) (Lp(a))* is a complex of LDL with apolipoprotein (a).

## 3.3.4 Nutritional determinants of lipid and lipoprotein levels in blood

The fact that plasma lipid and lipoprotein levels are important predictors of coronary heart disease risk (discussed in Chapter 19) has led to a great deal of research into nutritional and other lifestyle factors that interact with genetic factors to determine their concentration in the blood (Table 3.6).

The number of chylomicron particles produced and the fatty acid composition of chylomicron lipid are principally determined by the amount and type of fat eaten in the preceding meal. VLDL levels tend to be low in lean individuals and those who have regular physical activity. Obesity and an excessive intake of alcohol are associated with higher than average VLDL levels. An increased intake of carbohydrate (especially sugars and starches) is generally associated with an increase in VLDL as a result of increased hepatic synthesis of triacylglycerols, though adaptation may occur if the high carbohydrate intake is sustained over a prolonged period. Populations with habitual high carbohydrate intakes (e.g. Asian or African people who consume their traditional diets and are healthy weight) do not have particularly high plasma VLDL concentrations. Consumption of eicosapentaenoic (20:5ω3) and docosahexaenoic (22:6ω3) acids as fish or fish oils lowers plasma VLDL-triacylgylcerol levels. In routine clinical work, plasma triacylglycerol, rather than VLDL, is measured because the bulk of triacylglycerol levels in blood taken from fasting (10–12 hours) individuals tend to parallel levels of VLDL.

Levels of LDL and total plasma cholesterol are determined by an interaction of genetic factors and dietary characteristics. High intakes of saturated fatty acids, especially myristic and palmitic acids, and *trans*-fatty acids (e.g. elaidic acid) are associated with raised LDL cholesterol, while high intakes of linoleic acid, the major polyunsaturated acid in foods, and to a lesser extent *cis*-monounsaturated fatty acids tend to reduce cholesterol levels. The precise mechanism has not been established, but high intakes of saturated fatty acids appear to decrease the removal of plasma LDL by LDL receptors, whereas mono- and polyunsaturated fatty acids are associated with increased LDL receptor activity. Dietary cholesterol has a minor influence on plasma total and LDL cholesterol, but this is enhanced when saturated fatty acids comprise a high proportion of dietary lipid (greater than 15% of energy) and cholesterol intake exceeds 300 mg/day. It is less clear whether dietary cholesterol plays a major role over the relatively low range of intakes now seen in many countries and when saturated fatty acid intake is reduced. Plant sterols (e.g. β-sitosterol) are very poorly absorbed and interfere with the absorption of cholesterol. This property of plant sterols has been utilized by incorporating them into margarines.

**Table 3.6 Nutritional determinants of lipoprotein levels**

| Nutritional factor | VLDL | LDL | HDL |
|---|---|---|---|
| Obesity | ↑↑ | ↑↑ | ↓↓ |
| Replacement of MUFA or PUFA with a saturated fatty acid | | | |
| Lauric (12:0) | ↑ | ↑ | ↑ |
| Myristic (14:0) | ↑ | ↑↑ | ↑ |
| Palmitic (16:0) | ↑ | ↑↑ | ↑ |
| Stearic (18:0) | – | – | – |
| Replacement of SFA with MUFA | | | |
| Oleic (18:1 *cis*) | ↓ | ↓↓ | ↓ |
| 18:1 *trans* | – | ↓ | ↓ |
| Replacement of SFA with polyunsaturated fat (ω6) | | | |
| Linoleic (18:2ω6) | ↓ | ↓↓↓ | ↓ |
| Effect of consuming more polyunsaturated fat (ω3) | | | |
| Eicosapentaenoic (20:5ω3) | ↓↓ | (↑) | (↓) |
| Docosahexaenoic (22:6ω3) | ↓↓ | (↑) | (↓) |

↑, ↓, increase or decrease;
↑↑, ↓↓, appreciable increase or decrease;
(↑), (↓), indicating an effect only when fed in large quantities.
LDL, low-density lipoproteins; HDL, high-density lipoproteins; MUFA, monounsaturated fatty acid; PUFA, polyunsaturated fatty acid; SFA, saturated fatty acid; VLDL, very-low-density lipoproteins.

Consumption of these margarines (25 g/day containing roughly 2 g plant sterols) can lower plasma total and LDL cholesterol concentrations.

The ability of soluble forms of dietary fibre to reduce total and LDL cholesterol is small compared with the effect of altering the nature of dietary fat. Dietary protein may also influence plasma lipids and lipoproteins: soybean protein particularly has some cholesterol-lowering properties. Vegetarians have lower levels of total and LDL cholesterol in general than non-vegetarians, but it is not clear which characteristic of the vegetarian diet principally accounts for this effect.

Debate centres around whether saturated fatty acids should be replaced by whole grain fibre-rich carbohydrate-containing foods or by fats and oils with a more favourable fatty acid profile (i.e. monounsaturated or polyunsaturated fatty acids with a *cis* configuration). Both substitutions are appropriate; however, the greatest reduction in LDL cholesterol would be achieved by replacement of saturated fat with ω6 polyunsaturated fat.

Dietary factors have a small effect on HDL cholesterol concentration. HDL cholesterol can be slightly reduced when polyunsaturated fatty acids or monounsaturated fats replace saturated fat, but the reduction in LDL cholesterol is substantially larger, resulting in a considerable decrease in the LDL to HDL cholesterol ratio. HDL cholesterol is decreased when carbohydrate intake is increased from more usual levels consumed (less than 45% of energy) in affluent societies to 60% or more of total energy, or by increasing *trans*-unsaturated fatty acids. The HDL-lowering effect of a high-carbohydrate diet may be reduced or prevented if the carbohydrate is high in soluble forms of non-starch polysaccharide. HDL levels tend to be raised by diets relatively high in dietary cholesterol and saturated fatty acids, although this 'positive' effect is offset by the larger increases in LDL cholesterol caused by such diets. Increasing *cis* forms of monounsaturated fatty acids appears to be marginally better at maintaining HDL levels than polyunsaturated fatty acids when reducing saturated fat consumption. Most dietary studies have not included measurements of the subfractions of HDL. HDL cholesterol is raised in people who take substantial amounts of alcohol (see Chapter 6).

# 3.4 Essentials of lipid metabolism

## 3.4.1 Biosynthesis of fatty acids

Saturated and monounsaturated fatty acids can be synthesized in the body from carbohydrate and protein. In humans, this process predominantly occurs in the liver. This process of lipogenesis occurs especially in a well-fed person whose diet contains a high proportion of carbohydrate in the presence of an adequate energy intake. Insulin stimulates the biosynthesis of fatty acids. Lipogenesis is reduced during energy restriction or when the diet is high in fat or enriched with polyunsaturated fatty acids. Although the end-product of lipogenesis is often considered to be the saturated fatty acid C16:0, this can be further desaturated or elongated. Unsaturated fatty acids may be further elongated or desaturated by various enzyme systems (Fig. 3.9).

## 3.4.2 Essential fatty acids

Essential fatty acids are those that cannot be synthesized in the body and must be supplied in the diet to avoid deficiency symptoms. They include members of the ω6 (linoleic acid) and ω3 (α-linolenic acid) families of fatty acids. When the diet is deficient in linoleic acid, the most abundant unsaturated fatty acid in tissue, oleic acid, is desaturated and elongated to eicosatrienoic acid (20:3ω9), which is normally present in trace amounts. Increased plasma levels of this 20:3ω9 suggest a deficiency of essential fatty acids. Essential fatty acid deficiency is rare except in those with severe, untreated fat malabsorption or those suffering from famine. Symptoms include dry, cracked, scaly, and bleeding skin, excessive thirst due to high water

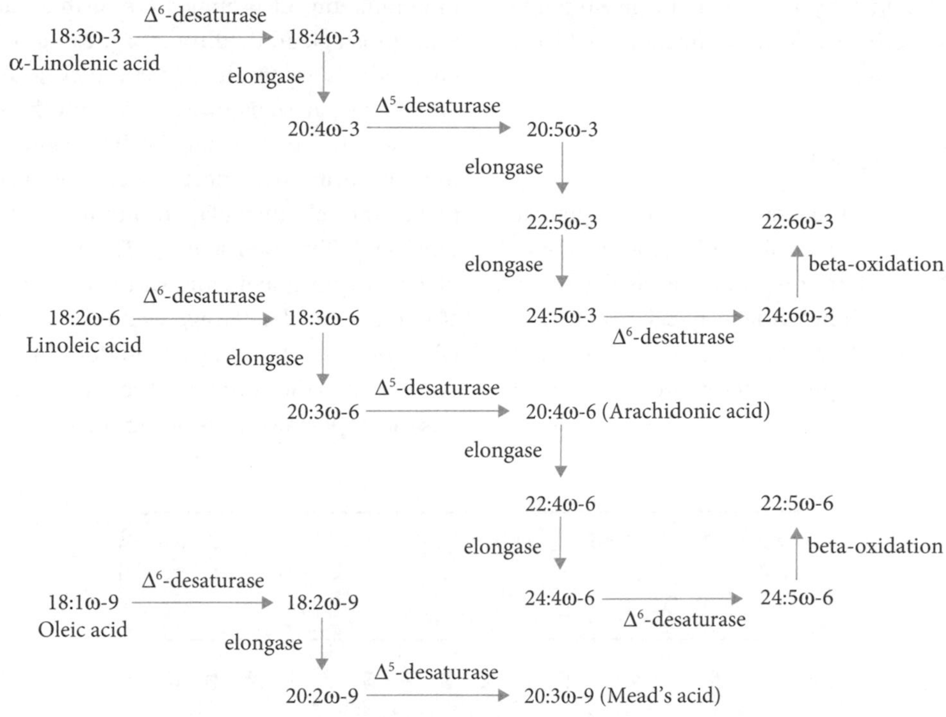


Fig. 3.9 Desaturation and elongation of polyunsaturated fatty acids.

loss from the skin, and impaired liver function resulting from the accumulation of lipid in the liver (i.e. fatty liver).

Linoleic acid and α-linolenic acid not only are required for the structural integrity of all cell membranes but also are elongated and desaturated, in limited amounts, into longer-chain, more polyunsaturated fatty acids that are the precursors to a group of hormone-like eicosanoid compounds, prostaglandins and leukotrienes (see Section 3.4.4). Linoleic acid (18:2ω6) is converted to arachidonic acid (20:4ω6), while α-linolenic acid (18:3ω3) is converted to eicosapentaenoic (20:5ω3) and docosahexaenoic (22:6ω3) acids. A high ratio of linoleic to α-linolenic acid in the diet tends to reduce the amount of α-linolenic acid converted to eicosapentaenoic and docosahexaenoic acids.

There is some question as to whether the arachidonic, eicosapentaenoic, and docosahexaenoic acids incorporated into the body's tissues come predominantly from endogenous desaturation and elongation of dietary essential fatty acids or are obtained from the diet as preformed fatty acids. Whatever the answer, the body appears to have a capacity to desaturate and elongate essential fatty acids, because individuals following strict vegan diets (no animal foods) ingest plenty of linoleic and α-linolenic acids but only negligible amounts of arachidonic, eicosapentaenoic, and docosahexaenoic acids, yet have adequate, albeit lower, levels of these latter fatty acids in their blood.

### 3.4.3 Membrane structure

Unsaturated fatty acids in membrane lipids play an important role in maintaining fluidity. The critically important metabolic functions of membranes, such as nutrient transport, receptor function, and ion channels, are affected by interactions between proteins and lipids. For example, the phosphoinositide cycle, which determines the responses of many cells to hormones, neurotransmitters, and cell growth factors and which controls processes of cell division, is influenced by the proportion of ω6 to ω3 fatty acids. Docosahexaenoic acid (22:6ω3) is uniquely abundant in brain tissue; it is the predominant polyunsaturated fatty acid in brain phospholipid, and it plays a critical role in the functions of the central nervous system.

### 3.4.4 Eicosanoids

Eicosanoids are biologically active, oxygenated metabolites of arachidonic acid, eicosapentaenoic acid (EPA), or dihomo-γ-linolenic acid (20:3ω6). They are produced in virtually all cells in the body, act locally, have short life spans, and act as modulators of numerous physiological processes, including reproduction, blood pressure, haemostasis, and inflammation. Eicosanoids are further categorized into prostaglandins/thromboxanes and leukotrienes, which are produced via the cyclo-oxygenase and lipoxygenase pathways, respectively (Fig. 3.10). Considerable recent interest has centred around the cardiovascular effects of eicosanoids, in particular the role they play in thrombosis (i.e. vessel blockage). Thromboxane $A_2$ ($TxA_2$), synthesized in platelets from arachidonic acid, stimulates vasoconstriction and platelet aggregation (i.e. clumping), while prostacyclin $I_2$ ($PGI_2$), produced from arachidonic acid in the endothelial cells of the vessel wall, has the opposing effects of stimulating vasodilation

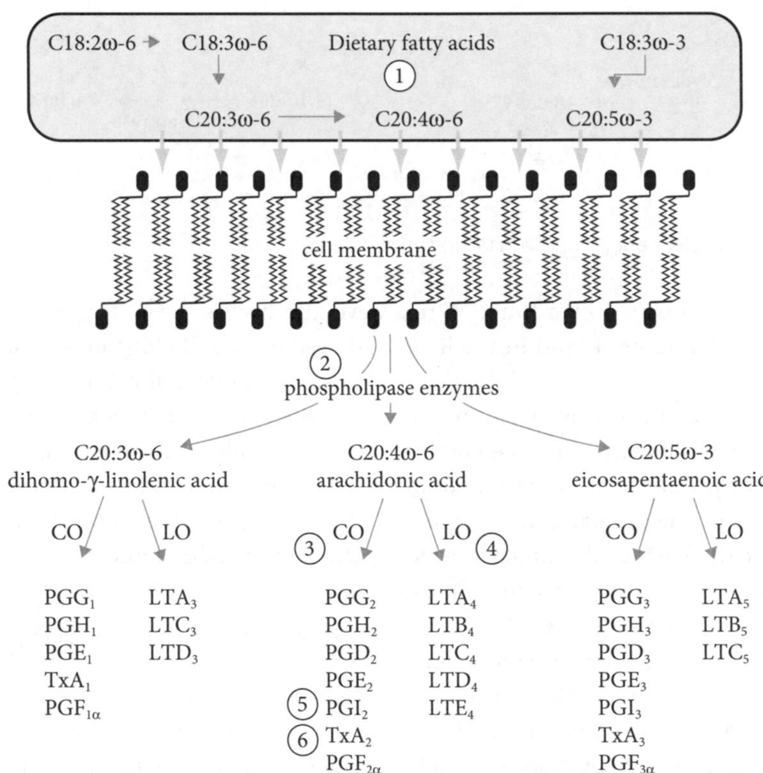


Fig. 3.10 Formation of eicosanoids: (1) Dihomo-γ-linolenic acid (DGLA), arachidonic acid (AA), and eicosapentaenoic (EPA) acid, obtained ready-made in the diet or via desaturation and elongation of their respective parent ω6 (linolenic acid) or ω3 (linolenic acid) essential fatty acids, are incorporated into the phospholipids of cell membranes. (2) When cells are stimulated by hormones or activating substances, phospholipase enzymes release DGLA, AA, and EPA into the interior of the cell. (3) Once released from the cell membranes, DGLA, AA, and EPA can be converted via the cyclo-oxygenase pathway (CO) into prostaglandins (PG) and thromboxanes (Tx) of the 1, 2, and 3 series, respectively; (4) or via the lipoxygenase pathway (LO) to leukotrienes (LT) of the 3, 4, and 5 series, respectively. (5) In platelets, AA is converted primarily to the 2 series thromboxane ($TxA_2$), which stimulates platelet clumping and increases blood pressure. (6) In the endothelial cells lining arterial vessels, AA is converted into the 2 series prostacyclin ($PGI_2$), which counters the action of $TxA_2$ by inhibiting platelet clumping and decreasing blood pressure.

and inhibiting platelet aggregation. The balance of these two counteracting eicosanoids affects overall thrombotic tendency.

Research, initially based on the observation that the Inuit (Eskimo) people of Greenland have very low rates of coronary heart disease, led to the demonstration that a high dietary intake of EPA (usually in fish oil) can profoundly influence the balance of thromboxanes and prostacyclins. Such diets lead to the substitution of EPA for arachidonic acid in platelet membranes. $TxA_2$ production decreases, not only because of lower levels of platelet arachidonic acid but also because the increased levels of EPA in platelets inhibit the conversion of arachidonic acid to $TxA_2$. On the other hand, $PGI_2$ production in the endothelial cells is only slightly reduced and there is a sharp rise in the production of $PGI_3$ from EPA, which has equal vasodilatory and platelet-inhibiting properties. The overall changes in eicosanoid production contribute to reducing thrombotic risk.

Leukotrienes are important in several diseases involving inflammatory or hypersensitivity reactions, including asthma, eczema, and rheumatoid arthritis. The effect of EPA consumption on leukotriene synthesis is to shift production from the more inflammatory 4-series leukotrienes, synthesized from arachidonic acid, to the less inflammatory 5-series leukotrienes synthesized from EPA. This metabolic effect helps to explain the improvements in some of the clinical symptoms experienced by rheumatoid arthritis sufferers who consume significant quantities of fish (i.e. EPA).

### 3.4.5 Effects of fatty acids on other metabolic processes

Fatty acids influence a range of other metabolic processes that have been less well studied in humans. Hydrolysis of some phospholipids results in the formation of biologically active compounds such as the platelet-activating factor (PAF) from 1-alkyl, 2-acyl phosphatidylcholine. Different polyunsaturated fatty acids in the precursor phospholipid can modify PAF formation. ω3 Fatty acids influence the production of cytokines, including the interleukins and tumour necrosis factors, which are involved in regulation of the immune system. An exciting area of current research is the study of the effects of fatty acids on the expression of genes encoding enzymes that are involved in lipid metabolism, as well as the expression of genes involved in cell growth regulation.

### 3.4.6 Oxidation of fatty acids

Those fatty acids not incorporated into tissues or used for synthesis of eicosanoids are oxidized for energy. Oxidation of fatty acids occurs predominantly in the mitochondria of cells and involves a multiple-step process by which the fatty acid is gradually broken down to molecules of acetyl CoA, which are available to enter the tricarboxylic acid cycle and so to generate energy. The rate of oxidation of fatty acids is highest at times of low energy intake and particularly during starvation. As the acetyl CoA splits off, adenosine triphosphate (ATP) is generated, which is also a source of energy. Fatty acids of different chain lengths and degrees of saturation are oxidized via slightly different pathways, but the ultimate purpose of each is the production of acetyl CoA and the generation of energy. Ketone bodies are produced by the liver and used peripherally as an energy source when glucose is not readily available, through the process of ketogenesis. In healthy individuals, the ketone bodies acetoacetate and 3-β-hydroxybutyrate are present in small amounts in the blood during fasting or prolonged exercise. Abnormally high concentrations of ketone bodies are found in the blood when there is an absolute insulin deficiency, such as is seen in severe uncontrolled insulin-dependent diabetes, which results in a very high rate of ketogenesis leading to ketoacidosis due to the accumulation of acetoacetic and β-hydroxybutyric acids. As the rate of ketogenesis depends upon the activity of enzymes that are controlled by the level of circulating insulin, in healthy individuals with a functioning pancreas, the ingestion of carbohydrate stimulates insulin secretion and thereby suppresses ketogenesis. In health, ketosis that is characterized by elevated serum levels of ketone bodies is nearly always a transient condition. Physiological ketosis is associated with prolonged fasting or exercise and ketogenic diets, and ketone body levels do not rise above 6–8 mol/L. (In severe diabetes, levels may be twice as high as this.) However, ketoacidosis does not occur in the absence of insulin deficiency.

### 3.4.7 Lipid storage

Energy intake in excess of requirements is converted to fat for storage. Stored fat in adipose tissue provides the human body with a source of energy when energy supplies are not immediately available from ingested carbohydrate, fat, protein, or stored glycogen stores. Triacylglycerols are the main storage form of lipids and most stored lipids are found in adipose tissue. The lipid is stored as single droplets in cells called adipocytes, which can expand as more fat needs to be stored. Most of the lipid in adipose tissue is derived from dietary lipid and the stored lipid reflects the composition of dietary fat. The triacylglycerol stores of adipose tissue are not static but are continually undergoing lipolysis and re-esterification.

### 3.4.8 Cholesterol synthesis and excretion

Cholesterol is present in tissues and in plasma lipoproteins as free cholesterol or combined with a fatty acid as cholesterol ester. About half the cholesterol in the body comes from endogenous synthesis and the remainder from the diet. It is synthesized in the body from acetyl CoA via a long metabolic pathway. Cholesterol synthesis in the liver is regulated near the beginning of the pathway by the dietary cholesterol delivered by chylomicron remnants. In the tissues, a cholesterol balance is maintained between factors causing a gain of cholesterol (synthesis, uptake into cells, hydrolysis of stored cholesterol esters) and factors causing loss of cholesterol (steroid hormone synthesis, cholesterol ester formation, bile acid synthesis, and reverse transport via HDL). The specific binding sites and receptors for LDL play a crucial role in cholesterol balance, since they constitute the principal means by which LDL cholesterol enters the cells. These receptors are defective in familial hypercholesterolaemia. Excess cholesterol is excreted from the liver in the bile either unchanged as cholesterol or converted to bile salts. A large proportion of the bile salts that are excreted from the liver into the gastrointestinal tract are absorbed back into the portal circulation and returned to the liver as part of the enterohepatic circulation, but some pass on to the colon and are excreted as faecal bile acids.

## 3.5 Health effects of dietary lipids

Most fatty acids can be made in the body, except for the essential fatty acids (EFAs) linoleic and α-linolenic acids, which must be obtained from the diet (see Section 3.4.2). The fact that specific deficiencies resulting from inadequate intakes of EFAs are very rare in adults, even in African and Asian countries where total dietary fat can provide as little as 10% of total energy, suggests that the minimum requirement is low. Amongst adults, EFA deficiency has only been reported when linoleic acid (18:2ω6) intakes are less than 2–5 g/day or less than 1–2% of total energy. Most adult Western diets provide at least 10 g/day of EFA and healthy people have a substantial reserve in adipose tissues. Clinical manifestations of α-linolenic acid deficiency are rare in humans.

Amongst adults in Western countries, the major health issues concerning intake of fat centre around the role of excessive dietary saturated fat in coronary heart disease (Chapter 19) and whether total fat intake plays a role in obesity (Chapter 17) and certain cancers (Chapter 20). There is concern too regarding the optimal amounts and balance of ω3 to ω6 fatty acids with regard to coronary heart disease risk, as well as the effect this balance may have on inflammatory and immunological responses.

Human milk provides 6% of total energy as essential fatty acids (linoleic and α-linolenic acids); it also contains small amounts of longer chain, more polyunsaturated fatty acids such as arachidonic acid (AA; 20:4ω6) and docosahexaenoic acids (DHA; 22:6ω3).

Commercial baby milk formulas contain comparable amounts of essential fatty acids and a number of brands—in some countries—contain AA and DHA. Infants fed exclusively with formula without AA and DHA have lower levels of these fatty acids in their plasma and red blood cells in comparison with breastfed infants. There is strong evidence that inadequate provision of AA and DHA to premature infants delays development of visual acuity (see Chapter 33).

Concern has been expressed that the desire to reduce total fat intake by some health-conscious parents in affluent societies might result in a diet high in complex carbohydrate and dietary fibre and containing insufficient energy for growth and development in childhood. These wide-ranging issues need to be taken into account when making nutritional and dietary recommendations.

## 3.6 Recommendations concerning fat intake

### 3.6.1 Minimum desirable intakes

In adults, it is necessary to ensure that dietary intake is adequate to meet energy needs and to meet the requirements for EFAs and fat-soluble vitamins. Adequate intakes are particularly important during pregnancy and lactation. Thus, for most adults, dietary fat should provide at least 15% of total energy, and 20% for women of reproductive age. World Health Organization (WHO) recommendations suggest that 2% of total daily energy is required from ω6, and at least 0.5% energy from ω3 polyunsaturated fatty acids. These levels are rather arbitrary and are based on the amounts required to cure EFA deficiency. Few countries recommend a specific ω3 to ω6 ratio, but a recommendation for eicosapentaenoic acids (EPA) and docosahexaenoic acids (DHA) is common. For example, the WHO recommends 0.25–2 g per day of EPA plus DHA. Particular attention must be paid to promoting adequate maternal intake of EFAs throughout pregnancy and lactation to meet the needs of the foetus and young infant in laying down lipids in their growing brains (which have a high content of DHA and AA).

For infants and young children, the amount and type of dietary fat are equally important. Breast milk fulfils all requirements (50–60% energy as fat, with appropriate balance of nutrients), and during weaning it is important to ensure that dietary fat intake does not fall too rapidly. At least until the age of 2 years, a child's diet should contain about 40% of energy from fat and provide similar levels of EFAs to breast milk. Infant formulas with AA and DHA in proportions similar to those found in breast milk are available in many countries.

It is necessary also to take into account associated substances, in particular several vitamins and antioxidants. These are considered in other chapters. In particular, vitamin E (Chapter 13) in edible oils is required to stabilize unsaturated fatty acids. Foods high in polyunsaturated fatty acids should contain at least 0.6 mg α-tocopherol equivalents per gram of polyunsaturated fatty acids. In countries where vitamin A deficiency is a public health problem, the use of red palm oil should be encouraged wherever it is available.

### 3.6.2 Upper limits of fat and oil intakes

In most Western countries, dietary recommendations concerning fat intake have focused primarily around desirable upper limits of intake. There is almost universal acceptance of the need to restrict intake of saturated fat to 10% or less of total energy (see Chapter 19). Limiting intake of total fat to no more than 30% total energy has been recommended in populations with high or increasing rates of overweight and obesity (see Chapter 17), though such advice has not been universally adopted. Some recent recommendations suggest that the upper limit may be as high as 40% total energy provided the fatty acid distribution is appropriate.

## Further reading

1. **Brown, M.S., Kovanen, P.T., and Goldstein, J.L.** (1981) Regulation of plasma cholesterol by lipoprotein receptors. *Science*, **212**, 628–35.

2. **Dyerberg, J., Bang, H.O., Stoffersen, E., et al.** (1978) Eicosapentaenoic acid and prevention of thrombosis and atherosclerosis? *The Lancet*, **15**, 117–9.

3. **Frayn, K.N. and Evans, R.D.** (2019) *Human metabolism. A regulatory perspective*, 4th edn. London: Wiley Blackwell.

4. **Gurr, M.I., Harwood, J.L., Frayn, K.N., et al.** (2016) *Lipids: biochemistry, biotechnology and health*, 6th edn. London: Wiley Blackwell.

5. **Hodson, L. and Gunn, P.J.** (2019) The regulation of hepatic fatty acid synthesis and partitioning: the effect of nutritional state. *Nat Rev Endocrinol*, **15**, 689–700.

6. **Laffel, L.** (1999) Ketone bodies: a review of physiology, pathophysiology and application of monitoring to diabetes. *Diabetes Metab Res Rev*, **15**, 412–26.

7. **Mensink, R.P., Zock, P.L., Kester, A.D.M., and Katan, M.B.** (2003) Effects of dietary fatty acids and carbohydrates on the ratio of serum total to HDL cholesterol and on serum lipids and apolipoproteins: a meta-analysis of 60 controlled trials. *Am J Clin Nutr*, **77**, 1146–55.

8. **Packard, C.J., Boren, J., and Taskinen, M.-R.** (2020) Causes and consequences of hypertriglyceridemia. *Front Endocrinol*, **11**, 252. doi.org/10.3389/fendo.2020.00252

# Protein

Colleen S. Deane, Daniel J. Wilkinson, and Philip J. Atherton



## 4.1 Normal growth and the maintenance of health

Proteins are fundamental structural and functional elements within every cell and undergo extensive metabolic interaction. This widespread metabolic interaction is intimately linked to the metabolism of energy and other nutrients (e.g. carbohydrates, fats). Following water, protein is the next most abundant chemical compound in the body. For an adult male who weighs 70 kg, about 16% will be protein, i.e. about 11 kg. A large proportion of this will be muscle (43%), with substantial proportions being present in skin (15%) and blood (16%). Half of the total is present in only four proteins: collagen, haemoglobin, myosin, and actin, with collagen comprising about 25% of the overall total.

Proteins fulfil a range of functions, and the amount of protein does not, in itself, provide any indication of the importance or relevance of the protein. Indeed, some of the most important functionally active proteins might only comprise a very small proportion of the total present, e.g. peptide hormones, such as insulin, growth factors, or cytokines. The biochemical activity of proteins is an attribute of their individual structure, shape, and size. This, in turn, is determined by the sequence of amino acids within the polypeptide chains, the characteristics of the individual amino acids (size, charge, hydrophobicity, or hydrophilicity), and the environment, which together determine the primary, secondary, tertiary, and quaternary structure of the protein.

Proteins taken in the diet are broken down into amino acids in the processes of digestion and absorption. Absorbed amino acids contribute to the amino acid pool of the body, from which all proteins are synthesized. The proteins of the body exist in a 'dynamic state' as they are constantly turning over through the processes of synthesis and degradation (see Fig. 4.1). On average, the rates of synthesis and degradation are similar in adults, so that the amount of protein in the body remains more or less constant over long periods of time, and nitrogen balance is achieved. During growth, synthesis exceeds degradation to enable net deposition of protein. For some proteins, it is clear that they undergo a process of turnover/renewal, e.g. shedding of skin, growth of hair, or replacement of gut mucosa, but this process is also true for plasma proteins, such as albumin or γ-globulins, and even structural proteins, such as muscle and bone. There are many thousands of proteins in the body, each of which is formed and degraded at a characteristic rate that may vary from minutes to days or even months.

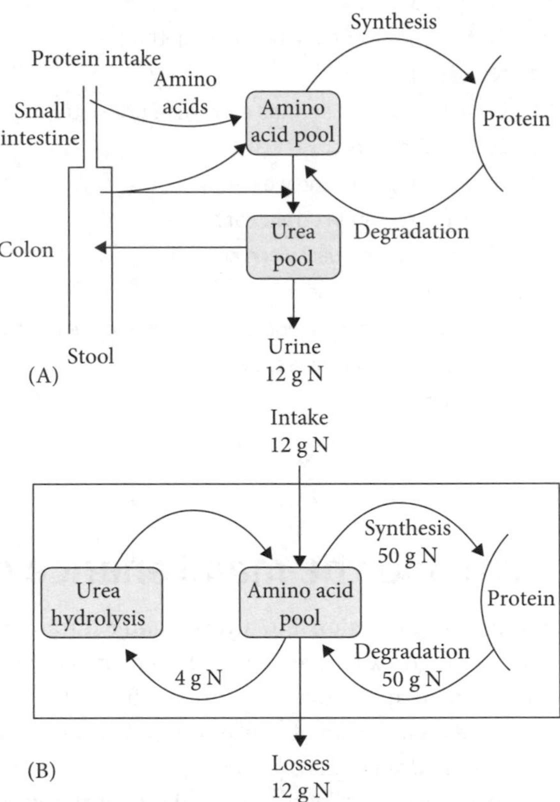


Fig. 4.1 (A) There is a dynamic interchange of protein, amino acids, and nitrogen in the body, which conceptually takes place through an amino acid pool. Amino acids are added to the pool from the degradation of body proteins and also from dietary protein, following digestion and absorption. There is the continuous breakdown of amino acids, with energy being made available when the carbon skeleton is oxidized to water and $CO_2$. The amino group goes to the formation of urea. Although urea is an end-product of mammalian metabolism, a significant proportion of the nitrogen is salvaged for further metabolic interaction following hydrolysis of urea by the flora resident in the colon. (B) The quantitative relations of these exchanges in normal adults expressed as grams of (amino) nitrogen (N) per day.

### 4.1.1 Growth

At the level of cells and tissues, growth can be characterized as an increase in the number of cells (hyperplasia), an increase in the size of cells (hypertrophy), or a combination of the two processes (mixed hypertrophy and hyperplasia), which then leads to differentiation and specialization of function.

Net protein deposition is required for growth to take place and therefore synthesis must exceed degradation. There is an increase in protein synthesis, but there is also an increase in protein degradation, so that overall, about 1.5–2.0 g protein is synthesized for every 1 g of net deposition. The apparent inefficiency of the system might be accounted for by a measure of flexibility to allow remodelling, transcriptional and translational errors, and/or wear and tear on the protein synthetic machinery.

## 4.2 Amino acids

The pattern and amounts of amino acids required to support protein synthesis are determined by the amount and pattern of proteins being formed. The habitual dietary intake of about 80 g protein/day in adults is only one-quarter of the protein being formed in the body each day, whereas the minimal dietary requirement for protein, about 35–40 g protein in adult men, is approximately one-tenth of the protein being formed in the body each day.

It has been presumed that the overall pattern is dominated by proteins of mixed composition similar to that seen in muscle, but this is not necessarily true for all situations, especially in some pathological states. For example, the amino acid profile of collagen is very rich in glycine and proline, but poor in leucine and the other branched-chain amino acids. During growth, where the demands for collagen formation are increased, the balance of amino acids needed is likely to be shifted towards a collagen pattern. During an inflammatory response, there is increased synthesis of glutamine to fuel immune cells and an increase in the antioxidant glutathione and the zinc-binding protein metallothionein, both of which are particularly rich in cysteine. The demand for the most appropriate pattern of amino acids might vary with different situations. The nature of the demand may change from one time to another, being determined by the physiological state, such as pregnancy, lactation, or growth, or the pathological state, such as infection, the response to trauma, or any other reason for an acute-phase response.

### 4.2.1 Amino acid structure

Amino acids constitute the building blocks of proteins. There are 20 amino acids required for protein synthesis and these are all 'metabolically essential' (Table 4.1). Of the 20 amino acids found in protein, nine have to be provided from the diet for adults and are identified as being 'indispensable' or 'essential'.

The other amino acids do not have to be provided from the diet, as long as they can be formed in the body from appropriate precursors in adequate amounts, and are identified as being 'dispensable' or 'non-essential' (Table 4.1). The non-essential amino acids are not necessarily of lesser biological importance. They have to be synthesized in adequate amounts endogenously. Their provision in the diet appears to spare additional quantities of other (essential) amino acids or sources of nitrogen that would be required for their synthesis.

In early childhood, a number of amino acids, which are not essential in adults, cannot be formed in adequate amounts, because the demand is high, the pathways for their formation are not matured, and/or the rate of endogenous formation is not adequate. These amino acids have been identified as being 'conditionally' essential, because of the limited ability of their endogenous formation relative to the magnitude of the demand. There may be disease situations during adult life when, for one reason or another, a particular amino acid, or group of amino acids, becomes conditionally essential (Table 4.1).

**Table 4.1 Amino acid classification, names and abbreviations**

| Classification | Amino acid | 3-letter abbreviation | 1-letter abbreviation |
|---|---|---|---|
| Indispensable (essential) amino acids | Histidine | His | H |
| | Isoleucine | Ile | I |
| | Leucine | Leu | L |
| | Lysine | Lys | K |
| | Methionine | Met | M |
| | Phenylalanine | Phe | F |
| | Threonine | Thr | T |
| | Tryptophan | Trp | W |
| | Valine | Val | V |
| Conditionally indispensable (essential) amino acids | Arginine | Arg | R |
| | Asparagine | Asn | N |
| | Cysteine | Cys | C |
| | Glutamine | Gln | Q |
| | Glycine | Gly | G |
| | Proline | Pro | P |
| | Serine | Ser | S |
| | Tyrosine | Tyr | Y |
| Dispensable (non-essential) amino acids | Alanine | Ala | A |
| | Aspartate | Asp | D |
| | Glutamate | Glu | E |

Although most of the amino acids are found in proteins, many amino acids also have metabolic activities that are not directly related to the formation of proteins. For example, some are major neurotransmitters (e.g. glutamate, *gamma*-aminobutyric acid) and others act as precursors for other important metabolically active compounds (Table 4.2). Additionally, not all amino acids are found in proteins and there are a number of metabolically important amino acids that play no direct part in the

**Table 4.2 Metabolic intermediates and products with amino acids as precursors**

| Product/function | Examples |
|---|---|
| Nucleotides | Formation of DNA, RNA, ATP, NAD |
| Energy transduction | ATP, NAD, creatine |
| Neurotransmitters | Serotonin, adrenaline, noradrenaline, acetylcholine |
| Membrane structures | Head groups of phospholipids: choline, ethanolamine |
| Porphyrin | Haem compounds, cytochromes |
| Cellular replication | Polyamines |
| Fat digestion | Taurine, glycine-conjugated bile acids |
| Fat metabolism | Carnitine |
| Hormones | Thyroid, pituitary hormones, insulin |

formation of proteins (e.g. citrulline). There is a relatively small pool of free amino acids in all tissues. This is the pool from which amino acids for protein formation are derived and to which amino acids coming from protein degradation contribute; therefore, the amino acid pools have a very high rate of turnover.

Structurally, each amino acid consists of a central carbon atom (α carbon) linked to a hydrogen atom, a carboxylic acid group, an amino group, and a variable side chain (also known as the R group) (see Fig. 4.2). The varying size, shape, charge, hydrophobic properties, and hydrogen-bonding capacity of the variable side chain renders each amino acid unique. At neutral pH, amino acids exist as dipolar ions (often referred to as zwitterions) whereby the carboxyl group is deprotonated ($-COO^-$) and the amino group is protonated ($-NH_3^+$). Therefore, the net charge of the amino acid is neutral unless the side chain carries charge. The ionization state of amino acids also varies based on pH.

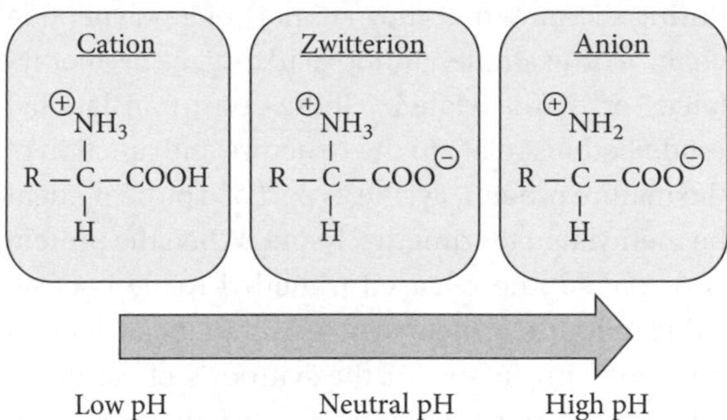


Fig. 4.2 Amino acid structure.

# 4.3 Amino acid turnover

The amino acid pool is the precursor pool from which all amino acids are drawn for protein synthesis and other pathways. It is helpful to identify the pool for each amino acid individually and to consider the general factors that are of importance (see Fig. 4.3). For any amino acid, there are three inflows to the pool and three outflows. The flows from the pool to protein synthesis and other metabolic pathways represent the metabolic demand for the amino acid. Flow to amino acid oxidation is determined either by the need to use the amino acid as a source of energy, or as a degradative pathway for amino acids in excess of what can be used effectively at that point in time. This demand has to be satisfied from:

- amino acids coming from protein degradation; or
- *de novo* synthesis of amino acids; or
- dietary amino acids.

It might be expected that, in a steady state, amino acids coming from protein degradation would represent a perfect fit for the proteins that are being synthesized, and hence there should be no general need for amino acids to be added to the system. However, this is not so. The amino acids released from protein degradation are different from those used in protein

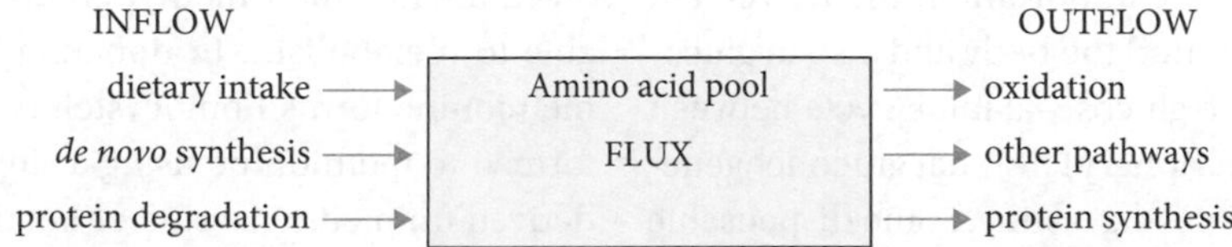


Fig. 4.3 The dynamic turnover of any amino acid in the body can be characterized by a model in which there is a single pool of an amino acid. Inflows to the pool are from three potential sources: protein degradation, the diet, and *de novo* synthesis. There are three outflows from the pool: to protein synthesis, to other metabolic pathways, and to oxidation with the breakdown of the amino acid.

synthesis, because some amino acids are altered during the time that they are part of a polypeptide chain. For example, amino acids might be methylated or carboxylated. These post-translational modifications relate to the structure and function of the mature protein. Lysine as part of a protein might be methylated to trimethyl lysine. When the protein is degraded, the released trimethyl lysine is of no value in future protein synthesis, but it can act as a metabolic precursor for the synthesis of carnitine. Carnitine plays a fundamental role in fatty acid metabolism. The endogenous formation of carnitine facilitates fatty acid oxidation and limits the need for a dietary source of carnitine, but an additional source of lysine is still needed for protein synthesis to be maintained. As lysine is an indispensable amino acid and cannot be synthesized endogenously in quantities sufficient to satisfy the metabolic need, lysine has to be obtained preformed in the diet.

Another example is that of 3-methylhistidine (3-MH), which is a post-translational modification of contractile protein histidine residues. Since 3-MH is not subject to re-incorporation into nascent peptide synthesis, due to no aminoacyl tRNA for 3-MH, 3-MH can be measured in the urine and is used as a marker of myofibrillar muscle protein breakdown. A limitation of this approach is the optimal requirement of a meat-free diet, since meat is an alternative source of 3-MH and may alter urinary 3-MH levels. Additionally, using 3-MH as a marker of myofibrillar breakdown assumes methylated-histidine is exclusively derived from skeletal muscle, which is incorrect since 3-MH is also found in other tissues such as cardiac muscle, smooth muscle, and the gut.

Hydroxylation (i.e. the addition of a hydroxyl group (-OH) to amino acids) is another post-translational modification. Hydroxylation occurs on several amino acids, chiefly proline and lysine because collagen proteins are the most abundant proteins in the human body. Hydroxylation of proline and lysine forms hydroxyproline and hydroxylysine, respectively, which are required for collagen strands to appropriately cross-link and thus are important for collagen synthesis.

## 4.4 Amino acid formation and oxidation

The *de novo* synthesis of an amino acid requires that its carbon skeleton can be made available from endogenous sources, and this skeleton then has an amino group effectively added in the right position. The sulphur moiety has to be added for the sulphur amino acids. In mammalian metabolism, some amino acids are readily formed from other metabolic intermediates, for example transaminating amino acids, alanine, and glutamic and aspartic acids, derived from intermediates of the citric acid cycle, pyruvate, α-ketoglutarate, and oxaloacetate. These amino acids are important in the movement of amino groups around the body and also in gluconeogenesis (e.g. the glucose–alanine cycle between the liver and the periphery) or renal gluconeogenesis from glutamine during fasting. Some dispensable amino acids derive directly from an indispensable amino acid (e.g. tyrosine from phenylalanine) and the endogenous formation of the amino acid is determined by the availability of the indispensable amino acid. Methionine and cysteine are amino acids that contain a sulfhydryl group, with considerable chemical activity. Although methionine can be formed in the body from homocysteine, this is part of a cycle (methionine cycle) that generates methyl groups for metabolism whilst also contributing methyl groups to other amino acids and other methylations, and in which there is no net gain of methionine (see Fig. 4.4). Methionine has a number of other important functions. In addition to acting as a signal for protein synthesis, it is the precursor for cysteine and other reactions in which methyl groups are made available to metabolism. In donating its methyl group, methionine forms homocysteine, which can be reformed to methionine using a single carbon group, derived from either serine or betaine (a breakdown product of choline). Homocysteine has an alternative metabolic fate, towards the formation of cysteine. In this pathway, the sulfhydryl group derived from methionine is made available to a molecule

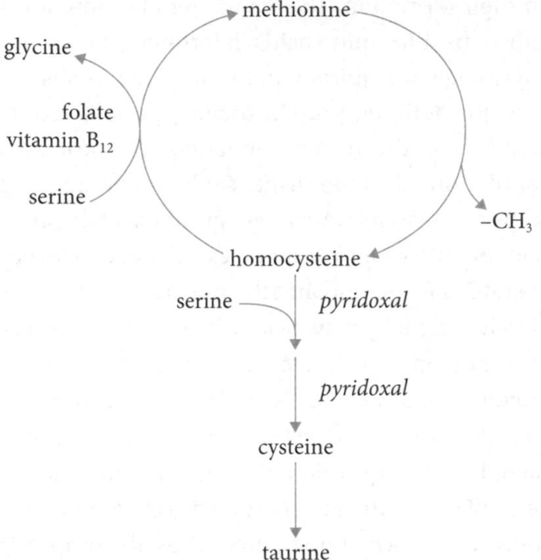


Fig. 4.4 The methionine cycle. Methionine has to be provided in the diet and, like other amino acids, is required for protein synthesis. In addition, it has a number of other important functions. Apart from acting as a signal for protein synthesis, it is the precursor for cysteine and other reactions in which methyl groups are made available to the metabolism. In donating its methyl group, methionine forms homocysteine, which can be reformed to methionine using a single carbon group, derived from either serine or betaine (a breakdown product of choline). Homocysteine is a branch point in metabolism as it also has another important fate, the formation of cysteine and taurine. In these pathways, the sulphydryl group is transferred to the carbon skeleton of serine. Thus, serine is used for the further metabolism of homocysteine, down either pathway. Increased amounts of homocysteine in the circulation are associated with increased risk of cardiovascular disease.

of serine with the formation of cysteine; the carbon skeleton derived from methionine is subsequently oxidized. Cysteine is the precursor for taurine, which with glycine is required for the formation of bile salts from bile acids. Thus, cysteine can be formed in the body provided there is sufficient methionine and serine available. However, the pathway for cysteine formation has not fully matured in the newborn, making cysteine a conditionally essential amino acid at this time.

For some amino acids, the pathway for their formation appears complex and tortuous, involving a complex pathway that is shared between a number of tissues. The renal formation of arginine is an example (see Fig. 4.5). Arginine is the precursor of nitric oxide, which has important metabolic functions as a neurotransmitter, in maintaining peripheral vascular tone, and is one of several oxidative radicals formed during the oxidative burst of leukocytes. Arginine is also the direct precursor of urea, the main form in which nitrogen is excreted from the body. Arginine is formed in two main sites, the kidney and the liver. Arginine formed in the kidney is available for the rest of the body, whereas that formed in the liver is cleaved to urea and ornithine in the urea cycle. In both locations the arginine is made from citrulline, but while in the liver the citrulline is generated locally in the mitochondria, in the kidney it is imported. The imported citrulline

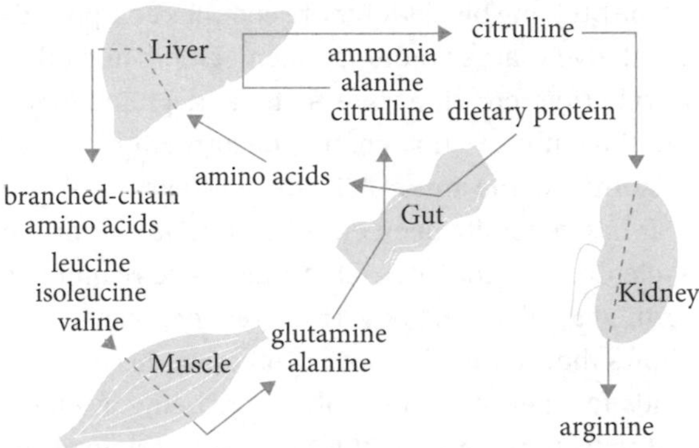


Fig. 4.5 There is a complex system of inter-organ cooperation for many aspects of amino acid metabolism and the partitioning of different functions between organs is of critical importance for effective metabolic control. Arginine is formed in significant amounts in both liver and kidney, but that formed in the liver goes to urea formation in general, whereas that formed in the kidney is exported and available for the needs of the rest of the body in, for example, the generation of nitric oxide. Arginine is formed from citrulline in the kidney, with the citrulline being derived from the gastrointestinal tract, where its formation is linked to the utilization of glutamine. Glutamine itself is formed in large amounts in muscle, in part from the breakdown of the branched-chain amino acids, leucine, isoleucine, and valine. The branched-chain amino acids coming from dietary protein are, as a group, handled in a special way; following absorption they tend to pass through the liver without being metabolized and are preferentially taken up by the muscles.

has been formed in the gastrointestinal tract, one of the end-products of the oxidation of glutamine. The glutamine itself has been generated in muscle from the branched-chain amino acids or through the liver, where a high level of glutamine synthetase compensates for a saturated urea cycle for ammonia nitrogen removal. During digestion and absorption of the amino acids consumed as protein in a meal, most are first taken up by the liver, but more of the branched-chain amino acids pass directly through the liver and are taken up preferentially by muscle, where they give rise to glutamine. This complex pattern of metabolic interchange enables control to be exerted and, in particular, creates a mechanism through which two important sets of metabolic interaction of arginine are separated and therefore can be controlled independently.

Other amino acids, such as glycine, are required as the building blocks for more complex compounds in relatively large amounts (haemoglobin and other porphyrins, creatine, bile salts, and glutathione), but the pathways that enable the formation of large quantities of the amino acid are not clear.

It has generally been considered that the carbon skeleton of the indispensable amino acids cannot be formed at all in the body. However, recent evidence shows that colonic bacteria synthesize these amino acids for their own use, with some being available to the host in amounts that may be physiologically useful.

In a number of inborn errors of metabolism, the enzymes involved in the oxidation of amino acids do not function normally, and there is considerable accumulation of an amino acid or its breakdown products. The extent to which the body can tolerate an excess of any single amino acid might vary, but almost always sustained, high, increased levels of amino acids in the body exert toxic effects. Therefore, the body goes to some lengths to maintain very low levels. The catabolic pathways for individual amino acids are active, although the activity may decrease on very low protein intakes. For the indispensable amino acids in particular, the amounts of individual amino acids usually found in diets are in excess of the usual requirements. This may be the reason why the body has not selected to maintain the pathways for their formation. The dietary requirement for the indispensable amino acids determines the minimal physiological requirement for protein and has been used for defining protein quality, protein requirements, and the recommendations for protein for populations. For the dispensable amino acids, the body is able to tolerate larger amounts of the amino acid, but also control can be exerted at two levels—the rate of formation and the rate of oxidation—and therefore toxicity may be less likely. On a daily basis, for a person in balance, an equivalent amount of protein or amino acids is oxidized as is taken in the diet. Therefore, for a person consuming 80 g protein each day, the equivalent of 80 g of amino acids will be oxidized to provide energy, with the amino group going to the formation of urea. Thus, the proportion of total energy derived from protein each day is similar to the relative contribution of protein to the energy in the diet.

### 4.4.1 Inborn errors of AA metabolism

There are about 100 uncommon or rare aminoacidopathies due to a genetic mutation that affects the function of an enzyme or transporter involved in the metabolism of one (or more) amino acid. Notable examples are phenylketonuria and maple syrup urine disease.

### 4.4.2 Phenylketonuria

The aromatic amino acid phenylalanine is essential for protein synthesis and over half of it is converted by phenylalanine hydroxylase (PAH) to tyrosine, another important amino acid. When PAH is not functioning, because of a mutation of its gene on chromosome 12, phenylalanine can only go down minor pathways to phenylpyruvic acid, which passes into the urine as phenylketonuria (PKU).

PKU was discovered by Følling in 1934. In an institution for people with a learning disability, the urine of a few cases gave a green colour with ferric chloride, due to phenyl ketone. PKU leads to irreversible, severe intellectual disability unless the elevated blood phenylalanine is brought down in the first few weeks of the infant's life and kept down throughout childhood and beyond. In developed countries, all newborns are screened routinely, with a heel prick for elevated plasma phenylalanine, above 250 μmol/L.

Management has to start with a low-phenylalanine milk, such as Lofenalac®, and go on to low-protein natural foods, plus special medical foods that are low in phenylalanine and contain extra tyrosine. The diet has to be titrated against plasma phenylalanine levels, weekly in early infancy. On the other hand, if the phenylalanine level is too low, deficiency of this essential amino acid (EAA) impairs growth from protein malnutrition. Dietetic treatment of PKU is specialized and demanding for parents and child.

PKU is a Mendelian recessive condition, occurring in about 1 in 10,000 infants.

### 4.4.3 Maple syrup urine disease (MSUD)

MSUD is characterized by the deficiency of the enzyme branched-chain alpha-keto acid dehydrogenase complex, which compromises the ability to break down the essential amino acids leucine, isoleucine, and valine. This leads to the harmful build-up of these amino acids and their toxic by-products (ketoacids), leading to the production of sweet-smelling urine, with a scent similar to that of maple syrup, giving rise to the name: MSUD.

The onset of MSUD symptoms typically occurs within the first few days to weeks after birth, which include sweet-smelling urine and sweat, weight loss, and loss of appetite/poor feeding. It is also possible for MSUD sufferers, especially babies, to experience 'metabolic crisis', particularly in response to infection (e.g. high temperature), which can be life-threatening. Signs of a 'metabolic crisis' include breathing difficulties, vomiting, lack of energy, and irritability, and require immediate medical attention.

Currently, there is no method to prevent MSUD. A key treatment method is diet control. This primarily consists of consuming a low-protein diet to carefully control the amount of leucine, isoleucine, and valine. It also includes consuming tailored metabolic formulas that aim to provide the necessary vitamins, minerals, trace elements, omega 3s, and other essential amino acids that are restricted due to the adoption of a low-protein diet. The only current treatment option that leads to the termination of dietary control and eradicates the occurrence of 'metabolic crisis' is a liver transplant. However, the risks of the associated surgery and potentially lifelong immunosuppressant treatment means that it must be considered with caution. A liver transplant is not considered a cure for MSUD as the patient will still be unable to produce the branched-chain alpha-keto acid dehydrogenase complex enzyme.

MSUD is an autosomal recessive disorder, which means that two copies of the defective gene have been inherited—one from each parent. The parents themselves are carriers of the defective gene but typically do not display symptoms of the condition. MSUD is rare, affecting 1 newborn out of ~185,000 live births worldwide.

### 4.4.4 Other AA disorders

**Cystinuria** is one of the earliest recognized inborn errors of metabolism. There is a defect in transport across membranes of cystine and dibasic amino acids (lysine, arginine). The only clinical effect is urinary tract stone formation, because the increased urinary cystine has low solubility. This is an uncommon cause of urinary tract stones. On microscopy of the urine, flat hexagonal crystals of cystine are diagnostic.

**Homocystinuria** is usually due to a defect in the enzyme that converts homocysteine, a metabolite of methionine, to cystathionine (and, hence, to cysteine). People with homocystinuria may have skeletal deformities, arachnodactyly, and dislocated lenses. They may also have premature vascular disease.

**Hartnup disease** has a pellagroid skin rash and intermittent neurological symptoms. The inherited defect is in renal transport of the neutral amino acids, so there is aminoaciduria. The most critical effect is loss of tryptophan. Symptoms are much improved with high-dose nicotinamide and a high-protein diet. Hartnup was the name of the first family described.

## 4.5 Protein structure

The structure and function of all proteins is related to their amino acid composition: the number and order of linkages, folding, intrachain linkages and the interaction with other groups to induce chemical change, e.g. phosphorylation/dephosphorylation, and oxidation and reduction of sulphhydryl groups. The amino acids are linked in chains through peptide bonds. Proteins may have critical requirements for either a tight or loose association with micronutrients and this is particularly likely for enzymes in which catalytic activity might require the presence of cofactors or prosthetic groups in close association with the active centres.

For each protein, the amino acid composition is characteristic. For a protein to be synthesized, it requires that all the amino acids needed are available at the point of synthesis. If one amino acid is in short supply, this will limit the process of protein synthesis; such an amino acid is defined as the 'limiting amino acid.'

### 4.5.1 Primary structure

Proteins in the most basic form are linear polymers formed by linking the α-amino group of one amino acid to the α-carboxyl group of another amino acid via a peptide bond (also known as an amide bond). When this peptide bond is produced, a water molecule is lost (synthesis condensation reaction) and when this bond is broken, a water molecule is used (hydrolysis reaction). While there are only 20 amino acids that are incorporated into proteins (i.e. proteinogenic amino acids), these exist in peptide chains in different sequences that give rise to thousands of different proteins, much like the alphabet only has 26 letters but can produce thousands of words by putting letters together in a different sequence. If one amino acid (or more) is not in the correct location, this can result in the production of a protein that is useless or harmful, again much like a single letter being in the wrong place leads to a misspelling and the wrong word being produced.

Table 4.3 **Amino acid-relevant prefixes**

| Prefix | Name | Number of amino acids |
|---|---|---|
| *Di* | Dipeptide | 2 |
| *Tri* | Tripeptide | 3 |
| *Tetra* | Tetrapeptide | 4 |
| *Penta* | Pentapeptide | 5 |
| *Hexa* | Hexapeptide | 6 |
| *Hepta* | Heptapeptide | 7 |
| *Octa* | Octapeptide | 8 |
| *Nona* | Nonapeptide | 9 |
| *Deca* | Decapeptide | 10 |
| *Oligo* | Oligopeptide | 10–20 |
| *Poly* | Polypeptide | 20+ |

Each amino acid joined into a peptide chain is referred to as an amino acid residue. The specific location of an amino acid unit within a peptide chain can be described by naming the amino acid and the location within the chain. For example, Thr389 refers to threonine, which is 389th in the amino acid sequence, where the amino acid at the N terminus is the first amino acid. Different prefixes can be used depending on the number of amino acids constituting a peptide chain (Table 4.3).

### 4.5.2 Secondary structure

The secondary structure of a protein refers to the stretch of a polypeptide chain, which can fold into two common conformations: alpha-helix or beta-pleated sheets. Alpha-helixes are amino acids in the spiral or coiled conformation, which allows hydrogen bonding to form between nitrogen and hydrogen. The alpha-helix conformation is found in proteins such as intermediate filaments and haemoglobin. Beta-pleated sheets are amino acids in a series of adjacent rows, which allows hydrogen bonds to form between the amino acid group and the carboxyl group of another amino acid. The beta-pleated sheet conformation occurs in proteins such as antibodies and fatty acid transport proteins.

### 4.5.3 Tertiary structure

The backbone of a polypeptide chain repeatedly folds, giving rise to a three-dimensional tertiary structure. Given the folding nature, this means that amino acids far apart in the primary structure may now be closer in proximity. Indeed, it is only when a protein has folded into the tertiary structure that it can function. Many types of chemical bonds give rise to the tertiary structure, including hydrophobic interactions, where hydrophobic fat-soluble amino acids will fold away from contact with water and towards the inside of the protein, and hydrophilic water-soluble amino acids will fold towards contact with water and away from the inside of the protein. Stronger covalent bonds (disulphide bonds) can also form at the tertiary level.

### 4.5.4 Quaternary structure

The quaternary structure of a protein is where more than one polypeptide chain, held together by hydrogen bonds, constitutes a protein. Each of these subunits has its own primary, secondary, and tertiary structure. An example of a protein with multiple polypeptide chains (referred to as subunit/s) is myosin, the contractile protein found in muscle cells, which is composed of two myosin heavy chains and four myosin light chains.

## 4.6 Protein function and status

Proteins participate in many different functions, which can be broadly classified into several categories, as described in Table 4.4. It should be noted, however, that many proteins can perform diverse roles and may not be confined to one single category/function.

### 4.6.1 Protein status of the body

There is no single way in which the protein status of an individual or the amount of protein contained in the body can be directly determined. However, most of the nitrogen in the body is present as amino acids in protein, and most protein is present as lean tissue. Thus, the measure of total body nitrogen using such techniques as *in vivo* neutron activation analysis (an expensive research procedure that requires special equipment and exposes the subject to radiation) can help estimate protein levels. Furthermore, on the assumption that the protein content of tissues remains fairly constant, the determination of the lean body mass (which approximates the fat-free mass) also gives an index of the total protein content. Indirect measures of lean body mass include bioelectrical impedance, assessment of total body water, total body potassium, underwater weighing, or fat-fold thickness. Organ and tissue size (and hence lean mass) can be determined by imaging techniques (e.g. magnetic resonance imaging, dual-energy X-ray absorptiometry). Muscle mass and changes in muscle content in clinical situations can be assessed from the urinary excretion of creatinine or the girth of limbs, such as the mid-upper arm circumference or thigh circumference. Determinations of the concentration of amino acids in plasma or the albumin concentration have been used as indirect indices of body protein status. However, each of these have their limitations.

Assessment of the function of the metabolically active tissues has also been used as an index of protein status, for example muscle function tests, liver function tests, and tests of immune function. Yet, the content of the body is not static, and therefore measurements of the rates at which proteins are formed and degraded in the body, and the rate at which nitrogen flows to the end-products of metabolism, provide one of the most important approaches to determining the mechanisms through which nitrogen equilibrium is maintained in health: positive balance is achieved during growth and negative balance is brought about during wasting conditions.

Table 4.4 **Types and functions of proteins**

| Type | Function | Working example | Example proteins |
|---|---|---|---|
| Structural | Maintain cell structure | Within skeletal muscle, the protein dystrophin provides key structural support by tethering the plasma membrane to the myofibrils. However, mutations to the dystrophin gene cause Duchenne's muscular dystrophy, which is characterized by progressive muscle wasting resulting in premature disability and death | Actin<br>Myosin<br>Collagen<br>Keratin |
| Enzymes | Accelerate biochemical reactions without being altered themselves | The enzyme glycogen phosphorylase breaks down muscle glycogen to provide ATP for muscle contraction | Amylase<br>Lipase<br>Lactase |
| Hormones | Chemical messengers secreted by endocrine glands to coordinate processes around the body | In response to elevated blood glucose levels (e.g. after a meal), the hormone insulin is secreted from the pancreas and binds to a receptor protein on the plasma membrane, therein initiating an intracellular signalling cascade that increases glucose uptake from the bloodstream into key organs (e.g. muscle), ultimately returning blood glucose levels back to baseline | Glucagon<br>Insulin |
| Transport | Carry substances into and out of cells, around the body, and into and out of intracellular organelles | The protein haemoglobin transports oxygen into muscles | Albumin<br>GLUT1<br>Protein channels/pumps |
| Contractile | Facilitate muscle contraction | Within muscle fibres, the protein myosin 'slides' past the protein actin to contract the muscle fibre, a process that requires ATP | Actin<br>Myosin |
| Immunological | Fight infection by recognizing and neutralizing foreign pathogens such as viruses and bacteria | Influenza vaccine provides a small dose of virus in the dead or inactive form so that our bodies can synthesize the necessary antibodies needed to produce an effective immune response | Immunoglobulin G<br>IL-6<br>TNF-α |
| Regulatory | Regulate the rate of transcription by binding to a specific DNA sequence (transcription factor) | In response to exercise, PGC-1α binds to the transcription factor TFAM, which increase mitochondrially-related gene expression | Tfam<br>MyoD<br>Mef2 |

## 4.7 Protein turnover and control in muscle

On average, about 50% of protein synthesis takes place in the visceral tissues, with liver predominating (25%), and 50% takes place in the carcass, with muscle predominating (25%). Although the mass of liver is much less than the mass of muscle, the intensity of turnover in liver (fractional turnover, 100% per day) is much greater than that in muscle (fractional turnover, 18% per day). In normal adults, about 4 g protein/kg body weight are synthesized each day: about 300 g protein/day in men and 250 g protein/day in women. In newborn infants, the rate is about 12 g protein/kg, falling to about 6 g/kg by 1 year of age. Basal metabolic rate is closely related to the size, shape, and body composition of an individual, and the same is true for protein synthesis. In adults, in a steady state, protein synthesis is matched by an equivalent rate of protein degradation, but in infancy and childhood, because of the net tissue and protein accretion associated with growth, protein synthesis exceeds protein degradation.

Muscle contains a large proportion of the protein in the body. Considerable interest has been shown in the growth of muscle, which is commercially important for the livestock industry, whilst also becoming a key aspect in maintenance of human health and in competitive sports. Muscle is one tissue most affected by wasting and, because it is relatively accessible, it has been the subject of a more detailed study *in vivo* in humans than any other tissue. Hormones, the availability of energy and nutrients, and muscle activity (loading) all contribute to the rate of muscle tissue deposition.

### 4.7.1 Diurnal regulation of muscle protein turnover by dietary protein and exercise

In healthy, young, recreationally active adults, diurnal muscle protein turnover exists in a dynamic equilibrium whereby muscle protein synthesis (MPS) exceeds muscle protein breakdown (MPB) in the fed state, and MPB exceeds MPS in the fasted state. This 'fasted-loss/fed-gain' cycle ensures skeletal muscle mass remains constant, exhibiting turnover rates of ~1.2% day$^{-1}$.

The provision of dietary protein is a principal determinant of adult skeletal muscle proteostasis, driven by the transfer and incorporation of dietary-derived amino acids into skeletal muscle proteins. This is demonstrated by the fact that the muscle protein synthetic response to a mixed meal (containing carbohydrate, protein, and fat) is entirely attributable to the essential amino acids. In isolation, as little as 10–20 g of essential amino acids (equating to ~20–40 g of high-quality protein) is needed to maximally stimulate MPS. In the rested non-exercised state, the MPS response is saturable and transient. Roughly 45–60 minutes after oral ingestion (time taken for digestion, absorption, and transport of amino acids into systemic circulation), MPS increases ~2-3-fold, peaks at around 1.5–2 h following ingestion, and thereafter returns to baseline at ~2–3 h, despite continued plasma and muscle amino acid availability and sustained intracellular anabolic (growth) signalling. At this point, muscle becomes refractory to MPS stimulation, whereby the muscle cells sense and divert essential amino acids towards oxidation, instead of incorporation into muscle proteins, a phenomenon termed 'muscle full'. This refractory response is why it is impossible to achieve muscle growth (hypertrophy) through consuming dietary protein alone (in healthy adults), even in excess. In addition to stimulating MPS, dietary protein stimulates the secretion of insulin similar to that of a mixed meal, causing a ~50% inhibition of MPB. So, dietary protein regulates protein turnover via both MPS and MPB, although the greater change in MPS renders MPS the driving force behind nutrient-induced anabolism.

Exercise is another principal determinant of adult skeletal muscle proteostasis. Following resistance exercise, both MPS and MPB transiently increase, resulting in an overall negative protein balance. As such, exercise must be coupled with adequate protein intake to achieve positive net protein balance. Exercise increases the anabolic sensitivity of the muscle, therein delaying the onset of 'muscle full', where increasing the availability of dietary EAA post-exercise enhances the magnitude and duration of the MPS response. This exercise-enhanced MPS response to protein intake can last for 24 hours. Despite exercise sensitizing the muscle to protein feeding, limits still exist whereby muscle full is still reached. In response to repeated bouts of resistance exercise performed over chronic periods of time (i.e. >3 weeks), the post-exercise positive net protein balance accrues, resulting in muscle growth (hypertrophy).

Protein turnover in muscle is also responsive to the hormonal state, with testosterone, growth hormone, and insulin-like growth factor 1 (IGF-I) all considered to have anabolic effects. So far, only testosterone has been demonstrated to directly stimulate MPS in humans, whereas direct robust evidence is currently lacking for growth hormone and IGF-I. Nonetheless, growth hormone and IGF-I may still be important for muscle turnover, as evidence in rodents shows IGF-I infusion or overexpression promotes muscle hypertrophy and IGF-I receptor knockout leads to muscle atrophy. Growth hormone also stimulates collagen synthesis, therein supporting extracellular matrix remodelling. In humans, resistance exercise transiently increases systemic hormones; however, associations between elevated individual hormones and ensuing muscle mass remain debated.

Corticosteroids produce a decrease in synthesis and an increase in degradation. Thyroid hormones increase protein synthesis and degradation. However, at normal physiological levels they have a greater effect on synthesis than on degradation, thus exerting a net anabolic effect. At hyperthyroid levels, the increase in degradation exceeds the increase in synthesis and thus the hormone exerts an overall catabolic effect. β-Adrenergic anabolic agents, such as clenbuterol, promote muscle growth by decreasing protein degradation.

Prostaglandins (PGs) have a direct effect on muscle protein synthesis and degradation, with $PGF_{2a}$ stimulating synthesis and $PGE_2$ stimulating degradation. The activity of phospholipase A2 makes arachidonic acid available for the synthesis of $PGF_{2a}$ (blocked by indomethacin). The activity of phospholipase A2 is enhanced by stretch (activation of calcium) or insulin (cyclic adenosine monophosphate (AMP)) and inhibited by glucocorticoids. Lipopolysaccharide exerts an influence on the release of arachidonic acid under the action of phospholipase A2, with the formation of $PGE_2$, which increases protein degradation through increased proteolysis within the lysosomes.

Stress, trauma, and surgery have all been shown to induce a negative nitrogen balance in rats and humans. In wasted individuals or people on low-protein diets, a catabolic response, with negative nitrogen balance, may not be evident. Those who do not show this response are more likely to die, giving the impression that the negative response is purposive and that under these circumstances, the catabolic response is more important than conserving body protein.

### 4.7.2 Molecular regulation

In addition to being substrates for MPS, certain EAA also act as signalling molecules regulating the MPS response. Following trans-sarcolemmal EAA transport, leucine (and perhaps other EAA) activates the mechanistic target of rapamycin complex-1 (mTORC1), independent of proximal insulin (e.g. PI3K/AKT) signalling. mTORC1 activation induces phosphorylation of 4E-binding protein and ribosomal protein S6 kinase (rps6), stimulating the binding of the eukaryotic initiation factor (eIF) 4A and eIF4E to eIF4G, forming the eIF4F complex. Thereafter, the eIF4F complex mediates mRNA binding to the 43S preinitiation complex, ultimately resulting in the formation of the 48S preinitiation complex. This relay system triggered by intracellular EAA accumulation results in an upregulation in rates of mRNA translation and, thus, MPS. Whilst the EAA-induced upstream regulation of mTORC1 is incompletely defined, recent evidence suggests leucine binding to Sestrin-2 (an intracellular leucine sensor of the mTOR pathway) and subsequent dissociation of Sestrin-2 and the GTPase-activating protein, GATOR2, leads to mTORC1 activation.

The molecular mechanisms regulating anabolic responses to exercise are more complex than those with nutrition alone, as resistance exercise stimulates multiple intracellular signalling networks associated with biochemical, mechanical, and metabolic stress. Like nutrition-induced MPS responses, mTORC1 is central to coordinating the anabolic response to exercise. Indeed, downstream mTORC1 substrates (e.g. rps6) are transiently upregulated in the hours following a bout of exercise. The precise regulation of mTORC1 by mechanical stimuli (termed 'mechanotransduction') remains to be fully determined, with evidence to suggest that a canonical pathway via IGF1-PI3K-Akt/PKB-mTOR exists. Emerging evidence also supports muscle intrinsic mechanosensitive signalling pathways, for example integrin-adhesome (also known as 'focal adhesions' or 'costameres') proteins such as focal adhesion kinase have been associated with activating mTORC1 post exercise, as has phospholipase D and its lipid second messenger, phosphatidic acid (PA).

MPB takes place through three major pathways—the calcium-protease, ubiquitin-proteosome, or lysosomal pathways.

### 4.7.3 Energy cost of protein turnover

Both protein synthesis and protein degradation consume energy, 4 kJ/g of protein of average composition. At the level of the whole body, the biochemical cost of peptide bond formation is estimated to be about 15–20% of the resting energy expenditure. There are additional energy costs to the body if protein turnover is to be sustained, as processes such as transport of amino acids into cells require energy. If the full energy costs of maintaining the system are included, then the physiological cost is probably in the region of 33% of resting energy expenditure.

There is a general interaction between the intake of dietary energy and nitrogen balance. Thus, although nitrogen balance and protein synthesis appear to be protected functions, modest increases in energy intake lead to positive nitrogen balance and decreasing energy intake results in a transient negative nitrogen balance. In general, about 2 mg nitrogen is retained or lost for a change in energy intake of 4 kJ (1 kcal).

## 4.8 Injury and trauma

Injury and trauma are characterized by an inflammatory or acute-phase response. This is a coordinated metabolic response by the body that appears to be designed to limit damage, remove foreign material, and repair damaged tissue. Under the influence of cytokines, there is a shift in the pattern of protein synthesis and degradation in the body. Substrate from endogenous sources is made available to support the activity of the immune system. In muscle, protein synthesis falls, and protein degradation might increase, resulting in net loss of protein from muscle, with wasting. The amino acids made available by muscle wasting may provide substrate for protein synthesis in the liver and the immune system. In the liver, there is a shift in the pattern of proteins synthesized, with a reduction in the formation of the usual secretory proteins, albumin, lipoproteins, transferrin, retinol-binding protein, and so on, and an increase in the formation of the acute-phase reactants, such as C-reactive protein, α-1-acid glycoprotein, and β-2-macroglobulin. In combination with a loss of appetite, the changes in protein turnover in liver and muscle result in a negative nitrogen balance. There is usually an increase in protein degradation overall, and the intensity of the increase is determined by the magnitude of the trauma. At the lower end of the scale, uncomplicated elective surgery is characterized by losses of less than 5 g nitrogen/day. At the upper end of the scale, burn injury can lead to losses in excess of 70 g nitrogen/day. The dietary intake appears to have an important influence on the extent to which protein synthesis can be maintained, and therefore the magnitude of the negative nitrogen balance and the severity of the consequent wasting.

## 4.9 Protein turnover measurement methods

Dynamic amino acid and protein metabolism can be safely measured in human muscle tissue using stable non-radioactive isotope tracers. Stable isotopes are species of an element that occupy the same position in the periodic table and are essentially chemically identical but differ in mass due to a different number of neutrons within the atomic nucleus. The difference in mass permits the measurement of these less common isotopes, such as those used as tracers (e.g. $^{2}H$, $^{13}C$, $^{15}N$, and $^{18}O$), from common isotopes, which make up most of our biological environment (e.g. $^{1}H$, $^{12}C$, $^{14}N$, and $^{16}O$). The distinction and measurement of stable isotopes from their more common counterparts mostly requires mass spectrometry equipment, which separates atoms or molecules based on their mass and/or charge.

The experimental use of traditional substrate-specific isotope tracers, such as amino acid tracers, allows us to measure the incorporation of isotopic motifs into biological tissues. For example, isotopically labelled amino acids can be used to measure muscle protein synthesis in protein obtained from muscle biopsy tissue. However, substrate-specific tracers used for protein turnover measurements require intravenous cannulation(s), sterile infusates, and multiple muscle tissue biopsies from volunteers held within a controlled clinical/laboratory environment. As such, measures of amino acid/protein metabolism are typically limited to <24 h, with most commonly captured over a ~2–8 h period, which affords insight into acute human metabolic responses to stress, nutrition, health, and disease. However, the invasive and time-restrictive nature of traditional tracers limits the application to free living environments and longer-term settings.

An alternative approach is the use of deuterium oxide (also referred to as $D_2O$, 'heavy water', or $^2H_2O$), which was one of the first stable isotope tracers to be used in metabolic research but is only re-emerging due to exponential advances in the sensitivity of mass spectrometry. $D_2O$ can be administered orally as a single bolus or as multiple regular doses, and it rapidly equilibrates with body water (~1–2 h in humans) creating a homogenous labelled, slowly turning over precursor pool (half-life ~9–11 days). Unlike substrate-specific AA tracers, deuterium from body water is exchanged through biological reduction during *de novo* (re)synthesis onto many metabolic substrates at stable C-H positions, including amino acids, therein allowing muscle protein synthesis rates to be measured from the incorporation of labelled amino acids into tissue protein over both acute and chronic time periods (please note, other substrate pools such as RNA/DNA synthesis, glycogenolysis, and triglyceride synthesis can also be measured using this $D_2O$ approach, due to its ability to ubiquitously label multiple substrate pools).

## 4.10 Nitrogen balance

About 16% of protein is amino nitrogen, and therefore by measuring this nitrogen (e.g. by the Kjeldahl method) and multiplying it by 6.25, the approximate protein content of a food or tissue can be obtained. Nitrogen balance identifies the overall relationship between the nitrogen removed from the environment for the body and the nitrogen returned to the environment. The intake is almost completely dietary, mainly as protein, but also in part as other nitrogen-containing compounds, such as nucleic acids and creatine in meat. Nitrogen can be lost from the body through several routes, but 85–90% is lost in urine and 5–10% in stool, with skin and hair or other losses making up the remainder. Nitrogen is lost as soluble molecules in urine as urea (85%), ammonia (5%), creatinine (5%), uric acids (2–5%), and traces of individual amino acids or proteins. There may be large losses of nitrogen through unusual routes in pathological situations (e.g. through the skin in burns, as haemorrhage, or via fistulae).

$$B = I - (U + F + \text{other})\ \text{g nitrogen/day}$$

where B = nitrogen balance, I = nitrogen intake, U = urinary nitrogen, and F = faecal nitrogen.

The achievement of nitrogen balance in response to a change in either intake or losses is brought about largely by a change in the rate at which urea is excreted in the urine. A reduction in the dietary intake of protein is matched by an equivalent reduction in the urinary excretion in urea, which returns nitrogen equilibrium within 3–5 days. Faecal losses of nitrogen on habitual intakes are usually 1–2 g nitrogen/day, or about one-tenth of the intake. However, faecal nitrogen may increase considerably on diets that are rich in non-starch polysaccharides or fibre. There is then a reduction in the excretion of urea in urine equivalent to the increase in faecal nitrogen. Increased cutaneous losses, through excessive sweating, exudation, or burns, are associated with a proportionate decrease in urinary urea.

In a system that is constantly turning over, the retention of body protein is a necessary condition for maintaining the integrity of the tissues and tissue protein. Any limitation in the availability of energy or a specific nutrient will lead to a net loss of tissue and negative nitrogen balance through increased losses of nitrogen. Thus, the major control over the protein content of the body is established by modifications in the rate at which nitrogen is lost from the body. Balance is re-established by a change in the rate of nitrogen excretion, which for the most part means a change in the rate at which urea is excreted. Therefore, it is important to consider in some detail factors that might exert control or influence over the formation and excretion of urea.

## 4.11 Urea metabolism and the salvage of urea nitrogen

Urea is formed in the liver (where the urea cycle takes place), in a cyclical process on a molecule of ornithine (see Fig. 4.6). Within the mitochondrion, carbamoyl phosphate (from ammonia and carbon dioxide) condenses with ornithine to form citrulline. The citrulline passes to the cytosol, where a further amino group is donated from aspartic acid, with the eventual formation of arginine (which has

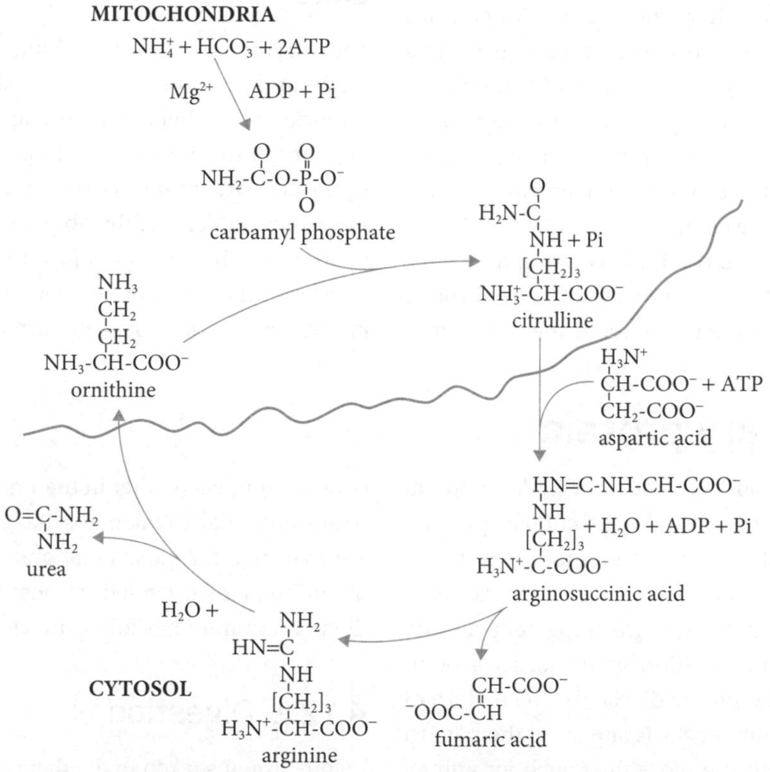

Fig. 4.6 The urea cycle. Urea is formed in the liver in a cyclical process between the mitochondria and the cytosol. A molecule of ornithine and the synthesis of a molecule of carbamyl phosphate from ammonia and $CO_2$ is the starting point with the ultimate formation of arginine, which is hydrolysed to reform ornithine and a molecule of urea. The cycle utilizes three molecules of ATP for each revolution.

three amino groups). This is hydrolysed with the formation of urea and the regeneration of ornithine. Urea is lost to the body by excretion through the kidney. In the kidney, urea fulfils an important physiological role in helping to generate and maintain the concentrating mechanism in the counter-current system of the loops of Henle. The rate of loss of urea through the kidney is influenced by the activity of the hormone vasopressin on the collecting ducts. Much of the urea is reabsorbed from urine in the collecting duct, so nitrogen is potentially retained in the system. Under all normal circumstances, more urea is formed in the liver than is excreted in the kidney. About one-third of the urea formed passes to the colon, where it is hydrolysed by the resident microflora. About one-third of the nitrogen from urea released in this way is returned directly to urea formation, but the other two-thirds are incorporated into the nitrogen pool of the body, presumably as amino acids. In other words, urea nitrogen has been salvaged. There are specific urea transporters in the collecting duct that are upregulated on low-protein diets. Similar transporters exist in the colonocyte, leading to coordinated regulation of increased retention of urea in the kidney with increased movement of urea into the colon.

In situations where the body is trying to economize on nitrogen, the proportion of urea-nitrogen lost in the urine is decreased, and the proportion salvaged through the colon is increased. This happens when the demand for nitrogen for protein synthesis increases, as in growth, or when the supply of nitrogen is reduced, as on a low-protein diet.

The optimal intake of protein is likely to be that which provides appropriate amounts of the different amino acids to satisfy the needs of the system. There is no evidence that the ingestion of large amounts of protein in itself confers any benefit. Indeed, the system may be stressed by the need to catabolize the excess amino acids that cannot be directed to synthetic pathways and have to be excreted as end-products. High protein intakes increase renal blood flow and glomerular filtration rate. In individuals with compromised renal function, this may increase the risk of further damage.

### 4.11.1 Low-protein therapeutic diets

The two clinical situations in which control of protein intake and metabolism are of considerable potential importance are hepatic failure and renal failure. In hepatic failure, there is a limitation in the liver's ability to detoxify ammonia through the formation of urea. In renal failure, the ability to excrete the urea is impaired. In each situation, reduction of the intake or modification of the metabolism of protein or amino acids is an important part of treatment.

## 4.12 Dietary protein

The majority of foods are made of cellular material and therefore in the natural state contain protein. Processing of foods may alter the amounts and relative proportions of some amino acids, for example the Maillard reaction and browning reduces the available lysine (its ε-amino group forms a bond with sugars that cannot be digested). The pattern of amino acids in animal cells is similar to the pattern in human cells and therefore the match for animal protein foods is good; plant materials may have very different patterns of amino acids. This difference has in the past led to the concept of first-class and second-class proteins, for animal and plant foods, respectively. However, diets are hardly ever made up of single foods. In most diets, different foods tend to complement each other in their amino acid pattern, so any potential imbalance is likely to be more apparent than real for most situations. Thus, the mixture of amino acids provided in most diets matches the dietary requirements for normal humans fairly well.

### 4.12.1 Digestion

Before proteins taken in the diet can be utilized, they have to be denatured (i.e. where the protein loses its shape and thus function), which can occur via the cooking process and/or via enzymes in the gut. Thereafter, the denatured proteins are broken down into the constituent amino acids through digestion. The catalytic breaking of the peptide bond is achieved

**Table 4.5 Human protein digestion**

| Organ | Activation | Enzyme | Substrate | Product |
|---|---|---|---|---|
| Stomach | pH <4 | Pepsin | Whole protein | Very large polypeptides with C-terminal Tyr, Phe, Trp, also Leu, Glu, Gln |
| Pancreas | pH 7.5 Enterokinase secreted by small intestinal mucosa | Endopeptidases | Bonds with peptide chain | Peptide with basic amino acid at C terminus (Arg, Lys) |
| | | Trypsin | Peptides | Peptide with neutral amino acid at C terminus |
| | | Chymotrypsin | Peptides | Amino acids |
| | | Exopeptidases | C-terminal bonds | Amino acids |
| | | Carboxypeptidase | Successive amino acids at C terminus | |
| | | Aminopeptidase | Successive amino acids at N terminus | |

*Note:* Enzymes that digest proteins are secreted as inactive precursors (e.g. pepsinogen, trypsinogen, chymotrypsinogen) and are activated under appropriate conditions by the removal of a small peptide from the parent molecule.

through enzymes, which act initially in the acid environment of the stomach, and the process is completed in the alkaline environment of the small intestine. There is a series of proteolytic enzymes that selectively attack specific bonds (Table 4.5). The products of digestion are presented for absorption as individual amino acids, dipeptides, or small oligopeptides. Absorption takes place in the small intestine as an energy-dependent process through specific transporters. There is evidence for the absorption of small amounts of intact protein—it is unlikely to be of great nutritional significance but may be of potential importance in the development of allergies. The extent of absorption of whole proteins is not clear, nor whether this can take place through intact bowel or requires mucosal lesions.

The absorptive capacity of the bowel for amino acids has to be greatly in excess of the dietary intake because there is a considerable net daily secretion of proteins into the bowel. The protein is contained in secretions associated with digestion and the enzymes contained therein, mucins, and sloughed cells. The amount varies, but estimates suggest a minimum of 70 g protein/day and possibly up to 200–300 g protein/day, i.e. the amounts are at least as great as the dietary intake. Therefore, dietary amino acids mix with and are diluted by endogenous amino acids. These dietary and endogenous amino acids are taken up into the circulation and distributed to cells around the body.

# 4.13 How much protein do we need?

## 4.13.1 Protein required from obligatory losses

One approach is to measure the losses of nitrogen on a protein-free diet adequate in energy. Urinary nitrogen (mostly urea) decreases rapidly for the first 5 days and then settles at a new low level. For adults, the 1973 committee of the Food and Agriculture Organization (FAO) and World Health Organization (WHO) estimated:

- *urine loss*: 37 mg/kg
- *faecal loss*: 12 mg/kg
- *skin loss*: 3 mg/kg
- *miscellaneous loss*: 2 mg/kg.

This gives a total of 54 mg nitrogen × 6.25, which is 0.34 g protein/kg/day.

This is an average requirement, so the recommended dietary intake (RDI) (+ 2 standard deviations) should be 0.44 g protein/kg or 29 g protein for a 65 kg adult.

However, this assumes an impossible 100% efficiency of metabolizing dietary protein and empirically it is not possible to achieve equilibrium nitrogen balance on 29 g protein/day. When protein intake is reduced, the many enzymes of amino acid catabolism rapidly decline in activity. Metabolic conditions for absorbed amino acids are far from normal in protein starvation. However, on an ordinary diet they are actively oxidizing and transaminating the amino acids. Thus, although the factorial method using obligatory losses is useful for some other essential nutrients with simple metabolism, it cannot provide us with realistic requirement numbers for protein.

### 4.13.2 From nitrogen balance

International and national recommendations for protein intake are based on this method. If protein intake is insufficient, the nitrogen balance is negative (below zero or equilibrium balance). However, above an adequate protein intake the balance becomes negative for a few days if protein intake is changed from high-adequate to moderate-adequate. This has to be allowed for by testing at intakes near the expected requirement and allowing about 5 days for adjustment at each intake level. Nitrogen balance experiments require attention to detail because nitrogen intakes tend to be overestimated since, for example, people do not swallow every last crumb, and nitrogen losses tend to be underestimated as some urine may be spilt and/or skin losses are very difficult to measure. The energy intake must be neither too much nor too little, because any deviation moves the balance more positive or more negative, respectively.

Rand et al. (2003) selected well-designed balance experiments in healthy adults in several different countries. The protein intake was at two or more levels and near the expected requirement; each intake was for 10–14 days and only the last 5 days were used for calculating the balance. Energy intake was based on the subjects' usual diets. The subjects needed to have a quiet and unvarying life during the experiments, and they lived in a special metabolic unit. The amount of protein for nitrogen equilibrium in 235 subjects in 10 studies was estimated by meta-analysis to be 105 mg nitrogen/kg/day (× 6.25 = 0.65 g protein/kg/day). However, this is an estimated average requirement (EAR). For the RDI, 2 standard deviations were added (and here the numbers were available, not estimated) giving 132 mg nitrogen/kg or (× 6.25) 0.83 g protein/kg. Importantly, the protein should be of good quality, i.e. not too low in any of the indispensable amino acids (see sections 4.15 and 4.16).

In USA and Canada, the RDA set by the Food and Nutrition Board, Institute of Medicine (2002) is 0.80 g protein/kg per day for all adults. FAO/WHO/UNU in 1985 had estimated 0.75 g protein/kg for men and women. Australia and New Zealand (2005) set 0.75 g/kg for women, 0.84 g/kg for men—all under 50 years of age, but for older adults they advise 25% more protein: 0.94 g/kg for women and 1.07 g/kg in men. More recently (2013), specialists in geriatric nutrition concluded that older people use protein less efficiently, and that there is increased splanchnic extraction and a declining anabolic response to ingested protein, perhaps suggesting older people require greater protein intake. Supporting this, Bauer et al. (2013) and the PROT-AGE Study Group recommend average dietary intakes of at least 1.0–1.2 g protein/kg body weight for people over 65 years of age. An associated challenge, however, is that older adults usually eat less calories, food, and thus protein (compared to younger counterparts) because their energy expenditure is lower. Thus, finding novel ways to increase protein intake in older adults is necessary to help offset sarcopenia.

### 4.13.3 Amino acid requirements

Amino acid requirements can vary since i) the indispensable (or essential) amino acids are not all present in the same amounts in body tissues and ii) individual tissues differ in amino acid pattern and in turnover rates.

The first estimates of amino acid requirements by W. C. Rose (in men) (1957), R.M. Leverton and others (in women), and S.E. Snyderman (in infants) in the 1950s used nitrogen balance. The adult subjects were usually college graduate students in the USA. They had to eat a diet of pure corn starch, sugar and syrup, butter, and oil, made into wafers or pudding, plus vitamin tablets and mineral salts. Instead of dietary protein they were given a mixture of pure L-amino acids. They were also allowed a little apple or grape juice (very low in protein), and lettuce and carrot. Average adult requirements from nitrogen balances are given in Table 4.6.

In subsequent years, researchers have reasoned that these requirements (from Rose, 1957) are too low. They add up to only 5.5 g, yet essential amino acids make up one-third to half the total amino acids in dietary proteins and 11–17 g protein would be quite inadequate. Indeed, there were technical limitations with the experiments. For example, Rose's subjects received more dietary energy than expected and in the women's experiments, balance did not always reach equilibrium and nitrogen losses in skin were not considered.

Young and Ajami (1999) estimated essential amino acid requirements by a different approach—biochemical evidence of increased oxidation of an indictor amino acid, labelled with the stable isotope $^{13}C$, when intake of the test amino acid is below requirement (the indicator amino acid oxidation method; Table 4.6). When one essential amino acid is inadequate, protein synthesis using all the amino acids is impaired, so there is both increased oxidation of all amino acids and increased urea production, increased urinary nitrogen, and negative nitrogen balance. In these stable isotope experiments, the diet also consists of protein-free starch, sugar, fat, and oil in cookies with multivitamins and minerals. The L-amino acids are given in a mixture, providing the expected requirement of all essential amino acids and also non-essential amino acids. Only the intake of the test essential amino acid is varied. Typically, each experiment lasts 7 days; each subject is tested at three levels of the test amino acid. On day 7, the $^{13}C$-labelled indicator amino acid (e.g. leucine) is given by constant intravenous infusion over 24 hours and $^{13}CO_2$ is collected and measured to indicate oxidation of the indicator amino acid. These experiments are very expensive and require gas chromatography mass spectrometry.

**Table 4.6 Early and recent estimates of essential amino acid requirements of healthy human adults (mg/kg/day)**

| | N balance method (1) | $^{13}C$ amino acid oxidation method (2) |
|---|---|---|
| Isoleucine | 10 | 19 |
| Leucine | 14 | 42 |
| Lysine | 12 | 38 |
| S-amino acids | 13 | 19 |
| Phenylalanine | 14 | 33 |
| Threonine | 7 | 20 |
| Tryptophan | 3.5 | 5 |
| Valine | 10 | 24 |
| Total | 83.5 | 200 |

*Adapted from:* (1) Joint FAO/WHO (1973/1985) *Energy and protein requirements*. Report of a Joint FAO/WHO Ad Hoc Expert Committee. WHO Tech Rep Ser, 522 724. Geneva: World Health Organization. (2) Food and Nutrition Board, Institute of Medicine (2002) *Dietary reference intakes for energy, carbohydrate, fiber, fat, fatty acids, cholesterol, protein and amino acids (macronutrients)*. Washington, DC: National Academy Press.

The Food and Nutrition Board, Institute of Medicine (2002) has accepted the amino acids in the right-hand column of Table 4.6 in its dietary reference intake report for macronutrients. To grade the amino acid pattern of dietary proteins, either this pattern can be used or the amino acid pattern of whole hen's egg (which has the highest biological value of all the dietary proteins). These two amino acid patterns are similar when compared at the same total of essential amino acids.

## 4.14 How much protein do we eat?

The amount of protein eaten each day is determined by the total food intake and the protein content of the food. In general, the proportion of energy derived from protein is between 11% and 15% of the total energy of the diet, which is generous for all normal purposes.

Fig. 4.7 shows the estimated amount of food available to different populations around the world, on

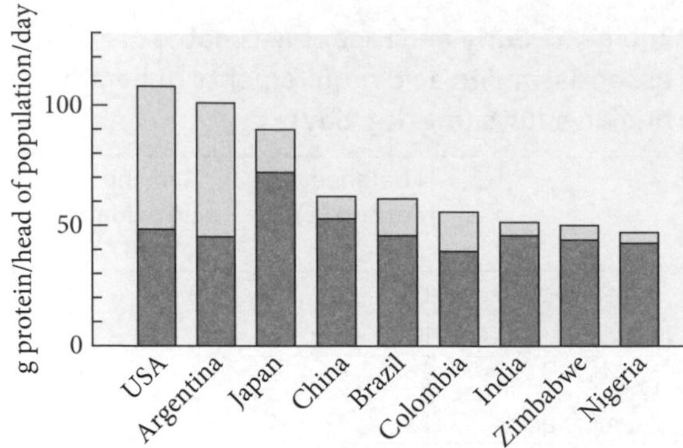


Fig. 4.7 The amount of protein available to the population can be assessed (g protein/head of population/day) and divided into meat protein (▢) and non-meat protein (■). For countries of widely different characteristics, the availability of non-meat protein falls within a narrow range (within ±10%) whereas the availability of meat protein varies widely, over a ten-fold range.

*Source:* Alan Jackson

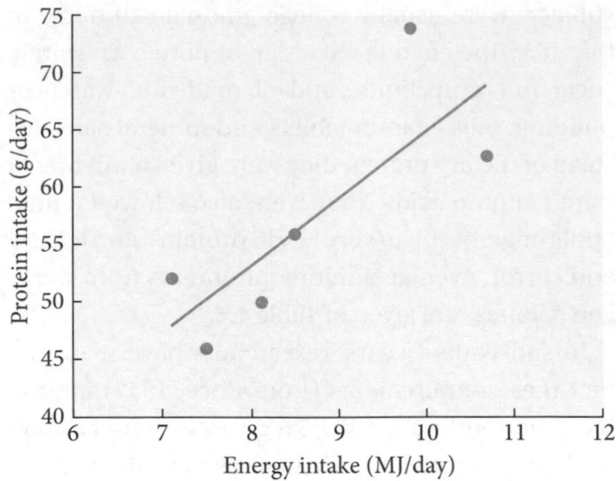


Fig. 4.8 Under most normal circumstances, the total food intake of a person is determined by their energy expenditure, which sets their energy requirement. The energy requirement consists of a reasonably fixed component, basal metabolic rate (BMR), and a variable component, which is determined primarily by the level of physical activity. The protein content of most diets provides between 10% and 15% of dietary energy and therefore the total protein intake is closely related to the total energy intake and hence the level of physical activity. MJ, megajoules.

*Source:* Alan Jackson

average. There is a two-fold difference overall between the technologically developed countries and the underdeveloped countries. The protein available from non-meat sources is very similar for all countries, around 50 g protein/head/day, varying by less than ±10% for the extremes. In contrast, the protein available from meat sources varied ten-fold between the extremes. Virtually all of the variation in protein availability between different countries was determined by the availability of meat protein. Within a population there are differences among individuals in the amount of protein taken. To a large extent this will be determined by the total food intake. Fig. 4.8 shows the relationship between protein intake and energy intake amongst a group of young women vegetarians. The proportion of energy derived from protein was similar for each woman. Those who were most active had the greatest intake of energy and the highest protein intake. Those who were relatively sedentary had a protein intake that approached the maintenance level.

In young children, energy expenditure per unit of body weight is high, and therefore energy intake per unit body weight is high. Consequently, as a proportion of total energy, the dietary protein requirement for normal growth is relatively low. For example, for an infant weighing 10 kg at 1 year of age and growing at a normal rate, for an energy intake of 95 kcal/kg/day and a protein requirement of 1.5 g/kg/day, the proportion of total energy coming from protein would be 6.3%. The highest relative requirement for protein is in sedentary individuals. For example, for a 70-year-old woman who weighs 80 kg, who is lying relatively immobile in bed, the proportion of total energy coming from protein would have to be 20%, which is not readily achievable on a normal diet.

## 4.15 Dietary protein quality and scoring

Protein quality is dependent upon the amino acid content and the subsequent bioavailability of these amino acids in the circulation, both factors that can influence their metabolism within various protein pools. As such, protein quality is typically based on protein digestibility scoring methods.

Traditionally, protein quality has been measured by the Protein Digestibility Corrected Amino Acid Score (PDCAAS), which evaluates protein quality by comparing the essential amino acid composition of a protein source to a reference, which is based on the essential amino acid requirements of a 2–5-year-old child. A protein source providing all of the essential amino acids required, such as whey, achieves the highest score of 1.0, whereas incomplete protein sources (i.e. those not containing all nine essential amino acids), such as nuts, achieve a lower score of ~0.5. PDCAAS has several notable limitations:

- It uses faecal protein digestibility, which can provide inaccurate values of true protein digestibility due to substantial bacterial (colonic) metabolism of amino acids.
- PDCAAS values are truncated at 1, suggesting that amino acids available over those contained in a reference protein provide no additional nutritional benefit, and so proteins of higher quality are not identified.

Due to the limitations associated with PDCAAS, the Food and Agriculture Organization of the United Nations now recommends the use of a newer scoring method called the Digestible Indispensable Amino Acid Score (DIAAS). DIAAS estimates protein digestibility based on true ileal digestibility, which is determined at the end of the small intestine where the amino acids are absorbed. DIAAS evaluates protein quality by determining the ratio of digestible amino acid content to the amino acid reference pattern taken from age-specific amino acid requirements.

As a primary example highlighting the differences between the scoring methods, whey and soy protein both score equally according to PDCAAS, whereas whey scores higher than soy protein according to DIAAS, reflecting the superior digestibility of whey. Because of these scoring attributes in addition to ileal digestibility consideration, DIAAS is the recommended scoring method. More examples of different protein sources and their score according to PDCAAS or DIAAS are shown in Table 4.7.

**Table 4.7 Examples of protein scoring for different protein sources using the DIAAS or PDCAAS method**

| Dietary protein | DIAAS | PDCAAS |
|---|---|---|
| Whey protein isolate | 1.18 | 1.00 |
| Whole milk | 1.14 | 1.00 |
| Egg (hard boiled) | 1.13 | 1.00 |
| Soy protein isolate | 0.90 | 0.98 |
| Almonds | 0.40 | 0.39 |
| Hydrolysed collagen | 0.00 | 0.00 |

DIAAS is not without limitation as it does not account for amino acid absorption kinetics, i.e. the rate at which amino acids are being absorbed; it only accounts for overall protein absorbability, i.e. cumulative absorption. Some, but not all, studies suggest that a faster rate of amino acid absorption is an independent factor that modulates the muscle protein synthetic response to feeding, and so this may need to be a consideration when thinking about the anabolic potential of a protein source.

In order to overcome the limitations associated with DIAAS (and PDCAAS), a minimally invasive dual-stable-isotope tracer method has recently been developed, which permits the simultaneous measurement of ileal digestibility of different indispensable amino acids from dietary protein sources. This method involves the feeding of an intrinsically isotope-labelled test protein alongside a different isotope-labelled standard protein or indispensable amino acid mixture (for which digestibility must already be known). Using this technique, Devi et al. (2018) found that feeding of an intrinsically $^{2}$H-labelled test protein together with $^{13}$C-labelled spirulina protein led to the simultaneous and accurate ileal digestibility measurement of the different $^{2}$H-labelled test protein-derived indispensable amino acids. Importantly, this method permits the evaluation of indispensable amino acid bioavailability from dietary protein, which can be used in different population groups, including vulnerable populations in which intestinal dysfunction may perturb indispensable amino acid bioavailability.

## 4.16 Dietary protein sources

The anabolic response to dietary protein is dependent on multiple factors, which are mainly: the amount of protein given, the digestion and absorption kinetics of the protein source, and the amino acid composition of the amino acid source. It stands that proteins containing more essential amino acids are more likely to stimulate a greater MPS response and are therefore generally considered higher quality proteins.

For many years, animal-derived protein sources have been the predominant dietary protein source and thus the focus of most research to date. Animal-animal-derived derived refers to proteins *directly* originating from animal sources such as meat, fish, poultry, eggs, and dairy (and the constituents whey and casein protein) and they are regarded as 'complete' proteins, meaning they provide sufficient amounts of all essential amino acids to meet human requirements. Animal-derived protein (~20–25 g whey) elicit a rapid postprandial increase in circulating plasma essential amino acids and subsequently a robust increase in MPS. Seminal investigations have shown that the essential amino acid leucine primarily drives the MPS response to protein feeding. Indeed, as little as 3 g essential amino acids containing 40% leucine can maximally stimulate MPS.

However, there is a global trend towards the consumption of non-animal-derived protein sources due to concerns surrounding environmental sustainability and/or in order to comply with vegetarian or vegan diets. As such, many are opting for plant-derived proteins (e.g. wheat, soy) as alternative or adjuvant dietary sources. In general, plant-derived proteins have a lower leucine content, poorer digestibility (possibly due to the tannins content), and lack one or more amino acids (particularly lysine and/or methionine) and are thus less anabolic, compared to animal-derived protein sources. There are, however, some exceptions. For example, the essential amino acid content of potato-derived protein is comparable to casein or egg protein, and the leucine content is in fact higher, so in theory, potato-derived protein could elicit similar or perhaps more robust MPS responses compared to animal-derived sources. Further studies are needed to confirm this theory. Nonetheless, certain strategies can be used to (at least partly) overcome the inherently inferior anabolic properties of plant-derived protein. For example, increasing the amount of plant protein consumed, thereby matching the leucine content of animal-derived protein, can elicit comparable MPS responses. As this strategy may not always be feasible (i.e. the amount of plant protein needed may be too high), fortifying plant-derived protein sources with the deficient amino acids may also enhance the anabolic potential. Further, extracting protein from plant-based sources can eradicate nutritionally limiting parts of the food matrix (e.g. tannins) and improve the digestibility so that it is similar to that of animal-derived sources.

Another topical protein source is collagen-derived proteins, which refers to proteins derived from gelatin and/or collagen hydrolysates that originate from animal sources (e.g. bone, pigskin, fish skin). Collagen-derived proteins are not regarded as 'complete' proteins because they are low in essential amino acids, high in non-essential amino acids (e.g. proline, glycine), and lack tryptophan, rendering a DIAAS of 0. Dietary collagen does, however, have superb digestibility and becomes rapidly bioavailable following consumption, and so may have some nutritional value.

To overcome the AA inadequacies of some dietary protein sources (i.e. plant/collagen sources), different sources/types of protein can be blended together to produce a viable and sustainable protein load. Most protein blends to date involve combining animal and plant blends to exploit the digestive properties of each protein source with the aim of maximizing amino acid availability and potentially enhancing and/or extending the MPS response.

The quest for sustainable protein sources continues, with other potentially viable and anabolic protein sources being investigated, including the breeding of insects, the growing of yeast, micro-algae, and fungi, and even the generation of lab-grown meat. While investigation into most of these sources is in its infancy, some emerging evidence indicates anabolic promise. For example, a protein source derived from the culture of fungus (named 'mycoprotein') has demonstrated good digestibility, contains a similar amino acid composition to dairy protein sources, and stimulates MPS at a similar level to that of leucine-matched milk protein.

## 4.17 Dietary protein deficiency and protein-deficient states

For the diets consumed by most populations, the intake of protein is adequate, provided that the overall intake of food is not limited (e.g. by inactivity or unavailability). However, for some diets in which the density of protein to energy is low, and/or where the quality of amino acids is low, there may be situations related to relative inactivity when the ability to satisfy the protein intake is marginal.

Protein-deficient states, where the content of protein in the body is reduced, are most likely to be the result of:

- an increase in demand (e.g. in infection or stress)
- an increase in losses (e.g. with haemorrhage, burns, or diarrhoea)
- a failure of the conservation systems (e.g. with impairment of urea salvage in the colon).

In correcting the deficient state, it is as important to remove the underlying cause as it is to provide adequate amounts of protein or amino acids in the diet.

### 4.17.1 Protein energy wasting

Protein energy wasting (PEW) is the progressive and simultaneous depletion of systemic protein and energy stores in people with chronic kidney disease (CKD), particularly in those with end-stage renal disease. PEW is highly prevalent, occurring in up to 50–75% of CKD patients, and is particularly undesirable, increasing the risk of morbidity and mortality whilst reducing quality of life. The pathogenesis of PEW is multifactorial, with loss of appetite and the ensuing reduction in nutrient intake just one of the key mediators, ultimately manifesting as reduced lean body mass leading to frailty.

Rigorous nutritional support remains a cornerstone in the management of PEW in order to preserve lean body mass and provide adequate energy supply. CKD patients undergoing dialysis (a medical procedure to remove excess fluid and waste products from the blood to compensate for inadequate kidney function) are at a high risk of PEW and so are advised to consume ≥1.2 g/kg/day protein, which is nearly double the recommended 0.6–0.75 g/kg/day protein intake for pre-dialysis CKD patients. The increased protein requirements of dialysis patients are primarily to compensate for protein and amino acid losses caused by dialysis, which can amass to a protein loss of 16 g per dialysis session. Pre-dialysis patients, on the other hand, are encouraged to consume low-protein diets (i.e. lower than the protein recommendation for healthy adults: 0.8 g/kg/day) in order to reduce uraemia, which is the accumulation in the blood of amino acid and protein metabolism end-products normally excreted in urine, ultimately slowing CKD progression and the need for dialysis.

Protein source is also a key consideration for prevention of PEW. Dialysis patients are advised to limit/avoid plant-derived protein sources due to the phosphorus and potassium content, instead consuming animal-derived protein, which is also characterized by a better bioavailability and higher quality protein profile.

### 4.17.2 Protein energy malnutrition

Protein energy malnutrition (PEM), also referred to as protein energy undernutrition, describes a range of malnutrition-related conditions that are characterized by a lack of dietary protein and/or energy (in the form of calories). In situations of PEW, to supply energy, the body initially breaks down adipose tissue. However, when these stores are depleted, the body breaks down muscle and organs (e.g. liver) to liberate protein for energy, resulting in a loss of body mass.

Primary PEW in children and adults is caused by inadequate nutrient intake. In children there are two common forms:

- **Marasmus**

  Marasmus is the most common form of PEW in children, which is caused by an unstable food supply leading to a loss of fat and muscle tissue, without the presence of tissue swelling (oedema).

- **Kwashiorkor**

  Kwashiorkor typically occurs in areas where the staple foods are high in carbohydrates but low in protein (e.g. sweet potatoes, green bananas). Ultimately, this leads to cell membranes leakage, which causes extravasation of intravascular protein and fluid, resulting in peripheral oedema.

Both marasmus and kwashiorkor compromise immunity, increasing susceptibility to bacterial infections (such as gastroenteritis, pneumonia), which can ultimately exacerbate muscle wasting and cause anorexia.

Secondary PEW results from disease or drug interventions that alter nutritional consumption. For example, muscle wasting associated with various cancers, chronic obstructive pulmonary disease, and AIDS, can reduce appetite and impair anabolic responsiveness to nutrition, resulting in PEW.

The main course of treatment for PEW is a balanced diet (possibly limiting/avoiding lactose), which may have to be controlled if PEW is severe. Concurrent infections and fluid/electrolyte abnormalities, if present, should also be managed. Treatment options must be considered carefully to minimize the risk and severity of complications associated with refeeding syndrome (e.g. cardiac arrhythmias, hyperglycaemia).

## 4.18 Ageing and physical inactivity

### 4.18.1 Ageing

Ageing is associated with the loss of skeletal muscle mass and function, termed sarcopenia, which increases the risk of falls, frailty, morbidity, and mortality and is of multifactorial aetiology. While post-absorptive MPS and MPB rates are similar across age, older adults display 'anabolic resistance', manifesting as a blunted increase in MPS and blunted decrease in MPB in response to the key growth signals regulating muscle homeostasis, i.e. nutrition and exercise. Over time, the insufficient replenishment of muscle protein during periods of inactivity, fasting, and/or remodelling leads to the progressive loss of muscle with age, and so 'anabolic resistance' is likely a major contributor to sarcopenic onset and progression.

Demonstrating age-related anabolic resistance to protein nutrition, provision of a 40 g EAA bolus fails to elicit a similar MPS response in older adults to that seen in young adults given a 10 g EAA bolus. This anabolic resistance, combined with the fact that older adults typically do not meet the protein RDA, likely a consequence of increased satiety after consuming a meal, means that other nutritional interventions have been sought. For example, fortifying protein feeds with leucine (the most anabolic essential amino acid) has been used to maximize the MPS response in older adults. However, it remains inconclusive whether standalone or adjuvant supplementation of leucine is effective to sufficiently stimulate MPS across populations, and it remains to be investigated whether submaximal doses of complete protein enriched with leucine lead to enhanced muscle anabolism in older adults. Demonstrating age-related anabolic resistance to exercise, following acute resistance exercise, older adults exhibit blunted MPS, and growth (mTORC1) signalling compared to younger adults. In response to resistance exercise training, younger adults exhibit increased MPS and gains in fat free mass, which are absent in older adults. Given the potent effects of protein feeding upon promoting sustainment of anabolic responses to resistance exercise, it has been tested whether anabolic resistance may be overcome by dietary manipulation. In this context, enrichment of low-dose essential amino acids with the addition of leucine after resistance exercise has shown anabolic efficacy.

### 4.18.2 Physical inactivity

Physical inactivity, such as sedentarism or immobilization due to bed rest, induces a premature muscle full state characterized by blunted fasted and fed MPS (i.e. anabolic resistance) leading to muscle wasting, which over time may contribute to sarcopenic onset and/or progression. Protein nutrition during periods of inactivity is a key strategy that can be used to offset such inactivity-induced declines in MPS.

Results on the efficacy of nutritional intervention so far have been mixed, and in general it seems that protein consumption of ≤1.6 g/kg/d during inactivity does not modulate muscle loss. Higher dose EAA/leucine supplementation shows more promising results. Middle-aged adults consuming leucine (0.06 g/kg/meal) during 14 days of bed rest showed attenuated reductions in post-absorptive MPS and reduced whole-body lean mass loss after 7 days, compared to alanine control. It is plausible that higher dose essential amino acids/leucine, as opposed to high-protein diets, may be a more effective intervention to overcome the premature onset of muscle full during inactivity. If so, it has been suggested that it is not the availability of AA *per se* that is limiting muscle anabolism during inactivity, but rather inactivity induces a significant increase in the threshold required for EAA/leucine to stimulate anabolism (e.g. intracellular growth signalling pathways).

Considering protein feeding in isolation may not fully counteract the premature onset of muscle full during inactivity in older adults, adjunct clinically relevant, feasible (i.e. financially), and effective interventions are needed. Neuromuscular electrical stimulation is a logical intervention that is clinically feasible even for those who are totally incapable of weight bearing. Work so far has shown that following 1 d bed rest, older adults subjected to unilateral neuromuscular electrical stimulation followed by 40 g casein ingestion prior to sleep had about an 18% greater increase in MPS overnight compared to the unstimulated leg, and this may therefore be a promising intervention.

## 4.19 Summary

Proteins are highly abundant structural and functional elements within every cell, which undergo extensive metabolic interaction and fulfil a range of functions (e.g. perform enzymatic, hormonal, contractile roles), rendering them fundamental for supporting life. At the most basic level, proteins are made from a combination of 20 different amino acids, which determines the structure and function of a protein. Of all 20 amino acids, 9 have to be obtained from the diet and are described as 'indispensable' or 'essential', with the remaining 11 synthesized *in vivo* and described as 'dispensable' or 'non-essential'. In healthy adults, diurnal protein turnover exists in a dynamic equilibrium whereby protein synthesis exceeds protein breakdown in the fed state, and protein breakdown exceeds protein synthesis in the fasted state. Dietary protein and exercise are the two key stimuli that promote positive protein turnover, which can be accurately and reliably measured using stable isotope methods (e.g. substrate-specific isotope tracers, deuterium oxide). The potency of the anabolic response to dietary protein is dependent upon the protein quality, which can be measured via the Digestible Indispensable Amino Acid Score, Protein Digestibility Corrected Amino Acid Score, and/or using stable isotope tracers. Protein quality depends on several factors, such as the amount of protein given, the digestion and absorption kinetics of the protein source, and the amino acid composition of the amino acid source. In situations of protein deficiency (e.g. protein energy wasting, protein energy malnutrition), significant body mass can be lost, drastically impairing health and quality of life, requiring dietary protein therapy to help overcome/manage these conditions. In situations of ageing and/or physical inactivity, the anabolic response to protein nutrition is blunted culminating in muscle mass loss, calling for new and novel dietary protein interventions to prevent such muscle decline.

## Further reading


1. **Bauer, J., Biolo, G., Cederholm, T., et al.** (2013) Evidence-based recommendations for optimal dietary protein intake in older people: a position paper from the PROT-AGE Study Group. *J Am Med Directors Assoc*, **14**, 542–59.
2. **Borgonha, S., Regan, M., Oh, S-H., Condon, M., and Young, V.R.** (2002) Threonine requirement of healthy adults, derived with a 24-hour indicator amino acid balance technique. *Am J Clin Nutr*, **75**, 698–704.
3. **Brosnan, J.T.** (2001) Amino acids, then and now—a reflection on Sir Hans Krebs' contribution to nitrogen metabolism. *IUBMB Life*, **52**, 265–70.
4. **Cochrane Injuries Group Albumin Reviewers** (1998) Human albumin administration in critically ill patients: systematic review of randomized controlled trials. *BMJ*, **317**, 235–40.
5. **Collins, S., Myatt, M., and Golden, B.** (1998) Dietary treatment of severe malnutrition in adults. *Am J Clin Nutr*, **68**, 193–9.
6. **Deane, C.S., Bass, J.J., Crossland, H., Phillips, B.E., and Atherton, P.J.** (2020) Animal, plant, collagen and blended dietary proteins: effects on musculoskeletal outcomes. *Nutrients*, **12**, 2670.
7. **Deane, C.S., Ely, I.A., Wilkinson, D.J., Smith, K., Phillips, B.E., and Atherton, P.J.** (2020) Dietary protein, exercise, ageing and physical inactivity: interactive influences on skeletal muscle proteostasis. *Proc Nutr Soc*, **80**, 106–117.
8. **Devi, S., Varkey, A., Sheshshayee, M.S., Preston, T., and Kurpad, A.V.** (2018) Measurement of protein digestibility in humans by a dual-tracer method. *Am J Clin Nutr*, **107**, 984–91.
9. **Food and Nutrition Board, Institute of Medicine** (2002) *Dietary reference intakes for energy, carbohydrate, fiber, fat, fatty acids, cholesterol, protein and amino acids (macronutrients)*. Washington, DC: National Academy Press.
10. **Joint FAO/WHO** (1985) *Energy and protein requirements*. Report of a Joint FAO/WHO Ad Hoc Expert Consultation. WHO Tech Rep Ser, **724**. Geneva: World Health Organization.
11. **Layman, D.K., Anthony, D.G., Rasmussen, B.B., et al**. (2015) Defining meal requirements for protein to optimize metabolic roles of amino acids. *Am J Clin Nutr*, **101**(Suppl.), 1330s–8s.
12. **Millward, D.J. and Jackson, A.A.** (2004) Protein/energy ratios of current diets in developed and developing countries compared with a safe protein/energy ratio: implications for recommended protein and amino acid intakes. *Public Health Nutr*, **7**, 387–405.
13. **Rand, W.M., Pellett, P.L., and Young, V.R.** (2003) Meta-analysis of nitrogen balance studies for estimating protein requirements in healthy adults. *Am J Clin Nutr*, **77**, 109–27.
14. **Rose, W.C.** (1957) The amino acid requirements of adult man. *Nutr Abst Rev*, **27**, 631–47.
15. **Stewart, G. and Smith, C.P.** (2005) Urea nitrogen salvage mechanisms and their relevance to ruminants, non-ruminants and man. *Nutr Res Rev*, **18**, 49–62.
16. **Young, V.R. and Ajami, A.** (1999) The Rudolph Schoenheimer Centenary Lecture. Isotopes in nutrition research. *Proc Nutr Soc*, **58**, 15–32.

# Energy

Andrew M. Prentice



Energy is the primary currency of nutrition. Mammals require energy to stay warm and to drive all the processes of life itself. All of this energy is derived from the chemical combustion of food, a process requiring oxygen, and producing carbon dioxide and water. It is the need to maintain an adequate supply of energy that is the major stimulus of food intake, and this appetite drive has an important influence on the intake of all other nutrients.

In humans, dietary energy is derived from four major food types—carbohydrate, fat, protein, and alcohol. These are termed macronutrients and each can be composed of numerous subtypes that have slightly different energy contents. Generation of energy from the various macronutrients requires different chemical processes, and for each of them there are optional pathways that can be used in different metabolic circumstances. For instance, glucose can initially be utilized by muscle without oxygen (anaerobically) when a short burst of movement is required, or with oxygen (aerobically) for longer periods of activity. Anaerobic metabolism incurs a temporary oxygen debt that must be repaid. Ultimately, all energy is derived through the process of oxidative phosphorylation that occurs in mitochondria. The biochemical pathways involved in these processes are summarized elsewhere (e.g. Cox, 2013). This chapter outlines the energy value of foods and how they may be calculated, and summarizes human energy needs and how these can be measured. It concludes by briefly summarizing the mechanisms by which energy balance is regulated.

## 5.1 The energy value of food

### 5.1.1 Chemical energy

The chemical energy of food is simply the total amount of energy that would be liberated by the food if it were combusted in oxygen (i.e. its heat of combustion). This can be measured directly in a bomb calorimeter, whereby the heat liberated by burning a small sample of the food is accurately recorded. This total chemical energy of food is also referred to as its *gross energy* (GE).

### 5.1.2 Metabolizable energy

A portion of the GE of food is unavailable for human metabolism for a variety of reasons. Firstly, not all of it can be digested and absorbed by the body (e.g. some components of dietary fibre, or the central parts of hard grains and nuts) and the energy will be lost in faeces. The proportion of GE that is actually absorbed across the digestive tract is termed *digestible energy* (DE). Secondly, even this DE is not fully available to the body because a number of the oxidative pathways are incapable of proceeding to completion. These are mostly confined to protein metabolism. For instance, amino acids are only oxidized as far as urea or ammonia. These compounds still contain energy, which is lost in the urine, and for which it is necessary to make a final adjustment in order to calculate the actual energy available for metabolism, known as *metabolizable energy* (ME).

### 5.1.3 Methods for assessing metabolizable energy intake

When very precise values are required in experimental studies of energy metabolism, it is possible to make direct measurements of ME, although this is extremely tiresome for the subjects and investigators alike. The method requires the accurate collection of duplicate portions of all foods consumed in proportion to the amount of each component of the diet that was eaten. Each food can then be analysed separately by bomb calorimetry, or the whole diet can be homogenized together and a small aliquot of the homogenate measured. The same process must be performed for all the faeces and urine collected throughout the period that the diet is eaten. Table 5.1 gives an example of the calculations used.

The ME of diets can also be estimated if the composition of the diet is accurately known. The composition can be obtained by chemical analysis of foods, or most frequently by reference to tables of food composition, which themselves have been derived from chemical analysis of a wide range of commonly consumed foods. Many countries have their own food tables reflecting their national diet. Once the macronutrient composition of a food or diet is known, the energy content is computed using standard conversion factors (Table 5.2). These were first derived by W. O. Atwater (and, hence, are frequently referred to as Atwater factors) in a series of experiments with human volunteers in which the conversion factors from GE to DE to ME were determined. It should be stressed that ME values of diets derived from food tables are somewhat imprecise, since there may be variations in the actual composition of listed foods and because the Atwater factors are approximate

**Table 5.1 Example of measuring the metabolizable energy of a diet**

| Subjects | |
|---|---|
| Young women | |
| **Methods** | |
| Provision of a diet of known composition over 7 days in a metabolic suite | |
| Bomb calorimetry of duplicate portions of the diet and of all faeces and urine | |
| **Measured values** | |
| Duplicate portions of diet | 9900 kJ/day |
| Faeces | 710 kJ/day |
| Urine | 420 kJ/day |
| **Derived values** | |
| Gross energy of food (a) | 9900 kJ/day |
| Digestible energy (a – b) | 9190 kJ/day |
| Digestibility coefficient [100(a – b)/a] | 92.8% |
| Metabolizable energy (a – b – c) | 8770 kJ/day |
| 4.184 kJ = 1 kcal | |

**Table 5.2 Example of calculating the metabolizable energy of a diet**

| Subjects | | |
|---|---|---|
| Young women | | |
| **Methods** | | |
| Weighed food records over 7 days | | |
| Food composition tables | | |
| Use of Atwater factors | | |
| Carbohydrate | | 17 kJ/g |
| Fat | | 37 kJ/g |
| Protein | | 17 kJ/g |
| Alcohol | | 29 kJ/g |
| **Derived values** | | |
| Carbohydrate | 245 g/day | 4165 kJ/day |
| Fat | 90 g/day | 3330 kJ/day |
| Protein | 75 g/day | 1275 kJ/day |
| Alcohol | 0 g/day | 0 kJ/day |
| Metabolizable energy | | 8770 kJ/day |

*Source:* Reprinted by permission from Macmillan Publishers Ltd. Adapted from Murgatroyd et al. 'Techniques for the measurement of human energy expenditure: a practical guide'. *Int J Obesity*, **17**(10), (1993).

values derived from people consuming a 'standard' Western diet. It should be also noted that the conversion factor for carbohydrates refers to the amount of carbohydrate in a food when expressed as available monosaccharides, since dietary fibre and non-starch polysaccharides have a lower digestibility.

## 5.2 Measurement of energy expenditure

The ability to measure human energy expenditure has been important in many aspects of nutritional science, ranging from very precise studies into how energy balance is regulated, to large-scale estimations of the energy needs of populations. There are numerous techniques available, each of which has advantages and disadvantages. It is important to match carefully the technique used to the situation at hand. Table 5.3 summarizes the major methods and lists their key features.

Table 5.3 **Methods for measuring human energy expenditure (EE)**

| Method | Measurement principle | Advantages | Disadvantages | Applications |
|---|---|---|---|---|
| Direct calorimetry | | | | |
| Whole-body chamber | Subject confined within a small-to-moderately sized chamber. Measures heat loss. | Historically had faster response time than indirect calorimetry (now no longer true). Good environment for strictly controlled studies. | Extreme technical complexity. Very high construction and running costs. Studies confined to small subject numbers. Artificial environment. | Initially used to validate the principle of indirect calorimetry. Some distinct applications in studying heat dynamics of exercise. Now largely obsolete. |
| Body suit | Subject wears an insulated metabolic suit. Measures heat loss. | As above. | As above. | As above. |
| Indirect calorimetry | | | | |
| Whole-body chamber | Subject confined within a small-to-moderately sized chamber. Measures oxygen consumption (and frequently also measures $CO_2$ production). Calculates EE from energy equivalence of oxygen consumed. Calculates macronutrient oxidation from RQ after adjustment for urinary nitrogen losses. | Very precise and repeatable. Provides minute-by-minute data. Measurements over 1–14 days. Good environment for strictly controlled studies. Measures macronutrient oxidation rates in addition to total EE. Represents gold standard. | Technically challenging. High production and running costs. Studies confined to relatively small subject numbers. Artificial environment. | Fundamental studies of the mechanisms regulating human energy balance. Includes effects of exercise, diet, physiological states such as pregnancy, and pharmacological effects of compounds intended to affect energy expenditure. |

*(Continued)*

Table 5.3 Methods for measuring human energy expenditure (EE) (*Continued*)

| Method | Measurement principle | Advantages | Disadvantages | Applications |
|---|---|---|---|---|
| Bedside methods | Supine subject has hood placed over their head or entire bed covered with a plastic tent. Measures oxygen consumption (and frequently also measures $CO_2$ production). Calculations as for whole-body indirect calorimetry. | Accurate and reliable data. Commercially available versions with integrated gas meters, computers, and display systems. User-friendly. Measurements possible over several hours. Good for measuring BMR. Can measure macronutrient oxidation rates in addition to total EE. | Relatively expensive. Requires periodic calibration. | Short-term studies of energy expenditure such as BMR or diet-induced thermogenesis. Can be used with hospital patients. |
| Ventilated hood methods | As for bedside methods, but subjects can be standing (e.g. on cycle ergometer). | As above. | As above. | As above, and for exercise studies. |
| Douglas bag method | Subject wears mouthpiece with one-way valve and nose clip. Collects expired air directly into an impermeable 'Douglas' bag then measures volume and gas concentrations of bag contents. Calculations as for whole-body indirect calorimetry. | Simple and robust. Provides reliable results. Inexpensive. | Prompt analysis required after gas collection to avoid gas leakages and diffusion. Only suitable for short periods of study. Inconvenient and uncomfortable for subjects. Interferes with normal activity. | Short-term studies of EE such as BMR or diet-induced thermogenesis. Can be used for ambulatory studies of work and exercise. Can be used with hospital patients. |
| Ambulatory methods | Subject wears mouthpiece with one-way valve and nose clip, or ventilated mask, and carries gas analysis 'respirometer' strapped to their back. Measures oxygen consumption. Calculations as for whole-body indirect calorimetry, but usually without $CO_2$ measurement and hence RQ. | Smaller and more compact than Douglas bag method. Relatively simple and robust. Yields reliable results. | Discomfort to subject after prolonged wearing of apparatus. Some versions require separate analysis of gas concentrations. Some versions are expensive. | Useful for studies of light-to-moderate physical activity in near-to-natural conditions. |

Table 5.3 Methods for measuring human energy expenditure (EE) (*Continued*)

| Method | Measurement principle | Advantages | Disadvantages | Applications |
|---|---|---|---|---|
| **Stable isotope methods** | | | | |
| Doubly labelled water | Assesses $CO_2$ turnover from differential rate of disappearance of $^2H$ and $^{18}O$.<br>Calculates EE from $CO_2$ production and an assumption about RQ. | Gold standard method for assessing habitual EE in free-living subjects.<br>Measurements over 10–20 days. | $^{18}O$ isotope is very expensive.<br>High capital investment for mass spectrometer.<br>Technically and mathematically challenging.<br>Requires certain assumptions that can lead to errors if not correctly applied. | Studies of free-living EE in all subjects.<br>Especially valuable for use in children as minimal subject cooperation is required. |
| Labelled bicarbonate | Assesses $CO_2$ turnover from rate of disappearance of constantly infused $^{13}C$ bicarbonate.<br>Calculates EE from $CO_2$ production and an assumption about RQ. | Requires no cumbersome apparatus except for a minipump to infuse the bicarbonate.<br>Can assess expenditure over a shorter time frame than doubly labelled water. | High capital investment for mass spectrometer.<br>Technically and mathematically challenging.<br>Requires certain assumptions that can lead to errors if not correctly applied. | Especially applicable in clinical studies where a shorter time frame is required. |
| Heart-rate monitoring | Electrodes collect minute-by-minute heart-rate data and store on computer chip.<br>EE can be calculated from individual calibration curves generated for each subject, or unconverted data can be used in a semi-quantitative manner. | Inexpensive and easy to use for both subject and investigator.<br>Provides minute-by-minute data over periods of 7 days or longer. | Calibration of each subject is a cumbersome and time-consuming process.<br>Can become tiresome to subjects if worn for long periods. | Generally used for large-scale epidemiological studies where comparative values are more important than absolute expenditure values (for instance, in studies of activity levels and health). |
| Movement sensors | Sensors collect minute-by-minute movements of the body and store on computer chip.<br>EE can be calculated from individual calibration curves generated for each subject, or unconverted data can be used in a semi-quantitative manner. | As above. | As above. | As above. |

(*Continued*)

Table 5.3 Methods for measuring human energy expenditure (EE) (*Continued*)

| Method | Measurement principle | Advantages | Disadvantages | Applications |
|---|---|---|---|---|
| Time-and-motion studies and factorial method | Daily activities are recorded by subjects themselves or by an observer. EE calculated by reference to standard tables for the energy cost of activities. | Very inexpensive if subjects record their own activities. Provides good data on types of activities. | High subject burden if self-recording. May cause alteration in activity patterns to simplify recording. | Many applications. For instance, has frequently been used (with fieldworkers doing the recording) to study work and activity patterns among farmers etc. in developing countries. |

EE, energy expenditure; RQ, respiratory quotient; BMR, basal metabolic rate.

Reprinted by permission from Macmillan Publishers Ltd: Adapted from Murgatroyd et al. 'Techniques for the measurement of human energy expenditure: A practical guide'. *Int J Obesity*, **17**(10) (1993).

## 5.2.1 Direct calorimetry

Principle Direct calorimetry, as the name implies, directly measures the *heat loss* from a subject. The first such study was conducted during a Parisian winter over two centuries ago by the father of energy metabolism, Antoine Lavoisier, when he measured the amount of water melted by a guinea pig kept in a small chamber surrounded by ice. By knowing the specific heat of melting ice, he was able to compute the heat liberated by all the metabolic processes within the animal.

Equipment Modern human direct calorimeters employ the same principle, but use complex thermocouple sensors and heat exchangers to measure the subject's radiative heat loss (heat lost through radiation as from any hot object), conductive heat loss (heat carried away by air passing over the skin and warming as it does so), and evaporative heat loss (heat lost as the specific heat of evaporation of perspiration). This can be achieved in a chamber ranging in size from just large enough to cover a man on a cycle ergometer to a moderately sized room. Attempts were also made to use an insulated body suit, but this proved too cumbersome. Direct calorimeters are extremely difficult and expensive to construct. Their main function has been to demonstrate that indirect calorimetry (which measures *heat production*; see section 5.2.2) gives precisely the same answers. There is a slight lag period between heat production and heat loss due to the temporary storage of heat in the body (for instance, when exercising a subject will become hot and some of the heat produced will take some minutes to exit the body as it cools down again). However, as long as time lag is accounted for, the two methods give precisely the same result. Because of the technical complexity of direct calorimeters, no more than a handful were ever constructed and it is doubtful if any are still in commission worldwide.

## 5.2.2 Indirect calorimetry

Principle Indirect calorimetry measures *heat production* by assessing oxygen consumption and (optionally) carbon dioxide production. Variations between different versions of the technique essentially come down to the method by which the exhaled respiratory gases are collected from the subject (Table 5.3). In one form or another, indirect calorimetry is now the usual method for measuring human energy expenditure. It is much easier to perform than direct calorimetry and is less expensive. It can also be used to estimate the relative contribution of each of the macronutrients to the total

**Table 5.4 Constants used in indirect calorimetry**

| | Fat | Carbohydrate[a] | Protein | Alcohol |
|---|---|---|---|---|
| Oxygen consumption (L/g) | 2.101 | 0.746 | 0.952 | 1.461 |
| Carbon dioxide production (L/g) | 1.492 | 0.746 | 0.795 | 0.974 |
| Respiratory quotient | 0.710 | 1.000 | 0.835 | 0.667 |
| Energy equivalence of oxygen (kJ/L) | 19.61 | 21.12 | 19.48 | 20.33 |
| Energy density (kJ/g) | 39.40 | 15.76 | 18.55 | 29.68 |

[a]As monosaccharide equivalents.

energy expenditure—a major advantage over direct calorimetry.

Table 5.4 lists the basic constants used to calculate energy expenditure by indirect calorimetry. The most important of these are the energy equivalence of each litre of oxygen consumed, which ranges between 19.48 kJ/L for protein and 21.12 kJ/L for carbohydrate—a difference of 8%. However, the body rarely combusts single macronutrients and for most practical purposes it can be assumed that it is combusting a mixture of fuels, similar to the diet that a person is consuming, and therefore a generalized value of 20.3 kJ/L oxygen is frequently used. In most circumstances, this assumption will lead to an error of less than 3%. To achieve a greater accuracy, as would be required for detailed physiological studies of human energy regulation, it is necessary also to measure carbon dioxide production to calculate the *respiratory quotient* (RQ = carbon dioxide produced/oxygen consumed) and the urinary nitrogen output in order to estimate protein oxidation. If alcohol has been consumed, it is further necessary to estimate its contribution to energy expenditure by making separate measurements of the rate of decline of blood alcohol levels. Once all these variables are known, it is possible to use the precise values for the energy equivalence of oxygen.

Indirect calorimetry can be used to assess the mixture of fuels oxidized as follows. First, since protein and alcohol oxidation can be assessed as described above, it is possible to calculate their contributions to a subject's total oxygen consumption and carbon dioxide production using the values in Table 5.4. The remaining gas exchange is due to carbohydrate and fat oxidation and can be expressed as the non-protein, non-alcohol RQ (frequently termed the non-protein RQ (NPRQ), since most measurements are made in the absence of alcohol consumption). The following equation shows that the RQ for carbohydrate (glucose) oxidation is precisely 1.0, since 1 mole of carbon dioxide is liberated for each mole of oxygen consumed:

$$C_6H_{12}O_6 + 6O_2 = 6CO_2 + 6H_2O + \text{heat}$$

A similar calculation can be performed for fatty acid oxidation (e.g. palmitic acid, $C_{16}H_{34}O_2$) and reveals that the average for different fats yields an RQ of 0.71 (Table 5.4). Thus, if the NPRQ is calculated to be 1.00, then only carbohydrate is being combusted, and if it is close to 0.7 then only fat is being combusted. Values between these indicate that a mixture of fat and carbohydrate is being combusted, the proportions of which can be calculated from nomograms or simultaneous equations. These are the basic principles. In practice, there are many complexities to the calculations and many special circumstances when adjustments must be made in order to obtain correct estimates. Readers interested in applying the techniques are referred to more advanced texts, such as that by Livesey and Elia (1988).

**Equipment** Table 5.3 lists the various types of equipment that can be used for indirect calorimetry, together with their major advantages and disadvantages.

The classic (now rarely used) method uses the Douglas bag (Fig. 5.1). This is a large bag (usually with a 100 L volume) that is impermeable to gases. The subject wears a nose clip and breathes through a mouthpiece connected to a one-way valve that

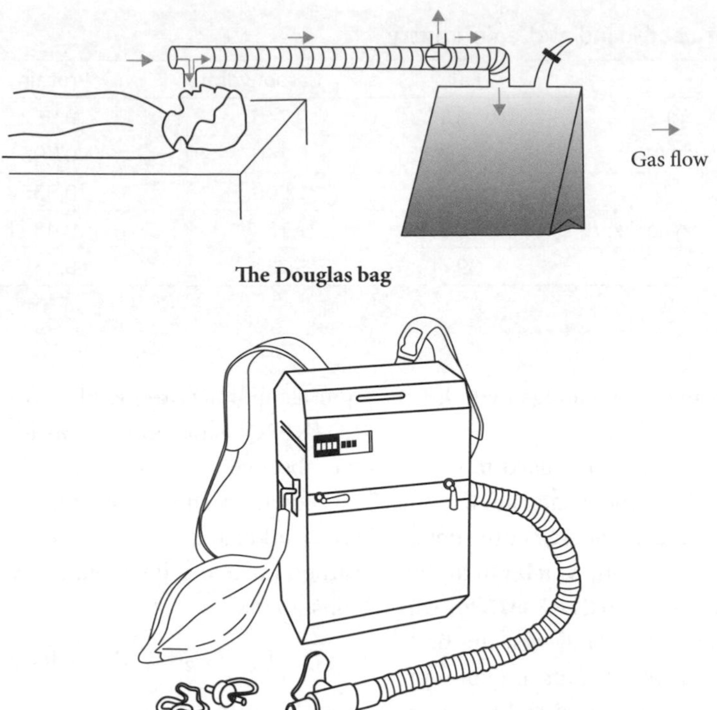


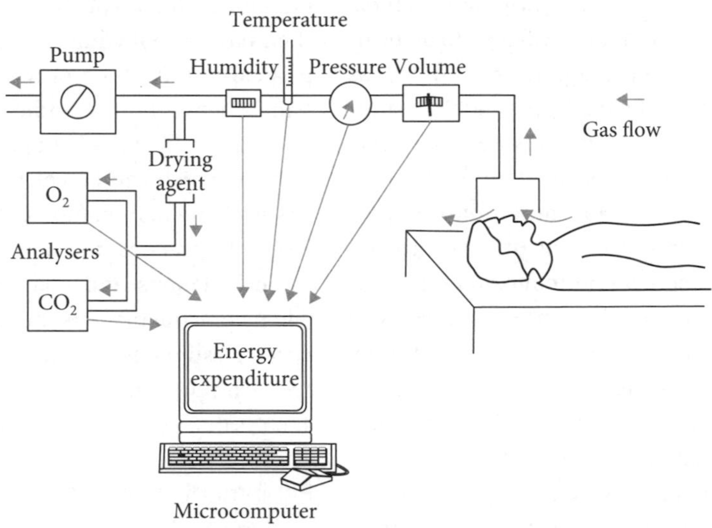


**A ventilated-hood indirect calorimeter**

Fig. 5.1 Some devices used for indirect calorimetry.

Reprinted from *Human nutrition and dietetics* 9e, Garrow, J.S. and James, W.P.T. (eds) (1993) with permission from Elsevier.

allows them to inhale fresh air and exhale into the bag. At the end of the experiment—which has to be short (up to 15 minutes) because of limited capacity in the bag—the volume of expired air is measured with a gas meter. The oxygen and carbon dioxide contents are analysed. Oxygen consumption is calculated from the difference between oxygen in the ambient (inspired) and expired air, multiplied by the ventilation rate.

For short-term measurements during exercise, the Kofrani–Michaelis (Max Plank) respirometer used to be used (see Fig. 5.1) because it measures expired air volume as it is produced, and therefore only a small sample of the expired air needs to be retained for subsequent gas analysis. There is now a wide selection of new generations of these machines with similar working principles. These portable respirometers were designed for short-term measurements at rest and during exercise in 'field' situations, but can also be used in laboratory settings.

Ventilated hood systems (Fig. 5.1) avoid the discomfort of a face mask. With this equipment, air flows over the subject's head (which is within a transparent Perspex or plastic hood) while they lie or sit quietly. The system is suited for situations in which the subject's gas exchange is measured in a laboratory or hospital setting for periods of between 30 minutes and several hours. There are a number of commercially available ventilated hood systems specially designed for use in a hospital setting.

Whole-body calorimeters (also called respiration chambers) represent the most sophisticated option and are chiefly used for detailed physiological experiments. Whole-body calorimeters are furnished airtight rooms (frequently about 10–15 $m^3$) in which subjects can live for periods of 1–14 days. Respiratory gas exchange is measured continuously using small samples of air drawn from the inlet and outlet vents of the chamber. Subjects can carry out all the activities of a relatively sedentary life, as well as controlled exercise on a cycle ergometer or treadmill. Food and drink are passed in through one airlock and waste products out through another. Whole-body calorimeters are extremely accurate and precise. Whole-body calorimeters are not commercially available and require specialized construction. Many of the critical research questions have now been answered and hence there are few whole-body calorimeters still in commission.

## 5.2.3 Isotopic tracer methods

### 5.2.3.1 Doubly labelled water method

**Principle** The disadvantage of the methods described above is that none of them can capture information on the total habitual energy expenditure of someone living their normal life. In the mid-1980s it became possible to do this using the doubly labelled water ($^2H_2{}^{18}O$) method (Speakman and Nagy, 1997).

The principle of the method is illustrated in Fig. 5.2. The subject drinks an accurately weighed amount of water labelled with the harmless, non-radioactive isotopes of deuterium ($^2H$) and oxygen-18 ($^{18}O$), and then provides a series of saliva or urine samples for the next 10–20 days (the optimal duration depends on their activity level). These are analysed by mass spectrometry to assess the disappearance of the isotopes from the body (Fig. 5.3). The $^2H$ labels the body's water pool, and its disappearance from the body ($k_2$) provides a measure of water turnover ($rH_2O$). The $^{18}O$ labels both the water and the bicar-

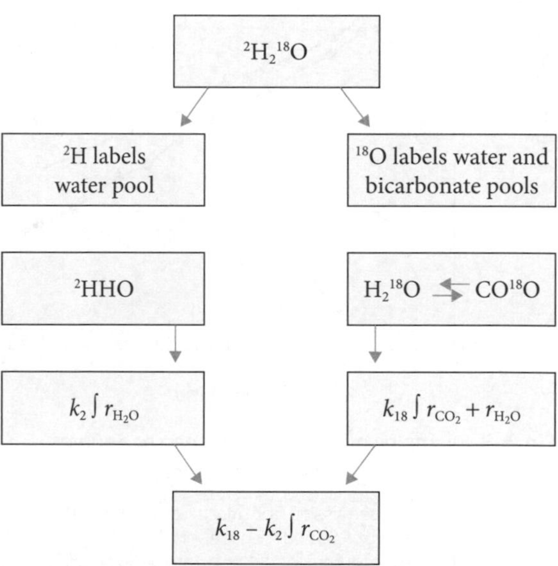


Fig. 5.2 Principle of the doubly labelled water method. Production rates are represented by *r* and rate constants (calculated from the slope of the isotope disappearance curves shown in Fig. 5.3) are represented by *k*.

*Source:* Reprinted by permission from Macmillan Publishers Ltd: Murgatroyd, P.R., Shetty, P.S., and Prentice, A.M. (1993) Techniques for the measurement of human energy expenditure: a practical guide. *Int J Obesity*, **17**(10), pp. 549–68.

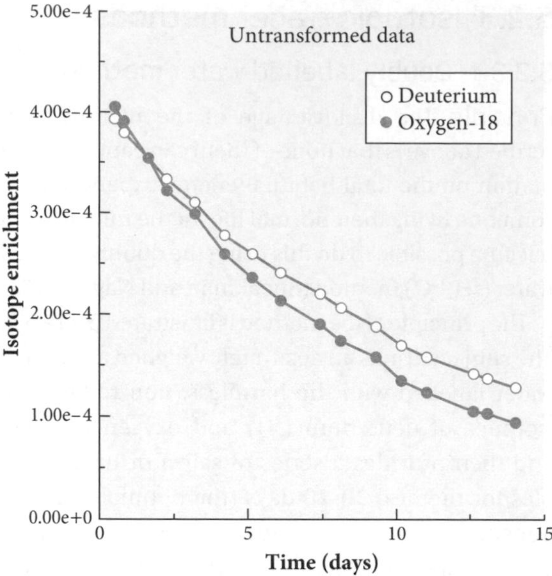


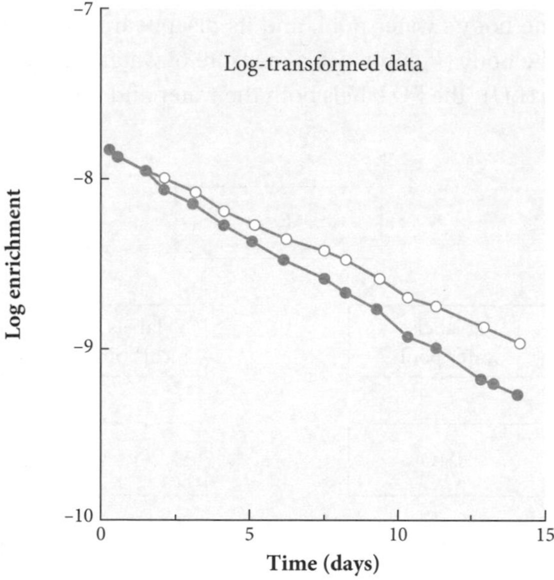


Fig. 5.3 Examples of isotope disappearance curves from the doubly labelled water method.

*Source:* Reprinted by permission from Macmillan Publishers Ltd: Murgatroyd, P.R., Shetty, P.S., and Prentice, A.M. (1993) Techniques for the measurement of human energy expenditure: a practical guide. *Int J Obesity*, **17**(10), pp. 549–68.

bonate pools, which are in rapid equilibrium with each other. The disappearance of $^{18}O$ ($k_{18}$) provides a measure of the combined turnover of water and bicarbonate ($rH_2O + rCO_2$). Therefore, bicarbonate turnover can be calculated by difference ($k_{18} - k_2$) and is equivalent to the subject's carbon dioxide production rate. The value $k_{18} - k_2$ is represented by the difference in slope between the two isotope disappearance curves shown in Fig. 5.3. Carbon dioxide production can be converted to energy expenditure using classical indirect calorimetric calculations.

In practice, there are many theoretical and mathematical complications in the technique, but these can be overcome by using appropriate procedures, and the method has been cross-validated against energy expenditure measured within a whole-body chamber. It is now recognized as the gold standard method for assessing total energy expenditure (TEE) in free-living people. By combining it with an assessment of basal metabolic rate (BMR), it becomes possible to estimate the energy costs of activity and thermogenesis (A&T = TEE – BMR) and, as thermogenesis represents a small and rather constant proportion of A&T, in practice it provides a good measure of a person's activity level.

**Equipment** Doubly labelled water measurements require a mass spectrometer to analyse the levels of $^{2}H$ and $^{18}O$. Because $^{18}O$ is so costly, it is necessary to use very small amounts that only raise the subject's levels a little above the natural background level. Hence, the mass spectrometer has to be extremely precise and the measurements are technically challenging. There are only a few specialist centres worldwide with the capacity to do the measurements and computations.

#### 5.2.3.2 Labelled bicarbonate method

**Principle** The principle of the labelled bicarbonate method is similar to the doubly labelled water method insofar as it assesses carbon dioxide production from the rate of turnover of the body's bicarbonate pool. In the labelled bicarbonate method, this is achieved through a constant subcutaneous infusion of $^{13}C$-labelled bicarbonate using a mini-pump strapped to the subject. As with doubly labelled water, samples of urine or saliva can be used to assess $^{13}C$ levels, which reach a steady state at a level proportional to the rate at which the subject's carbon dioxide production washes out the infused label. The method is suitable for relatively short-term assessments of up to about 24 hours and is ideally suited for use in hospitalized patients. In practice, however, this method has been rarely used.

Equipment The method requires a mass spectrometer capable of assessing $^{13}C$ enrichment.

### 5.2.4 Heart rate methods

Principle There is a linear relationship between energy expenditure and heart rate for all levels of expenditure above resting levels. The slope and intercept of this relationship vary among individuals according to age, sex, and fitness levels, so must be determined in each person by a calibration experiment using a treadmill or cycle ergometer in which heart rate can be measured at different work rates. Once this is known, it is possible to convert records of a person's heart rate into an estimate of TEE. The method has been validated within whole-body calorimeters and shown to have little bias, and hence is excellent for assessing group values. However, its precision is limited to about ±1 MJ/day. It should be noted that heart rate traces can provide very reliable assessments of activity patterns and of inter-individual differences without actually having to convert the output into energy expenditure. This is true also of the actometers described below.

Equipment There are various commercially available ambulatory heart rate monitors that can be worn by a subject for long periods and that collect minute-by-minute data on a computer chip. These data can be uploaded straight into software that computes energy expenditure.

### 5.2.5 Movement sensors (actometers)

Principle The principle of actometers is very similar to that for heart rate but a measure of movement in three dimensions (so called 'tri-axial') is substituted for heart rate. Modern actometers are an advance over simple instruments such as pedometers because they can capture movement in all directions and can assess the intensity of movement. As with heart rate, it is necessary to calibrate each individual against an alternative indirect calorimetric procedure in order to calculate energy expenditure.

Equipment There are a number of commercially available ambulatory actometers that operate in a similar way to heart rate monitors by accumulating minute-by-minute data over periods of up to a week.

### 5.2.6 Time-and-motion studies

Principle Frequently it is desirable just to obtain a relatively rough-and-ready estimate of people's energy expenditure, for instance to estimate a population's average energy requirements. Time-and-motion studies (often called activity diaries) can be used for this purpose. If subjects are literate, they can be asked to record their own activity patterns over short intervals (e.g. 5 or 15 minutes) throughout the day. It is necessary to have illiterate subjects followed and observed by trained field workers. Each activity is then ascribed an energy cost from standard tables. It will be appreciated that there are many approximations involved in such a process and it also carries a high burden for the subjects that can lead to inaccurate record keeping. In its most approximate form, this sort of calculation is used by the World Health Organization (WHO) and Food and Agriculture Organization (FAO) in their system for calculating the estimated average energy requirements of a population (FAO/WHO/UNU, 2004; James and Schofield, 1990).

Equipment The method only requires record sheets, a watch, and tables of the energy cost of different activities.

## 5.3 Human energy needs

The body's energy needs can be divided into three main components—basal metabolism, diet-induced thermogenesis, and physical activity. Children, or adults recovering from illness and weight loss, require additional energy for the growth of new tissue, and pregnant and lactating women require additional energy to sustain the growth of their offspring.

## 5.3.1 Basal metabolism

**Definition** Basal metabolism represents the energy required to sustain the basic processes of life, which include breathing, circulation, tissue repair and renewal, and ionic pumping. In humans, basal metabolism generally produces enough heat to maintain thermoregulation without any need to specifically generate additional heat. In most people (except the very active), basal metabolism is the largest component of daily energy expenditure, representing up to 70% of all energy used.

**Measurement** BMR is a specific term that describes energy expenditure measured under the following highly standardized conditions:

- subjects should be lying completely still and should be emotionally relaxed immediately after waking in the morning
- they must have fasted for the previous 12–14 hours and must not have performed heavy physical activity on the previous day
- they must be healthy and free from fever
- the measurement must be made at thermoneutral temperature.

The great advantage of applying such standardized conditions is that it permits comparisons among individuals and between different studies. Measurements made under similar conditions, but not quite meeting all of the above stipulations, are frequently referred to as resting metabolic rate. BMR is most frequently measured using a ventilated hood system but can also be assessed easily as part of a whole-body calorimetry protocol. The results as kJ/minute are frequently multiplied up over 24 hours and expressed as MJ/day. Because BMR is relatively predictable on the basis of a subject's age, sex, weight, and height, a number of predictive equations are available from which a reasonable approximation of a person's BMR can be obtained without having to measure it (Table 5.5) (FAO/WHO/UNU, 2004; James and Schofield, 1990).

**Table 5.5 Equations for estimating BMR from body weight**

| Age (years) | No. | BMR (MJ/day) | SEE | BMR (kcal/day) | SEE |
|---|---|---|---|---|---|
| *Males* | | | | | |
| <3 | 162 | 0.249 kg − 0.127 | 0.292 | 59.512 kg − 30.4 | 70 |
| 3–10 | 338 | 0.095 kg + 2.110 | 0.280 | 22.706 kg + 504.3 | 67 |
| 10–18 | 734 | 0.074 kg + 2.754 | 0.441 | 17.686 kg + 658.2 | 105 |
| 18–30 | 2879 | 0.063 kg + 2.896 | 0.641 | 15.057 kg + 692.2 | 153 |
| 30–60 | 646 | 0.048 kg + 3.653 | 0.700 | 11.472 kg + 873.1 | 167 |
| ≈60 | 50 | 0.049 kg + 2.459 | 0.686 | 11.711 kg + 587.7 | 164 |
| *Females* | | | | | |
| <3 | 137 | 0.244 kg − 0.130 | 0.246 | 58.317 kg − 31.1 | 59 |
| 3–10 | 413 | 0.085 kg + 2.033 | 0.292 | 20.315 kg + 485.9 | 70 |
| 10–18 | 575 | 0.056 kg + 2.898 | 0.466 | 13.384 kg + 692.6 | 111 |
| 18–30 | 829 | 0.062 kg + 2.036 | 0.497 | 14.818 kg + 486.6 | 119 |
| 30–60 | 372 | 0.034 kg + 3.538 | 0.465 | 8.126 kg + 845.6 | 111 |
| ≈60 | 38 | 0.038 kg + 2.755 | 0.451 | 9.082 kg + 658.5 | 108 |

Weight is expressed in kg.

SEE, standard error of the estimate.

*Source:* Food and Agriculture Organization of the United Nations (FAO) 2004. *Food and Nutrition Technical Report Series No 1. Human energy requirements.* Report of a Joint FAO/WHO/UNU Expert Consultation, Rome, October 2001. Available at: http://www.fao.org/home/en/ (accessed 5 October 2016). Reproduced with permission.

**Factors affecting basal metabolism** An individual's BMR is largely determined by their body size and body composition, and differences in these variables can explain the fact that women have a lower BMR than men and that BMR declines with age. The major determinant of BMR is the amount of lean tissue (often referred to as lean body mass (LBM)) or fat-free mass, since this is much more metabolically active than adipose tissue. BMR is therefore frequently expressed per kg LBM. Muscle is the major contributor to lean tissue, and in its resting basal state it has a moderately high energy expenditure. Visceral organs, especially the heart and liver, have an even higher metabolic rate and, hence, the total basal metabolism is determined also by the composition of the lean tissue in terms of the proportion of muscle to visceral organ mass. On average, women have a lower BMR than men because they are smaller and, even if matched for weight, they have a lower proportion of lean tissue than men. Older people have a lower BMR because ageing is associated with a gradual substitution of lean tissue by fat. BMR is high in children largely because they have a higher proportion of visceral organ mass to total mass.

BMR declines when people are in negative energy balance. There are two components to this decline. The first is a decrease (of up to 20%) in the metabolic rate per kg LBM; this is an adaptive mechanism to spare energy in starvation and is mediated by alterations in thyroid status. The second is due to the fact that LBM itself declines with longer-term energy deficiency. There has been considerable controversy for over a century as to whether BMR is increased when people over-consume energy, and whether this constitutes a homeostatic mechanism ('adaptive thermogenesis') for stabilizing body weight. The current consensus, based on detailed whole-body calorimeter studies and doubly labelled water measurements, is that such a mechanism does not exist, and that any increases in BMR can be accounted for by the increase in lean tissue mass that occurs with overfeeding.

BMR also changes as women pass through the various phases of reproduction and it even alters by a small percentage during the menstrual cycle. In pregnancy, BMR increases in proportion to the amount of new tissue accrued by the mother in the form of the foetus, placenta, and uterus. In thin women and in those short of food in pregnancy, BMR is suppressed in the early stages of gestation; this appears to be an adaptation to help women reproduce in marginal conditions.

Certain stimulants (e.g. caffeine) and pharmacological agents (e.g. ephedrine) also affect BMR. Drug companies have been trying to develop compounds that will increase metabolic rate as possible anti-obesity agents (e.g. through stimulating the β3-adrenoreceptors), but so far they either have had minimal effects in humans or have had unacceptable side effects on heart rate or blood pressure.

Certain clinical conditions also affect BMR, especially fevers. Alterations in thyroid function can have a pronounced effect on BMR: hypothyroidism decreases BMR and hyperthyroidism increases it. Before specific hormone assays were available, measurements of BMR were widely used in the diagnosis of thyroid disorders.

## 5.3.2 Diet-induced thermogenesis

**Definition** Diet-induced thermogenesis (DIT), often also called the thermic effect of food, represents the additional energy required to absorb, digest, transport, interconvert, and store the constituents of a meal. This is wasted energy that must be lost because no physiological process can be 100% efficient. It amounts to under 10% of total intake.

**Experimental approach** DIT is usually measured using a ventilated hood. With subjects resting, fasting energy expenditure is first measured for a short period in order to establish their baseline. Then they are fed a standard test meal and remain under the hood for a further 3–4 hours. Energy expenditure increases as the body processes the meal and then declines back to the initial baseline. DIT is assessed as the incremental area under the curve for energy expenditure.

**Factors affecting diet-induced thermogenesis** DIT is affected by the size and composition of the meal consumed. Protein tends to cause a higher

DIT than fat and carbohydrate, although in practice these differences are trivial within the normal range of the mixed diets consumed by humans.

### 5.3.3 Physical activity

**Definition** Energy expenditure caused by movements or performing physical work is generally classified under the overall heading of physical activity. This includes both conscious movements and subconscious ones (fidgeting).

**Experimental approach** The energy expended on standardized activities, such as walking at a fixed pace on a treadmill or cycling on a cycle ergometer, can be measured by Douglas bag, or a portable respirometer, or in a whole-body chamber. Everyday activities are usually measured using a portable respirometer and in the past these have been used for numerous studies of occupational physical activity, such as farming, factory work, and coal mining. Tables have been compiled that summarize these values, together with values for the energy costs of the everyday activities of life, and these can be used to make an approximation of a person's total energy expenditure (Durnin and Passmore, 1967; FAO/WHO/UNU, 2004). The energy cost of activities is usually expressed as kJ/minute, or as a multiple of BMR (termed the physical activity level (PAL)). The advantage of the latter is that it makes an automatic internal adjustment for differences between subjects of different weights, sexes, and ages, since these variables are already factored into the BMR.

**Factors affecting the energy costs of physical activity** Clearly, the total cost of physical activity is largely dependent on the amount of activity a person chooses to undertake. Within this, the specific cost of the individual activities will be influenced by the person's size, the speed of the activity, the times taken resting, the skill with which the activity is performed, and the efficiency of the muscles. Perhaps surprisingly, the efficiency with which muscles can convert food energy (glucose and fatty acids) into useful work is very constant among individuals and averages only about 25%.

The energy cost of weight-bearing activities, such as walking up stairs or uphill, is directly proportional to a person's body weight, but in non-weight-bearing activities, such as cycling, a person's body weight has less influence on the overall energy cost.

Differences in physical activity represent the major source of variability in the energy needs of different people. At the lowest end of the range are the bed-bound sick and the very elderly. These will have a PAL of about 1.35 × BMR. At the other end of the range are elite endurance athletes, such as Tour de France cyclists, who can sustain PAL values of almost 3 × BMR. People in the developing world who are engaged in hard physical labour, for instance at the peak of the farming season, sustain activity levels equivalent to about 2 × BMR. In modern society, where sedentary occupations are combined with very inactive leisure time pursuits, such as TV viewing, the average PAL is around 1.55 × BMR. Children tend to have spontaneously high levels of energy expenditure (often up at about 1.8 × BMR) that decline as they go through puberty, especially in girls.

### 5.3.4 Growth

Maintaining an adequate energy intake is essential at times of growth, and energy deficiency leads to stunting, wasting, and ultimately to severe malnutrition (see Chapter 18). In fact, in humans the marginal energy costs of growth (i.e. over and above the other daily energy needs) are surprisingly small because human growth is extraordinarily slow—an evolutionary adaptation that allows plenty of time for the growth, organization, and training of our large brain. Growth is fastest in the foetus, the very young neonate, and during the adolescent growth spurt, but even at these periods the energy required for normal growth rarely exceeds 5% of the daily energy need.

Faster tissue deposition rates than these can occur in people recovering from a severe illness, from severe childhood malnutrition, or from starvation. These very rapid rates are often accompanied by an inappropriate composition of new tissue with a higher proportion of fat tissue than is desirable. Generally, these deviations in body composition are corrected naturally after several months of weight stability. Growing children may also show episodic growth, particularly those in developing countries who are frequently affected by infections. During the recovery phase after an illness, children can have very high energy needs. These are often not met by energy-poor and protein-deficient diets in developing countries, leading to a gradual falling away from optimal growth rates and nutritional status.

### 5.3.5 Pregnancy and lactation

The marginal extra energy costs of pregnancy and lactation are also quite low in humans due to the slow growth of the offspring. In pregnancy, a mother requires only around 10% extra energy, and in lactation only about 25% (see Chapter 32 for more detail). There is good evidence that when women are short of energy, they display a range of energy-sparing mechanisms, both metabolic and behavioural, that can help ensure the success of reproduction.

## 5.4 Mechanisms for regulating energy balance

Energy balance is a dynamic state that constantly alters between positive deviations during meals and negative deviations during the intervals between meals. The challenge for the body is to ensure that these small deviations cancel out over time (except during periods of intentional growth when a slight positive energy balance is required). This regulation is achieved largely in the hypothalamus, which receives a wide range of neural and endocrine signals from the rest of the body; it integrates these through a complex network of interacting neural pathways. The hypothalamus then sends efferent neural signals to regulate appetite and energy expenditure. The short-term signals indicating energy sufficiency include blood glucose, amino acid, and fatty acid levels, together with stomach- and gut-derived hormones, and vagal signals from the liver. The long-term signals consist of hormones secreted by adipose tissue in proportion to the amount of fat that is stored there. Primary among these is leptin. Plasma concentrations of leptin are directly proportional to fat stores, but are also strongly influenced by the direction of change in the fat stores. Leptin acts as the body's fuel gauge, allowing the brain to assess its energy reserves and their rate of change; hence, it plays a vital role in regulating appetite and energy expenditure, tissue growth, reproduction, and various other physiological processes.

It used to be thought that differences in energy expenditure were major determinants of a person's energy balance. For instance, over several decades there was a popular theory that obese people must have extraordinarily efficient metabolisms and low energy requirements. In the reverse direction, it was thought that many weight-losing clinical conditions, such as AIDS and Alzheimer's disease, were caused by a hypermetabolism that raised energy needs. Studies using the doubly labelled water method have now shown that there is little truth in these theories and that most deviations in energy balance can be traced to differences in food intake. Most differences in energy expenditure can be adequately explained by differences in age and reproductive state, body size, body composition, and physical activity levels. There is, however, some flexibility in the efficiency of energy expenditure, particularly in times of weight loss or starvation, when metabolic rate can decline by about 20%.

The realization that most of the regulation of energy balance is achieved on the intake side of the energy balance equation has had a profound impact on research in the field, and there has been astonishing progress over the past decade in understanding the neurohormonal regulation of appetite. There is still much to be learnt about how these longer-term regulatory mechanisms modulate the influence of short-term internal appetite cues (such as low glucose levels or surges in the appetite-stimulating hormone, ghrelin) and external appetite cues (such as the sight and smell of food or advertising). Nonetheless, these advances hold promise for the development of therapeutic compounds to assist in the treatment of conditions such as obesity and anorexia nervosa.

## Further reading

1. **Blaxter, K.** (1989) *Energy metabolism in animals and man*. Cambridge: Cambridge University Press.
2. **Cox, S.** (2013) Energy: metabolism. In: Caballero, B., Allen, L.H., and Prentice, A.M. (eds) *Encyclopedia of human nutrition*, 3rd edn, pp. 177–185. London: Elsevier.
3. **Durnin, J.V.G.A. and Passmore, R.** (1967) *Energy, work and leisure*. London: Heinemann.
4. **FAO/WHO/UNU** (2004) *Human energy requirements. FAO Technical Report Series 1*. Rome: Food and Agricultural Organization.
5. **James, W.P.T. and Schofield, E.C.** (1990) *Human energy requirements: a manual for planners and nutritionists*. Oxford: Oxford University Press.
6. **Livesey, G. and Elia, M.** (1988) Estimation of energy expenditure, net carbohydrate utilization, and fat oxidation and synthesis by indirect calorimetry: evaluation of errors with special reference to the detailed composition of fuels. *Am J Clin Nutr*, **47**, 608–28.
7. **Murgatroyd, P.R., Shetty, P.S., and Prentice, A.M.** (1993) Techniques for the measurement of human energy expenditure: a practical guide. *Int J Obesity*, **17**, 549–68.
8. **Prentice, A.M., Black, A.E., Coward, W.A., and Cole, T.J. (1996)** Energy expenditure in overweight and obese adults in affluent societies: an analysis of 319 doubly-labelled water measurements. *Eur J Clin Nutr*, **50**, 93–7.
9. **Speakman, J.R. and Nagy, K.** (1997) *Doubly-labelled water: theory and practice*. Amsterdam: Kluwer Academic.

# Alcohol

## A. Stewart Truswell



Alcohol is the only substance that is both a drug affecting brain function and a nutrient (sometimes providing 5–10% of people's calorie intake). It is dispensed with food, not in a pharmacy. Alcohol is associated with happy times—weddings and celebrations—but it is also a cause of misery. The dose determines the effect!

Alcohol is normally consumed not pure ('neat'), but in aqueous solution in alcoholic beverages that were first developed thousands of years ago. Beer was first drunk by the Sumerians and Babylonians around 4000 years ago and has been brewed ever since. Wine is mentioned occasionally in the Old Testament (in Genesis 9, Noah planted a vineyard and got drunk), and was important in the life of classical Greece and Rome. It featured in Jesus' first miracle at the marriage feast in Cana and at his last supper, and passed into the central part of the Christian mass. Alcoholic beverages were also developed in prehistoric times in East Asia, e.g. sake fermented from rice, and in Africa beers from fermented millet or maize. Alcoholic beverages were thus independently discovered in different parts of the world by prehistoric sedentary agriculturalists who were growing barley, grapes, or rice. However, the indigenous peoples of Oceania (Polynesians and Australian Aborigines) and of America (American Indians) did not know of alcohol until the arrival of the Europeans, and had not established ways of using and controlling it. Until the development of piped water and sewage systems, beer was a healthier drink than water (it couldn't carry cholera).

From the basic fermented beverages, alcohol can be concentrated by the process of distillation (which was brought to Europe by the Arabs). Brandy and whisky first appeared in the 15th century.

## 6.1 Production of alcoholic beverages

Alcohol is produced by alcoholic fermentation of glucose. The specific enzymes are provided by certain yeasts, *Saccharomyces*. The biochemical pathway first follows the usual ten steps of anaerobic glycolysis to pyruvate, as in animal metabolism (see Chapter 2). Yeast contains the enzyme pyruvate decarboxylase, not present in animals. This converts pyruvate to acetaldehyde, then alcohol dehydrogenase converts acetaldehyde to ethanol. The overall reaction is:

$$C_6H_{12}O_6 + \text{cofactors} + \text{ATP} \rightarrow 2C_2H_5OH + 2CO_2$$

Grapes are unusual among fruits in containing a lot of sugar, nearly all glucose (around 16%), so providing an excellent substrate for alcoholic fermentation. Starch is a polymer of glucose. Before it can ferment to alcohol it has to be hydrolysed to its constituent glucose. Beers are made by malting the starch in barley. To do this, the barley is spread out, moist and warm, and allowed to germinate for several days. Enzymes are generated in the sprouting grain, which breaks down the stored starch into glucose. The barley is then heated and dried. This kills the embryo, which stops using sugar. For *sake* the starch is in rice. It is first treated with a mould, *Aspergillus oryzae*, which grows on the rice and secretes an amylase to yield glucose.

Table 6.1 **Volume that provides 10 g ethanol, a standard drink**

| Type of drink | Usual % ethanol[a] v/v (by volume) | Vol. that provides ~10 g ethanol |
|---|---|---|
| Low alcohol beer | 2–3% | 568 ml |
| Average beer | 4–5% | 285 ml |
| Average wine[b] | 10% | 120 ml = 4 oz |
| Fortified wine (e.g. sherry, port) | 20% | 60 ml = 2 oz |
| Spirits (e.g. whisky, gin, vodka, brandy) | 40% | 30 ml = 1 oz |

[a]*Note:* These are approximations. The exact percentage of alcohol should be on the label of the bottle. The specific gravity of ethanol is 0.790. To convert to g/100 mL, multiply by 0.79 (or 0.8).
[b]Wine bottles usually contain 750 mL = 6¼ standard drinks.

Beer contains around 5% alcohol (unless alcohol-reduced), wines contain around 10% alcohol (unless fortified), and spirits about 40% alcohol. Alcoholic beverages also contain variable amounts of unfermented sugars and dextrins (in beers), small amounts of alcohols other than ethyl (e.g. propyl alcohol), moderate amounts of potassium, almost no sodium, small amounts of riboflavin and niacin, but no thiamin, and sometimes vitamin C. They also contain a complex array of flavour compounds, colours (natural anthocyanins in red wines), phenolic compounds, a preservative (e.g. sodium metabisulphite), and sometimes additives. A standard drink (e.g. ½ pint beer, see Table 6.1) provides 10 g of ethanol.

## 6.2 Metabolism of alcohol

Ethanol is readily absorbed unchanged from the jejunum; it is one of the few substances that is also absorbed from the stomach. It is distributed throughout the total body water (moving easily through cell membranes), so that after having one drink, its 10 g of alcohol is diluted in about 40 L of water in an adult, giving a peak concentration of 0.025 g/dL in the blood and in the rest of the body water. For comparison, the permitted limit of blood alcohol for driving in many countries is double this—0.05 g/dL (11 mmol/L)—and the driving limit varies from 0.02 g/dL in Sweden to 0.08 g/dL in the British Isles (Fig. 6.1). Alcohol is nearly all metabolized in the liver, but a small amount is already metabolized as it passes through the stomach wall (first-pass metabolism). A small amount of alcohol passes unchanged into the urine and an even smaller (but diagnostically useful) amount is excreted in the breath.

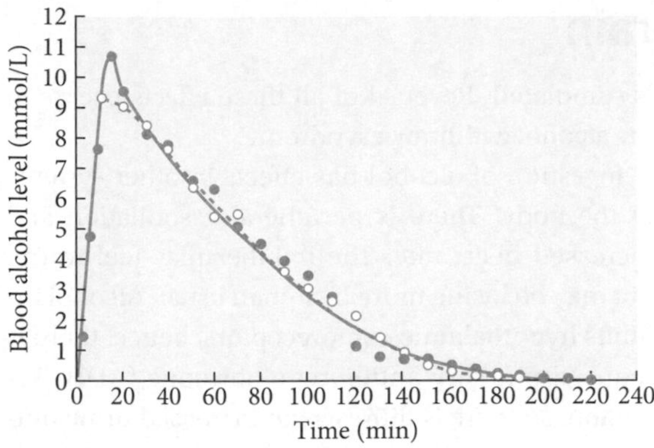


Fig. 6.1 Blood alcohol concentrations in a healthy young Caucasian man who took 0.3 g pure ethanol/kg body weight in orange juice, drunk rapidly and followed by a 4.4 MJ mixed meal (● and ○ are duplicate determinations). Ethanol was measured by gas chromatography.

There are three possible pathways for alcohol metabolism in humans. The major pathway in most people starts with alcohol dehydrogenase (ADH), a zinc-containing enzyme in the cytoplasm of the liver (Box 6.1). The ADH step is the rate-limiting step in alcohol metabolism. ADH occurs in slightly different forms, and some individuals have more active ADH than others.

It may seem surprising that humans naturally possess this enzyme for dealing with beer and wine, to which our hunter-gatherer ancestors were not exposed. However, some alcohols are produced naturally inside the body by fermentation in the large intestine (e.g. small amounts of methyl alcohol from pectin), and they occur in over-ripe fruits. The next step is conversion of acetaldehyde to acetate by aldehyde dehydrogenases (ALDHs), which are present in the cytoplasm and mitochondria (Box 6.2). In most people, there is no build-up of acetaldehyde, but nearly 50% of Chinese, Korean, and Japanese people have an inactive variant of one of the ALDHs, so after moderate intake of alcohol their blood acetaldehyde increases. This causes facial flushing and headache; sickness in homozygotes.

In long-term heavy drinkers, the microsomal ethanol oxidizing system (MEOS) with cytochrome P450 becomes a second important route for alcohol metabolism. The microsomes proliferate (are induced) in heavy drinkers. As with ADH, ethanol is converted to acetaldehyde. A third minor pathway for conversion of ethanol to acetaldehyde is via catalase in peroxisomes. The acetate that results from metabolism of alcohol goes into the tissues as a source of energy. Consequential changes in metabolism, which are important if the alcohol intake was large, are increased fatty acid synthesis, decrease in the Krebs cycle, and decreased gluconeogenesis, which can lead to hypoglycaemia after a heavy bout of drinking, and decreased excretion of urate.

On average, people can metabolize about 5 g of ethanol per hour (i.e. half a standard drink). The rate varies about two-fold between individuals. Alcohol absorption can be slowed by having a meal, or even milk, in the stomach, but there is no agent that increases the rate of alcohol metabolism. Smaller people are likely to have smaller livers and so metabolize less alcohol per hour. Women, on average, have smaller livers than men, a lower percentage of total body water (in which to distribute the alcohol), and also have less first-pass gastric alcohol dehydrogenase, so that they are less tolerant of alcohol than men. East Asian people may suffer from headaches and flushing at quite low intakes of alcohol because of acetaldehyde accumulation. This may limit their intake.

A drug used to control alcohol addiction, disulfiram (Antabuse) antagonizes ALDH. People taking it experience unpleasant symptoms (headache, nausea, flushing) as a result of acetaldehyde accumulation when they have a drink.

**Box 6.1**

$$CH_3CH_2OH^+ + NAD \rightarrow CH_3CHO + NADH + H^+$$

Ethanol ADH Acetaldehyde

**Box 6.2**

$$CH_3CH_2OH + +NAD + H_2O \rightarrow CH_3COOH + NADH + H^+$$

Acetaldehyde ALDH Acetic acid

## 6.3 Effect of alcohol on the brain

Pharmacologists classify ethanol as a central nervous system depressant, in the same group as volatile anaesthetic agents. With increasing levels of blood alcohol, people pass through successive stages of alcohol intoxication (Table 6.2). At the biochemical level alcohol affects a number of neurochemical processes simultaneously. γ-Aminobutyric (GABA) systems (inhibitory) are more active. Activity of the stimulatory glutamate *N*-methyl-D-aspartate receptor system is reduced. Dopamine is released and contributes to the reward effects of alcohol. The serotonergic system is stimulated. Reversal of all these effects occurs in the alcohol withdrawn syndrome.

Ingestion of alcohol has effects in other systems of the body. There is peripheral vasodilation and increased heart rate. The imbiber may feel warm, but may be losing more heat than usual. Alcohol inhibits hypothalamic osmoreceptors, hence, there is reduced pituitary antidiuretic hormone (ADH) secretion, so there is diuresis (an increased urine output), which can lead to dehydration, especially after drinking spirits.

## 6.4 Energy value of ethanol

The gross chemical energy of ethanol can be measured outside the body in a bomb calorimeter, and the value is between the energy value of carbohydrates and that of fat, about 30 kJ or 7.1 kcal/g.

However, in a metabolic ward, with food intakes strictly controlled, Lieber (1992) replaced 50% of subjects' energy (calorie) intake by isocaloric amounts of ethanol (they had been accustomed to high alcohol intakes). Instead of gaining weight, they lost weight. Free-living heavy drinkers are not usually overweight. It appears that above a certain intake, ethanol provides less than 7 kcal/g. Alcohol increases the basal metabolic rate (thermogenesis) and it is thought that metabolism of alcohol by liver microsomes yields less energy than the ADH route.

In heavy drinkers, 10–30% (or more) of dietary energy intake comes from alcohol, but alcoholic beverages contain no protein and very few micronutrients, so this nutrient-poor source of calories displaces other foods that normally provide essential nutrients. Appetite may be suppressed in heavy drinkers, either by alcoholic gastritis or by associated smoking. Alcohol dependency is an important cause of conditioned (or secondary) nutritional deficiency—the drinker may have access to enough foods and their nutrients, but is not eating them.

**Table 6.2 Successive stages of acute alcohol intoxication**

| Blood alcohol concentration (g/dL) | Stage | Effects |
|---|---|---|
| Up to 0.05 | Feeling of well-being | Relaxed, talks a lot |
| 0.05–0.08 | Risky state | Judgement and finer movements affected |
| 0.08–0.15 | Dangerous state | Slow speech, balance affected, eyesight blurred, wants to fall asleep, likely to vomit, needs help to walk |
| 0.2–0.4 | Drunken stupor | Dead drunk, no bladder control, heavy breathing, unconscious (e.g. deep anaesthesia) |
| 0.45–0.6 | Death | Shock and death |

Nutrients that are typically depleted in alcoholics include thiamin, folate, niacin, and several inorganic nutrients (see section 6.6).

Nutrition surveys that do not take alcohol intake into account cannot represent their subjects' full nutrient intake.

# 6.5 Direct consequences of alcohol intake

## 6.5.1 Acute intoxication

Acute intoxication can lead to road and other accidents, or domestic and other violence. Intoxicated people can suffer and inflict a range of injuries. Occasionally, people consume such a large dose of alcohol that they die with lethal blood levels. The breathalyser was developed to reduce road traffic accidents. In many countries, a driver stopped at random by a police check who has a breathalyser reading corresponding to a blood level of 0.05 g/dL (0.02–0.08 in different countries) will have his or her driver's licence suspended. This measure has reduced traffic accidents and contributed to the decline of alcohol consumption in a number of developed countries.

## 6.5.2 Hangovers

The excess intake of alcohol the night before may not yet have all been cleared from the blood. Dehydration may be present from diuresis and, with some drinks (e.g. brandy), toxic effects of small amounts of methanol and higher alcohols contribute to the symptoms.

## 6.5.3 Chronic alcoholism

Some people become dependent or addicted to alcohol and cannot face the world unless they have some alcohol in their blood throughout the day. Thus, they maintain an intake of alcohol per day larger than their liver's capacity to metabolize it.

## 6.5.4 Alcohol withdrawal syndrome

Alcohol addicts who have maintained some alcohol in their blood continuously for weeks or even longer, suffer withdrawal symptoms if an accident or illness abruptly removes them from their alcohol supply. There are tremors of the hands, anxiety, insomnia, and tachycardia. *Epileptic convulsions* can occur and, in severe cases, there is agitation, mental confusion, and hallucinations. This is *delirium tremens*, a severe illness.

## 6.5.5 Binge drinkers

One-night binge drinkers expect to get drunk (see Box 6.3). Men imbibe 80 g of alcohol (4 pints of beer) or more and women somewhat less.

**Box 6.3 Different patterns of alcohol consumption**

- The inexperienced drinker (e.g. an adolescent) who misjudges the dose and has an accident.
- The person who doesn't drink during the week, but drinks to excess and gets drunk on payday or Saturday night (one-night binge).
- The person who enjoys a controlled one or two drinks with dinner most days.
- The person who has too many drinks each day (mostly after work), but more or less maintains their (increasingly inefficient) usual life.
- The person who drinks very heavily for weeks.

The other pattern of alcohol excess is that a person drinks heavily for weeks. Consequently, as alcohol displaces much of the usual food intake, there can be an acute deficiency of micronutrients with the smallest reserve in the body, usually thiamin (see Wernicke-Korsakoff syndrome, Section 6.6.4 and Chapter 12).

## 6.6 Medical consequences of excess consumption

### 6.6.1 Liver disease

Alcohol causes liver damage in three stages. The least severe is fatty liver. Metabolism of large amounts of ethanol in the liver produce an increased ratio of NADH/NAD; this depresses the citric acid cycle and oxidation of fatty acids, and favours triglyceride synthesis in the liver cells. It used to be thought that the fatty liver was due to an associated nutritional deficiency, but fatty liver has been observed (using needle biopsy of the liver) in volunteers who took a moderately large intake of alcohol, but with all nutrients provided under strictly controlled conditions in hospital. The symptoms of fatty liver are not striking. On abdominal examination a doctor can feel that the liver is somewhat enlarged, and this shows with ultrasound; biochemical changes can be seen in a blood sample (see section 6.9).

*Alcoholic hepatitis* (inflammation of the liver) is more serious. This type is not caused by a virus, but by prolonged excess alcohol intake. There is loss of appetite, fevers, tender liver, jaundice, and elevation in the plasma of enzymes produced in the liver (e.g. aminotransferases (transaminases), γ-glutamyl transpeptidase, and alkaline phosphatase). If the patient continues drinking, this can progress to cirrhosis.

*Alcoholic cirrhosis* is associated with chronic alcoholism. When the liver has to metabolize large amounts of alcohol over a long time, membranes inside the cells become disordered; mitochondria show ballooning. In its fully developed form, irregular strands of fibrous tissue criss-cross the liver, replacing damaged liver parenchymal cells. These effects may be due to acetaldehyde or to free radical generation by neutrophil polymorph white cells in the liver. Cirrhosis seems to occur in people who have managed to consume large amounts of alcohol over many years, but carried on a reasonably regular life, and were able to eat and afford the alcohol. The amount of alcohol needed to cause cirrhosis is difficult to establish exactly because many people understate their alcohol consumption, especially heavy drinkers. It is greater than 40 g of ethanol per day in women and 50 g in men over years, usually much more. Most deaths from liver disease in the UK are due to alcohol—and they are increasing.

Not all cases of chronic hepatitis and cirrhosis are caused by alcohol excess. Some are caused by hepatitis viruses (B or C), non-alcoholic fatty liver disease (NAFLD) and non-alcoholic steatohepatitis are related to obesity (see Chapter 17), and there are other less common causes (see sections 9.6 and 10.2.8). Prolonged excessive alcohol intake is, however, the most common cause of liver cirrhosis and rates of mortality due to cirrhosis are an important indicator of population levels of harm from alcohol. In Western Europe, France had the highest mortality from cirrhosis in 1960, but by 2012 Austria and then Scotland had the highest rates.

### 6.6.2 Metabolic effects

Moderate regular drinkers who are apparently well may have increased plasma triglycerides (an overflow from the overproduction of fat in the liver). Plasma urate is raised because of reduced renal excretion probably due to increased blood lactate, which follows alcohol ingestion.

### 6.6.3 Foetal alcohol syndrome

Women who drink alcohol heavily during early pregnancy can give birth to a baby with an unusual facial appearance (small eyes, absent philtrum, thin upper lip), prenatal and postnatal growth impairment, central nervous system dysfunction, and often other physical abnormalities. Mothers of children with the foetal alcohol syndrome were heavy drinkers during their pregnancy and most were socially deprived. In a remote Aboriginal community, Fitzroy Crossing in north-west Australia, foetal alcohol syndrome affected about one in eight of the children. Female elders asked Sydney paediatricians to help tackle the problem, which had been under-recognized. Take-away sales of all alcohol drinks (except low alcohol beer) were prohibited and school teaching was modified for children with attention problems. Foetal alcohol syndrome and the broader foetal alcohol spectrum disorders (FASD) are a tragic legacy of society's careless relationship with alcohol. More moderate drinkers may have babies that are small for date, but otherwise normal. Some authorities insist that pregnant women should avoid all alcohol, but in a careful prospective study in Dundee, Scotland, Florey's group (Sulaiman et al., 1988) found that, after adjustment for the effect of smoking, social class, and mother's size, there was no detectable effect on pregnancy of alcohol consumption below 100 g/week (i.e. one standard drink a day).

### 6.6.4 Wernicke–Korsakoff syndrome

In heavy drinkers who consume large amounts of alcohol and virtually stop eating for three or more weeks, brain function can be affected by acute thiamin deficiency. Ethanol uses up thiamin for its metabolism, yet alcoholic beverages provide no thiamin; there is no rich food source of thiamin and body stores are very small (see Chapter 12). In Wernicke's encephalopathy, the patient is quietly confused—not an easy state to recognize in an alcoholic. The diagnostic feature, if the sufferer is brought to medical attention, is that the eyes cannot move properly (ophthalmoplegia). When Wernicke's encephalopathy is treated with thiamin, the ophthalmoplegia and confusion clear, but the patient may be left with a loss of recent memory, the inability to recall what has happened recently (Korsakoff's syndrome). It has been suggested that when an alcoholic has a partner who provides food, containing some thiamin, Wernicke–Korsakoff syndrome (WKS) is less likely. Korsakoff's psychosis can be permanent. It is one cause of alcohol-related brain damage. The incidence of WKS has been relatively high in Australia. As a preventive measure, bread in Australia has been fortified with thiamin since 1991, as it had been for a long time (for other reasons) in the USA, UK, and most other developed countries. WKS has become uncommon in Australia. Wernicke's

**Box 6.4 Similar drinks with low or zero alcohol**

At a special occasion, where everyone is expected to enjoy beer or wine, any substitute is likely to be a boring 'soft drink', and its drinker may seem to not fully participate in the celebration. Low alcohol beers and cider can help here (listed in McCance and Widdowson's British Food Tables), and there are now in industrial countries alcohol-free beers and wines that can have the traditional composition (and much of the taste) except the alcohol, e.g. 'alcohol removed Shiraz', and even 'Champagne'. These products can help drinkers who want to stop or reduce their alcohol intake.

It is good to see that some leading wine companies in Australia/New Zealand now have this advice on their label:

- Pregnancy warning
- Alcohol can cause lifelong harm to your baby

Some other wines advise:

- It is safer not to drink while you are pregnant

encephalopathy occasionally occurs in people who have not taken alcohol, e.g. with persistent vomiting of pregnancy, hyperemesis gravidarum.

### 6.6.5 Other nutritional deficiencies in alcoholics

In societies with adequate food supply, vitamin deficiencies are rare but do occur in heavy drinkers. Chronic thiamin or other B-vitamin deficiency may be responsible for a peripheral neuropathy in the legs, with reduced function of the motor and sensory nerves and diminished ankle jerks. Folate metabolism is commonly impaired in alcoholics, and megaloblastic anaemia may be seen. Vitamin A metabolism is abnormal where there is alcoholic liver disease. The liver does not store retinol normally or synthesize retinol-binding protein adequately. There can, consequently, be reduced plasma retinol and night blindness. Among inorganic nutrients, plasma magnesium and zinc can be subnormal in alcoholics.

### 6.6.6 Predisposition to some types of cancer

The risk of cancer of the mouth and pharynx is increased, especially when high alcohol intakes are combined with smoking. Other cancers associated with high alcohol consumption are those of the oesophagus, the liver (primary cancer of the liver is a complication of cirrhosis), the colon and rectum (in some beer drinkers), and breast cancer.

### 6.6.7 Gastrointestinal complications

Chronic gastritis and gastric or duodenal ulcers may be associated with excessive alcohol consumption. Acute pancreatitis is a severe complication.

### 6.6.8 Impaired immunity

Heavy alcohol consumption impairs immunity and increases susceptibility to pneumonia and tuberculosis.

### 6.6.9 Hypertension

The prevalence of raised arterial blood pressure increases with usual alcohol intakes above three or four drinks per day. Prompt falls of moderately elevated blood pressures have been well documented in heavy drinkers admitted to hospital for detoxication. Increased prevalence of hypertension explains the greater risk of stroke from cerebral haemorrhage in heavy alcohol drinkers. Heavy intakes of alcohol simulate secretion of corticotrophin-releasing hormone. Increased cortisol and sympathetic activity may explain the increased blood pressure. Limiting alcohol consumption is a standard part of lifestyle modification recommended for people with hypertension.

## 6.7 Alcohol and coronary heart disease

Opposed to the deleterious effect of alcohol on blood pressure is its apparent effect in reducing the risk of coronary heart disease (CHD), one of the major causes of death in affluent communities. Over 20 large prospective studies in several countries have all found that light-to-moderate alcohol consumption appears to protect against CHD (Table 6.3). At post-mortem examination, pathologists have long known to expect little or no atheroma in the arteries of people dying of alcoholic complications. However, the discovery that light-to-moderate drinking is negatively associated with CHD emerged in the 1990s first as a by-product of the classic prospective study of British doctors that established the health dangers of smoking (Doll et al., 2004). It is surprising that many subsequent epidemiological studies confirm this particular health benefit of drinking alcohol. Sceptics suggest confounding, but a large prospective study of twins reports the same lower risk of CHD in moderate drinkers, with no difference in all-cause mortality (Dai et al., 2015).

**Table 6.3 Relative risks of total mortality and mortality from coronary heart disease (CHD) in 276,802 men in the USA (aged 40–59 years at entry) in a 12-year follow-up**

| | Drinks per day | | | | | | | |
|---|---|---|---|---|---|---|---|---|
| | 0 | <1 | 1 | 2 | 3 | 4 | 5 | 6+ |
| Total death rate | 1.00 | 0.88 | 0.84 | 0.93 | 1.02 | 1.08 | 1.22 | 1.38 |
| CHD death rate | 1.00 | 0.86 | 0.79 | 0.80 | 0.83 | 0.74 | 0.85 | 0.92 |

*Adapted from*: Boffetta, P., and Garfinkel, L. (1990)

The longest established mechanism for this protective effect (known since 1969) is that alcohol consumption increases plasma high-density lipoprotein (HDL) cholesterol, a well-established protective factor for CHD (see Chapter 19). This increase of HDL in moderate drinkers is not sufficient to explain fully their lower risk of CHD.

There is also evidence that alcohol drinking reduces the tendency to thrombosis. It is not possible to test this directly, but alcohol reduces aggregation of platelets *in vitro*, in response to collagen and ADP.

Wines, especially red wines, contain flavonoid antioxidants, principally catechins and anthocyanins. These are better absorbed from alcoholic drinks than from vegetables and fruits. One of these polyphenols, resveratrol, has shown anti-inflammatory and cardioprotective effects in animals. But in a prospective study in Chianti, Italy, total resveratrol metabolites were measured in 783 older people: resveratrol level quartiles were not significantly associated with biomarkers of inflammation or incident cardiovascular disease or cancer (Semba et al., 2014). Sir Richard Doll (1997) suggested the extra benefit (in some countries) of wine over beer and spirits can be accounted for by differences in the pattern of drinking. No epidemiological study has conclusively shown a benefit of red over white wine (Klatsky et al., 1997).

This health benefit of alcohol consumption on CHD only applies to older people in developed countries. Even in this group, the benefit is almost balanced by deaths related to alcohol from other diseases. In the younger majority in developed countries and at all ages in developing countries, CHD hardly ever occurs: the effects of alcohol are almost all adverse (Fig. 6.2).

The cardio-protective effect of regular light-to-moderate drinking does not apply to episodic heavy drinking. A meta-analysis confirms that binge drinking confers no cardio-protective effects (Lieber, 1992). In Russian cities, excessive vodka drinking has been a major contributory cause of excessive deaths in men 15–54 years of age, some of which were certified as due to ischaemic heart disease (Zaridze et al.,

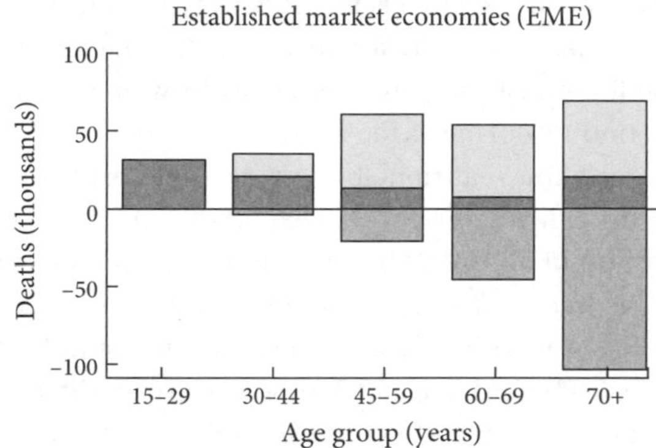


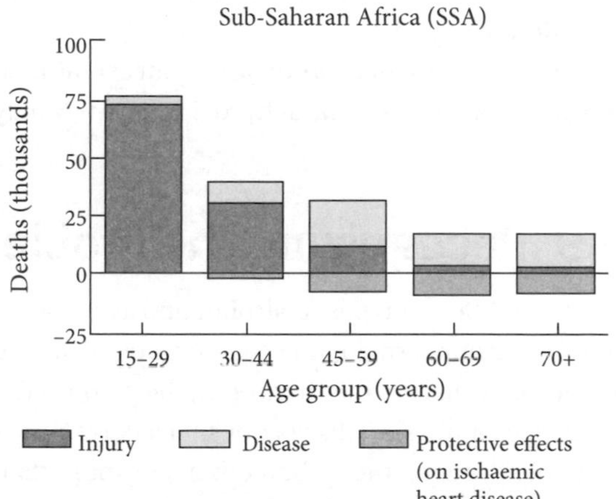


Fig. 6.2 Male deaths attributable to and averted by alcohol in established market economies (developed countries) compared with Sub-Saharan Africa.

*Source:* Murray, C.J.L. and Lopez, A.D. (eds) (1996) *The global burden of disease*. For WHO and World Bank. Cambridge, MA: Harvard University Press.

2014); Russia's death rate from CHD rose to be the highest in the world, while mortalities have been declining in Western countries.

### 6.7.1 Type 2 diabetes

In prospective cohort studies, it appears that small intakes of alcohol may also reduce the risk of developing type 2 diabetes (Baliunas et al., 2009). In a meta-analysis of 20 studies in 477,000 individuals followed over 10 years, those who stated their daily alcohol consumption as 20–30 g had a relative risk of diabetes averaging 0.60 in women and 0.87 in men. Higher intakes were not protective. In healthy subjects, postprandial glucose is lower if, alongside the test meal, beer, wine, or gin is taken in place of water (Brand-Miller et al., 2007).

## 6.8 Global burden of diseases related to alcohol

In the different WHO regions of the world, alcohol-attributable deaths (percentage of all deaths) in men (in 2004) were 11% in Europe, 9% in the Americas, and 8.4% in the West Pacific. They were 3.9% in South East Asia, 3.7% in Africa, and only 1.0% in the eastern Mediterranean region. Mortalities in women were around 17% of men's rates.

Disabilities attributable to alcohol, expressed as disability-adjusted life years (DALYs) as a proportion of all DALYs by sex and region, in men (in 2004) were 17.3% in Europe, 14.2% in the Americas, 11.8% in West Pacific countries, 4.7% in South East Asia, 3.4% in Africa, and 0.9% in eastern Mediterranean countries. DALYs from alcohol in women were around 18% of rates in men, i.e. 4.4% down to 0.1% in the different regions.

Thus, one component of our diet, of our overall food and drink—alcohol—although providing conviviality and social lubrication, is a serious cause of premature death and disease, except in Muslim countries. In poor developing countries, alcohol consumption has been increasing. For example, Thailand's alcohol consumption has increased by 33 times in the past 40 years and 8.1% of DALYs lost in that country are now attributable to alcohol (Gilmore, 2009).

It is estimated that in 200 three-digit disease codes in the International Classification of Diseases, alcohol is a component cause, and there are also 30 three- or four-digit codes that are alcohol-specific.

Health authorities in several countries are particularly concerned about increasing numbers of young people who go out binge drinking to excess on Friday or Saturday nights (Pincock, 2003). This includes teenage girls, some of whom mistakenly believe that they can tolerate the same excess intakes as their male companions (Frezza et al., 1990).

## 6.9 Recognizing the problem drinker

There are different types of alcohol abuse. An 'alcoholic' is a group term for any person whose drinking is leading to harm. This harm may be alcohol dependence, or physical disease, or social harm.

People drinking more than others, or more than they feel that they should, are very likely to underestimate their alcohol intake when asked. The spouse or other family member may give a very different answer. Health professionals are trained to suspect when someone is drinking too much and researchers use tactfully drafted questionnaires. Alcohol can be smelt on the breath and measured quantitatively in the breath, blood, or urine within hours of drinking. If a person has not been drinking recently, there are changes in the blood that are suggestive of long-term excessive alcohol intake:

- increased plasma $\gamma$-glutamyl transferase (GGT) activity
- increased plasma carbohydrate-deficient transferrin

- increased plasma (fasting) triglycerides (i.e. very low-density lipoproteins)
- increased plasma aminotransferases (transaminases)
- increased plasma urate
- increased red cell volume (mean corpuscular volume).

These vary in sensitivity and specificity; the first four findings listed above occur in the liver. Increased urate is a result of increased plasma lactate. Increased red cell volume is sometimes due to folate depletion; its cause in most cases is not yet clear.

## 6.10 Is alcoholism a disease or the top end of a normal distribution?

Alcoholism is a costly problem in most communities because of associated diseases, accidents, loss of earning, medical expenses, and social misery. There are two philosophical approaches. One is the medical model, which sees alcoholism as a disease in individuals who should be treated by the health professions. The other is the society model—the more alcohol sold and consumed, the larger will be the number of alcoholics. There are sections of society (e.g. some occupations, deprived minorities) who are at increased risk and there are social practices that contribute to alcohol abuse.

Clearly, primary care physicians cannot communicate with or control the majority of heavy drinkers, until they present with complications. At this stage, brief medical advice can be effective.

Ledermann (1956) put forward the hypothesis that, in a homogeneous population, the distribution of alcohol consumption is a logarithmic normal curve and that the number of people who drink a certain amount can be calculated if the average consumption is known (Fig. 6.3). This theory predicts that major complications of alcoholism in a country will be related to average national consumption. Governments rely on this principle in maintaining substantial taxes on alcohol, restricted outlets and hours, lower age limits, and other measures to reduce its free availability.

### 6.10.1 Government's responsibilities

Governments have a responsibility to work to limit excessive alcohol consumption. Here, public health can be in conflict with the great lobbying power of the alcohol industry—brewers, hotels, bars, restaurants, and wine growers.

Five well-established policies can be effective if governments are strong enough to implement them:

- taxation, making alcohol more expensive and which should be based on alcohol content (no discounts); there should be a minimum unit pricing
- drink-driving legislation—and enforcement
- banning advertising
- limiting availability, number of outlets, opening and closing times, minimum purchase age, preventing illicit alcohol production
- providing help for hazardous drinkers.

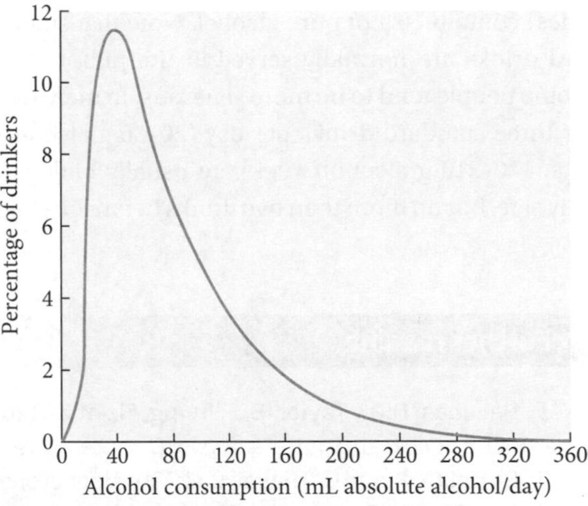


Fig. 6.3 Hypothetical curve proposed by Ledermann (1956). In a homogeneous population, alcohol consumption is distributed in a logarithmic normal curve.
*Source:* Smith, R. (1982) Alcohol problems: ABC of alcohol. In *Alcohol and alcoholism*, p. 29. London: BMJ.

Scotland, whose Department of Health is independent of England's, introduced minimum unit pricing of alcoholic drinks in 2018 after six years of legal fights with the drinks industry. One year later sales had fallen 4–5% in Scotland, while they continued to rise in England. Education campaigns alone are not effective.

## 6.11 Genetic liability to alcohol dependence?

Occurrence of alcoholism in families could be learnt, rather than genetic. From comparing monozygotic with dizygotic twins, the heritability of amount and frequency of alcohol drinking appears to be about 0.36 (i.e. one-third of the way along the scale from purely environmental to purely genetic). However, studies on twins cannot completely exclude environmental effects. Adoption studies have shown that the sons of alcoholic fathers are four times more likely to become alcoholics than the sons of fathers who were not alcoholics. The search is on to find one or more variations of brain metabolism that make people more likely to become alcohol-dependent. There have been several claims (e.g. abnormality of brain handling of dopamine or serotonin), but none has been convincingly confirmed.

## 6.12 Acceptable intakes of alcohol

The usual way in which alcohol is mentioned in national sets of dietary guidelines (see Chapter 27) is 'drink alcohol in moderation, if at all' or 'if you drink alcoholic beverages, do so in moderation'.

Because women have lower rates of metabolizing alcohol, advice on safe drinking levels has to be different for men and women. Recommendations are expressed in standard drinks that (in many countries) contain 10 g of pure alcohol. Note that standard drinks are normally served in the pub, but at home people tend to be more generous. In men, two or three standard drinks per day (20–30 g alcohol) (i.e. 140–210 g alcohol/week) are usually biologically safe, but no more than two drinks before driving, and a minority of men should not take this much, or even any alcohol (e.g. people with liver disease, taking other sedative drugs, or with a history of alcohol dependence). In women, the biologically safe intake is one or two drinks (10–20 g alcohol) per day (i.e. 70–140 g alcohol/week). In pregnancy, intake should be one drink or less; likewise if a woman thinks she might be pregnant. Children should not take alcohol, but in some cultures they are offered a small drink of wine with the family's main meal, and some believe this can be a good training in moderation in consumption of alcohol. The quantities of various alcoholic beverages providing 10 g ethanol are shown in Table 6.1.

### Further reading

1. **Baliunas, D.O., Taylor, B.J., Irving, H., et al.** (2009) Alcohol as a risk factor for type 2 diabetes: a systematic review and meta-analysis. *Diabetes Care*, **32**, 2123–32.
2. **Boffetta, P. and Garfinkel, L.** (1990) Alcohol drinking and mortality among men enrolled in an American Cancer Society Prospective Study. *Epidemiology*, **1**, 342–8.
3. **Brand-Miller, J.C., Fatima, K., Middlemiss, C., et al.** (2007) Effect of alcoholic beverages on postprandial glycemia and insulinemia in lean, young, healthy adults. *Am J Clin Nutr*, **85**, 1545–51.
4. **Chalmers, J.** (2014) Alcohol minimum unit pricing and socioeconomic status. *Lancet*, **383**, 1616–17.

5. **Christie, B.** (2020) Minimum pricing in Scotland leads to fall in alcohol sales. *BMJ,* **369,** m2324.
6. **Dai, J., Mukamal, K.J., Krasnow, R.E., Swan, G.E., and Reed, T.** (2015) Higher usual alcohol consumption was associated with a lower 41-y mortality risk from coronary artery disease in men independent of genetic and common environmental factors: the prospective NHLBI Twin Study. *Am J Clin Nutr*, **102**, 31–9.
7. **Doll, R.** (1997) One for the heart. *BMJ*, **315**, 1664–8.
8. **Doll, R.** (2004) Mortality in relation to smoking: 50 years' observations on male British doctors. *BMJ*, **328**, 1519.
9. **Frezza, M., di Padova, C., Pozzato, G., Teroin, M., Baraona, E., and Lieber, C.S.** (1990) High blood alcohol levels in women. The role of decreased gastric alcohol dehydrogenase activity and first-pass metabolism. *N Engl J Med*, **322**, 95–9.
10. **Gilmore, I.** (2009) Action needed to tackle a global drink problem. *Lancet*, **373**, 2174.
11. **Klatsky, A.L., Armstrong, M.A., and Friedman, G.D.** (1997) Red wine, white wine, liquor, beer, and risk for coronary artery disease hospitalization. *Am J Cardiol*, **80**, 416–20.
12. **Ledermann, S.** (1956) *Alcool, alcoolisme, alcoolisation.* Paris: Presse Universitaires de France.
13. **Lieber, L.S. (ed)** (1992) *Medical and nutritional complications of alcoholism. Mechanisms and management.* New York: Plenum.
14. **Pincock, S.** (2003) Binge drinking on rise in UK and elsewhere. *Lancet*, **365**, 1126–7.
15. **Popova, S., Lange, L., Shield, K., et al.** (2016) Comorbidity of fetal alcohol spectrum disorder: a systematic review and meta-analyses. *Lancet* **387**, 978–87.
16. **Semba, R.D., Ferrucci, L., Bartali, B., et al.** (2014) Resveratrol levels and all-cause mortality in older community-dwelling adults. *JAMA Intern Med*, **174**, 1077–84.
17. **Sulaiman, N.D., Florey, C. du V., Taylor, D.J., and Ogston, S.A.** (1988) Alcohol consumption in Dundee primigravidas and its effects on outcome of pregnancy. *BMJ*, **296**, 1500–3.
18. **Williams, R., Aspinall, R., Bellis, M., et al.** (2014) Addressing liver disease in the UK; a blueprint for attaining excellence in health care and reducing premature mortality from lifestyle issues of excess consumption of alcohol, obesity, and viral hepatitis. *Lancet*, **384**, 1953–97.
19. **Zaridze, D., Lewington, S., Boroda, A., et al.** (2014) Alcohol and mortality in Russia: prospective observational study of 151,000 adults. *Lancet*, **383**, 1465–73.

# Part 3

# Organic and Inorganic Essential Nutrients

# Water, Electrolytes, and Acid-base Balance

Lewis J. James and Gethin H. Evans



## 7.1 Body water

### 7.1.1 Importance of water

Water ($H_2O$) and human life are implicitly linked. This chapter will focus on the biological importance of water to humans, but beyond this, much of the human world is dependent on water. Proximity to water was fundamental for the development of human civilization, particularly urban civilization as we know it, which is why almost all major cities/towns are located on or very close to rivers, lakes, or springs. Clearly there was a need for easy availability of drinking water to support growing urban populations, but water was also vital for other reasons, including transport (human and material), cleaning, washing and waste removal, and the powering of early automated processes (mills etc.), amongst others. Indeed, water is now directly or indirectly used for almost all modern processes and the use (and abuse) of water is an important environmental consideration. With global warming, easily assessable clean water is only likely to become an increasingly important commodity.

From a biological perspective, water is the most abundant compound in the human body and is the nutrient both consumed and lost in the largest amount each day. Despite its abundance in the body, water stores are regulated within a narrow physiological range and even small deviations from normal body water stores exerts strong physiological, subjective, and behavioural responses that lead to the correction of the imbalance. Water can be considered the most essential of nutrients, as depending on the environment, a complete absence of water intake can result in death in hours or days, which contrasts with weeks or months in the complete absence of energy or other nutrients.

Water consists of two hydrogen atoms and one oxygen atom, which combine to form a polar molecule (the oxygen atom has a partial positive charge, whilst the hydrogen atoms have a partial negative charge). This polarity is responsible for much of water's biological importance, as it means neighbouring water molecules interact via hydrogen bonding. It is this hydrogen bonding that, as noted by Lawrence Henderson, in 1913, gives water its peculiar properties that make it an essential constituent of all known forms of life. Water is remarkably liquid at the range of temperatures in which biochemical reactions can occur, it has a high specific heat capacity that helps to moderate temperature gradients, a high latent heat that allows efficient cooling through evaporation, and a large dielectric constant that substantially reduces forces between charges. Water is also a good solvent for most organic compounds, except lipids and hydrocarbons, and, even when the active sites of enzymes are in clefts that exclude water molecules, reactants and products generally must arrive and leave in aqueous solution.

## 7.1.2 Body water volume and distribution

Water is the largest single constituent of the human body, with total body water (TBW) typically making up 40–70% of an adult human's body mass. Body water content as a percentage of body mass is highest in infants (up to 75% of body mass) and declines as we progress into adulthood, typically making up ~60% of body mass for an adult male and ~50% of body mass for an adult female. Much of the variability between individuals can be explained by differences in body mass and composition. The water content of blood (~83%), skin (~72%), organs (~68–79%) and muscle (~76%) are high, whilst the water content of the skeleton (~22%) and adipose tissue (~10%) are low. Therefore, the proportion of body mass made up by water will vary, mainly as a function of adipose tissue mass. As such, individuals that have a higher proportion of their body made up of adipose tissue will have a lower relative body water content, which explains much of variance commonly attributed to age, sex, or even fitness/athletic status.

The TBW is distributed into distinct compartments in the body (see Fig. 7.1), with approximately two-thirds contained inside cells in the intracellular fluid (ICF) and approximately one-third contained outside cells in the extracellular fluid (ECF). This ECF is further distributed between fluid contained in blood vessels (intra-vascular fluid or plasma; ~24% of ECF), fluid contained between blood vessels and cells (interstitial fluid; ~74% of ECF), and transcellular fluid (~2% of ECF), which includes ocular fluid, joint fluid, cerebrospinal fluid, and intestinal secretions.

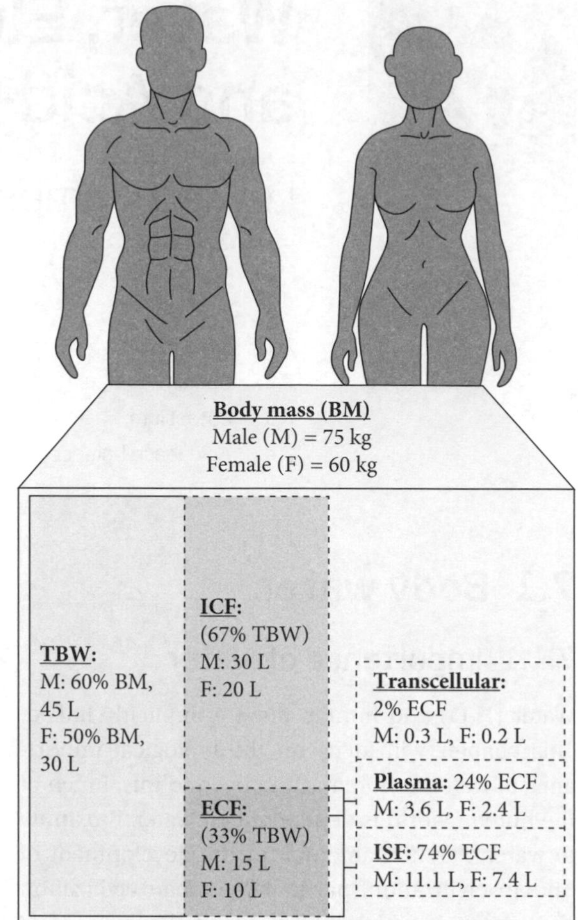


Fig. 7.1 Distribution of body water in a typical 75 kg male and 60 kg female.

BM, body mass; ECF, extracellular fluid; F, female; ICF, intracellular fluid; ISF, interstitial fluid; M, male; TBW, total body water.

## 7.1.3 Body fluid composition

In order to maintain a constant blood volume and pressure, water must be able to move easily between the different compartments. Cell membranes are normally impermeable to water, and water movement through cell membranes is facilitated by proteins connecting the inside and outside of a cell. These water channels or aquaporins (AQPs), the most common one being AQP1, help to balance out the osmolality caused by osmotically active substances, such as sodium, potassium, urea, or glucose, inside and outside the cells. Their discovery in 1986 by Benga and co-workers, confirmed in 1992 by Agre's group and awarded with the Nobel Prize, was as profound for our current understanding of life as the discovery of the structure of DNA in 1953 by Watson and Crick. AQPs are the answer to the old mystery of how water can pass lipid bilayers in order to maintain the flow of osmotically active substances, nutrients, and metabolites between the ECF and ICF compartments, and to perpetuate the loop of processes that ultimately defines life.

Although we refer to body water, the fluids of the body are not pure water, but are instead complex solutions consisting of many substances that contribute to the osmolality of the body fluids. The concentration of ions and substances (osmoles) dissolved in the ICF and ECF compartments differ due to differences in membrane permeability and expression of transporters, channels, and pumps for the individual solutes. If differences in concentration of solutes are maintained, then these solutes are regarded as 'effective osmoles', since they regulate the volume of the compartment. Osmolality is defined as the number of osmotically active particles dissolved in solution per kilogram of solvent (in this case water). This is similar, but not identical, to osmolarity, which is the number of osmotically active particles dissolved in solution per litre of solution. The osmolality of body fluids is tightly regulated (see section 7.1.5), with regulatory processes working to maintain this osmolality. The ICF and ECF have the same osmolality (~280–290 mosmol/kg), as water moves easily between the two fluid spaces through AQPs and down a concentration gradient from high to low water concentration (or from low osmolality to high osmolality). However, the composition of the two spaces differs considerably, with varying amounts of different cations (positively charged ions) and anions (negatively charged ions) in the ICF and ECF (see Table 7.1). The ECF contains mainly sodium ions ($Na^+$), small but important concentrations of potassium ($K^+$), calcium ($Ca^{2+}$), and magnesium ($Mg^{2+}$) ions, and the anions chloride ($Cl^-$) and bicarbonate ($HCO_3^-$). To maintain electroneutrality, cations must always be balanced by an equal number of anions. The ICF contains mainly potassium ions (mostly balanced by organic phosphate and the negatively charged groups on proteins), as well as small amounts of sodium, magnesium, bicarbonate, chloride, and calcium.

**Table 7.1 Composition of the intracellular (muscle cell fluid) and extracellular (blood plasma) fluids**

| Component | Muscle cell fluid (mmol/L) | Plasma (mmol/L) |
|---|---|---|
| $K^+$ | 150 | 5 |
| $Na^+$ | 15 | 140 |
| $Mg^{2+}$ | 18 | 1 |
| $Ca^{2+}$ | $10^{-4}$ | 2 |
| $Cl^-$ | 10 | 105 |
| $HCO_3^-$ | 16 | 24 |
| pH | 7.1 | 7.4 |
| Osmolality (mosmol/kg) | 280–290 | 280–290 |

*Note:* These values will vary from person to person and even within the same person over the day.

These large differences in ECF and ICF sodium and potassium concentrations are the result of a selective permeability of the cell membrane and the activity of the basolateral $Na^+/K^+$-ATPase, a transport protein that exchanges sodium against potassium by using adenosine triphosphate (ATP) as energy. Under resting conditions, cell membranes are only permeable for potassium facilitated by a basolateral potassium channel. Cells must therefore use a large part of the energy from their metabolism of nutrients such as glucose, fat, or protein to recover potassium that leaks out. The activity of the $Na^+/K^+$-ATPase also generates a large sodium gradient, which can be used by other transport proteins, such as the sodium-dependent glucose or amino acid transporter, to absorb nutrients, or by sodium channels mostly localized in different types of epithelia. Therefore, the provided sodium gradient is a form of energy for transporters and channels, which is vital for osmolyte balance and water homeostasis. This is further explained by the fact that the sodium-dependent glucose transporter is an important protein in the kidney and intestine to facilitate water absorption, as the absorption of sodium through cells (*trans*-cellular) or alongside cells (*para*-cellular) drives *trans*- or *para*-cellular water absorption facilitated by aquaporins along the osmotic gradient. Water absorption in the kidney and intestine is essential to maintain body water homeostasis; therefore, as a general principle: *water absorption is always driven by sodium absorption*. This explains why drugs such as diuretics are used to control blood pressure. By manipulating sodium transporters and channels in the kidney, water reabsorption can be reduced, and consequently blood volume and pressure lowered.

## 7.1.4 Body water balance

The balance of water in the human body is determined by gains to, and losses from, the body water pool (Fig. 7.2). This is a dynamic process, with continual losses and gains of water in different amounts through different routes over the day. A state of 'normal' body water is known as euhydration, which fluctuates slightly over the day and from day to day in response to different patterns of water intake and excretion (typically by less than ~1% of body mass). This fluctuation is because water losses are continuous, whilst water gains are usually episodic, occurring mainly around meals. Hyperhydration and hypohydration represent altered states of increased, and decreased, body water, respectively. Dehydration represents the process of losing water from the body and rehydration represents the process of regaining water from a hypohydrated state to a euhydrated state. Note that dehydration can mean losing water from hyperhydration to euhydration or from euhydration to hypohydration, but in common everyday terminology it is usually used to represent a reduced state of body water. It is important to note that dehydration (i.e. a loss of only water) is rare, and in most circumstances losses of water from the body are accompanied by losses of electrolytes and other substances contained in the lost body fluid.

Body water remains relatively stable, provided the individual is in a temperate environment and does not undertake substantial amounts of physical activity. Under such conditions, body water inputs (through drinks, foods, and metabolic water formation) balance body water outputs (urine, sweat, insensible water losses, and faeces), such that body water stores are maintained, at least on a day-by-day basis, with regulatory mechanisms controlling body water stores in most settings. See Fig. 7.2 for an overview of these routes of gain and loss.

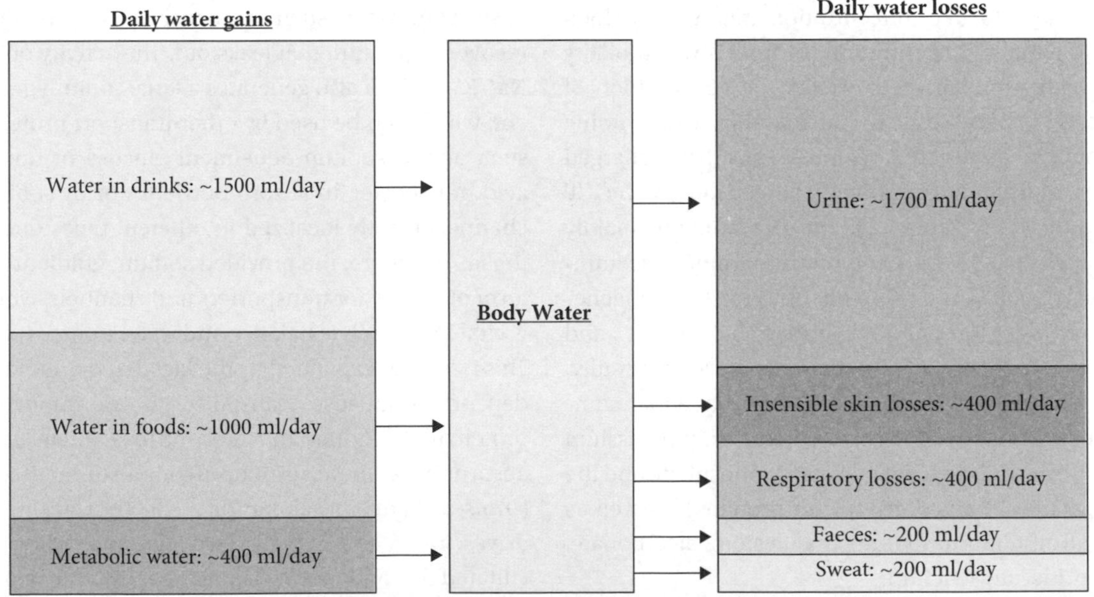


Fig. 7.2 Typical routes of water gain to and loss from the human body, with typical amounts.

*Note:* Amounts listed are typical values only and will vary substantially depending on the individual, the environment they are in, and their activity level or health.

#### 7.1.4.1 Body water outputs

Water is lost from the body through several routes, some of which are obvious and some less so, but most water losses are unavoidable, meaning that water must be ingested to replace losses. Routes of water loss from the body can be separated into two categories: sensible and insensible losses. Sensible water losses are defined as those that are perceived and can be measured (e.g. urine, faeces, and thermoregulatory sweat), whilst insensible water losses are defined as those that are not perceived or easily measured (e.g. transdermal and respiratory water losses). Insensible water losses result in the loss of pure water from the body and occur due to water diffusion through, and then evaporation from, the skin (~400 mL/day) and evaporation of water from the moist surfaces in the lung, which is then lost in expired breath (~400 mL/day). This respiratory water loss can vary substantially depending on the environmental conditions and energy expenditure (physical activity). For example, with low humidity (dry) environments or with increased ventilation/tidal volume during physical exercise, the losses can be up to 2 L/day. Urine output usually makes up the largest proportion of daily water losses and, whilst there is large inter- and intra-individual variability for urine losses (mainly due to differences in water intake and activity), it is typically in the range 500–2000 mL/day.

Urine output is the main mechanism used to excrete solute/waste from the body, meaning there is an obligatory requirement for urine production each day. The volume of urine will depend on the amount of solute that needs to be excreted and the individual's kidney function (i.e. how much the kidneys can concentrate the urine), as well as the requirement to excrete water to control body fluid volumes. Healthy kidneys can produce urine from close to 0 mosmol/kg up to 1200 mosmol/kg and the concentration of urine responds quickly to changes in fluid intake to increase or decrease water loss from the body. Typically, healthy functioning kidneys can concentrate urine ~4-times the concentration of plasma, although as kidney function declines with normal ageing or clinical conditions, the volume of water required to excrete the daily solute load must increase. For example, if we consider someone who must excrete the solute equivalent to 600 mosmol/day, if their kidneys can concentrate urine to 1200 mosmol/kg, they will be able to excrete this load in a minimum of 500 mL urine. If their kidneys are only able to produce urine up to 600 mosmol/kg, it would require a minimum of 1000 mL to excrete the same daily solute load. Commonly, the volume of urine produced is greater than the minimum requirement for excretion of the daily solute load.

For most inactive populations living in temperate climates (or with air conditioning), daily sweat losses are small (100–200 mL/day). Sweat is secreted onto the skin surface from eccrine sweat glands in response to changes in body temperature, and through the evaporation of this secreted sweat, heat is lost from the body (evaporative cooling). Therefore, daily water losses through sweating increase with physical activity and in hot environments, and some well-trained athletes have been reported to lose >4 L of sweat per hour of exercise.

The final route of water loss from the body is through faeces, which, like sweat production, usually produces a low daily loss of water (100–200 mL/day). However, infection with certain bacterial or viral pathogens can dramatically increase the loss of water through diarrhoea, with some severe cases resulting in >10 L of water loss per day.

#### 7.1.4.2 Body water inputs

As most routes of water loss are unavoidable, there is the essential requirement for water input to the body to balance these losses. The main route of body water input is through dietary intake, with a smaller but important source of body water input through metabolic water formation, as water (along with carbon dioxide) is one of the end-products of aerobic respiration. Most individuals consume more water than they need on a daily basis and thus the kidneys expel the additional water. There is still some scientific debate about the specific amount of water that should be consumed daily, but Table 7.2 gives suggested recommendations for water intakes for different age groups based on recommendations by the European Food Safety Authority.

In a sedentary individual, metabolic water formation from aerobic respiration contributes ~400 mL

Table 7.2 **Adequate intake recommendations for water in different age groups from the European Food Safety Authority**

| Age | Daily Adequate Intake | |
|---|---|---|
| 0–6 months | 100–190 ml/kg body mass | |
| 6–12 months | 800–1000 mL | |
| 12–24 months | 1100–1200 mL | |
| 2–3 years | 1300 mL | |
| 4–8 years | 1600 mL/day | |
| 9–13 years | Female: 1900 mL<br>Male: 2100 mL | |
| ≥14 years | Female: 2000 mL<br>Male: 2500 mL | Pregnant +300 mL<br>Lactating: +700 mL |

*Note:* These values include water from both food and fluids in the diet and apply to temperate environments with moderate levels of physical activity. Note that some individuals will survive on far less whilst some will need far more depending on the environment in which they live and their daily activities.

water a day, but if metabolic rate (energy expenditure) increases, so will metabolic water formation. Therefore, with physical activity, metabolic water formation may substantially increase and can reach more than 150 mL/h in some well-trained endurance athletes working at a high intensity for prolonged periods of time. However, the increase in metabolic rate also increases ventilation rate and tidal volume, which result in an increase in respiratory water loss that is comparable to the increase in metabolic water formation, meaning the two effectively cancel each other out.

Water intake through the diet is the major source of water input to the body and can increase or decrease regulatory mechanisms to influence physiology and behaviour, and consequently body water stores. Dietary water intake can be separated into two main sources: water in drinks and water in foods. Water in foods provides ~1000 mL of water a day but will vary substantially with the type of food consumed. Many foods have relatively low water contents, but some may contribute appreciable amounts to daily water intake. For example, many fruits, vegetables, and soups contain ≥90% water by weight, many staple foods (e.g. pasta, rice, etc.) accumulate water during the cooking process (e.g. dry pasta will accumulate more than its own weight in water during cooking), and it is common for milk to be added to certain foods (e.g. on breakfast cereals). Water intake through drinks typically makes up the largest component of water input (~1500 mL/day), but it is important to note that much of this water intake is not in response to physiological regulation/signals. Most of our daily water intake is driven by habit or context and therefore most individuals consume more than the minimum amount of water required to maintain TBW.

## 7.1.5 Body water regulation

Body water is tightly regulated, and it is estimated that the TBW usually varies by less than 1% over and between days. The regulation of TBW is controlled via mechanisms that act on both sides of the water balance equation. Coordinated homeostatic responses lead to changes in thirst sensation and urine output to help control the volume of fluid entering and leaving the body and restore any deviation from euhydration. It is important to note that, in normal circumstances, urine output via the kidney is the only physiologically regulated water loss process, but that the kidney can only limit water losses. Other routes of water loss are not generally under regulatory control and as such these losses must be replenished if

TBW is to be maintained. On the opposite side of the regulatory processes, thirst is initiated when sufficient water is lost from the body.

The main factors that are responsible for this regulatory process are the osmolality of body fluids (mainly the ECF) and blood volume. A change in plasma (blood) osmolality of as little as 1% (~3 mosmol/kg) initiates alterations in the factors that regulate the intake and excretion of body water. In contrast, blood volume is a far less sensitive regulator, with changes of more than 10% of blood volume required to prompt a homeostatic response. This is likely to be because changes in blood volume (e.g. with changes in posture/activity) are far more common than changes in EFC osmolality, which mainly responds to changes in water (and sometimes sodium or salt) intake or loss. The regulatory process is governed by mechanisms involving the hypothalamus, neurohypophysis (posterior pituitary gland), and the kidneys (Fig. 7.3), resulting in increases/decreases in urine output and increases/decreases in thirst and fluid-seeking behaviour. As water is lost

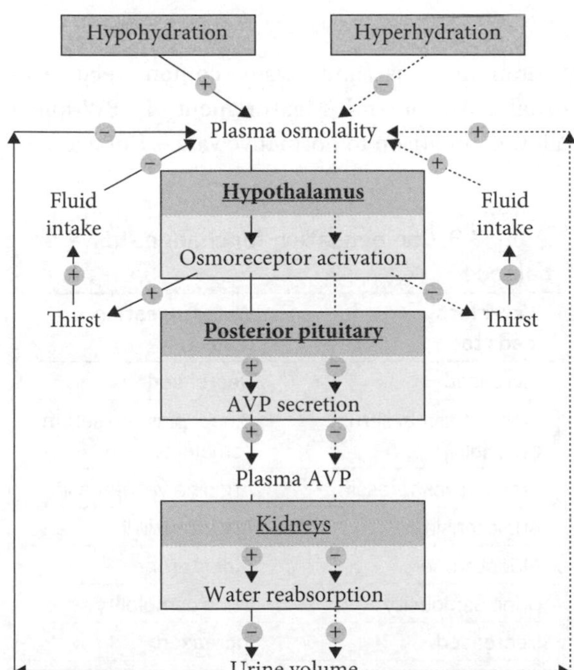


Fig. 7.3 Regulation of total body water involving the hypothalamus, neurohypophysis (posterior pituitary gland), and kidneys.

from the body, EFC osmolality increases, leading to a movement of water from the ICF to the ECF to balance osmolality. This means cells shrink, and the shrinkage of cells in regions of the hypothalamus of the brain containing osmoreceptors prompts two responses:

1. the secretion of arginine vasopressin (AVP)
2. the stimulation of thirst sensation.

AVP, better known as anti-diuretic hormone (ADH), is a hormone that is secreted from the posterior pituitary gland and acts on the principal cells in the kidney through the $V_2$ vasopressin receptor. Activation of $V_2$ receptors by AVP in the collecting duct, connecting tubule, distal convoluted tubule, and thick ascending limb of the loop of Henle in the kidney increases the reabsorption of solute-free water back into the circulation, concentrating the urine and reducing urine volume (antidiuresis). This reabsorption is orchestrated by an increased transcription and redistribution of AQP-2 (a water channel expressed in the collecting duct system) to the luminal membrane of the principal cells, consequently allowing more water to be reabsorbed from the filtrate. Conversely, if ECF osmolality is reduced, as happens when a large bolus of water is drunk, cell swelling of osmosensing cells reduces AVP secretion and AQP-2 channels are recycled, preventing water reabsorption in the kidney, leading to a larger volume of more dilute urine (diuresis) and the expulsion of water from the body. Nowhere is this latter process more obvious than when copious amounts of beer (or non-alcoholic drinks) are consumed in pubs. It is the rapid suppression of ECF osmolality and AVP that leads to the phenomenon known as 'breaking the seal' (see Fig. 7.4).

On the other side of the regulatory processes, the increase or decrease in ECF osmolality and subsequent osmosensing increases or decreases thirst sensation, respectively. If thirst increases, then this will prompt drinking when drink is available or prompt fluid-seeking behaviour if it is not. In much of the modern world this is simply turning on a tap or visiting a shop to purchase a drink, but in some settings, it may involve significantly more effort to quench the thirst. Interestingly, there are many

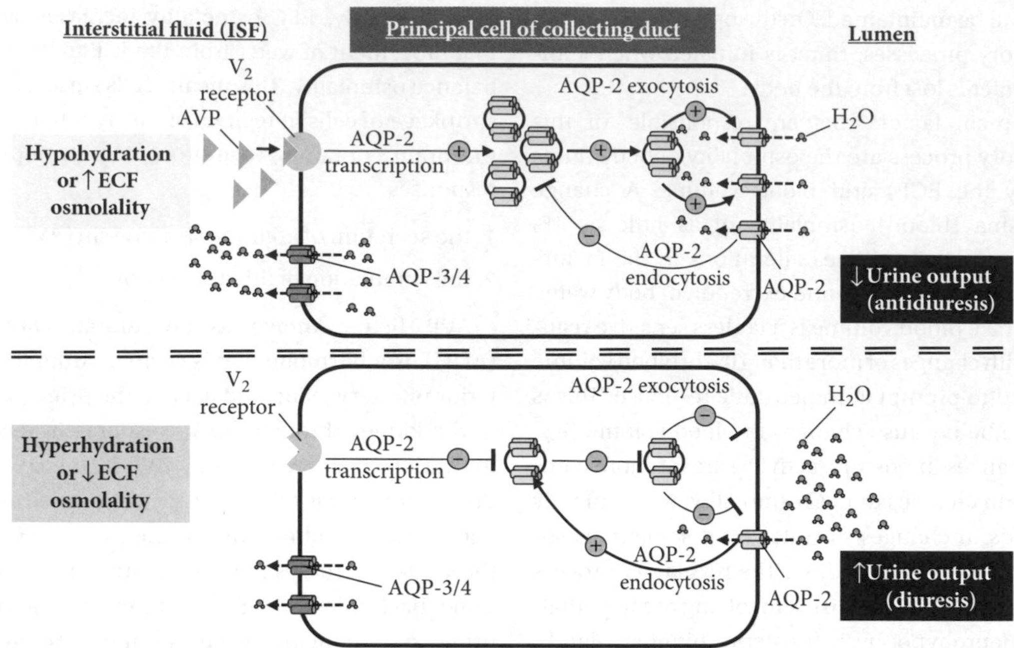


Fig. 7.4 A cell model of water reabsorption in the collecting duct of the kidneys. Changes in plasma vasopressin initiate changes in water reabsorption via interaction with the vasopressin $V_2$ receptor. Changes in binding of vasopressin to the $V_2$ receptor influence the transcription of aquaporin-2 (AQP-2) and its redistribution and insertion in/removal from the luminal (apical) membrane. This alters the permeability of the luminal membrane to water and consequently the amount of water that is reabsorbed. AVP, arginine vasopressin; ECF, extracellular fluid; $H_2O$, water.

reasons why humans feel thirsty. The thirst sensed because of ECF hyperosmolality is known as true physiological thirst, which is relatively rare in most healthy individuals. Most drinking is regulated by taste, habit, packaging size, and social circumstance, or the desire to ingest other nutrients (e.g. caffeine in coffee/tea and alcohol in alcoholic drinks). For example, when a friend offers to buy you a coffee, it is rarely the osmoreceptor-containing cells in your hypothalamus that decide if you say yes or no, or if you drink the whole cup or stop once your thirst is sated. Both excesses and deficiencies of body water can occur, and both may threaten life. The normal compensations for these changes are summarized in Table 7.3.

## 7.1.6 Measurement of body water

Measurements related to body water fall into two categories: those that measure or estimate TBW and those that measure or estimate an individual's hydration status (body water content relative to usual TBW content). Measurement of TBW might allow comparison to normative values (for age, sex,

**Table 7.3 Compensation for changes in water balance**

| Water loss leads to: | Water ingestion leads to: |
|---|---|
| Increased | Decreased |
| Blood/plasma/serum osmolality | Blood/plasma/serum osmolality |
| Arginine vasopressin | Arginine vasopressin |
| Angiotensin II | Angiotensin II |
| Aldosterone | Aldosterone |
| Urine osmolality | Urine osmolality |
| Decreased | Increased |
| Urine volume (oliguria, antidiuresis) | Urine volume (diuresis) |
| Urine sodium | Urine sodium |

fitness, etc.), but beyond this has relatively little value at an individual level, unless you know the individual's usual TBW. The precise measurement of TBW is possible through the administration and dilution of isotopic forms of water (usually the stable isotope deuterium oxide or $D_2O$) in the body water pool. A small, known amount of deuterium oxide (usually ~0.1 g/kg body mass or ~10 g in adults) is orally ingested and from the change in enrichment of a body water sample (urine, saliva, blood, etc.) before ingestion vs some hours later, it is possible to calculate the volume of water within which the isotope is diluted and thus TBW. Whilst precise, TBW measurement through isotope dilution takes time to yield results and requires specialist technical knowledge, consumables (e.g. the isotopically labelled water), and analytical equipment (e.g. an isotope ratio mass spectrometer), as well as the funds to cover the associated expenses. Less precise but far more practicable methods also exist to estimate TBW. These include methods that determine body composition and subsequently estimate TBW (e.g. bioelectrical impedance spectroscopy), but probably the simplest method is to use body mass as a proxy for body water. Whilst body mass is a relatively imprecise method for estimating the absolute amount of TBW, it can be very useful for determining changes in TBW in the short-term (a few hours or day-to-day). Provided energy balance is neutral (i.e. no negative or positive energy balance) and there are no substantial changes in body energy stores (glycogen, fat, or protein), then a change in body mass can be used to quite precisely estimate an acute change in TBW. Indeed, with athletes, this method is commonly used to determine water losses during exercise and make recommendations about water replacement during or after the exercise.

The second category of measurements assessing body water are those that assess an individual's hydration status (euhydration, hypohydration, or hyperhydration) from a spot sample, with these measurements typically revolving around the concentration of certain body fluids/secretions. These measures generally fall into two categories: urine-based measures and blood-based measures. Whilst a gold standard method to assess hydration status on the spot remains elusive, it is generally accepted that serum or plasma osmolality represent the best current measure. Although there is some inter-individual variation in serum/plasma osmolality, values of <280 mosmol/kg and >295 mosmol/kg are indicative of hyperhydration and hypohydration, respectively. However, this is technically challenging and time consuming, as it typically requires the collection of a venous blood sample, centrifugation of the blood to separate serum/plasma, and measurement with a freezing point or vapour pressure osmometer, making it unsuitable in most situations. Therefore, in many situations, urine-based measures are used to estimate hydration status, as urine collection is simple and relatively unintrusive. Osmolality or specific gravity of urine are commonly measured to estimate the hydration status and can be done relatively easily with the use of a handheld refractometer, with an osmolality of >900 mosmol/kg or a specific gravity of >1.025 indicative of hypohydration. Urine colour, which correlates with measures of urine osmolality/specific gravity, offers a simple measure that most people can use to track their hydration status. The darker the urine, the more water is being reabsorbed by the kidney, suggesting a raised blood osmolality. However, it must be noted that some foods/supplements (e.g. vitamin supplements high in beta-carotene, beetroot juice, etc.) and some medications can change the colour of urine, rendering this measurement impractical.

### 7.1.7 Deficiency of body water

Hypohydration (commonly termed dehydration—see section 7.1.4) describes a state of negative fluid balance that may be caused by numerous disease entities, athletic/occupational activities, or environmental conditions. Diarrhoeal illnesses are the most common causes, especially in children. Worldwide, hypohydration secondary to diarrhoeal illness is one of the leading causes of infant and child mortality. This results from the fact that, compared with adults, infants and children have much higher total body water content. This is required as they have a higher surface to mass ratio, causing more water loss via the skin, a higher respiratory and metabolic rate, and an immature kidney that cannot concentrate urine to the same extent as an adult, resulting in a much higher loss of body water. This also explains why the fluid requirements of infants and children are much higher than those of adults (infant: ~100–190 mL/kg, adult: ~40 mL/kg).

The effects of water loss depend on the electrolyte composition of the fluid lost, since this will affect the osmolality (also called tonicity) of the ECF. Loss of pure water only is relatively rare, with most water loss also being accompanied by at least some loss of solute, particularly electrolytes (mainly $Na^+$ and $Cl^-$). When pure water is lost or the fluid lost is hypotonic compared to the ECF (contains lower concentrations of electrolytes), this leads to an increase in the osmolality of the ECF (hypertonic dehydration/hypohydration). The increased ECF osmolality draws water into the ECF from the ICF to equalize osmolality between the fluid spaces. This shift helps attenuate volume losses from the ECF and circulating blood, and this takes precedence over the regulation of osmolality. This is sometimes referred to as intracellular dehydration, but really the water is drawn from both the ECF and the ICF. The elevated osmolality leads to AVP secretion and subsequent urine concentration to limit water losses. This response has survival value, as the increased osmolality maintains thirst (increasing drive to drink or find fluid) and minimizes urinary losses. This type of dehydration is common with inadequate fluid intake and exercise-induced dehydration, as the sweat gland reabsorbs sodium and chloride. Loss of fluid accompanied by isotonic loss of electrolytes (the same concentration as blood) will leave ECF tonicity unchanged (isotonic or isosmotic dehydration/hypohydration) and mean that much of the water loss is absorbed by the ECF. This is sometimes termed extracellular dehydration and is common with diuretic drug administration. Loss of fluid containing a higher electrolyte concentration than the ECF will reduce the osmolality of the ECF (hypotonic dehydration/hypohydration). Since the terms hyper- and hypotonic are sometimes confused in relation to loss of body water, they are probably best avoided, and the focus should be on describing exactly what is lost.

A note on 'dehydration' The distinctions in types of dehydration are obviously important in treatment, which involves replacing the deficit with appropriate fluid management and, in some settings, replacement of sodium and chloride losses. The term 'dehydration' is best avoided in dealing with altered sodium states (see section 7.2) and possibly best avoided altogether, since it refers to a loss of water alone.

Symptoms The cardinal manifestation of severe water deficiency is thirst; dry, cracking skin, confusion, seizures, and coma may follow. See Box 7.1 for information on the effects of water loss in an open, barren environment.

Fresh water may be as unavailable at sea as it is in deserts. The dry mouth, swollen tongue, and delirium with hallucinations are admirably described by Samuel Taylor Coleridge in *The Rime of the Ancient Mariner* (1798):

And every tongue through utter drought,
Was withered at the root;
We could not speak, no more than if
We had been choked with soot.
With throats unslaked, with black lips baked,
We could not laugh nor wail;

**Box 7.1 Survival without water**

**Amount of body water lost**

- *1–5% body weight loss*: thirst; vague discomfort; economy of movement; no appetite; flushed skin; impatience; increased pulse rate; nausea.
- *6–10% body weight loss*: dizziness; headache; laboured breathing; tingling in limbs; absence of saliva; blue body (cyanosis); indistinct speech; inability to walk.
- *11–12% body weight loss*: delirium; twitching; swollen tongue; inability to swallow; deafness; dim vision; shrivelled skin; numb skin.

Through utter drought all dumb we stood!
I bit my arm, I suck'd the blood,
And cried, 'A sail! a sail!'

During World War II, R.A. McCance (Professor of Experimental Medicine at Cambridge in England) was chairman of a subcommittee of the British Admiralty concerned with safeguarding the lives of hundreds of men cast adrift in lifeboats or rafts after their ships were sunk by enemy action. The reports showed that those who drank seawater had a mortality of 39% compared with 3.3% for those who did not. He studied volunteers in life rafts in temperate, arctic, and tropical seas, and concluded that seawater could not be used to supplement limited supplies of fresh water. Seawater has a higher salt concentration than the ECF, typically a 'salinity' of 3.5% (35 g/L), giving a sodium concentration of 469 mmol/L, which is more than human kidneys can achieve (about 270–300 mmol/L). As kidneys need to excrete more water to eliminate the salt than the water gained by drinking seawater, drinking seawater will generally make dehydration worse, as well as raising ECF osmolality. Any temporary improvement in circulation and feeling when water is withdrawn from cells to compensate for water loss is overshadowed by shortened survival. There is no ultimate advantage to feeling better on Monday and dying on Tuesday if it happens that your party is not going to be rescued until Wednesday.

Treatment The remedy for primary water deficiency is water, as the predominant symptom of thirst indicates. It is important to realize that this is not confined to oceans and deserts; it can occur in hospital wards if thirst fails or cannot be satisfied. Unconscious patients do not experience thirst; others may be too confused or too weak to drink water set beside them. Thirst, water diuresis, and the response to AVP tend to become attenuated with advancing years, highlighting that many older people are increasingly at risk of hypohydration.

### 7.1.8 Excess of body water

An excessive amount of water in the body (hyperhydration) is rarer than deficiency, because water diuresis generally protects against excess, and healthy functioning kidneys can usually excrete water as fast as the gut can absorb it (except after massive fluid ingestion). However, water diuresis can fail:

- *in anuria* (zero urine) or *oliguria* (reduced urine output) with impaired renal function, as the kidneys cannot respond to absence of AVP
- *with inappropriate release of AVP*: e.g. with head or chest tumours, concussions, trauma, or infection. AVP release is also stimulated by pain, some anaesthetics, and some drugs (including ecstasy), and AVP-like substances are secreted by some cancers, especially small-cell carcinoma of the lung.

If patients under these conditions are given much more than the ~1 L/day that they need to replace unavoidable losses, ECF osmolality and sodium concentration fall and water goes into cells, which swell. Weakness and cramps develop. Swelling of brain cells (cerebral oedema) leads to disturbances of consciousness and behaviour, and may progress to convulsions and death, so-called 'water intoxication.' This is known as hyponatraemia (low blood sodium concentration), which is the main method used to diagnose the condition. Hyponatraemia is defined as a blood (or serum/plasma) sodium concentration below 135 mmol/L and can be termed mild (blood sodium concentration of 130–135 mmol/L), moderate (blood sodium concentration of 125–129 mmol/L), or severe (blood sodium concentration <125 mmol/L). Whilst this is relatively rare in the modern world, cases in daily living or during or shortly after exercise (termed exercise-associated hyponatraemia) do occur and can, sadly, sometimes be fatal. Indeed, measurement of blood sodium during and after long-distance endurance events (e.g. marathons, long-distance triathlons, etc.) is often used to diagnose potential cases, where athletes that are smaller and/or complete the race more slowly (more time to overdrink) are more likely to develop hyponatraemia, although the vast majority of cases are asymptomatic.

On some rare occasions, hyponatraemia can occur without an excess of total body water (e.g. if the water but not the salt lost in profuse sweating is replaced). Muscle cramps can then be a major feature (previously known as miners' cramps and stokers' cramps, as they were common in English coal mines and ships). J.B.S. Haldane showed in 1928 that extra salt prevents or cures these cramps, as with miners in hot mines. Treatment of hyponatraemia depends on the cause and the severity. With mild cases, they are often treated by water restriction or ceasing medication (i.e. diuretics, etc.) that might be the route cause. Severe cases may require medical intervention, where the infusion of hypertonic saline can be used to pull water out of the ICF and reduce cerebral oedema.

## 7.2 Sodium

### 7.2.1 Significance and functions of sodium

Salt has long been a valuable commodity and part of the fabric of human life and culture. We pay salaries, derived from *salarium* (the allowance a Roman soldier was given to buy salt) and we ask whether someone is worth their salt, demonstrating the value placed on salt by humans. We often crave salt and some animals in arid regions travel vast distances to salt licks to get the sodium they need in order to excrete potassium from their high-potassium vegetable diet.

Common salt or table salt is sodium chloride (NaCl). Each gram contains 17.1 mmol of sodium, which is the principal cation in most extracellular fluids and is responsible for 95% of ECF osmolality. The ECF sodium concentration, at ~140 mmol/L in humans and most animals, helps determine the membrane potentials of most cells, and the action potentials underlying the transmission of nerve impulses and the contraction of muscles. Since ECF sodium concentration is tightly regulated, it may be argued that the sodium concentration maintains the ECF (at 1 L for every 140 mmol Na) that our cells live in. Consistent with this, disorders where plasma (ECF) sodium concentration is high or low (the dysnatraemias) are also disorders of water balance, not just sodium regulation.

### 7.2.2 Amount and distribution of sodium in the body

Adults contain about 5600 mmol of sodium (325 g NaCl). About half of that (2800 mmol) is dissolved in the extracellular fluids, with 300 mmol in cells and 2500 mmol in bone mineral. Half of the sodium in the bones is exchangeable with isotopically labelled Na; the rest is deeper and less accessible. Thus, in classical analyses of dissolved bodies, more sodium was found than with modern measurements, which are based on isotopic dilution during life. More recently, sodium stores have been found in skin, where sodium binds to glucose-aminoglycans (GAGs), although the metabolism/turnover of these stores is not well understood at present.

### 7.2.3 Sodium balance

#### Intake: diet

The diet typically provides 70–250 mmol/day, but this varies with habits, food preferences, and customs:

1. Natural foods typically contain 0–3 mmol sodium per 100 g: fruits typically <0.1 mmol/100 g; vegetables typically 0–3 mmol/100 g; meat, fish, and eggs typically ~3 mmol/100 g; animal milks typically ~2 mmol/100 g.
2. Processed foods contain far more salt: bread ~20 mmol/100 g; cheese ~30 mmol/100 g; salted butter ~40 mmol/100 g; raw lean bacon contains as much as 80 mmol/100 g.

Salt is added to processed foods as it is a flavour enhancer, but also for a variety of other reasons, including its effect as a preservative and on texture. It is usual for people to add salt during cooking or at the table, with the amount highly individual, but discretionary salt intake in the home is usually much lower than that obtained from manufactured or processed food.

### Output

1. *Faeces*: normally 5–10 mmol/day.
2. *Urine*: variable, but usually a little less than dietary intake (typically ~90% of dietary intake).
3. *Sweat:* extremely variable and dependent on physical activity and ambient conditions, but typically <10 mmol in a sedentary person living in a temperate climate.

The kidneys are capable of regulating sodium excretion in urine to less than 1 mmol/day, and to more than 500 mmol/day. They usually keep body sodium content constant by excreting intake in excess of the sum of other losses. Homer Smith's onetime remark that the composition of the body fluids depends 'not by what the mouth takes in but by what the kidneys keep' aptly describes their control of sodium balance.

As the kidneys are not actively secreting sodium to the urine, the amount excreted is entirely based on the amount of sodium filtered at any time by the glomerulus, defining the filtered load of sodium (glomerular filtration rate (GFR) × plasma concentration of sodium). The final rate of sodium excretion depends on the balance between GFR, which can be regulated according to plasma sodium and volume, and tubular reabsorption (usually greater than 99% of the filtered load). To balance out cases of low blood pressure or low blood/ECF volume, an initially reduced GFR is compensated by constriction of glomerular vessels (autoregulation), and reabsorption of sodium is increased by several factors, including angiotensin II and aldosterone. When blood pressure or volume is reduced, sympathetic nerves are activated by the input of different baro- and osmoreceptors, and renin is released from the kidney; renin is an enzyme that cleaves proteins and generates angiotensin I from its precursor angiotensinogen in the plasma. Angiotensin-converting enzyme (ACE) in the blood and some other tissues converts angiotensin I to angiotensin II, which stimulates production of aldosterone in the adrenal gland. In turn, aldosterone stimulates sodium and water reabsorption and simultaneous potassium loss in the principal cells of the distal nephron (see Fig. 7.5). This entire process controlling sodium homeostasis and blood pressure is called the 'renin–angiotensin–aldosterone system', or RAAS for short. Angiotensin II itself also stimulates sodium reabsorption. When blood volume increases again, these sodium-conserving mechanisms are inhibited, GFR increases, and tubular sodium reabsorption is reduced; sodium reabsorption is then also inhibited by atrial natriuretic peptide, which is released from the heart during volume expansion.

## 7.2.4 Sodium depletion

Deficiency of sodium does not simply result from deficient intake, as the kidneys can make the urine almost sodium-free. Abnormal losses causing depletion may arise from:

1. *Sweat*: up to as much as 15 L/day. Sweat is typically *hypotonic*, but the sodium concentration varies substantially from person to person, with an average of ~50 mmol/L (typical range 20–80 mmol/L). Thus, 15 L of sweat contains the same amount of salt as ~5 L of normal ECF, so even after osmolality is corrected by replacing water, the volume of the ECF will be reduced by 5 L unless the sodium is replaced. Note that the hypotonicity of sweat is important for the physiological regulation of water balance, as it effectively means that more water than solute is lost, leaving more solute in the ECF and raising ECF osmolality. It is this increase in ECF osmolality that initiates many of the counter regulatory responses (decreased urine output and increased water intake; see Fig. 7.3). This increase in ECF osmolality results in movement of water from the ICF, meaning the water loss is shared between fluid compartments. This type of fluid loss is also known as hypertonic hypovolaemia.
2. *Intestinal fluid*: 10 L/day of intestinal secretions are normally reabsorbed. However, with diarrhoeal disease, absorption is decreased, and intestinal fluid secretion often increased. Losses may reach 18 L/day with cholera (infection of the small intestine with the bacterium *Vibrio cholerae*). In this case, the fluid lost is almost isotonic and equivalent to its own volume of ECF.

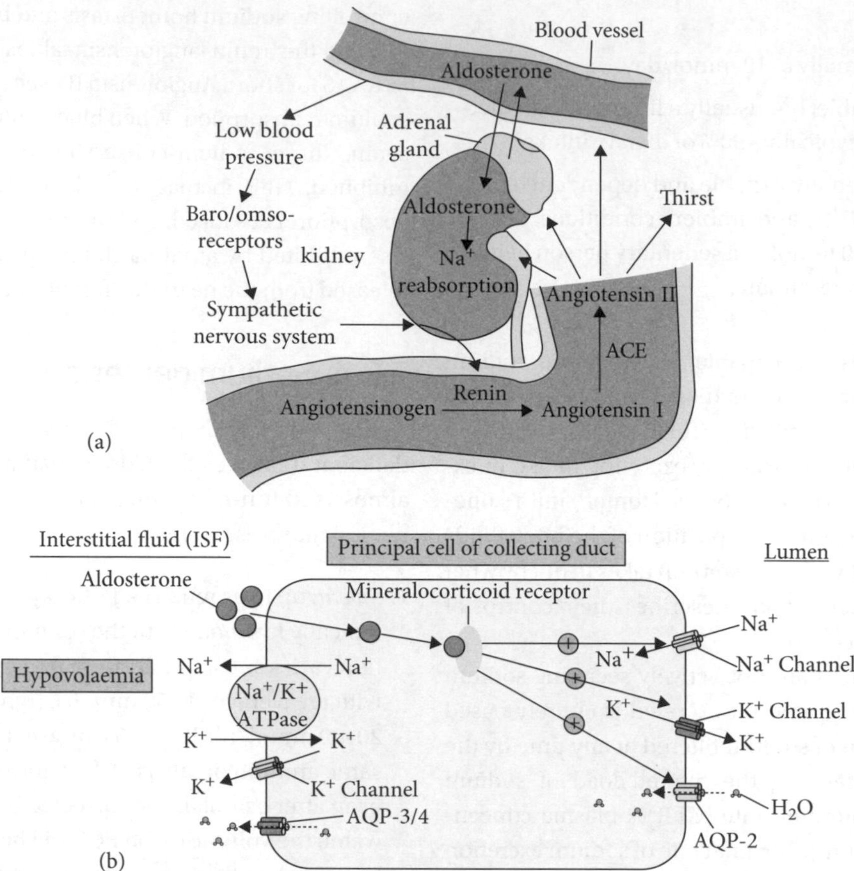


Fig. 7.5 The renin–angiotensin–aldosterone system (RAAS) and a model for the regulation of sodium reabsorption in the principal cell of the kidney by aldosterone. (a) Blood volume and pressure are tightly monitored and regulated, with the RAAS playing a central role. Baro- and osmoreceptors sense changes in blood pressure and osmolality and, via the sympathetic nervous system, alter the release of renin (an enzyme) from the kidney. Renin converts angiotensinogen into angiotensin I, which angiotensin-converting enzyme (ACE) converts to angiotensin II. Angiotensin II then initiates several processes to restore blood pressure/volume, including inducing thirst, constriction of arteries, increasing sodium reabsorption in the kidneys, and facilitating the release of aldosterone from the adrenal glands. (b) Aldosterone activates the mineralocorticoid receptor in the principal cell of the collecting duct, which results in increased sodium reabsorption and potassium excretion. AQP, aquaporin; ATP, adenine triphosphate; $H_2O$, water.

This type of fluid loss is also known as isotonic hypovolaemia.

3. *Urine*: the kidneys normally act to guard the body's stores of sodium. Diuretics and osmotic diuresis (e.g. with the load of glucose and ketone acids in diabetes) commonly remove large amounts of sodium in the urine. In adrenal insufficiency (e.g. in Addison's disease) the adrenal cortex fails to produce aldosterone, and the kidneys fail to conserve sodium (and retain potassium).

Most commonly (except perhaps with exercise) osmolality is maintained, and ~1 L of water is lost with every ~140 mmol of sodium, so that this volume depletion is largely confined to the ECF. As ECF volume depletion increases, blood pressure decreases and there may be a threat to life from circulatory failure, better known as 'shock'.

Symptoms Early symptoms of ECF volume depletion include dry mouth and tongue, loose skin that lacks turgor, and sunken eyes, and, eventually, a rapid, weak ('thready') pulse and low blood pressure complete the picture described as 'shock'. Packed cell volume, the concentrations of haemoglobin and plasma albumin, and blood viscosity all increase as the volume of plasma decreases. When oxygen transport to tissues is badly impaired, cells swell, taking up sodium and water; this further reduces the volume of the ECF and may set up a vicious cycle, further reducing blood pressure and oxygen delivery.

Treatment Salt as well as water are required to treat this desperate state. Water given alone or by infusing glucose solutions will disappear into the cells, with the risk of cerebral oedema (water intoxication). Isotonic saline ('normal', 9 g/L or 150 mmol/L NaCl, also known as 0.9% saline, and approximately equivalent to plasma sodium concentration) supplies sodium, chloride, and water in the proportions needed.

### 7.2.5 Experimental human salt deficiency

McCance subjected himself, and some other healthy people, to forced sweating for 2 hours each day while they ate a low-salt diet and drank only distilled water. They lost about 1 kg/day for 3 or 4 days, while plasma sodium concentrations remained normal. After that, the loss of sodium continued, but not the loss of weight; plasma sodium concentration fell and water moved into the cells. Their faces shrank, they lost their appetite and sense of taste, and they became very weak, weary, and muddle-headed, and experienced a general sense of exhaustion, reduced initiative, delayed excretion of water loads, and almost continuous muscle cramps. Many of McCance's symptoms resembled those of severe Addison's disease, although the adrenal glands were presumably overactive! They endured these miseries for a week or two, and then enjoyed a rapid cure by eating salty fried herrings and licking the salt out of the pan. McCance's fascinating description in the third of his Goulstonian lectures (McCance, 1936) is a nutritional classic.

### 7.2.6 Sodium excess

An increased intake of salt does not usually increase the amount in the body, because the kidneys normally excrete the extra sodium and keep the volume of the ECF constant. However, the kidneys can be misdirected or malfunction, and retain too much sodium. Examples (with some contributory factors) include:

1. *Patients with cirrhosis and cardiac failure* share the pathophysiology of decreased 'effective' arterial blood volume (EABV), resulting from hypoalbuminaemia (albumin is vital for the balance of glomerular filtration for the oncotic pressure of the capillaries) and splanchnic vasodilatation in cirrhosis, and from decreased cardiac output in cardiac failure. This decrease in EABV results in a constant stimulation of the renin–angiotensin–aldosterone system, and secondary sodium and water retention.
2. *Patients with a reduced plasma albumin*, e.g. from reduced retention within the plasma combined with loss in the urine in the nephrotic syndrome, or resulting from reduced synthesis in late cirrhosis, or from severe malnutrition, also develop a reduced EABV, with constant stimulation of the renin–angiotensin–aldosterone system. The reduction in albumin decreases plasma oncotic pressure, resulting in enhanced movement of plasma out of the intra- and into the extravascular (interstitial) compartment of the ECF. However, the plasma volume is not always low, indicating that there are, as of yet, incompletely understood factors involved.

The concept of EABV is important in understanding why and how the kidney retains sodium and water. EABV is an abstract term that refers to the adequacy of the arterial blood volume to 'fill' the capacity of the arterial vasculature. Elevated EABV can be reduced, therefore, by factors that reduce actual blood volume (hypohydration), or reduce cardiac output (cardiac failure). Whenever EABV falls, the kidney is triggered to retain sodium and water. The mechanisms involved are:

- reduced renal blood flow triggering renin release
- increased proximal tubular sodium and water reabsorption (angiotensin II)
- increased distal tubular sodium and water reabsorption (aldosterone)
- increased AVP activity.

Retained sodium increases osmolality in the ECF, inducing thirst and release of AVP, which immediately reduces renal 'free' water clearance (antidiuresis) until water is ingested/retained to normalize osmolality. With normal osmoregulation, 1 L of water is retained with each ~140 mmol of sodium. If just water is retained, it distributes throughout the total body water compartment and oedema will not usually form. However, if sodium is retained as well, it is confined to the extracellular spaces. The increased osmolality due to sodium retains water in the ECF. If the extra salt and water is not retained in the vasculature (e.g. if capillary pressure is raised or colloid oncotic pressure is low from lack of albumin as in cirrhosis), then it escapes into the interstitial spaces. Hence, blood volume is not expanded and the signals to retain sodium are not turned off. The excess sodium and water accumulate as generalized oedema, an excessive accumulation of fluid, usually beneath the skin, but sometimes in other compartments (see Fig. 7.1). Generalized oedema can occur with low, normal, or high blood (or serum/plasma) sodium concentration. Thus, blood sodium concentration does not alone reflect total body sodium. An increased total body sodium can occur with a low, normal, or high blood sodium concentration.

Excessive ECF volume can impair normal organ function. Oedema of the skin, particularly in the lower limbs, can be painful, interferes with normal circulation, impairs wound healing, and increases the likelihood of infection. Ascites (a condition in which fluid collects in spaces within your abdomen) can impair breathing, decrease venous blood return to the heart, and promote intraperitoneal infection. Pulmonary oedema interferes with respiratory gas exchange and is a major cause of morbidity and mortality. Furthermore, oedema is a sign of an underlying disease process that needs to be treated.

**Treatment** Apart from treating the underlying disease process, treatment involves decreasing salt and water intake and/or promoting salt and water excretion, usually with diuretics. Extremely low-salt diets are unpalatable, but some attempt to restrict salt should be made, including reducing intake of high-salt foods, including processed foods. Diuretics inhibit reabsorption of sodium by the renal tubules (remember the concept of sodium and water absorption—water absorption can be regulated by sodium absorption). To date, loop diuretics (e.g. furosemide) inhibit a sodium transporter in the loop of Henle, and thiazides reduce sodium reabsorption in the distal tubule by inhibiting a sodium transporter. See Table 7.4 for a summary of disturbances in the water content of the body.

Table 7.4 **Summary of disturbances in the amount of water in the body**

| Disturbance | Cause | Osmotic pressure (plasma [$Na^+$]) | Volume change | Manifestation |
|---|---|---|---|---|
| Excess water | Water only | Reduced | ICF ↑ ECF ↑ | Cerebral oedema |
| | Excess $Na^+$ | Normal | ECF ↑ | Generalized oedema |
| Dehydration | Water loss | Increased | ICF ↓ ECF ↓ | Thirst |
| | $Na^+$ loss | Normal | ECF ↓ | Circulatory failure |

ECF, extracellular fluid; ICF, intracellular fluid.

# 7.3 Potassium

## 7.3.1 Significance and functions of potassium

Potassium is the predominant cation in the cells of both animals and plants. Its salts, mainly organic, are responsible for most of the osmolality of animal cells and determine their volume. Intracellular enzymes have evolved to require an environment rich in potassium. Cell membranes are much more permeable to potassium than sodium. Hence, the ratio of the intracellular concentration to the extracellular concentration of potassium largely determines the resting potentials of cells and the transient action potentials, which transmit messages and activate nerve cells and muscle fibres.

An increase in the low extracellular concentration of potassium lowers the concentration ratio of intracellular potassium (150 mmol/L) to extracellular potassium (5 mmol/L); this depolarizes membranes and blocks transmission, whereas a decrease in extracellular potassium concentration increases the ratio, hyperpolarizes membranes, and raises the threshold for excitation. Consequently, the tight control of the low concentration of potassium in the ECF is critically important—large increases or decreases (2–3-fold) can paralyse muscles and even stop the heart from beating.

## 7.3.2 Amount and distribution in the body

An average adult human's body contains about 3800 mmol of potassium; most of this (about 3200 mmol) is in the cells. Indeed, a total body count of the natural isotope, potassium-40, can yield an estimate of cell mass. About 300 mmol is contained in the skeleton and only 80 mmol is in solution in the extracellular fluids. Hence, if the cells increase their content by only 1.25% (40 mmol) at the expense of the ECF, the external concentration would be halved, with potentially serious consequences for neuromuscular and cardiac function.

## 7.3.3 Potassium balance

### Intake: diet

Around 100 mmol/day:

1. Meat is animal muscle and vegetables contain plant cells, hence almost all foods contain potassium, except perhaps highly refined foods. There are no large differences between natural and processed foods, as there are for sodium.
2. *Wholemeal flour, meats, and fish*: 7–9 mmol/100 g
3. *Common vegetables*: 5–9 mmol/100 g
4. *Milk, eggs, and cheese*: 4–6 mmol/100 g
5. *Fruit*: 5–8 mmol/100 g (citrus fruits, bananas, and dry fruits have high potassium concentrations, e.g. oranges have 5 mmol/100 g, so orange juice is a useful source of potassium).

The lowest values are for salted butter at 0.5 mmol/100 g and apples at 0.3 mmol/100 g. Hence, ordinary mixed and vegetarian diets contain adequate amounts of potassium. It is difficult to devise a diet that is deficient, but it is easy to find one that is excessive for patients with kidney failure, who have trouble excreting potassium and may die from the consequences of uncontrolled hyperkalaemia (raised blood potassium concentration).

### Output

1. *Faeces*: about 10 mmol/day
2. *Urine*: 90 mmol/day

This can be varied widely to match alterations in intake. The kidneys hold the balance by adjusting the amount in the urine. They can excrete potassium rapidly if extracellular potassium concentration rises and can conserve it when scarce, though not as avidly or as briskly as sodium. The renal priority is to keep the critically important concentration of potassium in the ECF within its normal range of 3.5–5 mmol/L (see also section 7.2.3 about aldosterone effects). Note that unlike sodium, potassium is not lost in large amounts in sweat, as the concentration in sweat typically reflects the potassium concentration in the ECF.

### 7.3.4 Regulation of extracellular concentration

The most important factors are:

1. *Active uptake by cells*: this is maintained by ongoing metabolism, and is promoted by insulin and ECF acidosis. Potassium must be supplied to prevent a lethal fall in extracellular potassium concentration when patients with diabetic ketoacidosis treated with insulin and glucose begin to rebuild the severely depleted stores of glycogen and potassium in their muscles.
2. *Excretion by the kidneys*: the bulk of the potassium in the glomerular filtrate is reabsorbed from proximal tubules. What appears in the urine is mostly lost by distal tubular cells, as they preferentially reabsorb sodium in response to stimulation by aldosterone (see Fig. 7.5b). The rate of renal excretion is increased by:
   i. increased extracellular potassium concentration
   ii. aldosterone
   iii. *faster flow through distal tubules*—the transcellular diffusion gradient sets the concentration of potassium in the tubular fluid (the trans-tubular potassium gradient), so that excretion is proportional to the flow
   iv. increased sodium delivery or concentration in the distal tubular fluid. This increases sodium reabsorption via the sodium channel in the principal cells of the distal tubule and in the same way increases potassium release into the urine facilitated by the apical potassium channel (see Fig. 7.5b). This can be promoted by more proximally acting diuretics (loop diuretics or thiazides), which fundamentally work by inhibiting sodium reabsorption by various transporters in the LOH and DT.

### 7.3.5 Potassium depletion

Deficiency of potassium requires a failure of renal conservation, abnormal losses, or both:

1. Net absorption from the gastrointestinal tract is reduced or even negative (i.e. potassium is lost) if there is significant vomiting, aspiration of stomach contents, or diarrhoea, when lost fluid often contains more potassium than normal intestinal secretions. Colonic potassium loss will be further stimulated by aldosterone. Thus, the abuse or heavy use of laxatives can cause potassium depletion.
2. The kidneys usually excrete a certain amount of potassium depending on the diet and this excretion is increased by adrenal steroids (e.g. aldosterone) and diuretics, which increase the sodium load in the distal tubule, facilitating potassium secretion in principal cells.
3. Most disturbances of acid–base balance increase the rate of excretion of potassium. Cells accumulate hydrogen ions in preference to potassium (they effectively buffer the hydrogen ions). Thus, ECF acidosis will bring potassium out of cells, raising extracellular concentrations, so potassium removed from renal cells is excreted and lost. Alkalosis promotes uptake of potassium from plasma into cells, including renal tubular cells, from which potassium can be lost into the urine. Thus, alkalosis as well as acidosis can deplete the body of potassium.

**Symptoms** The symptoms of potassium deficiency are often mild, vague, and non-specific, and include fatigue and ill-defined malaise with weakness, especially of skeletal and intestinal muscles; the post-operative form of this is called an ileus. Cardiac arrhythmias are also common and characteristic changes may be visible in an electrocardiogram (ECG). Hypokalaemia can interfere with the kidney's maximum concentrating ability, which may promote further loss.

**Treatment** Oral potassium replacement is the usual treatment, except when serum concentration is dangerously reduced or when patients are receiving intravenous fluids. Intravenous potassium supplementation must be carried out cautiously because of the risk that a sudden high extracellular potassium concentration can produce cardiac asystole. Oral replacement with food (e.g. bananas) is safe in less urgent cases.

### 7.3.6 Excess of body potassium

Potassium is not stored in the body—there is no over-stocking of cells corresponding to oedema. Localized excesses in the form of high concentrations of potassium in the ECF are dangerous, but these are rare in the absence of renal failure (acute or chronic kidney injury). The kidneys may fail to protect against excessive concentrations in the following situations:

1. *In shock*: cells deprived of oxygen cannot retain potassium, and kidneys without adequate blood flow cannot excrete it.
2. *In crush injuries*: crushed muscles release potassium and also myoglobin, which can injure the kidneys, leading to acute failure.
3. *In anuria* (from any cause): excretion is impossible, and an increasing concentration of potassium in the plasma as cells break down may be a more pressing indication of the need for dialysis than a rising concentration of blood urea.
4. *In Addison's disease*: when adrenal mineralocorticoid secretion is deficient, extracellular potassium concentration may be moderately increased without an increase in total body potassium, because the kidneys fail to conserve sodium but retain potassium.

Note that high (like low) ECF potassium can produce changes in an ECG, producing large peaked T waves, broadening of the QRS complex, and flattening of the P wave. These changes are a sign that treatment of the hyperkalaemia is urgently required. This can be performed by some combination of driving potassium into cells with insulin and glucose administration or with alkali (as bicarbonate), and removal of potassium from the body using chelating agents (such as oral resonium), loop diuretics, or dialysis. The ECG changes can be temporarily stabilized by intravenous calcium gluconate.

## 7.4 Acid–base balance

The pH or concentration of $H^+$ in a solution of body fluids can vary according to their specific function. Some body fluids are alkaline. Saliva serves to neutralize acidity in the mouth resulting from the breakdown of sugars by bacteria. Pancreatic fluid neutralizes the highly acidic chyme coming from the stomach to set up an alkaline environment for several pancreatic enzymes required to digest food in the intestine. The acidic environment in the stomach is required to kill bacteria and digest protein. An imbalance in these pH-dependent environments can lead to a range of adverse health actions, for example dental caries as the bicarbonate in saliva does not protect teeth, or growth problems seen in cystic fibrosis, where the lack of water secretion impairs the release of a bicarbonate-rich fluid into the intestine. However, acid–base balance refers to plasma pH, which is closely monitored and kept in the range of pH 7.35–7.45, whilst the intracellular pH is about 7.1 due to metabolic processes. Small changes in plasma pH, also called acid–base disturbances, can have significant consequences, as many enzymes and processes require a certain optimum pH in which they can operate.

### 7.4.1 Regulation

The maintenance of acid–base balance implies keeping the plasma and other ECF at pH 7.35–7.45. This alkalinity is essential for cells in excitable tissues (nerves, muscles, and heart). The regulation of blood pH between the values of 7.35 and 7.45 is the result of a coordinated response including three mechanisms:

- buffering systems
- controlling excretion of carbon dioxide by the lungs (about 13,000 mmol/day); this removes the weak organic acid, carbonic acid, $H_2CO_3$
- excretion of smaller amounts of non-volatile acid (hydrogen ions) or regeneration of alkali (bicarbonate) by the kidneys.

Buffering systems can act within seconds to absorb hydrogen ions in response to disturbances in acid–base balance. These systems primarily consist of major proteins circulating in the blood, as they consist of amino acids that include positively and negatively charged groups. The major proteins that are involved in buffering hydrogen ions in the blood are haemoglobin and plasma proteins.

The concentration or partial pressure of $CO_2$ and the concentration of bicarbonate ($HCO_3^-$) in the blood are tightly controlled, and can be combined in a formula, from which the change in blood pH can be estimated if there is a change in either parameter. The kidneys control the numerator and the lungs the denominator of the Henderson–Hasselbalch equation:

$$\mathrm{pH} = 6.1 + \log\frac{[\mathrm{HCO_3^-}] \leftarrow \text{kidneys}}{0.03P_{\mathrm{CO_2}} \leftarrow \text{lungs and respiratory system}}$$

Normally, plasma bicarbonate ($HCO_3^-$) concentration is kept around 24 mmol/L and the partial pressure of carbon dioxide near 40 mmHg; hence, pH must be:

$$6.1 + \log(24/(0.03 \times 40)) = 6.1 + \log(24/1.2)$$
$$= 6.1 + \log 20 = 6.1 + 1.3 = 7.4.$$

Disturbances in acid–base balance are classed as being metabolic or respiratory in nature and either an acidosis (an increase in hydrogen ions and fall in pH below normal levels) or an alkalosis (a decrease in hydrogen ions and increase in pH above normal levels). Respiratory and renal diseases are common causes of altered acid–base balance. Considering only dietary factors, it has been known for more than 100 years that meat diets leave excess acid and vegetarian diets excess alkali in the body to be dealt with. Claude Bernard, in 1865, noticed that rabbits that happened to be starved produced acidic urine instead of the usual alkaline urine characteristic of herbivorous animals. He deduced that starvation made them temporarily carnivorous, living on their own flesh, and found that he could make their urine alkaline or acidic at will by giving them grass or meat to eat. He did the same with a horse. From a pathological perspective, a metabolic acidosis is usually associated with impaired renal acid excretion; however, undertaking relatively high intensity exercise can also temporarily cause this problem and can affect exercise performance, as the excess hydrogen ions produced as a result of increased metabolism can interfere with calcium handling in a sarcomere. A respiratory acidosis is commonly the result of a reduced respiratory rate and consequent $CO_2$ retention—this can be seen in conditions that lead to impaired respiratory function, such as chronic obstructive pulmonary disorder. Metabolic alkalosis is relatively uncommon but can occur in some digestive diseases that include repeated vomiting. Respiratory alkalosis from hyperventilation is quite common and can occur when an individual suffers from a panic attack. The important aspect to note here is that disturbances in acid–base balance can be acute or chronic and the lungs can compensate for metabolic disturbances by an increase or decrease of ventilation. The kidneys also contribute to this in chronic situations and provided the metabolic disturbance is not of renal origin. Respiratory disturbances are compensated for by the kidneys with an increase in bicarbonate reabsorption or acid secretion.

## 7.4.2 Dietary considerations

Meat diets yield sulphuric acid from sulphur-containing amino acids, and phosphoric acid from nucleoproteins and phospholipids. Mixed diets leave about 70 mmol/day of hydrogen ions to be excreted (the so-called non-volatile or titratable acids). Food faddists may label sour fruits as 'acid foods', but the organic acids they contain are either not absorbed or are mostly oxidized to water and carbon dioxide. Most of this is breathed out, although a little remains in the body as bicarbonate. Hence, these acids, taken in as potassium salts, leave an excess of potassium bicarbonate, which tends to make the blood more alkaline. Table 7.5 gives examples of food acids and their metabolic fates, and it can be seen that these dietary acids pose no threat to the body's mild alkalinity.

**Table 7.5 Acids in fruits and their metabolic fate**

| Food source | Acid | Fate |
|---|---|---|
| Citrus, pineapples, tomatoes, summer fruits | Citric | Oxidized to $CO_2$ and $H_2O$ |
| Apples, plums, tomatoes | Malic | Oxidized to $CO_2$ and $H_2O$ |
| Cranberries, blueberries | Benzoic | Excreted as hippuric acid |
| Grapes | Tartaric | Not absorbed |
| Strawberries, rhubarb, spinach | Oxalic | Not absorbed; forms calcium oxalate in the gut |

*Adapted from:* Passmore, R. and Eastwood, M.A. (1988) *Davidson and Passmore's human nutrition and dietetics*, 8th edn. Edinburgh: Churchill Livingstone.

Organic acids that can pose threats are:

- acetoacetic and other keto acids, particularly produced in diabetic ketoacidosis; also smaller, less important amounts during fasting.
- lactic acid produced in severe muscular exercise or from tissues inadequately supplied with oxygen (e.g. in shock when blood pressure is very low).

## 7.4.3 Renal bicarbonate regeneration

Filtered bicarbonate is reabsorbed by the secretion of hydrogen ions in the proximal and distal tubules. This reaction forms carbonic acid, which is converted to $CO_2$ and water. The $CO_2$ diffuses back into renal cells and forms intracellular bicarbonate, which is secreted into the plasma, maintaining buffer balance, and avoiding loss of the filtered bicarbonate in the urine. Additional bicarbonate can be generated by promoting ammonia secretion, which leads to excretion of further hydrogen ions. Since electroneutrality must be maintained, a molecule of bicarbonate is generated for every extra hydrogen ion excreted. The kidney thus makes bicarbonate by secreting acid (non-volatile/titratable acids), with 1 mmol of bicarbonate added to the plasma for each 1 mmol of hydrogen ions secreted into the urine. This restores the buffer balance that was used to neutralize non-volatile/titratable acids. Similarly, additional secreted hydrogen ions convert filtered buffers, especially phosphate, into their acid forms in acid urine. Hydrogen ions are also excreted as ammonium. Thus, 1 mmol of additional bicarbonate is added to the plasma for every 1 mmol of hydrogen ion excreted as acid buffer or ammonium. In alkalosis, the concentration of bicarbonate in the plasma may be so high that there is more in the glomerular filtrate than the total rate of hydrogen ion secretion can cope with. The excess bicarbonate then escapes in alkaline urine and lowers the concentration in the plasma.

## 7.4.4 Exercise and dietary supplements

As noted in section 7.4.1, undertaking relatively high intensity exercise results in an acute metabolic acidosis due to an increase in hydrogen ion production following anaerobic respiration. These hydrogen ions can interfere with calcium handling within a sarcomere, which can, ultimately, impact on exercise performance. As such, there has been much scientific and athlete interest in the use of pre-exercise dietary supplements that could attenuate the rise in hydrogen ions during exercise and potentially improve exercise performance. Two of the main supplements that have been investigated are sodium bicarbonate and beta-alanine.

The theory behind the use of sodium bicarbonate is reasonably straightforward. By providing an exogenous form of bicarbonate, there is an increase in the bicarbonate pool that can buffer any increase in hydrogen ions that may occur. Research in this area is conflicting, with some studies demonstrating that pre-exercise ingestion of sodium bicarbonate improves high intensity exercise performance in relatively short duration activity, whereas others demonstrate little or no effect. Several methodological issues in the literature prevent clear conclusions being drawn; however, of note is that a number of studies demonstrate that the use of sodium bicarbonate can lead to side effects including gastrointestinal discomfort.

Carnosine is an intracellular protein that is involved in buffering hydrogen ions during exercise and is present in relatively high concentrations within muscle cells. Given the role of carnosine in muscle buffering, it makes sense that attempting to increase carnosine concentration may be beneficial for exercise performance. Beta-alanine is the rate limiting precursor to carnosine synthesis and, as such, there is substantial scientific and commercial interest in beta-alanine supplements. Available evidence suggests that beta-alanine supplements result in improved exercise performance in events that last for relatively short durations and are undertaken at a high intensity, although there is some discussion on the best timing and dosage to use.

This chapter has been modified and updated from that originally written by Professor James Robinson (deceased), which appeared in the first three editions of *Essentials of Human Nutrition*; the chapter in the fourth edition was edited by Professor Zoltán Endre, and in the fifth edition it was edited by Dr Andrew Bahn.

## Further reading


1. **Bray, J.J., Cragg, P.A., Macknight, A.D.C., and Mills, R.G.** (1999) *Lecture notes on human physiology*, 4th edn. Oxford: Blackwell Scientific Publications.
2. **Danziger, J. and Zeidel, M.L.** (2015) Osmotic homeostasis. *Clin J Am Soc Nephrol*, **10**, 852–62.
3. **European Food Safety Authority (EFSA)** (2010) Scientific opinion on dietary reference values for water. *EFSA Journal*, **8**, 1459.
4. **Halperin, M.L., Kamel, K.S., and Goldstein M.B.** (2010) *Fluid, electrolyte, and acid-base physiology. A problem-based approach*, 4th edn. Philadelphia, PA: Saunders Elsevier.
5. **Knepper, M.A., Tae-Hwan Kwon, M.D., and Nielsen, S.** (2015) Molecular physiology of water balance. *The New England Journal of Medicine*, **372**, 14.
6. **McCance, R.A.** (1936) Medical problems in mineral metabolism. *Lancet*, **227**, 823–30.
7. **Passmore, R. and Eastwood, M.A.** (1988) *Davidson and Passmore's human nutrition and dietetics*, 8th edn. Edinburgh: Churchill Livingstone.
8. **Stanhewicz, A.E. and Kenney, W.L.** (2015) Determinants of water and sodium intake and output. *Nutr Rev*, **73**(Suppl. 2), 73–82. [Review.]
9. **Wrong, O.** (1993) Water and monovalent electrolytes. In: Garrow, J.S. and James, W.P.T. (eds) *Human nutrition and dietetics*, 9th edn, pp. 146–61. Edinburgh: Churchill Livingstone.

# 8 Major Minerals: Calcium and Magnesium

Helen Lambert, Ohood Hakim, and Susan A. Lanham-New



## 8.1 Calcium

Calcium (Ca) is a truly unique mineral and is an essential constituent of all forms of life. The skeleton contains approximately 99% of the body's total calcium and we need adequate dietary calcium intake, together with vitamin D, to facilitate normal growth and maintain healthy bones and teeth (Box 8.1). Calcium is a divalent cation, with an atomic weight of 40 (equivalents: 40 mg Ca = 1 mmol Ca). Calcium is the fifth most abundant element in our bodies. It is the main mineral in bone, being stored as hydroxyapatite, $Ca_{10}(OH)_2(PO_4)_6$. One of the key roles of our skeletal system is to protect the vital organs. Furthermore, the skeleton provides a store of minerals from which calcium (and phosphorus) may be continually withdrawn or deposited, according to physiological needs. The content of total body calcium differs considerably amongst individuals, at all ages, due to the fact that some people grow better skeletons than others, which is related to a combination of genetic/familial factors, as well as nutritional and other influences (Fig. 8.1).

In stark contrast, both intracellular and extracellular calcium concentrations are tightly controlled

**Box 8.1** Nutrition and bone

Essential for growth and maintenance of strong bones and teeth, healthy nerve and muscle function, blood clotting, and hormone release, calcium intake is needed daily to offset the obligatory losses seen in urine and faeces. When dietary supply is insufficient, bone calcium stores are resorbed, stimulated by parathyroid hormone (PTH). Food sources include dairy products, soy products, leafy green vegetables, bread, tap water in hard water areas, nuts and seeds, and dried fruits. Vitamin D is essential for alimentary calcium absorption. High calcium intakes and ensuring adequate vitamin D status are both important in slowing osteoporotic bone loss and reducing the risk of falling. Calcium balance is maintained by the calciotrophic hormones (PTH, calcitonin, and calcitriol (1,25-dihydroxyvitamin D)).

As a result, calcium is also released, and excreted in the urine. Supplementation with alkaline potassium salts (such as bicarbonate or citrate) significantly reduces calcium excretion, thus conserving calcium stores, and in a recent meta-analysis, this was shown to be associated with a decrease in bone resorption (Lambert et al., 2015). Fruit and vegetables are rich in alkaline potassium salts; increasing their consumption reduces the dietary acid load, and so could contribute to conserving bone mineral.

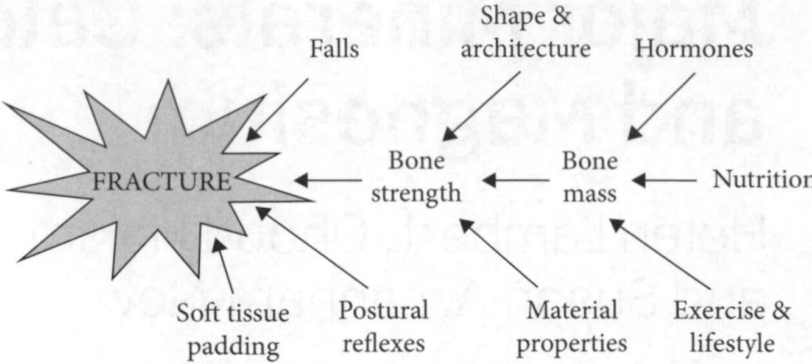


Fig. 8.1 Endogenous and exogenous factors affecting bone mass structure that can influence the risk of bone fracture. Bone health is determined by a combination of endogenous and exogenous factors. Genetics, or inherited traits, can play a role in most of these factors but are particularly important in determining peak bone mass.

*Source:* Heaney, R.P. (2003) Is the paradigm shifting? *Bone*, **33**, 457–65.

within narrow limits by the calciotrophic hormones. This is critical to human survival because interactions of calcium ions with proteins alter molecular activity. Ordered movement of ionic calcium plays an essential role in regulation of muscle contraction, nerve conductivity, ion transport, enzyme activation, blood clotting, and secretion of hormones and neurotransmitters. Life without calcium is not possible and small fluctuations in plasma calcium concentrations may have serious consequences. Hypocalcaemia and hypercalcaemia are common medical emergencies. Low blood calcium (hypocalcaemia) may cause seizures and tetany (musculoskeletal spasms and twitching, particularly in the fingers and face), and tingling and numbness due to increased neuromuscular activity. High blood calcium (hypercalcaemia) results in thirst, mild mental confusion and irritability, loss of appetite, and general fatigue and weakness. Polyuria and constipation are common. When the concentration of serum calcium is high, calcium salts may precipitate in soft tissues and kidney stones may subsequently form.

Another way in which diet influences calcium metabolism is through acid–base balance. The physiological buffering systems of the kidneys and lungs, as well as plasma protein buffers, keep serum acid–base balance under tight homeostatic control. However, if acid loading exceeds the capacity of these systems, bone mineral is mobilized to contribute bicarbonate and other anions to buffer the acid.

### 8.1.1 Bone metabolism

There are two types of bone: dense cortical bone (80% of the skeleton), often referred to as compact bone, and spongy trabecular bone (20% of the skeleton), often referred to as cancellous bone. The skeleton undergoes constant renovation, rather like a building site (see Fig. 8.2)! Old worn bits are being chiselled out or resorbed by multinucleate cells called osteoclasts, while teams of cells called osteoblasts busily rebuild excavation holes with strong new bone. These activities take place within a bone-remodelling unit. Bone cell activity can be followed using biochemical markers that are products of bone cell activity, or collagen synthesis or breakdown:

- blood levels of bone-specific alkaline phosphatase
- N-terminal propeptide of human procollagen type I
- osteocalcin reflects bone formation.

Urinary hydroxyproline, deoxypyridinoline and pyridinoline, C-terminal peptide (CTx), and N-terminal peptide (NTx) indicate resorption. Osteoblasts buried deep in bone mineral are called osteocytes; they sense weight-bearing and may help to regulate bone-remodelling responses to exercise. The different bone cells communicate actively. A complex, exquisitely sensitive 'internet' of chemical messages appears to control their differentiation and activity, but our understanding of this bone language remains limited. When osteoblast and osteoclast activity is matched, or coupled, bone mass is in a steady state. The amount of bone destroyed by osteoclasts

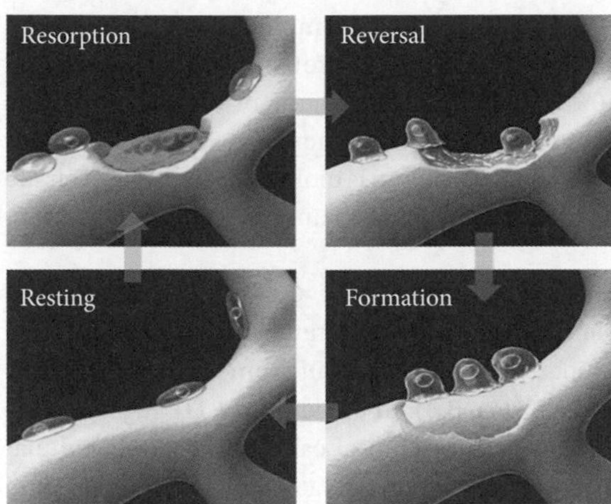


Fig. 8.2 Bone remodelling sequence: early in the remodelling sequence, osteoclasts (bone resorbing cells) are attracted to a quiescent bone surface and excavate an erosion cavity (*resorption*). Mononuclear cells then smooth off the erosion cavity and the site attracts osteoblasts (*reversal*). The osteoblasts (bone forming cells) synthesize an osteoid matrix and continuous new bone matrix synthesis is followed by calcification of the newly formed bone (*formation*). When complete, lining cells once more overlie the trabecular surface (*resting*).
*Source:* Coxon, J.P., Oades, G.M., Colston, K.W., and Kirby, R.S. (2004) Advances in the use of bisphosphonates in the prostate cancer setting. *Prostate Cancer Prostatic Dis*, **7**, 99–104. Available at: https://www.nature.com/articles/4500705/figures/1 (accessed 27 June 2022).

is replaced by an equal amount of new bone. When bone remodelling becomes uncoupled and resorption exceeds formation, bone is lost. Bone mass can be measured accurately *in vivo* using dual-energy X-ray absorptiometry (DEXA scanning, which provides an area density, units: g/cm$^2$) or computerized tomography (CT scanning, which provides a volumetric density, units: g/cm$^3$). Quantitative ultrasound is a useful tool for assessing risk of osteoporotic fracture, cheaper than DEXA and CT, and involves no ionizing radiation. However, unlike DEXA, it cannot be used to diagnose osteoporosis as defined by the World Health Organization (WHO, 1994), as there is no direct measure of bone mass.

## 8.1.2 Metabolic bone disorders

Children with vitamin D deficiency develop rickets, and adults, osteomalacia (see Chapter 14). Children with rickets do not calcify bone normally and their bones contain osteoid (unmineralized bone). Because this bone is weak, children with rickets often show bowed limbs.

Rickets is still seen in developing countries and amongst some populations in more affluent Western societies, although osteoporosis is the most frequently seen bone disease. Osteoporosis is defined by the WHO as a 'progressive systemic skeletal disease characterized by low bone mass and micro-architectural deterioration of bone tissue, with a consequent increase in bone fragility and susceptibility to fracture' (WHO, 1994). Osteoporosis is caused by substantial loss of bone. Despite there being too little bone, what remains is normally calcified. Osteoporotic bones are thin and break easily, especially in the wrist, spine, and hip. The lower the bone density, the higher the risk of fractures.

## 8.1.3 Calcium stores

**Bone stores** It is often observed in the clinic that when heavier people are being weighed they say 'I've got big bones'. However, bones are strong, rather than weighty. Bones of an average adult constitute 14% of body weight, and bone mineral 4%. Men accumulate more skeletal calcium (1200 g) than women (1000 g). Approximately one-fifth of this (21%) is in the skull, half (51%) in the arms and legs, and the remainder (28%) in the trunk (ribs 9%; pelvis 8%; spine 11%).

**Peak bone mass** The 'heaviest' bone mass that an individual achieves is called their peak bone mass (PBM). Although the rate of bone mineral accrual slows after puberty, consolidation continues well into the third decade in both men and women. The bone mass achieved at the end of this phase of consolidation is the PBM. The age at which this is attained varies between individuals, just as there are variations in the rate of bone loss in older age, and PBM occurs earlier in girls than in boys (see section 8.1.6). To achieve average PBM values, men require a positive daily calcium gain of 160 mg and women 130 mg for every single day of the first 20 years of their lives! Ethnic and familial studies and studies of monozygotic and dizygotic twins show that there are strong genetic influences on PBM attainment. These may account for up to 80% of the variability in adult bone density. The variance in PBM is wide, with values ranging between 20% higher and 20% lower than average. The variance does not change in the third and fourth decades of life, indicating that young adults with low density do not catch up bone density over time. Thus, if a good PBM is not attained by the late twenties, it is unlikely ever to be achieved.

**Calcium in extraskeletal stores and body fluids** These stores are small (15 g), comprising teeth (7 g), soft tissues (7 g), and plasma and intracellular fluids (1 g). Cytoplasm concentrations are 1000 times lower than those in plasma. Breast milk contains a high level of calcium (350 mg/L or 8.8 mmol/L) and a lactating mother transfers around 260 mg daily to her baby. Plasma calcium concentration is 2.2–2.6 mmol/L (8.8–10.4 mg/dL). Half of this is bound to protein (37% to albumin and 10% to globulin), 47% is free or ionized, and 6% is complexed to anions (phosphate, citrate, bicarbonate). Ionized calcium is biologically active. Levels are regulated by the PTH–vitamin D axis and calcitonin.

**Control of plasma calcium** When ionized plasma calcium concentrations fall, PTH is secreted to increase calcium input from the kidney, bone, and gut (Fig. 8.3). In the kidney, PTH augments the tubular reabsorption of calcium, decreases tubular reabsorption of phosphate and bicarbonate, and stimulates conversion of $25(OH)D_3$ to $1,25(OH)_2D_3$ (calcitriol). In bone, PTH promotes release of calcium and phosphate into blood. The effects of PTH on kidney and bone are direct and rapid and are

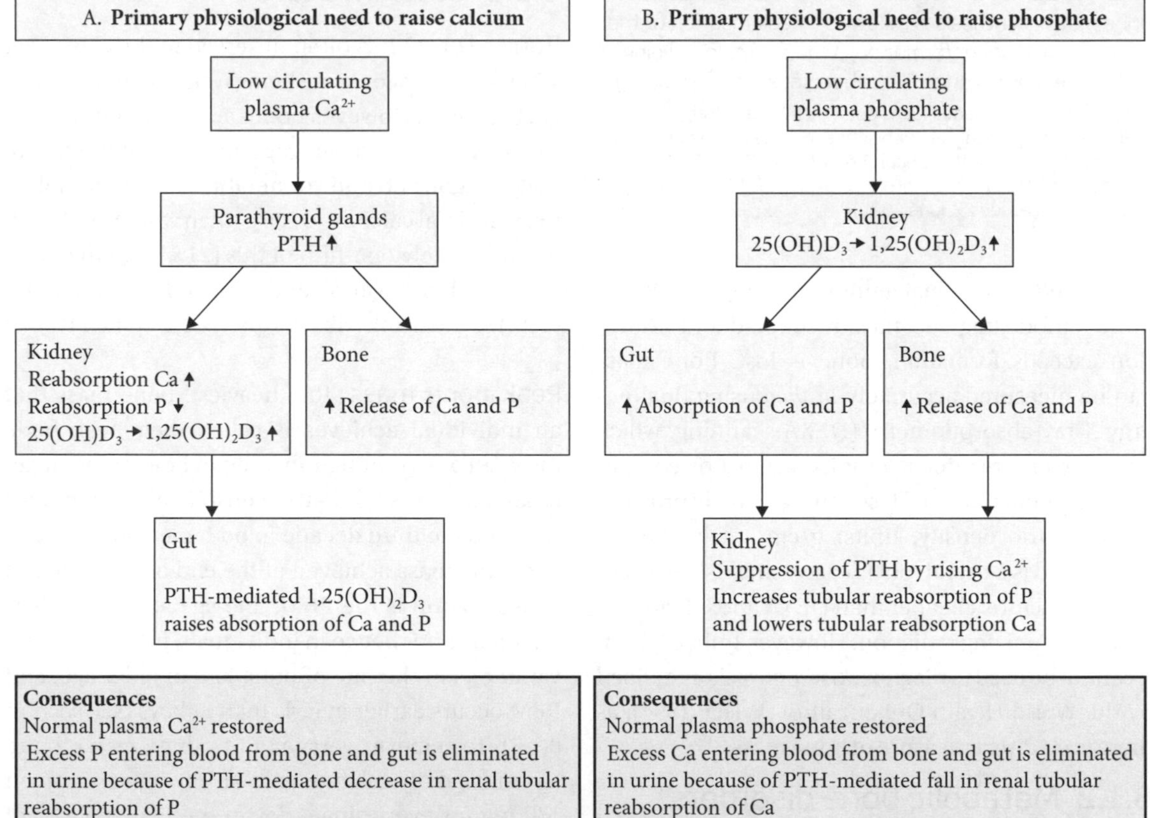


Fig. 8.3 Coordinated actions of parathyroid hormone and calcitriol in target organs regulate levels of calcium and phosphate in plasma. $Ca^{2+}$, ionized calcium; PTH, parathyroid hormone; $25(OH)D_3$, 25-hydroxyvitamin D3 (liver metabolite); $1,25(OH)_2D_3$, 1,25-dihydrocholecalciferol or calcitriol (kidney metabolite).

assisted by $1,25(OH)_2D_3$. The ability of PTH to raise alimentary calcium and phosphate absorption is mediated solely by calcitriol. When normal plasma calcium concentrations are restored, PTH secretion decreases, the flow of calcium from bone subsequently diminishes, urinary calcium rises, and $1,25(OH)_2D_3$ synthesis is closed down. This system is extremely robust and PTH and calcitriol influence each other's synthesis. The specific role of vitamin D is discussed in Chapter 14.

Calcitonin, a hormone secreted by the thyroid gland, helps to fine-tune plasma calcium regulation. It lowers serum calcium by inhibiting bone resorption and is secreted when ionized blood calcium levels rise above normal, helping to curb blood calcium fluctuations after meals. Calcitonin plays a smaller role in plasma calcium homeostasis than either PTH or calcitriol. Interestingly, patients who have had surgery on the thyroid gland maintain levels extremely well. In contrast, patients lacking either PTH or active vitamin D metabolites develop hypocalcaemia. Magnesium deficiency also causes hypocalcaemia because magnesium is a cofactor for PTH secretion (see section 8.2). Restoring magnesium corrects the problem of hypocalcaemia in alcoholics and patients with steatorrhoea. Total plasma calcium rarely falls below 1.25 mmol/L (5 mg/dL) because of this capacity of calcium ions from bone mineral to constantly exchange with extracellular fluid. This is why bone is referred to as a giant 'ion exchanger' facility in the body.

**Obligatory losses of calcium** Significant amounts of calcium 'leak' from the body. These unavoidable obligatory losses occur through the skin (generally less than 20 mg/day via epithelial cells and sweat), in the faeces (unabsorbed digestive juice may contain 80–120 mg calcium/day), and in urine. Obligatory urinary calcium loss varies between 40 and 200 mg/day, depending on how effectively calcium is reabsorbed from the glomerular filtrate. Renal tubules reabsorb more than 98% of the 10,000 mg calcium filtered daily by the glomeruli. Dietary salt (NaCl), protein, and caffeine are all known to increase obligatory urinary calcium loss.

### 8.1.4 Calcium balance and absorption

If more calcium is retained in the body than excreted, a person is said to be in positive calcium balance. Negative calcium balance occurs if more calcium is excreted than is ingested in the diet. Zero calcium balance implies that the amount of calcium absorbed daily from food exactly matches amounts lost in the faeces and urine and from the skin (see Fig. 8.4).

**Alimentary calcium absorption** Alimentary calcium absorption is not as efficient as renal tubular calcium reabsorption. Absorption is normally less than 70% (and often less than 30%) of calcium entering the gut. Net calcium absorption (the difference between calcium ingested by mouth and calcium excreted in the faeces) can be determined by traditional metabolic balance techniques. To ensure avoidance of negative calcium balance, net absorption must fully offset calcium losses from urine and skin. However, measurement of net calcium absorption considerably underestimates total calcium absorbed from the intestine into blood, because some faecal calcium (endogenous faecal) is derived from calcium resecreted into the intestine in the digestive juices, rather than from unabsorbed food calcium. True alimentary calcium absorption (amount actually absorbed from the gut) is measured with radioisotopes or stable isotopes, $^{42}Ca$ and $^{44}Ca$.

**Factors affecting the bioavailability of calcium in the intestine** Variations in the efficiency of absorption are predominantly determined by two factors:

1. vitamin D metabolites
2. rate of transit of gut contents through the intestine.

Calcitriol improves calcium absorption (see Chapter 14). However, some is absorbed even in vitamin D-deficient states, as some calcium is absorbed by passive concentration-dependent diffusion. The duodenum absorbs calcium most avidly, but larger quantities of calcium are absorbed by the ileum and

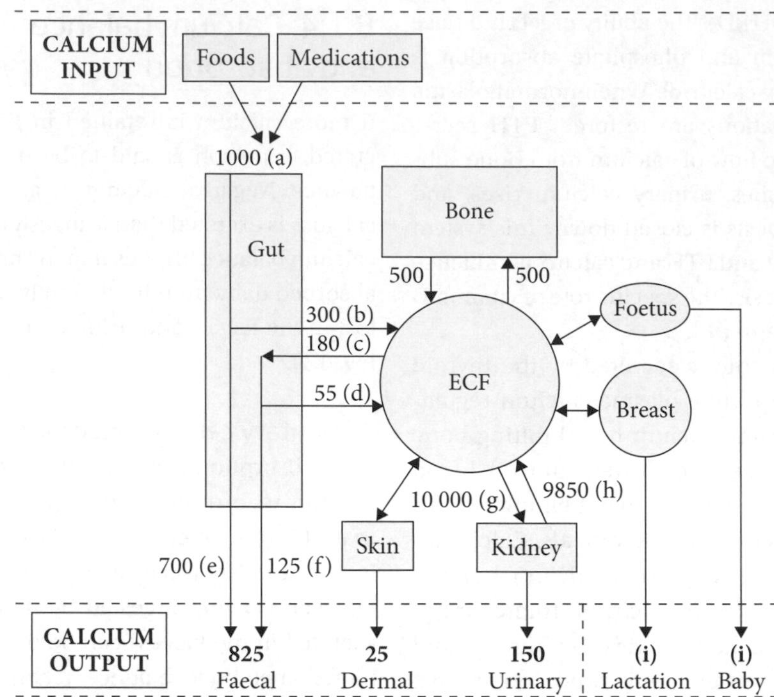


Fig. 8.4 Illustration of calcium influxes (mg/day) of subject in calcium balance (input = output). ECF, extracellular fluid; (a), total calcium ingested by mouth; (b), dietary calcium absorbed; (c), calcium in digestive secretions; (d), reabsorbed endogenous calcium; (e), unabsorbed dietary calcium; (f), endogenous faecal calcium excretion; (g), calcium load filtered at glomerulus; (h), calcium reabsorbed from glomerular filtrate (>98%); (i), losses only incurred in pregnancy (full-term baby has 25–30 g calcium) or lactation (160–300 mg/day in breast milk).

jejunum, mainly due to food components spending longer in these areas. Some calcium is also absorbed from the colon, and surgical resection can impair absorption. Carbohydrates, such as lactose, improve calcium absorption by augmenting its passive diffusion across villous membranes. Diets rich in oxalate, fibre, and phytic acid are reputed to depress alimentary absorption by complexing calcium in the gut. However, it appears that their overall effects seem small, possibly because bacterial breakdown of uronic acid and phytates in the colon frees calcium for absorption. Poor bioavailability of calcium from spinach is attributed to the high oxalate content. Dietary phosphorus increases the endogenous secretion of calcium into the gut. There is also evidence that calcium absorption diminishes in both sexes in the seventh decade of life because of lower renal synthesis of calcitriol and intestinal resistance to calcitriol, which contribute to the genesis of senile osteoporosis (see section 8.1.6).

### Factors influencing urinary calcium loss

Urinary calcium excretion rises when the filtered load of calcium increases or tubular reabsorption decreases. Acidifying agents, dietary sodium, protein, and caffeine raise excretion. Phosphorus, alkaline agents (bicarbonate, citrate), and thiazide diuretics lower excretion. Variations in salt intake explain much of the day-to-day fluctuation in urinary calcium. A useful pointer is that one teaspoonful of salt (100 mmol NaCl) raises urinary calcium by 40 mg calcium/day, and this occurs even on a low calcium intake. Purified sulphur-containing amino acids (methionine and cysteine) cause significant calciuria, but phosphate in whole proteins mitigates their calciuric effect when consumed in foods. There is evidence that vegetarians with an alkaline urine excrete less urinary calcium than meat-eaters, who have an acid urine.

## 8.1.5 Dietary calcium

The calcium stores in our bodies are built and maintained by extracting and retaining calcium from food. Our dietary needs vary according to gender, ethnicity, age, and magnitude of obligatory calcium loss. Table 8.1 provides details of the recommendations for calcium (and vitamin D) by the US Food and Nutrition Board of the Institute of Medicine (IOM) (2010). It is essential, at all stages, to consume and absorb enough dietary calcium to satisfy physiological calcium needs. This is because some bone will be mobilized to maintain blood calcium levels whenever losses of calcium exceed alimentary calcium absorption (see Fig. 8.5).

**Threshold concepts** Few individuals consume too much calcium from natural foods; however, many eat too little. This may subsequently affect their skeletons detrimentally, especially in childhood, when skeletal needs are high, and in later life, when alimentary absorption of calcium deteriorates. There is a threshold intake for calcium below which skeletal calcium accumulation is a function

**Table 8.1a US dietary recommendations for calcium and vitamin D**

| Life stage group | Age | Calcium (mg/day) | | Vitamin D (mg/day) | |
|---|---|---|---|---|---|
| | | RDA/AI | UIL | RDA/AI | UIL |
| Infants | (0–6 m) | 200 | 1000 | 400 | 1000 |
| | (6–12 m) | 260 | 1500 | 400 | 1500 |
| Children | (1–3 y) | 700 | 2500 | 600 | 2500 |
| | (4–8 y) | 1000 | 2500 | 600 | 3000 |
| Males | (9–13 y) | 1300 | 3000 | 600 | 4000 |
| | (14–18 y) | 1300 | 3000 | 600 | 4000 |
| | (19–30 y) | 1000 | 2500 | 600 | 4000 |
| | (31–50 y) | 1000 | 2500 | 600 | 4000 |
| | (51–70 y) | 1000 | 2000 | 600 | 4000 |
| | (>70 y) | 1200 | 2000 | 800 | 4000 |
| Females | (9–13 y) | 1300 | 3000 | 600 | 4000 |
| | (14–18 y) | 1300 | 3000 | 600 | 4000 |
| | (19–30 y) | 1000 | 2500 | 600 | 4000 |
| | (31–50 y) | 1000 | 2500 | 600 | 4000 |
| | (51–70 y) | 1200 | 2000 | 600 | 4000 |
| | (>70 y) | 1200 | 2000 | 800 | 4000 |
| Pregnancy | (14–18 y) | 1300 | 3000 | 600 | 4000 |
| | (19–30 y) | 1000 | 2500 | 600 | 4000 |
| | (31–50 y) | 1000 | 2500 | 600 | 4000 |
| Lactation | (14–18 y) | 1300 | 3000 | 600 | 4000 |
| | (19–30 y) | 1000 | 2500 | 600 | 4000 |
| | (31–50 y) | 1000 | 2500 | 600 | 4000 |

AI, adequate intake; RDA, recommended dietary allowance; UIL, tolerable upper intake level.

*Adapted from:* Institute of Medicine (IOM) (2010) *Dietary reference intakes for calcium and vitamin D*. Washington, DC: National Academy Press.

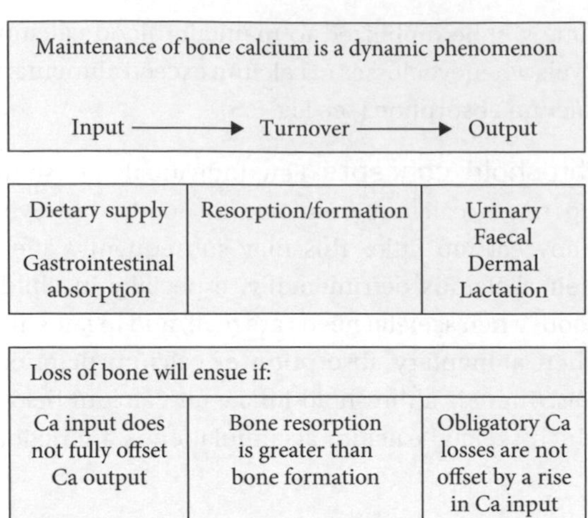


Fig. 8.5 Illustration of the ways in which bone loss can be caused. Maintenance of bone calcium is a dynamic phenomenon.

of intake, and above which skeletal accumulation does not further increase, irrespective of further increases in intake. In other words, 'enough' calcium is good, but 'extra' calcium will not increase bone formation.

**Recommended nutrient intake** Considerable controversy exists worldwide regarding the optimal daily dietary intake of calcium for individuals to:

- achieve optimal PBM attainment
- maintain adult bone mass
- prevent bone loss in later life.

There may be ethnic and genetic differences in calcium requirements. Those experts who argue for lower intakes point out that large sections of the population manage to grow and maintain bone on calcium intakes well below current US recommended dietary allowance (RDA). As shown in Table 8.1a, the US IOM recommendations for calcium and vitamin D published in 2010 maintain the levels set by the 1997 Food and Nutrition Board of the IOM. For the 1–3-year age group, calcium recommendations were increased from 500 to 700 mg/day, and for the 4–8-year age group, from 800 to 1000 mg/day. The recommended intakes in the UK, shown in Table 8.1b, are lower than those in the US. It is noteworthy that many people have difficulty in consuming more than 1000 mg calcium daily from natural foods. A possibly better way to boost the calcium economy would be to lower dietary salt intake, since this will reduce obligatory loss of calcium and improve calcium balance. Vitamin D supplementation, in addition to calcium, is essential for housebound elderly individuals.

**Table 8.1b UK dietary recommendations for calcium**

| Life stage group | Age | Calcium (mg/day) | | |
|---|---|---|---|---|
| | | LRNI | EAR | RNI |
| Infants | (0–12 m) | 240 | 400 | 525 |
| Children | (1–3 y) | 200 | 275 | 350 |
| | (4–6 y) | 275 | 350 | 450 |
| | (7–10 y) | 325 | 425 | 550 |
| Adolescents: males | (11–14 y) | 480 | 750 | 1000 |
| | (15–18 y) | 480 | 750 | 1000 |
| Adolescents: females | (11–14 y) | 450 | 625 | 800 |
| | (15–18 y) | 450 | 625 | 800 |
| Adults | (19–50 y) | 400 | 525 | 700 |
| | (50+ y) | 400 | 525 | 700 |
| Pregnancy | No increment | | | – |
| Lactation | | | | +550* |

EAR, estimated average requirement; LRNI, lower reference nutrient intake; RNI, reference nutrient intake.

*Reviewed in 1998—no longer thought necessary.

*Adapted from:* Department of Health Committee on Medical Aspects of Food Policy (1991) *Dietary reference values for food, energy and nutrients for the United Kingdom*. London: HMSO.

**Food sources of calcium** Foods vary greatly in their calcium content. Milk has an especially high calcium content (Table 8.2). Other excellent sources of calcium include cheeses and yoghurt. In many Western countries, dairy products supply up to two-thirds of the total daily intake of calcium. Soya milk substitutes are also excellent suppliers of calcium.

**Table 8.2 Calcium content of some common foods**

| Calcium sources | Serving size | Approximate g/serving | mg Ca/serving |
|---|---|---|---|
| **Dairy products** | | | |
| Milk, 2% fat | 1 cup | 250 ml | 300 |
| Milk, whole | 1 cup | 250 ml | 295 |
| Hard cheese (e.g. cheddar) | 1 oz | 30 | 240 |
| Yoghurt, plain dairy | 1 pot | 150 | 207 |
| Vanilla ice-cream | 1 cup | 150 | 132 |
| Mozzarella cheese | 1 oz | 30 | 121 |
| **Plant-based 'milk' alternatives*** | | | |
| Almond milk | 1 cup | 250 ml | 113 |
| Soy milk | 1 cup | 250 ml | 33 |
| Rice milk | 1 cup | 250 ml | 28 |
| Oat milk | 1 cup | 250 ml | 20 |
| **Other** | | | |
| Sardines, tinned | ½ can | 60 | 300 |
| Spinach (cooked) | 1 cup | 200 | 300 |
| Kale (raw) | 1 cup | 70 | 105 |
| Chick peas (cooked) | 1 cup | 170 | 84 |
| Almonds | 1 oz | 30 | 75 |
| Broccoli (raw) | 1 cup | 70 | 66 |
| Orange | 1 medium | - | 50 |
| Dried figs | 1 oz | 30 | 48 |

*Values are for unfortified products. Non-dairy milks may be enriched with calcium depending on the brand. Added calcium may be less well absorbed due to the presence of phytates in some plant-based milks.

*Adapted from:* USDA National Nutrient Database for Standard Reference, release 28, 2016 and McCance and Widdowson's Composition of Foods Integrated Dataset 2021.

Other good sources of calcium include nuts, canned fish with bones, leafy vegetables, and dried fruit. In some countries, foods are fortified with mineral calcium salts. An example of this is the mandatory fortification of white flour with calcium carbonate in the UK.

Dietary advice to increase calcium intake (while following nutritional guidelines for lowering fat intake) includes:

- have a serving of either yoghurt or milk daily for breakfast
- always have low-fat milk available in the fridge, but use the full-fat varieties for children
- choose low-fat dairy products when shopping
- add cheese chunks or a sprinkling of nuts/seeds to salads/vegetables
- eat pieces of cheese, nuts/seeds, or green vegetables as snacks
- add grated cheese or milk when serving soups and pasta
- use canned fish with bones in sandwich spreads
- try tofu chunks with salads and casseroles
- add a little skimmed milk powder to recipes when baking
- serve vegetables in white sauces made with milk

- use yoghurt in place of cream with desserts
- ensure that non-dairy milks, yoghurt, and cheese are enriched with calcium to at least the equivalent value of dairy products.

When individuals choose the calcium-rich foods they like, a satisfactory intake of calcium may be sustained throughout all the life stages.

**Calcium supplementation and food fortification** Individuals who find it difficult to eat enough calcium may benefit from mineral supplements. At risk are those with very low calorie intakes, milk allergies, or symptomatic lactose malabsorption. They may need to consume foods fortified with calcium (soybean and citrus drinks, breakfast cereals) or take supplements, which are as well absorbed as food calcium.

**Calcium toxicity** Ingestion of large amounts of alkaline calcium salts (the US IOM reference nutrient intakes for calcium and vitamin D recommend an upper limit for Ca of between 2000 and 3000 mg/day from age 1 year to >70 years) can override the ability of the kidney to excrete unwanted calcium, causing hypercalcaemia and metastatic calcification of the cornea, kidneys, and blood vessels. People consuming huge quantities of calcium carbonate in antacids are prone to this effect (milk–alkali syndrome). Patients taking vitamin D, or its metabolites, may suffer similar symptoms. Very large amounts of vitamin D are known to be toxic (see section 14.1.11). The US IOM upper intake level for vitamin D from age 9 years to >70 years has therefore been set at 4000 IU.

## 8.1.6 Factors affecting bone growth and attrition

**Normal growth** Both boys and girls display similar linear gains in calcium up to the age of about 10 years (Fig. 8.6). Their total body calcium then averages 400 g, indicating a daily increment of 110 mg over this period. Skeletal growth is accelerated at puberty. Spinal density matures earlier in girls, who go through puberty earlier than boys (UK average is 12.2 years for girls and 14.1 years for boys).

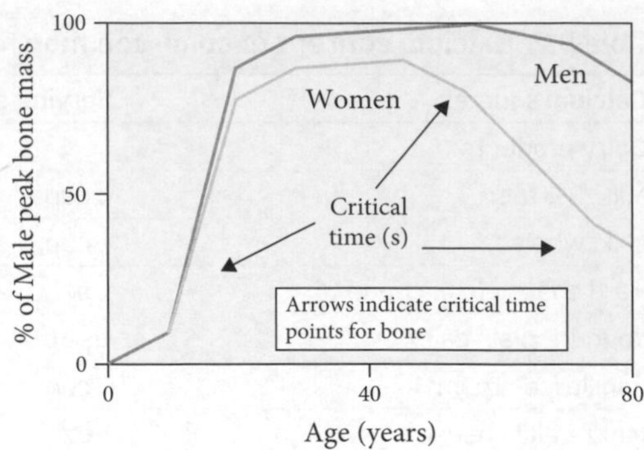


Fig. 8.6 Changes in skeletal mass throughout the life cycle.

Total body calcium doubles in girls between the ages of 10 and 15 years (an average gain of 200 mg daily). Boys have two extra years of pre-pubertal bone gain before their pubertal growth spurt, when they deposit over 400 mg calcium daily in bone. Children do not have higher calcium absorption than adults. Obligatory losses are also high and there are concerns that many children consume too little calcium to meet their skeletal needs. Indeed, in the UK, the National Diet and Nutrition Surveys show that 15% of boys and girls aged 11–18 years have intakes below the lower reference nutrient intake for calcium (400 mg/day). In teenagers, bone mass can be increased by calcium supplementation. However, the gain may be temporary and catch-up may occur in children who consume less calcium (Lambert et al., 2008). It may just take children longer to attain their skeletal potential on moderate calcium intakes than on very high intakes. However, there is evidence that if supplementation is achieved using milk or dairy products early enough in puberty, the beneficial effects on bone mass may be more permanent (Cadogan et al., 1997; Bonjour et al., 2001).

Exogenous factors, such as calcium intake, physical activity, and sex steroid status, influence bone accrual. Cigarette smoking and excess alcohol affect bone mass adversely. High levels of weight-bearing physical activity, adequate dietary calcium, and sex steroids favour bone accrual. Regular moderate exercise should be recommended for children and teenagers. Children with greater lean body mass

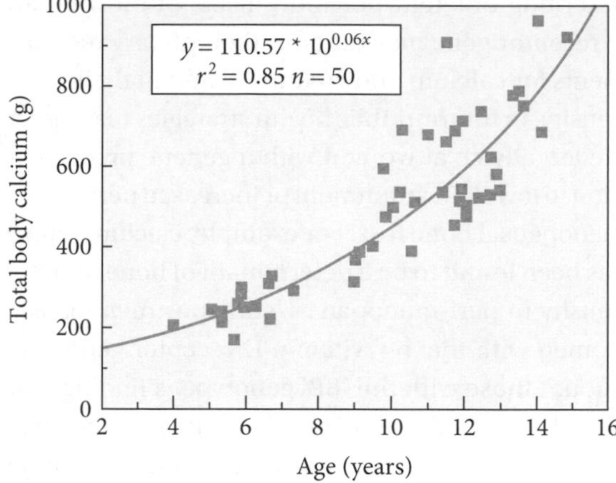


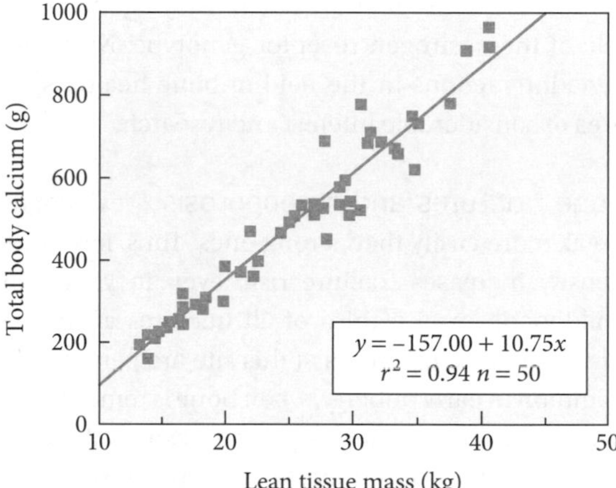


Fig. 8.7 Total body calcium changes in 50 growing girls aged 3–15 years, in relation to age (r = 0.92) and lean tissue mass (r = 0.97).

*Source:* Previous author's data (Assoc. Professor Ailsa Goulding, University of Otago, Dunedin, NZ).

have the best bone mass (see Fig. 8.7). In teenagers with anorexia nervosa and athletic amenorrhoea, low oestrogen status results in low PBM. Girls who recover from these conditions continue to show low spinal bone density years after plasma oestrogen levels have returned to normal.

**Fit but fragile phenomenon** More than a century ago, a German scientist, Julius Woolf, stated the theory now called 'Woolf's Law'—'bone accommodates the forces applied to it by altering its amount and distribution of mass'. More recently, this concept has been refined to a general theory of bone mass regulation, known as the mechanostat model. It is well known that in the absence of weight-bearing exercise, bone loss will occur at both axial and appendicular skeletal sites. Whilst bone mass has been shown to be higher in athletes involved in various sports, there is increasing concern for the bone health of women engaged in high-intensity physical training, for whom amenorrhoea is a common characteristic. Some such sports also demand extremely low body weights and there is high reported incidence of anorexia nervosa amongst participants. The combination of amenorrhoea and anorexia is of detriment to bone mass, and there is now good evidence to show that they 'under-achieve' their PBM potential and thus are at considerably increased risk of osteoporosis. This picture of undernutrition, amenorrhoea, and osteoporosis is defined as the 'female athletic triad' and, in 1997, the American College of Sports Medicine published a position stand to 'encourage the prevention, recognition and management of this syndrome' (Otis et al., 1997) (Fig. 8.8). The exact mechanisms involved in PBM reduction are unclear, but some data suggest suppression of osteoblasts, rather than increased osteoclastic activity. Of further interest is the finding that in gymnasts, despite a high prevalence of oligo- and amenorrhoea, bone mass shows a higher than predicted value. In a 3-year longitudinal study undertaken at the University of Surrey, UK, in gymnasts and controls, despite evidence of late age of menarche and/or amenorrhoea in the gymnast group, bone density was as much as 25% higher in these girls compared with age-matched and pubertal age-matched controls (Fig. 8.9), probably because extreme weight-bearing in gymnastics has a compensatory effect.

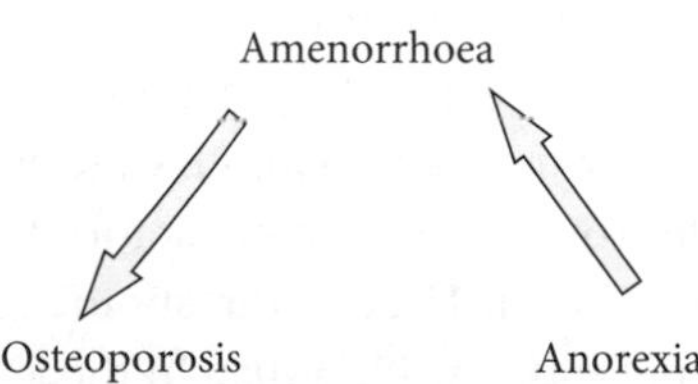


Fig. 8.8 Detrimental effects of exercise on the skeleton—the female athletic triad.

*Adapted from:* Otis, C.L., Drinkwater, B., Johnson, M., Loucks, A., and Wilmore, J. (1997) American College of Sports Medicine position stand. The female athletic triad. *Med Sci Sport Exer*, **29**, i–ix.

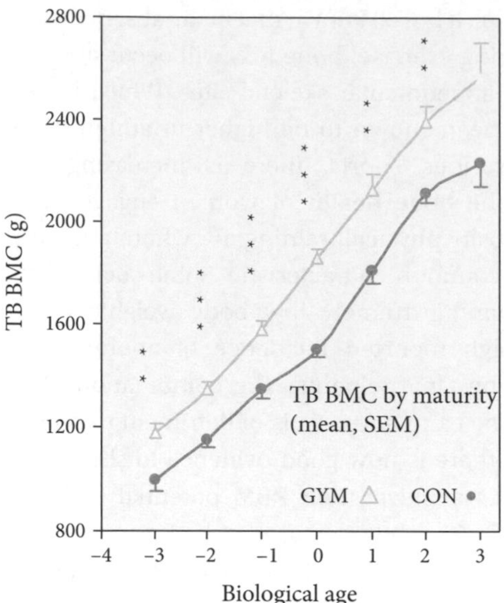


Fig. 8.9 Positive effect of impact loading exercise on PBM attainment. Values are mean ± standard error of the mean adjusted for height and weight. **, P < 0.01; ***, P < 0.001; BMC, bone mineral content; GYM, gymnasts; CON, controls.

*Source:* Nurmi-Lawton, J.A., Baxter-Jones, A.D., Mirwald, R.L., et al. (2004) Evidence of sustained skeletal benefits from impact-loading exercise in young females: a 3-year longitudinal study. *J Bone Miner Res*, **19**, 314–22. With permission from John Wiley and Sons.

**Bone attrition** Bone mineral density declines after middle life. Falling levels of sex steroids cause trabecular loss, while calcium deprivation speeds cortical loss. Effects of oestrogen deprivation are particularly pronounced at the menopause, with annual bone loss averaging 1–2% of bone over a 5–10-year period, although we know that there are fast and slow bone losers (Fig. 8.7). Bone losses are considerable: from youth to old age, women lose half their trabecular bone and a third of their cortical bone, while men lose a third of their trabecular bone and a fifth of their cortical bone. Low bone mass in the elderly may be due to poor PBM, to subsequent excessive loss of bone, or to both factors. Bone density will fall below the fracture threshold at a younger age in people with low PBM than in those with high PBM. Variations in the genetic inheritance of factors affecting mineral metabolism and bone cells probably affect PBM, rate of bone loss, and, hence, their susceptibility to bone fracture. Individuals inheriting different polymorphisms of the vitamin D receptor gene may differ in their dietary requirements for calcium and vitamin D, and in their bone density. In future, public health strategies may target dietary advice at women with a genetic predisposition to low PBM attainment or increased peri-/postmenopausal bone loss. For example, calcium intake has been found to be a determinant of bone mineral density in peri-menopausal/early postmenopausal women with the 'bb' vitamin D receptor genotype, but not those with the 'BB' genotype, a finding that is exclusive to those women not taking exogenous oestrogen. Modest alcohol intake (1–2 units/day) has been shown to be associated with reduced bone loss in peri-menopausal women carrying the '*p*' allele of the oestrogen receptor genotype. Nutrient–gene interactions in the field of bone health is an area of considerable interest and research.

**Bone fractures and osteoporosis** Weak bones break more easily than strong ones. Thus, low bone density increases fracture risk, even in youth. In childhood, up to a third of all fractures affect the distal forearm. Fractures at this site are particularly common in early puberty, when bone is remodelling fast and calcium requirements are especially high. Children who break their wrists have lower bone density than those without fractures. Increasing calcium intake and physical activity during growth helps to strengthen their skeletons and avoid fractures (see Box 8.2).

Osteoporosis is a serious and expensive public health problem, particularly in women. It causes significant pain and morbidity among elderly people (see section 8.1.2). The incidence of osteoporotic fractures is expected to increase in future because people are living longer. In 1990, there were 1.16 million estimated hip fractures worldwide. By the year 2050, this figure is projected to reach 6 million.

At present, the best way to avoid osteoporotic fractures in later life is to grow a good skeleton, to achieve optimal genetic skeletal mass, and then to retain this as long as possible. Every effort should therefore be made to ensure life-long consumption, absorption, and retention of sufficient calcium to do this (see Box 8.3).

**Box 8.2 Case 1**

- A boy who habitually avoided milk products because he disliked the taste was seen at 6.5 years and, again, 2 years later. At 7.8 years he broke his forearm falling less than standing height on carpet, playing with his sister. Although his total body calcium at 8.5 years was 582 g and his DXA bone mineral density (BMD) values were slightly above average (111% of expected values), he had experienced rapid, inappropriate weight gain, possibly because he ate almond croissants daily for breakfast and exercised little. In 2 years he had gained 14.1 kg (11.2 g fat per day) and his body mass index (BMI) had risen from the 75th to >95th percentile for age. At 8.5 years he had the average weight of a 12-year-old (41.2 kg), but distal radius BMD suitable only for a 10-year-old. He thus fell heavily on insufficient bone. Obesity, milk avoidance, and inactivity can increase fracture risk in growing children.

**Box 8.3 Case 2**

- A 73-year-old woman (height 152 cm, weight 48 kg, BMI 20.8 kg/m$^2$) referred for investigation had been diagnosed lactose intolerant at age 7 years. She then avoided dairy products all her life without consuming other calcium-rich foods, but was active and a keen gardener. DXA scans taken 65 years after diagnosis of lactose intolerance confirmed severe osteoporosis and a low total body calcium content of only 488 g (half the young adult value of 960 g). Her BMD T scores were: −5.63 (wrist); −3.10 (hip); −5.71 (spine); −3.89 (total body). She was promptly placed on bone-conserving medication to reduce fracture risk. This case shows that sustained dietary calcium deprivation is detrimental to bone health. T score compares BMD with normal young adult.

## 8.1.7 Other possible health effects (positive and negative) of calcium

Relatively high intakes of calcium appear to be protective against hypertension, especially when consumed in low-fat dairy products in a diet high in fruit and vegetables, and low in salt. The series of DASH (dietary approaches to stop hypertension) trials showed that consumption of low-fat dairy products resulted in further reductions in blood pressure when added to a diet with the other blood pressure-lowering attributes. Furthermore, such a diet is associated with decreased urinary calcium excretion and reduced bone resorption.

The consumption or absorption of other substances may be influenced by calcium intakes. Some dairy products are high in saturated fat, but lower-fat products contain as much, and often more, calcium than full-fat versions. There are concerns that high dietary calcium may lower iron absorption in some individuals, but few data confirm that this effect is found in the general population. High levels of calcium also aggravate the inhibitory effect of phytic acid on zinc absorption. Calcium binds gut oxalate so supplements do not induce kidney stones. Calcium may lower absorption of tetracyclines.

Population subgroups listed below, who have special nutritional needs and who may be vulnerable to calcium deprivation, should be targeted to improve calcium economy and hence safeguard bone health. Such groups include:

- people with habitually low dietary calcium intakes
- people with food allergies or lactose malabsorption
- adolescents building maximal bone
- girls with anorexia nervosa or athletic amenorrhoea
- those on weight-loss diets who avoid dairy products
- people with very low dietary energy intakes
- pregnant women (last trimester) and lactating women
- people with high intakes of common salt

- people with heavy alcohol consumption and smokers
- patients with malabsorption syndromes
- patients taking corticosteroid medication
- patients with renal disease
- elderly people
- people confined indoors who get no vitamin D from sunlight.

There has been recent interest in the possibility that calcium supplements may increase risk of myocardial infarction and other cardiovascular events. A meta-analysis published in 2010 showed that use of calcium supplements (without vitamin D) was associated with an increased risk of myocardial infarction and the authors suggest a reassessment of calcium supplementation in the management of osteoporosis (Bolland et al., 2008, 2010). These conclusions have been criticized on the grounds that data from randomized controlled trials were misrepresented (Sabbagh and Vatanparast, 2009) and the Women's Health Initiative (WHI) study, criticized widely for poor study design and compliance, was overemphasized. A subsequent meta-analysis, which also included studies of calcium with vitamin D, but excluded studies in men, concluded that these supplements did not increase the risk of cardiovascular outcomes (Lewis, 2014). Thus, definitive conclusions require more research. It is, however, key to point out that the link between increased calcium intake from supplements and risk of myocardial infarction applies to calcium supplements and not calcium from the diet.

## 8.2 Magnesium

Since the early 1930s when McCollum first observed magnesium deficiency in rats and dogs, magnesium has been an intriguing mineral to research. It has both physiological and biochemical functions, and important interrelationships, especially those with the cations calcium, potassium, and sodium. Magnesium is also involved with second messengers, PTH secretion, vitamin D metabolism, and bone functions.

### 8.2.1 Distribution and functions

Approximately 60–65% of the body content, i.e. 1 mol (25 g), of magnesium in an adult person is found in the skeleton. Like calcium, it is an integral part of the inorganic structure of bones and teeth. However, unlike calcium, magnesium is the major divalent cation in the cells, accounting for most of the remaining magnesium, with 27% in the muscles and 7% in other cells. Intracellular magnesium is involved in energy metabolism, acting mainly as a metal activator or cofactor for enzymes requiring adenosine triphosphate, in the replication of DNA, and the synthesis of RNA and protein. Hence it appears to be essential for all phosphate-transferring systems.

Several magnesium-activated enzymes are inhibited by calcium, while in others, magnesium can be replaced by manganese. The remaining 1% of the body content of magnesium is in the extracellular fluids. Plasma concentrations of magnesium are about 1 mmol/L, of which, as with calcium, about one-third is protein bound. Magnesium and calcium have somewhat similar effects on the excitability of muscle and nerve cells, but calcium has a further important function in signalling, which requires its concentration in the cells to be kept extremely low.

### 8.2.2 Metabolism

Magnesium is predominantly absorbed from the small intestine, both by a facilitated process and by simple diffusion. The relative importance of these two processes depends on the magnesium levels in the gastrointestinal tract, as well as gut transit time. Absorption can vary widely and depends on other dietary factors, including calcium intakes, the presence of chelators (such as fluoride), and the pH of the gastrointestinal tract (the latter is affected by medication, including antacids and some antibiotics). On average about 40–60% of dietary intake is absorbed.

Excretion is mainly through the kidneys and increases with dietary magnesium intake, as well as caffeine and alcohol intakes. The kidney is extremely efficient in conserving magnesium; when the intake decreases, the urine can become almost magnesium-free. The intestinal and renal conservation and excretory mechanisms in normal individuals permit homeostasis over a wide range of intakes.

### 8.2.3 Dietary sources of magnesium

Similar to potassium, magnesium is present in both animal and plant cells and is also the mineral in chlorophyll. Green vegetables, unrefined cereals, legumes, seeds and nuts, and animal products are all good sources (Table 8.3). In contrast to calcium, dairy products tend to be low in magnesium, with cow's milk containing 120 mg Mg/L (5 mmol/L) compared with 1200 mg Ca/L (30 mmol/L). The calcium, phosphate, and protein in meat and other animal products reduce the bioavailability of magnesium from these sources. Average magnesium intake is about 320 mg (13 mmol) per day for males and 230 mg (10 mmol) per day for females.

**Table 8.3 Magnesium sources of some common foods**

| Magnesium sources | Serving size | g/serving | mg Mg/ serving |
|---|---|---|---|
| Pumpkin seeds | 1 cup | 130 | 592 |
| Brazil nuts | 1 cup | 130 | 376 |
| Dark chocolate (70–85% cocoa) | 1 oz | 30 | 228 |
| Quinoa (raw) | 1 cup | 170 | 197 |
| Oats | 1 cup | 160 | 177 |
| Whole grain wheat flour | 1 cup | 150 | 176 |
| Red kidney beans | 1 cup | 180 | 160 |
| Mackerel | 1 fillet | 90 | 97 |
| Kale (raw) | 1 cup | 70 | 88 |
| Spinach (cooked) | 1 cup | 200 | 87 |

*Adapted from:* USDA National Nutrient Database for Standard Reference, release 28, 2016 and McCance and Widdowson's Composition of Foods Integrated Dataset 2021.

### 8.2.4 Magnesium deficiency

Since magnesium is the second most abundant cation in cells after potassium, dietary deficiency is unlikely to occur in people eating a normal varied diet. In his classical studies in the 1960s, Shils found it difficult to produce experimental magnesium deficiency (Shils, 1969). This research demonstrated the interrelationships between magnesium and the other principal cations, calcium, potassium, and sodium. The plasma concentration of magnesium decreased progressively, as did serum potassium and calcium, whereas the serum sodium remained normal even though sodium was being retained (see Table 8.4). Functional effects, including personality changes, abnormal neuromuscular function, and gastrointestinal symptoms, were restored to normal only by repletion of magnesium. Hypomagnesaemia can precipitate hypocalcaemia because magnesium is required for the secretion of PTH.

Several clinical conditions are associated with magnesium deficiency (see Table 8.5) and most frequently result from losses of magnesium from the kidney or gastrointestinal tract. Magnesium depletion induces neuromuscular excitability and may increase the risk of cardiac arrhythmias and cardiac arrest.

### 8.2.5 Magnesium excess

Excessive dietary intakes of magnesium appear unharmful to humans with normal renal function. Hypermagnesaemia is uncommon and is almost impossible to achieve from food sources alone.

**Table 8.4 Magnesium depletion and accompanying changes**

| Cation | Blood chemistry | Metabolic balances |
|---|---|---|
| Magnesium | Plasma Mg ↓ | Mg negative |
| Potassium | Serum K ↓ | K negative |
| Calcium | Serum Ca ↓ | Ca positive |
| Sodium | Serum Na no change | Na positive |

**Table 8.5 Clinical conditions associated with occurrence of magnesium deficiency**

- Habitual or sustained low dietary supply/intake (<250 mg/day)
- Poor alimentary absorption—malabsorption syndromes (such as Crohn's disease), short bowel syndrome, laxative abuse
- Excessive body losses—via sweat or urine (genetic disorders, diabetes, alcohol abuse, diuretics)
- Losses from high-output stomas
- Increased requirement—pregnancy or lactation
- Hospitalized patients—65% of intensive care patients are hypomagnesaemic
- Endocrine disorders, such as parathyroid disorders and hyperaldosteronism

Magnesium supplementation is useful clinically to treat pregnancy-induced hypertension (pre-eclampsia and eclampsia). Many people also take oral magnesium supplements (such as Epsom salts) to prevent constipation, because they appear to have a cathartic effect.

## 8.2.6 Magnesium status

Serum magnesium concentration is the most frequently used index of magnesium status, with the normal range defined as 0.7–1.0 mmol/L. Plasma is not used because anticoagulants may be contaminated with magnesium.

The US Food and Nutrition Board of the IOM recommended intakes of magnesium within the different age groups are shown in Table 8.6a. The recommendations for the UK (Table 8.6b) are 300 mg/day for men and 270 mg/day for women, as determined from balance studies using typical UK diets. The values for infants are based on the magnesium content of breast milk.

**Table 8.6a US dietary recommendations for magnesium**

| Life stage group | | Magnesium (mg/day) | |
|---|---|---|---|
| | | RDA/AI | UIL[a] |
| Infants | (0–6 m) | 30* | ND |
| | (6–12 m) | 75* | ND |
| Children | (1–3 y) | 80 | 65 |
| | (4–8 y) | 130 | 110 |
| Males | (9–13 y) | 240 | 350 |
| | (14–18 y) | 410 | 350 |
| | (19–30 y) | 400 | 350 |
| | (31–50 y) | 420 | 350 |
| | (51–70 y) | 420 | 350 |
| | (>70 y) | 420 | 350 |
| Females | (9–13 y) | 240 | 350 |
| | (14–18 y) | 360 | 350 |
| | (19–30 y) | 310 | 350 |
| | (31–50 y) | 320 | 350 |
| | (51–70 y) | 320 | 350 |
| | (>70 y) | 320 | 350 |
| Pregnancy | (14–18 y) | 400 | 350 |
| | (19–30 y) | 350 | 350 |
| | (31–50 y) | 360 | 350 |
| Lactation | (14–18 y) | 360 | 350 |
| | (19–30 y) | 310 | 350 |
| | (31–50 y) | 320 | 350 |

AI, adequate intake; ND, not determined; RDA, recommended dietary allowance; UIL, tolerable upper intake level.

[a]UIL for magnesium represents intake from pharmacological agents only, and does not include intake from food or water.

*Adapted from:* Food and Nutrition Board, Institute of Medicine (1997) *Dietary reference intakes for calcium, phosphorus, magnesium, vitamin D and fluoride*. Washington, DC: National Academy Press.

## 8.2.7 Magnesium effects on bone

Magnesium deficiency has been shown to cause cessation of bone growth, decreased osteoblastic and osteoclastic activity, osteopenia, and increased bone fragility. It is necessary for vitamin D synthesis, transport, and activation, and also affects mineralization, both by acting directly on bone cells, and indirectly via PTH secretion. There is conflicting evidence from epidemiological studies on the role of magnesium in preventing fractures. Veronese et al. (2018) demonstrated an increase in fracture risk with

**Table 8.6b UK dietary recommendations for magnesium**

| Life stage group | Age | Magnesium (mg/day) | | |
|---|---|---|---|---|
| | | LRNI | EAR | RNI |
| Infants | (0–3 m) | 30 | 40 | 55 |
| | (4–6 m) | 40 | 50 | 60 |
| | (7–9 m) | 45 | 60 | 75 |
| | (10–12 m) | 45 | 60 | 80 |
| Children | (1–3 y) | 50 | 65 | 85 |
| | (4–6 y) | 70 | 90 | 120 |
| | (7–10 y) | 115 | 150 | 200 |
| Adolescents | (11–14 y) | 180 | 230 | 280 |
| | (15–18 y) | 190 | 250 | 300 |
| Adults: males | (19+ y) | 190 | 250 | 300 |
| Adults: females | (19+ y) | 150 | 200 | 270 |
| | | | | |
| Pregnancy | | | | No increment |
| Lactation | | | | +50 |

EAR, estimated average requirement; LRNI, lower reference nutrient intake; RNI, reference nutrient intake.

*Adapted from:* Department of Health Committee on Medical Aspects of Food Policy (1991) *Dietary reference values for food, energy and nutrients for the United Kingdom*. London: HMSO.

decreasing magnesium intakes in men and women, whereas an increased risk of hip fracture had previously been reported in a case-control study of Swedish women (Michaëlsson 1995). However, there is evidence from population-based, cross-sectional studies of a positive association between magnesium and bone mass/markers of bone metabolism, including the Aberdeen Prospective Osteoporosis Screening Study (APOSS) cohort and the Framingham population cohort. Carpenter et al. (2006), in a well-designed randomized controlled trial (RCT), showed that magnesium supplementation (300 mg elemental Mg per day in two divided doses) versus placebo given orally for 12 months significantly increased hip bone mineral content (BMC) in peri-adolescent girls (aged 8–14 years) who habitually consume a low magnesium intake (<220 mg/day). In postmenopausal women given a magnesium supplement for 30 days, Aydin et al. (2010) found that bone formation increased, and bone resorption decreased, along with PTH. Few other good Mg supplementation studies exist and it is an area for urgent further research given the growing recognition that some population groups have suboptimal intakes of magnesium. (In the 2020 National Diet and Nutrition Survey, 33% of boys and 47% of girls aged 11–18 years in the UK had dietary intakes below the LRNI of 180–190 mg/day).

## Further reading

1. **Bolland, M.J., Barber, P.A., Doughty, R.N., et al.** (2008) Vascular events in healthy older women receiving calcium supplementation: randomized controlled trial. *BMJ*, **336**, 262–6.
2. **Bolland, M.J., Avenell, A., Baron, J.A., et al.** (2010) Effect of calcium supplements on risk of myocardial infarction and cardiovascular events: meta-analysis. *BMJ*, **341**, c3691.
3. **Bonjour, J.P., Chevalley, T., Ammann, P., Slosman, D., and Rizzoli, R.** (2001) Gain in bone mineral mass in prepubertal girls 3.5 years after discontinuation of calcium supplementation: a follow-up study. *Lancet*, **358**, 1208–12.
4. **Cadogan, J., Eastell, R., Jones, N., and Barker, M.E.** (1997) Milk intake and bone mineral acquisition in adolescent girls: randomized, controlled intervention trial. *BMJ*, **315**, 1255–60.
5. **Darling, A.L., Millward, D.J., Torgerson, D.J., Hewitt, C.E., and Lanham-New, S.A.** (2009) Dietary protein and bone health: a systematic review and meta-analysis. *Am J Clin Nutr*, **90**, 1674–92; editorial **90**, 1142–3.

6. **Department of Health** (1991) *Dietary reference values for food, energy and nutrients for the United Kingdom*. London: HMSO.

7. **Goulding, A.** (2001) Bone mineral density and body composition in boys with distal forearm fractures: a dual energy X-ray absorptiometry study. *J Pediatr*, **139**, 509–15.

8. **Holick, M. and Dawson-Hughes, B.** (2004) *Nutrition and bone health*. New York, NY: Humana Press.

9. **Lambert, H.L., Eastel, R., Karnik, K., et al.** (2008) Calcium supplementation and bone mineral accretion in adolescent girls: an 18-month randomized controlled trial with 2-y follow-up. *Am J Clin Nutr*, **87**, 455–62.

10. **Lambert, H., Frassetto, L., Moore, J.B., et al.** (2015) The effect of supplementation with alkaline potassium salts on bone metabolism: a meta-analysis. *Osteoporosis Int*, **26**, 1311–18.

11. **Lanham-New, S.A.** (2008) Importance of vitamin D and calcium in the maintenance of bone health and prevention of osteoporosis. *Proc Nutr Soc*, **67**, 163–76.

12. **Lanham-New, S.A.** (2008) The balance of bone health: tipping the scales in favour of the potassium case. *J Nutr*, **138**, 172S–7S.

13. **Lanham-New, S.A. and Gannon, R.H.T.** (2007) Fruit and vegetable link to bone health. In: Lanham-New, S.A., O'Neill, T., Sutcliffe, A., and Morris, R. (eds) *Prevention and treatment of osteoporosis*, pp. 35–49. Oxford: Clinical Publishing.

14. **Lanham-New, S.A., Thompson, R.L., More, J., et al.** (2007) Importance of vitamin D, calcium and exercise to bone health with specific reference to children and adolescents. *Nutr Bull*, **32**, 364–77.

15. **Matkovic, V.** (2005) Calcium supplementation and bone mineral density in females from childhood to young adulthood: a randomized controlled trial. *Am J Clin Nutr*, **81**, 175–88.

16. **Nurmi-Lawton, J.A., Baxter-Jones, A.D., Mirwald, R.L., et al.** (2004) Evidence of sustained skeletal benefits from impact-loading exercise in young females: a 3-year longitudinal study. *J Bone Min Res*, **19**, 314–22.

17. **Otis, C.L., Drinkwater, B., Johnson, M., et al.** (1997) American College of Sports Medicine position stand. The Female Athlete Triad. *Med Sci Sport Exerc*, **29**, i–ix.

18. **Sabbagh, Z. and Vatanparast, H.** (2009) Is calcium supplementation a risk factor for cardiovascular disease in older women? *Nutr Rev*, **67**, 105–8.

19. **Shils, M.E.** (1969) Experimental production of magnesium deficiency in man. *Ann NY Acad Sci*, **162**, 847–55.

20. **USDA** (2009) National Nutrient Database for Standard Reference, Release 22.

21. **Walsh, J.S., Henry, Y.M., Fatayerji, D., and Eastell, R.** (2009) Lumbar spine peak bone mass and bone turnover in men and women: a longitudinal study. *Osteoporosis Int*, **20**, 355–62.

22. **WHO** (1994) *Study group on assessment of fracture risk and its application to screening and postmenopausal osteoporosis*, Report of a WHO Study Group, Technical Report, Series 84. Geneva: World Health Organization.

# Iron

Patrick MacPhail



Iron deficiency is the most frequently encountered nutritional deficiency in humans. The World Health Organization (WHO) has ranked it very high among the preventable risks of disability in the world. Paradoxically, iron overload is also a major clinical problem in some populations. Both iron deficiency and iron overload have serious consequences and are major causes of human morbidity.

Iron owes its importance in biology to its remarkable reactivity (see Box 9.1). Of paramount importance is the reversible one-electron oxidation–reduction reaction that allows iron to shuttle between ferrous ($Fe^{2+}$) and ferric ($Fe^{3+}$) forms. This reaction is exploited by most iron-dependent enzyme systems involving electron transport, oxygen carriage, and iron transport across cell membranes. It is also responsible for the toxicity seen in acute and chronic iron overload. These contradictory properties are managed by highly specialized and conserved proteins involved in the storage and transport of iron and in regulating the concentration of intracellular iron.

In recent years there has been an explosion of knowledge about the proteins of iron metabolism and the genes that produce them, filling many gaps in our understanding and, at the same time, creating new questions that have still to be answered.

### Box 9.1 Important themes in iron metabolism

- Iron has the ability to accept and donate electrons. This reactivity is exploited in numerous physiologic processes such as oxygen transport (haemoglobin), electron transfer (cytochromes), and the many enzymes with iron-sulphur clusters. It is also dangerous, producing free radicals and oxidative stress. Specialized proteins have evolved to protect organisms from damage when iron is in transit (transferrin), within cells (iron chaperones), and when stored (ferritin).
- Ferrous iron ($Fe^{2+}$) crosses cell membranes, but ferric iron ($Fe^{3+}$) is stored and transported. Iron transporters are associated with ferric reductases (Dcytb and DMT1) and ferroxidases (hephaestin and ferroportin) to swap between these forms.
- Iron itself controls the expression of many proteins essential for its distribution and storage (hepcidin, ferritin, DMT1, transferrin receptor, ferroportin).

## 9.1 Basic iron metabolism

The total body iron content is about 50 mg/kg (3–4 g per person). It is convenient to think of body iron as distributed between four interconnected compartments. Over 60% is in the red cell compartment, mainly in haemoglobin, while about 25% is in the storage compartment, mainly in the liver. The remainder is distributed between myoglobin in muscles (8%) and in enzymes (5%). A small amount (about 3 mg) is in transit in the circulation bound to the plasma transport protein, transferrin (see Fig. 9.4).

### 9.1.1 Iron absorption

The mechanism by which iron is absorbed from the gut is not clearly understood, but recent work has uncovered the existence of a number of genes coding for proteins involved in the control of iron absorption and transport of iron across membranes. Four phases of iron absorption are recognized. In the *luminal phase*, food iron is solubilized, largely by acid secreted by the stomach, and is presented to the proximal duodenum, where most iron absorption takes place. Factors that maintain the solubility of iron in the face of rising pH, such as valency (ferrous iron is better absorbed), mucin secreted by the cells lining the gut (mucosa), and chelators (ascorbic acid, secreted by the stomach), appear to be important in this phase. In addition, other foods present in the meal may promote or inhibit iron absorption (see section 9.3).

The second phase, *mucosal uptake*, depends on iron binding to the brush border of the apical cells of the duodenal mucosa and the transport of iron into the cell (Fig. 9.1). Two forms of dietary iron, haem and non-haem iron, need to be accommodated. The mechanism by which haem iron, derived mainly from myoglobin and haemoglobin in food, is transported into the cell is not clear (1 in Fig. 9.1). Non-haem iron, derived from a wide variety of foods, has to be in the ferrous form ($Fe^{2+}$) before it can be transported across the cell membrane by the *divalent metal transporter* (DTM1) (3 in Fig. 9.1). The reduction from ferric to ferrous iron is probably achieved by a membrane-bound ferrireductase called *duodenal cytochrome b* (Dcytb) (2 in Fig. 9.1). Protons, required by DMT1, are supplied by gastric acid flowing into the proximal duodenum, where DMT1 is most highly expressed. This explains why proton pump inhibitors reduce non-haem iron absorption.

In the third *intracellular phase*, iron, whether derived from haem or non-haem sources, is either stored in the storage protein, *ferritin*, or transported directly to the opposite (inner) side of the mucosal cell and released. In the *release phase*, iron is oxidized to the ferric form by a membrane-bound ferroxidase, *hephaestin* (5 in Fig. 9.1), and released by a specialized iron transporter, *ferroportin 1* (4 in Fig. 9.1) into the portal circulation, where it is bound to the transport protein, *transferrin*. Both iron uptake and particularly iron release by the mucosal cell are inversely related to the amount of iron stored in the body and directly related to the rate of erythropoiesis. The mechanism by which this is achieved is discussed in sections 9.1.3 and 9.1.4.

### 9.1.2 Internal iron exchange

Once released from the mucosal cell, iron enters the portal circulation and is bound to the transport protein, *transferrin* (Fig. 9.1 and Box 9.2), which keeps iron non-reactive in the circulation. While transferrin is able to distribute iron to most tissues, 80% of the iron in circulation is transported directly to the bone marrow, where it is incorporated in haemoglobin. The mechanism of transferrin uptake by the young red cells, and all active cells, involves a specific *transferrin receptor* (TfR1) (7 in Fig. 9.1) expressed on the surface of the cell. The iron–transferrin–TfR1 complex is taken into the cell by receptor-mediated endocytosis. A fall in the pH within the endosome causes the iron to be released from transferrin. This $Fe^{3+}$ is reduced to $Fe^{2+}$ by another ferrireductase, STEAP3, and then transported through the vesicle membrane into the cell by DMT1 (3 in Fig. 9.1). The transferrin–TfR1 complex, now devoid of iron, is cycled back to the cell surface, where the apotransferrin (transferrin without iron) is released back into the circulation and is available to transport

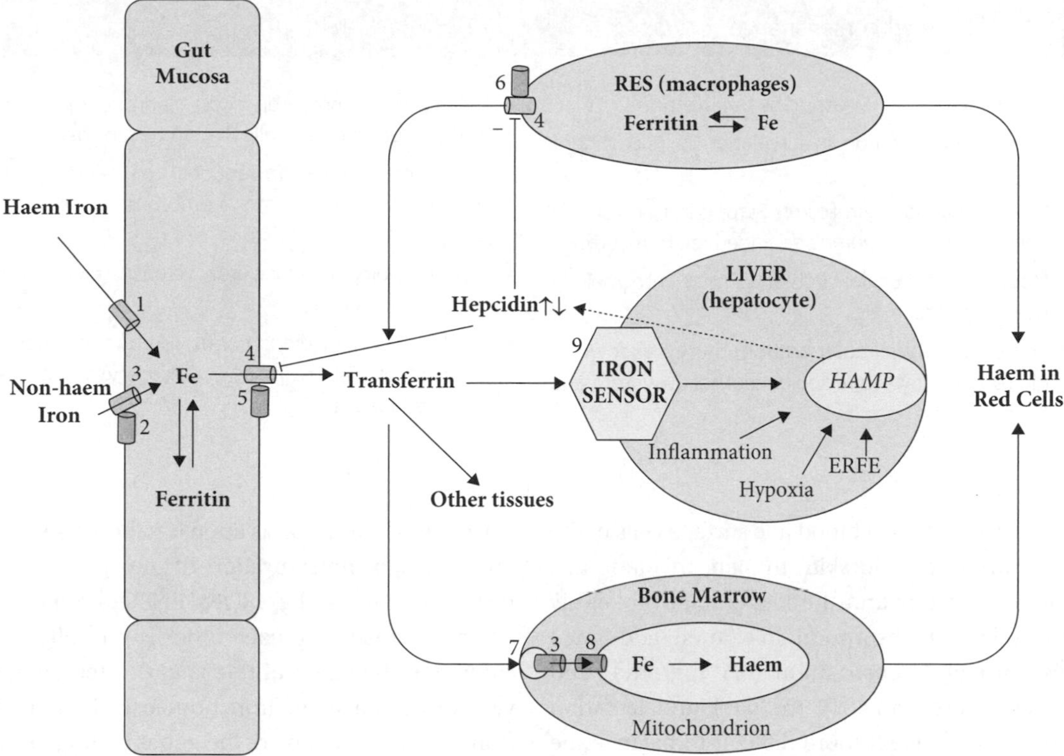


Fig. 9.1 The iron circuit and the central role of the liver. Iron, absorbed from the lumen of the small intestine (left), passes through the mucosal cell and enters the circulation, where it is bound to transferrin and redistributed to tissues. Most goes to the bone marrow for the production of haemoglobin in red cells, which circulate and, after 120 days, are engulfed by macrophages of the reticuloendothelial system (RES), located mainly in the liver, spleen, and bone marrow. Iron in the RES is either stored in ferritin or redistributed back to transferrin. Most of the iron in circulation comes from the recycling of haemoglobin iron via the RES. Control of iron release by ferroportin (4) from both absorption and recycling is mediated by hepcidin, which binds to ferroportin, inhibiting iron release. Several factors modulate the transcription of the *HAMP* gene to increase (high iron, inflammation) or decrease (low iron, anaemia, hypoxia, erythropoiesis via erythroferrone (ERFE)) hepcidin expression. (**1**) haem-iron transporter; (**2**) duodenal cytochrome b (Dcytb), a ferrireductase, converts $Fe^{3+}$ to $Fe^{2+}$ to facilitate transfer by (**3**) divalent metal transporter (DMT1), which transports non-haem iron ($Fe^{2+}$) into cells; (**4**) ferroportin 1, iron transporter on the basolateral surface of the mucosal cell and the macrophage RES, transports $Fe^{2+}$ out of cells and is regulated by hepcidin; (**5**) hephaestin and (**6**) ceruloplasmin are ferroxidases, converting $Fe^{2+}$ to $Fe^{3+}$ for binding to transferrin; (**7**) transferrin receptor 1 (TfR1) binds with iron-loaded transferrin and transfers both into cells by endocytosis; (**8**) mitoferrin transports iron into the mitochondrion, the site of haem and Fe-S cluster synthesis; (**9**) 'IRON SENSOR' represents the complex interaction of at least ten proteins (including transferrin, HFE, transferrin receptors 1 and 2, and haemojuvalin (HJV)) in and on the surface of the hepatocyte, which sense iron status and modulate the expression of hepcidin and hence the function of ferroportin.

more iron. The incorporation of iron into protoporphyrin IX by ferrochelatase, the first step in haemoglobin synthesis, takes place in mitochondria of the erythroid cell, which means that iron must cross the mitochondrial membrane. A specific transporter, *mitoferrin* (8 in Fig. 9.1), plays a crucial role in achieving this.

It should be noted that most of the iron entering the circulation comes from recycled red cells via the reticuloendothelial system (RES) and not from iron absorption (ratio about 20:1). There is normally only one way into the iron circuit and there is no way out except through blood loss or, in pregnancy, to the foetus. In reality, a small amount of iron is lost,

**Box 9.2 Internal iron exchange**

- Iron in circulation is carried by *transferrin*.
- The amount of iron in circulation is under tight control.
- Humans cannot excrete iron. Surplus iron is stored in cells in *ferritin* or *haemosiderin* (derived from ferritin).
- Most of the iron in circulation goes into making haemoglobin.
- Iron enters the circulation from the breakdown of old red cells by the reticuloendothelial system and from iron absorption.
- Twenty times more iron in circulation comes from the breakdown of red cells than from iron absorption.
- The hormone *hepcidin* reduces iron release from cells by binding to the membrane transporter, *ferroportin*.
- Low hepcidin increases iron release; high levels inhibit iron release.
- The level of hepcidin is controlled by the amount of iron in circulation and by hypoxia, erythropoiesis, and cytokines (IL-6).

mainly through loss of blood and surface cells of the gut, urinary tract, and skin. In men, this amounts to about 1 mg/day and the loss is relatively easily balanced by iron absorption. In women, additional losses through menstruation (0.5 mg/day), and the cost of pregnancy (2 mg/day) and lactation (0.5 mg/day), make it more difficult to balance the loss through iron absorption (see section 9.1.5).

### 9.1.3 Control of iron recycling: circulating iron and hepcidin

Elegant experiments using radio-iron in the 1950s by Bothwell and Finch showed that the body responds to changes in iron requirements by increasing or decreasing iron absorption and iron release from the reticuloendothelial system (RES). The discovery of the iron transporter *ferroportin* and its regulator *hepcidin* has greatly enhanced our understanding of how this is achieved (Fig. 9.1). Hepcidin, a small, 25-amino-acid peptide, is secreted by hepatocytes under the influence of several known mechanisms, and its expression is modulated by changes in the transcription of the *HAMP* gene. Importantly, the production of hepcidin is regulated by the level of iron in circulation, so that when iron is abundant more hepcidin is secreted by the hepatocyte, so limiting further iron absorption and release from stores. When iron is in short supply, the secretion of hepcidin is suppressed, allowing more iron to enter the plasma.

Changes in iron status appear to be sensed, probably through the saturation of transferrin, by two distinct pathways (the 'IRON SENSOR', 9 in Fig. 9.1). Their relationship to each other is not fully understood, but their central role is illustrated by the severe disturbances of iron homeostasis that result from mutations in any of the genes coding for these proteins. The pathway that has been best characterized involves haemojuvalin (HJV), the stability of which is influenced by transferrin saturation. HJV is a co-receptor for a group of growth factors known as bone morphogenic proteins (BMPs). Binding of BMP sets off a cascade of events that results in phosphorylation of intracellular signalling molecules (SMAD), which bind to the *HAMP* promoter to increase hepcidin expression. Disturbances in this pathway lead to severe iron overload early in life.

The other pathway involves the HFE protein, and mutations in the gene (*HFE*) that codes for this protein are responsible for the majority of cases of hereditary haemochromatosis (HH) (see section 9.6). HFE, transferrin receptor 1 (TfR1), and its homologue TfR2 are known to be closely associated on the cell surface and, through binding of circulating transferrin to TfR2, appear to be able to sense iron status and modulate *HAMP* transcription accordingly by an unknown mechanism. The result of both systems is a change in the rate of transcription of *HAMP*, the gene responsible for encoding hepcidin. Hepcidin, released into the circulation, in turn binds to ferroportin, causing the iron transporter to be internalized and degraded, thus inhibiting iron release. This feedback mechanism causes increased iron release when iron is scarce or in high demand and switches off iron release when it is plentiful.

It has long been known that inflammation has a profound effect on iron metabolism. Again, Bothwell and Finch showed that inflammation induced by injection of turpentine resulted in hypoferraemia (low plasma iron), impaired iron absorption, and impaired iron transport to the foetus. It is now understood that these effects are mediated through increased secretion of hepcidin and thus impaired iron release by ferroportin. As part of the immune response several cytokines, particularly IL-6, increase transcription of *HAMP*, independently of the iron sensor mediated pathway. Injection of IL-6 results in a rapid rise in hepcidin and a profound fall in plasma iron. If this is sustained, the resulting iron starvation inhibits erythropoiesis (red cell production) and leads to anaemia. While other factors, such as shortened red cell survival and direct inhibition of the bone marrow, contribute to the anaemia of inflammation (anaemia of chronic disorders), hepcidin-induced hypoferraemia is the central cause. The advantage to the body of the hypoferraemia induced by inflammation appears to be the limitation of iron supply to an invading organism.

Other mechanisms have been shown to inhibit *HAMP* transcription, and therefore hepcidin expression, leading to increased iron release by ferroportin (Fig. 9.1). These operate in situations where erythropoiesis would benefit from an increase in iron supply. Factors operating in this manner include hypoxia operating directly through hypoxia inducible transcription factor (HIF) and erythropoiesis through the recently identified hormone *erythroferrone* (ERFE).

### 9.1.4 Control of iron recycling: intracellular iron

While hepcidin controls the entry of iron into the circulation, and hence its distribution, the level of intracellular iron within the cytosol of the cell controls the expression of some important proteins involved in the movement of iron in and out of cells. Control is largely at the level of the translation of messenger RNA (mRNA) to protein, although iron-related changes in transcription also occur. Control of translation is mediated by iron-responsive proteins (IRP1 and IRP2), which bind to specific stem loop structures, called iron-responsive elements (IRE) of the mRNA. In the iron-deficient state, the IRP binds to an IRE in the 5′ untranslated region of some mRNAs (e.g. ferritin and ferroportin), inhibiting translation, and to IREs in the 3′ untranslated region of other mRNAs (e.g. transferrin receptor and DMT1), stabilizing the mRNA. In the presence of iron, the binding is lost, and ferritin and ferroportin mRNA are translated, while the mRNA of transferrin receptor and DMT1 becomes unstable and cannot be translated. The way in which the two IRPs sense iron levels differs. IRP1 contains an iron sulphur cluster (Fe-S) that, when iron is plentiful, changes the shape of the protein, facilitating binding to the IRE while IRP2, which lacks an Fe-S cluster, is rapidly degraded. When iron is scarce, IRP1 opens and is no longer able to bind to the IRE, while IRP2 becomes stable. The result of these complex reciprocal arrangements is that when the level of iron is low, more transferrin receptor and more DMT1 are translated, facilitating the movement of iron into the cell, while when the level of iron in the cell rises, the need for iron storage is met by increased translation of ferritin mRNA, and iron export is facilitated by increased ferroportin translation.

### 9.1.5 Iron balance

Iron requirements must be balanced by iron supply if iron deficiency or iron overload are to be avoided. Several factors combine to influence iron balance (Fig. 9.2). Obligatory iron losses, the requirements of growth and pregnancy, as well as pathological losses due to excessive menstrual and other bleeding must be balanced against iron supply. Iron supply is influenced by the amount and type of iron in food and the combination of various inhibitors and promoters of iron bioavailability. These requirements are buffered by iron, which can be mobilized from stores. The body's iron requirements and the ability of the diet to meet the demand vary during life (Fig. 9.3).

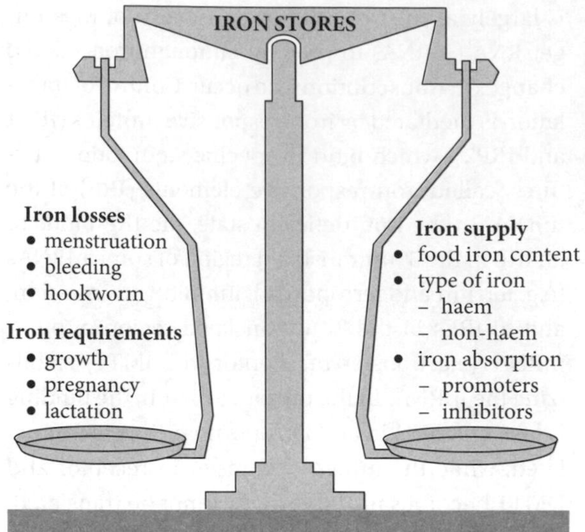


**Fig. 9.2** Iron losses and requirements for growth (left) are balanced by iron supplied in the diet (right). Surplus iron is stored and can be drawn upon to supplement increased losses or requirements.

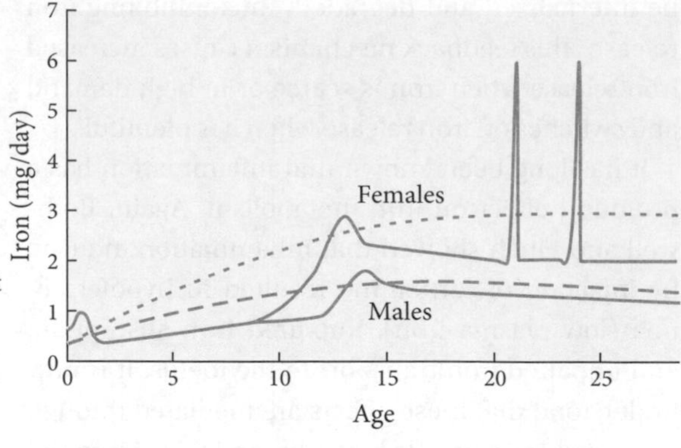


**Fig. 9.3** Iron requirements for males and females vary during life. A Western diet, rich in meat and iron promoters (----), is able to meet iron requirements of the majority of females at all ages except in infancy, at the peak of pubertal growth spurt and onset of menstruation, and during pregnancies. In contrast, a cereal-based diet, without meat or iron promoters and with excess inhibitors of iron absorption (- - -), is not able to meet the requirements of most childbearing women and some men.

In infancy, during the pre-adolescent growth spurt, and in women during reproductive life, particularly during pregnancy, the iron requirements may exceed the iron supply, making iron deficiency more common during these periods. Individuals consuming a diet of low iron bioavailability are even more at risk.

## 9.2 Iron in food

Iron supply is greatly influenced by the composition of the diet. Two broad categories of iron are present in food: *haem iron* derived mainly from haemoglobin and myoglobin in meat, and *non-haem iron* in the form of iron salts, iron in other proteins (e.g. ferritin), and iron derived from processing or storage methods. Haem iron enters the mucosal cells by a different mechanism and is better absorbed than non-haem iron. It is also less influenced by the body's iron status and, because the iron is protected by the haem molecule, it is not affected by other constituents in the diet. Not surprisingly, the relative proportion of these two categories in the diet has a profound influence on iron status.

Non-haem iron compounds are found in a wide variety of foods of both plant and animal origin. The iron is present in metalloproteins (e.g. ferritin, haemosiderin, and lactoferrin), soluble iron, iron bound to phytates in plants, and contaminant iron such as ferric oxides and hydroxides introduced in the preparation and storage of food and by contamination from soil. The bioavailability of these forms of non-haem iron, unlike haem iron, is influenced by other constituents of the diet. Forms of iron that are similarly affected are said to enter a 'common pool'. The importance of this concept is that fortificant iron added to food will be subjected to the same inhibitory and promotive influences as the intrinsic food iron and will, therefore, have similar bioavailability. This concept, however, does not hold true for all forms of added iron. For example, most ferric salts, whether contaminant or added as fortificants, do not enter the common pool and have very low bioavailability, while chelated iron (e.g. NaFeEDTA) is protected from inhibitors and has a higher availability than the native iron.

## 9.3 Promoters and inhibitors of iron absorption

The relative concentrations of promoters and inhibitors of iron absorption in foods are responsible for the wide range of bioavailability that has been demonstrated (Table 9.1). The most important *promoters* of non-haem iron absorption are ascorbic acid (vitamin C) and meat. Other organic acids (e.g. citric acid) and some spices have also been shown to enhance iron absorption. The major *inhibitors* of iron absorption are phytates and polyphenols, common constituents of cereals and many vegetables.

Ascorbic acid is thought to enhance iron absorption by converting ferric to ferrous iron and by chelating iron in the lumen of the gut. This keeps iron in a more soluble and absorbable form and prevents binding to inhibitory ligands. It follows that the bioavailability of non-haem iron from foods with significant ascorbic acid content is high. Moreover, the addition of ascorbic acid to meals with potent inhibitors increases non-haem iron absorption.

The factor in meat responsible for enhancing non-haem iron absorption has not been identified. The enhancing effect is not shared by proteins derived from plants or other animal proteins such as milk, cheese, and eggs. For example, substitution of beef for egg albumin as a source of protein in a test meal resulted in a five-fold increase in iron absorption. Present evidence suggests that peptides rich in the amino acid cysteine may play a role in the enhancement of non-haem iron absorption.

Polyphenols commonly found in many vegetables and in some grains are potent inhibitors of non-haem iron absorption. There is a strong inverse relationship between the concentration of polyphenols in foods and the absorption of iron from them. Many of the foods with low iron bioavailability listed in Table 9.1 are rich in polyphenols. Among the best-known polyphenols is tannin, found in tea and other beverages, which has a profound inhibitory effect on iron absorption. Polyphenols also form strongly coloured compounds with iron, which is a major problem in food fortification. This phenomenon can be illustrated by dropping a few crystals of ferrous sulphate into a cup of tea.

Phytates, found mainly in the husks of grains, are also major inhibitors of iron absorption. In this regard, iron absorption from unpolished rice is significantly worse than from polished rice, while increasing the bran content of a meal produces a dose-related depression in iron bioavailability. Both meat and ascorbic acid are able to overcome this inhibitory effect.

**Table 9.1 Relative bioavailability of non-haem iron in foods**

| Food | Bioavailability | | |
|---|---|---|---|
| | **Low (5%)** | **Intermediate (10%)** | **High (15%)** |
| Cereals | Most cereals<br>Whole wheat flour | Corn flour<br>White flour | |
| Fruits | Apple<br>Banana<br>Peach<br>Strawberry | Cantaloupe<br>Mango | Guava<br>Citrus<br>Pawpaw<br>Tomato |
| Vegetables | Aubergine<br>Legumes | Carrot<br>Potato | Beetroot<br>Broccoli<br>Pumpkin |
| Beverages | Tea<br>Coffee | Red wine | White wine |
| Nuts | All | | |
| Animal proteins | Cheese<br>Egg<br>Cow's milk | | All meat<br>Poultry<br>Breast milk |

## 9.4 Recommended dietary intake

The concept of a recommended dietary intake of iron is difficult to reconcile with the wide range of bioavailability (Table 9.1 and Box 9.3). The total iron content of a diet is a meaningless, although commonly employed, measure of its nutritional adequacy, and may provide a false sense of nutritional security. For example, foods with high iron content due to large quantities of contaminant iron, or inappropriate fortificant iron, may be nutritionally worthless because of the low bioavailability of the iron. On the other hand, haem iron, making up only 10–15% of the total ingested iron, may account for a third of the iron actually absorbed. Similarly, the iron absorbed from a meal containing non-haem iron may be doubled if the meal is taken with a glass of orange juice (30 mg ascorbic acid) or reduced to a third if taken with tea. However, it is possible to divide diets into ones of low, intermediate, and high iron bioavailability. These correspond to iron absorption of about 5%, 10%, and 15% in subjects with depleted iron stores.

A diet of low bioavailability (<5%) with a high inhibitor content, negligible amounts of enhancers, and little haem iron is based largely on unrefined cereals and legumes. Such diets are typical of many developing countries and supply about 0.7 mg of available iron daily, which is insufficient to meet the needs of most women, growing children, and some men. A diet of intermediate bioavailability (about 10%) includes limited amounts of foods that promote iron absorption and supplies enough absorbed iron (about 1.5 mg) to meet the needs of 50% of women. A diet of high iron bioavailability (>15%) contains generous amounts of food rich in promoters and haem iron. The inhibitor content is low as cereals are often highly refined. Such diets, typical of many developed countries, supply sufficient iron (>2.1 mg/day) for most adults, but still cannot match the daily amounts of absorbed iron required in the second half of pregnancy (5 mg/day).

The Food and Agriculture Organization/World Health Organization (FAO/WHO) recommendations (Table 9.2), based on estimates that apply to 97.5% of the population, are an attempt to take variations in bioavailability and requirements into account.

## 9.5 Iron deficiency

In the past, iron deficiency was thought to be due largely to blood loss, rather than insufficient iron supply. Credence for this view was given by the obvious effects of pathological blood loss and the high prevalence of iron deficiency in the developing world, where hookworm infestation is endemic. However, it is now apparent that the poor bioavailability of iron in largely unrefined, cereal-based diets is a major cause of iron deficiency in most developing countries. The impact of such diets is obviously enhanced when pathological blood loss or increased physiological iron demand is also present. These factors explain the geographical and gender variation in the prevalence of iron-deficiency anaemia (IDA), which is most common in Asia, where up to 60% of women and over 30% of men are anaemic, while in

**Box 9.3 The milk paradox: breast is best**

- Although human milk contains less iron than cow's milk, the iron in human milk is much more bioavailable (estimated absorption up to 50%). The US Institute of Medicine considers that the iron provided by exclusive breastfeeding is adequate to meet the needs of infants up to the age of 6 months.

Table 9.2 **FAO/WHO recommended daily iron intake for individuals consuming diets of low, intermediate, and high iron bioavailability**

| Group | Age (years) | Total absolute requirements (median) (mg/day) | Recommended iron intake for diets of different bioavailability (mean + 2 SD) (mg/day) | | |
|---|---|---|---|---|---|
| | | | Low (5%) | Intermediate (10%) | High (15%) |
| Children | 0.5–1 | 0.72 | 19 | 8 | 6 |
| | 1–3 | 0.46 | 12 | 6 | 4 |
| | 4–6 | 0.50 | 13 | 6 | 4 |
| | 7–10 | 0.71 | 18 | 9 | 6 |
| Males | 11–14 | 1.17 | 29 | 15 | 10 |
| | 15–17 | 1.50 | 38 | 19 | 13 |
| | 18+ | 1.05 | 27 | 14 | 9 |
| Females | 11–14[a] | 1.20 | 28 | 14 | 9 |
| | 11–14 | 1.68 | 65 | 32 | 22 |
| | 15–17 | 1.62 | 62 | 31 | 21 |
| | 18+ | 1.46 | 59 | 29 | 20 |
| Post-menopausal | | 0.87 | 23 | 11 | 8 |
| Lactating | | 1.15 | 30 | 15 | 10 |

[a]Premenstrual.

*Source:* Modified from FAO/WHO (2004) *Vitamin and mineral requirements in human nutrition*, 2nd edn.

Europe and North America less than 5% of females and 2% of males are anaemic. The preponderance of females can be explained by increased physiological loss of iron in menstruation and pregnancy, and by their lower food and, therefore, iron intake.

The development of iron deficiency is characterized by sequential changes in the amount of storage iron in the various iron compartments of the body. The measurement of iron in these compartments is discussed in section 9.8. In the first stage, iron stores become depleted, but there is enough iron to meet the needs of red cell production. When iron stores are exhausted, the amount of iron in the circulation starts to fall, and red cell production becomes compromised (iron-deficient erythropoiesis). In the final stage, iron stores are exhausted, the amount of iron in the circulation is very low, red cell production is drastically reduced, and anaemia develops. The point at which the function of iron-containing enzymes becomes impaired is uncertain, but probably depends on the rate of renewal of the enzymes and the growth of the tissues involved.

While blood loss and the diet are of fundamental importance in causing iron deficiency, other factors need to be considered. Gastrointestinal diseases, not necessarily causing bleeding, may hinder iron absorption and are important causes of IDA refractory to oral iron therapy. Atrophic gastritis and gastrectomy reduce acid output vital for iron solubilization and functioning of DMT1 (see section 9.1.1). Coeliac disease, associated with antibodies directed against small intestinal epithelium, is a cause of iron malabsorption. Recent work has shown that eradication of *Helicobacter pylori* infection improves response to oral iron. As the genetics of more proteins of iron metabolism are unravelled, inherited causes of iron deficiency in humans are beginning to emerge. Hereditary iron-refractory IDA is due to mutations in the gene coding for matripase-2 (*TMPRSS6*), part of the HJV pathway (see section 9.1.3), which result in increased hepcidin and inhibition of iron recycling. Mutations in DMT1 causing severe congenital iron deficiency because of impaired membrane iron transfer, previously described in mice and rats, has now been found in humans (3 in Fig. 9.1).

Iron deficiency has been associated with a number of pathological consequences of which anaemia is the most obvious. Severe anaemia is associated with weakness, impaired effort tolerance, and, eventually, heart failure. There is no doubt that even mild IDA limits work performance and studies in Indonesia and Sri Lanka have linked it to reduced productivity. In children, IDA, particularly in infancy, is associated with impaired psychomotor development, the effects of which may be irreversible. Although well-designed trials are lacking, the effect of iron supplementation in infancy on cognitive function is uncertain. In pregnancy, IDA is associated with prematurity, low birth weight, and increased perinatal mortality, as well as an increased risk of iron deficiency in the infant after 4 months of age. Changes in the gastrointestinal tract (atrophy of the mucosa of the mouth, oesophagus, and stomach) and the skin and nails (spoon-shaped nails) are well described but infrequent. Other, less well-recognized abnormalities include inability to adapt to cold and impaired immunity.

## 9.6 Iron overload

Excessive amounts of iron may accumulate in the body and result in organ damage. The *acute* ingestion of a large amount of bioavailable iron, usually in the form of ferrous sulphate tablets, will exceed both the ability of the mucosa to control iron absorption and the capacity of transferrin to bind iron in the circulation. The acute iron toxicity that results is thought to be due to the generation of free radicals by free iron, both in the gut and in the circulation. Most of the victims are children, who develop severe abdominal pain, vomiting, metabolic acidosis, and cardiovascular collapse. Severe poisoning, requiring urgent chelation therapy, may follow ingestion of as little as 30 mg of iron/kg.

*Chronic iron overload* develops insidiously and the recent discovery of novel genes coding for proteins involved in iron sensing and transport (see section 9.1) has greatly increased our understanding of iron overload. *Haemochromatosis* is the name given to inherited forms of iron overload, in which abnormalities of hepcidin and ferroportin function play a central role (Table 9.3). Type 1 is the most common form and is found almost exclusively in people of north-western European origin, reaching a homozygous prevalence of 1.2–1.4% in Ireland and Denmark. Types 1–3 share common features, but differ in severity and age of onset. In these types, iron floods the circulation because of uninhibited release of iron by ferroportin due to the absence of hepcidin (see section 9.1; Fig. 9.1). Transferrin becomes saturated with iron and free, toxic, non-transferrin-bound iron (NTBI) is rapidly taken up by the liver, heart, pancreas, and other organs, where excessive iron accumulates over years. The damage to these organs is due to free radicals induced by iron and can result in liver cirrhosis, liver cancer, heart failure, arthritis, and endocrine disease (diabetes and impotence). In addition, large deposits of ferric iron cause oxidation of vitamin C, which can lead to scurvy and osteoporosis. Removal of haemoglobin by repeated bleeding is an effective treatment because iron stores have to be mobilized to replace the lost haemoglobin. Interestingly, recent epidemiological studies have shown that many people homozygous for the common mutation of type 1 haemochromatosis (C282Y) do not go on to develop the full-blown clinical features, possibly because more than one iron-sensing pathway is available (see section 9.1.3). Also of interest is the possible selective advantage of increased iron absorption and good iron stores, seen even in heterozygous C282Y carriers, at the start of pregnancy (see section 9.7).

Two other clinically important forms of iron overload, sometimes called secondary, should be mentioned. *African dietary iron overload* is caused by the ingestion, over many years, of large amounts of

Table 9.3 **Classification of haemochromatosis**

| Type | Protein (see Fig. 9.1) | Gene (common mutation) | Clinical picture |
|---|---|---|---|
| 1 Classic (adult) | HFE (9, iron sensor) | *HFE* (C282Y in >80%) | Hepcidin deficiency. Recessive inheritance. Age usually over 40. Severe iron overload with organ damage (liver, heart, pituitary). Variable expression. Parenchymal iron distribution (hepatocytes, cardiac muscle, endocrine tissues) with little or no iron in reticuloendothelial tissues (spleen, bone marrow). High transferrin saturation (>50%), high serum ferritin (can be thousands μg/L). |
| 2 Juvenile | A Haemojuvelin (9, iron sensor) | *HJV* | Recessive inheritance. Presents before age 30 with massive iron overload, cardiac failure, and central hypogonadism. Rare. |
| | B Hepcidin | *HAMP* | |
| 3 Non-HFE (adult) | Transferrin receptor 2 (9, iron sensor) | *TFR2* | Recessive inheritance. Clinical picture between types 1 and 2. |
| 4 Dominant | Ferroportin (4) | *SLC40A1* | Dominant inheritance. Variable effects from severe to trivial. Type A: ferroportin cannot export iron. Normal to low transferrin saturation. Iron 'trapped' in reticuloendothelial cells (spleen, Kupfer cells, bone marrow). Type B: impaired ferroportin binding to hepcidin leads to picture similar to type 1. |

highly bioavailable iron in low-alcohol traditional beer brewed in iron containers. Recent evidence suggests that there is also a genetic predisposition, possibly related to the ferroportin gene, since the distribution of iron is the same as that seen in haemochromatosis type 4A. In severe cases, the toxic effects of iron are similar to those seen in the other forms of haemochromatosis. Scurvy and osteoporosis due to oxidation of vitamin C by large deposits of iron was first described in this condition. *Secondary iron overload* occurs in the so-called 'iron-loading anaemias,' of which thalassaemia major is the most common. Excessive amounts of iron are absorbed over a relatively short period because of the inhibition of hepcidin expression caused by the increased turnover of red cells (ERFE in Fig. 9.1). This effect can be suppressed by regular blood transfusion, while the iron burden can be reduced by iron chelation therapy augmented by vitamin C supplementation. The iron overload occurs rapidly and most victims die from iron-induced heart failure if not treated. Patients with bone marrow failure, kept alive by repeated blood transfusions, may similarly develop iron overload.

## 9.7 Iron in pregnancy

The cost of pregnancy, in terms of iron, is considerable. The iron transferred to the foetus (about 270 mg) and the placenta (about 90 mg) is greater than that found in the iron stores of an adult woman. To this must be added the cost of the expansion of the mother's red cell mass (about 450 mg) and the normal obligatory iron loss during pregnancy (about 230 mg). In all, this is more than the body iron stores of an adult male. To keep pace with this there is an increase in iron absorption but, in the second and third trimesters, the daily iron requirement (5–6 mg/day) greatly exceeds the amount that can be absorbed even from a high-iron diet (see Fig. 9.3). The balance must come from the mother's iron stores.

It follows that a woman with no or low iron stores at the onset of pregnancy will be iron deficient at term. The consequences of iron deficiency during pregnancy are serious for both mother and child. The risk of maternal haemorrhage is greater, as is the risk of foetal loss, prematurity, small infant size, and abnormal cognitive development. While iron supplementation during pregnancy may be avoided in iron-replete women in the developed world consuming diets of high iron bioavailability, there is little room for such complacency in the developing world. The WHO guidelines (2012) recommend daily oral iron supplementation but the UK NICE guidelines (2016) actively discourage routine iron supplementation, reserving it for women with established anaemia. This is likely to be too little iron too late and a more sensible trigger for supplementation would be the detection of low iron stores (serum ferritin <20 μg/L) early in pregnancy. The possibility that iron supplementation may increase susceptibility to malaria in mother and neonate is probably overstated and the consensus is that the known benefits of supplementation outweigh the risks.

The changes in iron metabolism during pregnancy can be largely anticipated from the general model described in section 9.1, although some aspects remain poorly understood. In humans, and most other mammals with a haemochorial placenta, maternal iron, from both stores and absorption, is delivered to the placenta via the transferrin-transferrin receptor mechanism seen in erythropoiesis (section 9.1.2). Pregnancy can be described as an 'iron deficient state' so it is not surprising that maternal transferrin saturation and plasma hepcidin are low, so encouraging ferroportin iron release and facilitating iron absorption and mobilization of iron stores. As in other iron-exporting tissues, iron leaves the placenta via ferroportin to enter the foetal circulation and is rapidly taken up by the liver, the site of foetal and neonatal erythropoiesis. Hepcidin is present in the foetal circulation but appears to have no effect on placental ferroportin, explaining why foetal and neonatal transferrin saturation is commonly greater than 60%, as in haemochromatosis. Foetal hepcidin does, however, appear to facilitate the rapid build-up of foetal liver iron stores in the last trimester, thus ensuring sufficient iron in early neonatal life when iron demand is high and supply is limited.

## 9.8 Assessment of iron status

The iron status of an individual, or a population, can be gauged by measuring the amount of iron in each of the body iron compartments. The progression from normal through to IDA can be measured by sequential changes in the biochemical markers that reflect the iron status of each compartment. No single biochemical index can assess all the stages of iron depletion. Fig. 9.4 shows how different measurements of iron status relate to these stages and the movement of iron between major compartments.

- During *iron depletion* (as with blood loss), storage iron in the liver and in the reticuloendothelial cells of the spleen and bone marrow is progressively reduced and this can be detected by a parallel fall in serum ferritin concentration. The function of circulating ferritin, which carries almost no iron, is not known, but there is a near-linear relationship between iron stores and serum ferritin. The serum ferritin concentration falls below 12 μg/L (12 ng/mL) when the iron stores are exhausted. Iron depletion is the only cause of a serum ferritin below this level. Other measurements of iron status are normal at this stage.
- In *iron-deficient erythropoiesis*, iron stores are exhausted (serum ferritin <12 μg/L) and iron supply to the marrow is insufficient to meet the needs of haemoglobin production. This stage is detected by low serum iron and serum ferritin concentrations and a transferrin saturation below 16%, although the haemoglobin concentration is still within the normal range.

| Iron Compartments | Normal | Depleted Stores | Iron-deficient Erythropoiesis | Iron Deficiency Anaemia (IDA) | Anaemia of Chronic Disorders (ACD) | Haemochromatosis Types 1–3 |
|---|---|---|---|---|---|---|
| IRON STORES CIRCULATING IRON RED CELLS | | | | | | |
| **Measurements** | | | | | | |
| Bone marrow iron | Normal | Absent | Absent | Absent | Increased | Normal |
| Serum ferritin | 12–400(300[a]) μg/L | Low | Low | Low | Normal or increased | Greatly increased |
| Serum iron | 10–30 μmol/L | Normal | Low | Low | Low | Greatly increased |
| Transferrin saturation | 16–45% | Normal | Low | Low | Normal or low | Greatly increased |
| Haemoglobin | >130(>120[a]) g/L | Normal | Normal | Low | Low | Normal |
| Serum transferrin receptor (sTfR) | 1.8–4.6 mg/L | Normal | Increased | Increased | Normal or low | Normal |
| Free erythrocyte protoporphyrin (FEP) | 0.28–0.64 μmol/L RBC | Normal | Increased | Increased | Increased | Normal |
| Serum hepcidin | Assay dependent[b] | Normal | Low | Low | Increased | Low |

Fig. 9.4 Changes in measurements of iron status in various conditions.
*Adapted from: Iron metabolism in man.* Bothwell, T.H., Charlton, R.W., Cook, J.D., and Finch, C.A. (Oxford: Blackwell Scientific Publications, 1979).

- In *iron-deficiency anaemia*, the supply of iron to the marrow is so reduced that the concentration of haemoglobin falls below normal. The cut-off value below which anaemia is diagnosed varies according to age and gender (<110 g/L in children younger than 6 years and in pregnant women; below 120 g/L in women and adolescents under 15 years; and <130 g/L in adult men). There are obviously many other causes of anaemia besides iron deficiency (e.g. vitamin $B_{12}$ and folate deficiency, chronic infection, and intrinsic diseases of the bone marrow). The diagnosis of anaemia due to iron deficiency therefore requires that other measurements of iron status are also in the iron-deficient range (serum ferritin <12 μg/L and transferrin saturation <16%). In addition, in established IDA, the red cells become small (microcytosis) and pale (hypochromia). These changes can be detected by examination of a blood film or by a fall in the mean cell volume (MCV) below 85 fL and in the mean cell haemoglobin concentration (MCHC) below 27 pg.
- The diagnosis of *iron overload* is also measured by a combination of measurements of iron status.

A raised serum ferritin concentration (>400 μg/L) and transferrin saturation greater than 60% are highly suggestive of types 1–3 haemochromatosis (Table 9.3). As a rough guide, 100 μg/L of serum ferritin is equivalent to a gram of storage iron and the normal range is usually given as 50–400 μg/L in men or 50–300 μg/L in women. In haemochromatosis, serum ferritin levels greater than 1000 μg/L (iron stores ten times normal) are associated with organ damage. Unfortunately, circulating ferritin is an acute phase reactant and increases when inflammation is present, making serum ferritin an unreliable estimate of iron stores in this situation. Some of this increase is a reflection of iron trapped in stores through the action of increased levels of hepcidin on ferroportin (see section 9.1.3). In addition, tissue destruction is associated with high levels of serum ferritin because of the release of tissue ferritin into the circulation. Iron stores can also be measured by chemical estimation of iron in biopsies of liver or bone marrow or by electromagnetic resonance. Very high levels of serum iron are seen in situations where hepcidin is low or absent (haemochromatosis) and in acute iron poisoning, in which case the capacity for transferrin to hold iron becomes impaired (at 60% to over 100%) and free non-transferrin-bound iron (NTBI) becomes detectable in the plasma (see section 9.6).

- The changes in the measurements of iron status that often accompany inflammation (see section 9.1.3) are often confused with iron deficiency. In fact, these changes represent a shift of iron from the red cell compartment to stores caused by inhibition of iron release by hepcidin, while the total amount of iron in the body is actually unchanged. The serum iron is low (*hypoferraemia of inflammation*), while the transferrin saturation may be low or normal. In prolonged inflammation, the haemoglobin falls and the red cells may develop changes suggestive of IDA. Because iron is trapped in stores, the serum ferritin is high and is out of keeping with the low serum iron. This high serum ferritin is not indicative of iron overload. Similar changes have been described in obesity (*hypoferraemia of obesity*), but whether this is due to inflammation alone is not clear. The diagnosis of iron deficiency in the presence of inflammation is often difficult. A hint that iron deficiency may also be present is provided by a serum ferritin in the low normal range, since the total iron is diminished. The only sure way to resolve this is to assess iron stores and erythropoiesis in the bone marrow.

Other tests that may be used to infer iron status rely on the levels of proteins of iron metabolism that are controlled by iron itself. A neglected and simple test that can be done on a drop of blood is the concentration of *free erythrocyte proto-porphyrin* (FEP), which gives information similar to the transferrin saturation. The production of haem is dependent on the supply of iron to the marrow. When iron supply is restricted, free protoporphyrin, a precursor of haem, accumulates in red cells. A transferrin saturation lower than 16% is associated with an FEP greater than 1.24 μmol/L red cells. The FEP is, however, also elevated in lead poisoning and in inflammation.

Tissue iron depletion can also be measured by the concentration of soluble *transferrin receptor* in plasma or serum. The expression of transferrin receptors on the surface of all cells is determined by the level of intracellular iron (see section 9.1.4). In cellular iron depletion, the concentration of soluble transferrin receptor in the plasma rises but, unlike the serum ferritin concentration, the level is less affected by inflammation. The ratio of serum transferrin receptor to log serum ferritin may be useful in distinguishing iron deficiency from inflammation as the cause of anaemia. The recent development of methods to measure the concentration of hepcidin in serum, which is high in inflammation and low in iron deficiency (section 9.1.3), may prove helpful in this regard. Serum hepcidin is undetectable in haemochromatosis types 1 to 3 (Table 9.3).

## 9.9 Treatment and prevention of iron deficiency

Iron-deficiency anaemia in individuals is best treated by the oral administration of ferrous iron salts. The cheapest and most effective is ferrous sulphate, which is usually given in a dose of one tablet (65 mg of iron) two or three times a day. The increase in haemoglobin concentration that can be expected with optimal doses of ferrous sulphate is about 2 g/L/day. Gastrointestinal side effects of oral iron therapy are common, which has led to a plethora of different oral iron compounds being available. Most differ in their formulation in an attempt to limit side effects, the most popular being slow-release preparations. None has been shown to be convincingly better than ferrous sulphate and all will correct iron deficiency in time. The addition of ascorbic acid, while enhancing food iron availability, does little to improve therapeutic efficacy and probably increases the side effects. In persons intolerant of, or refractory to, oral iron therapy (see section 9.5), it is possible to give iron by intravenous injection.

The treatment and prevention of iron deficiency in high-risk groups and the population at large presents many problems. *Supplementation* refers to the administration of iron compounds in the form of tablets or syrups usually targeted to high-risk groups such as children and pregnant women. Again, the major problem encountered is the high prevalence of side effects and its impact on compliance. Giving smaller or less frequent (weekly) doses prolongs the time to response considerably, while supply and distribution limit efficacy. Food *fortification* offers a more cost-effective approach through the fortification of staples such as cereals or condiments (salt, fish sauce, or soy sauce). Target groups (such as infants and children) can be reached by iron fortification of infant formulas and cereals. While there is abundant evidence that iron fortification of food is effective, the process is fraught with practical difficulties. Apart from the logistic problems in the manufacture of iron-fortified food and its distribution, so important in the developing world, the major dilemma is that bioavailable iron compounds are highly reactive and cause unacceptable changes in the taste and colour of the food vehicle. Insoluble salts, such as ferric orthophosphate and elemental iron powders, while producing fewer changes in food, are poorly absorbed. Strategies that have been employed to overcome these barriers include encapsulating ferrous salts, reducing the phytate content of cereals, and adding a promoter to cereals fortified with a ferric salt. This last has been used successfully in the fortification of infant formulas and cereals with iron and ascorbic acid, but is only effective when oxidation of the ascorbic acid can be prevented by keeping the product dry, avoiding heat and exposure to air until shortly before use.

The iron chelate, sodium ferric EDTA (NaFeEDTA), escapes the effects of inhibitors, particularly phytates, and, in this setting, the iron is two to three times better absorbed than from ferrous sulphate. Judicious choice of the food vehicle can avoid unwanted colour changes. Field trials have shown NaFeEDTA to be an effective fortificant. In one study, the prevalence of iron deficiency in women was reduced from 22% to 5% over a 2-year period. Meta-analyses of fortification programmes using NaFeEDTA fortified condiments, particularly in China, have confirmed its efficacy. Similar claims have been made regarding the amino acid chelate ferrous bisglycinate, which has the advantage of being considered a natural compound and as 'generally recognized as safe' (GRAS) by the FDA, but is more costly. Despite these limitations, fortification of food with iron is commonplace in the industrialized world. Paradoxically, these fortification programmes have often been carried out in a haphazard way, with little attention being paid to the bioavailability of the fortificant or the efficacy of the programme. Elemental iron, widely used as a fortificant in Western countries, has a very low bioavailability that varies with particle size. In the developing world, iron fortification programmes face additional problems. Foods are seldom centrally processed, making fortification a difficult logistical problem. Furthermore, the diets are largely cereal-based and lack natural promoters of iron absorption, which means that the added iron will be poorly absorbed. These factors have been extensively reviewed by SUSTAIN (Bothwell and Lynch, 2004), while the Micronutrient Forum (Raiten et al., 2016) has addressed their increasingly complex interaction with disease burden, and social and cultural determinants.

## Further reading

1. **Anderson, G.J. and Frazer, D.M.** (2017) Current understanding of iron homeostasis. *Am J Clinical Nutr*, **106**(Suppl), 1559S–66S.
2. **Anderson, G.J., Frazer, D.M., and McLaren, G.D.** (2009) Iron absorption and metabolism. *Curr Opin Gastroenterol*, **25**, 129–35.
3. **Anderson, C.P., Shen, M., Eisenstein, R.S., and Leibold, E.A.** (2012) Mammalian iron metabolism and its control by iron regulatory proteins. *Biochim Biophys Acta*, **1823**(9), 1468–83.

4. **Bothwell, T.H.** (2000) Iron requirements in pregnancy and strategies to meet them. *Am J Clin Nutr*, **72**, 257S–64S.
5. **Bothwell, T.H. and Lynch, S.** (Guest Editors) (2004) Special issue: Innovative ingredient technologies to enhance iron absorption. Proceedings of a SUSTAIN workshop, Washington D.C., March 9–12, 2003. *Int J Vitam Nutr Res*, **74**, 385–466.
6. **Bothwell, T.H and MacPhail, A.P.** (2004) The potential role of NaFeEDTA as an iron fortificant. *Int J Vitam Nutr Res*, **74**, 421–34.
7. **Camaschella, C.** (2015) Iron-deficiency anemia. *N Engl J Med*, **372**, 1832–43.
8. **Camaschella, C., Nai, A., and Silvestri, L**. (2020) Iron metabolism and iron disorders revisited in the hepcidin era. *Haematologica*, **105**(2), 260–72.
9. **Ganz, T. and Nemeth, E.** (2012) Hepcidin and iron homeostasis. *Biochim Biophys Acta*, **1823**, 1434–43.
10. **Keats, E.C., Neufeld, L.M., Garrett, G.S., Mbuya, M.N.N., and Bhutta, Z.A.** (2019) Improved micronutrient status and health outcomes in low- and middle-income countries following large-scale fortification: evidence from a systematic review and meta analysis. *Am J Clin Nutr*, **109**(6), 1696–708.
11. **Lane, D.J.R. and Richardson, D.R.** (2014) The active role of vitamin C in mammalian iron metabolism: much more than just enhanced iron absorption. *Free Radical Biology and Medicine*, **75**, 69–83.
12. **Raiten, D.J., Neufeld, L.M., De-Regil, L., et al**. (2016) Integration to implementation and the Micronutrient Forum: A coordinated approach for global nutrition. Case study. Application: safety and effectiveness of iron interventions. *Adv Nutr*, **7**, 135–48.
13. **Weiss, G., Ganz, T., and Goodnough, L.T.** (2019) Anemia of inflammation. *Blood*, **133**(1), 40–50.
14. **World Health Organization; Allen, L., de Benoist, B., Dary, O., and Hurrell, R. (eds)** (2006) *Guidelines on food fortification with micronutrients.* Geneva: World Health Organization. Available at: https://apps.who.int/iris/handle/10665/43412 (accessed 11 April 2022).

# Trace Elements



## 10.1 Zinc

Samir Samman

### 10.1.1 Historical perspective

Experimental zinc deficiency was demonstrated in laboratory animals, but the likelihood of deficiency in humans was considered remote because of the ubiquitous nature of zinc in the food supply and the relative difficulty in creating zinc-deficient animal models. Zinc deficiency was first recognized in humans in 1958.

### 10.1.2 Distribution and function

Zinc is one of the IIB series of metals, with a molecular weight of 65.4 (see Fig. 10.1). It is the most common catalytic metal ion in the cytoplasm of cells. Adult humans contain between 1.2 and 2.3 g of zinc, which is distributed in all tissues. The highest concentrations of zinc are observed in the choroid of the eye and the prostate gland, but most of the body zinc is in bones and muscles. In liver cells, zinc is associated with all sub-cellular fractions. The plasma concentration of zinc is approximately 15 μmol/L, of which a third is bound to $\alpha_2$ macroglobulin and the rest to albumin. However, only 10–20% of the zinc in blood is found in the plasma; the rest is in red blood cells, associated mainly with carbonic anhydrase. The red cell membrane contains some zinc. Semen has 100-fold the zinc concentration of plasma.

A number of genes are regulated by zinc through the presence of short DNA sequence motifs known as metal-responsive elements. Some genes are positively affected, others negatively, and some are affected only by extremes of zinc status (deficiency or excess). Some of the genes identified include those involved in the regulation of redox signalling, fatty acid synthesis and degradation, platelet activation, and the regulation of homocysteine concentrations.

Zinc 'fingers' have been identified in the human genome. These are small proteins that have a zinc ion coordinated with a combination of cysteine and histidine residues. The proteins, which may contain up to 30 fingers, interact with DNA, RNA, and other cellular proteins. Thus, they have a widespread role in cellular metabolism.

Zinc also plays a role in stabilizing macromolecules and cellular membranes, and it can function as a site-specific antioxidant. It can bind to or in close proximity to thiol groups of proteins and reduce their reactivity. Zinc is a constituent of a large number of mammalian enzymes (more than 150), where it functions at the active site, or as a structural component, or both. Carbonic anhydrase was the first discovered zinc metalloenzyme; other enzymes include: carboxypeptidase, alkaline phosphatase, transferases, ligases, lyases, isomerases, DNA/RNA polymerase, reverse transcriptase, and superoxide dismutase.

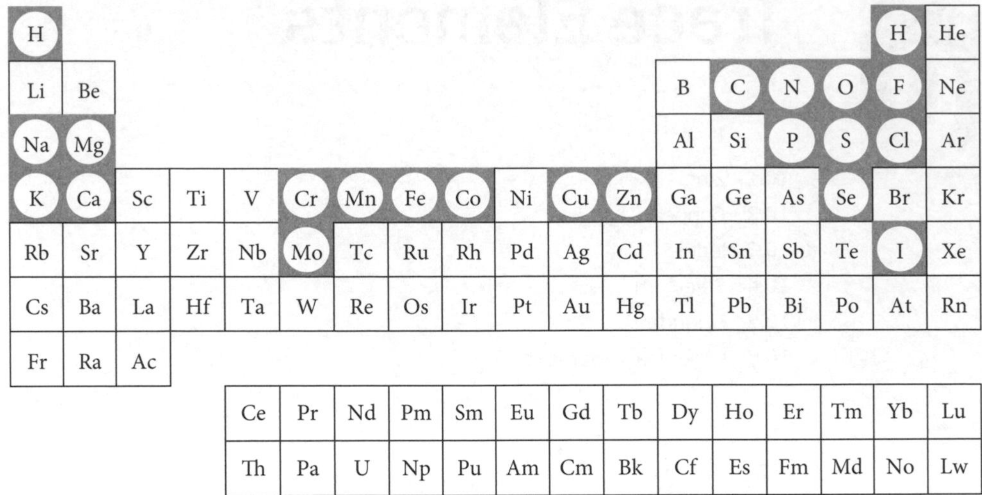


Fig. 10.1 Periodic table of the elements. Those essential for humans are encircled in red. In addition, boron, silicon, nickel, arsenic, and vanadium are still under consideration as ultratrace elements.

Therefore, zinc is involved in a number of major metabolic processes, including protein and nucleic acid synthesis. Zinc is essential for the synthesis and action of insulin. It also helps to stabilize the proinsulin and insulin hexamers by forming complexes with them.

### 10.1.3 Absorption and excretion

Zinc is absorbed mainly from the duodenum, but some is absorbed lower down the small intestine. The mode of absorption involves both saturable and passive mechanisms. A number of zinc transporters have been identified: ZiPs (at least 15 different ones) promote zinc influx into cells, while ZnTs (ZnT1–ZnT9) promote efflux across membranes. The distribution of different ZnTs and ZiPs is tissue specific. The exact site of zinc absorption depends on the form of zinc and the presence or absence of other nutrients that may form complexes with zinc or impact on intestinal transit time. Once absorbed, zinc is transported to the liver bound to albumin.

The major route of zinc excretion is by the intestine, followed by the kidneys and the skin. Faecal zinc originates from unabsorbed dietary sources, as well as zinc that is excreted into the intestine along with the digestive juices (endogenous excretion). Smaller amounts of zinc are excreted in the urine or shed in skin cells. In addition, sexual activity in males contributes to zinc losses. In well-controlled metabolic ward studies, it has been shown that each ejaculate can contain up to 0.5 mg of zinc, probably derived from secretions of the prostate gland. Although conservation of zinc occurs during experimental zinc deficiency, the amount of zinc in ejaculates remains relatively high, thus representing a significant loss of zinc, particularly in people with low intakes. Hence, the role of zinc in men is analogous to iron in women, in that it is lost as part of normal sexual function.

### 10.1.4 Deficiency

Zinc deficiency was first observed in adolescents in Iran and Egypt. The first case was a 21-year-old male subject who had the stature of a 10-year-old boy. His main food was unleavened bread from unrefined wheat flour and he ate a considerable amount of clay. Other cases had hookworm infections, and ate mostly unleavened wheat bread and beans. Unleavened bread prepared from unrefined wheat flour has a high phytate content, which interferes with zinc absorption. Further investigations identified zinc as the limiting nutrient, which was responsible for numerous symptoms, including growth retardation, hypogonadism, and delayed sexual maturation. Other manifestations of zinc deficiency that have been reported subsequently include diverse

**Box 10.1 Some symptoms of zinc deficiency**

| Mild deficiency | Severe deficiency |
|---|---|
| • Stunted growth in children | • Dwarfism |
| • Decreased taste sensation | • Hypogonadism and delayed sexual maturation |
| • Impaired immune function | • Hypopigmented hair |

forms of skin lesions, impaired wound healing, loss of taste (hypogeusia), behavioural disturbances, night blindness, and immune deficiency (Box 10.1). These symptoms do not always occur together and seem to depend on the setting. For instance, in patients on total parenteral nutrition (if it lacks zinc), there is mental confusion, depression, eczema, and alopecia. In young children, zinc deficiency is expressed as a reduction in appetite, poor taste acuity, and poor growth.

The involvement of zinc in the immune system is well-established, with an initial consequence of zinc deficiency being an impairment of immunological functions. Zinc is crucial for the normal development and function of cells, mediating both innate and acquired immunity. The acute phase response includes a decline in the plasma zinc concentration as a result of the redistribution of zinc, which is thought to promote protein synthesis and prevention of microbial invasion. Patients with head trauma exhibit hypozincaemia along with a prominent cytokine and acute phase response. In infected adults, lower plasma zinc concentrations are associated with higher illness scores and increased cytokine production.

Zinc deficiency is estimated by the WHO to be one of the 10 biggest factors that contribute to the burden of disease in developing countries. In children, zinc deficiency contributes to up to 15% of diarrhoea deaths, 10% of malaria deaths, and 7% of pneumonia deaths. Zinc supplementation was evaluated for the treatment of children with diarrhoea. In a Cochrane review on the relationship between zinc and diarrhoea, it was shown that in children aged greater than 6 months, zinc supplementation shortened the duration of diarrhoea, with the improvement being greatest in malnourished children, especially those who are greater than 12 months of age. Despite these important findings in developing countries, there is no strong evidence that zinc lozenges are effective in treating symptoms of the common cold in developed countries.

Night blindness, a significant symptom of vitamin A deficiency, can be due to deficiency of zinc, which is a coenzyme for the conversion of retinol to retinaldehyde by retinol dehydrogenase. Zinc deficiency has been observed in association with protein energy malnutrition in infants with marasmus or during their recovery. Zinc supplements supplied to malnourished children during recovery promote weight gain and synthesis of lean tissue, and reduce common comorbidities. Anorexia nervosa in some ways resembles marasmus; however, the role of zinc in anorexia nervosa is unclear. It is likely that zinc deficiency develops in some anorexic patients through a generally inadequate diet, which sustains the disorder and prevents adequate weight gain. However, zinc does not appear to play a causal role.

In pregnancy, plasma zinc has been reported to be low and although this can be attributed partly to physiological changes unrelated to zinc depletion, the intakes of pregnant women are often below the recommendations. Adaptations in pregnancy, such as increased absorption and reduced endogenous losses, may help meet the requirement. Apart from possible reduction in induction of labour, caesarean section, and pre-term delivery, and in some cases increases in birth weight and head circumference, zinc supplementation does not appear to have a significant or consistent effect on pregnancy outcome. Such inconsistencies may be related to small sample sizes, differing zinc status of pregnant women, and inadequate study design.

Biochemical abnormalities of zinc deficiency include a reduction in plasma zinc concentrations, protein synthesis, activity of metalloproteins, resistance to infection, collagen synthesis, and platelet aggregation. In view of the large number of zinc finger proteins, and the interaction between zinc and DNA, it has been hypothesized that zinc primarily restricts gene expression, rather than the enzyme activities.

Conditions that predispose to deficiency are related to:

- decreased intake, possibly associated with an eating disorder
- decreased absorption and/or bioavailability due to a high intake of an inhibitor (e.g. phytate), as noted in the first reported case of human zinc deficiency
- decreased utilization secondary to other conditions, such as alcoholism
- increased losses in conditions such as diarrhoea and excessive vomiting, which may also be associated with an eating disorder
- increased requirement associated with growth, pregnancy, and lactation. This is recognized by a small increase in the recommended dietary intake in some countries.

## 10.1.5 Bioavailability and food sources

Zinc is available widely in the food supply, but its bioavailability from different foods is highly variable. Zinc in animal products, crustacea, and molluscs is more readily absorbed than from plant foods. Rich sources of zinc include oysters, red meat, lamb's liver, and cheese. Cereal grains, legumes, and nuts are rich in phytate, which reduces zinc absorption. The zinc content of refined cereals is lower than unrefined cereals, but because the bran, which contains most of the phytate, has been removed, the bioavailability is greater. Although a number of factors are known to influence the bioavailability of zinc, a reliable algorithm to calculate available zinc remains to be worked out. The molar ratio of phytate:zinc has been proposed as a predictor of zinc bioavailability and ratios greater than 15 have been associated with suboptimal zinc status. The phytate × calcium:zinc in the diet has been suggested as a marker of zinc bioavailability, although there is limited data on the phytate content of foods. The WHO has put forward three categories of bioavailability (Table 10.1).

**Table 10.1 Dietary determinants of zinc bioavailability (WHO, 2004)**

| Estimated absorption | Type of diet |
|---|---|
| Low (<15%) | Diets high in unrefined cereal grain<br>Phytate:Zn molar ratio >15<br>Calcium >1 g/day |
| Moderate (15–35%) | Mixed diet containing animal or fish protein<br>Phytate:Zn molar ratio <10 |
| High (35–55%) | Refined diets, low in cereal fibre<br>Phytate:Zn molar ratio <5<br>Dietary protein primarily from animal foods |

*Adapted from:* World Health Organization (2004) *Vitamin and mineral requirements in human nutrition.* Geneva: WHO.

The extent of adaptation to foods with low bioavailability of zinc is not fully understood and is confounded by the interaction with other nutrients. Current methods used for studying zinc bioavailability in humans include metabolic balance studies, and radioisotope and stable isotope techniques. The radioisotope techniques are limited by ethical considerations, such as long radioactive half-lives and the amount of radiation exposure to subjects. Use of stable isotopes circumvents this issue, but this technique requires costly instrumentation and demanding analytical procedures.

## 10.1.6 Nutrient reference values

The recommended dietary intake for zinc is 11–14 mg/day for men in different committee reports, 8 mg for women, with 3 and 4 mg extra for pregnancy and lactation, respectively. The upper intake level (UIL) is 40 mg/day.

### 10.1.7 Biochemical tests for status

The plasma zinc concentration represents less than 1% of the body pool of zinc and, hence, its measurement provides a limited amount of information about the zinc status of individuals. Zinc from the plasma is taken up by the liver in response to cytokines released during stress and infection. In addition, plasma zinc concentrations fall in pregnancy, with injuries, and in diseases such as liver cirrhosis and pernicious anaemia. Plasma zinc also undergoes diurnal variation, with a U-shaped curve over a 24-hour period. Peak concentrations are found in the mornings and trough concentrations in the mid-evening. Despite its limitations, the concentration of zinc in plasma is the most commonly used diagnostic indicator, and the balance of evidence shows that the concentration falls in deficiency and rises in sufficiency (or with supplementation).

Under the EURopean micronutrient RECommendations Aligned (EURRECA) consortium, meta-analyses that examined the usefulness of biomarkers of zinc status in humans showed that the plasma zinc concentration responded in a dose-dependent manner to dietary manipulation in a range of population groups such that, for every doubling in zinc intake, the difference in zinc serum or plasma concentration is 6%. Data on urinary zinc excretion, although limited, appeared to respond in the same manner. The small magnitude of this relationship places further emphasis on the technical aspects of sample collection and analytical processing.

### 10.1.8 Toxicity

The ingestion of very high doses (>1 g zinc) results in a metallic taste in the mouth, nausea, fever, lethargy, and gastric distress. This acute response occurs with deliberate supplementation, occupational exposure, or food poisoning. Rapid infusion of intravenous feeding solutions containing zinc can cause similar symptoms. Very large doses have resulted in death.

Zinc supplements (50–150 mg zinc) decrease plasma HDL-c concentrations, thus increasing the risk of cardiovascular disease in normolipidaemic individuals. The major effect, however, is due to the adverse interaction between zinc and copper absorption. Zinc induces the synthesis of metallothionein, a sulphur-rich protein that binds copper with high affinity. In chronic toxicity, copper status is compromised, resulting in a decrease in copper-related functions including the reduction in copper metalloenzyme activity and anaemia. The decrease in copper absorption is advantageous under some circumstances. It is required in the treatment of patients with (Kinnear) Wilson's disease and zinc supplementation is part of the management strategy.

## Further reading

1. **Fischer Walker, C.L., Ezzati, M., and Black, R.E.** (2009) Global and regional child mortality and burden of disease attributable to zinc deficiency. *Eur J Clin Nutr*, **63**, 591–7.
2. **Gibson, R.S.** (2005) *Principles of nutritional assessment*, 2nd edn. Oxford: Oxford University Press.
3. **International Zinc Nutrition Consultative Group; Brown, K.H., and Hess, S.Y. (guest eds)** (2009) Technical Document #2: Systematic reviews of zinc interventions strategies. *Food Nutr Bull*, **30**, S5–S184.
4. **King, J.C., Brown, K.H., Gibson, R.S., et al.** (2016) Biomarkers of Nutrition for Development (BOND)—Zinc Review. *J Nutr*, doi: 10.3945/jn.115.220079.
5. **Lazzerini, M. and Ronfani, L.** (2013) Oral zinc for treating diarrhoea in children. *Cochrane Database Syst Rev*, **1**, CD005436.
6. **Lowe, N.M., Medina, M.W., Stammers, A.L., et al.** (2012) The relationship between zinc intake and serum/plasma zinc concentration in adults: a systematic review and dose-response meta-analysis by the EURRECA Network. *Br J Nutr*, **108**, 1962–71.
7. **World Health Organization** (2004) *Vitamin and mineral requirements in human nutrition*. Geneva: WHO.

# 10.2 Copper

Samir Samman

## 10.2.1 Historical perspective

The essential role of copper was realized in 1926, and soon after it was shown that it is required for the synthesis of haemoglobin in rats. In 1962, copper deficiency was reported in humans.

## 10.2.2 Distribution and function

Copper is one of the IB series of metals, with a molecular weight of 63.5. It is one of the most effective cations for binding to organic molecules. It is commonly used in biological reactions that involve electron transfer.

Adult humans contain about 100 mg Cu, which is distributed in concentrations of about 1.5 μg/g in the skin, skeletal muscle, bone marrow, liver, and brain. Studies in animals suggest that the copper content may decrease with age. The plasma concentration is 15 μmol/L (similar to zinc); up to 90% of this is associated with caeruloplasmin. Other Cu proteins include many of the oxidases, metallothionein, α-fetoglobulin, superoxide dismutase, and transcuprein.

Copper has diverse functions including erythropoiesis, connective tissue synthesis (via lysyl oxidase), oxidative phosphorylation, thermogenesis, and superoxide dismutation. As well as transporting copper, caeruloplasmin is one of the acute phase proteins and, via its ferroxidase activity, catalyses the oxidation of ferrous iron. This latter reaction is essential for the mobilization of iron as a complex with transferrin. It is believed to be the mechanism by which copper is able to regulate the homeostasis of iron.

## 10.2.3 Absorption and excretion

Dietary copper is reduced to $Cu^{1+}$ and transported across the apical membrane of the enterocyte by a specific transporter known as copper transporter 1 (Ctr1). Copper is incorporated into cellular protein (including enzymes), but the majority is released from the basolateral membrane of the intestinal cell to portal blood. Copper is transported to the liver by albumin and transcuprein. In the liver copper is incorporated into caeruloplasmin and subsequently circulates to other tissues.

The efficiency of copper absorption depends on the individual's copper status. The efficiency of absorption increases in cases of deficiency or when dietary copper intake is low. Copper is excreted mainly via the gastrointestinal tract. Less is excreted in the urine and from the skin.

## 10.2.4 Bioavailability and food sources

Copper has a wide distribution in the food supply, but in particular it is found in foods of animal sources, legumes, nuts, and the water supply (copper pipes). Copper absorption is enhanced by organic nutrients, such as amino acids and, in particular, histidine. Conversely, absorption is inhibited by excesses of other divalent cations such as zinc and iron. Studies in animals suggest that vitamin C may have an adverse effect on copper absorption, but the results of trials in humans are not conclusive. Phytic acid and dietary fibre do not appear to inhibit copper absorption.

## 10.2.5 Deficiency

Copper deficiency is relatively rare. It has been observed in protein energy malnutrition, in patients on long-term copper-free total parenteral nutrition, and in premature infants fed cow's milk or unfortified formula. Symptoms include anaemia, neutropenia, skeletal demineralization, decreased skin tone, connective tissue aneurysms, hypothermia, neurological symptoms, and depigmented hair.

A decreased copper status has been identified in obese patients who have undergone bariatric surgery. Bariatric surgery is an effective treatment for obese individuals, and is associated with reduced comorbidities. However, complications result primarily from micronutrient deficiencies, including copper. Results of a meta-analysis to explore copper status 6 months post-bariatric surgery show a significant reduction in the plasma/serum copper concentration. Copper status also decreased over time and differences between pre- and post-operative values were statistically significant. Similarly, caloric restriction combined with exercise induces a decrease in plasma concentration of caeruloplasmin. This copper transport protein was inversely correlated with insulin sensitivity. These data suggest that copper status needs to be monitored in obese patients who are undergoing surgery, particularly as this is becoming a more common procedure.

Defects in copper metabolism have been identified. Menkes' disease, an X-linked progressive brain disease, was established as related to copper following the recognition by Australian researchers that patients with the disease have kinks in their hair, which was similar to the kinks in the wool of sheep grazing on copper-deficient soils. The characteristics of the hair together with low concentrations of plasma copper and caeruloplasmin are features of the disease. Intestinal absorption of copper is defective.

Patients who lack plasma caeruloplasmin have been identified. Recent findings in patients with acaeruloplasminaemia have confirmed the essential role of this copper protein in iron metabolism. Symptoms associated with acaeruloplasminaemia include decreased copper and iron in plasma, increased iron concentrations in tissues, and impaired copper absorption.

### 10.2.6 Nutrient reference values

No estimated average requirement (EAR) or recommended dietary intake (RDI) has been estimated for copper. An adequate intake for adults is approximately 1.2 (for women) to 1.7 (for men) mg/day, with another 0.3 mg recommended for lactation. The UIL should not exceed 10 mg/day.

### 10.2.7 Assessment of copper status

Serum copper is the most useful biomarker of copper status. The assessment of marginal deficiency remains a challenge. However, it appears that the activity of some copper-dependent enzymes (e.g. serum diamine oxidase) respond to increases in dietary or supplemental copper, and have the potential to reflect copper status. Frank copper deficiency can be determined by the measurement of plasma copper concentrations and plasma caeruloplasmin. The haematocrit decreases and there is microcytic hypochromic anaemia.

### 10.2.8 Toxicity

The amounts of copper in the diet (including water) are not likely to be toxic because of the ability to maintain homeostasis by decreasing absorption and increasing excretion. Acute toxicity has been reported as a result of accidental ingestion of large doses of copper or in industrial accidents. The symptoms of small doses include vomiting and nausea, while larger doses induce hepatic necrosis and haemolytic anaemia. Chronic toxicity is relatively rare.

Wilson's disease, Indian childhood cirrhosis, and idiopathic copper toxicosis are disorders that predispose individuals to copper overload. Wilson's disease is a rare inborn error of metabolism with a reported incidence of 1:30,000. It is an autosomal recessive disease that gives rise to hepatolenticular degeneration (juvenile cirrhosis, coarse tremor, browning around the cornea). Less well quantified are the incidences of Indian childhood cirrhosis, reported initially in India, but also in other parts of the world in non-Indian children, and idiopathic copper toxicity. There is little evidence to support the efficacy of copper restriction for the management of Wilson's disease and other copper storage diseases. The primary intervention has to be pharmacological (chelation) therapy to increase urinary copper excretion.

## Further reading

1. **Fairweather-Tait, S.J., Harvey, L.J., and Collings, R.** (2011) Risk-benefit analysis of mineral intakes: case studies on copper and iron. *Proc Nutr Soc*, **70**, 1–9.
2. **Freeland-Graves, J.H., Lee, J.J., Mousa, T.Y., and Elizondo, J.J.** (2014) Patients at risk for trace element deficiencies: bariatric surgery. *J Trace Elements Med Biol*, **28**, 495–503.
3. **Prohaska, J.R.** (2014) Impact of copper deficiency in humans. *Ann NY Acad Sci*, **1314**, 1–5.

# 10.3 Iodine

Sheila Skeaff and Christine D. Thomson

Iodine was one of the first trace elements to be identified as essential. As early as 2700 BC, the Chinese were treating goitre by feeding seaweed, marine animal preparations, and burnt sponge (all rich in iodine). In the first half of the nineteenth century, the incidence of goitre was linked with low iodine content of food and drinking water, and by the late nineteenth century the geographical distribution of endemic goitre was recognized to extend around the world. In the 1920s, iodine was shown to be an integral component of the thyroid hormone thyroxine, which is required for normal growth and metabolism, and later, in 1952, of triiodothyronine.

## 10.3.1 Chemical structure and functions of iodine

Iodine functions as an integral part of the thyroid hormones, the prohormone thyroxine ($T_4$), and the active form 3,5,3′-triiodothyronine ($T_3$), which is the key regulator of important cell processes. The thyroid hormones are required for normal growth and development of individual tissues, such as the brain and central nervous system, and maturation of the whole body, as well as for energy production and oxygen consumption in cells, thereby maintaining the body's metabolic rate.

If thyroid hormone secretion is inadequate (i.e. hypothyroidism), the basal metabolic rate is reduced, and the general level of activity of the individual is decreased. Normal growth and development will also be impaired.

The regulation of thyroid hormone synthesis, release, and action is a complex process involving the thyroid, the pituitary, the brain, and peripheral tissues. The hypothalamus regulates the plasma concentrations of the thyroid hormones by controlling the release from the pituitary of thyroid-stimulating hormone (TSH, thyrotropin) through a negative feedback mechanism related to the level of $T_4$ in the blood. If blood $T_4$ falls, the secretion of TSH is increased, which enhances the activity of the thyroid and, consequently, the output of $T_4$ into the circulation. This fine control of $T_4$ secretion is important, as either an excess or a deficit in the hormone is detrimental to normal function. If the level of circulating $T_4$ hormone is not maintained because of severe iodine deficiency, TSH remains elevated, and both these measures are used for diagnosis of hypothyroidism due to iodine deficiency.

## 10.3.2 Body content

Iodine occurs in the tissues in both inorganic (iodide) and organically bound forms. The adult human body contains about 15–50 mg iodine, with 70–80% of this found in the thyroid gland, which has a remarkable concentrating power for iodine, and the remainder mainly in the circulating blood.

## 10.3.3 Metabolism

The metabolism of iodine is closely linked to thyroid function, since the only known function for iodine is in the synthesis of thyroid hormones. Iodine is an anionic trace element that is rapidly absorbed in the form of iodide, and taken up immediately by the thyroid gland. The thyroid gland is a globular, butterfly-shaped gland located at the base of the front of the neck. It is composed of spherical follicles lined with thyroid cells filled with colloid. A sodium-iodide symporter protein located on the basal membrane of the thyroid cell actively transports iodide into the thyroid cell. Iodide then migrates to the apical membrane of the thyrocyte and crosses into the follicular lumen. Two enzymes, thyroperoxidase and hydrogen peroxidase, oxidize iodide, which is attached to tyrosyl residues of thyroglobulin (Tg) to form mono-iodotyrosine (MIT) and di-iodotyrosine (D-IT). MIT and D-IT are coupled to make $T_3$ or $T_4$ within the Tg molecule. Tg enters the thyrocyte by endocytosis of the colloid and is proteolysed, releasing $T_3$ and $T_4$, which subsequently enter the circulation. Within the blood, more than 99% of the thyroid hormones are bound to plasma proteins. At receptors on the surfaces of target cells in organs around the body, $T_4$ is converted to $T_3$ by various iodothyronine 5′ deiodinase enzymes. $T_3$ is the main physiologically active form of the hormone and binds to nuclear receptors. The half-life of $T_4$ is approximately 7 days, but only 24 hours for $T_3$. The majority (~80%) of $T_3$ is formed extrathyroidally from deiodination of $T_4$. The thyroid gland needs to trap around 60 μg iodide per day to maintain an adequate supply of $T_4$. Excess inorganic iodine is readily excreted in the urine.

### 10.3.4 Deficiency

Iodine deficiency is recognized as a major international public health problem because of the large number of populations living in iodine-deficient environments, characterized primarily by iodine-deficient soils. Remarkable progress has been made in improving iodine status over the last 20 years through the implementation of salt iodization programmes. Only 67 countries had adequate iodine intake in 2003, which almost doubled to 111 countries in 2021. However, 19 countries are still iodine deficient, 11 countries have an excessive iodine intake, and there is no data on 53 countries. Iodine deficiency is still found in all regions worldwide and in populations at every stage of economic development. Basil Hetzel coined the term iodine-deficiency disorders (IDD), which refers to the wide spectrum of adverse effects that iodine deficiency can have on growth and development (Table 10.2). Goitre, a swelling of the thyroid gland (as shown in Fig. 10.2), is the most obvious and familiar feature of iodine deficiency, and is the body's adaptive response to inadequate dietary iodine. A number of changes take place in the thyroid gland, including hyperplasia of the thyroid cells, resulting in an increase in the size of the gland, and a more efficient use of available iodine to produce thyroid hormones.

The most damaging consequences of iodine deficiency are on foetal and infant development. Thyroid hormones (therefore iodine) are essential for normal development of the brain, and insufficient levels may result in permanent intellectual disability of the foetus or newborn child. Iodine deficiency has been regarded as the world's greatest single cause of

**Table 10.2 Spectrum of the iodine-deficiency disorders (IDD)**

| Foetus | Abortions |
|---|---|
| | Stillbirths |
| | Congenital anomalies |
| | Increased perinatal mortality |
| | Increased infant mortality |
| | Neurological cretinism: intellectual disability, hearing impairment, spastic diplegia, squint |
| | Myxoedematous cretinism: intellectual disability, dwarfism, hypothyroidism |
| | Psychomotor defects |
| Neonate | Neonatal hypothyroidism |
| Child and adolescent | Impaired mental and physical development |
| Adult | Goitre and its complications |
| | Iodine-induced hyperthyroidism |
| All ages | Goitre |
| | Hypothyroidism |
| | Impaired mental function |
| | Increased susceptibility to nuclear radiation |

Adapted from WHO/NHD (2001) *Assessment of iodine deficiency disorders and monitoring their elimination: a guide for programme managers*, 2nd edn. Geneva: World Health Organization.

Fig. 10.2 Three women of the Himalayas with stage II goitres.

With permission, Editor, *Thyroid Manager*, Chapter 20 by Professor Creswell Eastman.

Fig. 10.3 Myxoedematous endemic cretinism in the Democratic Republic of Congo. Four inhabitants aged 15–20 years: a normal male and three females with severe longstanding hypothyroidism with dwarfism, retarded sexual development, puffy features, dry skin, and severe intellectual disability.

With permission, Editor, *Thyroid Manager*, Chapter 20 by Professor Creswell Eastman.

preventable brain damage and intellectual disability. The most severe effect of foetal iodine deficiency is congenital iodine-deficiency syndrome, historically referred to as endemic cretinism, which occurs when a pregnant woman is severely iodine deficient, particularly in the first trimester. In general, people with congenital iodine-deficiency syndrome have significantly impaired cognition, as well as physical abnormalities. Clinical manifestations differ with geographical location, and two types have been observed. In *myxoedematous cretinism*, features of hypothyroidism are present (dry skin, hoarse voice) with stunted growth and intellectual impairment (Fig. 10.3). In the *neurological type of cretinism*, intellectual impairment is present, as well as hearing and speech defects, and characteristic disorders of stance and gait, while hypothyroidism is absent. In myxoedematous cretinism, thyroid hormone treatment can lead to some improvement (iodine cannot help—the thyroid gland is atrophic). With the widespread implementation of salt iodization programmes, reports of congenital iodine-deficiency syndrome are now rare.

**Mild to moderate iodine deficiency** There are also detrimental effects of less obvious iodine deficiency on mental performance of school children, which may have considerable social consequences that could affect national development. A meta-analysis of 18 studies estimated that the mean IQ scores of children living in moderately to severely iodine-deficient areas were 13.5 points lower than those of children living in iodine sufficient areas. Another meta-analysis of 37 studies in China confirmed this, finding a mean difference of 10 IQ points. Both meta-analyses primarily used data from cross-sectional studies. Only two well-conducted randomized controlled trials have been undertaken in this area. Albanian schoolchildren with moderate iodine deficiency given a bolus dose of iodized poppy seed oil showed an improvement in four out of seven cognitive tests after 24 weeks. A similar study in New Zealand schoolchildren with mild iodine deficiency who took a daily iodine supplement for 28 weeks showed an improvement in two out of four cognitive tests compared with placebo children. The cerebral effect was not mirrored by total thyroid hormone levels.

The major cause of IDD is inadequate dietary intake of iodine from foods grown in soils from which iodine has been leached by glaciation, high rainfall, or flooding. Goitre is usually seen where the intake is less than 50 μg/day, and congenital iodine-deficiency syndrome where the mother's intake is 30 μg/day or less. However, thyroid function may also be impaired after exposure to antithyroid compounds in foods and drugs—called goitrogens (e.g. thiocyanate)—which prevent the uptake of iodine into the thyroid gland. Selenium has the potential to play a part in the outcome of iodine deficiency in two ways. Firstly, the selenium-containing iodothyronine 5′-deiodinases regulate the synthesis and degradation of $T_3$. Secondly, selenoperoxidases protect the thyroid gland from hydrogen peroxide produced during the synthesis of thyroid hormones.

In countries where iodine deficiency is not endemic, hypothyroidism is typically due to autoimmune disease.

### 10.3.5 Measures to prevent iodine deficiency

Iodization of salt has been the primary method for combating iodine deficiency since the 1920s, when it was first introduced in Switzerland. In 2020, salt iodization was mandatory in 124 countries and voluntary in 21 countries, and it is estimated that, globally, 88% of the population uses iodized salt. WHO recommends universal salt iodization (USI), which means that all salt used by the food industry and livestock salt is iodized at 20–40 mg/kg, and adequately iodized salt (≥15 mg/kg) is consumed in >90% of households. This recommendation is based on the assumption that 10 g salt/day iodized at 15 mg/kg will provide 150 μg of iodine, the recommended dietary intake for adults. UNICEF estimates that approximately 1 billion people do not have access to adequately iodized salt. Despite WHO's recommendation, USI is not common; the exception is China, which has successfully used USI since 1993 with a consequent reduction in goitre rate in children from 20% to 6%. Indeed, USI has been so successful that in 2010 the Chinese government called for a lowering of the amount of iodine in salt from 20–60 mg/kg to 20–30 mg/kg in response to a study showing many Chinese people now have excessive iodine intakes.

In the past, iodized vegetable oil that contains 200–400 mg of iodine, given either as an injection or orally, was successfully used to improve iodine status in areas of severe iodine deficiency. In 2007, the WHO and UNICEF recommended that pregnant and lactating women and children <2 years, who do not have access to iodized salt and live in areas of moderate to severe iodine deficiency, should be given oral doses of iodized oil annually. In countries where iodine intakes are higher, but still not adequate, daily iodine supplements are recommended, containing 150–250 μg.

### 10.3.6 Towards elimination of iodine deficiency: global action

Iodine deficiency is recognized as a major international public health problem because of the large populations at risk in iodine-deficient environments. The International Council for the Control of Iodine Deficiency Disorders (ICCIDD) (http://www.iccidd.org) formed in 1986 and worked closely with other international organizations such as UNICEF, the Micronutrient Initiative, and the Global Alliance for Improved Nutrition (GAIN) to develop national programmes to prevent and control IDD. In 1990, the UN adopted a global action plan to eliminate IDD as a major public health problem by the year 2000. In 2002, the Network for Sustained Elimination of Iodine Deficiency was launched at the UN Special Session for Children. In 2012, the ICCIDD Global Network was formed from these two organizations and in 2014, the organization was renamed and is now called the Iodine Global Network (IGN). Every year, the IGN produces the Global Scorecard of Iodine Nutrition and a corresponding Global Map, which is freely available on the IGN website. As stated earlier, substantial progress has been made towards the elimination of iodine deficiency; however, continued efforts are needed to monitor at-risk populations, and to strengthen and maintain salt iodization programmes. Australia and New Zealand are a case in point, where a lack in surveillance systems and changes in food patterns resulted in the re-emergence of mild iodine deficiency from the early 1990s, necessitating the mandatory fortification of commercial breads with iodized salt in 2009.

### 10.3.7 Assessment of iodine status

The assessment of nutritional status of iodine is important in relation to a population or group living in an area or region that is suspected to be iodine deficient. To date, the most important information comes from measurement of the urinary iodide and the prevalence of goitre. Other indices of iodine status, such as thyroglobulin, newborn thyroid-stimulating hormone, and breast milk iodine, are also recommended, for selected groups of the population, such as children, pregnant women, and lactating women, respectively.

**Table 10.3 Epidemiological criteria for assessing iodine nutrition based on median urinary iodine concentrations in a group**

| Median urinary iodine (μg/L) | Iodine intake | Iodine nutrition |
|---|---|---|
| <20 | Insufficient | Severe iodine deficiency |
| 20–49 | Insufficient | Moderate iodine deficiency |
| 50–99 | Insufficient | Mild iodine deficiency |
| 100–199 | Adequate | Optimal for children and adults, including breastfeeding women |
| >150 | Adequate for pregnant women | Optimal for pregnant women |
| 200–299 | More than adequate | Risk of iodine-induced hyperthyroidism within 5–10 years following introduction of iodized salt in susceptible groups |
| >300 | Excessive | Risk of adverse health consequences (iodine-induced hyperthyroidism, autoimmune thyroid diseases) |

*Adapted from:* WHO/UNICEF/ICCIDD (2007) *Assessment of iodine deficiency disorders and monitoring their elimination: a guide for programme managers*, 3rd edn. Geneva: WHO.

**Urinary iodine excretion** Approximately 90% of iodine intake is excreted in the urine and therefore 24-hour excretion of iodine reflects dietary intake and may be used for estimating the intake. However, 24-hour urine samples are difficult to collect in the field, and non-fasting casual or spot urine specimens are usually collected. The ease of obtaining a casual urine sample from subjects is offset by the large variability of such samples at the level of the individual. Urinary iodine cannot be used to assess iodine status in an individual and should not be used to categorize an individual as iodine deficient. Thus, the median urinary iodine concentration of a group is used as a biomarker for the assessment of population iodine nutrition. Although urinary iodine can be expressed as μg/L, μg/day, or μg/g creatinine, the epidemiological criteria are reported for μg/L. An optimal median urinary iodine concentration between 100 and 200 μg/L corresponds to an intake of approximately 150–200 μg/day. Table 10.3 gives median urinary iodine concentrations associated with levels of iodine nutrition. Urinary iodine, however, is a sensitive indicator of recent iodine intake (i.e. days), but not of thyroid function. Furthermore, where goitrogens are preventing the uptake of iodine into the thyroid gland and synthesis of thyroid hormones, urinary iodine may be normal and, therefore, not a suitable marker for iodine status.

**Thyroid hormones and thyroglobulin** The level of thyroid hormones provides an indirect measure of iodine nutritional status. When iodine in the diet is limited, TSH increases, while $T_4$ concentration decreases; however, these changes are relatively transient because the iodine gland can respond to low iodine intakes. Typically, alterations in TSH and $T_4$ concentration outside the reference range are only observed in moderate iodine deficiency, while $T_3$ concentrations only decline in severe deficiency. Thus, the assessment of TSH, $T_4$, or $T_3$ are not particularly useful or sensitive for determining iodine status in adults, especially in areas of moderate to mild iodine deficiency. The WHO has advocated the use of neonatal TSH to assess iodine status for many years; however, technical issues have hindered its widespread use. Serum thyroglobulin is a more sensitive indicator of mild iodine deficiency and iodine repletion than TSH or $T_4$, as levels are elevated in subjects with mild iodine deficiency. Increased thyroglobulin concentration indicates increased thyroid activity and is a marker of intermediate iodine status (i.e. weeks to months).

For population studies, both TSH and thyroglobulin are recommended surveillance measures and can be determined in blood spots on filter paper or serum samples. The percentage of neonates with TSH values greater than 5 μIU/mL whole blood defines the level of deficiency—mild (3–19%), moderate (20–39%), and severe (≥40%). Similarly, serum thyroglobulin values of 10–19, 20–39, or ≥40 ng/mL serum in groups of school-age children represent mild, moderate, and severe deficiency, respectively; values for adults and pregnant women have yet to be firmly established.

**Assessment of thyroid size and goitre rate** The prevalence of goitre reflects a population's history of iodine nutrition (i.e. months to years), but it does not properly reflect its present iodine status, because thyroid size decreases only slowly after iodine repletion. Goitre assessment is made by inspection, palpation, or more recently by ultrasonography. The recommended target group for monitoring goitre rate is school children. Normative values proposed by the WHO and the ICCIDD for thyroid volume by ultrasonography are based on data obtained from a large international sample of iodine-replete school-age children. The percentage of children with thyroid glands greater than the 97th percentile of normative values characterizes mild (5–19%), moderate (20–29%), and severe (≥30%) deficiency.

## 10.3.8 Dietary intakes of iodine

Iodine intakes vary considerably depending on geographical location, dietary habits, and salt iodization. Adequate dietary intakes of iodine are around 100–150 μg/day, with intakes of 220–270 μg/day recommended in pregnancy and lactation.

Foods of marine origin, such as sea fish, shellfish, seameal, and seaweeds, are rich in iodine and reflect the greater iodine concentration of seawater compared with fresh water. The iodine content of plants and animals depends on the environment in which they grow; generally, vegetables, fruit, and cereals grown on soils with low iodine content are poor sources of iodine.

Because the mammary gland concentrates iodine, dairy products are usually a good source, but only if the cows get enough iodine. Iodine contamination of dairy products from iodophors and bread from iodates (used as bread improvers) have made major contributions to the daily intake. The use of iodophors as sanitizers in the dairy industry, now declining, has resulted in variable, but considerable, amounts of residual iodine in milk, cheese, and other milk products in the past. Because many foods that are good sources of iodine are of animal origin, people consuming plant-based diets often have low iodine intakes, which can be mitigated by the use of iodized salt at the table and in cooking. Other adventitious sources of iodine include kelp tablets and drugs, and beverages or foods containing the iodine-containing colouring erythrosine.

Iodized salt is another source of iodine and has been the most primary means of improving iodine nutrition. Salt is considered to have an adequate iodine concentration if ≥15 mg/kg; however, the amount of iodine added to salt varies widely. In Canada, salt is iodized to a concentration of 77 mg/kg iodine as potassium iodide, so that the daily recommended intake might be obtained from 2 g salt. In the USA, iodized salt contains 45 mg/kg, whereas in New Zealand and Australia the permitted range is 24–65 mg/kg. In some countries both iodized salt and non-iodized salt is available. In developed countries, much of the salt intake now comes from processed foods, which often does not contain iodized salt. Sea salt, rock salt, and pink salt are increasing in popularity but, unless iodized, typically contain negligible amounts of iodine, and are not recommended.

## 10.3.9 Interactions

The utilization of absorbed iodine is influenced by goitrogens, which interfere with the biosynthesis of the hormones. Goitrogens are found in vegetables of the genus *Brassica*—cabbage, turnip, swede, Brussels sprouts, and broccoli; in some staple foods such as cassava, maize, millet, and lima beans that are used in developing countries; and in some parts of the world, goitrogens can be found in the water. Goitrogens can become a problem where people whose iodine intakes are only marginal eat these staple foods, particularly if they are not well cooked. Most goitrogens are inactivated by heat, but not when milk is pasteurized.

### 10.3.10 Requirement and recommended dietary intakes

Goitre occurs when iodine intakes are less than about 50 μg/day, and congenital iodine-deficiency syndrome when maternal intake is 30 μg/day or less. Minimum requirement to prevent goitre, based on the urinary excretion associated with a high incidence of goitre in a population, is approximately 1 μg/kg body weight/day. However, recommended dietary intakes are based on physiological requirements, which are, in turn, based on a number of indicators, including thyroidal radio-iodine accumulation and turnover, iodine balance studies, urinary iodide excretion, thyroid hormone measures, and thyroid volume. From these data, a physiological requirement of around 100 μg/day is indicated, and a rather large safety margin is generally advised to ensure an adequate intake. In most countries, the recommended intake is in the range 150–200 μg/day. This is adequate to maintain normal thyroid function that is essential for growth and development. Because of the increased requirements for thyroid hormones during pregnancy and the importance of adequate iodine for the foetus and neonate, recommended intakes for pregnant and lactating women are considerably higher at 200–230 μg/day for pregnancy and 200–290 μg/day for lactation.

### 10.3.11 Toxicity

Intakes between 50 and 1100 μg/day are considered safe. The effects of high iodine intake on thyroid function are variable and depend on the health of the thyroid gland. Dietary intakes of up to 1100 μg/day have few long-term effects when the thyroid is healthy. Daily intakes of 2000 μg iodine should be regarded as excessive or potentially harmful. Such intakes are unlikely to be obtained from diets of natural foods except where they are exceptionally high in marine fish or seaweed, such as in Japan, or where foods are contaminated with iodine from iodophors. Other large sources of iodine include iodine-containing drugs, radiographic contrast media, and the use of kelp supplements that can contain very high, but variable, amounts of iodine.

Excess intakes of iodide can cause enlargement of the thyroid gland, just as deficiency can, as well as hypothyroidism and elevated TSH and increased incidence of autoimmune thyroid disease. People who have underlying autoimmune disease, such as Graves' disease or Hashimoto's thyroiditis, or who have lived in areas of previous severe iodine deficiency but still have nodular goitres, may be more sensitive to an increase in iodine. Iodine-induced thyrotoxicosis (Jod–Basedow syndrome), characterized by a high pulse rate, weight loss, perspiration, and tremor, has been observed following iodization programmes. However, cases of thyrotoxicosis disappear alongside the disappearance of goitre as the iodine status of the population improves.

#### Further reading

1. **Bath, S.C.** (2019) The effect of iodine deficiency during pregnancy on child development. *Proc Nutr Soc*, **78**(2), 150–60.
2. **Eveleigh, E.R., Coneyworth, J.L., Avery, A., and Welham, S.J.M.** (2020) Vegans, vegetarians, and omnivores: how does dietary choice influence iodine intake? A systematic review. *Nutrients*, **12**(6), 1606.
3. **Chen, Z.-P. and Hetzel, B.S.** (2010) Cretinism revisited. *Best Pract Res Clin Endocrinol Metab*, **24**, 39–50.

4. **Gorstein, J.L., Bagriansky, J., Pearce, E.N., Kupka, R., and Zimmermann, M.B.** (2020) Estimating the health and economic benefits of universal salt iodization programmes to correct iodine deficiency disorders. *Thyroid*, **30**(12), 1802–09.
5. **Hetzel, B.S.** (1989) *The story of iodine deficiency: an international challenge in nutrition*. Oxford: Oxford University Press.
6. **Preedy, V.R., Burrow, G.N., and Watson, R.R. (eds)** (2009) *The comprehensive handbook of iodine*. London: Academic Press.
7. **WHO/UNICEF/ICCIDD** (2007) *Assessment of iodine deficiency disorders and monitoring their elimination: a guide for programme managers*, 3rd ed. Geneva: WHO.
8. **Zimmermann, M.B.** (2009) Iodine deficiency. *Endocrine Rev*, **30**, 376–408.
9. **Zimmermann, M.B. and Andersson, M.** (2021) Global perspectives in endocrinology: coverage of iodized salt programs and iodine status in 2020. *Eur J Endocrinol*, **185**, R13–R21.
The **IDD newsletter** is very interesting, with illustrated stories about iodine deficiency and work to reduce it, in different countries around the world. It is published by the IGN and appears four times a year: https://www.ign.org/home.htm (accessed 16 December 2021).

## 10.4 Selenium

Christine D. Thomson

Selenium first attracted interest in the 1930s as a toxic trace element that caused loss of hair and blind staggers in livestock that consumed high-selenium plants in South Dakota. In 1957, selenium was shown to be essential for mammalian life, when traces of this mineral prevented liver necrosis in vitamin E-deficient rats, and later to prevent a variety of economically important diseases in domestic animals, such as white muscle disease in cattle and sheep, hepatosis dietetica in swine, and exudative diathesis in poultry. The demonstration in 1973 of a biochemical function for selenium as an integral component of the selenoenzyme glutathione peroxidase (GPx) was followed by identification of several other selenoproteins, all of which were found to contain selenocysteine. The importance of selenium in human nutrition was highlighted in reports in 1979 of selenium deficiency in a patient in New Zealand on total parenteral nutrition and of the selenium-responsive condition Keshan disease in China. Considerable research during the past four decades has provided information on the metabolism and importance of selenium in human nutrition, leading to the establishment of recommended dietary intakes based on amounts required to maximize plasma selenoproteins. More recently, the focus has turned to possible benefits from higher levels of selenium intake in maintaining human health, by protecting against certain types of cancer and cardiovascular disease (CVD), and through the maintenance of a healthy immune system. Genomics is now rapidly advancing our knowledge of functions of and requirements for selenium.

### 10.4.1 Functions of selenium

Selenium exerts its biological effects as a constituent of selenoproteins, of which there are 25 in humans. These selenoproteins are involved in a wide variety of processes in the body, including antioxidant defence and redox metabolism, thyroid metabolism, immune function, reproductive function, and many others, with implications of clinical importance in many diseases, such as cancer, autoimmune thyroid disease, and neurological disorders. The following have been purified and studied:

- Glutathione peroxidase (GPx):
  - cytosolic, cellular (GPx1)
  - gastrointestinal (GPx2)

    - plasma (GPx3)
    - phospholipid hydroperoxide (GPx4)
    - embryonic and olfactory (GPx6)
- Selenoprotein P (SEPP)
- Iodothyronine 5′-deiodinases (DI1, DI2, DI3)
- Thioredoxin reductase (TR1, TR2, TR3)
- Also selenophosphate synthetase 2, Sep15 (15 kDa selenoprotein), selenoprotein W, selenoprotein R (methionine-R-sulphoxide reductase; MsrB1), selenoproteins H, I, K, M, N, O, P, S, T, and V.

The selenium is present in all selenoproteins as selenocysteine at the active site. Selenocysteine, the twenty-first amino acid, is inserted into proteins cotranslationally in response to the UGA codon, which, in addition to selenocysteine insertion, functions to terminate protein synthesis.

The first of the selenoproteins to be characterized was GPx, which consists of four identical subunits, each containing selenocysteine at the active site. GPx is present in at least five different forms, all of which use glutathione to catalyse the reduction of hydrogen peroxide and/or phospholipid hydroperoxides. In cells, including erythrocytes (GPx1), the gastrointestinal tract (GPx2), and plasma (GPx3), this enzyme may function *in vivo* to remove hydrogen peroxide, thereby preventing the initiation of peroxidation of membranes and oxidative damage. These GPxs may have more specific functions in arachidonic acid metabolism in platelets, microbiocidal activity in leukocytes, thyroid hormone synthesis, immune function, and cancer prevention. GPx1 is one of the selenoproteins more highly sensitive to selenium deficiency and changes in selenium status.

Another selenium-containing enzyme, phospholipid hydroperoxide GPx (GPx4), differs from other GPxs in several ways. It can metabolize phospholipid hydroperoxides in cell membranes, and thus may play a role as an antioxidant in protecting biomembranes. In addition, GPx4 acts as a structural protein, which is required for sperm maturation.

Selenoprotein P (SEPP), a glycoprotein containing multiple selenocysteine residues, is the major selenoprotein in plasma, providing 40–50% of plasma selenium. Its concentration in rat plasma falls to 10% in selenium deficiency, and so is a useful biomarker of selenium status. Its function is still unclear, but there is evidence for both an antioxidant role and a transport role in the testis and in the brain.

Three iodothyronine 5′-deiodinases (types 1, 2, and 3) are selenoproteins. These enzymes catalyse the conversion of thyroxine ($T_4$) to its active metabolite triiodothyronine ($T_3$). Severe selenium deficiency results in an increase in levels of plasma $T_4$ and a corresponding decrease in levels of $T_3$. The interactions of selenium and iodine deficiencies may have implications for both human health and livestock production. In humans, selenium deficiency has been suggested to exacerbate effects of concurrent iodine deficiency and autoimmune thyroid disorders.

Another family of selenoproteins is the thioredoxinreductases, NADPH-dependent flavoprotein oxidoreductases that reduce the disulphide of thioredoxin. Activity of thioredoxin reductase declines in selenium deficiency. In humans there are three distinct thioredoxin reductases, which support cell proliferation, antioxidant defence, and redox-regulated signalling cascades and may be involved in spermatogenesis and embryonic development.

Selenoprotein W is found in muscle and other tissues, and derived its name because it is one of the missing selenoproteins in heart and muscle of lambs suffering from white muscle disease. Its function, however, is still uncertain.

Selenophosphate synthetase 2 is a selenoenzyme required for the formation of selenocysteine in selenoprotein synthesis.

Several other selenoproteins, including selenoproteins R and N, are emerging as important in various processes, while selenium-containing enzymes have been identified in micro-organisms and other selenoproteins have been found in animal tissues, suggesting further functions for selenium. There is growing evidence that mutations or single-nucleotide polymorphisms in selenoproteins most likely affect selenocysteine incorporation efficiency, which in turn may contribute to the aetiology of diseases such as cancer, CVD, autoimmune thyroid diseases, and neuromuscular disorders.

## 10.4.2 Metabolism of selenium

Selenium metabolism is dependent mainly on the chemical form ingested. Selenoamino acids are the main dietary forms of selenium, with selenium replacing sulphur in selenomethionine in general proteins in plant and animal foods and selenocysteine in selenoenzymes in animal foods. Inorganic forms of selenium, such as selenite and selenate, are used in experimental diets and as supplements.

**Absorption** Selenium is absorbed mainly from the duodenum. Selenomethionine and methionine share the same active transport mechanism, and selenocysteine probably shares the cysteine transporter. Absorption of inorganic forms such as selenite and selenate is via a passive mechanism. Absorption of selenium is generally high in humans, probably about 80% from most dietary sources; selenomethionine appears to be better absorbed than selenite. Absorption is unaffected by selenium status, suggesting that there is no homeostatic regulation of absorption.

**Bioavailability** Animal studies show a wide variation in the bioavailability of selenium from different foods. In rats, the bioavailability from mushrooms, tuna, wheat, beef kidney, and Brazil nuts is 5, 57, 83, 97, and 124%, respectively, in comparison with sodium selenite. Human studies also show differences among various forms such as selenate, wheat, and yeast.

**Transport** Selenium appears to be transported in the body bound to plasma proteins: selenomethionine bound to albumin and inorganic selenium to very-low-density lipoproteins. SEPP also appears to be a transport protein, involved particularly in transporting selenium to the brain and testis.

**Metabolism and distribution** An outline of selenium metabolism is shown in Fig. 10.4. Selenium in animal tissues is present in two main forms: selenocysteine, which is present as the active form of selenium in selenoproteins; and selenomethionine, which is non-specifically incorporated in place of methionine in a variety of proteins, unregulated by the selenium status of the animal.

Selenium levels in tissues are influenced by dietary intake, as reflected in the wide variation in blood selenium concentrations of residents of countries with differing soil selenium levels (Fig. 10.5). The

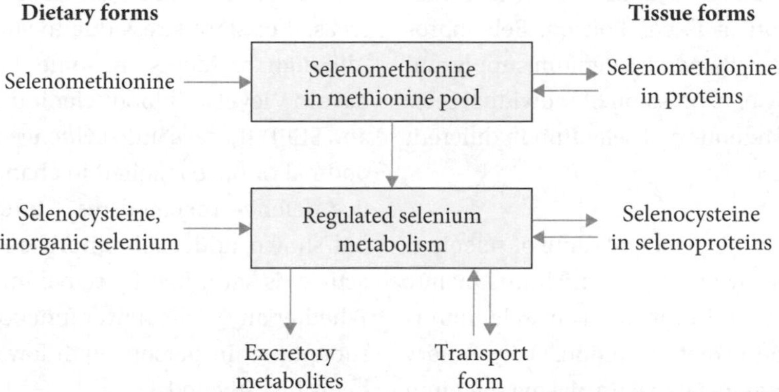


Fig. 10.4 Outline of selenium metabolism. Two compartments of selenium metabolism. Entry points for dietary forms of selenium and the low molecular weight (MW) and protein forms of selenium are shown for the selenomethionine (top) and selenocysteine (bottom) compartments. Common compounds in these pools include selenomethionine ((Se) Met), selenocysteine (Sec), selenite ($SeO_3^{2-}$), selenate ($SeO_4^{2-}$), selenide ($HSe^-$), and selenophosphate ($HSePO_3^{2-}$). Excretory forms are 1 β-methylseleno-N-acetyl-galactosamine, methyl selenol ($CH_3SeH$), dimethyl selenide ($(CH_3)_2Se$), and tri-methyl selenonium ion ($(CH_3)_3Se^+$).

*Source:* Levander, O.A. and Burk, R.F. (1996) (2006) Selenium. In: Ziegler, E.E. and Filer, L.J. (eds). *Present knowledge in nutrition*, 7th edn. Washington, DC: ILSI Press, pp. 320–8.

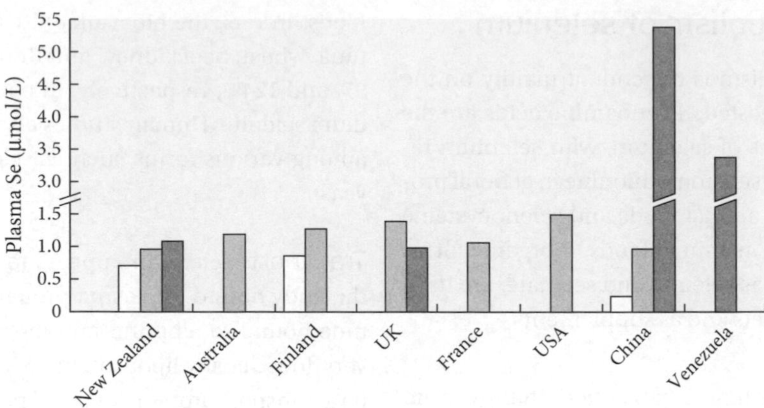


Fig. 10.5 Blood selenium values (μmol/L). New Zealand, pre-1990 (□), 1995–99 (■), post-2000 (■); Australia, (■); Finland, pre-1984 (□), post-1984 (■); UK, pre-1990 (■), post-1990 (■); France, (■); USA, (■); China, low Se (□), medium Se (■), high Se (■); Venezuela, (■).

form administered also influences retention of selenium, with selenomethionine more effective in raising blood selenium levels than sodium selenite or selenate. Both inorganic and organic forms of selenium are transformed to selenide. Selenide (the –2 oxidation state) is transformed to selenocysteine on tRNA and the selenocysteinyl residue is incorporated into the active site of selenoproteins by the UGA codon. The non-specific incorporation of selenomethionine into protein contributes to tissue selenium, which is not immediately available for synthesis of selenoproteins, until it is catabolized. Selenoprotein expression is regulated by selenium supply and there is a hierarchy of expression of individual selenoproteins and of retention of selenium in different organs and tissues.

**Excretion** Urine is the principal route of selenium excretion, followed by faeces, in which it is mainly unabsorbed selenium. Homeostasis of selenium is achieved by regulation of its excretion. Daily urinary excretion is closely associated with plasma selenium and dietary intake, and accounts for approximately 50–60% of the total amount excreted.

Surplus selenium is methylated to methylated selenium metabolites from the common intermediate selenide. 1β-Methylseleno-*N*-acetyl-D-galactosamine (selenosugar) is the major urinary metabolite. Trimethylselenonium ion is excreted in response to very high intakes of selenium and may be used as a biological marker for excessive doses. Excess selenium is also excreted in expired air as dimethyl selenide.

### 10.4.3 Deficiency

Interaction between selenium and vitamin E is observed in the aetiology of many deficiency diseases in animals and pure selenium deficiency is, in fact, rare. Thus, selenium deficiency may only occur when low selenium status is linked with an additional stress such as chemical exposure or increased oxidant stress due to vitamin E deficiency. Although residents in some low-selenium areas have low levels of blood selenium and of GPx activity and SEPP, there is little evidence that these are suboptimal or have resulted in changes in other oxidative defence mechanisms. Moreover, people have not shown noticeably improved health when GPx activity is saturated by selenium supplementation. Whether any of the newer functions of selenium are suboptimal in persons with low selenium status is being investigated.

**Selenium-responsive diseases in humans—Keshan disease** An endemic cardiomyopathy occurring in low-selenium areas of north-east China was reported in 1979 to be responsive to supplementation with sodium selenite. The disease is associated with low selenium intake and low blood and hair levels, and affects mainly children and women of childbearing age. Because some features of Keshan disease (e.g. seasonal variation) cannot be explained solely on the basis of very low selenium status, it is thought likely that a virus was able to cause heart damage in the presence of selenium deficiency, as has been shown with selenium-deficient mice (Ge and Yang, 1993).

Kashin-Beck disease An endemic osteoarthritis of ankles, knees, wrists, and elbows that occurs during pre-adolescent or adolescent years found in rural areas of Tibet, China, and Siberia has also been associated with severe selenium deficiency, although other factors such as iodine deficiency or presence of mycotoxins on barley may be equally important (Ge and Yang, 1993; Moreno-Reyes et al., 2003).

Selenium deficiency in combination with inadequate iodine status contributes to the pathogenesis of myxodematous cretinism. Even mild to moderate selenium deficiency might be responsible for autoimmune thyroid disorders.

Selenium deficiency has been associated with long-term intravenous nutrition, because of previously low levels of selenium in the fluids. Clinical symptoms of cardiomyopathy, muscle pain, and muscular weakness are responsive to selenium supplementation, but are not seen in all patients with extremely low selenium status, indicating that there may be other interacting factors.

### 10.4.4 Selenium and human health

Selenium and immune function Selenium is essential for optimal function of many aspects of the immune system, influencing both the innate and the acquired immune system, and has a role to play in the defence system of animals against bacteria and other infections, including viral infection. The mechanisms for the involvement of selenium in the immune system are likely to be related to its antioxidant function through the antioxidant selenoproteins GPx, thioredoxin reductases, or SEPP.

Studies of host response to myocarditic and non-myocarditic strains of Coxsackie virus B3 in mice showed that selenium deficiency and vitamin E deficiency potentiated cardiotoxicity of myocarditic strains, but, in addition, the non-myocarditic strain caused heart lesions in selenium-deficient mice, apparently as a result of a change in the viral genome. This observation is relevant to the aetiology of Keshan disease, which has been attributed in part to an endemic Coxsackie virus. Mutational changes as a result of selenium deficiency also enhance the intensity of infection of another RNA virus, influenza A, and the protozoan parasite *Trypanosoma cruzi*. There is evidence that both selenoproteins and redox-active selenium species use different mechanisms to attenuate oxidative stress triggered by viruses, including human immunodeficiency virus 1 and SARS-CoV-2, excessive inflammatory responses, and immune system dysfunction, which may explain the improved outcomes observed in people with higher selenium intakes and status; clinical trials are needed to confirm if selenium supplements are both safe and beneficial.

Selenium and cancer Several lines of scientific enquiry suggest an association of cancer with low levels of selenium in the diet. Evidence for the role of selenium as an anticarcinogenic agent comes from *in vitro* and animal studies that suggest that selenium is protective against tumorigenesis at high levels of intake. Evidence from prospective studies linking low selenium status with increased incidence of cancer at various cancer sites has been conflicting, but the strongest evidence is available for prostate and breast cancer. There have been several intervention trials in humans of the effect of selenium, alone or with other nutrients, on the incidence of cancer or concentrations of biomarkers. Again, the results of these trials are conflicting. The strongest evidence comes from The Nutritional Prevention of Cancer Trial, which examined the efficacy of high selenium yeast in preventing skin cancer. There was no effect of selenium on skin cancer, but there was a statistically significant reduction in total cancer (50%) and cancer of the prostate, lung, and colorectum. The effects were strongest in subjects with the lowest selenium status. On the other hand, the more recent Selenium and Vitamin E Cancer Prevention Trial (SELECT) was discontinued early because of a lack of beneficial effect on prostate cancer incidence and an increase in the incidence of diabetes in the selenium group and prostate cancer in the vitamin E group.

The level of selenium intake required for the protective effect appears to be higher than that required to maximize selenoproteins, suggesting that other processes, such as the involvement of anticarcinogenic methylated selenium metabolites such as Se-methyl selenocysteine, might be involved. However, the association of selenium with a reduction in DNA damage and oxidative stress, and recent evidence of an effect of selenoprotein polymorphisms on cancer risk, suggest that selenoproteins are also involved.

**Selenium and cardiovascular disease** Lack of dietary selenium has also been implicated in the aetiology of cardiovascular disease, but the evidence is less convincing than for cancer. A meta-analysis of 25 observational studies found a moderate inverse relationship between plasma selenium and coronary heart disease. However, the few clinical trials have found no evidence of selenium supplementation providing cardiovascular protection in selenium-replete populations, and this was confirmed in the SELECT trial. In addition, higher selenium status has been shown to be associated with increased total and low-density lipoprotein cholesterol, higher fasting plasma glucose and glycosylated haemoglobin levels, and higher prevalence of diabetes and hypertension. On the other hand, a low level of erythrocyte GPx1 has been shown to be a predictor of cardiovascular events in patients with coronary artery disease. GPxs protect against processes relevant to atherosclerosis, in particular in individuals, such as smokers, at risk from increased oxidant stress. These processes include the inhibition of low-density lipoprotein oxidation.

These confusing observations on the effects of selenium on chronic disease clearly indicate that a U-curve represents the effects of selenium status. Because of the adverse effects of higher-than-recommended selenium intakes—diabetes, hypertension, skin cancer, and hypercholesterolaemia—we should be cautious about recommending selenium supplementation for chronic disease prevention.

### 10.4.5 Assessment of selenium status

Blood selenium concentration is a useful measure of selenium status and intake, but other tissues are often assessed as well. Plasma or serum selenium reflects short-term status and red cell selenium reflects longer-term status. Toenail selenium is also used, as toenail concentrations provide a stable assessment of longer-term dietary intake, but selenium-containing shampoos restrict the use of hair. Urinary excretion can also be used to assess selenium status, and total dietary intake is estimated as twice the average daily urinary excretion.

The close relationship between plasma GPx3 or red cell GPx1 activity and selenium concentrations (Fig. 10.6) is useful for assessment in people with relatively low status, but not once the saturating activity of the enzyme is reached at blood selenium concentrations above 80 μg/L (1.00 μmol/L). SEPP, which accounts for more than 50% of selenium in blood, has been shown to be a reliable marker of selenium status in populations with low-to-moderate selenium status, but like GPx, not with higher selenium status. The conclusions drawn from measurement of one selenoprotein may not apply

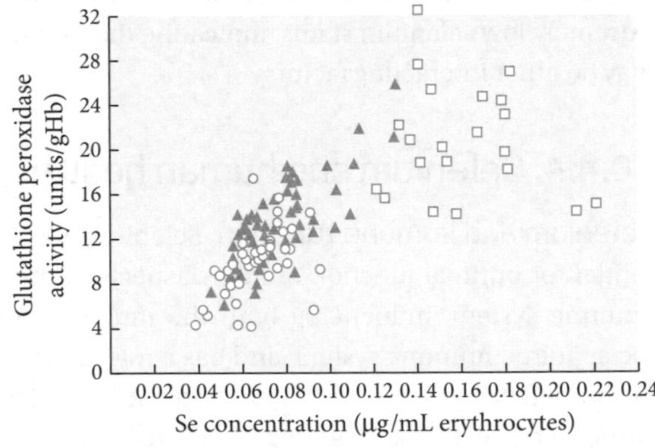


Fig. 10.6 Relationship between selenium concentration of erythrocytes (μg/mL erythrocytes) and glutathione peroxidase (GPx) activity for Otago patients (○), Otago blood donors (▲), and overseas subjects (□). (To convert μg selenium/mL to μmol/litre, multiply by 12.66.)

*Source:* Rea, H.M., Thomson, C.D., Campbell, D.R., and Robinson, M.F. (1979) Relation between erythrocyte selenium concentrations and glutathione peroxidase (EC 1.11.1.9.) activities in New Zealand residents and visitors to New Zealand. *Br J Nutr*, **42**, 201–8, reproduced with permission.

to all biological functions of selenium because of differences in responses of tissues and these proteins to deficient, adequate, or high levels of selenium.

### 10.4.6 Dietary intake

Dietary intake of selenium varies with the geographical source of foods and eating habits of the people. Plant food concentrations reflect selenium content of soils and availability for uptake, as plants generally do not require selenium for growth; the selenium content of cereals and grains grown in soils poor or rich in selenium may vary 100-fold. However, some plants, such as garlic, mushrooms, and broccoli, have developed the ability to accumulate selenium from the soil and therefore may contain high levels of selenium. Brazil nuts are also exceptionally good sources of selenium, but the content varies greatly, depending on where they are grown. Animal foods vary less because animals have an absolute requirement for selenium, which they must get through feed or supplements. Fish and organ meats are the richest sources, followed by muscle meats, cereals and grains, and dairy products, with fruits and vegetables mostly poor sources. Average daily dietary intakes vary considerably depending on the soil selenium levels (Table 10.4), ranging from about 10 μg selenium in low-soil-selenium areas of China where Keshan disease is endemic, to median intakes in New Zealand of 67 and 47 μg/day for males and females, respectively, and up to over 200 μg or more in seleniferous areas in Venezuela and parts of USA. In 1985, selenium was added to fertilizers in Finland as a way of increasing selenium intake throughout the population, and the daily intake rose from 40 μg to close to 100 μg/day, resulting in a significant increase in serum selenium in healthy individuals of 0.85 μmol/L in 1985 to 1.52 μmol/L in 1989–91. In New Zealand, intakes are higher in the North Island because of importation of Australian wheat, but intakes have also increased as a result of increases in selenium concentrations in animal foods due to supplementation of commercial fertilizers and animal feeds and greater consumption of imported foods.

**Table 10.4 Daily dietary intakes and whole blood values of selenium**

| Country | Selenium intake (μg/day) | Plasma or serum selenium[a] (μmol/L) |
|---|---|---|
| **China** | | |
| Keshan disease area | 7–11 | 0.20–0.30 |
| Non-Keshan disease area | 40–120 | 0.49–1.41 |
| Seleniferous area | 750–4990 | 4.52–6.25 |
| **New Zealand** | | |
| Before 1990 | 28–32 | 0.56–0.87 |
| After 1990 | 30–60 | 0.89–1.17 |
| **Finland** | | |
| Before 1984 | 25 | 0.70–1.05 |
| After 1984 | 67–110 | 0.92–1.60 |
| **Great Britain** | | |
| Before 1990 | 60 | 1.25–1.52 |
| After 1990 | 29–39 | 0.78–1.00 |
| **North and South America** | | |
| USA | 60–220 | 1.11–1.88 |
| Venezuela | 200–350 | 2.73–3.99 |

[a]Mean values.

*Adapted from:* Thomson, C.D. (2004) Selenium and iodine intakes and status in New Zealand and Australia. *Br J Nutr*, **91**, 661–72; Combs, G.F., Jr (2001) Selenium in global food systems. *Br J Nutr*, **85**, 517–47.

The reason for the reduction in selenium intake in Great Britain is the change from importation of wheat from North America to European wheat.

### 10.4.7 Requirements and recommended dietary intakes

The minimum requirement to prevent selenium deficiency (20 μg) is based on comparison of intakes in endemic and non-endemic Keshan disease areas of China. However, most countries base their recommended dietary intakes of selenium on estimates of intakes at which saturation of plasma GPx activity occurs, obtained from studies in China and New Zealand. Recommended intakes of the USA and Canada, the UK, Australia/New Zealand, and European countries are summarized in Table 27.2.

Whether optimal health depends upon saturation of GPx activity has yet to be resolved. Recommended dietary intakes of the USA/Canada and most countries are based on desirability of full activity of GPx, whereas a WHO group concluded that only two-thirds of the maximal activity was needed. Recent studies using SEPP as an endpoint for determining selenium requirements may result in higher dietary reference values in the future.

Intakes of selenium that may reduce risk of the chronic diseases cancer and coronary heart disease are likely to be higher than those for maintenance of maximal selenoprotein levels.

### 10.4.8 Toxicity

Overexposure or selenosis may occur from consuming high-selenium foods grown in seleniferous areas in Venezuela, Colombia, northern USA, and Enshi county in China. The most common sign of poisoning is loss of hair and nails, but the skin, nervous system, and teeth may also be involved. Garlic odour on the breath is an indication of excessive selenium exposure (from breathing out dimethylselenide). Sensitive biochemical techniques are lacking for selenium toxicity, which is at present diagnosed from hair loss and nail changes. Some effects of selenium toxicity are seen in individuals with dietary intakes as low as 900 μg, and the UIL has been set at 400 μg per day.

### Further reading

1. **Burk, R.F. and Hill, K.E.** (2015) Regulation of selenium metabolism and transport. *Annual Reviews of Nutrition*, **35**, 109–34.
2. **Cardoso, B.R., Roberts, B.R., Bush, A.I., and Hare, D.J.** (2015) Selenium, selenoproteins and neurogenerative diseases. *Metallomics*, **7**, 1213–28.
3. **Coombs, G.F. Jr** (2015) Biomarkers of selenium status. *Nutrients*, **17**, 2209–36.
4. **Davis, C.D., Tsuji, P.A., and Milner, J.A.** (2012) Selenoproteins and cancer prevention. *Annu Rev Nutr*, **32**, 73–95.
5. **Duntas, L.H.** (2015) Selenium: an element for life. *Endocrine*, **48**, 756–75.
6. **Fairweather-Tait, S.J., Bao, Y., Broadley, M.R., et al.** (2011) Selenium in human health and disease. *Antioxid Redox Signal*, **14**, 1337–83.
7. **Ge, K. and Yang, G.** (1993) The epidemiology of selenium deficiency in the etiological study of endemic diseases in China. *Am J Clin Nutr*, **57**, 259S–63S.
8. **Hurst, R., Collings, R., Harvey, L.J., et al.** (2013) EURRECA–Estimating selenium requirements for deriving dietary reference values. *Crit Rev Food Sci Nutr*, **53**, 1077–96.
9. **Moreno-Reyes, R., Mattieu, F., Boelaert, M., et al.** (2003) Selenium and iodine supplementation of rural Tibetan children affected by Kashan–Beck osteoartopathy. *Am J Clin Nutr*, **78**, 137–44.
10. **Rayman, M.P.** (2020) Selenium intake, status and health: a complex relationship. *Hormones*, **19**, 9–14.
11. **Roman, M., Jitaru, P., and Barbante, C.** (2014) Selenium biochemistry and its role for human health. *Metallomics*, **6**, 25–54.
12. **Zhang, J., Saad, R., Taylor, E.W., and Rayman, M.P.** (2020) Selenium and selenoproteins in viral infection with potential relevance to COVID-19. *Redox Biology*, **37**, 101715.

## 10.5 Fluoride

A. Stewart Truswell

The fluoride ion ($F^-$) is generally regarded as a beneficial nutrient at low dosage because intakes of 1-4 mg/day reduce the prevalence of dental caries (decay). Nutritional authorities have until recently hesitated to classify fluoride as an essential nutrient. The dental benefit is not life-saving. However, many millions of people in the USA, Australia, and New Zealand, and parts of other countries—some

400 million worldwide—have fluoride added to their drinking water at the water works, at the controlled level of 1 mg/L (1 ppm). The compounds used are sodium fluoride [NaF] or sodium fluorosilicate [$Na_2SiF_6$].

As well as this, most toothpastes contain added fluoride, and dentists periodically paint children's teeth with fluoride solution as a preventive measure against tooth decay.

In 1997, the US Institute of Medicine, in collaboration with Canada, decided to set a recommended intake for fluoride; Australia and New Zealand followed suit in 2005; and in Germany, Austria, and Switzerland 'guiding values' for fluoride were published in 2001 and 2002.

Natural water supplies commonly range from 0.1 to 5 ppm in fluoride content and in some places (e.g. bore holes) the fluoride concentration is much higher. From many studies of US children, Dean et al. (1942) showed an inverse relation between the natural fluoride concentration of communal water supplies in the range of 0–2 ppm (μg/g) fluoride and the prevalence of dental caries (see Fig. 10.7). Earlier, a direct association had been found between the fluoride content of water supplies (0–6 ppm) and the occurrence of enamel fluorosis (or mottled enamel), from barely noticeable white flecks affecting a small percentage of the enamel to brown-stained or pitted enamel in the most severe cases (see Fig. 10.8). This is a cosmetic effect and milder forms are not readily apparent to the affected individual or casual observer. Figures 10.7 and 10.8 show little increase in benefit with water containing fluoride above 1 ppm, while mottling only became apparent above 2 ppm. This concentration therefore offered maximal protection against dental caries with minimal risk of dental fluorosis. Moreover, in communities where the natural water supplies were unusually low in fluoride, less than 0.3 ppm, addition of fluoride or fluoridation of the water supply to achieve 1 ppm has been followed by a remarkable decline in prevalence of dental decay in the children, of up to 60%. Indeed, the impact of fluoridation of water supplies and more recently of toothpastes containing fluoride has been a cost-effective triumph of public health.

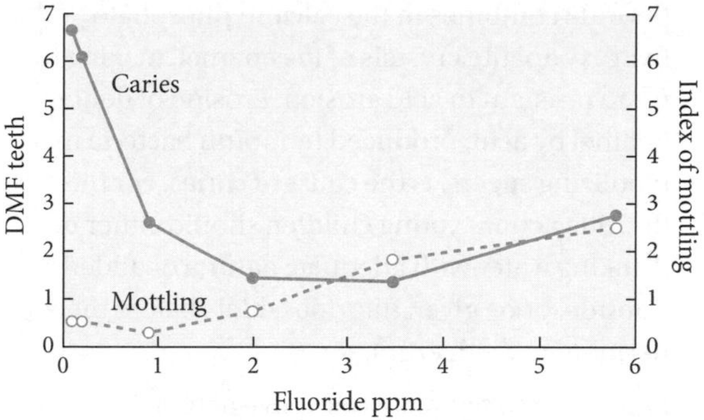


Fig. 10.8 Caries and enamel mottling according to fluoride level in drinking water for UK children aged 12–14 years. DMF, decayed, missing, and filled teeth.

*Source:* Forrest, J.R. (1956) Caries incidence and enamel defects in areas with different levels of fluoride in the drinking water. *Br Dental J*, **100**, 195–200.

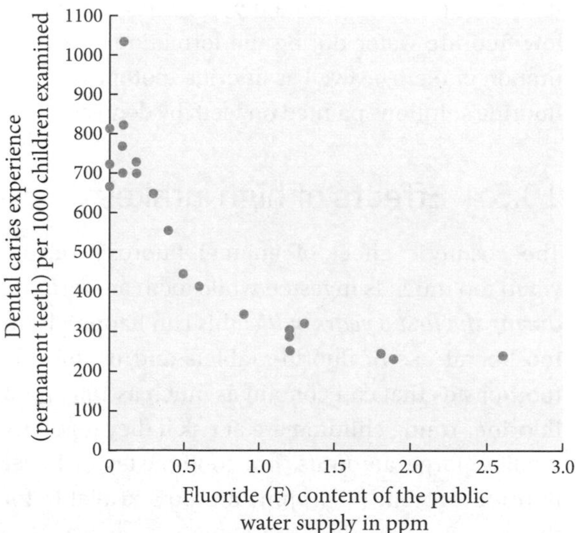


Fig. 10.7 Relation between the amount of dental caries (permanent teeth) observed in 7257 selected 12–14-year-old white schoolchildren from 21 US cities (in four states) and the fluoride content of the public water supply.

*Source:* Dean, H.T., Arnold, F.A., and Elvove, E. (1942) *Pub Health Rep*, **57**, 1155–79.

## 10.5.1 Function

Fluoride acts to reduce dental caries in two ways:

1 When ingested by young children while the permanent (second set of) teeth are forming inside the jaw, before they erupt, the blood-borne

fluoride combines in the calcium phosphate hydroxy apatite crystals of the enamel, making it more resistant to acid erosion. Erosion of dental enamel by acid, produced by mouth bacteria metabolizing sugars, is the cause of caries. For this fluoride action, young children should either be drinking water with adequate natural or added fluoride, or be given fluoride tablets before their permanent teeth erupt.

2 Fluoride also has a post-eruptive action. In solution in the saliva or in contact with the teeth via toothpaste, it inhibits bacterial enzymes that produce acid in plaques on the teeth, and it increases remineralization of incipient enamel lesions. This means that fluoride continues to have cariostatic action in adult life.

## 10.5.2 Metabolism of fluoride

The fluoride ion occurs in water, and both ionic and non-ionic (or bound) forms occur in food and beverages. Ionic fluoride is rapidly and almost completely absorbed, whereas organic or protein-bound forms are less well absorbed (about 75%), and inorganic bone fluoride as in bonemeal even less (<50%). There is a transitory rise in plasma fluoride following fluoride ingestion, after which it returns to about 0.1 ppm. Some fluoride is taken up by the bones and retained for a long time, but most is rapidly excreted in the urine, with small amounts in sweat and faeces. The urinary output gives a good indication of the daily fluoride intake, whereas the bone fluoride content reflects the long-term intake. There is minimal transfer of fluoride across the placenta. Fluoride content of breast milk is little affected by small supplements of fluoride, such as 1.5 mg/day.

## 10.5.3 Sources of fluoride

Dietary sources Beverages are the principal sources of fluoride but their contribution depends on the fluoride concentration of the water supply. Tea leaves, and hence tea infusions, are also major sources (about 1–2 ppm), depending on the water fluoride content and strength of the infusion. Thus, beverages can give as little as 0.2 mg/day for non-tea drinkers drinking unfluoridated or low fluoride water, and up to 2–4 mg/day or even more for frequent drinkers of strong tea prepared with fluoridated water.

Bottled water The fluoride content of bottled water (and carbonated beverages) depends on that of its source, usually a natural spring, and it is usually absent. If people drink most of their day's water from bottles, they are likely to by-pass the dental benefit of tap water.

Foods contain traces of fluoride, contributing for an adult about 0.5 mg/day; plant foods (1 ppm) generally contain more fluoride than animal foods (0.1 ppm), apart from marine fish (1–3 ppm). The fluoride content of processed food comes mainly from the fluoride content of the water used in processing or in the home. The fluoride content of infant formulas reflects the processing of the powdered formula and also the water used to make it up. Milk formulas usually contain more fluoride than human milk.

Non-dietary sources These include fluoride tablets or drops given mainly to children drinking low-fluoride water during the formation and maturation of teeth, as well as fluoride toothpastes and fluoride solutions painted on teeth by dentists.

## 10.5.4 Effects of high intakes

The cosmetic effect of enamel fluorosis occurs when too much is ingested while teeth are forming *during the first 8 years of life*; this can happen from too liberal use of fluoride tablets and/or fluoride toothpastes that can contain as much as 1000 ppm fluoride. Young children are at risk if they regularly swallow large amounts, but toothpastes of lower fluoride content (400 ppm) are now available for them.

The skeleton is affected by chronic high levels of fluoride intake, as from long-term drinking water with 20 ppm fluoride. This may cause dense bones and joint abnormalities; this skeletal fluorosis occurs in parts of India, China, and South Africa.

The acute lethal dose of sodium fluoride is 5 g (2300 mg F). It would be impossible for water containing 1 mg/L to cause this.

Large doses of fluoride have been used in the treatment of osteoporosis, but the bone quality tends to be poor, fractures may increase, and there are doubts about the safety of such treatment.

### 10.5.5 Fluoride and dental health

The first public water supply to be fluoridated was in Grand Rapids, Michigan, in 1945. Hastings in 1954 was first in New Zealand; Canberra and Hobart in Australia followed in 1964. The following countries have some of their water supply fluoridated: USA [the most; rest in alphabetical order]: *Argentina, Australia, Brazil, Brunei, Canada, Chile, Colombia, Guyana, Hong Kong, Ireland, Israel, Korea, Malaysia, New Zealand, Philippines, Singapore, Serbia, Spain and Vietnam.*

Rates of decayed, missing, and filled teeth (DMFT) indicating dental caries fell significantly in children with water fluoridation.

Yet in other developed countries, in Europe and elsewhere, with only some or no water fluoridated (10% in the UK), caries rates are also much lower than in earlier generations. The explanation is thought to be fluoride in toothpastes, universal school dental services, use of topical fluoride by dentists, in some cases fluoride tablets, even fluoridated salt in France, Germany, Sweden, Jamaica, and Colombia, and fluoridated milk in some countries.

While dental caries is controlled in affluent communities, it is increasing in poor people in the rest of the world (Bagramian et al., 2009), who cannot afford toothpaste, let alone dentists. A 2006 conference of dental experts from 30 countries 'expressed their deep concern about growing disparities in dental health and the lack of progress in tackling the worldwide burden of tooth decay, particularly in disadvantaged populations.' They called for action to promote dental health by using fluoride.

Water fluoridation has the great advantage of reaching all members of the community, particularly those with poor dental hygiene and no access to dentists. The US Centers for Disease Control and Prevention listed water fluoridation as one of the ten great public health achievements of the twentieth century. It is endorsed by the WHO, the European Academy of Paediatric Dentistry, and national dental associations of USA, Canada, Australia, and other countries. When it is proposed that an area's water be fluoridated, the majority of the public usually agree, but people who oppose it (similar to those against immunization) tend to be intense. Note that 1.0 ppm F in water is lower than the average F content of seawater. So the decision should ideally be made by the Ministry of Health, not the local council. Where water is supplied by a private company, objection is because of the (modest) cost to the company, which is very much less than the cost of dealing with dental caries in the community.

### 10.5.6 Recommended dietary intake

Most national and international health organizations recommend a water supply in the range 0.7–1.0 ppm fluoride for temperate climates, using the average local maximum temperature as a predictor of water intake; the lower concentrations are for warmer climates, where more beverages are drunk. The US Institute of Medicine (1997) suggests 'adequate intakes' of fluoride are between 0.5 and 1.0 mg/day for young children from 6 months to 8 years of age, and in adults 4 mg/day in men and 3 mg/day in women. The same amounts are recommended for Australia and New Zealand.

### 10.5.7 Risks of water fluoridation?

It is widely agreed that fluoridation of drinking water to a level of 1 ppm has no known adverse health effects. There is no good evidence that fluoridated water is associated with allergic reactions or hypersensitivity, sudden infant death syndrome, stomach or intestinal problems, birth defects, Down syndrome, or genetic mutations. No association of cancer with exposure to fluoridated water has been found (Doll and Kinlen, 1977; Smith, 1980).

## Further reading

1. **Bagramian, R.A., Garcia-Godoy, F., and Volpe, A.R.** (2009) The global increase in dental caries. A pending public health crisis. *Am J Dent*, **22**, 3–8.
2. **British Fluoridation Society** (2004) *One in a million: the facts about fluoridation*, 2nd edn. Liverpool: British Fluoridation Society.
3. **Dean, H.T., Arnold, F.A., and Elvove, E.** (1942) Domestic water and dental caries. V. Additional studies: experience in 4,425 white children, aged 12 to 14 years, of 13 cities in 4 states. *Publ Health Rep*, **57**, 1155–79.
4. **Doll, R. and Kinlen, L.** (1977) Fluoridation of water and cancer mortality in the USA. *Lancet*, **1**, 1300–2.
5. **Gluckman, P. and Skegg, D.** (2014) *Health effects of water fluoridation: a review of the scientific evidence*. Wellington: Royal Society of New Zealand/Office of the Prime Minister's Chief Science Advisor.
6. **Institute of Medicine** (1997) *Dietary reference intakes for calcium, phosphorus, magnesium, vitamin D and fluoride*. Washington, DC: National Academy Press.
7. **McDonagh, M.S., Whiting, P.F., and Wilson, P.M.** (2000) Systematic review of water fluoridation. *BMJ*, **321**, 855–9.
8. **Rugg-Gunn, A., Spencer, A.J., Whelton, H.P., et al.** (2016) Critique of the review of 'Water fluoridation for the prevention of dental caries' published by the Cochrane Collaboration in 2015. *Br Dent J*, **220**(7), 335–40.
9. **Smith, A.H.** (1980) An examination of the relationship between fluoridation of water and cancer mortality in 20 large US cities. *NZ Med J*, **91**, 413–16.

# 10.6 Other trace elements

A. Stewart Truswell

### 10.6.1 Chromium

Chromium (Cr) has been found in some situations to potentiate the action of insulin. No mammalian enzyme is known that requires chromium. Cr is widely distributed in trace amounts in the food supply, but is poorly absorbed. A small number of cases on long-term total parenteral nutrition developed symptoms—weight loss and high blood glucose—that responded to small amounts of chromium salt. Chromium at 10–15 μg/day is now routinely included in the fluids for long-term parenteral nutrition. The Institute of Medicine set the adequate intake (AI) at 25 μg/day for women and 35 μg/day for men.

### 10.6.2 Manganese

Manganese (Mn) is involved in the formation of bone and in the function of several enzymes, particularly superoxide dismutase and arginase. Mn deficiency can be demonstrated in animals and is a practical problem in the poultry and pig industries. However, the few reports that describe human deficiency features have shortcomings. Tea is a good dietary source. One child on long-term total parenteral nutrition had poor growth and had bone demineralization corrected by manganese supplementation. Manganese at 60–100 μg/day is included in fluids for total parenteral nutrition. The Institute of Medicine set the AI at 5 mg/day for women and 5.5 mg/day for men.

### 10.6.3 Molybdenum

Molybdenum (Mo) is essential for growing vegetables. In animals, including humans, it is a cofactor (bound to protein) for three enzymes: sulphite oxidase, xanthine oxidase, and aldehyde oxidase. The main case for essentiality of molybdenum is a severe genetic defect that prevents synthesis of sulphite oxidase, but this cannot be corrected by diet. A single case with biochemical features of Mo deficiency occurred in a patient on long-term total parenteral nutrition, corrected with ammonium molybdate. The best dietary source is legumes. Molybdenum is well absorbed. The Institute of Medicine set the RDI at 45 μg/day for men and women.

# Vitamin A and Carotenoids

David I. Thurnham



## 11.1 History

Vitamin A deficiency was known to the ancient Egyptians and to Hippocrates. Liver was prescribed for night blindness. More recently, Snell demonstrated in 1880 that cod liver oil was effective in curing not only night blindness but also Bitôt's spots. Poor growth was a common feature of young rats fed on diets of pure protein, starch, sugar, lard, and salts, and McCollum and Davis (1912) reported there was an essential fat-soluble factor in butter, egg yolk, and cod liver oil that would overcome this growth inhibition. Osborn and Mendel reported similar results, but it was McCollum and Davis who gave the name 'fat-soluble factor A' to the component in these foods to distinguish it from the 'water-soluble factor B' they found in whey, yeast, and rice polishings. Rosenheim and Drummond reported in 1920 that the vitamin A activity in plant foods was related to their content of carotene, a pigment isolated from carrots 100 years earlier. The ultimate confirmation that carotene is the source of plant vitamin A activity came from Thomas Moore in Cambridge in 1957.

## 11.2 Units, terminology, nomenclature, and chemical structures

Vitamin A is the generic term used to include retinol and related structures with 20 carbon atoms, and the pro-vitamin A carotenoids with 40 carbon atoms. The term 'retinoids' is used for the group of naturally occurring compounds that have a structure similar to retinol as well as others that have been synthesized in the search for new therapeutic compounds.

Pre-formed vitamin A structures include all-*trans* retinol (vitamin $A_1$, alcohol form), all-*trans* retinal (aldehyde form), and 3-dehydroretinol (vitamin $A_2$) (Fig. 11.1). Vitamin $A_2$ is found in freshwater fish. In addition, there are various oxidized forms of retinol with vitamin A activity (e.g. all-*trans* retinoic acid and 9-*cis* retinoic acid), which are important in the genetic control of metabolic functions. Small amounts of both retinol and retinoic acid can also circulate as β-glucuronide conjugates. The basic unit of activity of vitamin A is the retinol equivalent (RE) by which 1 μg RE is the same as 3.33 IU or 3.5 nmol of retinol. In foods and pharmaceutical preparations, vitamin $A_1$ occurs mainly as vitamin A palmitate. To overcome the need to use different weights if different esters are present, vitamin A activity is usually expressed as international units (IU).

The common carotenoids found in the blood are shown in Fig. 11.2. Of these, only α-carotene,

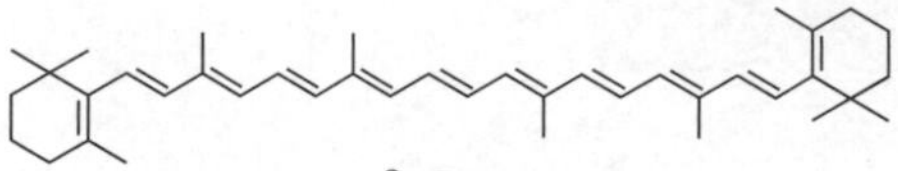

β-Carotene

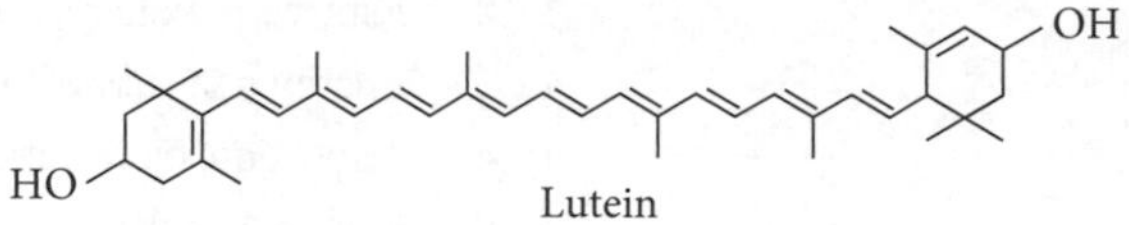

HO OH

Lutein

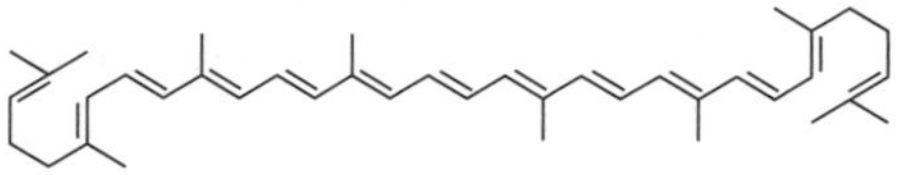

Lycopene

OH

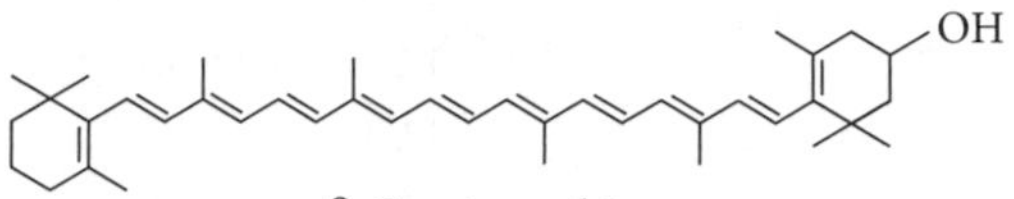

β-Cryptoxanthin

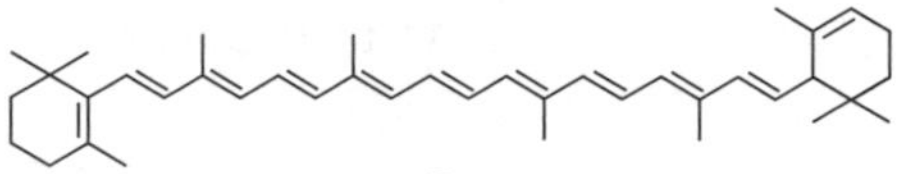

α-Carotene

Fig. 11.2 Carotenoids found commonly in human blood.

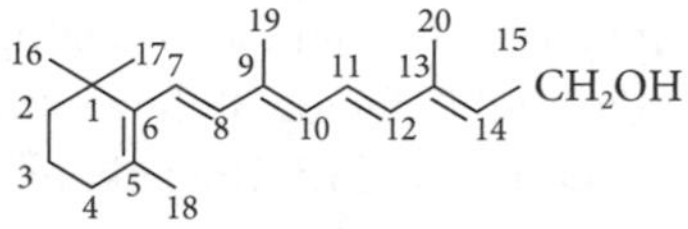

16 17 7 9 19 11 13 20 15 $CH_2OH$ 2 1 6 8 10 12 14 3 4 5 18

All-*trans* retinol (vitamin $A_1$)

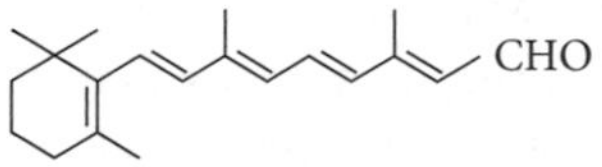

CHO

All-*trans* retinal

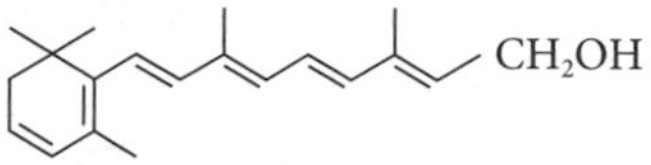

$CH_2OH$

3-Dehydroretinol (vitamin $A_2$)

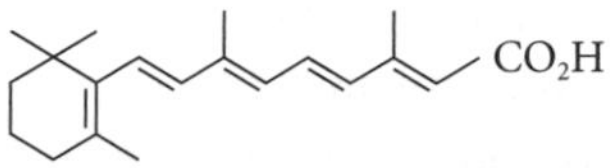

$CO_2H$

All-*trans* retinoic acid

9 $CO_2H$

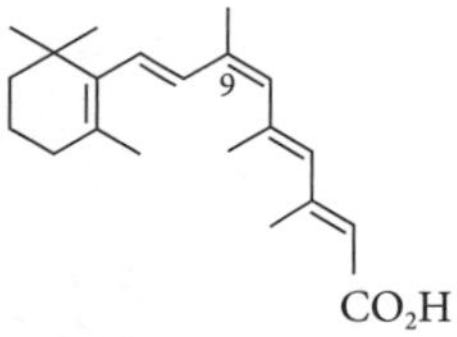

9-*cis* Retinoic acid

Fig. 11.1 Structures of the common retinoids.

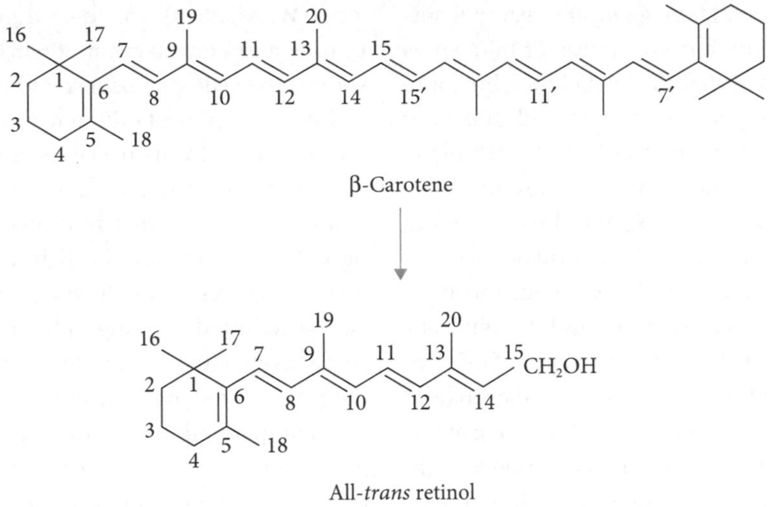

β-Carotene

All-*trans* retinol

Fig. 11.3 β-carotene conversion to retinol.

β-carotene, and β-cryptoxanthin are pro-vitamin A carotenoids, as they contain at least one β-ionone structure with no functional groups attached. The enzyme required to produce vitamin $A_1$ from pro-vitamin A carotenes is β-carotene 15, 15′-monooxygenase (EC 1.13.11.21) that is found in the small intestine and liver. It is believed to split carotene down the middle of the molecule (Fig. 11.3). β-Carotene has two β-ionone structures, at each end of the molecule, so theoretically should form two molecules of vitamin $A_1$, whereas all other carotenes have only one β-ionone structure and can only form one molecule of vitamin $A_1$.

Units of β-carotene are determined by their bioequivalence to vitamin $A_1$. WHO/FAO (FAO/WHO, 1967) recommends 1 μg RE is equivalent to 1 μg retinol, 6 μg of β-carotene, or 12 μg of carotenoids containing only one β-ionone ring. The equivalency was revised by the US Institute of Medicine in 2000 to 12 and 24 μg for β-carotene and other pro-vitamin A carotenoids, respectively; however, the revision is not universally accepted yet (FAO/WHO, 2004). IU are not used to quantify carotenes in current literature.

## 11.3 Functions of retinol: physiology, biochemistry, and molecular biology

Retinoic acid supports many of the important functions of vitamin A—cellular differentiation, embryogenesis, synthesis of glycoproteins, immunity, and growth—but it cannot be reduced back to retinol in the body; therefore, an animal that is maintained only on retinoic acid will be blind and also will fail to reproduce successfully. Only retinol can fully support vision and reproduction.

### 11.3.1 Vision

Both retinol and retinoic acid are needed to maintain a healthy eye. Retinol is needed for the visual process and one of the earliest signs of vitamin A deficiency is a failure to see in dim light. This is known as night blindness and is due to impaired ability to regenerate 11-*cis*-retinal following bleaching of the visual pigment layer. Damage on the external surface of the eye in the form of xerosis or Bitôt spots are also early signs of vitamin A deficiency, but more likely due to inadequacy of retinoic acid. The latter is required to maintain the surface epithelium of the eye and, in vitamin A deficiency, tear production is impaired, debris accumulates, and the eye is more vulnerable to bacterial attachment and disease.

The visual process in the retina is dependent on the ability to synthesize 11-*cis* retinal and its behaviour on exposure to light. There are two types of light receptor cell in the retina of the human eye, the rods and the cones. The rods are responsible for seeing at low light intensities, whereas the cones are used for light of higher intensities and colour vision. The chromophore (11-*cis* retinal) is the same in both, but the proteins attached to it are different. In the rods, the chromophore is attached to rhodopsin (a guanosine triphosphate (GTP)-binding protein, or G-protein), while in the cones, there are three very similar proteins that are sensitive to short (blue), middle (green), or long (red) wavelengths. Light detection by these G-proteins is due to the covalent binding of the chromophore to the apo-opsin proteins. The specific amino acid sequence of each opsin coupled to 11-*cis* retinal determines its spectral sensitivity. Rhodopsin contains the protein opsin, an ethanolamine-containing phospholipid, and the chromophore. The basic mechanism of light excitation is common to both systems, but has been studied in far more detail in the rods as there are about 100 million rods compared with 3 million cones in the human eye.

The retina of the mammalian eye comprises ten layers and the photoreceptors form the layer underneath the retinal pigment epithelium (the outermost layer). This means that before the process of phototransduction can begin, light has to pass through all nine layers. The chromophore receptors are linked to G-proteins and the latter regulate specific plasma membrane enzymes or, in the case of rhodopsin, ion channels in response to receptor binding. Before light excitation, the chromophore locks the receptor protein opsin by a Schiff-base linkage in its inactive form.

The primary event in visual excitation of rhodopsin is the photoisomerization of the 11-*cis* retinal to all-*trans* retinal within a few picoseconds, leaving the rhodopsin in the activated, metarhodopsin II (Meta II) form (Weiss, 2020). The Meta II forms of the opsins in rods and cones activate their G-proteins, which stimulates cyclic guanosine monophosphate (cGMP) hydrolysis by phosphodiesterase 6. This reduction in cGMP causes closure of cGMP-gated $Na^+/Ca^{2+}$ channels, hyperpolarization of the photoreceptor cell membrane, and signal transmission via the optic nerve to the brain. Note that light activation will cause graded changes in membrane potential. Each opsin can be activated by a single photon but the more opsins activated, the greater the hyperpolarization and the greater the signal transmitted to the brain.

Accompanying the above changes, all-*trans* retinal is released from the opsin molecule, reduced to all-*trans* retinol, and taken up by the retinal pigment epithelium. Within the outermost layer of the eye and in the absence of light, all-*trans* retinol is firstly esterified, then isomerized to 11-*cis* retinol, and subsequently oxidized to 11-*cis* retinal when it re-enters the photoreceptor outer segment ready to commence the cycle with another apo-opsin (Pepperberg and Crouch, 2001).

The action of light on rhodopsin is to alter its colour from magenta through orange to yellow and ultimately white 'bleached'. In this form, it is apo-opsin and unattached to all-*trans* retinal, although the latter is still bound to interstitial retinoid binding protein.

### 11.3.2 Reproduction

Vitamin A deficiency results in infertility in males, while in females there are low rates of conception and high rates of stillbirths. Both retinol and retinoic acid are needed for successful reproduction and retinoic acid has been shown to have a profound influence on embryonic development, being responsible for formation of limb buds, hind brain, spinal cord, and the eye. β-Carotene may also function in reproduction, as it is deposited in substantial amounts in the corpus luteum of the ovaries.

### 11.3.3 Cellular functions

Almost all the other functions of vitamin A are under genetic control and are mediated by retinoic acid derivatives binding to nuclear receptors that mediate retinoid action on the genome.

There are retinoic acid receptors (RAR) and retinoid X receptors (RXR), and each family has three major subtypes, α, β, and γ. The subtypes have different functions, their distribution in cells is different, and each subtype can also have several different isoforms. The nuclear retinoid receptors generally act as hetero-dimers, of which the most common is RAR–RXR. To be active, RAR must bind to retinoic acid (either 9-*cis* or all-*trans*), whereas RXR can bind to 9-*cis* retinoic acid, a synthetic ligand, or it can also form heterodimers with nuclear receptors for triiodothyronine, 1,25-hydroxyvitamin $D_3$, and the peroxisome proliferator-activated receptor (PPAR, involving essential fatty acid metabolites), farnesoid receptor (bile acid and lipids), pregnane X receptor (steroid and xenobiotic-sensing metabolites), adrostane receptor (drug metabolites), and others. When activated, the dimeric nuclear receptors bind to 'response elements' in specific genes to change the level of expression of that gene. Most interactions produce positive responses but some are repressors of target genes, e.g. some thyroid hormone interactions. In this way, vitamin A regulates the synthesis of a large number of proteins vital to maintaining a great many physiological functions in the body.

**Cellular differentiation** This is the series of morphological changes that take place to produce mature epithelia. The outer skin is characterized by keratin-producing cells, whereas the gut is characterized by mucus-secreting tissue containing many goblet cells. Most of the functions of vitamin A in cell differentiation are regulated by retinoic acid. When vitamin A is lacking, keratin-producing cells replace mucus-secreting cells in the intestinal and respiratory tracts. In the eye there is xerosis and drying of the conjunctiva and cornea.

**Embryogenesis** Retinoic acid isomers play important roles in embryogenesis through their control of genes linked to development and growth (homeobox genes). Both deficiency and excess can have adverse effects. Implants containing all-*trans* retinoic acid placed in the anterior part of a developing chick limb bud mimic the activity of the naturally occurring zone of polarizing activity. Retinoic acid has been shown to have a profound influence on embryonic development, being responsible for formation of the limb buds, hind brain, spinal cord, and eyes. Correct morphological development depends on the concentration of all-*trans* retinoic acid. Experimental evidence and tragedies from overexposure to synthetic retinoids highlight the particular importance of avoiding excess vitamin A during pregnancy.

**Synthesis of glycoproteins and glycosaminoglycans (GAGs)** Glycoproteins are polypeptides with short chains of carbohydrates. They are important components in mucus. Many glycoproteins on the surface of the cell are receptors for other glycoproteins (e.g. growth factors). GAGs are long, unbranched polysaccharide chains that provide a viscous extracellular matrix on the cell surface. They are important in connective tissue and can provide a passageway for cell migration or lubrication between joints. Retinoids have been shown to be involved in the synthesis of some of these compounds. Sulphate is an important component of several members of the GAG family, e.g. chondroitin sulphate and heparin, and all-*trans* retinoic acid has been shown to induce several sulphotransferase enzymes in cellular and animal experiments (see Box 11.1).

**Immunity and host defence** Vitamin A is generally believed to be important for resistance

**Box 11.1 Effects produced in vitamin A deficiency by impairment of mucopolysaccharide synthesis**

- Reduced wettability of the eye surface.
- Reduced tear production contributing to the xerosis (dry and rough) of the eye surface.
- Reduced mucus production by mucous membranes with increased susceptibility to bacterial attachment and infection.
- Reduced ability to taste through changes in the taste buds.
- Changes in the skin giving rise to follicular keratitis: seen more in adults than children.
- Changes in the ground substance of bone, cartilage, and teeth resulting in defective formation of these substances during growth.

to infections—hence the term 'anti-infective vitamin'. Measles is a very serious infection in vitamin A-deficient children and there is a strong protective effect of supplemental vitamin A (Hussey and Klein, 1990). Vitamin A deficiency impairs immunity by impeding normal regeneration of mucosal barriers damaged by infection and by diminishing the function of neutrophils, macrophages, and natural killer cells. Thus, mortality from infections is higher in communities where vitamin A deficiency is found.

Pathogens that infect mucosal surfaces trigger adaptive immune responses by initially interacting with antigen-presenting cells such as dendritic cells. Antigen-loaded dendritic cells interact with antigen-specific B- and T-lymphocytes in subepithelial aggregates, such as Peyers patches, resulting in proliferation and differentiation of naive T cells into memory/effector T cell subsets, potentially including T helper (Th) 1, 2, and 17 cells. Dendritic cells in the intestine produce retinoic acid, which is not produced by dendritic cells at other sites (Benson et al., 2007). Retinoic acid potentially steers the development of Th cells in the direction of Th2 and Treg subsets. However, the effector functions of mucosal adaptive immunity will only be effective if the IgA-secreting plasma cells, memory B cells, and memory T cells return to the lamina propria underlying the mucosal epithelium in the intestine. Evidence suggests that retinoic acid produced by dendritic cells at the time of antigen presentation promotes the development of specific mucosal cell adhesion molecules to assist return and activity of previously stimulated immune cells.

**Growth** Vitamin A influences bone growth by modulating the growth of bones through remodelling. The vitamin is necessary for the normal cycle of growth, maturation, and degeneration of cells in the epiphyseal cartridge. However, in only one study did preschool children who received extra vitamin A in the condiment monosodium glutamate show higher rates of growth by comparison with children from control villages. Observations on the effect of vitamin A supplements on early child growth in areas where vitamin A deficiency is a risk are inconsistent, probably because growth is dependent on many nutrients and just replacing vitamin A in a child's diet may do no more than allow more vitamin A storage, until such time as the correct balance of nutrients in the diet is restored.

**Haemopoiesis** Vitamin A deficiency in humans and in experimental animals has been consistently associated with anaemia, and studies have shown that both vitamin A and iron are required to promote a full haematological response. Vitamin A may assist iron homeostasis by influencing iron regulatory proteins (IRP), (Schroeder et al., 2007). The anti-inflammatory effects of vitamin A supplements may also stimulate re-utilization and absorption of iron indirectly by reducing infection and inflammation (Hess et al., 2005).

# 11.4 Absorption, distribution, and transport

## 11.4.1 Digestion

A schematic representation of vitamin A metabolism and the role of retinol-binding protein and transthyretin is given in Fig. 11.4. Vitamin A and its precursors are ingested in the food matrix. Proteolysis in the stomach may release some of the vitamin A and carotenoids from foods. However, to release carotenoids from vegetables, they should be thoroughly cooked and masticated, otherwise the carotenoids will remain within the cellulose structures. Released vitamin A and carotenoids aggregate with lipids into globules and pass into the upper part of the small intestine. Here, pancreatic lipase and other esterases hydrolyse lipids (triglycerides, etc.), retinyl esters, and any esters of carotenoids. Bile salts assist in emulsifying the contents of the gut lumen and lipid micelles are formed.

## 11.4.2 Absorption

**Retinol** Lipid micelles are taken up by the cells lining the intestine and as much as 90% of retinol in foods is absorbed and utilized. The high efficiency

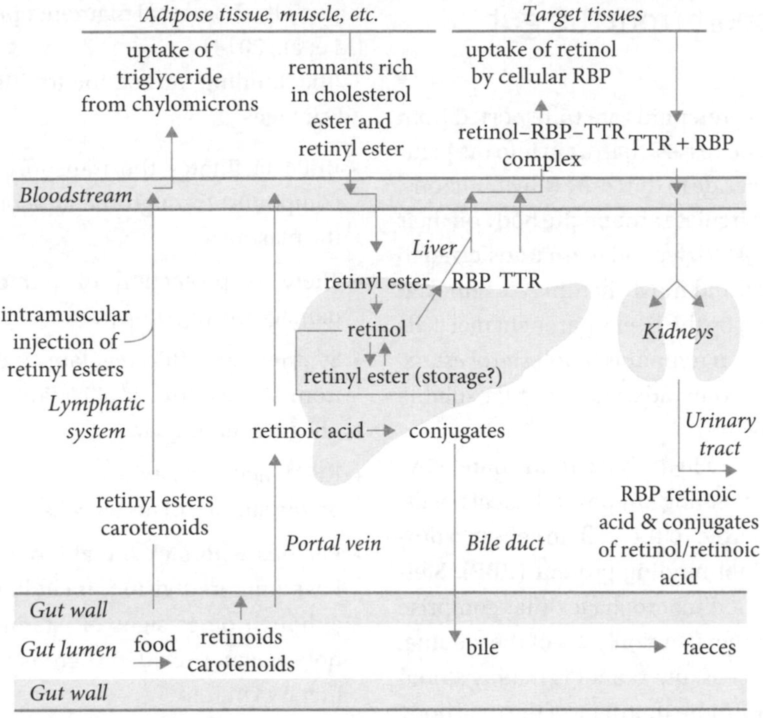


Fig. 11.4 Schematic representation of vitamin A metabolism and role of retinol-binding protein (RBP) and transthyretin (TTR). In the gut wall some β-carotene and retinol is oxidized to all-*trans* retinoic acid, where it may have local activity. Additionally, in target tissues the all-*trans* and 9-*cis* isomers of retinoic acid are formed from retinol to interact with the retinoic acid receptors (RAR) and retinoid X receptors, respectively, which are found in every cell type (see section 11.3.3).

of this process may be due to the specific cellular binding protein (CRBPII) in the mucosal cell that carries retinol to the enzyme lecithin:retinol acyltransferase (LRAT), the main intestinal enzyme that esterifies retinol and delivers it to the chylomicrons. Very little retinyl ester is absorbed, but hydrolysis of the vitamin A esters in the gut is fairly efficient, so more than 50% of the vitamin A in large (pharmaceutical) doses is also absorbed. Within the enterocyte, absorbed retinol is re-esterified to retinol palmitate and, together with triglycerides and other fat-soluble nutrients, is packaged into chylomicrons for transport to the liver.

**Carotenoids** Carotenoids are also fairly efficiently absorbed at low doses (<5 mg), but the amount absorbed falls off steeply as the dose rises. There are three potential fates for the carotenes absorbed. Some is metabolized by β-carotene 15,15′-monooxygenase to form first retinal, then retinol, and finally retinol palmitate. Some carotene is taken up by the chylomicrons unchanged, while the epithelial cell retains the remainder, which, if not converted to retinol, is lost through cell turnover in the faeces. The amount of carotene converted to retinol is variable. A number of factors may be responsible. Two common non-synonymous single-nucleotide poor-converter polymorphisms with variant allele frequencies of 42% and 24%, respectively, have been identified. Carriers of these variant alleles had 32–69% lower ability to convert β-carotene to retinol (Leung et al., 2009). There is also evidence that other carotenoids in the diet, like lutein, may inhibit conversion of β-carotene to retinol if the molar ratio of lutein:β-carotene is greater than 3:1.

## 11.4.3 Transport from the gut to the liver

Retinol esters and carotenoids are transported from the gut, via lymphatic vessels that drain into the jugular vein, with triglyceride in the core of chylomicrons. The chylomicrons circulate around the body on their way to the liver. Most triglycerides are transferred to extrahepatic tissues and most vitamin A is removed from the circulation by the liver's parenchymal cells when the chylomicron remnants (cholesterol esters, retinol palmitate, carotenoids, and other fat-soluble vitamins) reach the liver.

The retinyl esters are hydrolysed in the parenchymal cells and, after meeting any physiological needs, the retinol is transferred to the stellate cells in a process involving retinol-binding protein (RBP). Stellate cells are modified macrophages that comprise 7% of liver cell numbers, but only 2% of the volume. Within the stellate cells, the retinol is mainly stored as palmitate (>90%). More than 80% of the total body vitamin A is stored in the liver and some in the kidney. Generally, vitamin A in the liver increases with age. On average, a 70-kg man with a liver weighing 1.8 kg would have 150–300 mg of stored vitamin A, enough to last for a year or more of no intake.

## 11.4.4 Mobilization of vitamin A from the liver and serum transport

Retinol is released from the liver bound to RBP4. Following the completion of the Human Genome Project, the gene encoding RBP, the circulating transport protein for retinol, was given the name *RBP4*. RBP4 has a molecular weight of 21,000 and one binding site for retinol. Holo-RBP (the RBP4–retinol complex) is released from the liver bound to another protein, transthyretin (TTR). TTR has a molecular weight of 55,000 and was previously known as thyroxine-binding pre-albumin; it also has one binding site, so the whole complex is a 1:1:1 structure. In plasma, 95% of the retinol is bound in the retinol–RBP4–TTR complex. In 2007, a specific cell surface receptor for RBP4 was reported, STRA6. The receptor is expressed in a number of tissues with a high demand for vitamin A, especially the retinal pigment epithelium of the eye (Li et al., 2014).

The binding of retinol to RBP4 confers some advantages:

- RBP4 facilitates the transport of a lipid-soluble compound through the aqueous environment of the plasma.
- There is protection of retinol from oxidative damage during transport.
- Synthesis of RBP4 regulates the release of retinol from storage for mobilization when dietary supplies are inadequate.
- RBP4 facilitates delivery of retinol to specific sites on the surface of target cells.
- Persons who lack the ability to synthesize RBP4 have impaired vision, including night blindness, although other signs of vitamin A deficiency do not appear, provided there is a regular supply of dietary vitamin A.

Holo-RBP4 is taken up by specific cell-surface receptors. Once the retinol is transferred within the cell, the apo-RBP4 (the free protein) is released from the receptor and can be recycled. Some is excreted by the kidney. Inside cells, all-*trans* retinol and all-*trans* retinal bind to cellular retinol-binding protein (CRBP).

Some cells can also take up retinol palmitate from circulating chylomicrons, very-low-density lipoproteins (VLDL), and low-density lipoprotein (LDL) via lipoprotein receptors. Within the cell, the ester is hydrolysed and retinol is bound to CRBP for further metabolism. The lack of clinical evidence of vitamin A deficiency (except impaired vision) in persons unable to synthesize RBP4 suggests that tissue uptake of retinol esters from lipoproteins may be the main source of vitamin A for most tissues and that RBP4 was developed to enable stores of vitamin A to be utilized at times of deficiency (Li et al., 2014).

Inside the cell, retinol is oxidized to retinal and then to all-*trans* retinoic acid by local expression of retinoic acid-synthesizing aldehyde dehydrogenase. Some of the all-*trans* retinoic acid is converted to 9-*cis* retinoic acid. The two forms, all-*trans* and 9-*cis* retinoic acid, interact with the nuclear receptors RAR and RXR, respectively (see section 11.3.3). The cytochrome P450 enzyme CYP26, which has specific retinoic acid 4-hydroxylase activity, may regulate steady state levels of the active retinoids in target tissues.

Plasma carotenes are transported mainly in the low-density lipoproteins, while the more water-soluble xanthophylls are more concentrated in the high-density lipoproteins. Depletion studies suggest that half-lives of β- and α-carotene and β-cryptoxanthin are <2 weeks, of lycopene is 2–4 weeks, and lutein and zeaxanthin are 4–8 weeks. The shorter half-lives of the pro-vitamin A carotenoids may be evidence of their conversion to vitamin A in the tissues, but >80% of retinol synthesis from carotenes takes place in the gut during absorption.

### 11.4.5 Vitamin A excretion

In a healthy person no vitamin A is excreted per se. Oxidized metabolites can be found in the urine and any conjugated vitamin A products that might be formed by vitamin A excess would be secreted into the bile and then lost in the faeces. During illness, particularly with fever, retinol is lost in the urine, together with RBP; amounts can be as high as 500 μg retinol/day.

## 11.5 Vitamin A deficiency

### 11.5.1 Experimental deficiencies in animals

In 1928, Green and Mellanby showed that, when animals were placed on vitamin A-deficient diets, practically all died with infective lesions. The most useful models for such work are rats and mice, and the chicken is probably the next most useful species. Chickens are also useful to study xanthophyll absorption and metabolism. Viral diseases in the hen provide a useful model to study the influence of infection on both retinol and carotenoid metabolism, and interaction with deficiencies.

### 11.5.2 Experimental human deficiencies

The first important deficiency study in humans was the 'Sheffield' study (Hume and Krebs, 1949). Twenty-three men and three women, conscientious objectors to military service, volunteered to receive a diet containing no vitamin $A_1$ and <7 μg RE of carotene. By 18 months only three of the volunteers actually showed early signs of vitamin A deficiency, but intervention with carefully chosen amounts of retinol or β-carotene in arachis oil provided the information that established vitamin A requirements in the human adult to reverse signs of vitamin A deficiency to be 750 μg vitamin $A_1$ or 1800 μg β-carotene, and the bioequivalence of β-carotene to retinol (6 μg β-carotene 1 μg retinol).

### 11.5.3 Natural human deficiencies

For many years, the prevalence of vitamin A deficiency was defined by the number of persons showing signs of xerophthalmia (meaning dry eyes). Using such criteria, Sommer and West (1996) estimated that over 40 million children under the age of 6 years annually have mild to moderate xerophthalmia, 1% go blind annually, and 50–75% of these will die within the year.

Work with vitamin A showed that children at risk of xerophthalmia also have a high risk of infection. The term 'anti-infective vitamin' was attached to vitamin A. As techniques to measure plasma retinol concentrations improved, this has been used as the preferred biomarker of vitamin A status and current estimates of vitamin A deficiency in the world are of the order of 500 million. An estimated 250,000 to 500,000 vitamin A-deficient children become blind every year, half of them dying within 12 months of losing their sight (WHO, 2015). Vitamin A deficiency is a particular problem in groups with the highest vitamin A requirements: infants, preschool children, and pregnant and lactating women. Vitamin A status is commonly assessed in these groups using plasma retinol concentrations. Infection is particularly common in infants and children, and plasma retinol concentrations are depressed by inflammation. So, current estimates of global vitamin A deficiency are probably overestimates. Nevertheless, vitamin A deficiency is a major risk to human health in many parts of Asia, East Asia, Africa, Middle Eastern countries, and South and Central America.

## 11.5.4 Features of deficiency disease in humans

Changes in the eye People with marginal vitamin A deficiency become less able to see in dim light. This is **night blindness** and in communities where vitamin A deficiency exists, there is usually a specific word to describe the condition. The word frequently compares the person's behaviour to that of a chicken. Chickens have no rod cells in their retina and cannot see in the dark. If disturbed after nightfall, they bump into things when moving around. Night blindness correlates with low plasma levels of vitamin A, but retinol concentrations depressed by inflammation will not necessarily correlate with night blindness. Night blindness can occur in children and adults.

Another clinical indicator of early or marginal vitamin A deficiency is Bitôt's spots. These are foamy deposits on the surface of the conjunctiva (Fig. 11.5). They are found more frequently in pre- and school-age children than in adults. Both night blindness and Bitôt's spots disappear on treatment with vitamin A, with no lasting damage. In prolonged or more severe vitamin A deficiency, a series of changes can take place in the cornea, some of which are irreversible, as summarized in Box 11.2.

Changes to other epithelial tissues The influence of vitamin A on cellular differentiation is reflected in the widespread effects of vitamin A deficiency on other epithelial tissues. Follicular hyperkeratosis seen in vitamin A-deficient adults is due to skin keratinization blocking sebaceous glands with horny plugs, though this may not be a specific vitamin A effect. Vitamin A deficiency also affects the epithelial cell lining of the respiratory,

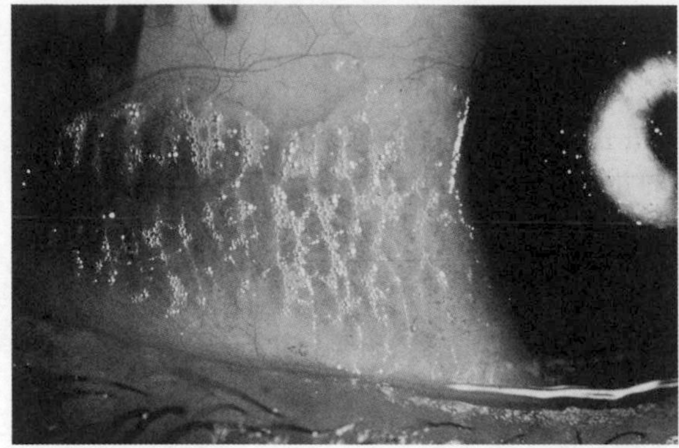

Fig. 11.5 Photograph of a Bitôt's spot.
*Source:* © David Thurnham.

gastrointestinal, and genitourinary tracts, as well as immune cell maturation. The different epithelia lose their characteristic structure and, hence, specialized function. The tracheal lining, for example, loses the cilia (which sweep foreign material up and out) and, in severe cases, the columnar epithelium is replaced by squamous epithelium in the intestine, villi are flattened, and mucous glands reduced.

Morbidity and mortality Sommer and colleagues in Indonesia showed that the death rate of children with mild xerophthalmia (night blindness and Bitôt's spots) was, on average, four times higher than in those with no xerophthalmia. Meta-analysis of intervention studies with vitamin A in countries where xerophthalmia was occurring showed reduced overall mortality by a highly significant 23%. However, a recent large vitamin A supplementation trial in Indian children, DEVTA (Deworming and Vitamin A supplementation Trial; Awasthi et al. (2013)), did not confirm the earlier results (Sommer et al., 2013).

Vitamin A and measles One particular infectious disease, measles, is much more severe, with about a 12% fatality rate in communities where xerophthalmia is seen. Even in developed countries the benefit of vitamin A treatment in cases of measles is striking. WHO and UNICEF recommend that all children with measles in developing countries should be given a massive dose of vitamin A. It is also advised for severe cases in developed countries (see Hussey and Klein, 1990).

**Box 11.2 Stages of xerophthalmia**

- *Conjunctival xerosis* (X1A) is one or more patches of dry, non-wettable conjunctiva described as 'emerging like sand banks at a receding tide at the sea-shore' when a child ceases to cry. Due to changes in the epithelium of the conjunctiva and lack of tear production.
- *Bitôt's spots* (X1B) are found in association with X1A (see Fig. 11.5).
- *Corneal xerosis* (X2) is an extension of conjunctival xerosis to the cornea. The corneal surface also begins to lose its transparent appearance. A light shone at an angle on the surface of the eye will often reveal the ripple-like surface and irregular reflection of the source of illumination.
- *Corneal ulceration* (X3A) usually begins first at the edge of the cornea and is characterized by small holes, 1–3 mm in diameter, with steep sides. If treatment with vitamin A is initiated at this stage, it may be possible to reverse the lesion and retain some sight.
- *More extensive corneal ulceration* (X3B) is characterized by larger defects that result in blindness. In Africa, it is reported that children with measles can develop X3B quickly without the appearance of the intermediate stages.
- *Corneal scars* (XS) result from the healing of the irreversible changes described above and may appear as white scar-like tissue in the cornea. On the other hand, if the cornea ruptures and the eye contents escape, a shrunken eyeball results.

*Source:* WHO (1982) *Control of vitamin A deficiency and xerophthalmia.* Report of a Joint WHO/USAID/Helen Keller International/IVACG Meeting, WHO Technical Report Series, No. 672. Geneva: World Health Organization.

**Nutritional anaemia** Human intervention studies with vitamin A have shown that there is a specific effect of vitamin A on haemoglobin synthesis in the absence of additional dietary iron (see section 11.3.3) (Hess et al., 2005; Schroeder et al., 2007).

## 11.6 Influence of disease/trauma on plasma retinol concentrations and mobilization of vitamin A from the liver

Infection or inflammation decreases plasma retinol concentrations and reduces mobilization of retinol from liver stores, part of the acute phase response to stress. RBP is a negative acute phase protein. Stimulation of the acute phase response by infection, surgery, or other stresses induces production of cytokines like interleukin-6 (IL-6) by macrophages that induce a wide variety of responses in the liver and elsewhere in the body. One effect is to reduce transcription of the messenger RNA for RBP synthesis. There is also increased vascular permeability and some plasma retinol moves into extracellular fluid compartments. As much as 500 μg/day of retinol and RBP can be lost in the urine. Figure 11.6 shows plasma retinol concentrations following surgery.

## 11.7 Biochemical tests for vitamin A deficiency

Plasma retinol concentration is the most widely used method of assessing vitamin A status, but it is influenced strongly by age (Fig. 11.7), female sex hormones, and inflammation. Infection or inflammation decrease plasma retinol concentrations and reduce mobilization of retinol from liver stores. These effects can have a major impact on several methods of measurement, namely

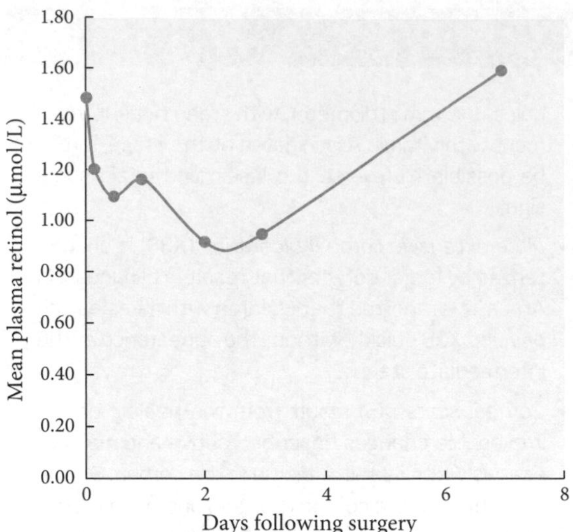


**Fig. 11.6** Influence of surgery on plasma retinol concentrations in South African women.

*Source:* Adapted from Louw, J.A., Werbeck, A., Louw, M.E.J., et al. (1992) Blood vitamin concentrations during the acute-phase response. *Crit Care Med*, **20**, 934–41.

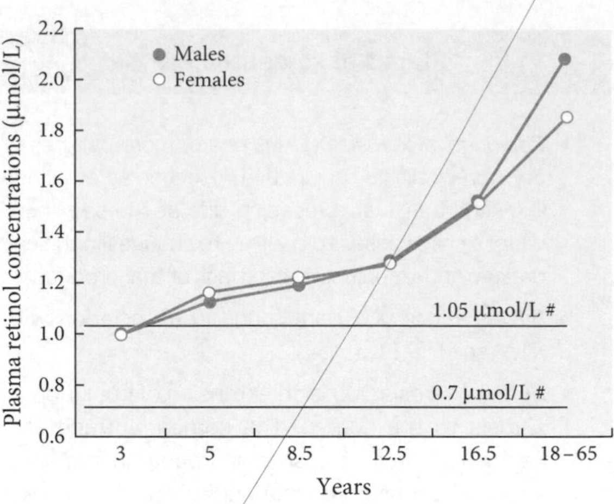


**Fig. 11.7** Influence of age on plasma retinol concentrations from three national surveys in Great Britain. #Thresholds of biochemical deficiency in children (0.7 μmol/L) and adults (1.05 μmol/L). Reproduced with permission.

*Source:* Thurnham, D.I., Mburu, A.S.W., Mwaniki, D.L., and de Wagt, A. (2005) Micronutrients in childhood and the influence of subclinical inflammation. *Proc Nutr Soc*, **64**, 502–9.

plasma retinol and RBP4 concentrations, and dose-response tests.

Recently, a method of correcting plasma retinol has been suggested, utilizing plasma concentration of two acute-phase proteins, C-reactive protein (CRP) and α1-acid glycoprotein (AGP). CRP is a marker of the early phase of inflammation and is associated with large depressions in plasma retinol concentrations. In contrast, AGP is associated with chronic inflammation or late convalescence when plasma retinol concentrations are less depressed. The use of these two biomarkers enables plasma retinol concentrations to be corrected for the influence of inflammation (Thurnham et al., 2003) or regression equations (Namaste et al., 2020) to adjust plasma retinol. In the absence of measurements of inflammation, the presence of β-carotene in plasma is a clue that a low plasma retinol is depressed by inflammation and not poor vitamin A status.

Probably the best (research) method of measuring total body vitamin A reserves is the radioisotope dilution assays. These tests may be more or less unaffected by inflammation, as 3 weeks is allowed for equilibration of the labelled tracer with body stores of vitamin A. This period should allow sufficient time for any inflammation at the time of treatment to subside and equilibration with body reserves of vitamin A to take place.

# 11.8 Functions of carotenoids

## 11.8.1 Bioavailability and conversion of β-carotene to retinol

Carotenoids are fat-soluble and the presence of adequate fat in the diet (5–10 g) at the time of consumption is necessary for optimal absorption. In plant leaves, carotenoids are present within pigment-protein complexes in the cell chloroplasts, and for the satisfactory utilization of plant carotenoids, the cellulose structure of cell walls has to be ruptured to release carotenoids. Cooking and chewing assist rupture of the cell walls of leaves during ingestion of food. However, in fruit, cell wall structure is usually much weaker than in leaves, and carotenes are found in the lipid droplets in chromoplasts. Hence, carotenoids from fruit are more easily bioavailable than those in leaves.

On the basis of data available to early committees, it was decided that, on average, only one-third of dietary plant carotene was absorbed. Furthermore, conversion of dietary β-carotene to retinol was also poorly efficient. Even when small amounts of pure β-carotene dissolved in oil (<2 mg) were fed to vitamin A-deficient volunteers, only 50% was converted to retinol. Hence, for many years the assumption was made that 6 μg β-carotene from plant sources was bioequivalent to 1 μg retinol. More recently, work suggested that far less β-carotene in vegetable foods was available for absorption. Hence, it was proposed (Institute of Medicine, 2000) that the bioequivalence was 12 μg β-carotene in plant food equal to 1 μg retinol, and for β-cryptoxanthin and α-carotene (with only one β-ionone ring per molecule) the retinol activity equivalent in plant foods was 24:1. The FAO/WHO (2004) considered the evidence but concluded that the 1:6 bioconversion factor should be retained until more information was available.

## 11.8.2 Macular pigments: lutein, zeaxanthin, and *meso*-zeaxanthin

The retinal epithelium in the human body contains almost exclusively only the three related xanthophyll carotenoids—zeaxanthin, *meso*-zeaxanthin (MZ), and lutein (see Fig. 11.8). They are concentrated in the **macula lutea** in the centre of the retina and provide its yellow colour. Humans consume 1–3 mg lutein per day and the ratio of lutein to zeaxanthin in the diet is about 5:1. Lutein and zeaxanthin occur in the blood in roughly the same proportions as found in food, but no MZ is usually found. The xanthophyll pigments occur widely in vegetables and fruits, but MZ is found in only a few foods. In spite of the different amounts of the three xanthophylls in the diet, lutein, zeaxanthin, and MZ occur in approximately equal amounts in the macula, but the two zeaxanthin isomers predominate in the centre. In addition, a binding protein that specifically binds zeaxanthin and MZ and not

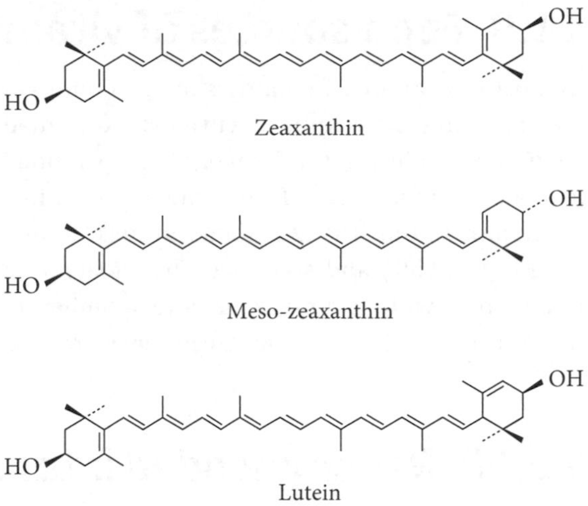

HO ... OH
Zeaxanthin

HO ... OH
Meso-zeaxanthin

HO ... OH
Lutein

Fig. 11.8 Main macular pigments showing zeaxanthin, meso-zeaxanthin, and lutein.

lutein has been isolated from optical tissues. This protein may enable the conversion of lutein to MZ in the eye. Experiments in monkeys suggest that lutein is the main source of the MZ in the macula.

Macular pigment optical density (MPOD) is a measure of the pigment density in the macula of the eye. The concentration of pigment extracted from the eyes of patients with macula disease at autopsy is lower than that of patients without disease. MPOD can also be measured in living subjects using physiological methods and can be shown to increase following supplementation with xanthophyll supplements and with vegetables like spinach. The increase was particularly rapid with supplements containing MZ (Connolly et al., 2010). Once increased, MPOD remains elevated for several weeks to months after withdrawal of the supplement. Green, leafy vegetables are the richest source of dietary lutein, but lutein and zeaxanthin in eggs yolks are more bioavailable. Orange peppers (capsicum) and yellow maize are sources of zeaxanthin.

**Lycopene** is the red pigment of tomatoes. It is often the most abundant of the carotenoids in human plasma. Having no β-ionone rings (see Fig. 11.2), it is not a pro-vitamin A. It is, however, an antioxidant (as is β-carotene). Early reports of a negative association of tomato sauce consumption and prostate cancer await further substantiation.

## 11.9 Food sources of vitamin A and carotenoids

In most industrialized countries the predominant dietary source of vitamin A activity is pre-formed vitamin $A_1$, mainly in the form of retinol palmitate in foods of animal origin. Liver is the richest source of vitamin A, but it is also found in milk, butter, cheese, egg yolk, and some fatty fish. Margarine is enriched with vitamin A to levels similar to those in butter. Pro-vitamin A carotenes are also obtained from plant foods, the main ones being dark green leafy vegetables, and some yellow- or orange-coloured fruits (e.g. mango, papaya, apricots, oranges).

In developing countries, plant sources of vitamin A are predominant in the diet. Certain foods like red palm oil, papayas, mangoes, and carrots are particularly rich in β-carotene.

## 11.10 Recommended intakes

Vitamin A requirements vary with age and the requirement for growth is a major determinant. Thus, requirements in infancy and childhood are higher per kg body weight than in adults. Infants are born with almost no stores of vitamin A, so it is critical that newborn infants obtain sufficient vitamin A to meet the needs of growth and of a developing immune system, and to reduce the risk of vitamin A deficiency-related severe morbidity, as well as to accumulate body stores. Mothers' colostrum and early milk is high in vitamin A, and neonatal mortality was significantly less in Nepal and Ghana when infants were breastfed within the first 24 hours after birth (see Edmond et al., 2006; Mullany et al., 2007). The needs of the infant also increase the vitamin A requirements of lactating mothers.

## 11.11 Toxic effects of vitamin A

The most serious toxic effects of vitamin A are teratogenic as a result of overdose during the first trimester of pregnancy. Such effects include spontaneous abortions or foetal abnormalities, including those of the cranium (microcephaly), face (hairlip), heart, kidney, thymus, and central nervous system (deafness and lowered learning ability). Since embryogenesis is under the control of retinoic acid isomers, short-term increases in these compounds are probably responsible. Normal concentrations of plasma retinoic acid are 1–2 nmol/L. In one experiment, large doses of vitamin A (>300,000 IU, >100 mg) given to 10 women caused 10–100-fold increases in plasma retinoic acid concentrations at 4 hours. The same amount of vitamin A given as liver-only increased plasma retinoic acid concentrations 10-fold at 4 hours. Thus, women who are pregnant or who could become pregnant should not be exposed to retinoid therapy either for ointments applied to the skin for skin conditions or as supplements. Daily intakes should not exceed 10,000 IU (3 mg RE).

Acute and chronic toxic effects of vitamin A overdose can also occur in all individuals. Very high single doses can cause transient symptoms that may include bulging fontanelles in infants,

**Table 11.1 Summary of RDIs**

| | μg retinol equivalents/day |
|---|---|
| Infants | 350–500 |
| 1–6 years | 400 |
| 7–12 years | 500 |
| Adolescents | 600 |
| Men | 600–900 |
| Women | 600–900 |
| Pregnant | 700–800 |
| Lactating | 850, 1100, 1300 |

headaches in older children and adults, and vomiting, diarrhoea, and loss of appetite in all age groups. It is rare for toxicity to occur from ingestion of food sources of vitamin A. When it does, it is usually due to the consumption of a large amount of liver as, for example, in Arctic and Antarctic explorers who consumed polar bear, seal, or dog liver. In these extreme circumstances additional symptoms included blurred or double vision, vertigo, uncoordinated movements, elevated cerebrospinal pressure, and skin exfoliation. Deaths have also occurred.

Single large doses of vitamin A in infancy and childhood have been reported to cause transient toxic effects, but these are usually avoided if the dose is not more than 50,000 IU for infants below 6 months, 100,000 IU between 6 and 12 months, and 200,000 IU for children over 1 year. Doses are not given more frequently than once every 3 months.

Chronic toxicity is induced by consuming for a month or more at least 10 times the recommended daily allowance (e.g. 10 mg RE per day or 33,300 IU). A wide range of symptoms have been reported including headache, bone and muscle pain, ataxia, visual impairment, skin disorders, alopecia, liver toxicity, and hyperlipidaemia, and there is concern in developed countries that osteoporosis, bone fractures, and respiratory tract infections may be linked to retinoid toxicity (Li et al., 2014). Toxicity may be associated with high concentrations of retinol palmitate in the blood (3–8 μmol/L).

High intakes of carotenoids, e.g. from tomato or carrot juice or red palm oil, can lead to hypercarotenaemia and yellow colouration of the skin, especially the palms of the hands or soles of the feet, and the nasolabial folds (not the eyes), but this is not associated with toxic effects. High doses of β-carotene (180 mg/day) are used in the treatment of erythropoietic protoporphyria and have never been found to cause harm. The only worrying effects with carotenoids are prolonged use of high-dose β-carotene in smokers. Doses of 25–30 mg/day were associated with a higher incidence of lung cancer in two large intervention studies.

# 11.12 Measures to prevent vitamin A deficiency

## 11.12.1 Using available foods

Deficiencies rarely occur alone, and the clinical features in a community are those of the most seriously deficient nutrient. In developing countries, vitamin A deficiency is a particular problem because dietary sources of pre-formed vitamin A (animal food products) are expensive and people are reliant on vegetable sources. The bioavailability of carotene in plant food is poor (section 11.8.1). Fruit carotene is more bioavailable than vegetables, but is often seasonal. In many places, there is a greater variety of plant foods to overcome seasonal shortages, but thorough cooking of vegetables and fat (which improves carotene absorption) optimize bioavailability. Cooking methods are part of the culture of a community and not changed easily, and fat is very often an animal product and in short supply. Where vitamin A deficiencies exist, people need to be made aware of the diversity of foods that can provide vitamin A. Growing mangos, papaya (yellow and orange varieties), red sweet potato, and red palm oil need to be encouraged, as well as the introduction of chickens for their eggs, and fish where appropriate. As poverty is often the root cause of the lack of food, such schemes will need assistance and take time to be effective. High-dose supplementation of vulnerable groups, food fortification, and introduction of nutrient-enriched foods are short-term measures to rectify specific nutrient deficiencies; some may become part of the long-term solution to vitamin A deficiency.

Prevention—ideally eradication—of vitamin A deficiency is a priority for WHO and UNICEF. The deficiency is widespread, and it causes death or blindness in many thousands of young children. Measures of prevention are known, they are safe and practical (see 11.12.3), and are being generally applied by national governments and NGOs.

### 11.12.2 Breastfeeding

Infants are born with very little stored vitamin A. They rely on vitamin A in their mothers' milk to supply their needs for growth, immune function, and storage. The colostrum is particularly rich in vitamin A and early introduction of infants to the breast promotes mother–child bonding, more successful breastfeeding, and a continued supply of vitamin A through infancy. Introduction straight after birth has been shown to lower neonatal mortality (Edmond et al., 2006; Mullany et al., 2007). Delayed introduction is common in some African and Asian countries through cultural taboos, and may play a role in the unacceptably high neonatal mortalities in these continents. Breastfeeding promotion programmes should emphasize early initiation, as well as exclusive breastfeeding. The latter should provide adequate vitamin A for infants for the first 6 months of life (FAO/WHO, 2004).

### 11.12.3 Massive dosing and supplementation

Massive dosing of communities with vitamin A to overcome vitamin A deficiency was first tried in the 1960s in India. Doses of 200,000–300,000 IU (60–90 mg RE) were given to preschool children and any children showing side effects were noted. Specific problems included headaches, nausea, vomiting, and bulging fontanelles, but none of the side effects caused permanent damage, and the prevalence of signs like night blindness, Bitôt's spots, and xerosis was reduced. Night blindness is cured first. Oral and intramuscular administration of vitamin A dissolved in oil was tested, and oral administration by capsule or spoon was found to be highly effective. Use of the method has extended to many countries and clear benefits in reduced mortality have been demonstrated (see section 11.5.4). Currently recommended treatments are shown in Box 11.3. Vitamin A, 200,000 IU, is given therapeutically to children admitted to hospital with measles, xerophthalmia, or malnutrition.

The high dose for a child provides sufficient vitamin A for a period of 4–6 months, so where vitamin A deficiency is a constant problem, treatment should be provided every 6 months. Women of childbearing age must not be given high doses except immediately post-partum. The highest dose for a non-pregnant woman is 10,000 IU (3 mg RE) or 1 mg RE daily.

**Box 11.3 Massive dosing 6 months apart**

- Under 6 months, 50,000 IU orally (15 mg RE)
- 6 months to 1 year, 100,000 IU (30 mg RE)
- Over 1 year, 200,000 IU (60 mg RE)
- Women 1–8 weeks post-partum, 2 × 200,000 IU in separate doses

## 11.13 Influence of micronutrient deficiencies and drugs on vitamin A status

### 11.13.1 Influence of other nutrients on vitamin A status

A balanced and adequate diet is necessary for optimal vitamin A status. Specific deficiencies of protein and zinc may adversely affect status. Experimental protein deficiency reduces the activity of β-carotene 15,15′-dioxygenase activity, so may impair the conversion of β-carotene to retinal. Absorption of retinol is impaired in kwashiorkor and serum retinol is low. It improves with refeeding before vitamin A is given. Zinc-dependent enzymes convert retinol to retinal, and in one study, night blindness in a group of alcoholics was explained by this. Zinc is also involved in the synthesis of RBP4, which is particularly rich in zinc. Plasma retinol concentrations frequently correlate with plasma α-tocopherol concentrations. Vitamin E may protect free retinol from oxidation.

### 11.13.2 Interactions with drugs

Several drugs are known to affect the absorption of vitamin A, e.g. cholestyramine colchicine mineral oil (laxative), neomycin (antibiotic that inactivates bile salts), olestra, and phytostanols and phytosterols, but unless their use is prolonged over many months, the large liver stores in adults will maintain vitamin A status.

Alcohol abuse results in a striking depletion of hepatic vitamin A. Increased ethanol-oxidizing capacity, as a result of alcohol abuse, may increase the oxidation and loss of retinol from the liver and cause the poor vitamin A status of many alcoholics. In contrast, oestrogen-containing drugs (oral contraceptives, hormone replacement therapy) can have a stimulatory effect on the concentration of plasma retinol.

## 11.14 Pharmaceutical uses of vitamin A

Several synthetic retinoids influence proliferation and differentiation of the skin epidermis, inhibit keratinization, reduce production of sebum, and influence immune response, especially cell-mediated immunity. These properties have been used in medicine, particularly to treat such skin disorders as acne, seborrhoea (overproduction of sebum), and psoriasis. One of the more effective drugs is 13-*cis* retinoic acid (isotretinoin, 'Roaccutane'), but even for topical applications, its use is restricted in women of childbearing age because of the risk of teratogenesis. Topical use of all-*trans* retinoic acid can also reduce wrinkling and hyperpigmentation caused by photo-ageing. All-*trans* retinoic acid and 13-*cis* retinoic acid have been used to treat certain acute myeloid leukaemias (AML), where the retinoids stop proliferation of the cells and induce terminal differentiation to the granulocyte. Unfortunately, there are a number of different AMLs and only a limited number of patients respond.

### Further reading


1. **Awasthi, S., Peto, R., Read, S. et al.** (2013) Vitamin A supplementation every 6 months with retinol in 1 million pre-school children in north India: DEVTA, a cluster-randomised trial. *Lancet*, **381**, 1469–77.
2. **Benson, M.J., Pino-Lagos, K., Rosemblatt, M., et al.** (2007) All-trans retinoic acid mediates enhanced T reg cell growth, differentiation, and gut homing in the face of high levels of co-stimulation. *J Exp Med*, **204**, 1765–74.
3. **Connolly, E.E., Beatty, S., Thurnham, D.I., et al.** (2010) Augmentation of macular pigment following supplementation with all three macular carotenoids: an exploratory study. *Current Eye Research*, **35**, 335–51.
4. **Edmond, K.M., Zandoh, C., Quigley, M.A., et al.** (2006) Delayed breastfeeding initiation increases risk of neonatal mortality. *Pediatrics*, **117**, e380–6.
5. **Hess, S.Y., Thurnham, D.I., and Hurrell, R.F.** (2005) *Influence of provitamin A carotenoids on iron, zinc and vitamin A status*, Harvest Plus Technical Monographs 6. Washington: International Food Policy Research Institute
6. **Hume, E.M. and Krebs, H.A.** (1949) *Vitamin A requirements of human adults.* A report of the vitamin A sub-committee of the accessory food factors committee. MRC Report No. 264. London: His Majesty's Stationery Office.
7. **Hussey, G.D. and Klein, M.** (1990) A randomized, controlled trial of vitamin A in children with severe measles. *New Engl J Med*, **323**, 160–4.

8. **Leung, W.C., Hessel, S., Méplan, C., et al.** (2009) Two common single nucleotide polymorphisms in the gene encoding beta-carotene 15,15'-monoxygenase alter beta-carotene metabolism in female volunteers. *FASEB J*, **23**, 1041–53.
9. **Li, Y., Wongsiriroj, N., and Blaner, W.S.** (2014) The multifaceted nature of retinoid transport and metabolism. *Hepatobiliary Surg Nutr*, **3**, 126–39.
10. **Mullany, L.C., Katz, J., Li, Y.M., et al.** (2007) Breast-feeding patterns, time to initiation, and mortality risk among newborns in southern Nepal. *J Nutr*, **138**, 599–603.
11. **Namaste, S.M., Ou, J., Williams, AM., et al.** (2020) Adjusting iron and vitamin A status in settings of inflammation: a sensitivity analysis of the Biomarkers Reflecting Inflammation and Nutritional Determinants of Anemia. *Am J Clin Nutr*, **112**(Suppl), 458s–467s.
12. **Pepperberg, D.R. and Crouch, R.K.** (2001) An illuminating new step in visual-pigment regeneration. *The Lancet*, **358**, 2098–9.
13. **Schroeder, S.E., Reddy, M.B., and Schalinske, K.L.** (2007) Retinoic acid modulates hepatic iron homeostasis in rats by attenuating the RNA-binding activity of iron regulatory proteins. *J Nutr*, **137**, 2686–90.
14. **Sommer, A. and West, K.P.** (1996) *Vitamin A deficiency: health, survival and vision*. New York: Oxford University Press.
15. **Sommer, A., West, K.P., Jr, and Martorell, R.** (2013) Vitamin A supplementation in Indian children. *Lancet*, **382**, 591–6. (Comment on Awasti et al. (2013) *Lancet*, **381**, 1469–77.).
16. **Thurnham, D.I., McCabe, G.P., Northrop-Clewes, C.A., et al.** (2003) Effect of subclinical infection on plasma retinol concentrations and assessment of prevalence of vitamin A deficiency: meta-analysis. *Lancet*, **362**, 2052–8.
17. **Weiss, E.** (2020) Shedding light on dark adaptation. *The Biochemist*, **42**, 44–50.
18. **WHO** (2015) *Micronutrient deficiencies; vitamin A deficiency*. Available at: http:/who.int/nutrition/topics/vad/en/ (accessed 22 March 2023).

# The B Vitamins

A. Stewart Truswell



All B vitamins are water soluble. The water-soluble B vitamin that McCollum differentiated from fat-soluble A was later shown to consist of several distinct essential nutrients, although often found in the same foods, like liver and yeast. As they were separately isolated, they were named vitamin $B_1$, $B_2$, $B_3$, etc., at first, but some of these were the same as someone else had found in another laboratory and fell by the wayside. So, there are now eight well-established vitamins from the original B complex. Most of them are now given their chemical name, but $B_1$ is sometimes still used, and $B_6$ and $B_{12}$ persist in common use.

## 12.1 Thiamin (vitamin $B_1$)

In 1897, Eijkmann, a Dutch medical officer stationed in Java, discovered that a polyneuritis resembling beriberi (which was very common in South East Asia at that time) could be produced in chickens fed on polished rice. Subsequently, he and his successor, Grijns, showed that this polyneuritis could be cured with rice bran or polishings. This contained $B_1$, the first of the vitamins to be identified, but it was not until 1936 that R.R. Williams finally elucidated the unusual structure and synthesized thiamin.

### 12.1.1 Functions

Thiamin (Fig. 12.1) as the diphosphate (or 'pyrophosphate'), thiamin pyrophosphate (TPP), is a co-enzyme for the following major decarboxylation steps in carbohydrate metabolism:

1. Pyruvate → acetyl CoA (pyruvate dehydrogenase complex) at the entry to the citric acid cycle. Hence, in thiamin deficiency, pyruvate and lactate accumulate.



Fig. 12.1 Structure of thiamin.

2. α-Ketoglutarate → succinyl CoA (α-ketoglutarate dehydrogenase), halfway round the citric acid cycle.
3. Transketolase reactions in the hexose monophosphate shunt, alternative pathway for oxidation of glucose. Hence, in thiamin deficiency, oxidation of glucose is impaired with no alternative route.
4. The second step in catabolism of the branched-chain amino acids leucine, isoleucine, and valine.

### 12.1.2 Absorption and metabolism

Thiamin is readily absorbed by active transport at low concentrations in the small intestine and by passive diffusion at high concentrations. Total body content is only 25–30 mg, mostly in the form of TPP in the tissues. There is another co-enzyme form, thiamin triphosphate, in the brain. Thiamin is excreted both unchanged and as metabolites in the urine. It has a relatively high turnover rate in the body; there is really no store anywhere in the body. On a diet lacking in thiamin, signs of deficiency can occur after 25–30 days.

### 12.1.3 Deficiency in animals

Pigeons and chickens are more susceptible than mammals. The characteristic effect is head retraction, called opisthotonus, from neurological dysfunction. In mammals with experimental deficiency there is incoordination of muscle movements, progressing to paralyses, convulsions, and death. The brain is dependent on glucose oxidation for its energy, but the decrease in its pyruvate dehydrogenase and α-ketoglutarate dehydrogenase activities does not seem sufficient to explain the severe neurological dysfunction. Reduced formation of the neurotransmitter, acetylcholine (because acetyl CoA is not being formed), and a role of thiamin triphosphate in nerve transmission are other possible mechanisms. Loss of appetite, cardiac enlargement, oedema, and increased pyruvate and lactate are also seen.

### 12.1.4 Deficiency in humans

There are two distinct major deficiency diseases: beriberi and Wernicke-Korsakoff syndrome. They do not usually occur together.

*Beriberi* is now rare in the countries where it was originally described—Japan, Indonesia, and Malaysia (the name comes from the Singhalese language of Sri Lanka). In Western countries, occasional cases are seen in alcoholics. In *acute* beriberi, there is a high output cardiac failure, with warm extremities, bounding pulse, oedema, and cardiac enlargement. These features appear to be the result of intense vasodilatation from accumulation of pyruvate and lactate in blood and tissues. There are few electrocardiographic abnormalities. Response to thiamin treatment is prompt, with diuresis and usually a full recovery. In *chronic* beriberi, the peripheral nerves are affected, rather than the cardiovascular system. There is an inability to lift the foot up (foot drop), loss of sensation in the feet, and absent ankle jerk reflexes.

*Wernicke's encephalopathy* is usually seen in people who have been drinking alcohol heavily for some weeks and have eaten very little. Alcohol requires thiamin for its metabolism and alcoholic beverages do not contain it. Alcohol may also interfere with thiamin absorption. Occasional cases are seen in people on a prolonged fast (such as hunger strikers) or with persistent vomiting (as in Wernicke's first described case). Cases occurred in malnourished soldiers in Japanese prisoner-of-war camps in World War II. Clinically, there is a state of quiet confusion, lowered level of consciousness, and incoordination (fairly non-specific signs in an alcoholic). The characteristic feature is paralysis of one or more of the external movements of the eyes (ophthalmoplegia). This needs urgent injection of a large dose of thiamin, 350 mg intramuscularly. Thiamin is not actively transported across the blood-brain barrier; if treatment is delayed the memory may never recover. The memory disorder that is a sequel of Wernicke's encephalopathy is called Korsakoff's psychosis after the Russian psychologist who first described it. There is an inability to retain new memories and sometimes confabulation.

In people who die of Wernicke-Korsakoff syndrome, lesions are found in the mamillary bodies, mid-brain, and cerebellum. It is not clear why one deficient person develops beriberi and another develops Wernicke–Korsakoff syndrome, and why the two diseases seldom coincide. Possibly, the cardiac disease occurs in people who use their muscles for heavy work and so accumulate large amounts of pyruvate, producing vasodilatation and increasing cardiac work, while encephalopathy is the first manifestation in inactive people.

### 12.1.5 Biochemical test

Red cell transketolase activity, with and without TPP added *in vitro*, is a good test. However, heparinized whole blood must be used, it must be analysed fresh (or specially preserved), and the test will be normal if thiamin treatment has been already started. If the transketolase activity is increased more than 30% in the test tube with added TPP, this indicates at least some degree of biochemical thiamin deficiency. In Wernicke's encephalopathy, this 'TPP effect' can be higher than this, around 70% or even 100%. (*Note*: in this test, reported as 'TPP effect,' high values are abnormal.)

### 12.1.6 Interactions: nutrients

The requirement for thiamin is proportional to the intake of carbohydrates + alcohol + protein. In homogeneous societies, where proportions of fat and carbohydrate do not greatly differ, the thiamin requirement is proportional to the total energy intake.

There are no real stores of thiamin and the body runs out of it after about 3 weeks of starvation. When a malnourished person is given food (that uses thiamin), there is a danger of precipitating Wernicke's encephalopathy, e.g. with an intravenous glucose/water infusion. Thiamin should always be given with refeeding (see Chapter 43).

### 12.1.7 Food sources

There are no rich food sources of thiamin. The best natural sources in descending order are wheatgerm, whole wheat and products, yeast and yeast extracts, pulses, nuts, pork, duck, oatmeal, fortified breakfast cereals, cod's roe, and other meats. In many industrial countries (UK, North America, etc.), bread flour is enriched with thiamin and so are most breakfast cereals. Australia introduced mandatory fortification of bread flour with thiamin in 1991 and it has reduced that country's previously high rate of Wernicke–Korsakoff syndrome. Thiamin is readily destroyed by heat, and by sulphite and thiaminase (present in raw fish).

The recommended dietary intake (RDI) of thiamin is 0.4 mg per 1000 kcal (0.1 mg/kJ) (i.e. about 1.0 mg per day in adults). The toxicity of thiamin is very low.

#### Further reading

1. **Carpenter, K.J.** (2000) *Beri beri, white rice, and vitamin B.* Berkeley, CA: University of California Press.
2. **Truswell, A.S.** (2000) Australian experience with the Wernicke–Korsakoff syndrome. *Addiction*, **95**, 829–31.
3. **Victor, M., Adams R.D., and Collins, G.H.** (1989) *The Wernicke–Korsakoff syndrome and related neurologic disorders due to alcoholism and malnutrition*, 2nd edn. Philadelphia, PA: FA Davis.

## 12.2 Riboflavin

Vitamin B was originally considered to have two components, heat-labile $B_1$ thiamin and heat-stable $B_2$. In the 1930s, it was discovered that a yellow growth factor (riboflavin) in this latter fraction is distinct from the pellagra-preventing substance (niacin).

### 12.2.1 Structure

Riboflavin, or 7,8-dimethyl-10-(1′D-ribityl) isoallaxazine, comprises an alloxazine ring connected to a ribose alcohol—the ribityl side chain is required for full vitamin activity. It is a yellow-green fluorescent compound.

### 12.2.2 Functions

Riboflavin (Fig. 12.2) is part of two important co-enzymes, flavin mononucleotide (FMN) and flavin adenine dinucleotide (FAD), which are oxidizing agents. They participate in flavoproteins in the oxidation chain in mitochondria. They are also cofactors for several enzymes, e.g. NADH dehydrogenase, xanthine oxidase, l-amino acid oxidase, glutathione reductase, l-gulonolactone oxidase, and methylene tetrahydrofolate reductase (MTHFR).

### 12.2.3 Absorption and metabolism

Absorption is by a specialized carrier system in the proximal small intestine, which is saturated at levels above 25 mg. The vitamin is transported in plasma as free riboflavin and FMN, or bound to albumin. The body contains only about 1 g of riboflavin, mostly found in the muscle as FAD. Riboflavin is excreted primarily in urine; urinary excretion tends to reflect dietary intake.

### 12.2.4 Deficiency in animals

The most common effects in animals are cessation of growth, dermatitis, hyperkeratosis, alopecia, and vascularization of the cornea. Abortion or skeletal malformations of the foetus may occur. In some species, anaemia, fatty liver, and neurological changes have also been reported.

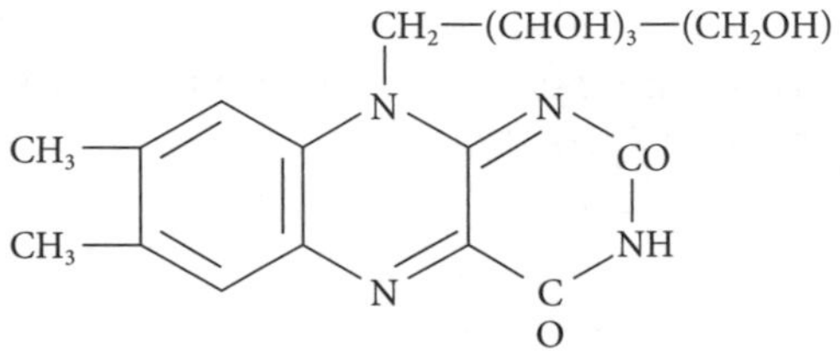

$CH_2$—$(CHOH)_3$—$(CH_2OH)$

$CH_3$ $CH_3$ N N N CO NH C O

Riboflavin

Fig. 12.2 Structure of riboflavin. Ribitol is the alcohol form of the 5-carbon sugar, ribose. In FMN two phosphates are attached to the end of the ribitol. In FAD this is extended further with adenylate (ribose-adenine).

### 12.2.5 Deficiency in humans

The clinical symptoms of deficiency—angular stomatitis, cheilosis, atrophy of the tongue papillae, nasolabial dyssebacea, and anaemia—are surprisingly minor, presumably due to the body's ability to conserve riboflavin, and the high affinity of the co-enzymes for their respective enzymes. Riboflavin deficiency (ariboflavinosis) is most commonly seen alongside other nutrient deficiencies (e.g. pellagra).

### 12.2.6 Biochemical tests

1. Erythrocyte glutathione reductase activity (EGRA) coefficient: FAD is a cofactor for this enzyme and its activity correlates with riboflavin status. The activity coefficient (or FAD effect): (EGRA with added FAD *in vitro*)/(EGRA without FAD *in vitro*). Values of <1.2 are considered to be acceptable, but values of 1.3–1.7 indicate inadequate riboflavin status. However, some doubts have been raised about the validity of the FAD effect, as it is elevated during exercise and pregnancy.
2. Measurement of urinary excretion of riboflavin: non-vitamin flavins (from foods) can be excreted, so a high-performance liquid chromatography–fluorometric method should be used to separate the riboflavin. Levels below 100 μg riboflavin/day are low.

### 12.2.7 Interactions: drugs

Phenothiazine derivatives (e.g. chlorpromazine) and tricyclic antidepressants have similar structures, and can interfere with riboflavin metabolism. Reduced riboflavin status is observed in alcoholics, but is due more to decreased dietary intake and absorption than to a direct effect of alcohol.

### 12.2.8 Food sources

Riboflavin is present in most foods, although the best sources are milk and milk products, eggs, liver, kidney, yeast extracts, and fortified breakfast cereals. Dairy products contribute significantly to riboflavin intake in Western diets. However, riboflavin is unstable in ultraviolet light, and after milk has been exposed to sunlight for 4 hours, up to 70% of riboflavin is lost.

The RDI for adults is about 1.3 mg/day. The requirement is less in people with small energy intakes and more in those with large energy intakes.

### 12.2.9 Toxic effects

The toxicity is very low. The gastrointestinal tract cannot absorb more than about 20–25 mg of riboflavin in a single dose.

## 12.3 Niacin

Niacin (Fig. 12.3) is a generic term for the related compounds that have activity as pellagra-preventing vitamins; the two that occur in foods are nicotinic acid (pyridine 3-carboxylic acid) and its amide, nicotinamide. They have apparently equal vitamin activity. Nicotinic acid is a fairly simple chemical (molecular weight 123) that was known long before its nutritional role was established. It was first isolated as an oxidation product of the natural alkaloid, nicotine, from which its name is derived. However, nicotinic acid and amide have very different physiological properties from nicotine (which is α-*N*-methyl-D-β-pyridyl pyrrolidine).

### 12.3.1 Functions

Nicotinamide is part of the co-enzymes nicotinamide-adenine dinucleotide (NAD) and nicotinamide-adenine-dinucleotide phosphate (NADP), the pyridine nucleotides. NAD has the structure: adenine–ribose–$PO_4$–$PO_4$–ribose–nicotinamide. It plays a central role in metabolism: it functions as the first hydrogen receptor in the electron chain during oxidative phosphorylation in the mitochondria. The pyridine ring of the nicotinamide is the part of the molecule that takes up a hydrogen (NAD ↔ NADH). NADP has an extra $PO_4$ (phosphate) attached to the ribose adjacent to adenine. It has a more specialized function as a hydrogen donor in fatty acid synthesis.

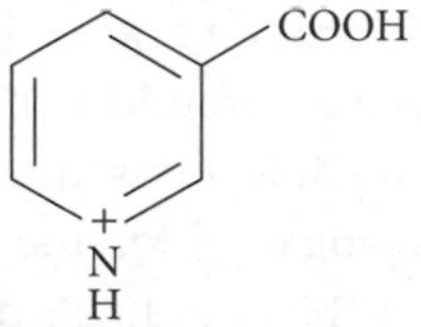

COOH
+
N
H
Nicotinic acid

Fig. 12.3 Structure of nicotinic acid (niacin). Nicotinamide is the corresponding amide, with $CONH_2$ in the side chain.

### 12.3.2 Absorption and metabolism

Nicotinic acid and its amide are water-soluble, and well absorbed from the stomach and small intestine, and are transported in solution in the plasma. Stores of niacin and its co-enzymes are only small, and early features of pellagra can occur in human subjects after 45 days of depletion.

### 12.3.3 Synthesis from tryptophan

A special feature of niacin is that in most conditions only about half of what is in the body is absorbed as preformed nicotinic acid or amide from the diet. About the same amount is synthesized in the liver from tryptophan, part of dietary protein, in a sequence of seven enzyme steps down the kynurenine pathway.

Most tryptophan in the body is used for protein synthesis—it is the least abundant in foods of all the essential amino acids—and some also goes to serotonin. The rest goes down the kynurenine pathway (Fig. 12.4). Approximately 60 mg tryptophan has been shown in humans to convert to 1 mg of niacin. The first enzyme in the kynurenine pathway, hepatic

Tryptophan → N-formylkynurenine → kynurenine → 3OH kynurenine → 3OH anthranilic acid → 2-amino-3-carboxy-muconaldehydate → quinolinic acid → nicotinic acid → NAD

Fig. 12.4 Tryptophan conversion to NAD.

tryptophan oxygenase, is under hormonal control, and the amount of niacin formed appears to be increased in pregnancy. It is downregulated when the protein intake is inadequate.

## 12.3.4 Deficiency in animals

The classic animal model for pellagra is 'black tongue' in dogs. Puppies lose their appetite and have inflamed gums, dark tongue, and diarrhoea with blood. Elvehjem's group at Wisconsin tested different components of liver for their ability to cure black tongue and in 1937 found that the 'pellagra-preventing factor' was nicotinamide.

## 12.3.5 Deficiency in humans

There is one deficiency disease, *pellagra* (the name means 'sour skin' in Italian). The skin is inflamed where it is exposed to sunlight, resembling severe sunburn, but the affected skin is sharply demarcated. The skin lesions progress to pigmentation, cracking, and peeling. Often the skin of the neck is involved (Casal's collar) (Fig. 12.5). Students are taught that pellagra is the disease of three Ds: dermatitis, diarrhoea, and delirium or dementia. As well as diarrhoea there is likely to be headache and insomnia, and an inflamed tongue (glossitis). In mild chronic cases, mental symptoms (the third 'D') are not prominent. It is hard to explain the clinical manifestations by the known biochemical functions of niacin. Because some niacin is formed from tryptophan, pellagra can be cured by giving either niacin or a generous intake of easily assimilated protein. Some patients have low serum albumin.

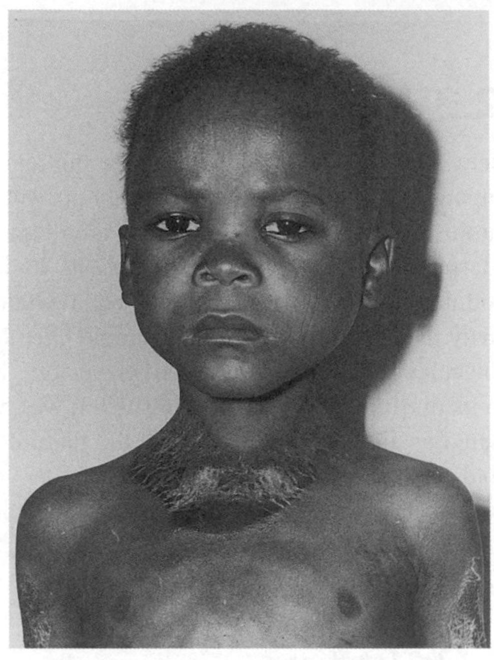

Fig. 12.5 A girl suffering from pellagra. Note the Casal's collar and lesions on outer arms.

*Source:* Reproduced with permission, *Am J Clin Nutr*, 1968; **21**, 1314–20.

Pellagra appeared in Europe after maize was introduced as a cereal crop from the New World after 1500, but the Mayas, Aztecs, and indigenous North Americans do not seem to have suffered from pellagra.

There was a major epidemic of pellagra in poorer communities (share croppers) in the southern states of the USA from around 1905. It was generally thought to be an infectious disease, but Joseph Goldberger, investigating for the Federal government between 1914 and 1929, demonstrated by epidemiology and crucial human experiments that it was due to a diet of maize grits and little else. He identified an animal model, black tongue in dogs (rats are not susceptible), cured by yeast in which the pellagra-preventing factor eventually turned out to be nicotinamide, already known in tissue culture biochemistry.

The niacin in cereals is in a complex, 'niacytin,' which humans cannot absorb, so that, although it appears in food tables (because the complex is split during extraction), it is not biologically available. If subsistence farmers eat a diet predominantly of maize, with few other foods, their niacin has to come from tryptophan in the protein of the cereal. However, the protein of maize is deficient in tryptophan, unlike other cereals, so little or no niacin can be made in the body via the kynurenine pathway. All cereals are low in lysine, but maize has less tryptophan than wheat and rice. In pre-Columban America, the ground maize was steeped in warm lime water (calcium hydroxide), which liberates the niacin (making it biologically available), and then made into tortillas, flat cakes—as it still is today in Mexico and Guatemala.

Pellagra is rare in developed countries. It occurs in parts of Africa, where maize is the staple cereal.

## 12.3.6 Biochemical tests

1. Urinary $N'$-methylnicotinamide (and/or its 2-pyridone) is the best-known test, but tests that require a 24-hour urine are inconvenient.
2. Red cell NAD concentration
3. Fasting plasma tryptophan

## 12.3.7 Interactions: nutrients

The most important is that tryptophan and hence dietary proteins (except in maize) provide niacin. Tryptophan makes up about 1% of mixed dietary proteins, so 6 g protein (60 mg tryptophan) = 1 mg niacin, hence a protein intake of 70 g/day is equivalent to about 12 mg niacin. The niacin requirement is thought to be proportional to the energy expenditure or energy intake.

Two of the enzymes in the kynurenine pathway are vitamin $B_6$ dependent, so vitamin $B_6$ deficiency is likely to reduce niacin synthesis from tryptophan.

## 12.3.8 Food sources

Good sources of preformed niacin, in descending order, are liver and kidney (the richest sources), other meat, poultry, fish, brewer's yeast and yeast extracts, peanuts, bran, pulses, wholemeal wheat, and (surprisingly) coffee, including instant coffee. Other foods that are rich in protein provide tryptophan. Maize meal and bread are now enriched with niacin in South Africa. If food tables give values for milligram niacin equivalents (NE), this is preformed niacin, mg + (tryptophan ÷ 60). The British food tables (McCance and Widdowson) have separate columns for 'niacin' and for 'niacin, mg tryptophan ÷ 60'. From this one can see, for example, that 100 g fresh whole cow's milk provides 0.2 mg preformed niacin, but 0.6 mg potential niacin from tryptophan.

The RDI for niacin in adults, expressed as NE, is 6.6 mg NE per 1000 kcal (1.6 mg NE per 1000 kJ): in absolute numbers, about 14 mg for women and 16 mg for men.

## 12.3.9 Pharmacological doses

Well above the nutrient dose, *nicotinic acid* (but not the amide) produces cutaneous flushing from histamine release, at doses of 100 mg/day or more, and has been used to treat chilblains. At doses of 3 g/day or more (200 × RDI), nicotinic acid inhibits lipolysis in adipose tissue, and lowers plasma triglyceride and cholesterol. It is in the pharmacopoeia as a second-line drug for combined hyperlipidaemia (i.e. high plasma cholesterol with raised triglycerides). Side effects, as well as flushing, include gastric irritation, impaired glucose tolerance, and disturbed liver function tests.

In 2015, Chen et al. in Sydney reported that in a phase 3 randomized trial, nicotinamide 500 mg twice daily over 18 months reduced recurrence of (non-melanoma) skin cancer by 23% ($p = 0.02$) in half of 386 middle-aged people who had had recurrent basal cell or squamous skin cancers (the other half took placebo tablets). There were no significant differences in adverse effects between nicotinamide and placebo subjects.

## Further reading

1. **Chen, A.C., Martin, A.J., Choy, B., et al.** (2015) A phase 3 randomized trial of nicotinamide for skin-cancer chemoprotection. *N Engl J Med*, **373**, 1618–26.
2. **Roe, D.A.** (1973) *A plague of corn. The social history of pellagra*. Ithaca: Cornell University Press.
3. **Seal, A.J., Creeke, P.I., Dibari, F., et al.** (2007) Low and deficient niacin status and pellagra are endemic in postwar Angola. *Am J Clin Nutr*, **85**, 218–24.


# 12.4 Vitamin $B_6$

## 12.4.1 Structure

Vitamin $B_6$ (Fig. 12.6) occurs in nature in three forms—pyridoxine, pyridoxal, and pyridoxamine—which are interconvertible within the body. Each form (vitamer) also occurs as a phosphorylated compound: the principal one in the body and in food is pyridoxal 5′-phosphate (PLP).

## 12.4.2 Functions

The major co-enzyme form in the body is PLP. It functions in practically all the reactions involved in amino acid metabolism, including:

- transaminations and synthesis of non-essential amino acids
- deamination of serine and threonine
- metabolism of sulphur-containing amino acids including homocysteine
- *decarboxylations*:
  - o formation of neurotransmitters adrenalin, noradrenalin, serotonin, and γ-amino butyric acid (GABA)
  - o formation of γ-aminolaevulinic acid, which is the first step in porphyrin synthesis, making haemoglobin
  - o synthesis of sphingomyelin and phosphatidyl choline (lecithin)
  - o synthesis of taurine, a conjugator of bile acids and important in eye and brain function
- *kynureninase*: for the conversion of tryptophan to niacin. When this reaction is impaired, xanthurenic acid (major metabolite of 3(OH) kynurenine) accumulates in the urine, which is used as biochemical marker for $B_6$ status.

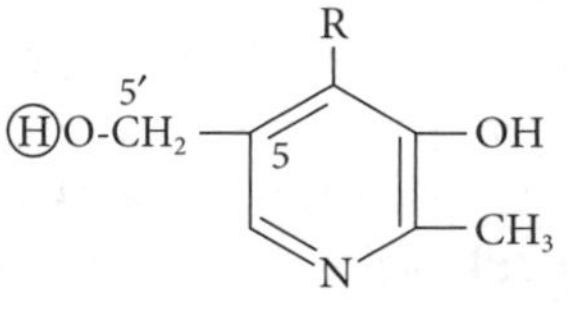

R
5′
(H)O-$CH_2$
5
OH
$CH_3$
N
Vitamin $B_6$

Fig. 12.6 Structure of vitamin $B_6$. In pyridoxine, R = $CH_2OH$; in pyridoxal, R = CHO; in pyridoxamine, R = $CH_2NH_2$. In co-enzyme forms, phosphate replaces the ringed H.

However, the role of vitamin $B_6$ is not restricted to protein metabolism. Over half of total body $B_6$ is associated with glycogen phosphorylase enzyme in the muscles, which releases glucose as glucose 1-phosphate from glycogen stores. PLP may also have a role in modulating steroid hormone receptors.

## 12.4.3 Absorption and metabolism

In the small intestine vitamin $B_6$ is absorbed by passive diffusion, mainly in the unphosphorylated form. Even large doses are well absorbed. The different forms of the vitamin are rapidly converted to pyridoxal in the intestinal cell, by the FMN-requiring enzyme pyridoxal phosphate oxidase. Pyridoxal is transported in the circulation largely bound to albumin and haemoglobin, and after diffusion into cells, pyridoxal is rephosphorylated by pyridoxal kinase, which maintains it within the cells.

The total body content of vitamin $B_6$ is estimated to be between 50 and 150 mg in adults. Most of this (90%) is tightly bound in tissues. Vitamin $B_6$ in the liver, brain, kidney, spleen, and muscle is bound to protein, which protects it from hydrolysis. The major metabolite of vitamin $B_6$ is 4-pyridoxic acid, which is inactive as a vitamin and excreted in the urine.

### 12.4.4 Deficiency in animals

Dermatological and neurological changes are commonly observed in animals when vitamin $B_6$ is deficient. In rats, impaired growth, muscular weakness, irritability, dermatitis, anaemia, fatty liver, impaired immune function, hypertension, and insulin insufficiency have all been observed. Neurological changes include convulsions.

### 12.4.5 Deficiency in humans

The symptoms of deficiency in humans are general weakness, sleeplessness, peripheral neuropathy, personality changes, dermatitis, cheilosis and glossitis (as in riboflavin deficiency), anaemia, and impaired immunity. Deficiency on its own is rare; it is most often seen with deficiencies of other vitamins, or with protein deficiency. In 1953, a minor epidemic of convulsions in infants in the USA was traced to a milk formula that contained no vitamin $B_6$ because of a manufacturing error. Convulsions in pyridoxine deficiency are probably due to impaired synthesis of γ-aminobutyric acid (GABA), the major inhibitory neurotransmitter in the brain.

### 12.4.6 Deficiency: secondary

A number of inborn errors of amino acid metabolism may respond to supranutritional doses of pyridoxine. Hyperhomocysteinaemia, a condition that may increase the risk for cardiovascular disease, responds to supplements of folate, vitamin $B_{12}$, and sometimes $B_6$.

Low vitamin $B_6$ status is common in chronic alcoholics, who may have impaired absorption. Acetaldehyde (oxidation product of ethanol) can inhibit the conversion of pyridoxine to PLP.

Pregnant women have a decrease in plasma PLP levels. It is unclear whether this indicates a deficiency or is a normal physiological change.

### 12.4.7 Biochemical tests

1. Measurement of plasma PLP. Normal levels are above about 30 nmol/L.
2. Increased urinary xanthurenic acid after a load of the amino acid tryptophan.
3. Activity of erythrocyte alanine aminotransferase, with and without *in vitro* PLP.
4. Urinary 4-pyridoxic acid.

### 12.4.8 Interactions: other nutrients

High protein intakes increase metabolic demand for vitamin $B_6$.

### 12.4.9 Interactions: drugs

Isoniazid (used to treat tuberculosis) increases urinary excretion of vitamin $B_6$. Several drugs, including cycloserine, gentamicin, penicillamine, L-DOPA, and phenelzine, are vitamin $B_6$ antagonists. Some biochemical indices of vitamin $B_6$ state may be abnormal in a proportion of women taking oral contraceptives, but these are indirect indices (e.g. alanine aminotransferase).

### 12.4.10 Food sources

The vitamin is distributed in a wide range of unprocessed (or lightly processed) foods. Major food sources in the Western diet are meats, whole grain products, vegetables, bananas, and nuts. Refined cereal products, such as white bread and white rice, are not significant sources of vitamin $B_6$ due to milling losses.

The RDI is 0.02 mg vitamin $B_6$ per gram of protein intake, which works out to about 1.5 mg/day in average adults. In pregnancy and lactation, 1.9 mg/day and 2.0 mg/day are recommended.

### 12.4.11 Pharmacological doses

Pharmaceutical preparations, tablets of pyridoxine HCl, are indicated for several rare inborn errors of metabolism. They are used for radiation sickness and for premenstrual syndrome. The few controlled trials for the latter condition are unimpressive. Vitamin $B_6$ can contribute to lowering raised plasma homocysteine (though folic acid usually has more effect). Whether it has value in reducing the risk of cardiovascular disease is not clear at present.

### 12.4.12 Toxic effects

Vitamin $B_6$ toxicity was first reported in women taking supplements of very large doses of pyridoxine (2000–6000 mg/day). These supplements were taken for premenstrual syndrome or carpal tunnel syndrome, and the women developed peripheral neuropathy and lost sensation in their feet. Intakes of supplements down to 200 mg/day (133 × RDI) have been associated with neuropathy. The upper intake level (UIL) set by the US Institute of Medicine is 100 mg/day. This can only be obtained from supplements. The amount of vitamin $B_6$ obtainable from foods is far below this.

#### Further reading

1. **Schaumberg, H., Kaplan, J., Windebank, A., et al.** (1983) Sensory neuropathy from pyridoxine abuse. *N Engl J Med*, **309**, 445–8.

## 12.5 Biotin

### 12.5.1 Functions

Biotin is a co-enzyme for several carboxylase enzymes—pyruvate carboxylase (formation of oxaloacetate for the tricarboxylic acid cycle), acetyl co-enzyme A (CoA) carboxylase (fatty acid synthesis), propionyl CoA carboxylase (catabolism of odd-chain fatty acids and some amino acids), and 3-methylcrotonyl CoA carboxylase (catabolism of the ketogenic amino acid leucine).

### 12.5.2 Deficiency: animals and humans

Biotin deficiency is very rare as biotin is found in a wide range of foods, and bacterial production in the large intestine appears to supplement dietary intake. Deficiency can, however, be produced when animals or humans eat large amounts of uncooked egg white, which contains avidin. This tightly binds biotin in the gut, preventing absorption. Avidin is destroyed by heating. 'Egg white injury' (e.g. biotin deficiency) impairs lipid and energy metabolism in animals. It produces seborrhoeic dermatitis, alopecia, and paralysis of the hind limbs in rats and mice.

In humans, cases of biotin deficiency have been associated with a red, scaly skin rash (altered fatty acid metabolism may contribute to this skin condition), glossitis, loss of hair, anorexia, depression, and hypercholesterolaemia. Some cases of seborrhoeic dermatitis in young breastfed infants have responded to administration of biotin to the mother. Human milk contains much less biotin than cows' milk. Biotin deficiency has been reported in patients on total parenteral nutrition whose infusions did not contain biotin. The human requirement is estimated to be about 30 μg/day. In experimental human biotin deficiency (feeding raw egg whites), the biochemical indicators have been reduced urinary biotin and increased 3-hydroxyisovaleric acid (that should normally be metabolized by 3-methylcrotonyl CoA carboxylase).

### 12.5.3 Food sources

As well as being synthesized by the gut microflora, biotin is also found in a wide range of foods, the richest including liver, kidney, egg yolks, meat, whole grain cereals, cheese, and milk. In contrast, green vegetables (e.g. spinach) contain very little biotin. See The Health and Food Supplements information Service (HSIS). Available at https://www.hsis.org/a-z-food-supplements/biotin/(accessed10 August 2022).

## 12.6 Pantothenic acid

Co-enzymes often contain unusual structures—unusual in the sense that higher animals have lost the ability to form them and they must be supplied in the diet. For co-enzyme A it is pantothenic acid.

### 12.6.1 Functions

Pantothenic acid is part of co-enzyme A (CoA) and acyl carrier protein (ACP). CoA and ACP are both carriers of acyl groups. Acetyl-CoA participates in the tricarboxylic acid cycle in the disposal of carbohydrates and ketogenic amino acids. CoA is also involved in the synthesis of lipids—fatty acids, glycerides, cholesterol, ketone bodies, and sphingosine—and in acylation of proteins. ACP is involved in chain elongation during fatty acid synthesis.

Pantothenic acid is transported primarily in the CoA form by red cells in the blood, and is taken up into cells by a specific carrier protein. The highest concentrations of the vitamin are found in the liver, adrenals, kidney, brain, heart, and testes. Most is in the CoA form.

All tissues are able to synthesize CoA from pantothenic acid. ACP is synthesized from a 4-phosphopantetheine residue transferred from CoA. These metabolically active forms can be degraded to free pantothenic acid, which is the major form of excretion in the urine. Urinary pantothenic acid reflects dietary intake, ranging from 2 to 7 mg/day in adults.

### 12.6.2 Deficiency: animals

In most species, pantothenic acid deficiency is associated with dermatitis, changes to hair or feathers, anaemia, infertility, irritability, ataxia, paralysis, convulsions, and even death. In rats, a condition called 'bloody whiskers' is caused by release of protoporphyrin via the nose and tear ducts.

### 12.6.3 Deficiency: humans

Spontaneous human deficiency has never been described. As pantothenic acid is so widely distributed in foods, any dietary deficiency in humans is usually associated with other nutrient deficiencies. The word 'pantothen' means 'from everywhere' (Greek), but highly refined foods do not contain pantothenic acid.

Subjects given the antagonist ω-methylpantothenic acid developed a deficiency with symptoms of depression, fatigue, insomnia, vomiting, muscle weakness, and a burning sensation in the feet. Changes in glucose tolerance, an increase in insulin sensitivity, postural hypotension, and decreased antibody production were also noted.

During World War II, malnourished prisoners of war in the Far East developed 'burning feet syndrome', which appeared to respond to large doses of Ca-pantothenate, but not to other B complex vitamins.

There is insufficient evidence to derive a precise requirement figure for pantothenate. The US/Canadian adequate intake is 5 mg/day, based on estimated usual intakes and urinary pantothenate excretion. It must be provided in total parenteral nutrition.

## 12.7 Folate

Folate is used as the generic name for compounds chemically related to pteroyl glutamic acid, folic acid. Deficiency of folate is quite common in hospital patients, secondary to diseases, especially intestinal, neoplastic, and haematological. Requirements are notably increased in pregnancy. The word 'folic' is from the Latin 'folia' (leaf), coined in 1941 for an early preparation of this vitamin from spinach leaves.

### 12.7.1 Structure

Folic acid (pteroyl glutamic acid) is the primary vitamin from the chemical point of view, and it is the pharmaceutical form (and used for food fortification) because of its stability. However, it is rare (naturally occurring) in foods and in the body. Most folates are in the reduced form, tetrahydrofolate (THF); they also have 1-carbon components (methyl or formyl)

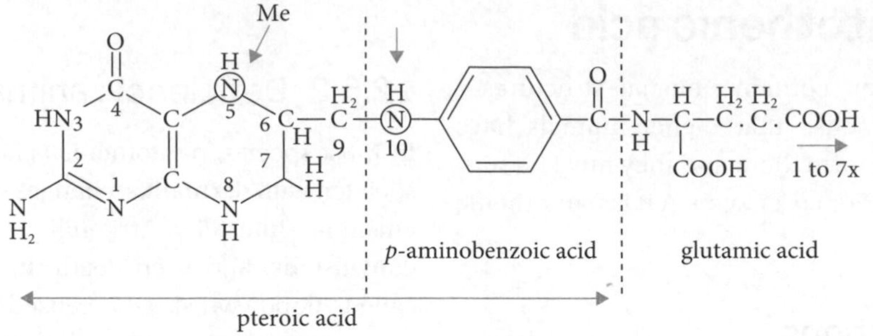

*p*-aminobenzoic acid

glutamic acid

pteroic acid

Tetrahydrofolate

Fig. 12.7 Tetrahydrofolate (pteroyl glutamic acid) monoglutamate: folic acid with extra hydrogens at 5, 6, 7, and 8.

attached to nitrogen atom 5 or 10, or bridging between them (5,10-methylene THF) (Fig. 12.7). In addition, they have up to seven glutamic acid residues in a row (at the right in Fig. 12.7).

## 12.7.2 Functions

Tetrahydrofolate plays an essential role in 1-carbon transfers in the body. It receives 1-carbon radicals from, for example, serine, glycine, histidine, and tryptophan, and donates them at two steps in purine synthesis and one important step in pyrimidine synthesis. In synthesis of the adenine and guanine in RNA and DNA, carbons at their 2 and 8 positions are provided by N10 formyl THF at different stages of their development. And in DNA synthesis, 5,10-methylene THF adds a methyl group at the 5 position to desoxyuridylic acid to make thymidylic acid. Thus, three of the four bases in DNA require folate for their *de novo* synthesis.

5-Methyl THF cooperates with vitamin $B_{12}$ in the action of methionine synthase, which adds a methyl group to homocysteine and forms methionine and THF (from which 5,10-methylene THF can be formed). When vitamin $B_{12}$ is deficient, folate is trapped as the 5-methyl compound and 5,10-methylene THF is not available to form thymidylate for DNA synthesis (Fig. 12.8).

## 12.7.3 Absorption and metabolism

Most food folates are in polyglutamate form. These are hydrolysed in the gut by folate conjugase (γ-glutamyl-hydrolase) and absorbed as monoglutamate. Most food folates are in reduced form as THF. Folic acid is reduced to THF in the process of absorption. Folate is present in plasma mainly as 5-methyl THF. Inside cells it is usually in polyglutamate form. Total body folate has been estimated at about 10 mg, half being in the liver. Very little is excreted in the urine. More is excreted in the bile and about half of this is reabsorbed.

In the colon, microorganisms produce folate compounds. In a human study, 13C-labelled 5-formyl THFA was given in a special caplet whose outside stays intact until reaching the colon. Venous blood showed that the folate released in the colon had been absorbed. Thus, about 350 micrograms of folate may be added to intake from food and supplements.

## 12.7.4 Deficiency in animals

Chicks show reduced growth, anaemia, and impaired feather growth. In guinea pigs there is a low white blood cell count and growth failure.

## 12.7.5 Deficiency in humans

Folate is the final anti-megaloblastic substance. In deficiency, the basic abnormality is reduced ability of cells to double their nuclear DNA in order to divide, because of impaired synthesis of thymidylate.

There is megaloblastic anaemia (cells are enlarged, their nuclei large, but with reduced density of chromatin) and similar changes in leukocytes, platelets, and epithelial cells. There is also infertility and may be diarrhoea.

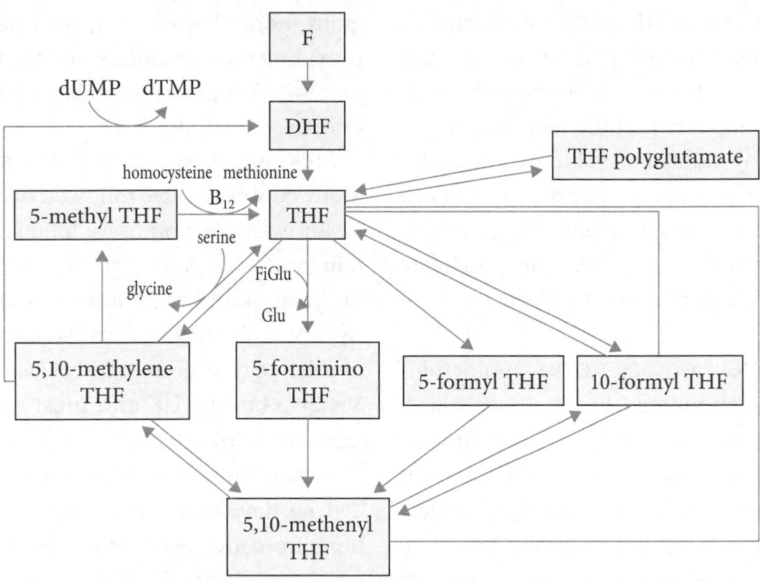


Fig. 12.8 Folate metabolism. Note that 5-methyltetrahydrofolate can only return to the pool for other functions if there is enough vitamin $B_{12}$ for the homocysteine → methionine reaction, which takes up the methyl group. In the reaction of dUMP → dTMP, dTMP is thymidylate, one of the four essential bases of DNA. If there is insufficient 5,10-methylene tetrahydrofolate (the specific cofactor), DNA synthesis is reduced or stops.

DHF, dihydrofolate; F, folate; FiGlu, formiminoglutamic acid; THF, tetrahydrofolate.

Pure dietary deficiency is seen occasionally. In a previously healthy physician, Victor Herbert, experimental depletion resulted in anaemia after 125 days, with biochemical and histological changes before this. However, when there is increased cell proliferation or interference with folate metabolism, features of deficiency appear earlier. Secondary deficiency is common in late pregnancy, in alcoholics, in haemolytic anaemias, uraemia, intensive hospital therapy, and people with malabsorption.

## 12.7.6 Benefits of extra folate

Folate is important at both ends of pregnancy. In late pregnancy megaloblastic anaemia can occur, especially in women in developing countries. This was first described in India by Dr Lucy Wills in 1931, before folic acid was isolated. This anaemia can sometimes be first noticed in the early weeks after childbirth. One method used as a preventative measure to avoid folate deficiency in infants is delayed cord clamping. In areas with limited resources, delayed clamping is recommended for 60–180 seconds after birth. This has many benefits, including lowered incidence of anaemia and improvements in heart rate, cardiac output, and cerebral oxidation (Bruckner et al., 2021).

In affluent countries, too low serum folate and megaloblastic anaemia is possible and is prevented with routine folic acid tablets, sometimes combined with iron.

**Neural tube defects** In early pregnancy, folic acid supplements have more recently been found to reduce the risk of serious foetal malformations, spina bifida, and anencephaly, together termed neural tube defects.

In human embryology the neural tube closes at days 24–28 after conception and the extra folate must be taken before this (i.e. often before the woman may be sure she is pregnant). In the UK trial by the Medical Research Council Vitamin Study Group (1991), women in several countries who had previously had a baby affected by neural tube defect agreed to take one of four different nutritional supplements periconceptionally, i.e. before as well as during pregnancy. It was found that those taking supplements containing folic acid had 70% fewer babies with neural tube defects. Other trials and cohort studies have been supportive. From studies of serum folate of women who were subsequently found to have an abnormal baby, it appears that the neural tube defect is not caused by a deficiency, but by a need for extra folate in some women at this time of extra cell division in embryonic development. There is also evidence based on multivitamin use that extra folate reduces the risk of congenital heart defects.

**Increased plasma homocysteine** Epidemiological evidence accumulated in the 1990s that a raised plasma homocysteine is a risk factor for cardiovascular diseases in older adults (Chapter 19) and also for the development of dementia (Seshadri et al., 2002). Homocysteine is derived from the amino acid methionine—it is methionine minus its methyl group—and folic acid supplements increase the activity of methionine synthase, which remethylates homocysteine back to methionine. Extra vitamins $B_{12}$ and $B_6$ augment this effect.

### 12.7.7 Folate fortification

Neural tube defect (NTD) is a very serious deformity and relatively common. With the evidence that folic acid can largely prevent it, there are three options:

1. Nutrition education—encourage women aiming to become pregnant to eat more foods with higher folate content.
2. Advise such women to take a tablet of 0.4 or 0.5 mg folic acid daily.
3. Fortify the national food supply with folic acid.

Option 1 is least efficient. The extra folate has to be taken periconceptionally before the woman is sure she is pregnant. Option 2 will only help a woman who is both planning pregnancy and has good health information.

On 1 January 1998 the USA and Canada took the major step of enacting mandatory fortification of all cereal grains (bread, flour, pasta, breakfast cereals, and rice) with folic acid at 140 μg per 100 g of grain food. This was a giant nutrition experiment providing the extra folate needed for pregnancy, but possibly giving the rest of the population more folate than they needed.

Chile and Costa Rica followed North America. Many countries have followed over time. In Australia, for example, voluntary folate fortification of certain foods was permitted from 1995 and mandatory fortification of bread flour was introduced in 2009. By now over 50 countries have mandatory addition of folic acid to flour, and NTDs have decreased by 30–60%; but the UK and most countries in Europe have not followed (relying on options 1 and 2 above).

In North America, folate intake went up by nearly 200 μg/day, more than had been intended, perhaps from overages added by cereal manufacturers. Serum folate doubled (Jacques et al., 1999) and neural tube defects were roughly halved in North America (Ray, 2004). Homocysteine levels declined (Jacques et al., 1999). There was no change in serum vitamin $B_{12}$ and no masking reported of vitamin $B_{12}$ deficiency.

Have there been other benefits or adverse effects? With increasing pharmaceutical and surgical treatment of cardiovascular diseases (CVD), it is not possible to distinguish any national folic acid effect, and so far the randomized controlled trials of folic acid, with or without vitamin $B_{12}$ and $B_6$, have not shown significant reduction of CVD. National statistics on incidence of dementia are probably impossible to obtain. There is substantial epidemiological evidence that raised plasma homocysteine predisposes to dementia but randomized controlled trials with B vitamins (including folic acid) together have not found significant cognitive benefit (Clarke et al., 2014).

In the big picture there was a small 3-year upswing in the declining trend of incidence of colorectal cancer in both USA and (separately) in Canada (Mason et al., 2007). In one of the trials of folic acid and vitamin $B_{12}$, in Norway (where grain foods are not fortified), those on extra vitamins had developed more cancer (Ebbing et al., 2010). After this, colorectal cancer incidence continued to decrease in the USA and is now lower than when folate fortification started.

In the USA, on top of folate fortification of grains, about a third of adults take multivitamin supplements that include folic acid, so that 5% of older men now consume more than the recommended UIL (see section 12.7.11). It appears that if people on high folate intakes have very low serum $B_{12}$, they have more severe clinical and biochemical effects of the $B_{12}$ deficiency (Selhub et al., 2009).

Another change that has been observed is that there is now more free folic acid in serum than there was in sera collected before fortification and stored. The significance of this is not clear and it does not seem to affect breast milk.

### 12.7.8 Interactions

Folate interacts with vitamin $B_{12}$ (see section 12.8.7). Vitamin C in foods reduces loss of folate in cooking. Several drugs interfere with folate metabolism, most of them by antagonizing dihydrofolate reductase (which converts 2H folate to 4H folate, i.e. THF): methotrexate, aminopterin, amethopterin (used for chemotherapy for cancer), pyrimethamine (antimalarial), and cotrimoxazole (antibacterial). Most of the antiepileptic drugs (carbamazepine, phenytoin, valproate), if taken through early pregnancy, increase the risk of neural tube defect in the foetus. When they are prescribed, folic acid, at the higher dose of 5 mg/day, should be taken periconceptionally.

### 12.7.9 Genetics

A fairly common single-nucleotide polymorphism in the gene for MTHFR, the 667 C → T mutation, affects the activity of this enzyme. Overall, around 10% of people are homozygous for the *TT* allele (with a range in different communities of 1–20%). They have reduced activity of MTHFR, so that availability of methyl THF, and hence THF, is lower (Fig. 12.8). They have been found to have lower blood folates (hence, higher requirement) and also increased risk of neural tube defects and higher plasma homocysteines, but not more cardiovascular disease.

### 12.7.10 Biochemical tests

Serum folate reflects recent intake, but not yesterday's. Levels continue upwards for about 3 weeks when intake is moved to a fixed higher level (Truswell and Kounnavong, 1997). Levels below 7 nmol/L (3 ng/mL) reflect deficiency. Since fortification, levels in North America are around 24 nmol/L and higher in those taking supplements. Red cell folate reflects cellular status. It is normally above 225 nmol/L (100 ng/mL).

### 12.7.11 Recommended intakes

The Institute of Medicine (Food and Nutrition Board, Institute of Medicine, 1998) for the USA and Canada recommends (RDI) 400 μg per day of dietary folate equivalents (DFEs) for men and women, and for the latter 600 μg/day in pregnancy and 500 μg/day in lactation. The upper level is 1000 μg/day as folic acid or fortified food for adults. Australia and New Zealand followed in 2005 with the same recommendations.

DFEs adjust for better absorption of free folic acid compared with folate naturally in foods (Bailey, 1998). 1 μg of food folate = 0.6 μg of folic acid added to foods or taken with food = 0.5 μg of folic acid supplement taken on an empty stomach.

### 12.7.12 Food sources

Although the name comes from the Latin 'folia' (leaf), and it does occur in leafy vegetables (spinach, broccoli, cabbage, lettuce), folate occurs also in other foods: liver, kidney, beans, beetroot, bran, peanuts, yeast extract, avocadoes, bananas, wholemeal bread, eggs, and some fish. There is even a little folate in beer and tea. In the USA and Canada, all cereal products have to be fortified with folate. In other countries, voluntary fortification is permitted, so that food tables may be out of date in their folate content of foods. Many breakfast cereals and some breads are now fortified with extra folic acid.

Pure folic acid as the supplement is about twice as bioavailable as most food folates. Also, the folic acid added in fortifying food is more available than the intrinsic folate in most foods. The US Institute of Medicine proposes the use of DFEs.

Analysis of folate in foods is difficult because of the multiple compounds. Early analyses used microbiological assay, but human cells may not respond to different folate compounds in the same way as bacteria. Some recent methods are based on the major peaks on HPLC (high-performance liquid chromatography). Folate is destroyed in foods by prolonged boiling.

### 12.7.13 Toxicity

The main concern is that if someone with vitamin $B_{12}$ deficiency (pernicious anaemia) is treated with a fairly high (supranutritional) dose of folic acid (5 mg/day), the anaemia may improve but the biochemical basis for the neurological symptoms of vitamin $B_{12}$ deficiency is not corrected. Therefore, correct biochemical diagnosis with serum $B_{12}$ and folate is essential before anyone is treated for anaemia with folic acid. Otherwise, the toxicity of folic acid is low.

## Further reading


1. **Bailey, L.B.** (1998) Dietary reference intakes for folate: the debut of dietary folate equivalents. *Nutr Rev*, **56**, 294–9.
2. **Bruckner, M., Katheria, A.C., and Schmölzer, G.M.** (2021) Delayed cord clamping in healthy term infants: more harm or good? *Semin in Fetal and Neonatal Med*, **26**(2), 101221.
3. **Clarke, R., Bennett. D., Parish, S., et al.** (2014) Effects of homocysteine lowering with B vitamins on cognitive aging: meta-analysis of 11 trials with cognitive data on 22,000 individuals. *Am J Clin Nutr*, **100**, 657–66.
4. **Ebbing, M., Bønaa, K.H., Arnesen, E., et al.** (2010) Combined analyses and extended follow-up of two randomized controlled homocysteine-lowering B-vitamin trials. *JIM*, **268**, 367–82.
5. **Food and Nutrition Board, Institute of Medicine** (1998) *Dietary reference intakes for thiamin, riboflavin, niacin, vitamin B-6, pantothenic acid, biotin and choline*. Washington DC: National Academy Press.
6. **Jacques, P.F., Selhub, J., Bostom, A.G., et al.** (1999) The effect of folic acid fortification on plasma folate and total homocysteine concentrations. *N Engl J Med*, **340**(19), 1449–54.
7. **Mason, J.B., Dickstein, A., Jacques, P.F., et al.** (2007) A temporal association between folic acid fortification and an increase in colorectal cancer rates may be illuminating important biological principles: a hypothesis. *Cancer Epidemiol Biomarkers Prev*, **16**, 1325–9.
8. **Medical Research Council Vitamin Study Group** (1991) Prevention of neural tube defects: results of the MRC vitamin Study Research Group. *Lancet*, **338**, 131–7.
9. **Morris, J.K., Rankin, J., Draper, E.S., et al.** (2015) Prevention of neural tube defects in the UK: a missed opportunity. http://dx.doi.org/10.1136/archdischild-2015-309226
10. **National Health and Medical Research Council (NHMRC)** (1994) Folate fortification: report of the expert panel on folate fortification. Canberra: Australian Government Publishing Service.
11. **Ray, J.G.** (2004) Folic acid food fortification in Canada. *Nutr Rev*, **62**, 535–9.
12. **Selhub, J., Morris, M.S., Jacques, P.F., and Rosenberg, I.H.** (2009) Folate–vitamin B12 interaction in relation to cognitive impairment, anaemia and biochemical indicators of vitamin B-12 deficiency. *Am J Clin Nutr*, **89**(Suppl.), 702S–6S.
13. **Seshadri, S., Beiser, A., Selhub, J., et al.** (2002) Plasma homocysteine as a risk factor for dementia and Alzheimer's disease. *N. Engl J Med*, **346**, 476–83.
14. **Smithells, R.W., Shepard, S., Schorah, J., et al. (1990)** Possible prevention of neural-tube defects by periconceptual vitamin supplements. *Lancet*, **1**, 339–40.
15. **Truswell, A.S. and Kounnavong, S.** (1997) Quantitative responses of serum folate to increasing intakes of folic acid in healthy women. *Eur J Clin Nutr*, **51**, 839–45.

# 12.8 Vitamin $B_{12}$

In the late 1920s it was postulated that human gastric juice contained an 'intrinsic factor', which combined with an 'extrinsic factor' in animal protein foods (notably raw liver), and that the combination would cure a type of anaemia that was until then untreatable, pernicious anaemia. In 1948 the extrinsic factor, vitamin $B_{12}$, was identified and human intrinsic factor was isolated in the 1960s.

## 12.8.1 Structure

Vitamin $B_{12}$ or cobalamin is a red compound containing a corrinoid ring (four pyrrole rings) with an atom of cobalt in its centre. It is only synthesized by bacteria. Vitamin $B_{12}$ has the largest molecule of the vitamins, with a molecular weight of 1355. The structure is large and three-dimensional—to view it, refer to a good textbook of biochemistry. Dorothy Hodgkin was awarded the Nobel Prize for Chemistry (1964) for elucidating vitamin $B_{12}$'s structure by X-ray crystallography.

## 12.8.2 Functions

The co-enzyme forms of vitamin $B_{12}$ are methylcobalamin and deoxyadenosylcobalamin. Only two $B_{12}$-dependent enzymes have been identified in humans: methylmalonyl-CoA mutase (which requires deoxyadenosylcobalamin) and methionine synthase (which requires methylcobalamin).

Methylmalonyl-CoA mutase is involved in the conversion of methylmalonyl-CoA to succinyl-CoA in the catabolism of propionate, in the mitochondria.

Methionine synthase, found in the cytosol, transfers a methyl group from the donor 5-methyl THF to homocysteine to produce methionine (Fig. 12.8). Increased plasma homocysteine, a condition that may increase the risk for vascular disease, responds to supplements of vitamins $B_{12}$, folate, and $B_6$.

## 12.8.3 Absorption and metabolism

Absorption of vitamin $B_{12}$ is by a highly specific mechanism. It first has to be released from its binding to animal proteins in food. This requires pepsin and acid in the stomach. When gastric acid is secreted after a meal, the parietal cells at the same time release the specific glycoprotein, intrinsic factor (IF). This doesn't attach to vitamin $B_{12}$ until they are both in the duodenum. (While in the stomach, $B_{12}$ is attached to R-binder (haptocorrin) from the saliva, which is digested in the duodenum.) The $B_{12}$/IF complex passes down the small intestine and in the terminal ileum there are specific receptors for IF. The $B_{12}$/IF complex is absorbed and after 2–3 hours, vitamin $B_{12}$ appears in the bloodstream carried on transcobalamin II (TCII), the main $B_{12}$ transport protein.

Vitamin $B_{12}$ is concentrated, stored in the liver. It is excreted in the bile into the duodenum, where much of it combines with IF (secreted after a meal) and also gets absorbed in the terminal ileum. This enterohepatic cycle helps to conserve the vitamin.

Some anaerobic bacteria in the large intestine can synthesize vitamin $B_{12}$. But this is formed below the ileal receptor site, so it is not likely to be absorbed.

The total body store is estimated to range from only 3–5 mg, but enough for several years! Only about 0.25% of total body stores (2–5 μg) are excreted daily. The requirement is only 2 μg/day.

## 12.8.4 Deficiency in animals

The most common signs of deficiency in animals are lack of growth and reduced food intake. Alterations in lipid metabolism occur—fatty liver, and increase in triglycerides and free fatty acids. In pigs, a mild anaemia is observed. Neurological changes have been observed in monkeys after 3–5 years on a vitamin $B_{12}$-deficient diet, and in fruit bats after 7 months.

## 12.8.5 Deficiency in humans

Vegans: Very strict vegetarian (vegan) diets containing no fish, poultry, eggs, or dairy products, and without vitamin supplements, contain practically no vitamin $B_{12}$. Vegans have low circulating levels of vitamin $B_{12}$, but clinical symptoms are surprisingly uncommon. Normal body stores of the vitamin are sufficient to last for 2–5 years. Bacteria in the intestine produce some vitamin $B_{12}$ that might perhaps be absorbed from the caecum, but the bioavailability of such $B_{12}$ is uncertain. However, infants breastfed by strict vegan mothers are at serious risk of impaired neurological development, anaemia, and even severe encephalopathy (Von Shenck et al., 1997). Vegans are advised to take vitamin $B_{12}$ tablets or foods fortified with $B_{12}$ (produced microbiologically, i.e. not from an animal source). However, for vegan breastfed infants, vitamin $B_{12}$ supplements are essential.

Pernicious anaemia and vitamin $B^{12}$ neuropathy: Inadequate dietary intake is not the usual cause of clinical vitamin $B_{12}$ deficiency. Most common is malabsorption due to an autoimmune atrophy of the gastric mucosa, so there is failure to produce IF. Other less common causes include total gastrectomy and disease of the terminal ileum. Severe vitamin $B_{12}$ deficiency from gastric atrophy is called pernicious anaemia because it used to be untreatable. There are two effects of vitamin $B_{12}$ deficiency—megaloblastic anaemia and/or neurological dysfunction. Anaemia usually precedes neurological symptoms, but not always. Anaemia is megaloblastic, so called because blood cells are characteristically large with reduced nuclear density, and white cells, platelets, and epithelial cells are also affected the same way. Vitamin $B_{12}$ deficiency interrupts normal nuclear division by 'trapping' folate, leading to a reduction in the synthesis of DNA (see section 12.7). The anaemia is morphologically the same in folate and vitamin $B_{12}$ deficiency. Biochemical tests have to be used to distinguish between them. An injection of 100 μg/month will successfully treat pernicious anaemia but high oral doses (working by passive absorption) are also used. The modern pharmaceutical preparation is hydroxocobalamin.

The characteristic *neuropathy* of vitamin $B_{12}$ deficiency is subacute combined degeneration of the spinal cord. There is loss of position sense and spastic weakness in the lower limbs due to demyelination of the posterior and lateral (pyramidal) tracts (upper motor neurone). Sometimes the nervous system can be affected in other ways, e.g. with neuropsychiatric disorders. There is not always an accompanying anaemia. Serum vitamin $B_{12}$ is subnormal and symptoms should respond to vitamin $B_{12}$ treatment, although more slowly than the anaemia.

Spinal cord disease is not seen in folate deficiency. The biochemical basis of the pernicious anaemia involves an interaction with folate; the neurological disease does not. The most likely explanation of the neuropathy is impaired methylation of myelin basic protein from deficient methionine synthase.

Subclinical vitamin $B_{12}$ deficiency: Pernicious anaemia, being a serious clinical disease, has been known since it was first described by Addison in 1894. However, in recent years, since serum vitamin $B_{12}$ assays have been generally available, it has been found that subnormal levels of serum $B_{12}$ occur in 10% or more of old people without anaemia in developed countries. Serum methylmalonate is raised; absorption of crystalline vitamin $B_{12}$ is normal. In a proportion of cases, loss of gastric acid and failure to free $B_{12}$ from protein binding in food is the probable cause. Apart from raised homocysteine (a risk factor for cardiovascular disease and dementia), no consistent serious effect has been found. The Institute of Medicine recommends for older people that part of the intake of vitamin $B_{12}$ should be in crystalline form, e.g. fortified food or in multivitamins. Other RDI committees (e.g. Australia and New Zealand) have not yet followed this advice. Antagonists to vitamin $B_{12}$: prolonged nitrous oxide anaesthesia, metformin, and stomach acid blockers.

### 12.8.6 Biochemical tests

Serum vitamin $B_{12}$ can be assessed by radioligand-binding assay or microbiological assay. Normal levels range from 200 to 900 pg/mL (pg = $10^{-12}$ g) or over 150 pmol/L. Deficiency is indicated by values below this. Concentrations of vitamin $B_{12}$ in serum are exceedingly small; different methods are available. Serum $B_{12}$ results are not always reliable, so two other biochemical tests are useful.

Elevated serum or urinary excretion of methylmalonate and raised plasma homocysteine are the other biochemical tests indicating low $B_{12}$ status. Methylmalonate is more specific because homocysteine is also elevated with folate deficiency. The Schilling test is used to confirm the diagnosis of pernicious anaemia. It measures absorption of oral vitamin $B_{12}$ labelled with radioactive cobalt on two occasions, the first without and the second test with IF.

## 12.8.7 Interactions: nutrients

In vitamin $B_{12}$ deficiency, 5-methyl THF accumulates due to decreased activity of methionine synthase, thus holding folate in what is termed the 'methyl-folate trap'. The ultimate cause of megaloblastic anaemia is impaired conversion of deoxyuridylic acid to thymidylic acid and so DNA synthesis is impaired because of lack of 5,10-methylene THF.

## 12.8.8 Food sources

As vitamin $B_{12}$ is synthesized by micro-organisms, the vitamin is only found in bacterially fermented foods, or meat and offal from ruminant animals in which the vitamin is synthesized by ruminal microflora. The richest source of the vitamin is liver. Other sources include shellfish, fish, meat, eggs, milk, cheeses, and yoghurt.

Not all corrinoids exhibit vitamin $B_{12}$ activity. In spirulina—a type of algae often promoted as a source of $B_{12}$ for vegetarians—80% of the corrinoids do not have vitamin $B_{12}$ activity. Up to 30% of the $B_{12}$ in supplement pills may be analogue(s) of $B_{12}$ with little or no activity.

The RDI of vitamin $B_{12}$ is very small—in adults it is 2.5 μg/day.

## 12.8.9 Toxic effects

Oral intakes of several hundred times the nutritional requirement are safe, as intestinal absorption is specific and limited. Vitamin $B_{12}$ injections (a nice red colour) are used in medicine as a placebo.

### Further reading

1. **Bates, C.J., Schneeds, J., Mishra, G., Prentice, A., and Mansoor, M.A.** (2003) Relationship between methylmalonic acid, homocysteine, vitamin $B_{12}$ intake and status and social-economic indices, in a subset of participants in the British National Diet and Nutrition Survey of people aged 65 y and over. *Eur J Clin Nutr*, **57**, 349–59.

2. **Von Shenck, U., Bender-Gotze, C., and Koletzo, B.** (1997) Persistence of neurological damage induced by dietary vitamin $B_{12}$ deficiency in infancy. *Arch Dis Child*, **77**, 137–9.

# Vitamins C and E

A. Stewart Truswell and Jim Mann



## 13.1 Vitamin C

Also called ascorbic acid, vitamin C is the antiscorbutic vitamin that prevents scurvy (Latin *scorbutus*).

### 13.1.1 History

Scurvy became very important after Columbus connected Europe with America (1492). West European explorers and navies ventured ever longer and longer distances in sailing ships that took months to cross oceans and round continents. When Vasco da Gama sailed from Portugal round the Cape of Good Hope in 1490, he lost 60% of his crew of 160 sailors to scurvy. The sailors had food, but not vegetables and fruit. It took centuries before the concept was accepted that prolonged lack of certain foods can cause severe illness without people getting thinner.

As early as 1601, Sir James Lancaster had avoided scurvy on an expedition round Africa to Sumatra to buy spices. He gave the sailors lemon juice and put in to ports to buy oranges, etc., but this was an exception. Scurvy was responsible for more deaths at sea than shipwreck, warfare, and all other illnesses combined. There was probably much more land scurvy, with less dramatic presentation than at sea. Drummond and Wilbraham, in their history of *The Englishman's food*, from knowledge of what foods were available, estimate that scorbutic conditions must have been common in mediaeval and Tudor England, particularly in early spring. Some degree of scurvy must have been prevalent in prisons and in boarding schools. Florence Nightingale thought that scurvy had caused much loss of life in British forces in the Crimean War. The introduction of the potato in the 17th century reduced the incidence of land scurvy.

In 1747 James Lind, surgeon on *HMS Salisbury*, performed the first controlled therapeutic trial and cured scurvy in two men with two oranges and a lemon daily; five other treatments had no effect on pairs of sailors in the same sickbay. Captain James Cook sailed to and around the Antipodes on two very long voyages (1768–71 and 1772–75) and returned to England without losing a single sailor from scurvy. He was strict with cleanliness, stopped for fresh vegetables wherever possible, and took with him a number of supposed remedies for scurvy, including wort of malt (inexpensive and favoured by the Admiralty), sauerkraut, and small quantities of lemon juice concentrate. Cook himself was not clear which had been the effective preventive.

No official action was taken to prevent scurvy on ships until Dr Gilbert Blane, an aristocrat with experience in the 1782 naval war against the French in the West Indies, persuaded the Lords of the British Admiralty to issue lemon juice as a daily ration on all navy ships from 1804. This was at the beginning of the Napoleonic war. The stronger manpower on British ships must have contributed to naval victories like Trafalgar. At one stage, West Indian lime replaced Mediterranean lemon juice to save money, but was a less effective preventive (and later shown to contain only a quarter of the vitamin C).

Confusion persisted about prevention of scurvy into the twentieth century, until the antiscorbutic substance was identified and could be quantified in different foods. Scott's tragic South Pole expedition walked 1700 miles under severe conditions with no vitamin C in their rations in 1912.

In 1907, Holst and Frölich in Oslo, Norway, had been trying to produce an animal model of beriberi. They found that guinea pigs developed scurvy on a cereal diet with no green food ('one-sided diet'). Common laboratory animals, such as pigeons, rats, mice, and dogs, do not develop scurvy on deficient diets.

A substance, first called hexuronic acid, was isolated from adrenal glands, oranges, and cabbage by Szent-György in 1928. Glen King and Szent-György in 1932 independently showed that it prevented guinea pig scurvy. Vitamins A and B had been identified earlier and Drummond proposed 'vitamin C' for the antiscorbutic vitamin. Its glucose-like structure was worked out by Haworth and Hirst, and ascorbic acid was the first vitamin to be synthesized, by Reichstein (1935). Szent-György and Haworth were awarded Nobel Prizes for physiology/medicine and chemistry, respectively, both partly for their research on vitamin C.

## 13.1.2 Deficiency disease

Scurvy in adults is rare today in developed countries and should be diagnosed before the patient is dangerously ill, but recognition of this rare disease requires doctors to be alert to its distinctive features (Box 13.1).

In the old sailing ship days, people with scurvy were listless and weak, with bleeding gums, loss of teeth, foul breath, painful legs, and generalized haemorrhages. Old wounds broke down. Severe cases were fatal, sometimes suddenly.

There are three types of haemorrhage in scurvy:

- small skin haemorrhages around hyperkeratotic hair follicles with coiled hairs (diagnostic)
- larger bruises in muscles
- internal haemorrhages, which could be into the brain or pericardium.

The spongy bleeding gums have to be distinguished from common gingivitis and do not appear in edentulous people. Most of the features can be explained

### Box 13.1 A recent case of scurvy

A 42-year-old man presented to the emergency department with a 1-month history of lethargy, and significant spontaneous bruising and pain over his thighs and bleeding from his gums. There was no known trauma and he had not had any significant bleeding history. His health was otherwise unremarkable.

His social history was significant in that he had severe agoraphobia (extreme fear of public places) and was living with his mother in a retirement village. She had recently been moved to a hostel because of severe dementia. His diet over the past 5 years had consisted almost entirely of sausage rolls, meat pies, wheat cereal, biscuits, peanut butter, chocolate cake, and milk. He had not had any vegetables or fresh fruit in over 5 years and was unable to cook for himself. Recently, his shopping had been done by community-support staff and they had previously tried to encourage a better diet, but he was non-compliant.

On examination, he appeared malnourished and had extreme ecchymoses (bruising) and petechiae (tiny skin haemorrhages) over the lower limbs, and marked gingivitis.

His blood tests on admission showed an undetectable ascorbic acid (vitamin C) level. He had normal electrolytes and renal function, but decreased albumin of 29 g/L. His vitamin B12, thiamin, and folate levels were normal. He was anaemic, with a haemoglobin level of 81 g/L, with normal white cells, platelets, and coagulation studies.

#### Two questions

Why did the doctor suspect scurvy, and request plasma vitamin C measurement (not a routine investigation)?

How do you think the patient was treated?

*Source:* Dean, M.G. (2006) Extensive bruising secondary to scurvy. *Intern Med J*, **36**, 393.

by impaired synthesis (and repair) of collagen and capillary fragility.

In young children, Barlow (1894) described infants whose legs were too sore to move. They had lost weight and were anaemic. This was not rickets; it was vitamin C deficiency with haemorrhages under the periosteum of bones or into joints. If the child was too young to have teeth through, there were no bleeding gums. The cause was feeds of condensed or sterilized cows' milk, which lacked any vitamin C (unlike breast milk or modern infant formulas).

Experimental vitamin C deficiency was produced in human volunteers in Britain during World War II (supervised by Professor Krebs) and by Hodges et al. (1971) in the USA. In the later study, the first signs of scurvy appeared after 2–3 months of zero intake. Both studies found that 10 mg of pure ascorbic acid was sufficient to prevent scorbutic features.

### 13.1.3 Chemistry and biosynthesis

Ascorbic acid is a 6-carbon compound, derived from glucose, but it is $C_6H_8O_6$, not a carbohydrate. It is a weak acid, soluble in water, and a strong reducing agent (antioxidant). X-ray crystallography shows the molecule is flat.

It has two asymmetric carbon atoms, so there are four stereoisomers:

- L-ascorbic
- D-ascorbic
- L-isoascorbic
- D-isoascorbic.

Only the first is antiscorbutic; all four are antioxidants and D-isoascorbic is used for this in food processing.

Vitamin C in aqueous solution is unusually sensitive to heat, especially with oxidizing agents, at alkaline pH, and in the presence of copper. Care has to be taken, therefore, with vegetables to keep cooking time short, and not to add sodium bicarbonate, or use a copper pan.

**Biosynthesis** For most animals, ascorbic acid is not a vitamin: they make what they require from glucose. The steps via glucuronic acid, gulonic acid, and L-gulonolactone are shown in Fig. 13.1.

The enzyme for the final step, L-gulonolactone oxidase, is lacking only in primates (including humans), fruit-eating bats, guinea pigs, and some passerine birds. All these susceptible animals eat fruit as a natural part of their diet.

### 13.1.4 Functions

Ascorbic acid is a cofactor for eight mammalian enzymes. It is required to convert iron on the enzyme from the ferric state ($Fe^{3+}$) to ferrous ($Fe^{2+}$).

The best-established function of vitamin C is its role in synthesis of collagen, the major material in all connective tissue, including bones. Collagen is

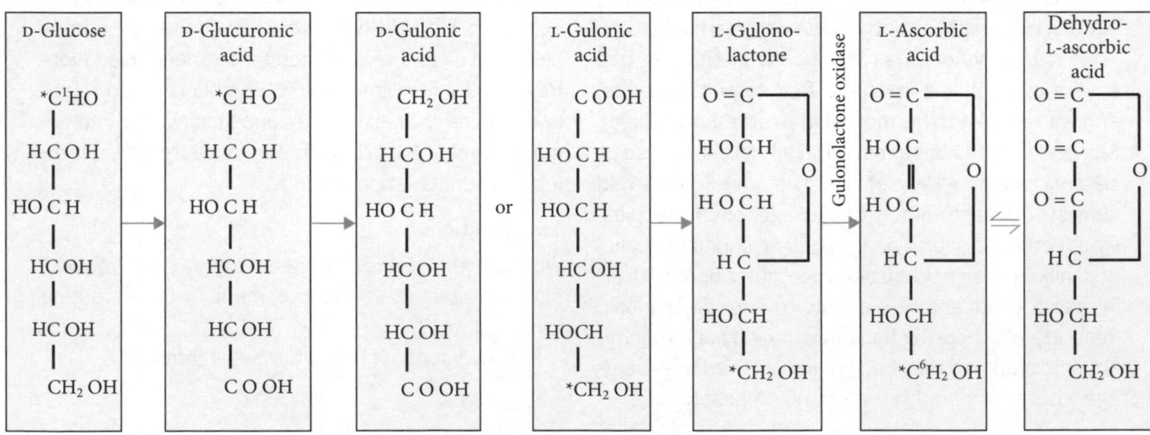


Fig. 13.1 Biosynthesis of vitamin C.

first synthesized as procollagen polypeptides, rich in proline. Some of the proline and lysine is then hydroxylated post-translation. The three hydroxylase enzymes require ascorbate to function. Hydroxyproline is necessary for the formation of collagen's triple helix structure. Normal collagen cannot be made in the absence of vitamin C.

The adrenal gland has a very high ascorbate concentration. It plays a role in conversion of dopamine to noradrenaline and is also involved in the effect of ACTH on cortisol production. When ACTH stimulates secretion of cortisol, this is preceded by increased ascorbate in the adrenal vein. Two other organs with very high ascorbate are the pituitary and the lens of the eye, where it must play a role in preventing photo-oxidative damage of the transparent lens protein.

Ascorbate is also required for conversion of lysine to carnitine, which transfers long-chain fatty acids to the inner mitochondria for conversion to energy by way of β-oxidation.

In muscles and the heart, fat is a significant source of energy and function. This can explain muscle weakness in severe scurvy.

White blood cells contain higher concentrations of ascorbate than plasma and lymphocytes appear to have the highest concentration. Detailed mechanisms of how it protects proteins in these cells from free radical damage during phagocytosis are not yet fully elucidated.

### 13.1.5 Absorption, metabolism, and excretion

Ascorbic acid is absorbed as itself in the small intestine by an active transport mechanism that gets saturated at higher intakes. On ordinary dietary intakes nearly all the vitamin C is absorbed, but above 1 g/day, i.e. with supplements, the law of diminishing returns applies and much of the vitamin is not absorbed. People with the solute carrier family 23 SLC23Airs 33972 313 variant G have plasma ascorbate about 20% higher than average (Kobylecki et al., 2015).

Ascorbate is carried free in the plasma. Its plasma concentration increases to a plateau of 70–80 mmol/L at intakes around 400 mg/day (Levine et al., 1996). Urinary excretion starts before this. This means that if people take more than 500 mg/day, e.g. as supplement, plasma concentration will not increase. Progressively larger proportions are not absorbed and larger proportions are lost in the urine. (Higher plasma ascorbate can only occur if there is renal failure or the vitamin is given intravenously.)

In the organs, ascorbate concentrations are higher than in plasma. It is transported into the cells by sodium-dependent vitamin C transporters SVCT1 and SVCT2. When ascorbate is oxidized, the product is dehydroascorbate. In neutrophils this can be taken into the cell with glucose transporters GLUT1 and GLUT3, where it is rapidly reduced to ascorbate by the protein glutaredoxin.

The normal total body pool of ascorbate is 1500–2500 mg. There are no defined stores.

Ascorbate is excreted unchanged in the urine. It is also oxidized reversibly to dehydroascorbatic acid, which is further hydrolysed (irreversibly) to diketogulonic acid, whose metabolic products include oxalic acid.

### 13.1.6 Interactions

Vitamin C enhances intestinal absorption of non-haem iron. It facilitates the conversion of ferric to ferrous iron, which is more soluble. This happens with foods rich in vitamin C.

Smokers have significantly lower serum ascorbate: 40% lower with 20 or more cigarettes/day. This is only partly explained by smokers' lower consumption of fruits and vegetables; radioisotope-labelled ascorbate studies show that the turnover rate is almost doubled. Ascorbate is lost *in vitro* in plasma exposed to cigarette smoke.

Several drugs have been reported to lower plasma ascorbate, including corticosteroids, aspirin, indomethacin, phenylbutazone, and tetracycline.

### 13.1.7 Biochemical tests

Serum ascorbic acid is the standard laboratory test. Blood taken for measuring vitamin C must be specially preserved with metaphosphoric acid or trichloracetic acid, and then frozen if not analysed fresh. Older methods used 2,6-dichlorophenolindophenol for analysis, but this is not completely specific and the method of choice is high-performance liquid chromatography with electrochemical detection.

Women have somewhat higher levels than men at the same vitamin C intakes. Serum ascorbate cannot be used to identify people taking very high doses of vitamin C supplements (see section 13.1.5). Very low levels, between 11 μmol/L and zero, are found in scurvy. Leukocyte (buffy layer) ascorbate indicates cell content rather than recent intakes. The expected concentration is higher in mononuclear cells than in neutrophils.

## 13.1.8 Food sources

Potatoes, *Brassica* vegetables, and orange juice are the major contributors in Europe and North America. There is about a 250-fold range in the vitamin C content of fruits (Table 13.1).

Vitamin C is heat labile and lost in cooking or warm holding, whether in the food factory, kitchen, or bain-marie. Alkaline conditions, e.g. sodium bicarbonate, and copper containers accelerate cooking losses. So, fresh fruits and salads have more reliable vitamin C than cooked plant foods. Dried fruits have lost their original vitamin C.

In animal foods, there is no vitamin C in usual meats, i.e. muscle, but liver and kidney contain around 10 g–100 g (depending on cooking method). Cows' milk (unwarmed) contains a little, but human milk averages 4 mg vitamin C/100 mL, which is sufficient for the first 6 months of infancy.

**Table 13.1 Ascorbic acid content of foods**

| Content (mg/100 g edible portion) | Food |
|---|---|
| <10 | Apricot, pear, apple, plum, grape, banana, cherries |
| 10 | Potatoes (boiled) |
| 50 | Citrus fruit and strawberries |
| 100 | Broccoli (raw), parsley, Brussels sprouts (raw), cauliflower (raw), kiwi fruit |
| 200–300 | Blackcurrants, guavas, peppers |
| 1250 | Rosehips (*Rosa canina*) |
| 1000–2300 | Acerola fruit |
| 2300–3150 | *Terminalia ferdinandiana* (Kakadu plum) |

## 13.1.9 Recommended nutrient intakes

Some Expert Committees, e.g. FAO/WHO, UK, and Australia/New Zealand, recommend 40–45 mg/day for adults as sufficient to prevent scurvy (10 mg/day), with generous safety factors, including for smokers. USA/Canada and Germany/Austria/Switzerland recommend 90–100 mg vitamin C/day as the amount to achieve saturation of white blood cells without wastage from urinary secretion.

Average vitamin C intakes from food in the 2000–01 Diet and Nutrition Survey of British Adults were 83 mg/day in men and 81 mg/day in women. Intakes are higher in the USA and Australia. In Australia in 1995, intakes were 116 mg in men and 98 mg in women. In addition, a minority of people consume extra vitamin C in multivitamin supplements.

Nutrient recommendations are for healthy people and based on physiological research. In sickness, more vitamin C may be desirable (see Box 13.2). Serum or leukocyte ascorbate have been reported low with trauma and during recovery from surgery (which involves a big increase in collagen synthesis) and in people with severe infections.

## 13.1.10 Toxic effects

The US Institute of Medicine (2000) set the upper intake level for adults at 2000 mg/day. The more recent Australian/New Zealand committee were not able to set a definite upper level but considered 1000 mg/day would be a prudent limit. There is no point in megadosage of vitamin C above an intake of 200 mg/day. At that point, neutrophil (i.e. cellular) ascorbate reaches a plateau and any further intake only goes into the urine.

**Box 13.2** Case study: does extra vitamin C have health benefits?

The problem is the intake distance between prevention of scurvy and a possible optimal intake. As ascorbate is concentrated in immune reactive cells and in the adrenal, pituitary, and anterior eye, it is reasonable to suppose that more than enough to prevent scurvy may help achieve optimal immune function, response to stress, and prevention of cataract. The antioxidant function was stressed in health education until randomized controlled trials with vitamin C or E proved very disappointing. The British MRC/BHF trial in over 20,000 people did not find any benefit in total mortality, coronary heart disease, strokes, or cancer mortality in those who took 250 mg/day vitamin C plus vitamin E and β-carotene (Heart Protection Study Collaborative Group, 2002). Cohort studies such as EPIC-Norfolk have shown an inverse relation of plasma ascorbate and deaths from all causes of cardiovascular diseases (Khaw et al., 2001), but the authors think it more likely that plasma vitamin C is a marker for intake of fruits and vegetables, which contain mixtures of potentially protective substances. Two large Mendelian randomization studies (Kobylecki et al., 2015; Wade et al., 2015) did not show less coronary or other heart disease in those with the gene *SLC23A1* associated with high plasma vitamin C due to a variant of the sodium-dependent vitamin C transporter. Protective effects of fruit and vegetables must have other explanations.

Ascorbic acid is secreted in the gastric juice and can reduce formation of nitrosamines, which are carcinogenic. An international panel made a thorough review of all the epidemiological evidence about diet and cancer in 1997 for the World Cancer Research Foundation. They concluded that vitamin C 'possibly' protected against stomach cancer. This extensive review was repeated in 2007 with stricter evidence-based criteria, and now vitamin C has largely disappeared from the conclusion tables. For oesophagus cancer, 'foods containing vitamin C' are probably protective; for stomach cancer, 'non-starchy vegetables, allium vegetables, and fruits are probably protective' (i.e. vitamin C is not specifically named).

There is still a popular belief that if you take vitamin C tablets they will prevent colds. Randomized controlled trials to prove this are easy to organize; many have been done and the findings have nearly all shown no reduced incidence of colds; they may possibly be of shorter duration (Truswell, 1986; Douglas et al., 2001).

There have been several reports of low plasma ascorbate in people with diabetes. Doctors in the diabetic clinic might justifiably advise more fruit and vegetables (or small doses of vitamin C tablets).

With cataract, the largest cohort study found that women who took vitamin C tablets in the long term had a significantly lower incidence of cataract operations (Hankinson et al., 1992). There remains the possibility of confounding, e.g. people who take vitamin supplements may use sunglasses.

Convalescence after a major illness, surgery, or trauma, and old people with poor appetite, especially if bedsores threaten, are situations where intake is close to the EAR or requirements may be increased. Such patients should be encouraged or helped to increase their vitamin C intake by eating more fruit or vegetables: if they cannot manage this, with 100 mg vitamin C tablets.

Vitamin C tablets are provided by the pharmaceutical industry at doses of 50, 100, 250, 500, and 1000 mg, partly as sodium ascorbate. Indications in prescribers' guides are 'general well being, antioxidant, colds, flu, assist wound healing'. It seems very unlikely that there could be any possible benefit from doses above 500 mg/day, because plasma ascorbate is tightly controlled and cannot be pushed higher with oral administration, as Levine et al. (1996) showed.

**Urinary oxalate stones** Ascorbic acid is partly metabolized to oxalic acid, which is a common ingredient of urinary tract stones (it is not only derived from ascorbate).

For years this was considered a serious side effect of vitamin C supplements, but more recently it was found that: (a) an important proportion of the oxalate found in urine from people taking vitamin C tablets is formed from chemical change after the urine has been passed; and (b) in the large Harvard Health Professional cohort (over 45,000), those who suffered urinary tract stones had not had higher vitamin C intakes (Curham et al., 1996).

**Iron absorption** Because ascorbic acid increases absorption of non-haem iron, people with haemochromatosis (see Chapter 9) or with the gene for it should be advised to avoid vitamin C supplements.

## Further reading

1. **Brown, S.R.** (2003) *Scurvy. How a surgeon, a mariner and a gentleman solved the greatest medical mystery of the age of sail.* Camberwell, Victoria: Penguin Group.
2. **Carpenter, K.J.** (1986) *The history of scurvy and vitamin C.* New York: Cambridge University Press.
3. **Curham, G.C., Willett, W.C., Rimm, E.B., and Stampfer, M.J.** (1996) A prospective study of the intake of vitamins C and B6, and the risk of kidney stones in men. *J Urol*, **755**, 1847–51.
4. **Dean, M.G.** (2006) Extensive bruising secondary to scurvy. *Intern Med J*, **36**, 393.
5. **Douglas, R.M., Chalker, E.B., and Treacy, B.** (2001) *Vitamin C for preventing and treating the common cold.* Cochrane Review, The Cochrane Library 3, 2001. Oxford: Oxford Update Software.
6. **Drummond, J.C. and Wilbraham, A.** (1939) *The Englishman's food. A history of five centuries of English diet.* London: J. Cape, Ltd.
7. **Hankinson, S.E., Stampfer, M.J., Seddon, M.J., et al.** (1992) Nutrient intake and cataract extraction in women: a prospective study. *BMJ*, **305**, 335–9.
8. **Heart Protection Study Collaborative Group** (2002) MRC/BHF Heart Protection Study of antioxidant vitamin supplementation in 20,536 high-risk individuals: a randomized placebo-controlled trial. *Lancet*, **360**, 23–33.
9. **Holley, A.D., Oxland, E., Barnes, A., et al.** (2011) Scurvy: historically a plague of the sailor that remains a consideration into the modern intensive care unit. *Intern Med J*, **41**, 283–5.
10. **Kobylecki, C.J., Afzal, S., Davey Smith, G., and Nordestgaard, B.G.** (2015) Genetically high plasma vitamin C, intake of fruit and vegetables, and risk of ischaemic heart disease and all-cause mortality: a Mendelian randomization study. *Am J Clin Nutr*, **101**, 1135–43.
11. **Levine, M.** (1986) New concepts in the biology and biochemistry of ascorbic acid. *N Engl J Med*, **314**, 892–902.
12. **Levine, M., Cowry-Cantilena, C., Wang, Y., et al.** (1996) Vitamin C pharmacokinetics in healthy volunteers: evidence for a recommended dietary allowance. *Proc Natl Acad Sci USA*, **193**, 3704–9.
13. **Truswell, A.S.** (1986) Ascorbic acid. *N Engl J Med*, **315**, 709.
14. **Wade, K.H., Forouhi, N.G., Barnes, A., et al.** (2015) Variation in the SLC23A1 gene does not influence cardiometabolic outcomes to the extent expected given its association with L-ascorbic acid. *Am J Clin Nutr*, **101**, 201–9.

# 13.2 Vitamin E

## 13.2.1 History

In 1922, two researchers, Evans and Bishop, found that a lipid-soluble factor was essential for reproduction in rats. Foetal resorption occurred when female rats were fed on a diet including rancid fats, but was found to be preventable by the addition to the diet of wheatgerm, dried alfalfa leaves, or fresh lettuce. This lipid-soluble factor was named vitamin E and, because of its role in reproduction, tocopherol (from the Greek: to bring forth offspring; the 'ol' indicates the alcohol in its structure). The vitamin was isolated and its chemical structure identified in the 1930s.

## 13.2.2 Sources, structure, and bioavailability

Vegetable oils (especially wheat germ oil), nuts, sweet potatoes, and, to a variable extent, other vegetables and fruits are the richest dietary sources of vitamin E. There are eight naturally occurring forms of vitamin E—four tocopherols and four tocotrienols. All forms consist of a chromanol ring with an isoprenoid 16-carbon side chain, saturated in the case of tocopherols and unsaturated in the case of tocotrienols. The α, β, γ, and δ forms of the tocopherols and tocotrienols differ with regard

Tocopherol

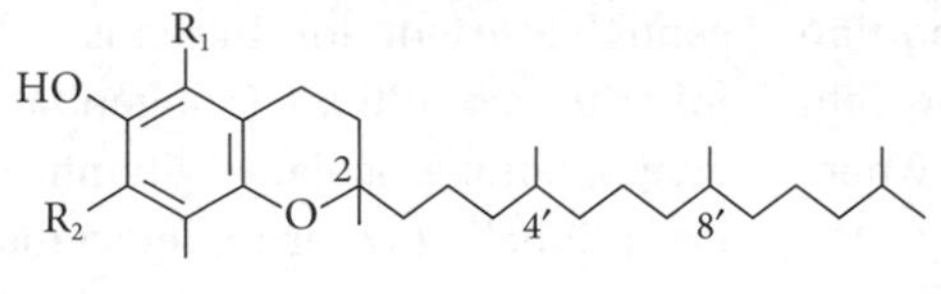

| $R_1$ | $R_2$ | |
|---|---|---|
| $CH_3$ | $CH_3$ | α-tocopherol |
| $CH_3$ | H | β-tocopherol |
| H | $CH_3$ | γ-tocopherol |
| H | H | δ-tocopherol |

Tocotrienol

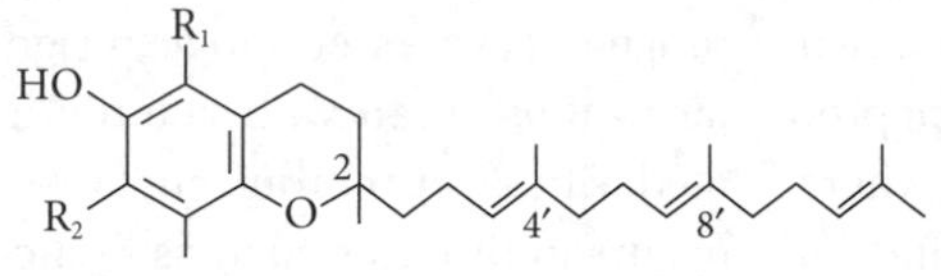

| $R_1$ | $R_2$ | |
|---|---|---|
| $CH_3$ | $CH_3$ | α-tocotrienol |
| $CH_3$ | H | β-tocotrienol |
| H | $CH_3$ | γ-tocotrienol |
| H | H | δ-tocotrienol |

**Fig. 13.2** Structure of tocopherols and tocotrienols.

*Source:* Reprinted by permission from Macmillan Publishers Ltd: Surh, Y.J. (2003) Cancer chemoprevention with dietary phytochemicals. *Nature Rev Cancer*, **3**, 768–80.

to the position and number of the methyl groups (Fig. 13.2). Furthermore, there are potentially eight stereoisomers for each of the vitamin E compounds. Tocopherols occur in foods as the free alcohols, and tocotrienols as free alcohols and esters. Acetate and succinate esters are used by the pharmaceutical industry since they are more stable.

Biological activity of these compounds is expressed by comparing potency in an animal model system with the potency of a synthetic tocopherol (all-*rac*-a-tocopherol acetate), the activity of which is set at 1.00 IU/mg. The most potent naturally occurring form of vitamin E is 2R,4′R,8′R-α-tocopherol. The vitamin E content of the diet is typically expressed as α-tocopherol equivalents (α-TE), where 1 α-TE is the activity of 1 mg RRR-α-tocopherol. The eight stereoisomers of α-tocopherol (known collectively as all-*rac*-α-tocopherol), when considered as a group, have relatively high biological activity and make the major contribution to the vitamin E content of the diet. The β, γ, and δ tocopherols and the entire group of tocotrienols make a relatively small contribution to total vitamin E activity.

### 13.2.3 Functions

Vitamin E is one of the principal antioxidants in the body. Free radicals are generated as the body's cells use oxygen as an energy source. Free radicals react with the polyunsaturated fatty acids of cell membranes and protein in a process known as lipid peroxidation, which can influence and impair membrane fluidity and function. Oxidation of apolipoprotein B results in the accumulation of oxidized low-density lipoprotein in the arterial walls, thus promoting the development of atherosclerotic plaques and increasing the risk of cardiovascular disease. Free radical damage plays a role in other disease processes, including cancer, arthritis, and cataracts.

Vitamin E is able to neutralize free radicals by transferring an electron from the hydroxyl group on the chromanol ring to the free radicals, thus making it less reactive. Following this transfer, the remaining vitamin E (now α-tocopheroxyl radical) has an impaired electron, which can become permanently inactivated by reacting with another free radical or it can react with vitamin C or glutathione, and be regenerated to active vitamin E. Although there are several other important antioxidant systems and there are close relationships between these and vitamin E, vitamin E appears to have a uniquely important role in preventing the peroxidation of the polyunsaturated fatty acids of cell membranes.

A range of additional functions of vitamin E, independent of its antioxidant properties, has been reported. α-Tocopherol has a role in modulating gene transcription, inhibits platelet aggregation and vascular smooth muscle proliferation, and may have a further signalling role in the immune system.

### 13.2.4 Absorption, transport, and metabolism

Vitamin E is, like other dietary lipids, absorbed in micelles into the cells lining the small intestine.

Absorption is fairly efficient; generally at least half of the dietary intake is absorbed, but the proportion is reduced at the high intake levels associated with pharmacological doses and in any situation where absorption of fat is reduced.

In the mucosal cells, all forms of tocopherols and tocotrienols present in the diet are incorporated into chylomicrons, which are broken down to chylomicron remnants by the action of lipoprotein lipase. Most are transported to the liver where α-tocopherol, especially RRR α-tocopherol, binds to liver α-tocopherol transport protein and is then exported in very-low-density lipoprotein to various tissues, especially to those organs with a high fatty acid content (e.g. liver, brain, and adipose tissue). Most of the other vitamin E stereoisomers, which do not bind well to α-tocopherol transport protein, are metabolized in the liver and excreted in bile. The preferential binding of RRR α-tocopherol to the transport protein, incorporating it into very-low-density lipoprotein (VLDL) and transporting it to the tissues, explain its high biological activity.

There is some transfer of vitamin E directly to adipose tissue and muscle via chylomicron remnants, which do not reach the liver. Generally, little vitamin E is released from adipose tissue. However, in experimentally depleted subjects, plasma tocopherol levels do not fall for many months because under such circumstances vitamin E is released from the substantial tissue reserves.

## 13.2.5 Deficiency states

Soon after the first discovery by Evans and Bishop, deficiency of the lipid-soluble factor was found in experimental animals to be associated with other conditions, including sterility in male rats, muscle wasting in guinea pigs and rabbits, and haemolytic anaemia and macular degeneration in monkeys. No symptoms or signs of experimental vitamin E deficiency have been found in humans and deficiency appears not to be a problem, even among people consuming relatively poor diets. However, vitamin E status is low in the newborn, due to poor placental transfer, and premature infants are given vitamin E. Furthermore, deficiency in some uncommon clinical situations shows that vitamin E is an essential nutrient for humans. Children with rare inherited conditions (congenital abetalipoproteinaemia, familial isolated vitamin E deficiency) have exceptionally low or undetectable levels of plasma vitamin E, resulting in vitamin E not reaching the tissues. Patients with these conditions develop a range of severe neurological signs, which are similar to those seen with vitamin E depletion associated with severe chronic malabsorption of fat, which occurs in diseases such as cystic fibrosis and cholecystatic liver disease. These conditions require treatment with very high doses of vitamin E, which can halt progression and may result in improvement of the neurological signs. If vitamin E deficiency is detected and treated at the early stage, these neurological consequences can be avoided.

## 13.2.6 Assessment of vitamin E status

Vitamin E status is typically assessed by measuring plasma vitamin E, of which over 90% is generally α-tocopherol. The reference range is 12–30 μmol/L. Because raised levels of plasma lipoproteins are associated with high levels of vitamin E, vitamin E status is also often expressed per unit of plasma cholesterol; an index of greater than 2.0 μmol/mmol cholesterol is regarded as an indication of adequate status.

## 13.2.7 Recommended nutrient intakes

Because clinical deficiency of vitamin E is so rare in humans, even when diets are poor, recommended intakes are somewhat arbitrary. Average requirements are considered to be around 10–12 mg of α-tocopherol equivalents per day, which is relatively easy to achieve, given the high content in polyunsaturated oils. Since, at least in theory, vitamin E requirements are increased if intake of polyunsaturated fats is high, requirements are sometimes expressed in terms of amount of the fatty acids in the diet; a ratio of 0.4 mg α-tocopherol/g polyunsaturated fat intake is considered to be an adequate intake of vitamin E.

The antioxidant properties of vitamin E and the findings of some prospective studies have led to the suggestion that intakes higher than physiological requirements might protect against heart disease, some cancers, and other chronic diseases. However, most randomized controlled trials and systematic literature reviews relating to vitamin E supplementation have not confirmed these earlier suggestions (see Chapter 19). Indeed, an increased risk of heart failure and increased all-cause mortality has been reported. A single trial has suggested that a subgroup of people with haptoglobin genotype 2-2 and who are at high risk of cardiovascular disease may benefit from vitamin E supplementation. Nevertheless, despite very low toxicity up to intakes of around 3000 μg or more/day, there is no justification for recommending extremely high doses as preventive or therapeutic measures. The Institute of Medicine in the USA has suggested an adult upper limit of 1000 mg/day for adults.

Some 40 years ago, it was suggested that 'Vitamin E is one of those embarrassing vitamins that have been identified, isolated and synthesized by physiologists and biochemists and then handed to the medical profession with the suggestion that a use should be found for it, without any satisfactory evidence to show that human beings are ever deficient of it or even that it is a necessary nutrient for man' (Davidson and Passmore, 1969). While it does appear to be an essential nutrient for humans, its role is still not fully understood.

## Further reading


1. **Brigelius-Flohe, R., Kelly, F.J., Salonen, J.T., Neuzil, J., Zingg, J.-M., and Azzi, A.** (2002) The European perspective on vitamin E: current knowledge and future research. *Am J Clin Nutr*, **76**, 703–16.
2. **Davidson, S. and Passmore, R.** (1969) *Human nutrition and dietetics*. Baltimore: Williams & Wilkins.
3. **Guallar, E., Hanley, D.F., and Miller, E.R.** (2005) An editorial update: annus horribilis for vitamin E. *Ann Intern Med*, **143**, 143–6.
4. **Horwitt, M.K.** (1960) Vitamin E and lipid metabolism in man. *Am J Clin Nutr*, **8**, 451–61.
5. **Kayden, H.T. and Traber, M.G.** (1993) Absorption, lipoprotein transport, and regulation of plasma concentrations of vitamin E in humans. *J Lipid Res*, **34**, 345–58.
6. **Leth, T. and Sondergaard, H.** (1977) Biological activity of vitamin E compounds and natural materials by the resorption-gestation test, and chemical determination of the vitamin E activity in foods and feeds. *J Nutr*, **107**, 2236–43.
7. **Meydani, M.** (1995) Vitamin E. *Lancet*, **345**, 170–5.
8. **Traber, M.G. and Arai, H.** (1999) Molecular mechanisms of vitamin E transport. *Ann Rev Nutr*, **19**, 343–55.
9. **Vardi, M., Levy, N.S., and Levy, A.P.** (2013) Vitamin E in the prevention of cardiovascular disease: the importance of proper patient selection. *J Lipid Res*, **54**, 2307–14.

# Vitamins D and K

A. Stewart Truswell



## 14.1 Vitamin D

### 14.1.1 History

In the first step towards identifying individual vitamins, E.V. McCollum at the University of Wisconsin postulated (1915) two essential dietary factors, as well as macronutrients and minerals—'fat-soluble A' and 'water-soluble B'. Fat-soluble A prevented growth failure and the eye disease xerophthalmia in animals fed purified diets. In 1919, Edward Mellanby in London found that some fats would cure experimental dietary rickets in puppies kept indoors, but others would not. Cod liver oil was very active against the bone disease rickets and against xerophthalmia (Chapter 11), but when heated, with oxygen bubbled through, its antixerophthalmia activity was lost, though not its antirachitic (antirickets) activity, and McCollum realized in 1922 that there were two nutritional factors in cod liver oil. He designated the antirachitic factor vitamin D, because 'water-soluble C' had been proposed for the antiscorbutic (antiscurvy) factor in 1919. Meanwhile, Harriette Chick, a British scientist, proved that rickets in children in Vienna after World War I could be cured either by cod liver oil or by exposure to an ultraviolet (UV) light lamp. Pure vitamin $D_2$ was first obtained by irradiating ergosterol with UV light in 1927. Its chemical structure was established by A. Windaus in Germany and F. Askew in England in 1932. In 1936, Windaus published the structure of vitamin $D_3$, the natural form of the vitamin made by UV light in the skin and present in cod liver oil.

### 14.1.2 Chemistry and metabolism

Vitamin $D_3$, cholecalciferol, is derived by the effect of UVB irradiation (wavelength 290–315 nm) on 7-dehydrocholesterol (cholesterol with a double bond at carbon 7), a minor companion of cholesterol, in the skin. There is a rearrangement of the molecule, with opening of the B ring of the steroid nucleus (Fig. 14.1). Cholecalciferol is the naturally occurring form of the vitamin in humans and animals, for example in cod liver oil, fatty fish, butter, and animal liver.

Because vitamin D was discovered in an era when trace nutrients were being found, it was classified as a food factor, later vitamin, although few foods contain it. The production of this substance in the skin was discovered about the same time, and this origin of the vitamin is the predominant source of vitamin D for humans. The majority of people around the world do not eat the limited number of foods that contain vitamin D. What is in their bodies has come largely by the skin route.

Vitamin $D_2$ is derived from ergosterol (a fungal sterol) by irradiating it with UV light via the same sequence of chemical changes and is called ergocalciferol. It is used as a pharmaceutical (also called calciferol) and in some of the foods fortified with vitamin D (e.g. milk in North America, margarine). Ergosterol and ergocalciferol differ from 7-dehydrocholesterol and cholecalciferol only in having an extra double bond at carbon 22 and a

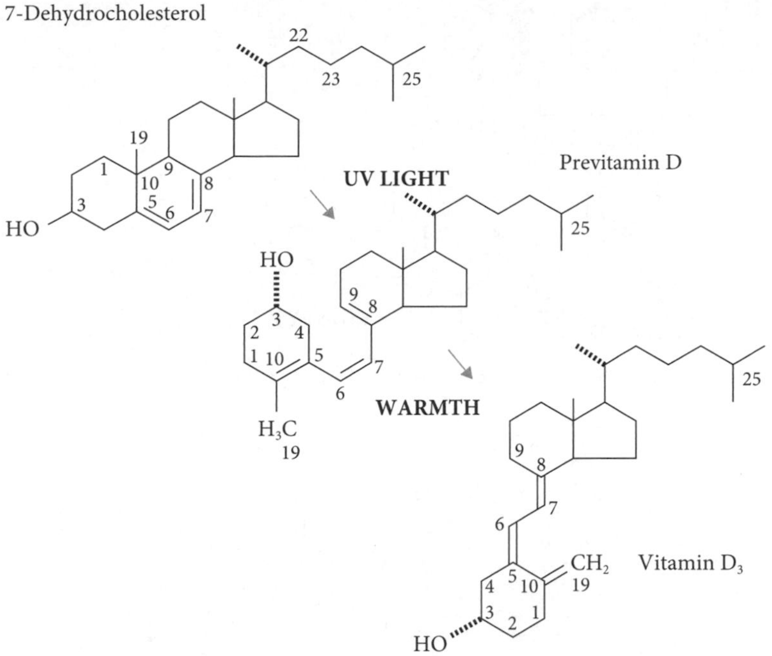

Fig. 14.1 Formation of vitamin $D_3$ in the skin. 7-Dehydrocholesterol is present in the skin as a minor companion of cholesterol. Under the influence of short-wavelength UV light (290–315 nm) from sunlight, the B ring of the sterol opens to form a secosterol, previtamin $D_3$. The first step takes place rapidly. The second stage is a rearrangement of the secosterol to make vitamin $D_3$ (cholecalciferol). It takes place more slowly, under the influence of warmth.

methyl group at carbon 24 in the side chain (Fig. 14.1). The original vitamin $D_1$ turned out to be an impure mixture of sterols. Using the older quantitative unit for vitamin D, one international unit (IU) = 0.025 µg of cholecalciferol (so 1 µg = 40 IU).

In the tropical regions of the world the intensity of UV radiation from the sun is similar across the seasons, so unless people avoid the sun (e.g. housebound or skin completely covered) enough vitamin D is made in the skin to meet entirely the body's needs. However, at higher latitudes large seasonal variations in UV radiation occur, with a peak in summer and a low (nadir) in winter. Additionally, lower temperatures in winter usually force people to be heavily clothed and to spend more of their time indoors. This lowering of UV exposure in winter means the skin may not make enough vitamin D and intake of cholecalciferol (vitamin D) present in foods—or supplements—is needed to meet the body's requirement. Since cholecalciferol is formed in one organ of the body (the skin) and transported by the blood to act on other organs (the bones, gut, kidneys), it is a unique nutrient in that it shares the properties of a hormone.

Inside the body, vitamin D itself is not biologically active until it has been chemically modified (hydroxylated) twice. The first clue to this was the observation of a lag period of 8 hours before one could see an effect of administered vitamin D in experimental animals. Vitamin D, whether of cutaneous origin or absorbed ($D_3$ or $D_2$), is carried in the plasma on a specific $\alpha_2$-globulin, vitamin D-binding protein (DBP). In liver microsomes, the end of the side chain is hydroxylated to form 25-hydroxy-vitamin D (25(OH)D). This compound has a more stable concentration in the blood than that of vitamin D, which rises temporarily as some is absorbed or synthesized in the skin.

25(OH)D is still not the active metabolite. It has to have a third hydroxyl (OH) group put on at carbon 1. This is done by an enzyme, 1α-hydroxylase, in the kidneys (in the mitochondria of the proximal convoluted tubule) to make 1,25-dihydroxy-vitamin D (1,25$(OH)_2$D), also called calcitriol (Fig. 14.2).

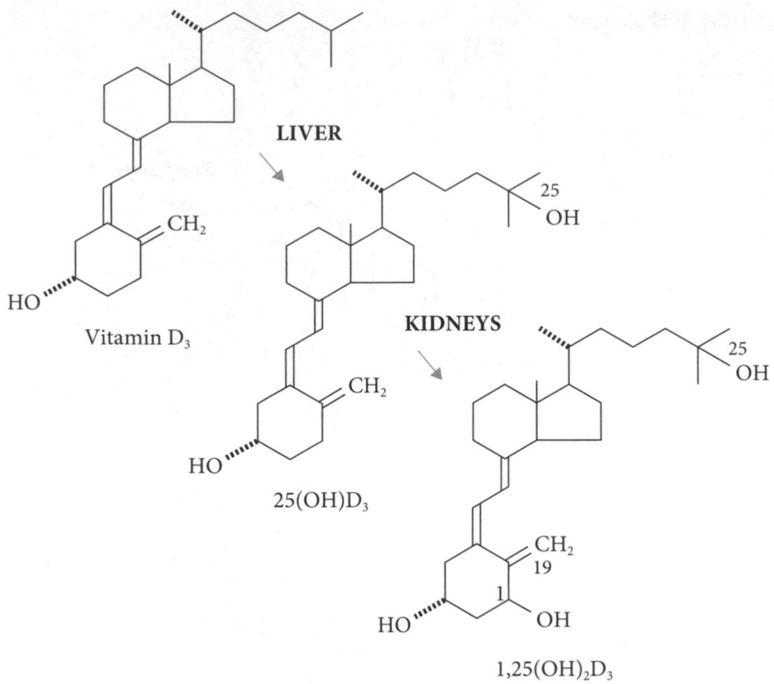

LIVER

KIDNEYS

Vitamin $D_3$

25(OH)$D_3$

1,25(OH)$_2$$D_3$

Fig. 14.2 Activation of vitamin D. In the liver parenchymal cells, vitamin $D_3$ (or $D_2$) is hydroxylated to 25-hydroxyvitamin D (25(OH)D), which circulates in the blood. A small proportion of the available 25(OH)D is further hydroxylated by a specific 25(OH)D-1α-hydroxylase in the kidneys to the active form, 1,25(OH)$_2$ vitamin D (calcitriol). During pregnancy, some 1α-hydroxylation also takes place in the placenta.

The plasma concentration of 1,25(OH)$_2$D is about 1000 times smaller than that of 25(OH)D. The activity of renal 1α-hydroxylase is tightly controlled, so the rate of production of 1,25(OH)$_2$D is increased by any fall in plasma calcium or rise in parathyroid hormone level. The half-life of 1,25(OH)$_2$D is about 15 hours. The half-life of 25(OH)D is much longer. It averages 15–50 days, but can be as long as 89 days (Datta et al., 2017). Adipose tissue contains vitamin D, but this is not a functional store because it will only be released in states of negative energy balance, when some fat tissue is lost. The liver does not act as a functional store of vitamin D.

1,25(OH)$_2$D is one of the three hormones that normally act together to maintain the extracellular calcium concentration constant; the other two are parathyroid hormone and calcitonin (see Chapter 8). There are about 30 other known metabolites of vitamin D, probably all inactive.

1,25(OH)$_2$D acts in a similar manner to steroid hormones. There is a specific vitamin D receptor (VDR) protein in the cell nuclei, which has great affinity for 1,25(OH)$_2$D. It also has a DNA-binding domain. This receptor, when activated, switches on the gene that induces synthesis of a calcium-transport protein (calbindin) in the epithelium of the small intestine. VDR has actually been found in a range of tissues (see section 14.1.6), but normally has its main effect in the small intestinal epithelium and the cells in bone, osteoblasts (that form new bone) and osteoclasts (that break bone down).

The vitamin D status of newborn infants tends to reflect that of the mother because during pregnancy vitamin D is delivered across the placenta from the mother's blood to the foetus. Breast milk is not a good source of vitamin D, so breast milk alone does not provide enough to meet the infant's need. Therefore, until the newborn begins to get sufficient safe sun exposure, breastfed infants are advised to be supplemented daily with 10 micrograms of vitamin D. For infants who are formula fed, infant formulas normally include added vitamin D.

The skin pigment melanin absorbs UV light, thus reducing the amount available to produce vitamin D in skin. For this reason, less vitamin D is synthesized in people with darker compared with lighter coloured skin. Older people also make less vitamin D after exposure to short-wave UV light; their skin contains less of the starting material, 7-dehydrocholesterol. Vitamin D taken by mouth is digested and absorbed, then transported from the upper small intestine in chylomicrons, like other lipids. Like other lipids, its absorption can be impaired in chronic biliary or intestinal disease with malabsorption. Excretion of vitamin D is in the bile, principally as more polar metabolites.

## 14.1.3 Deficiency diseases

In *rickets*, there is reduced calcification of the growing ends (epiphyses) of bones. Thick seams of uncalcified osteoid cartilage are seen histologically. Rickets only occurs in young people, whose bones are still growing. *Osteomalacia* is the corresponding decalcifying bone disease in adults, whose epiphyses have fused so that the bones are no longer growing. Bone density is reduced. This is because the bones contain less calcium: the ratio of calcium to organic bone is reduced. Rickets can occur in premature infants. It can also occur in full-term infants but, with few exceptions, only when a number of factors align: the infant's mother has very low vitamin D status during pregnancy—more likely in women with darker skin and who get little sun exposure, the infant also gets little exposure to sunlight, and the infant is exclusively breastfed. In affluent countries, osteomalacia is possible in elderly people confined indoors. Malabsorption increases the risk. Muscular weakness and susceptibility to infections in rickets or osteomalacia may reflect roles for VDR in the muscles and the immune system. In chronic kidney failure, 1α-hydroxylation is impaired. Renal osteomalacia does not respond to vitamin D (or sunlight), only to administration of $1,25(OH)_2D$ (pharmaceutical name calcitriol) or to 1α(OH)D (pharmaceutical name alphacalcidol). This shows the critical importance of 1α-hydroxylation to normal vitamin D function.

## 14.1.4 Biochemical tests of vitamin D status

Plasma calcium and phosphate levels fall in severe vitamin D-deficient states. Plasma alkaline phosphatase (the isoenzyme originating in bone) is increased in mild as well as in severe rickets and osteomalacia, but also can be elevated in some other bone diseases unrelated to vitamin D. Therefore, vitamin D status is best assessed by measuring serum 25(OH)D concentration (SI unit, nmol/L; alternative units, ng/ml [divide nmol/L by 2.6]). Methods for analysis use either automated commercial immunoassay (e.g. DiaSorin®) or liquid chromatography followed by mass spectrometry. Plasma 25(OH)D levels above 50 nmol/L are considered normal and sufficient to meet the needs of almost all in the population. Levels below 30 nmol/L, when appearance of symptoms of poor bone health become more likely, indicate deficiency. Levels between 30 and 50 nmol/L are considered to be suboptimal. In tropical regions of the world, the mean serum 25(OH)D concentrations in the population tend to vary little across the seasons because UV radiation is similar all year round. For example, Masai tribespeople in Tanzania who spend most of the day outdoors have 25(OH)D levels of 115 (range 58–167) nmol/L (Luxwolda et al., 2012). However, in populations living at higher latitudes the mean population levels can vary by as much as 30 to 40 nmol/L across the year. An unresolved question when assessing vitamin D status is whether serum 25(OH)D concentration should be above 50 nmol/L at all times of the year or the annual average should be above 50 nmol/L.

Plasma $1,25(OH)_2D$ can also be measured but is a specialized investigation.

## 14.1.5 Osteoporosis as well as osteomalacia

It used to be thought that vitamin D deficiency causes osteomalacia, not osteoporosis. (In osteoporosis, total bone is reduced—organic as well as calcium.) Now that vitamin D status can be quantified with serum 25(OH)D, it is becoming clear that as levels fall below 50 nmol/L, the risk increases of compensatory secondary increases of parathyroid hormone and increased mobilization of bone. However, the results of randomized controlled trials of vitamin D supplementation, most involving participants with vitamin D status above 30 nmol/L, have shown little meaningful effect on rates of osteoporosis.

### 14.1.6 Other possible roles for vitamin D beyond bones

Plausible mechanisms and associations between vitamin D and several disease states observed in prospective cohort studies have suggested a possible protective or therapeutic role for vitamin D other than in relation to bone health (see Table 14.1).

**Table 14.1 Potential health benefits of vitamin D beyond the effect on bones**

| Condition | Observational data | Clinical trials of supplementation | Comment |
|---|---|---|---|
| Coronary heart disease (CHD) | ++ | – | No reduction in CHD risk, even in those with low 25(OH)D at baseline |
| CHD risk factors | ± | ± | Limited evidence of increased incidence of hypertension in those starting with low 25(OH)D, and variable effects of supplementation on systolic and diastolic blood pressure. Limited evidence that supplementation may reduce total and LDL cholesterol |
| Cancer | + | ± | Some prospective studies report associations between low 25(OH)D and total cancer incidence and deaths. Inconsistent findings for individual cancers. Some suggestion of benefit of supplementation in terms of total cancer mortality but findings not consistent |
| Type 2 diabetes | ++ | – | Strong observational evidence may be confounded by the fact that many people with T2DM are overweight or obese, conditions that are associated with low 25(OH)D. Clinical trials provide no evidence that vitamin D supplements reduce progression from prediabetes to T2DM or is beneficial in treatment |
| Type 1 diabetes | + | ± | There is some evidence that vitamin D supplementation may reduce the risk of developing T1DM in childhood and improve blood glucose levels when given during the 'honeymoon phase' following the diagnosis of T1DM |
| Multiple sclerosis | ++ | – | Strongly suggestive observational data. However, no benefit of supplemental vitamin D in trials that have examined the effects in terms of relapse or increasing disability |
| Acute respiratory infections | ++ | ± | Meta-analyses suggest low serum 25(OH)D concentrations are associated with acute respiratory infections. However, other meta-analyses and large trials of supplementation show no benefit, small reduction of duration of symptoms, or benefit from low but not high levels of supplementation |
| Tuberculosis | ++ | – | Vitamin D deficiency predicts risk of tuberculosis with a concentration-dependent effect but a large RCT in children with low 25(OH)D found no benefit of supplementation |

*Key:* – no significant effect or no data; ± inconsistent or inconclusive findings; + modest benefit; ++ moderate to strong benefit

Although not entirely consistent, observational epidemiological data suggest that relatively high vitamin D status (as measured by 25(OH)D levels) may be protective against some important non-communicable diseases (notably coronary heart disease, diabetes, and some cancers), acute respiratory infections, and tuberculosis. However, a growing number of well conducted randomized controlled trials of vitamin D supplementation with or without calcium have shown that epidemiological associations do not necessarily translate into clinical benefit. For example, despite cohort studies showing an increasing risk of type 2 diabetes with lower levels of 25(OH)D, supplementation with vitamin D does not appear to reduce risk of progression from pre-diabetes to diabetes. Failure to demonstrate benefit of supplementation in published clinical trials does not exclude the possibility of benefit in those who are deficient in vitamin D, since most of the trials have been carried out in populations where the majority are vitamin D replete. However, in some instances there is evidence that vitamin D supplementation does not confer benefit even in populations with relatively low levels of 25(OH)D. It is also conceivable that many of the epidemiological associations might be explained by confounding, or that supplementation may be of limited or no value if initiated when the disease process has progressed. The information in Table 14.1 suggests that the expectation of widespread benefit of vitamin D beyond its effect on bone health may have been overly optimistic. However, further research in populations or groups with relatively low average levels of 25(OH)D, or who live in countries where there is limited exposure to sunlight for prolonged periods, may alter this conclusion.

### 14.1.7 Sources

Fish liver oils (e.g. cod and halibut) and some fish and marine animals' livers are rich sources. Mushrooms can be a good source (of ergocalciferol) if grown with UVB light. Moderate sources are fatty fish (herring, sardine, salmon, etc.), margarines (which in most countries are fortified with vitamin D), infant milk formulas, eggs, red meat, and liver. Milk is fortified with vitamin D in North America and Scandinavia. Human milk contains little vitamin D unless the mother takes a vitamin D supplement (moderate exposure to sunlight is good for babies). For those prescribing or taking vitamin D tablets, these should be $D_3$, cholecalciferol. Ergocalciferol (vitamin $D_2$) is less efficient in raising 25(OH)D.

### 14.1.8 Interactions

Long-term use of anticonvulsants (e.g. for epilepsy), by inducing liver microsomes, increases metabolic losses of vitamin D. Glucocorticoids (e.g. prednisone) inhibit vitamin D-dependent intestinal calcium absorption; patients on long-term steroids may benefit from additional vitamin D to maintain serum 25(OH)D in the mid-normal range (above 50 nmol/L).

### 14.1.9 Genetics

Variants in three genes affect vitamin D status. *DHCR7* encodes the enzyme 7-dehydrocholesterol (7-DHC) reductase, which converts 7-DHC to cholesterol, thereby reducing the starting substance for the action of sunlight. *CYPZRI* probably encodes the enzyme for 25-hydroxylation of vitamin D in the liver. The third gene, *GC*, encodes vitamin D-binding protein, which binds and transports in the plasma vitamin D, 25(OH)D, and 1,25(OH)$_2$D.

### 14.1.10 Recommended nutrient intake

The US Institute of Medicine (IOM) estimates adequate intakes (AI) of vitamin D. The recommendations are based on the amount of vitamin D needed in the diet for 97.5% of the population to have a plasma 25(OH)D concentration above 50 nmol/L, assuming minimal vitamin D synthesis in the skin, such as would occur during the winter in higher Northern and Southern latitudes. The IOM revised its recommendations upwards in 2010. The RDA is now 15 μg (600 IU) per day for ages 1–70 years, and 20 μg (800 IU) per day for infants and adults over 70 years. Older people make less vitamin D in their skin, tend to avoid sun exposure, and use UV light-blocking sunscreen. If they are housebound and cannot get in the sun, they probably need vitamin D supplements because most diets provide under 5 μg vitamin D/day unless the milk is fortified.

### 14.1.11 Toxicity

Exposure of the skin to sunlight, if excessive, causes sunburn and brings up the serum 25(OH)D level if it was low, but does not lead to vitamin D toxicity because with the excessive UV light, previtamin $D_3$ (7-dehydrocholesterol) is photoisomerized to biologically inert products (lumisterol and tachysterol). Vitamin $D_3$ is also photodegraded if it is not taken inside the body by vitamin D-binding protein. However, with oral intake, the margin between the upper limit of the nutritional dose and the lower limit of the toxic dose is quite narrow. Over-dosage causes raised plasma calcium (hypercalcaemia), with thirst, anorexia, raised plasma levels of 25(OH)D, and risk of calcification of soft tissues and of urinary calcium stones. Infants are most at risk of hypervitaminosis D; a few children have developed hypercalcaemia on intakes of only 50 μg/day. The upper level of 50 μg (2000 IU) per day for adults set by the US IOM in 1997 has been shown to be unnecessarily cautious (Vieth et al., 2007). The upper intake level was increased to 100 μg (4000 IU) by the IOM in 2010. In some conditions, people are unusually sensitive to vitamin D (e.g. in sarcoidosis and a rare condition in infants with elfin facial appearance, Williams syndrome).

### Further reading

1. **Amrein, K., Scherkl, M., Neuwersch-Sommeregger, S., et al.** (2020) Vitamin D deficiency 2.0: an update on the current status worldwide. *Eur J Clin Nutr*, **74**(11), 1498–1513.

2. **Bouillon, R., Manousaki, D., Rosen, C., Trajanoska, K., Rivadeneira, F., and Richards, J.B.** (2022) The health effects of vitamin D supplementation: evidence from human studies. *Nat Rev Endocrinol*, **18**(2), 96–110. Doi: 10.1038/s41574-021-00593-z.

3. **Chang, S-W. and Lee, H-C.** (2019) Vitamin D and health—The missing vitamin in humans. *Pediatrics and Neonatology*, **60**, 237–44.

4. **Datta, P., Philipsen, P.A., Olsen, P., et al.** (2017) The half-life of 25(OH)D after UVB exposure depends on gender and vitamin D receptor polymorphism but mainly on the start level. *Photochem Photobiol Sci*, **16**(6), 985–95. doi: 10.1039/c6pp00258g.PMID: 28485745

5. **Gallagher, J.C.** (2021) Vitamin D and respiratory infections. *Lancet Diabetes Endocrinol* **9**(2), 54–6. doi: 10.1016/S2213-8587(20)30403-4.

6. **Ganmaa, D., Uyanga, B., Zhou, X., et al.** (2020) Vitamin D supplements for prevention of tuberculosis infection and disease. *N Engl J Med*, **383**, 359–68.

7. **Hilger, J., Friedel, A., Herr, R., et al.** (2014) A systematic review of vitamin D status in populations worldwide. *Brit J Nutr*, **111**(1), 23–45.

8. **Itkonen, S.T., Andersen, R., Björk, A.K., et al.** (2021) Vitamin D status and current policies to achieve adequate vitamin D intake in the Nordic countries. *Scand J Public Health*, **49**(6), 616–27. doi.org/10.1177/1403494819896878.

9. **Kawahara, T., Suzuki, G., Mizuno, S., et al.** (2022) Effect of active vitamin D treatment on development of type 2 diabetes. DPVD randomized controlled trial in Japanese population. *BMJ*, **377**, e066222.

10. **Keum, N., Lee, D.H., Greenwood, D.C., Manson, J.E., and Giovannucci, E.** (2019) Vitamin D supplementation and total cancer incidence and mortality: a meta-analysis of randomized controlled trials. *Ann Oncol*, **30**, 733–43.

11. **LeBoff, M.S., Chou, S., Ratliff, K.A., et al.** (2022) Supplemental vitamin D and incident fractures in midlife and older adults. *N Eng J Med*, **387**, 299–309.

12. **Luxwolda, M.F., Kuipers, R.S., Kema, I.P., et al.** (2012) Traditionally living populations in East Africa have a mean serum 25-hydroxyvitamin D concentration of 115 nmol/l. *Br J Nutr*, **108**(9), 1557–61. doi: 10.1017/S0007114511007161. PMID: 22264449

13. **Manson, J.E., Cook, N.R., Lee, I-M.,** et al. (2019) Vitamin D supplements and prevention of cancer and cardiovascular disease. *N Engl J Med*, **380**, 33–44.
14. **Office of Dietary Supplements, National Institutes of Health (NIH)** (2022) *Vitamin D fact sheet for health professionals*. Available at: https://ods.od.nih.gov/factsheets/VitaminD-HealthProfessional/ (accessed 1 September 2022).
15. **Saraf, R., Morton, S.M.B., Camargo, C.A., and Grant, C.C.** (2016) Global summary of maternal and newborn vitamin D status—a systematic review. *Matern Child Nutr*, **12**(4), 647–68.
16. **Vieth, R., Bischoff-Ferrari, H., Boucher, B.J., et al.** (2007) The urgent need to recommend an intake of vitamin D that is effective. *Am J Clin Nutr*, **85**(3), 649–50. doi: 10.1093/ajcn/85.3.649
17. **Virtanen, J.K., Nurmi, T., Aro, A., et al.** (2022) Vitamin D supplementation and prevention of cardiovascular disease and cancer in the Finnish vitamin D trial: a randomized controlled trial. *Am J Clin*, **115**, 1300–10. doi: 10.1093/ajcn/nqab419.

# 14.2 Vitamin K

## 14.2.1 History

The name 'vitamin K' was proposed by Henrik Dam of Denmark in 1935. K was the next letter of the alphabet not already used for a vitamin at that time. It is also the first letter of the German word *koagulation*, which refers to its best-known function. While investigating the essentiality of cholesterol in the diet of chickens (their eggs, of course, are rich in cholesterol), Dam fed them rations from which the lipid had been extracted with organic solvents. They developed haemorrhages and their blood was slow to clot. This bleeding tendency could be corrected with alfalfa or with decayed fishmeal. The alfalfa was soon shown to provide vitamin $K_1$; bacteria in the fishmeal were responsible for producing vitamin $K_2$.

## 14.2.2 Chemistry

H.J. Almquist solved the chemical search in 1939, reporting that a lipid from the sheath of tubercle bacilli, phthiocol, had vitamin K activity (Fig. 14.3a).

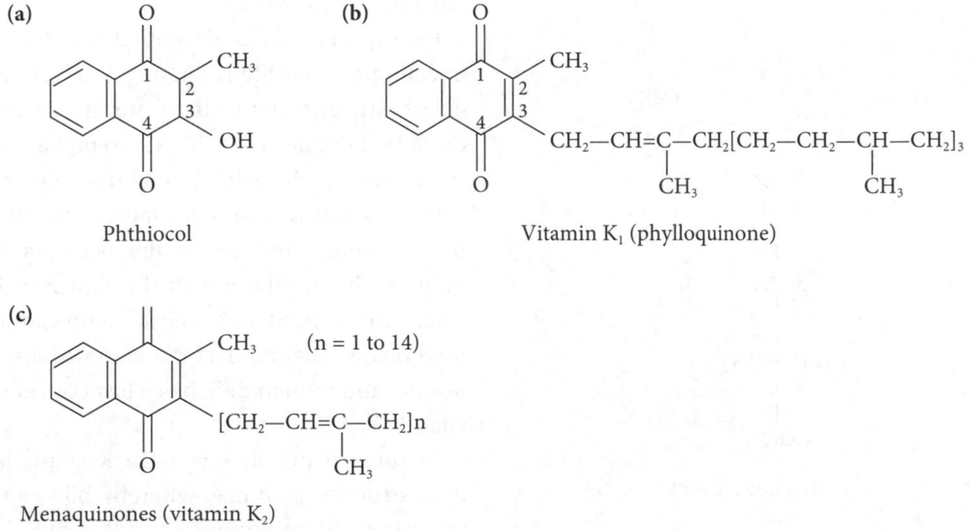

(a) O, $CH_3$, 1, 2, 3, 4, OH, O

Phthiocol

(b) O, $CH_3$, 1, 2, 3, 4, O, $CH_2—CH{=}C—CH_2[CH_2—CH_2—CH—CH_2]_3$, $CH_3$, $CH_3$

Vitamin $K_1$ (phylloquinone)

(c) $CH_3$ (n = 1 to 14), $[CH_2—CH{=}C—CH_2]n$, $CH_3$, O

Menaquinones (vitamin $K_2$)

Fig. 14.3 Structures of (a) phthiocol, (b) phylloquinone (vitamin $K_1$), and (c) menaquinones (vitamin $K_2$).

Vitamin $K_1$ and the $K_2$ series are all based on this 2-methyl-3-hydroxy-1,4-naphthoquinone. They have side chains in place of the 3-hydroxyl group. In vitamin $K_1$ (phylloquinone), the side chain is a 20-carbon terpenoid alcohol (four-fifths of the phytol side chain of chlorophyll) (Fig. 14.3b). It is found in green leaves.

Vitamin $K_2$ comprises a family of compounds, called menaquinones, whose side chain consists of repeated (5-carbon) isoprene units, from 1 to 14 of them (Fig. 14.3c). Depending on the number of isoprene units, they are referred to as MK-1 to MK-14. The menaquinones are synthesized by several bacterial species (*Bacteriodes*, *Enterobacteria*, etc.), some of which occur naturally in the large intestine of animals, including humans.

## 14.2.3 Functions

In the liver, vitamin K promotes the synthesis of a special amino acid with three carboxylic acid groups, γ-carboxyglutamic acid (Gla) (Fig. 14.4). The enzyme responsible for putting another carboxylic acid on to glutamic acid requires vitamin K as a cofactor.

Gla is an essential part of four of the coagulation factors, all proteins: prothrombin (factor II) and factors VII, IX, and X. Factors II and VII contain 10 Gla residues per molecule; factors IX and X each contain 12. The Gla residues confer on these coagulation proteins the capacity to bind to phospholipid surfaces in the presence of calcium ions.

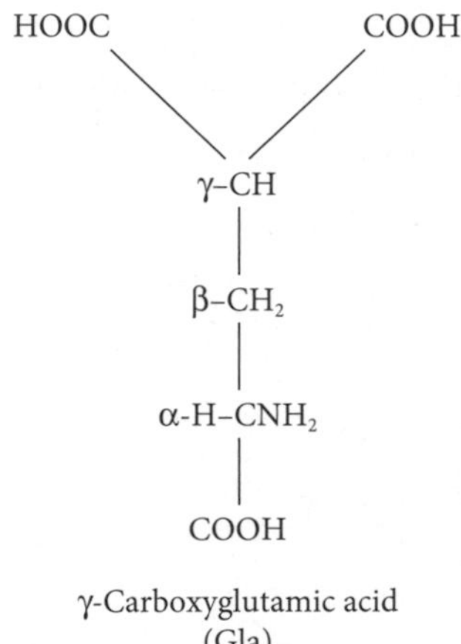


Fig. 14.4 γ-Carboxyglutamic acid (Gla).

Five other Gla-containing and vitamin K-dependent proteins were discovered more recently. Proteins C, S, and Z all have anticoagulant activities. They seem to act to limit clot formation. In bone are osteocalcin and matrix Gla protein (MGP). MGP is also expressed in soft tissues and appears to inhibit arterial calcification. Osteocalcin can be assayed in plasma.

Vitamin K, being fat-soluble, requires bile for its absorption. It is transported in the plasma on triglyceride-rich lipoproteins. Analysis of vitamin K in tissues has been difficult because the several compounds are present in very tiny (nanomolar) amounts. Adipose tissue has much higher vitamin $K_1$ than plasma. In adult liver, there is not only vitamin $K_1$ derived from green leaves in the diet but also significant amounts of menaquinones ($K_2$), some presumably synthesized by anaerobic bacteria in the large intestine and absorbed. MK-4 is apparently derived from phylloquinone ($K_1$).

## 14.2.4 Deficiency disease

In vitamin K deficiency, there is a bleeding disorder, characterized by low plasma prothrombin activity (hypoprothrombinaemia). Vitamin K deficiency can occur in obstetric or paediatric practice, in surgical and medical patients.

For most people in developed countries, the first injection of their life is vitamin $K_1$ given intramuscularly straight after birth to prevent haemorrhagic disease of the newborn. This is to prevent the small risk of haemorrhage in the first days after birth, because vitamin K, like other fat-soluble vitamins, is poorly transported across the placenta from the mother's blood; the gut of the newborn is sterile (there are no resident bacteria, so no vitamin $K_2$). It is some days before bacteria colonize the large intestine, and human milk has a low concentration of vitamin K ($K_1$).

In surgical practice, vitamin K status is critical in obstructive jaundice, whereby bile cannot flow into the small intestine so that vitamin $K_1$ is not absorbed. It is, of course, very dangerous to operate on someone with a coagulation defect, so before surgery on the bile duct, the prothrombin activity must be checked and vitamin $K_1$ given as a precaution. Vitamin K deficiency also occurs in patients with malabsorption, and sometimes following prolonged use of broad-spectrum antibiotics by mouth (which can destroy the colonic bacteria). In serious liver disease, the coagulation factors may not be adequately synthesized, but this is because of poor liver function, rather than vitamin K deficiency.

In internal medicine, the most common cause of vitamin K deficiency is the use of anticoagulant drugs, given to prevent clotting in veins—warfarin and dicoumarol, which owe their therapeutic action to blocking one of the enzymes that recycle vitamin K in the liver.

Dosage is controlled by following the prothrombin time in patients' blood. This is usually expressed as international normalized ratio, which adjusts for the thromboplastin used in the test. This is the usual way to make the diagnosis of vitamin K deficiency, because plasma concentrations of $K_1$ are minute, only around 1 nmol/L. A more specific method is to measure under γ-carboxylated prothrombin, PIVKA-II. Recently introduced anticoagulants, such as dabigatran ('Pradaxa'), do not work by antagonizing vitamin K.

There are reports of low circulating vitamin K in elderly patients with femoral neck fractures or spinal crush fractures, suggesting that suboptimal vitamin K status may play a role in osteoporosis, presumably because osteocalcin is not made adequately.

### 14.2.5 Dietary sources

Vitamin $K_1$ is present in dark-green leaves eaten as foods. Some contain more than others. It is associated with the photosynthetic tissues. Kale, spinach, Brussels sprouts, broccoli, parsley, coriander, mint, cabbage, and lettuce are good sources (in descending order). Other good sources are some vegetable oils (soybean and canola). Absorption is more efficient from these oils than from the leafy vegetables. Small amounts are present in beef liver, apples, and green tea, and there is some vitamin $K_2$ in cheese and fermented soybeans.

There are no experimentally derived reference values yet for vitamin K, but it is agreed that 1 μg/kg body weight is a safe and adequate intake (i.e. around 80 μg/day in men and 65 μg/day in women). In the United States, the Institute of Medicine suggests 120 μg/day in men and 90 μg/day in women based on median intakes in NHANES III.

Bone Gla protein may be more sensitive to low intakes than the Gla haemostatic proteins. Usual adult intakes are variously estimated at 60–200 μg/day. However, the synthetic water-soluble pharmaceutical with vitamin K activity, menadione, can cause haemolytic anaemia and jaundice in newborn babies. It is now obsolete, superseded by vitamin $K_1$.

## Further reading

1. **Booth, S.L., Sadowski, J.A., Weikrauch, J.L., and Ferland, G.** (1998) Vitamin K (phylloquinone) content of foods: a provisional table. *J Food Comp Anal*, **6**, 109–20.
2. **Duggan, P., Cashman, K.D., Flynn, A., Bolton-Smith, C., and Kiely, M.** (2004) Phylloquinone (vitamin $K_1$) intakes and food sources in 18–64-year-old Irish adults. *Br J Nutr*, **92**, 151–8.
3. **Feskanich, D., Weber, P., Willett, W.C., Rockett, H., Booth, S.L., and Colditz, G.A.** (1999) Vitamin K intake and hip fractures in women: a prospective study. *Am J Clin Nutr*, **69**, 74–9.
4. **Johnson, M.A.** (2005) Influence of vitamin K on anticoagulant therapy depends on vitamin K status and the source and chemical forms of vitamin K. *Nutr Rev*, **63**, 91–7.
5. **Shea, M.K., Booth, S.L., Gundberg, C.M., et al.** (2010) Adulthood obesity is positively associated with adipose tissue concentrations of vitamin K and inversely associated with circulating indicators of vitamin K status in men and women. *J Nutr*, **140**, 1029–34.
6. **Shearer, M.J. and Newman, P.** (2014) Recent trends in the metabolism and cell biology of vitamin K with special reference to vitamin K cycling and MK-4 biosynthesis. *J Lipid Res*, **55**, 345–62.

# Other Biologically Active Substances in Plant Foods: Phytochemicals

Bernhard Watzl



Plants contain a wide range of secondary metabolites of low molecular weight that, in the broadest sense, are biologically active molecules that have evolved in the interaction between the plant and its environment, including ultraviolet light. When plant foods are consumed, a diverse range of secondary plant metabolites is ingested. In the past, emphasis was placed on secondary metabolites as natural toxicants that are present in plant foods (i.e. glycoalkaloids in potatoes and tomatoes, cyanogenic glycosides in cassava) and their potential hazardous effects on humans. However, for the past three decades there has been increasing recognition of the health benefits of consuming diets rich in vegetables, fruits, whole grain, and legumes. Therefore, a resurgence of interest in secondary plant metabolites has evolved due to the epidemiological evidence that has demonstrated a protective effect of plant food intake against chronic illnesses, such as cardiovascular disease and cancer.

Primary plant metabolites are substances that mainly contribute to energy metabolism and to the structure of the plant cell, i.e. carbohydrates, including dietary fibre, proteins, and fats. Secondary plant metabolites are non-nutritive dietary components (excluding vitamins) that have been referred to as *phytochemicals*. Secondary plant metabolites have various functions in the plant, such as serving as a defence against destructive weeds, insects, and micro-organisms, as growth regulators, and as pigments. They are essential for the plant's interactions with its environment. Chemically, these secondary plant metabolites are quite diverse compounds and are found only in minute amounts, in contrast to the primary plant metabolites. Secondary plant metabolites have potential pharmacological effects on humans, a thought that was already discussed in the 1950s. More recently, nutrition scientists have systematically begun to investigate the health-promoting effects of these plant substances.

The total number of naturally occurring phytochemicals is not known. Present assumptions vary from 60,000 to 100,000 substances. With a mixed diet, around 1.5 g of phytochemicals are ingested daily. On a vegetarian diet regimen, the intake of phytochemicals can be distinctly higher.

Phytochemicals can have beneficial, as well as detrimental, health effects. Until a few years ago, they were merely seen under the aspect of toxicity and some were described as antinutritive, or even toxic, metabolites because they restrict the availability of nutrients and increase the permeability of the intestinal wall. However, under the usual conditions of food consumption, nearly all natural components—

with a few exceptions, such as solanine—are harmless. Many phytochemicals that were previously regarded as having health-adverse effects may have a variety of health-promoting effects. This is exemplified by protease inhibitors in legumes and glucosinolates of various *Cruciferae* species.

In the following, the health-promoting potential of phytochemicals is illustrated on the basis of experimental findings of their biological activity. The evidence found in epidemiological studies helps to estimate the importance that phytochemicals may have for human health.

## 15.1 Classification of phytochemicals

Phytochemicals are classified according to their chemical structure and their functional characteristics. The main groups of phytochemicals and their physiological effects show their great diversity (see Table 15.1).

### 15.1.1 Carotenoids

Carotenoids are widespread phytochemicals in fruits and vegetables, and one of their main functions in plants is to provide the red and yellow pigments essential for photosynthesis. They can be divided into oxygen-free and oxygen-containing (xanthophylls) carotenoids. Of the about 700 natural carotenoids, only around 40–50 are of significance in human nutrition. Depending on the carotenoid structure, several carotenoids possess provitamin A activity (Chapter 11). Human serum mainly contains the oxygen-free carotenoids α- and β-carotene, and lycopene, as well as the xanthophylls lutein, zeaxanthin, and β-cryptoxanthin in varying proportions, depending on the individual diet. β-Carotene accounts for 15–30%. Oxygen-free and oxygen-containing carotenoids differ mainly in their thermal stability. β-Carotene in carrots and lycopene in tomatoes, for example, are heat stable; xanthophylls (mainly in green vegetables) are sensitive to thermal processing. The total daily intake of carotenoids on a Western diet is about 6 mg. The absorption of carotenoids differs between raw and heat-treated vegetables and fruits (see Table 15.2).

**Table 15.1 Classification of phytochemicals and their main effects**

| Phytochemical | Evidence for the following effects | | | | | | | | | |
|---|---|---|---|---|---|---|---|---|---|---|
| | A | B | C | D | E | F | G | H | I | J |
| Carotenoids | X | | X | | X | X | | X | | X |
| Phytosterols | X | | | | | | | X | | X |
| Saponins | X | X | | | X | | | X | | |
| Glucosinolates | X | X | | | | | | X | | |
| Polyphenols | X | X | X | X | X | X | X | | X | X |
| Protease inhibitors | X | | X | | | | | | X | X |
| Monoterpenes | X | X | | | | X | | X | | |
| Phyto-oestrogens | X | | X | | X | | | | | |
| Sulphides | X | X | X | X | X | X | X | X | | |

A = anticarcinogenic; B = antimicrobial; C = antioxidative; D = anti-thrombotic; E = immunomodulatory; F = anti-inflammatory; G = influence on blood pressure; H = cholesterol-lowering; I = modulates blood glucose levels; J = influence on cognitive function.
*Source:* Modified from Watzl, B. and Leitzmann, C. (2005) *Bioaktive Substanzen in Lebensmitteln*, 3rd edn. Stuttgart: Hippokrates.

Table 15.2 **Absorption of phytochemicals in humans**

| High (>15%) | Medium (3–15%) | Low (<3%) |
|---|---|---|
| Carotenoids* | Phytosterols | Carotenoids** |
| Glucosinolates | Phenolic acids | Saponins |
| Flavonoids*** | Protease inhibitors | Anthocyanins |
| Phyto-oestrogens | | Flavones |
| Monoterpenes | | |
| Sulphides | | |

*From heat-treated food.
**From raw food.
***Flavonoids excluding anthocyanins and flavones.

## 15.1.2 Phytosterols

Phytosterols such as β-sitosterol, stigmasterol, and campesterol are mainly found in plant seeds, nuts, and oils. Chemically, phytosterols differ from cholesterol by an additional side chain only. The daily phytosterol intake amounts to 100–500 mg. In humans, absorption of phytosterols is low (0–10%) compared with the > 40% for cholesterol. Absorbed phytosterols in the enterocytes are actively transported back to the intestinal lumen, contributing to the low bioavailability of phytosterols. The cholesterol-lowering effect of phytosterols (see Chapter 23) has been known since the 1950s and is, in part, due to their property of inhibiting cholesterol absorption. This activity of phytosterols resulted in the generation of one of the first functional foods, a margarine enriched with phytosterol or phytostanol.

## 15.1.3 Saponins

Saponins are bitter-tasting, surface-active compounds that form complexes with proteins and lipids, such as cholesterol. Saponins are particularly abundant in legumes. The daily intake of saponins may be higher than 200 mg, depending on the dietary habits, but usually averages around 15 mg/day. Saponins have a low absorption rate (see Table 15.2) and, therefore, are primarily active in the intestinal tract. Due to their haemolytic properties, saponins were solely considered to be detrimental to health. In studies conducted with humans, however, this could not be confirmed.

## 15.1.4 Glucosinolates

Glucosinolates (thioglucosides) are found in food plants belonging to the family of Cruciferae. Their degradation products contribute to the typical flavour of mustard, horseradish, and broccoli. In these plants, glucosinolates are associated with, but sterically separated from, the enzyme myrosinase. Mechanical damage of the plant tissue (cutting, chewing) eliminates the sterical separation between the enzyme and its substrates. Myrosinase then hydrolyses the glucosinolate to its active metabolites: isothiocyanates, thiocyanates, and indoles. Heating of cabbage reduces its glucosinolate content by 30–60% and inactivates its myrosinase activity. However, microbial myrosinase activity in the large intestine contributes to glucosinolate degradation. The total daily intake of glucosinolates is in the range of 10–50 mg. With vegetarian diets, the total daily intake can be as high as 100 mg. The glucosinolate metabolites, such as isothiocyanates, are completely absorbed in the small intestine (see Table 15.2). An isothiocyanate that has been intensely studied is sulphoraphane, which has an anticarcinogenic effect and induces antioxidant responses (see Table 15.1).

## 15.1.5 Polyphenols

The term polyphenol is used for all substances that are made up of phenol derivatives. Polyphenols mainly include phenolic acids (including hydroxycinnamic acids) and flavonoids (including flavonols, flavones, flavanols, flavanones, and anthocyanins; see Fig. 15.1). Polyphenols normally occur bound to sugars and are rarely found as aglycones in plant foods. Besides monomeric flavonoids, oligomeric flavonoids (with two or more linked molecules), such as the flavanols (procyanidins), occur in red wine, tea, dark chocolate, and apples. Fresh plant foods contain up to 0.1% polyphenols located almost exclusively in the outer layer of fruits (e.g. apple skin) and vegetables (green leaves of lettuce). The flavonoid content of green leafy vegetables is highest with

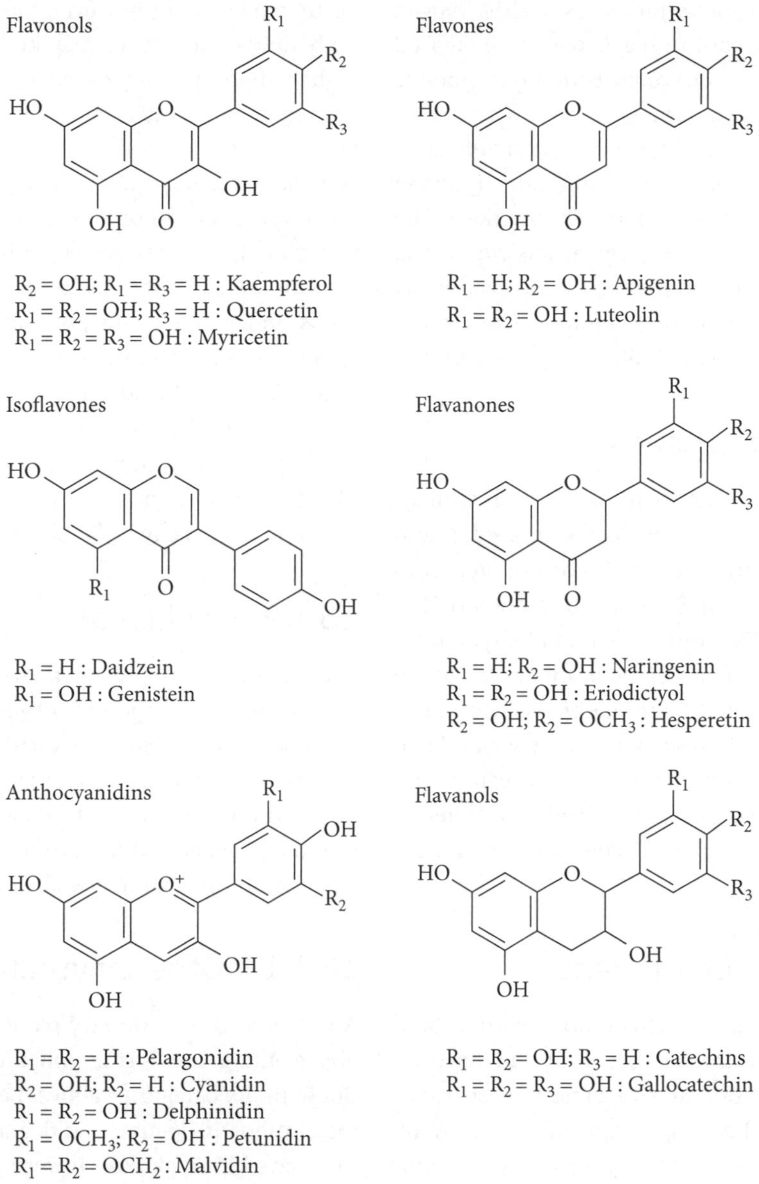

Flavonols

$R_2$ = OH; $R_1$ = $R_3$ = H : Kaempferol
$R_1$ = $R_2$ = OH; $R_3$ = H : Quercetin
$R_1$ = $R_2$ = $R_3$ = OH : Myricetin

Flavones

$R_1$ = H; $R_2$ = OH : Apigenin
$R_1$ = $R_2$ = OH : Luteolin

Isoflavones

$R_1$ = H : Daidzein
$R_1$ = OH : Genistein

Flavanones

$R_1$ = H; $R_2$ = OH : Naringenin
$R_1$ = $R_2$ = OH : Eriodictyol
$R_2$ = OH; $R_2$ = $OCH_3$ : Hesperetin

Anthocyanidins

$R_1$ = $R_2$ = H : Pelargonidin
$R_2$ = OH; $R_2$ = H : Cyanidin
$R_1$ = $R_2$ = OH : Delphinidin
$R_1$ = $OCH_3$; $R_2$ = OH : Petunidin
$R_1$ = $R_2$ = $OCH_2$ : Malvidin

Flavanols

$R_1$ = $R_2$ = OH; $R_3$ = H : Catechins
$R_1$ = $R_2$ = $R_3$ = OH : Gallocatechin

**Fig. 15.1** Chemical structures of flavonoids.
*Adapted from:* Manach, C., Scalbert, A., Morand, C., Remesy, C., and Jimenez, L. (2004) Polyphenols: food sources and bioavailability, with permission from: *Am J Clin Nutr*, **79**, 727–47.

increasing ripeness. Field-grown vegetables have a higher flavonoid content than greenhouse produce. The most commonly found flavonoid is the flavonol quercetin, with a daily intake of about 25 mg. Recent findings suggest that some flavonoids, such as quercetin, can be absorbed by humans in quantities relevant to human health, while other flavonoids, such as the anthocyanins, are hardly absorbed in the small intestine (see Table 15.2). The intestinal microbiota contributes to the metabolism of polyphenols. Polyphenols exert a variety of physiological effects (see Table 15.1).

### 15.1.6 Protease inhibitors

Protease inhibitors are found especially in plant seeds (legumes, grains). Dependent on the mammalian species, protease inhibitors in the gut hamper the activity of endogenous proteases, such as trypsin. In response, the organism reacts with an increased synthesis of digestive enzymes. Humans synthesize a specific trypsin form, among others, which is resistant to protease inhibitors. Cooking significantly reduces the activity of protease inhibitors. Protease inhibitor intake averages about 300 mg daily. The protease inhibitor intake of vegetarians with a diet high in grains and legumes can be considerably higher. Absorbed protease inhibitors can be detected in various tissues in biologically active form.

### 15.1.7 Monoterpenes

Active substances in herbs and spices such as menthol (peppermint), carvone (caraway seeds), and limonene (citrus oil) are examples of monoterpenes in food. The average daily intake of monoterpenes is up to 200 mg. Due to their fat solubility, monoterpenes display a high degree of bioavailability in humans (see Table 15.2). Limonene has been studied in animal models as an anticarcinogen and had been undergoing preliminary trials in cancer patients. However, its rapid metabolization resulted in a low efficiency and in the discontinuation of these human trials.

### 15.1.8 Phyto-oestrogens

Phyto-oestrogens are plant components that bind to mammalian oestrogen receptors and have effects similar to those of endogenous oestrogens. Isoflavones and lignans, chemically both polyphenols, are the two major groups of the phyto-oestrogens in plant foods. Phyto-oestrogens have only about 0.1% of the efficacy that human oestrogens exhibit; however, their concentration in body fluids and tissues may be 100- to 10,000-fold higher. Therefore, phyto-oestrogens can act both as oestrogens and anti-oestrogens, depending on the amount and concentration of endogenous oestrogens. Isoflavones are almost exclusively found in soybeans and soybean products. The major isoflavones in soy are the glycosides of genistein and daidzein. Lignans are present in higher concentrations in flax seeds and whole grain products. With traditional Asian diets and vegetarian diets, the phyto-oestrogen intake is high (15–40 mg/day), but Western diets provide little phyto-oestrogen (<2 mg/day). Phyto-oestrogens have a high absorption rate resulting in blood concentrations associated with various *in vitro* and *in vivo* effects (Tables 15.1 and 15.2).

### 15.1.9 Sulphides

The sulphides among phytochemicals include all organosulphur compounds of garlic and other bulbous plants. The main active substance of garlic is oxidized diallyl disulphide or allicin. Damage to the tissue in the garlic clove leads to the release of the enzyme alliinase, which produces allicin from the basic compound, alliin or S-allylcysteine sulphoxide.

### 15.1.10 Other phytochemicals

Apart from the secondary plant metabolites listed above, there are further phytochemicals that do not fit in the categories above. Lectins, for example, are present in legumes and grain products. They may have blood glucose-lowering effects. Other examples are glucarates, phthalides, chlorophyll, and tocotrienols, as well as phytic acid.

## 15.2 Physiological effects of phytochemicals

The following is a short overview of the major effects of phytochemicals. For detailed information and references see Watzl and Leitzmann (2005), Liu (2004), Williamson et al. (2018), Parmenter et al. (2020), and Wang et al. (2021).

### 15.2.1 Anticarcinogenic effects

Cancer is the second-most frequent cause of death in industrialized countries. Nutrition is the major exogenous factor that modulates cancer risk, and contributes to about one-third of all types of cancer. There are dietary factors that may promote carcinogenesis, but also others that may lower cancer risk. Evidence from epidemiological and animal experimental studies, as well as information from biomarker and mechanistic studies, indicate that a higher intake of vegetables and fruits is associated with a lower risk of various types of cancer. For all classes of phytochemicals occurring in vegetables and fruits, anti-cancer effects have been described. Based on the potential cancer-preventative activity of plant foods, it has been recommended to increase the consumption of vegetables and fruits to at least five servings per day.

Phytochemicals may interfere with and inhibit carcinogenesis at almost any stage in the multistep process: tumour initiation, promotion, and progression (Fig. 15.2). Knowledge of anticarcinogenic effects of vegetables and fruits, and of isolated phytochemicals, has been obtained from different experimental systems (*in vitro*, animal, human). Animal experiments yield direct information about the extent of suppression of spontaneous and chemically induced tumours by ingestion of certain plant foods or isolated phytochemicals (dose–effect studies). However, human studies, especially epidemiological, intervention, and biomarker-related studies, are of particular relevance.

Carcinogens (e.g. nitrosamines) are usually ingested in their inactive form. Their endogenous activation by phase I enzymes (e.g. cytochrome P-450-dependent monooxygenases) is a prerequisite for the interaction with the DNA and genotoxic activity. Phase II enzymes (e.g. glutathione S-transferase, GST) usually detoxify activated carcinogens. In general, phytochemicals (glucosinolates, polyphenols, monoterpenes, sulphides) can inhibit carcinogenesis by inhibiting phase I enzymes and inducing phase II enzymes in cell cultures and in animal experiments, thereby acting as blocking agents (Fig. 15.2). In this manner, the risk of DNA damage and tumour initiation is reduced. For example, the isothiocyanate sulphoraphane, which can be isolated from broccoli, activates the phase II detoxifying enzyme quinone reductase in cell culture systems. In human studies, 300 g Brussels sprouts/day led to increased levels of alpha-GST (phase II enzyme) in male subjects, but not in female subjects.

Recent data on genetic polymorphisms in humans contribute to a better understanding of the cancer-preventative activity of phytochemicals and of vegetables and fruits. According to new studies, the potential effects of phytochemicals such as carotenoids and isothiocyanates on cancer prevention, for example, depend highly on GST genotypes. In subjects with deletions of certain GST isotypes, a high intake of these phyto-chemicals was associated with a lower cancer risk.

An influence on hormone metabolism has been demonstrated by phyto-oestrogens. This is an example of cancer-suppressive activities of phytochemicals during tumour promotion and progression (Fig. 15.2). Animal experiments have shown that phyto-oestrogens and the glucosinolate metabolite indole-3-carbinol influence oestrogen metabolism in such a manner that oestrogens known for their low-promoting tumour-growth properties are produced (i.e. catechol oestrogen). Furthermore, phyto-oestrogens induce the synthesis of the sex hormone-binding globulin (SHBG) in the human liver, leading to an increase of oestrogens bound to this transport protein, making them less active. In prospective studies, however, phyto-oestrogens were not associated with lower cancer risk. Genistein, an isoflavone and phyto-oestrogen present in soybeans that can be detected in humans after soybean consumption, inhibits the growth of blood vessels *in vitro*, possibly having an effect on the growth and metastasis of tumours.

A further point of attack for anticarcinogenic phytochemicals is the regulation of cell growth (proliferation) and programmed cell death (apoptosis). Tumour cells distinguish themselves by a distorted regulation of cell proliferation and apoptosis. Phytochemicals such as isothiocyanates and carotenoids could intervene in such a twisted regulation by modulating the endogenous formation of cell growth-promoting substances and the process of intracellular signal transduction.

### 15.2.2 Antioxidative effects

The pathogenesis of cancer and other diseases has been associated with the presence of reactive oxygen molecules and free radicals. The human body

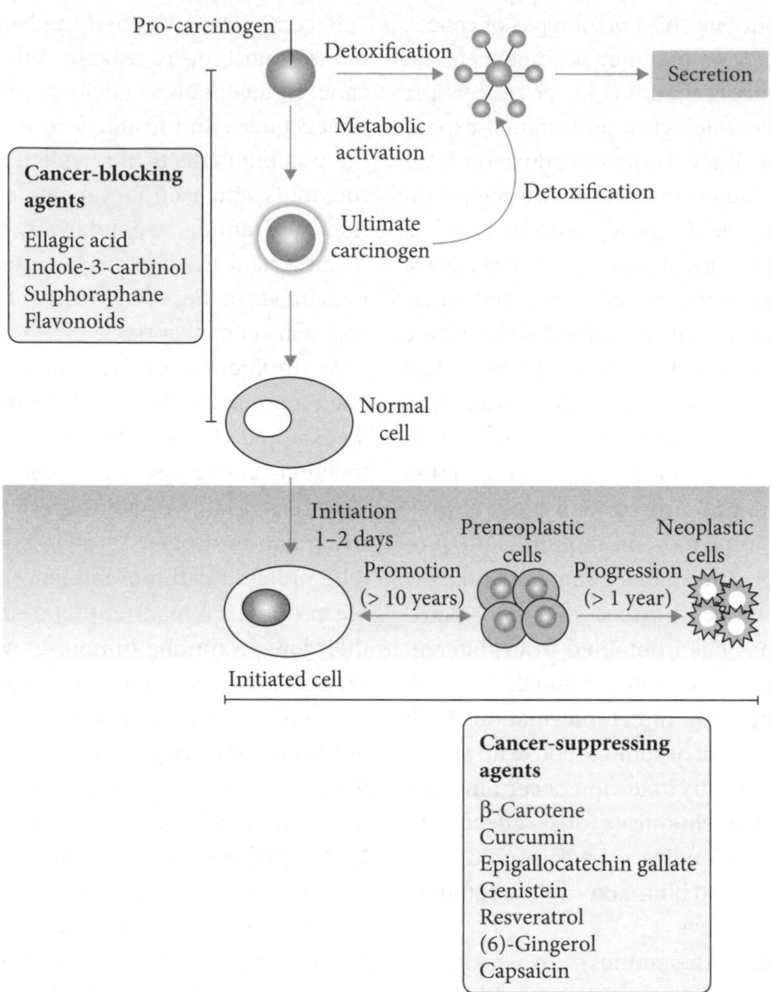


**Fig. 15.2** Dietary phytochemicals that block or suppress multistage carcinogenesis.

*Source:* Reprinted by permission from Macmillan Publishers Ltd: Surh, Y.J. (2003) Cancer chemoprevention with dietary phytochemicals. *Nature Rev Cancer*, **3**, 768–80.

is equipped with several protective mechanisms against these reactive substances, including enzyme systems, superoxide dismutase, and glutathione peroxidase, as well as endogenous antioxidants (uric acid, glutathione, α-lipoic acid, co-enzyme $Q_{10}$). Essential nutrients with antioxidant activity include vitamins E and C. Additionally, phytochemicals (carotenoids, polyphenols, protease inhibitors, phyto-oestrogens, and sulphides) exhibit antioxidative effects.

Certain carotenoids, such as β-carotene, lycopene, and canthaxanthin, provide effective protection against singlet oxygen or oxygen radicals *in vitro*. However, human studies have failed to demonstrate an antioxidative effect of carotenoid supplements in well-nourished individuals. Of all antioxidants in plant foods, polyphenols *in vitro* have the highest potential in terms of quantity and effect. While most human studies could not demonstrate an antioxidative effect of polyphenols *in vivo*, recent studies using dark chocolate or tea showed an effect on plasma antioxidant capacity of polyphenol-rich foods.

Certain fruit and vegetable species influence the level of naturally occurring oxidative DNA damage. For example, the daily consumption of apples or Brussels sprouts over a 3–4-week period led to a significant decrease of oxidative DNA damage as assessed in blood lymphocytes or urine (reduced excretion of 8-oxo-7,8-dihydro-2-deoxyguanosine). Essential nutrients with antioxidative potential (vitamins C and E) are ingested in amounts of around 100 mg/day. In contrast, the daily ingested amount of phytochemicals with antioxidant potential may exceed 1 g. This emphasizes the potential physiological importance of phytochemicals as antioxidants and as agents for other mechanisms that reduce the risk of chronic diseases by consumption of vegetables and fruits.

### 15.2.3 Immunomodulatory effects

The immune system, primarily responsible for the defence against pathogens and transformed cells, is also involved in pathophysiological processes that lead to cancer and cardiovascular disease. Adequate nutrition is the basis for an optimally functioning immune system. Of special importance in this respect are total energy intake, quantity and quality of fats, and certain micronutrients.

Immunomodulatory activities of phytochemicals are a further mechanism through which plant foods may reduce disease risk. The immunomodulatory effects of phytochemicals have so far only been investigated to a very small extent, except for carotenoids and polyphenols. The stimulating effects of various carotenoids and carotenoid-rich foods on the immune system have been demonstrated in numerous animal experiments and human intervention studies.

Flavonoids, in contrast to the carotenoids, have been almost exclusively studied *in vitro*. Most studies demonstrate immunosuppressive activity of the flavonoids. Recent human studies suggest that dietary flavonoids can inhibit the inflammatory processes that are associated with obesity. Saponins, sulphides, phyto-oestrogens, and phytic acid show immunomodulatory effects. As human studies with pure phytochemicals are lacking, with the exception of certain carotenoids and polyphenols, the evaluation of their immunomodulatory effects in humans is not yet possible. For some phytochemicals, however, such an effect seems likely.

### 15.2.4 Antimicrobial effects

Vegetables, fruits, and spices have been used to treat infections since antiquity. The discovery of sulphonamides and microbial antibiotics and their successful use in treating infections resulted in a decline of interest in antimicrobial constituents in food. Only recently is there a renewed interest in plant foods with antimicrobial effects.

Earlier studies clearly verified the antimicrobial action of sulphides from bulbous plants. Recent studies also demonstrated an inhibitory effect of garlic sulphides against *Helicobacter pylori in vitro*. Eliminating allicin from garlic prevented this effect. The glucosinolate metabolites isothiocyanate and thiocyanate have likewise been shown to have antimicrobial activity. They are present in bacteriostatic concentrations in the urinary tract after consumption of garden cress, nasturtium, and horseradish roots. However, the consumption of these plants alone will not lead to successful therapy.

In naturopathy, some berries, such as cranberries and blueberries, are frequently used for the prevention and therapy of infectious diseases. Data from several human intervention studies indicate that the daily intake of 300 mL of cranberry juice, as well as of other berries, significantly reduces the risk of urinary tract infections, mainly in women. Juice consumption increased the intake of polyphenols (procyanidins) that inhibit the adherence of infectious bacteria to the urinary tract epithelium.

### 15.2.5 Cholesterol-lowering effects

Phytosterols, saponins, sulphides, flavonoids, and lycopene have been found to lower serum cholesterol levels in both animal experiments and in clinical studies. The extent to which cholesterol levels are reduced depends on the cholesterol and fat content of the diet. The minimal effective dose of phytosterol is 1 g/day, while 170–440 mg/day is consumed in a typical Western diet. Margarines (a portion of 20 g/day) enriched with the esters of phytosterol or phytostanol provide 1.5–3.0 g/day, resulting in a 10–15% reduction of low-density lipoprotein cholesterol (see Chapter 19).

Although the cholesterol-lowering effect of phytochemicals has been known for over 50 years, the underlying mechanism is still not clear. Several mechanisms may be responsible for this. Saponins bind to primary bile acids in the gut and form micelles. These micelles are too large to pass the intestinal wall, thus leading to reduced absorption of bile acid and, in turn, to their excretion. As a consequence, an increased synthesis of primary bile acids in the liver from the endogenous cholesterol pool is initiated, leading to a decrease of the serum cholesterol level. Phytosterols probably also retard cholesterol absorption by driving cholesterol out of the micelles that normally help to absorb cholesterol from the gut.

Phytochemicals can also inhibit key enzymes of the cholesterol synthesis in the liver. Of these key enzymes, the most important is 3-hydroxy-3-methylglutaryl-CoA-reductase, which is inhibited by monoterpenes and sulphides in animals.

### 15.2.6 Phytochemicals affecting cognitive function

The role of nutrition in cognitive decline and dementia is currently not clear. Preliminary results from human intervention studies as well as from prospective observational studies suggest that flavonoid-rich foods, such as cocoa, blueberries, and orange juice, could improve cognitive functions. Together with flavonoids, carotenoids, phytosterols, and monoterpenes may exert antineuroinflammatory effects. In addition, flavonoids and metabolites produced by the intestinal microbiota have been shown to influence the central nervous system via the gut–brain axis.

### 15.2.7 Phytochemicals affecting drug metabolism

Phytochemicals in grapefruit juice interact with the metabolism of a variety of drugs. Drinking a single glass of grapefruit juice before administration of these drugs can severely affect drug bioavailability and pharmacokinetics of the drug. The mechanism for this effect is the post-transcriptional inhibition of the cytochrome P-450 3A4 enzyme in the small intestine, without affecting the same enzyme system in the large intestine or in the liver. This results in a reduction of pre-systemic metabolism of the drug followed by enhanced drug effects. Human experimental data suggest that furanocoumarins contribute to this grapefruit effect.

Other health-promoting effects of phytochemicals include regulation of blood pressure, blood glucose level, and blood coagulation and inhibition of inflammatory processes (Table 15.1). Further, carotenoids may be involved in the prevention of macular degeneration in the retina, as well as other diseases of the eye (see Chapter 11).

## 15.3 Epidemiological evidence for the protective effects of vegetables, fruits, whole grains, and phytochemicals

A number of epidemiological studies analysed by the World Cancer Research Fund, the American Institute for Cancer Research, and the International Agency for Research on Cancer of the WHO suggest that a high intake of plant foods is inversely associated with the risk for some cancers. While the outcome of recent prospective cohort studies was less supportive for a general cancer-protective effect, overall the totality of evidence still suggests that constituents in plant foods protect against cancer. In addition, data from prospective studies analysing associations at the level of phytochemical intakes and cancer risk suggest that phytochemicals are related to the reduction of cancer risk. As well as evidence about cancer, a number of prospective cohort studies have reported a reduction of risk of cardiovascular disease by 30% in subjects consuming high amounts of vegetables, fruits, and whole grains compared with subjects with a low intake. Similar trends for the intake of specific phytochemicals in prospective, as well as in human intervention, studies were observed. For example, the increased intake of specific flavonoids reduced blood pressure in human intervention studies.

With the present state of knowledge, it is hard to differentiate to what degree the various components in plant foods (essential nutrients, dietary fibre, and phytochemicals) contribute to the observed reduction in disease risk. Human intervention studies are needed to prove that the health-promoting effects of plant foods observed in epidemiological studies are causally related to the intake of phytochemicals.

### 15.3.1 Summary

Present knowledge of the effects of phytochemicals allows us to conclude that these non-nutritive dietary compounds of plant foods can have health-promoting effects. Phytochemicals, along with vitamins, minerals, trace elements, fatty acids, and dietary fibre, are responsible for the protective effects of vegetables and fruits, nuts, whole grains, and legumes against cancer and cardiovascular disease. Clearly, there is no evidence that a single phytochemical is especially effective in the prevention of cancer or cardiovascular disease. The most protective effect is observed when a high number of different phytochemicals are consumed with plant foods, which presumably exert cumulative or synergistic effects. For many phytochemicals, detection methods in foods and body fluids have been established. Although the determination of phytochemicals in terms of content, bioavailability, and biokinetics is now possible, only key phytochemicals of the individual classes have been carefully studied. Further epidemiological and experimental studies should elucidate the links between the ingestion of certain phytochemicals and the incidence of specific diseases, including mechanisms of protection. In short-term intervention studies, biomarkers need to be identified that yield indications for long-term preventative effects of phytochemicals in humans.

The toxic potential of phytochemicals is negligible, as long as consumption habits are restricted to whole food and avoid extracts or isolates from food. So far no adverse effects of phytochemicals as part of wholesome foods have been reported, even in subjects on predominantly vegetarian diets.

Nutritional recommendations do not need to be modified in the light of the latest understanding of the health benefits of phytochemicals. Recommended dietary allowances for certain plant foods for prevention or therapy of certain diseases cannot be given at this point in time. However, most nutritional recommendations include an increase in the consumption of plant-derived foods (to five a day for vegetables and fruit) based on epidemiological evidence that phytochemicals have beneficial effects on the health and well-being of humans. In summary:

- plant foods contain phytochemicals of different chemical classes
- intake of dietary phytochemicals modulates physiological processes in humans
- a high dietary intake of phytochemicals with vegetables, fruits, nuts, pulses, and whole grains is associated with a reduced risk for cardiovascular disease and other diseases.

### Further reading


1. **Bohn, T., McDougall, J. Alegria, A., et al.** (2015) Mind the gap—deficits in our knowledge of aspects impacting the bioavailability of phytochemicals and their metabolites—a position paper focusing on carotenoids and polyphenols. *Mol Nutr Food Res*, **59**, 1307–23.
2. **Dominguez-Lopez, I., Yago-Aragon, M., Salas-Huetos, A., et al.** (2020) Effects of dietary phytoestrogens on hormones throughout a human lifespan: a review. *Nutrients*, **12**, 2456.

3. **Howes, M.R., Perry, N.S.L., Vasquez-Londono, C., et al.** (2020) Role of phytochemicals as nutraceuticals for cognitive functions affected in ageing. *Br J Pharmacol*, **177**, 1294–315.
4. **Liu, R.H.** (2004) Potential synergy of phytochemicals in cancer prevention: mechanism of action. *J Nutr*, **134**, 3479S–85S.
5. **Marelli, M., Conforti, F., Araniti, F., et al.** (2016) Effects of saponins on lipid metabolism: a review of potential health benefits in the treatment of obesity. *Molecules*, **21**, 1404.
6. **Miekus, N., Marszalek, K., Podlacha, M., et al.** (2020) Health benefits of plant-derived sulfur compounds, glucosinolates, and organosulfur compounds. *Molecules*, **25**, 3804.
7. **Parmenter, B.H., Croft, K.D., Hodgson, J.M., et al.** (2020) An overview and update on the epidemiology of flavonoid intake and cardiovascular disease risk. *Food & Function*, **11**, 6777–806.
8. **Plat, J., Baumgartner, S., Vanmierlo, T., et al.** (2019) Plant-based sterols and stanols in health and disease: 'consequences of human development in a plant-based environment?' *Progress Lipid Research*, **74**, 87–102.
9. **Rowles III, J.L. and Erdman Jr, J.W.** (2020) Carotenoids and their role in cancer prevention. *BBA—Mol Cell Biol Lipids*, **1865**(11), 158613.
10. **Seden, K., Dickinson, L., Khoo, S., et al.** (2010) Grapefruit–drug interactions. *Drugs*, **70**, 2373–407.
11. **Surh, Y.J.** (2003) Cancer chemoprevention with dietary phytochemicals. *Nat Rev Cancer*, **3**, 768–80.
12. **Wang, Y., Lim, Y.Y., He, Z., et al.** (2021) Dietary phytochemicals that influence gut microbiota: roles and actions as Alzheimer agents. *Crit Rev Food Sci* Nutr, **62**(19), 5140–66. https://doi.org/10.1080/10408398.2021.1882381
13. **Watzl, B. and Leitzmann, C.** (2005) *Bioaktive Substanzen in Lebensmitteln*, 3rd edn. Stuttgart: Hippokrates.
14. **Williamson, G., Kay, C.D., and Crozier, A.** (2018) The bioavailability, transport, and bioactivity of dietary flavonoids: a review from a historical perspective. *Compr Rev Food Sci Food Saf*, **17**, 1054–1112.
15. **World Cancer Research Fund/American Institute for Cancer Research** (2018) *Diet, nutrition, physical activity, and cancer: a global perspective*, 3rd edn. Washington, DC: AICR.

# Nutrition, Genetics, and Personalized Nutrition

Anne-Marie Minihane



## 16.1 Introduction

There is often confusion within the scientific community between the terms nutrigenomics and nutrigenetics. Nutrigenomics is an umbrella term that refers to the impact of dietary components on physiological processes and health by altering the gene expression, epigenetic, protein, or metabolite profiles of an individual, and also includes nutrigenetics (Daniel et al., 2008; Ordovas et al., 2018). Nutrigenetics specifically relates to the impact of genetic variation (genotype) on food intake and the response to dietary change (Ordovas et al., 2018). Nutrigenetics will be the focus of the current chapter (see Box 16.1).

The influence of genetic variation on the metabolism of nutrients (nutrigenetics) has been known for over a hundred years, with the first report of an inborn error of metabolism in 1908. Inborn errors of metabolism are a mixed grouping of over 500 genetic disorders, many of which are in genes that code for enzymes involved in carbohydrate, protein, and fatty acid metabolism. Examples are glycogen storage disease, phenylketonuria, and medium-chain acyl-co-enzyme A dehydrogenase deficiency. Such conditions are typically attributable to a single gene defect (monogenic disorder), with a prevalence of approximately 1 in 2000 live births, and treated with strict dietary regimes. Although of high consequence for the individual, such genetic variation does not affect the nutrition status of the majority of the population. The modern field of nutrigenetics emerged less than 50 years ago and is focused on the impact of common genetic variability on food intake and preferences, and the absorption, metabolism, physiological response, and impact on health of a particular dietary pattern, food, nutrient, or other food-derived non-nutrient (such as fibre or polyphenols, see Fig. 16.1).

In recent years there has been an emerging nutrigenetic interest in the ability of dietary strategies to mitigate the effect of an overall 'at-risk' genotype, as identified by having a high genetic risk score (GRS) for a particular condition (see section 16.7) (Lourida et al., 2019).

Current UK and global dietary guidelines take a general approach, with some differences in

**Box 16.1 Genetic terms**

Allele: An alternative form of a gene.

Chromosome: A single strand of DNA, wrapped around proteins called histones, which carry the genetic information of an organism. Humans have 23 pairs of chromosomes, with one of each pair received from the biological father and one from the biological mother.

Copy number variation (CNV): Also known as a structural variation. It is a large segment of DNA > 1000 bases (1 kb), which may be deleted, duplicated, inserted, inverted, or translocated, and therefore present in a variable copy number relative to a reference genome.

DNA methylation: A biological process by which methyl groups are added to the DNA molecule, which typically influences gene expression.

Epigenetics: Changes in metabolic regulation caused by modification of gene expression rather than alteration of the genetic code itself. DNA methylation is an example.

Exon: A segment of DNA containing the 'blueprint' for protein synthesis. Remarkably, only 1–2% of DNA is coding, with the purpose of the remaining 98–99% non-coding DNA incompletely understood.

Gene: Unit of DNA that encodes for a protein. A gene is made up of exons, introns, and regulatory sequences, which alter gene expression.

Genome: The complete set of genes or genetic material present in a cell.

Genome-wide association studies (GWAS): An observational study of a genome-wide set of thousands of genetic variants (typically SNPs, see below) in individuals to see if any gene variant is associated with a trait (e.g. disease).

Genotype: The genetic make-up of an organism.

Heterozygous: Having two different alleles of a particular gene.

Homozygous: Having two identical alleles of a particular gene.

Intron: Non-coding region of DNA, which constitute 98–99% of the total.

mRNA: The molecules that convey genetic information from DNA within the nucleus of the cell to the ribosome where proteins are made.

Mutation: Change to the DNA sequence.

Phenotype: The observable traits of an organism (e.g. body weight, plasma cholesterol levels), which are the result of the interaction between its genotype and environment.

Polymorphism: One or more variants of a particular DNA/gene sequence within a population.

Single nucleotide polymorphism (SNP): A variation in DNA in a single nucleotide (A, T, C, G) between individuals or paired chromosomes in an individual. It is the most common form of genetic variation, accounting for > 90% of the total.

recommended intakes between children and adults, males and females, for pregnancy and lactation, and in those with a disease diagnosis, such as type 2 diabetes. These generic recommendations are based on population estimates of optimal intakes for reducing risk of chronic disease, and the prevention of nutrient deficiency in most of the population. However, it is widely recognized that dietary change and the physiological response to dietary change is highly heterogeneous due to a range of personal factors, including personal food preferences, facilitators or barriers to behaviour change, habitual diet and nutrition status, age and stage of life (pregnant, lactating, etc.), microbiome composition, disease status, BMI, medication use, and genotype. The ethos of personalized (individual recommendations) or stratified (population subgroup) nutrition is that dietary advice based on characteristics of the individual or subpopulation will increase adherence, result in greater health benefits, or both (Ordovas et al., 2018). Nutrigenetics is evolving from examining the impact and potential of single variants with single dietary components, to taking a whole genome approach and considering how it interacts with dietary patterns or multiple lifestyle behaviours to affect well-being.

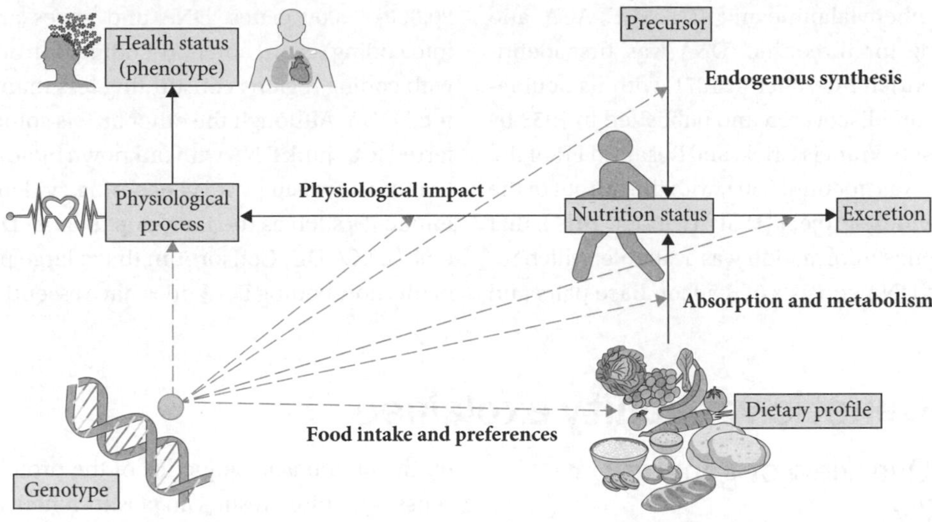


Fig. 16.1 Overview of the potential effect of genotype on food intake and metabolism.

# 16.2 The history of DNA

In humans, DNA is organized into 23 pairs of chromosomes within the nucleus of the cell. One member of each pair is inherited from the biological mother and one from the father. DNA molecules are double-stranded helices of two long chains of repeating units called nucleotides. Each nucleotide consists of a sugar and phosphate group (which makes up the backbone of the chain) and a base attached to the sugar; see Fig. 16.2.

The bases are adenine (A), thymine (T), guanine (G), and cytosine (C). A series of three bases encodes for an amino acid, with, for example, TTT and TTC

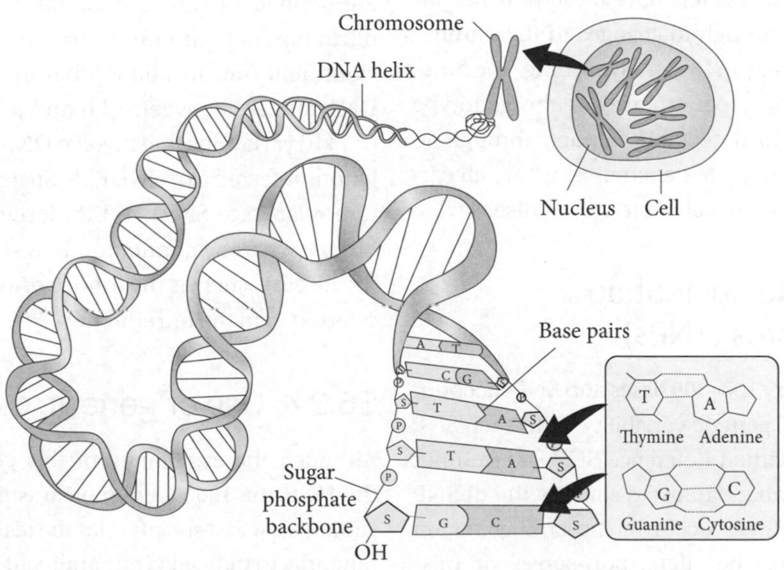


Fig. 16.2 DNA and nucleotide structures.

coding for phenylalanine and ACT, ACC, ACA, and ACG coding for threonine. DNA was first identified by Friedrich Miescher in 1871, with its double-helix structure discovered and published in 1953 by James Watson, Francis Crick, and Rosalind Franklin. However, it was not until 2004, with the output of the Human Genome Project (HGP) (Lander, 2011), that DNA sequence information was available, which revealed that DNA consists of 3 billion base pairs and 20,000–25,000 genes. DNA and genes are divided into coding (exon) and non-coding (intron) regions, with coding regions constituting less than 2% of the total DNA. Although the other 98% is sometimes referred to as junk DNA with unknown biological function, unsurprisingly it is becoming evident through initiatives such as the Encyclopaedia of DNA Elements (ENCODE) Consortium that a large proportion of this non-coding DNA does have essential roles.

## 16.3 Genetic variability explained

### 16.3.1 Overview of genetic variability

Only identical twins have identical DNA. In the general population, 99.5% of the DNA is common between individuals, with the remaining 0.5% variable and defining an individual's phenotype (observable characteristics) and response to their environment, including diet. In 2015, output from the 1000 Genome Consortium indicated that there are typically 88 million variants in a human genome (Auton et al., 2015).

Genetic variability is a change in the base sequence and structure of DNA. Such mutations can be minor, with a change to only a single base in the DNA referred to as a single nucleotide polymorphism (SNP), right through to changes in the number of chromosomes, e.g. in Down's syndrome. The functional consequence (penetrance) of the genotype varies between no detectable impact, through to mutations with complete penetrance, where all carriers of the mutation develop the trait or disease.

### 16.3.2 Single nucleotide polymorphisms (SNPs)

A SNP occurs every 100–300 bases and SNPs account for over 90% of all genetic variability, with each SNP assigned an individual Reference SNP (rs) number on publicly available databases such as the dbSNP database. When SNPs occur in the coding region of DNA, they may be silent, non-sense, or missense. Silent, as the term suggests, has no impact on the amino acid sequence of the protein. A non-sense mutation results in a shortened and often non-functional protein. A missense SNP results in an amino acid change in the protein sequence, with the functional consequences dependent on where in the protein the amino acid change has occurred and how similar the replacement amino acid is to the original amino acid. SNPs rarely cause disease in isolation, but may increase the risk of and response to prescribed medications or lifestyle behaviours, such as eating and physical activity, as will be discussed.

### 16.3.3 Larger structural variants

Structural variants, such as copy-number variants, are defined as genomic alterations (deletions, duplications, copy-number variants, insertions, inversions, and translocations) that involve segments of DNA that are between 1 kb and 3 Mb, with smaller (< 1 kb) variations that involve DNA insertions or deletions referred to as INDELS. Structural variants are rare relative to SNPs and underlie monogenic disorders, whereby a mutation in a single gene causes the disease, such as the inborn errors of metabolism referred to in the introduction.

### 16.3.4 Use of genetic data

Although the availability of this genetic data from the HGP, the 1000 Genome Consortium, and other, similar, disease-specific databases has tremendous potential to radicalize our ability to diagnose disease risk, and move towards a greater personalization of dietary advice and interventions, establishing the individual and combined impact of these variants represents a huge research challenge. Unsurprisingly, therefore, the application of the field of nutrigenetics in public health and personalized nutrition is an emerging field. At present there are only a limited number of diet*genotype*health interactions that have been consistently observed, as will be discussed.

# 16.4 *APOE* genotype, health, and response to dietary change

## 16.4.1 Role of apolipoprotein E and the basis and prevalence of *APOE* genotypes

Apolipoprotein E (apoE) is mainly produced by the liver but also by macrophages and by glial cells in the brain (Minihane, 2013). It is pleiotropic, with apoE being an integral part of lipoproteins, mediating lipid transport throughout the body and in particular in the brain. In addition, apoE regulates immune function, inflammation, oxidative status, and neurotransmission. The gene for apoE is on chromosome 19. *APOE* genotype is comprised of two common SNPs (rs429358 and rs7412) resulting in three alleles, ε2, ε3, and ε4, and six possible genotype combinations, *APOE2/E2*, *APOE2/E3*, *APOE2/E4*, *APOE3/E3*, *APOE3/E4*, and *APOE4/E4* (see Fig. 16.3; Liu et al., 2013).

The variants are missense SNPs at positions 112 and 158 of the apoE protein. About 60–65% of typical populations are the wildtype *APOE3/E3*. The ε4 is the ancestral allele, with a higher prevalence in northern latitudes (low solar irradiation) and global regions with high rates of infectious diseases, where it is thought to offer an evolutionary advantage, as it has been associated with a higher vitamin D synthesis and higher inflammatory responses. However, being an ε4 carrier is associated with a modest increase in cardiovascular diseases (CVD) due to higher total and LDL-cholesterol and inflammation (Minihane, 2013). *APOE* genotype is the strongest common genetic determinant of dementia and

| APOE | SNP rs429358 | Amino acid 112 | SNP rs7412 | Amino acid 158 |
|---|---|---|---|---|
| ε2 | TGC | Cystine | TGC | **Cystine** |
| ε3 | TGC | Cystine | CGC | Arginine |
| ε4 | CGC | **Arginine** | CGC | Arginine |

| Genotype (% typical general population) | | | | | |
|---|---|---|---|---|---|
| *E2/E2* | *E2/E3* | *E2/E4* | *E3/E3* | *E3/E4* | *E4/E4* |
| 1% | 15% | 2% | 60% | 20% | 2% |

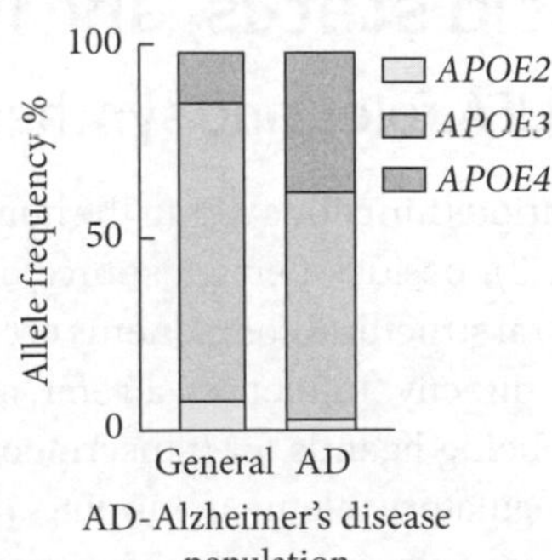


Fig. 16.3 *APOE* genotype.

Alzheimer's disease (AD the main form of dementia), with a 2–3-fold and 12–15-fold higher risk in *APOE3/E4* and *APOE4/E4,* respectively, and typically a decade earlier onset (Pontifex et al., 2018). The impact is further evidenced by the fact that the ε4 allele frequency is three times higher in AD relative to the general population (see Fig. 16.3). It appears that the ε4 allele has a greater impact on AD risk in females compared to males, which may partly explain the fact that two-thirds of AD patients are females. The higher AD and CVD risk are likely to be major contributors to the fact that *APOE4* carrier status is one of the few common genetic predictors of a shorter life expectancy.

### 16.4.2 APOE genotype and the response to lifestyle behaviours

*APOE4* genotype is also more responsive to lifestyle behaviours, including diet (Minihane, 2013). In the Cardiovascular Risk Factors, Aging, and Dementia (CAIDE) study, where participants were followed up for 21 years, *APOE4* carriers were more susceptible to the effects of physical inactivity, alcohol drinking and smoking, and moderate to high intake of saturated fats on AD risk. For saturated fat, quartile 4 (highest intake) had a 7-fold increased risk of AD in *APOE4* carriers compared to quartile 1 (lowest intake) in the *APOE4* non-carrier reference group, who had only a 2-fold higher AD risk in quartile 4 versus quartile 1 of saturated fat (Kivipelto et al., 2008).

In addition to saturated fat, those with an *APOE4* genotype are more sensitive to dietary cholesterol and the long chain n-3 PUFA, docosahexaenoic acid (DHA). Using the robust prospective recruitment according to *APOE* genotype approach, with *APOE* genotype-matched groups, high dose fish oil supplementation (providing 3.45 g DHA per day) resulted in greater triglyceride reductions in *APOE4* carriers, which was consistent with an earlier study (Carvalho-Wells et al., 2012; Minihane, 2013). However, such high DHA doses also modestly increase LDL-cholesterol in *APOE4* carriers, which is likely offset by an overall beneficial increase in LDL particle size. This increase in LDL-cholesterol is not observed at lower, more physiologically relevant eicosapentaenoic (EPA) + DHA intakes (< 2 g per day). Furthermore, emerging evidence indicates that older *APOE4* females are particularly susceptible to low brain DHA levels and would benefit from meeting the current recommended intakes of 500 mg EPA + DHA per day, or potentially higher DHA intakes (Pontifex et al., 2018).

Overall, *APOE4* carriers represent a group with a higher risk of certain prevalent chronic disease who would benefit from targeted dietary fatty acid intake advice to moderate saturated fat intake and ensure adequate dietary DHA in order to delay or prevent disease onset.

## 16.5 Fatty acid desaturase (FADS 1 and 2) genotype, fatty acid status, and response to intake

### 16.5.1 PUFA roles and synthesis

PUFA fulfil various functions within the human body (see Chapter 3), besides being a source of energy. They are central structural components of cell membranes, can directly influence several metabolic pathways by being ligands for transcription factors like sterol regulatory element binding protein 1 (SREBP-1) and peroxisome proliferator-activated receptors (PPARs), and are the parent compounds for the formation of various oxylipin metabolites that are highly bioactive (Schulze et al., 2020). Apart from linoleic acid (LA, n-6 PUFA) and α-linolenic acid (ALA, n-3 PUFA), which are essential and need to be derived from the diet, PUFA can be synthesized endogenously, mainly in the liver (see Fig. 16.4). Delta-6 and delta-5-desaturases are the key enzymes in this process, with delta-6-desaturase considered the rate limiting step of conversion of LA and ALA to longer and more unsaturated fatty acids

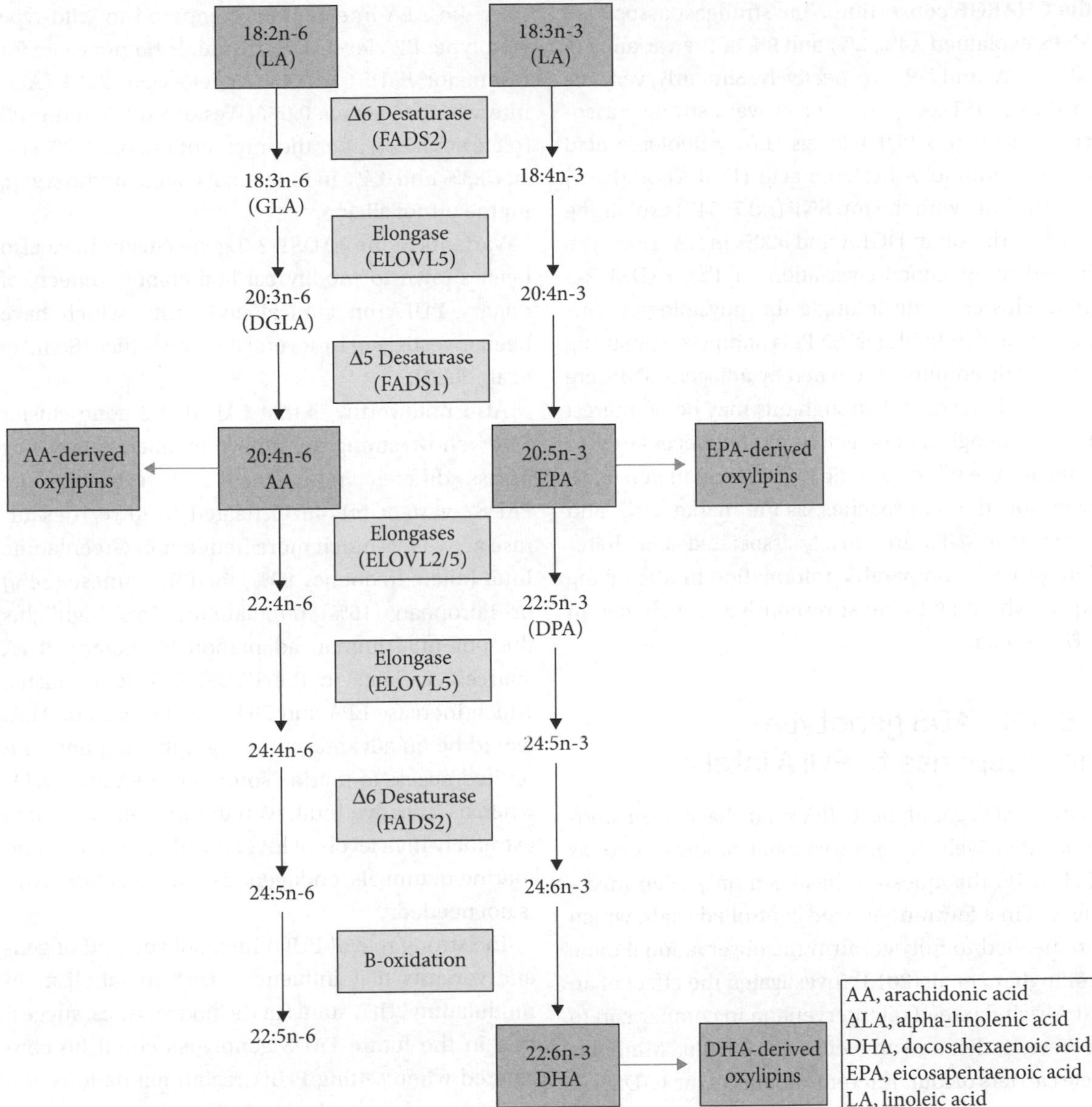


Fig. 16.4 Polyunsaturated fatty acid (PUFA) metabolism.

such as arachidonic acid (AA), EPA, and DHA; for an overview see Schulze et al. (2020).

### 16.5.2 FADS genotype and PUFA status

In candidate gene studies (specifically focusing on the FADS gene region), variations in FADS1 and FADS2—the genes adjacent to each other on chromosome 11, encoding the delta-5- and delta-6-desaturases—have been related to blood PUFA status and to the relationship between PUFA intake and health outcomes. Furthermore, GWAS have identified this FADS region to have the strongest genetic link to blood PUFA levels.

For example, SNPs in the FADS1-2-3 gene cluster were associated with higher levels of ALA and lower levels of EPA and docosapentaenoic acid (DPA) in the CHARGE consortium. The strongest associated SNPs explained ~4%, 2%, and 9% of the variance of ALA, EPA, and DPA, respectively. Similarly, variants in the FADS1-2-3 gene cluster were strongly associated with n-6 PUFA levels (LA, γ-linolenic acid (GLA), dihomo-γ-linolenic acid (DGLA), and AA) in CHARGE, with the top SNP (rs174547) explaining ~10% variation in DGLA and >20% in AA. However, investigating genetic variation in the FADS1-2-3 gene cluster to disentangle the physiological importance of individuals' SNPs is hampered by strong linkage disequilibrium, whereby adjacent SNPs are inherited together. Although this may be of interest from a biological perspective, it is not necessarily an issue for the use of genetic information in personalized nutrition approaches, as the marker SNP and functional SNPs are closely associated and therefore provide comparable information in identifying those who may be most responsive to a change in PUFA intake.

### 16.5.3 FADS genotype and response to PUFA intake

While FADS genotype*PUFA status have been studied extensively in observational studies such as CHARGE, this question has been only been investigated in a few randomized controlled trials, which are needed to fully confirm the observational data. Gillingham et al. (2013) investigated the effect of an ALA-rich flaxseed oil intervention in comparison to a Western diet and an oleic acid-rich diet. Minor allele carriers of four different variants in the FADS1-2-3 had a lower increase in EPA plasma concentrations after the ALA intervention compared to wild-type genotype: EPA levels in individuals homozygote for the major FADS1 rs174561 allele were 2.2% (ALA intervention) versus 0.6% (Western diet) and 0.7% (oleic acid diet) after the intervention, but 0.9% versus 0.3% and 0.4% in individuals being homozygote for the minor allele.

Variants in the FADS1-2-3 gene cluster have also been shown to modify cardiometabolic effects of dietary PUFA on clinical endpoints, which have been investigated in several cohort studies (Schulze et al., 2020).

Also noteworthy is that FADS1-2-3 gene cluster SNPs show strong variability in allele frequency across different populations. For example, the FADS2 variant rs174570, related to lower desaturase activity, is much more frequent in Greenlandic Inuit (allele frequency 99%) than in Chinese (34%) or European (16%) populations. This highlights the potential human adaptation to dietary PUFA sources. Variants in the FADS1-2-3 gene cluster, which increase EPA and DHA synthesis from ALA, would be an advantage in geographic regions with limited access to marine sources of EPA and DHA, whereas in native Inuit, who traditionally consume extremely high levels of EPA and DHA from fish and marine mammals, endogenous synthesis from ALA is not needed.

The strong role of PUFA metabolism, and of genetic variants that influence PUFA metabolism in modulating PUFA status in the body tissues, suggest that in the future FADS genotypes could be considered when setting PUFA recommendations at a population or individual level.

## 16.6 *MTHFR* genotype, riboflavin, and blood pressure

### 16.6.1 *MTHFR* genotype explained

Perhaps the most consistently observed nutrigenetic example, with wide public health application, is the role of folate and riboflavin in mitigating the increased risk of high blood pressure and CVD associated with the *MTHFR* genotype (McAuley et al., 2016). Methylenetetrahydrofolate reductase (MTHFR) is an enzyme that breaks down the amino acid homocysteine (see Fig. 16.5). Homocysteine is an independent risk factor for CVD, with homocysteine levels being inversely associated with folate levels. The common *MTHFR C677T* (rs1801133) genotype results in a cysteine to threonine amino acid change at position 677 in the protein and an enzyme with reduced activity (~35% of the

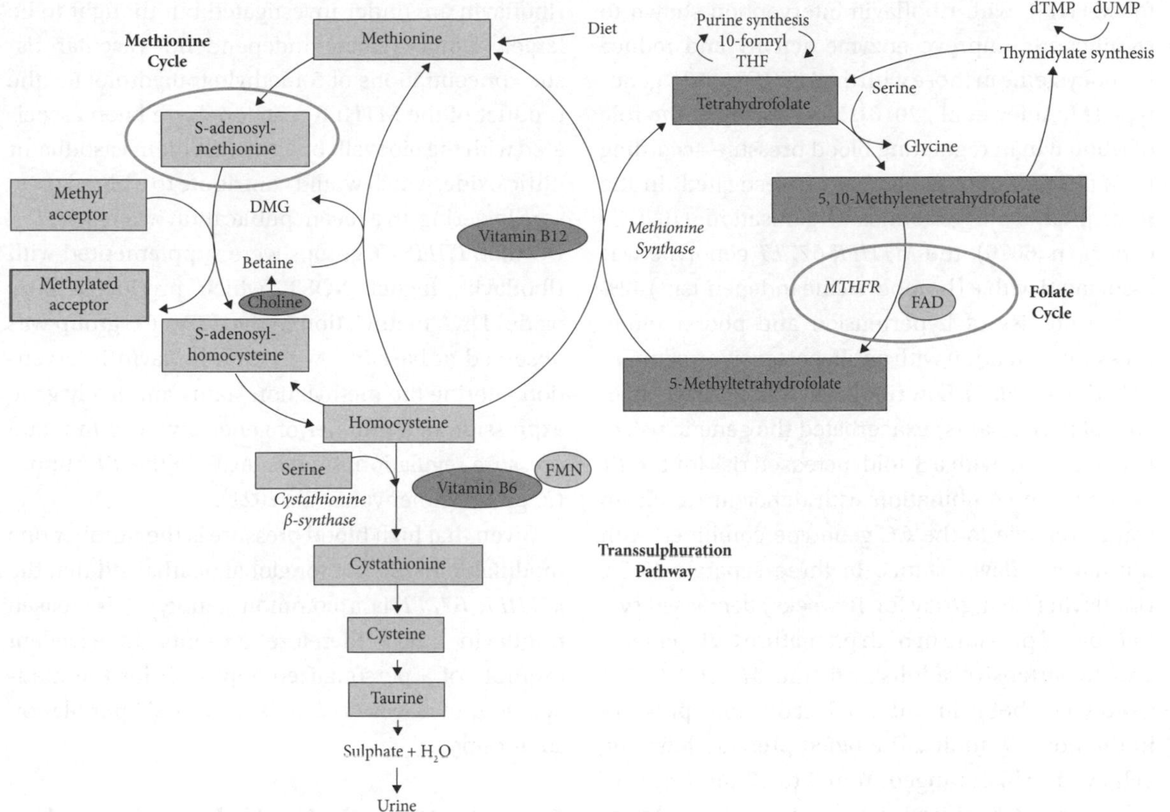


**Fig. 16.5** One-carbon metabolism.

*Source:* Clarke et al. *Proceedings of the Nutrition Society* (2014).

MTHFR-677C). Homozygosity (*MTHFR 677TT*) affects over 10% of the UK and Irish populations and up to 32% of other global populations and has been associated with an increased risk of CVD (especially stroke) and hypertension (McAuley et al., 2016). Although *MTHFR 677TT* can increase the risk of CVD and hypertension by up to 40% and over 50%, a wide geographical variation in the extent of excess risk indicates that the penetrance is affected by behavioural or other environmental variables.

## 16.6.2 *MTHFR* genotype and response to dietary vitamin B status

Low folate status impacts on homocysteine levels more severely in individuals with the *MTHFR 677TT* genotype, and the standard recommendations for folate intake (with green vegetables, legumes, liver, and fortified breakfast cereal providing good dietary sources) have been shown to be insufficient to maintain homocysteine levels below the risk level in this population. It is accepted that increasing folate intake (from 200 μg/day to 400–600 μg/day) reduces the risk for hyperhomocysteinemia in most *MTHFR 677TT* individuals.

In a 2009 GWAS, which tested 2.5 million SNPs, eight regions including *MTHFR* were associated with blood pressure. Subsequent targeted genotyping identified the *MTHFR C677T* as the functional variant. Accumulating evidence indicates that dietary riboflavin (vitamin $B_2$), provided by sources such as milk, eggs, fortified breakfast cereals, lean beef and pork, organ meats (such as beef liver), and yoghurt, contributes to the effect of *T* allele carrier status. Riboflavin in the form of flavin adenine dinucleotide (FAD) is required as a cofactor for MTHFR, with riboflavin intervention shown to significantly improve enzyme activity and reduce homocysteine in those with the *MTHFR 677TT* genotype (McAuley et al., 2016). More recently, the role of riboflavin in regulating blood pressure according to *MTHFR* genotype has been investigated. In the Joint Irish Nutrigenomics Organisation (JINGO) cohort (n=6076), the *MTHFR 677TT* genotype was associated with a (homocysteine independent) 42% increased risk of hypertension and poorer blood pressure regulation with antihypertensive treatment (Ward et al., 2020). Low riboflavin status (observed in 30% of participants) exacerbated the genetic risk of hypertension, with a 3-fold increased risk for the *TT* genotype in combination with deficient riboflavin status, relative to the *CC* genotype combined with normal riboflavin status. In three separate RCTs, riboflavin (1.6 mg/day for 16 weeks) decreased systolic blood pressure in both premature CVD patients and hypertensive adults with the *MTHFR 677TT* genotype; riboflavin had no effect on blood pressure in the non-*TT* adults. The blood pressure lowering achieved (which ranged from 5 to 13 mmHg) is of large clinical significance given that a 2 mmHg decrease in BP confers a 10% reduction in stroke risk. Deficiency of riboflavin is highly prevalent in low income countries but is also a widespread problem in high income countries in population subgroups. For example, in the UK, older adults (over 65 years) are particularly susceptible, with a significant proportion not reaching the recommended nutrient intake (RNI) and 8% below the lower RNI, with the proportion increasing with age.

The mechanism underlying the effect of *MTHFR* genotype on blood pressure and its modulation by riboflavin are under-investigated but thought to be largely homocysteine independent. Vascular tissue concentrations of 5-methyltetrahydrofolate, the product of the MTHFR reaction, have been associated with the bioavailability of the potent vasodilator nitric oxide, which would contribute to a blood pressure lowering. In a recent publication where *MTHFR CC* vs *MTHFR TT* groups were supplemented with riboflavin, higher NOS3 (which produces nitric oxide) DNA methylation in the *TT* vs *CC* group was observed at baseline, with the riboflavin intervention altering the methylation status and likely gene expression of a number of genes involved in blood pressure regulation, such as ACE, in the *TT* but not *CC* group (Amenyah et al., 2021).

Given that high blood pressure is the number one modifiable risk factor for global deaths and that the *MTHFR 677TT* is a common genotype, increased riboflavin intake therefore presents an excellent example of a personalized approach for the management of disease risk in large 'at-risk' population subgroups.

### 16.6.3 Other potential nutrigenetic targets

A number of other genes and common genotypes, such as *FTO* (rs9939609), *CD36* (rs1527483), *TCF7L2* (rs7903146), *NOS3* (also called *eNOS*, rs1799983), *TAS2R38*, *PPARG*, and the *CYP*-family, have been extensively studied for their effect on health outcomes, and response to dietary and pharmaceutical intervention, with potential in future personalized/stratified nutrition approaches, which need further verification.

## 16.7 Genetic risk score approach

Instead of considering each variant/SNP in isolation, a genetic risk score (GRS) combines multiple SNPs. For example, in an analysis of 480,000 individuals looking at coronary artery disease (CAD) risk, 1.7 million variants were used (Inouye et al., 2018). Those in the highest 20% of GRS data had a 4-fold increased risk of CAD relative to the lowest 20%, decreasing to 2.8-fold for those on lipid lowering or antihypertensive medications. Looking at a GRS for dementia, and how its effect can be influenced by lifestyle behaviours (eating, physical activity, smoking, and alcohol consumption), Lourida and colleagues conducted a UK Biobank analysis, in 196,383 individuals (Lourida et al., 2019). Of the participants with a high GRS, an unfavourable lifestyle increased the risk of dementia by 2.8-fold. This indicates that the penetrance of genotype is highly modifiable by dietary and other behaviour changes.

The advantage of using a GRS rather than conventional risk factor profile to target recommendations is that we are born with our genotypes, whereas other risk factors such as excess weight and high blood pressure develop over time, and are often not detected until significant irreversible physiological damage has occurred. Therefore, genetics, such as a GRS approach, can be used to target intervention early, affording the maximum lifetime benefits.

## 16.8 Translating nutrigenetic research into public health and clinical care: what needs to be considered

When the HGP was first made available in 2004, it was naively considered by many to be the panacea and likely to rapidly deliver a step change in disease risk assessment and the delivery and uptake of personalized, effective guidelines and interventions. Although there are robust examples outlined above as to how genetic information may be used to deliver this, the consensus from researchers working in the field is that the widespread application of nutrigenetics is not ready for routine use, with much more research needed before personalized nutrition can deliver these expected results.

The majority of nutrigenetics research to date is either from observational studies or from RCT using retrospective genotyping (at the end of the trial) approaches, with often small group sizes in the rare allele subgroups and genotype groups unmatched for important variables such as age and sex, which runs the risk of erroneous and often false negative results. More fit-for-purpose RCTs with prospective genotype approaches are needed to confirm promising findings from observational or retrospective RCT approaches.

More recently, new technologies (such as wearables, microbiome sequencing, metabolomics, etc.) and associated data analytical capacity, along with artificial intelligence and machine learning, has enabled multiple endogenous (physiological and genetic) and exogenous factors to be studied repeatedly and in real time and used to predict risk of disease and the response to intervention. Much of this technology may be prohibitively expensive if deployed in large groups and may increase health inequalities, by only being available to those who can afford it. Genotyping has become a rapid and relatively cheap technology, if conducted at cost and available through public healthcare systems. However, at present it is only accessible through direct-to-consumer commercial organizations, where the inflated costs make it inaccessible to many and the information associated with the genotype (its impact on health and supporting personalized nutrition advice) is of variable and unregulated quality. The challenge for research will be to define the minimum set of affordable genetic and other robust measurements that predicts individual response to personalized nutrition in large numbers of people and to incorporate these into existing primary and secondary healthcare systems.

Before we invest further in a move to a more personalized approach to health, a key question is whether personalized dietary advice is more effective than general recommendations. The Food4Me Study (Celis-Morales et al., 2017) is the largest RCT to test the efficacy of personalized nutrition by implementing an internet-based intervention completed by 1269 adults in seven European countries over 6 months. Participants who followed the personalized nutrition approach consumed less red meat, salt, and saturated fat, and had increased folate intake and a higher overall diet quality, as assessed by a Healthy Eating Index score, compared to the control, general dietary advice arm. The core personalized nutrition intervention was based on an individual's habitual diet. There was no evidence that adding more expensive personal phenotypic (disease risk biomarker levels) and phenotypic plus genotypic information enhanced the effectiveness of the personalized nutrition advice. In line with the finding of Food4Me, a recent systematic review, which included 11 studies that included between 57 and 1488 participants and follow-up durations from 1 to 12 months, concluded that dietary intake is improved to a greater extent in participants randomly assigned to receive personalized nutrition advice compared with generalized dietary advice (Jinnette et al., 2021). Although many of the growing number of studies that personalized on the basis of genetic information do show evidence of greater uptake of dietary advice compared to the general advice control group, the Food4Me (Celis-Morales et al., 2017) or other individual studies as reviewed by Jinnette et al. (2021) do not provide a direct comparison of the efficacy of personalized nutrition based on habitual diet versus phenotype versus genotype in isolation.

## 16.9 Final message

Poor diet is a top risk factor for disease globally, responsible for 1 in 5 deaths. As current public health campaigns using general dietary recommendations are not achieving the changes in dietary behaviour needed to shift dietary intake towards healthier dietary patterns that reduce population risk of disease, there is a strong justification and need for more personalized and stratified nutrition approaches to complement the population-based dietary guidelines. Personalized nutrition can be applied in two broad areas: firstly, for the dietary management of people with specific diseases or other groups who need special nutritional support; or secondly, for the development of more effective interventions for improving public health. For the latter in particular, such strategies should be cost effective and easy to implement at scale to ensure they are widely available, and should include emerging gene variants that are known to have a large effect on nutrition status or response to dietary change. The use of a GRS approach can be used to target interventions and support to high-risk groups prior to the development of disease or even an 'at-risk' physiological profile. All personalized dietary interventions should be based on behaviour-change techniques in order to optimize their likelihood of success.

### Further reading


1. **Amenyah, S.D., Ward, M., McMahon, A., et al.** (2021) DNA methylation of hypertension-related genes and effect of riboflavin supplementation in adults stratified by genotype for the MTHFR C677T polymorphism. *Int J Cardiol*, **322**, 233–39.
2. **Auton, A., Brooks, L.D., Durbin, R.M., et al.** (2015) A global reference for human genetic variation. *Nature*, **526**, 68–74.
3. **Carvalho-Wells, A.L., Jackson, K.G., Lockyer, S., et al.** (2012) APOE genotype influences triglyceride and C-reactive protein responses to altered dietary fat intake in UK adults. *Am J Clin Nutr*, **96**, 1447–53.
4. **Celis-Morales, C., Livingstone, K.M., Marsaux, C.F., et al.** (2017) Effect of personalized nutrition on health-related behaviour change: evidence from the Food4Me European randomized controlled trial. *Int J Epidemiol*, **46**, 578–88.
5. **Daniel, H., Drevon, C.A., Klein, U.I., et al.** (2008) The challenges for molecular nutrition research 3: comparative nutrigenomics research as a basis for entering the systems level. *Genes Nutr*, **3**, 101–6.
6. **Gillingham, L.G., Harding, S.V., Rideout, T.C., et al.** (2013) Dietary oils and FADS1-FADS2 genetic variants modulate [13C]α-linolenic acid metabolism and plasma fatty acid composition. *Am J Clin Nutr*, **97**(1), 195–207. doi: 10.3945/ajcn.112.043117

7. **Igo Jr, R.P., Kinzy, T.G., and Cooke Bailey, J.N.** (2019) Genetic Risk Scores. *Curr Protoc Hum Genet*, **104,** e95. doi:10.1002/cphg.95.
8. **Inouye, M., Abraham, G., Nelson, C.P., et al.** (2018) Genomic risk prediction of coronary artery disease in 480,000 adults: implications for primary prevention. *J Am Coll Cardiol*, **72**, 1883–93.
9. **Jinnette, R., Narita, A., Manning, B., et al.** (2021) Does personalized nutrition advice improve dietary intake in healthy adults? A systematic review of randomized controlled trials. *Adv Nutr*, **12**, 657–69.
10. **Kivipelto, M., Rovio, S., Ngandu, T., et al.** (2008) Apolipoprotein E epsilon4 magnifies lifestyle risks for dementia: a population-based study. *J Cell Mol Med*, **12**, 2762–71.
11. **Lander, E.S.** (2011) Initial impact of the sequencing of the human genome. *Nature*, **470**, 187–97.
12. **Liu, C.C., Liu, C.C., Kanekiyo, T., et al.** (2013) Apolipoprotein E and Alzheimer disease: risk, mechanisms and therapy. *Nat Rev Neurol*, **9**, 106–18.
13. **Lourida, I., Hannon, E., Littlejohns, T.J., et al.** (2019) Association of lifestyle and genetic risk with incidence of dementia. *JAMA*, **322**, 430–37.
14. **McAuley, E., McNulty, H., Hughes, C., et al.** (2016) Riboflavin status, MTHFR genotype and blood pressure: current evidence and implications for personalised nutrition. *Proc Nutr Soc*, **75**, 405–14.
15. **Minihane, A.M.** (2013) The genetic contribution to disease risk and variability in response to diet: where is the hidden heritability? *Proc Nutr Soc*, **72**, 40–7.
16. **Ordovas, J.M., Ferguson, L.R., Tai, E.S., and Mathers, J.C.** (2018) Personalised nutrition and health. *BMJ*, **361**, bmj k2173.
17. **Pontifex, M., Vauzour, D., and Minihane, A.M.** (2018) The effect of APOE genotype on Alzheimer's disease risk is influenced by sex and docosahexaenoic acid status. *Neurobiol Aging*, **69**, 209–20.
18. **Schulze, M.B., Minihane, A.M., Saleh, R.N.M., and Riserus, U.** (2020) Intake and metabolism of omega-3 and omega-6 polyunsaturated fatty acids: nutritional implications for cardiometabolic diseases. *Lancet Diabetes Endocrinol*, **8**, 915–30.
19. **Ward, M., Hughes, C.F., Strain, J.J., et al.** (2020) Impact of the common MTHFR 677C→T polymorphism on blood pressure in adulthood and role of riboflavin in modifying the genetic risk of hypertension: evidence from the JINGO project. *BMC Med*, **18**, 318.

# Part 4

# Nutrition-related Disorders

# Evidence-based nutrition advice

Nutrition recommendations for individuals and populations are typically made by international organizations (e.g. the World Health Organization (WHO)), national governments, and professional and other non-governmental organizations (e.g. National Heart Foundations, World Cancer Research Fund (WCRF)). Prior to the realization that carbohydrates and fats had direct relevance to human health, interest centred around optimal intake of amino acids, vitamins, minerals, and trace elements. Recommended intakes were largely based on a range of absorption and metabolic experiments and the levels of intake shown to be associated with clinically detectable deficiencies. Potentially toxic levels may also be specified. Carbohydrate and fats were principally regarded as energy sources.

While energy and nutrient deficiencies continue to occur in many parts of the world and in some sections of the populations in affluent countries (e.g. in the elderly and disadvantaged), there has been a rapid increase of many chronic non-communicable diseases in almost all countries. While the causes of such diseases are multifactorial, in many there is now clear evidence that nutritional factors are important contributors because of excessive or inappropriate intakes of some fats and carbohydrates, and/or micronutrients. However, the nutrient–disease relationships are typically more complex than is the case for single nutrient deficiencies and a different approach is required for recommendations regarding optimal intakes.

Early nutritional recommendations for reducing the risk of the most important of these diseases (cardiovascular disease and some cancers) were based on the 'expert' opinions of researchers, clinicians, and public health experts appointed to advisory committees by relevant international or national bodies. Recommendations that emanated from such bodies were based on the experts' assessment of existing scientific literature and clinical experience. Given the absence of a clearly defined process for developing recommendations, there was clearly the potential for strongly held opinions, vested interests, and a range of possible biases to influence the decisions of expert committees.

The now accepted concept of 'evidence-based medicine' (EBM) is the stimulus to the development of a comparable more robust approach to nutrition recommendations. The underpinning principle of EBM is that RCTs with meaningful clinical endpoints are required to establish the benefit of any medical, surgical, or preventive treatments. Systematic literature reviews (SLRs) of all the RCTs (with meta-analyses of the trials where appropriate) form the basis of recommendations. An early attempt to establish clear guidelines for developing nutrition recommendations aimed at reducing the risk of chronic non-communicable diseases was made by the WCRF in order to develop advice aimed at reducing the risk of nutrition-related cancers. The approach acknowledged the considerable difficulties involved in the conduct of RCTs, examining the potential of dietary modification to reduce cancer risk. Such trials would inevitably involve very large numbers of participants who would need to follow prescribed diets over prolonged periods. Thus, the WCRF recommended alternative criteria for describing the relationships between dietary factors (foods or individual nutrients) as '*convincingly causal*' or '*probably causal*'. Such associations were regarded as sufficient to warrant recommendations for dietary change, but those based on lesser degrees of evidence were not. In addition to RCTs that showed that modifying a dietary factor could influence cancer risk, consistent findings from several prospective cohort studies could also form the basis of establishing that a diet–cancer association was convincingly causal, provided there was clear evidence that the associations could not be explained by chance or confounding. For an association to be regarded as convincing, there should also be confirmatory evidence from experimental studies in humans or animals (e.g. the effect of dietary manipulation on a well-established disease risk factor), as well as a biological gradient or dose–response between the degree or level of exposure to the dietary factor and the disease risk—meaning the greater the exposure to the food or nutrient, the greater the risk. Associations may be defined as '*probably causal*' if several cohort and/or case-control studies consistently demonstrate biologically plausible associations that cannot be explained by chance, bias, or confounding. The demonstration of a dose–response effect is not an essential prerequisite.

Initially, this approach was also used by the WHO. However, the WHO, and many other national and international organizations, have now espoused an even more prescribed method for formulating questions, selecting outcomes, developing systematic reviews, and evaluating the quality of the evidence using the GRADE (WHO, 2014) approach in order to make '*strong*' or '*conditional*' recommendations. Typically, the strong recommendations are based on '*high*' or '*moderate*' quality evidence derived from RCTs; but '*low*' quality evidence based principally on observational studies may be upgraded if specified criteria are met (e.g. a dose–response gradient). GRADE recommendations are also required to take into account issues beyond the immediate health consequences of the advice. In the context of nutritional guidelines, such a consideration would include availability, acceptability, and environmental consequences of any required change in the food supply in order to meet the recommendations, as well as potential consequences on equity of health outcomes. While several other methods are used by authoritative organizations worldwide, all depend upon systematic literature reviews, which attempt to synthesize the evidence, and a clearly defined method for grading quality and developing recommendations. Clearly, discussion will continue regarding standardized approaches to developing nutrition recommendations. There is a need to clarify how best to incorporate experimental data into the total body of evidence, since current methods rely largely on RCTs and prospective observational studies. Furthermore, there may be instances when natural experimental studies may be used to guide public health action.

Current evidence based nutrition guidelines now cover foods, food groups, patterns, and lifestyle factors as well as nutrients. While most evidence relates to nutrient intakes, this translates readily into advice on foods and dietary patterns relevant to a wide range of cultural practices.

## Reference

WHO (2014) *Handbook for guideline development*, 2nd edn. Geneva: World Health Organization.

# Overweight and Obesity

Mike Lean



While uncommon through most of our evolution, now 80–90% of all adults in affluent societies will accumulate excess body fat, to be classified as clinically overweight or obese. This pandemic has developed over 40–50 years, most rapidly in lower- and middle-income countries, in step with highly effective global social marketing to normalize increased food-energy consumption and snacking, while occupational physical activity has greatly reduced. Depending on individual and racial susceptibility, many chronic diseases are mediated by the *disease-process of obesity*, with its combined genetic and environmental aetiology. The disabilities and premature mortality attributable to overweight and obesity have profound personal, social, and economic effects.

## 17.1 The disease-process of obesity: definitions and measurements

In the past, obesity was considered simply as a *state* in which fat stores are visibly excessive, impairing health and quality of life. It was viewed largely as the result of excessive food consumption (gluttony) and lack of exercise (sloth), reflecting poor decision-making and lack of self-control. This ill-informed victim-blaming stance induces discrimination and stigmatization, and still exists, even within some healthcare sectors.

Body fat, at a degree that is excessive for an individual's age, sex, and race, impairs every aspect of health—physical, mental, social, and genomic. Obesity has become an exciting field of biomedical research, with more complete understanding of its biological basis. Both biological and environmental conditions are necessary to perturb the complex regulation of energy balance: in order to avoid weight gain and/or to lose weight, an individual must negotiate a powerful obesogenic environment. Appreciating the underlying neuroendocrine biology and genetics means that no person should ever be blamed for causing their disease. Healthcare should offer sympathy and evidence-based support.

Body fat can be measured in several ways (Box 17.1), but the most useful clinical measure to detect and measure changes in body fat is weight change. For epidemiology, obesity is still usually defined in terms of body mass index (BMI), weight in kilograms divided by height in metres squared (kg/m$^2$), correlating with adiposity, at least at higher levels. At lower levels, and in younger people, BMI is considerably influenced by variations in muscle mass. The World Health Organization (WHO) has established cut-offs, mainly for categorizing populations or groups of adults, with a BMI of 25–30 kg/m$^2$ categorized as 'overweight', and BMI >30 kg/m$^2$ as 'obese' (Table 17.1). For children, whose body compositions are very different from adults and change with age, special charts assess BMI adjusted for age (Chapter 35).

The conventional WHO BMI cut-offs were based mainly on American 1960s life assurance data that showed elevated cardiovascular risks with BMI above 25, and greater risks above BMI 30. Those data may not be truly representative of the whole population, and improved medical care has reduced cardiovascular mortality. More recent studies confirmed the association between BMI and total mortality, but with lower risk persisting close to BMI 30. However, emphasis has shifted towards other health outcomes mediated by obesity, importantly type 2 diabetes and hypertension, for which the original WHO cut-offs remain appropriate. For people of Asian origin, who have more body fat and less muscle at any given BMI, secondary health risks emerge at lower levels of BMI, so the accepted cut-offs are BMI above 23 for 'overweight', and above 27 for 'obesity' (Table 17.1). People of Polynesian descent have more lean body mass than Europeans for a given BMI, but a strong predisposition to diabetes and other metabolic complications of obesity, so European BMI ranges are probably appropriate.

The site of fat deposition is important to identify individuals at high risk of metabolic diseases. Specifically, excess abdominal (visceral) adipose tissue and intrahepatic fat are associated with high risks of diabetes, hypertension, dyslipidaemia, cardiovascular diseases, and premature mortality ('metabolic syndrome': see section 17.8 and Chapter 21). To quantify whole body, intra-abdominal, and intra-organ fat accurately, the gold-standard method, magnetic resonance imaging (MRI), is necessary, but prohibitively expensive. Dual-energy X-ray absorptiometry (DEXA) remains useful because radiation dose is very small, and it can quantify changes in body fat over time, as well as estimate abdominal

### Box 17.1 Techniques for measuring body composition, to identify and characterize obesity

Routine clinical practice and field methods:

- Body mass index (BMI routine in clinical and epidemiological use)
- Waist circumference (correlates slightly better than BMI with body fat and metabolic outcomes)
- Body fat prediction equation from simple anthropometry (better estimation of body fat and prediction of chronic diseases than simple BMI or waist circumference)

Specialist methods:

- Skinfold thicknesses (cheap but complicated method, no better than waist circumference)
- Bioelectrical impedance (expensive 'black-box' method: a crude estimate of body water, not better than waist circumference to estimate body fat)
- Air displacement (whole body plethysmography; very expensive and difficult to standardize)
- Dual-energy X-ray absorptiometry (DEXA, very reproducible so good for detecting change in body fat. Can make crude estimates of abdominal and limb components)

Reference methods (to validate other methods):

- Hydrodensitometry (underwater weighing: the original 2-compartment gold standard)
- Deuterium dilution (stable isotope method to measure non-fat body water space)
- Whole body computerized tomography (CT: cheaper than MRI but involves radiation exposure. Can quantify total and intra-abdominal fat)
- Whole body magnetic resonance imaging (MRI: the current gold standard. Can quantify total and intra-abdominal fat, and intra-organ fat by spectroscopy)

**Table 17.1 Epidemiological definitions of obesity according to the WHO (2000) for European populations and WHO (Western Pacific Region) IOTF (2000) for Asian populations**

| Classification | BMI (kg/m²) | | Risk of comorbidities |
|---|---|---|---|
| | Caucasian | Asian | |
| Underweight | <18.5 | <18.5 | Low (but risk of other clinical problems increased) |
| Normal range | 18.5–24.9 | 18.5–22.9 | Average |
| Overweight | ≥25.0 | ≥23.0 | |
| Pre-obese | 25–29.9 | 23.0–24.9 | Mildly increased |
| Obese | ≥30.0 | ≥25.0 | |
| Class I | 30.0–34.9 | 25.0–29.9 | Moderate |
| Class II | 35.0–39.9 | ≥30.0 | High |
| Class III | ≥40.0 | | Very high |

fat. The simple measure, for clinical practice and epidemiology, which gives some indication of intra-abdominal fat, is waist circumference. Measurements >102 cm in men and >88 cm in women reflect increased intra-abdominal fat mass and indicate a greatly increased risk of metabolic disease. Waist circumference >94 cm in men and >80 cm in women indicate increased risk. People of Asian extraction have more abdominal fat for a given BMI, so waist circumferences >90 cm in men and >80 cm in women signify high cardiometabolic risk (Table 17.2).

Waist circumference is measured in a horizontal plane, midway between the iliac crest and the lowest rib margin. Clothing should be removed to ensure correct positioning of the tape-measure, and the subject should be standing, at a normal minimal respiration.

While BMI and waist circumference provide reasonably reliable field methods to identify overweight and obesity, and more complicated indices such as waist/height may have marginally greater correlations with body fat in some situations, these indices do not themselves have any biological meaning. They are used as proxy measures for the biologically important variables, whole body fat mass, or fat-free mass as an indication of muscle mass. It is more appropriate and makes better use of the anthropometric measurements being collected routinely in surveys to use them in equations to estimate whole body fat mass and skeletal muscle mass separately (Box 17.2). These estimates for body fat used underwater weighing as the reference method, validated in a large separate population. Equations developed to estimate *whole body adipose tissue*, and separately *muscle mass*, again validated in separate studies, correlate highly with MRI as the reference measurement. The estimates are slightly better for men than for women. Applying these equations to data

**Table 17.2 Waist circumference criteria for action and prevention, including cut-offs for Europeans and Asians**

| | Action Level 1 | Action Level 2 |
|---|---|---|
| | **Health risks rising**<br>**Take personal action** | **Health risks high**<br>**Seek professional help** |
| **European ancestry** | Men 94–102 cm<br>Women 80–88 cm | Men >102 cm<br>Women >88 cm |
| **Asian ancestry** | Men 78–90 cm<br>Women 72–80 cm | Men >90 cm<br>Women >80 cm |

**Box 17.2 Estimation of whole body fat and of muscle mass, using anthropometric measurements collected routinely in national surveys**

Total body fat (%)
Men = 0.567 waist (cm) + 0.101 age (years) − 31.8
Women = 0.439 waist (cm) + 0.221 age (years) − 9.4

Whole body adipose tissue mass (kg) (multiply by 0.8 to estimate whole body fat mass)
Men = 0.198 weight (kg) + 0.478 waist (cm) − 0.147 height (cm) − 12.8
Women = 0.789 weight (kg) + 0.0786 age (years) − 0.342 height (cm) + 24.5

Whole body skeletal muscle mass (kg)
Men = 39.5 + 0.665 body weight (kg) − 0.185 waist circumference (cm) − 0.418 hip circumference (cm) − 0.08 age (y)
Women = 2.89 + 0.255 body weight (kg) − 0.175 hip circumference (cm) − 0.038 age (y) + 0.118 height (cm)

See references: Lean et al. (1996), Al-Gindan et al. (2014; 2015)

collected in large population surveys has shown how, for example, risks of type 2 diabetes, or of hypertension, rise with greater body fat, but fall with greater muscle mass. The simplicity of these equations, from measurements already usually made in surveys, questions the continuing practice of reporting national surveys with BMI alone.

## 17.2 Obesity prevalence: patterns over time and across geography

Obesity emerges as a problem in developing countries as soon as populations cease subsistence farming and adopt cash-based economies and 'Western' lifestyles. In the developed world, prevalence, particularly of severe obesity (e.g. BMI >40), is greater amongst lower socioeconomic groups than amongst the more affluent and better educated. In both environments, obesity prevalence has been increasing steadily over the past few decades, and at almost every level of income or education. Already, the prevalence of BMI >30 in adults has reached 30% in the United States, and it is close to this figure in much of Europe and Australasia. If this trend continues, by 2025, global prevalence of BMI >30 will reach 18% in men and surpass 21% in women (NCF Risk Factor Collaboration (NCD-RisC); *Lancet*, 2017). Examples of the increasing prevalence data for some countries are illustrated in Fig. 17.1. Global prevalence data are continuously updated by World Obesity Federation.

Because weight gain progresses with age, and diseases mediated by obesity depend on duration at high adiposity, the trend for younger onset of weight gain has increased the attributable clinical burdens. Prevalences of BMI >30 approached 10% in developed countries 30 years ago; now a similar proportion have BMI levels >35. Extreme prevalences of BMI >30, up to 50% or more of all adults, are found in some island states, and in Middle Eastern countries, where unifying features include high reliance on imported manufactured foods, and relative wealth without need or facilities for physical activity. The rise in prevalence of BMI >30 is of course preceded by a rise in the BMI 25–30 category, when health risks begin to accumulate. Among the American, European, Middle Eastern, and Australasian populations, where BMI >30 is now common, the total prevalence of BMI >25 is about 70%.

The radical changes in body composition over just one or two generations introduces a new problem in the efforts to define the optimal BMI range. In the UK, by age 65–70, only 10–15% of people still have a 'normal' BMI, below 25. With so few remaining with BMI in the conventional 'normal' range (18.5–25), a greater proportion have reached this range through

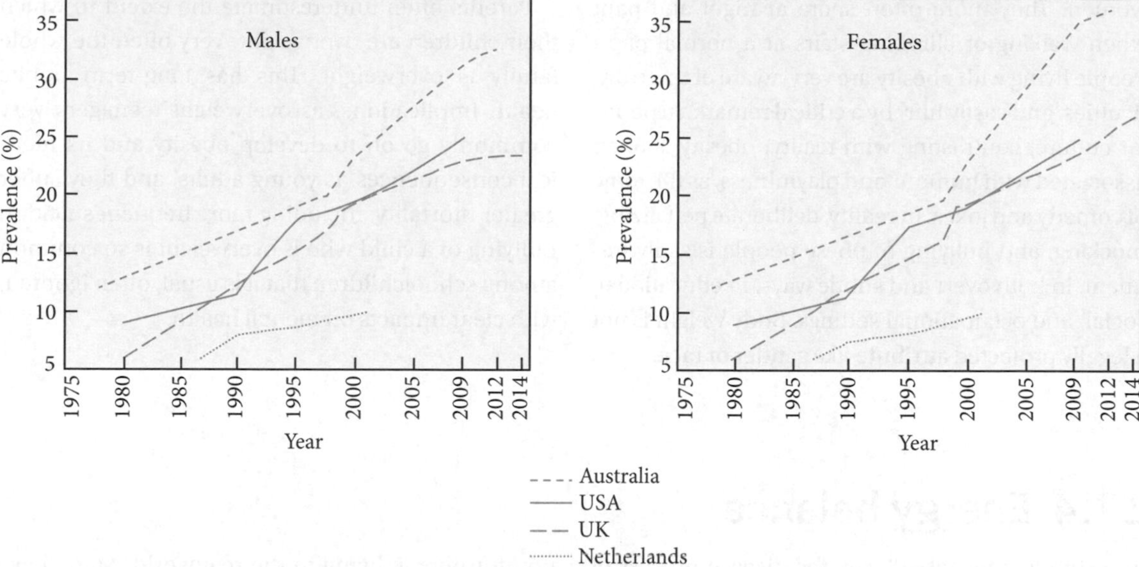


Fig. 17.1 Changing obesity prevalence rates in four countries.
*Adapted from:* http://www.worldobesity.org/ (accessed 20 October 2016).

weight loss associated with illness, and have reduced life expectancy. Thus, while it may still be *optimal* to maintain a BMI 20–22 or BMI <25 through a well-balanced diet and physical activity, this is no longer reflected by the overall health and mortality data of everyone in this category.

## 17.3 Perceptions: how it is to live with obesity

Perceptions of body size differ from country to country, and influence public health and clinical management of obesity. In most Western countries, those who are obese may be perceived poorly by the community and health professionals. It is generally accepted that negative attitudes and discrimination by others at least aggravate the problem with low self-esteem, resulting in depression and 'motivational collapse' with prolonged failure to control a weight problem. Treatment failures are common, often through underestimation of the task within complicated lives, and unrealistic expectations of the weight loss that can be achieved. People lose less than they deem a success, and with little to show for a big effort, the simplicity of returning to their previously 'normal' eating habits overcomes the domestic and social disruption necessary to adopt a new eating habit and retain a lower weight.

Women seeking help greatly outnumber men, although the health consequences of the same degree of obesity are more profound for men. As a very broad generalization, women more often try to lose weight for appearance and for rather ill-defined 'health' improvement. Men commonly only present when persuaded to do so, and often as a result of developing one or more diseases secondary to established obesity. In some cultures, overweight and obesity are still regarded as desirable attributes.

Adults living with obesity suffer daily discrimination. Seats in buses and cinemas do not fit them, chairs and beds (even in hospital) may be too small, and shops do not stock their sizes in attractive lines. Negative, mocking, and abusive comments are common. People with obesity have high metabolic rates, so tend to get hot and to sweat more than others, and stress incontinence is common among women. They more often snore at night and pant when walking or climbing stairs at a normal pace. People living with obesity are very aware of these difficulties, and easily hurt by a critical remark. In popular culture, contrasting with reality, obesity is often associated with humour and playfulness, as the topic of comedy and jokes. In reality, deliberate penalizing, mocking, and bullying of obese people is sadly frequent, in both overt and subtle ways, in educational, social, and occupational settings. Body weight is not a legally protected attribute like gender or race.

Parents often underestimate the extent to which their children are overweight. Very often the whole family is overweight. This has long-term public health implications, as overweight teenagers very commonly go on to develop obesity and its medical consequences as young adults, and they suffer greater mortality, including more frequent suicides. Bullying of a child who is overweight is so common among schoolchildren that it is usual, often ignored, with clear impacts on mental health.

## 17.4 Energy balance

At a simple, inevitable level, the disease-process of obesity entails an accumulation of excess body fat, which represents storage of excess energy (measured as calories or kilojoules). To become overweight, calories eaten have exceeded calories used over a long period.

Each of us has a metabolic rate, expressed as calories or kilojoules used per 24 hours. If this is matched precisely by energy intake, the result is a steady weight. The metabolic rate varies a little between individuals of the same weight but rises with weight gain. Thus, if a person is to gain weight, they must consume enough food for the extra energy stored as fat, plus a daily extra amount in order to match their increased metabolic rate, to prevent weight from falling.

Conversely, if a person is to lose weight, they must under-eat by the necessary amount to lose weight at 7000 kcal per kilo, and additionally adjust their daily energy intake downwards to match the fall in metabolic rate. It follows that if a person under-eats to a level below their metabolic rate, they will inevitably be using up stored energy in adipose tissue to make up the difference. So weight loss is inevitable, until energy intake once again equals expenditure. These principles reflect the laws of Newtonian physics. They are absolutely guaranteed, but do not of course capture the complexity of influences on energy expenditure, or on appetite, eating pattern, and preferences, which influence energy consumption for people living in the real world. Many have eating habits that are, or appear to be, within the ranges eaten by people of normal weight. However, measuring the components of energy balance is fraught with difficulty and errors. Firstly, there are major difficulties in assessing people's food habits outside controlled experimental conditions (see Chapter 38). Secondly, people with weight problems tend to underestimate and/or under-report their true food consumptions, often consistently and to a large degree. Thirdly, most people (overweight or not) tend to exhibit 'optimistic bias' in reporting their habitual physical activity. People who are significantly overweight find regular physical activity difficult, tiring, and unpleasant, and overestimation is usually revealed when objective methods are used. Each kilo of weight gain entails net storage of about 7000 kcal. Thus, someone who is 100 kg, who has stored about an extra 30–40 kg, has accumulated about 200,000–300,000 kcal from food intake above requirement for a steady weight. Because we eat intermittently, and during our evolution we had to have a way to store energy in order to withstand famines and winters, our physiology has evolved in order to consume extra energy when it is available, and to avoid unnecessary physical activity. There are wide differences between individuals.

Despite energy balance oscillating from meal to meal, and fluctuating day to day, under 'normal' conditions in non-pregnant adults there are no persistent or cumulative changes in body stores. It is amazing that anybody, eating 40–50 tonnes of food over a lifespan, can maintain a reasonably stable body weight: long-term weight regulation is extremely well balanced within the neuroendocrine systems involved. However, when energy balance is consistently positive, with calories consumed exceeding calories expended even to a very small degree for longer periods, body weight will increase. If a 1% failure of the physiological regulation of energy balance results in intake exceeding expenditure by 20 kcal every day (5 g of sugar or 2 g of cooking oil) above the requirement for energy balance, energy stores in adipose tissue have to enlarge to accommodate 365 × 20 = 7300 kcal per year. That will result in weight gain of about 1 kilo/year.

Energy expenditure includes basal (resting) metabolic rate, dietary thermogenesis (meal-induced heat production), and a variable amount for physical activity, which is also influenced by numerous domestic and social factors (Chapter 5). The physiological regulation of appetite involves integrated functions of over 30 endocrine and paracrine hormones, centrally (brain) and peripherally (adipose tissue and gut). In humans, while appetite is under the same regulatory controls as experimental animals, eating may also be initiated, and sustained, by a variety of environmental factors, some specific to humans. Psychological and social influences, and learned beliefs and attitudes towards foods and eating, interact with olfactory and visual stimuli from the food itself. Successful food marketing, especially for 'added-value' foods (whose actual value is commonly less than its packaging), has found that sales can be increased by use of colour, music, role models, and of course by adding an addictive substance like caffeine. Eating for pleasure or reward can assume features very similar to an addiction.

The 'satiety cascade' describes a continually repeated cycle of the interactions between food intake and hunger signals (Fig. 17.2). Following *hunger*, food intake promotes *satiation*, ultimately bringing eating to an end. This is followed by a period of *satiety*, a warm feeling of fullness that theoretically keeps us satisfied between meals, until the next hunger feeling develops. The appetite-stimulating pathway is regulated in a tiny brain centre, in the ventromedial part of the arcuate nucleus of the hypothalamus. The hormone leptin, one of many released from adipose tissue, promotes satiety by suppressing neurones that express two neuropeptides that increase appetite: neuropeptide Y (NPY) and agouti-related protein (AGRP). When energy balance is negative, leptin secretion from adipose tissue decreases, so NPY and AGRP activity rises to stimulate appetite. During fasting, appetite is further

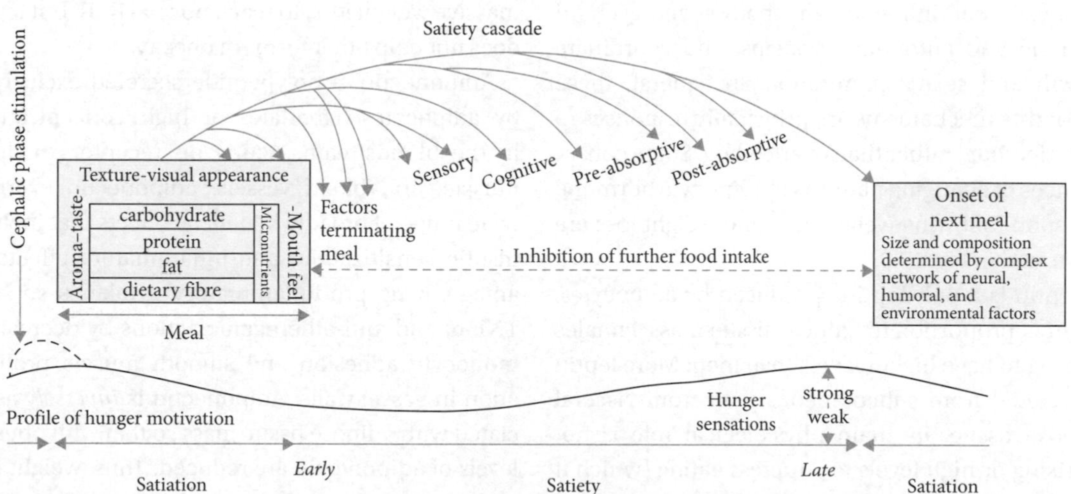


Fig. 17.2 Postprandial satiety cascade involving mechanical, social, hormonal, and neuronal regulatory functions.
*Source:* Blundell, J. Pharmacological approaches to appetite suppression. *Trends Pharmacol Sci* 1991; **12**, 147–57. With permission from Elsevier.

increased by declining concentrations of ghrelin, a hormone secreted by cells in the fed gastrointestinal tract that suppresses further eating. The central appetite-suppressing pathway is stimulated by leptin and insulin via expression of melanocortins and corticotrophins. Cholecystokinin (CCK) and peptide YY (PYY) are other gut-derived factors released after eating that also contribute to appetite suppression. Thus, leptin, which circulates in proportion to adipose tissue mass, may be regarded as a long-term signal to appetite control, whereas insulin, CCK, and PYY provide more acute responses to meal ingestion.

In theory, these complex biochemical interactions should ensure energy balance and weight stability. However, the evolutionary need to avoid or correct negative energy balance and weight loss was always more critical than any innate demand to avoid positive energy balance and weight gain. So, psycho-social factors, and the presentation and energy density of food (high fat, high sugar), coupled with sustained persuasive marketing, can override the relatively weak control mechanisms, and lead to the small positive energy balance needed for weight gain.

## 17.5 Hormonal and adipocyte factors

Adipose tissue was conventionally regarded as an inert reservoir for energy storage. However, it is now known that there are several adipose organs with rather different functions. The internal and perivascular adipose tissue, developmentally brown adipose tissue responsible for heat generation in infancy, is very distinct from subcutaneous adipose tissue, the principal site for energy storage. It has also emerged that adipose tissue is a very active endocrine organ, playing key roles in energy and metabolic homeostasis, via endocrine, paracrine, and autocrine functions, secreting hormones and cytokines ('adipokines') that influence the body's energy, and immune and autonomic systems, and coordinate growth and sexual maturation. In general, these endocrine mechanisms are principally responses to underfeeding, rather than overfeeding. Some consequences of changing patterns of adipocyte hormone secretion following weight gain and weight loss are illustrated in Fig. 17.3.

Leptin is a polypeptide produced by adipocytes, in direct proportion to adipose tissue mass, females tending to have higher levels than men. More leptin is secreted from subcutaneous than from visceral adipose tissue. Its main physiological role is not for rising or high levels to suppress eating (which it does, weakly), but for low levels to stimulate eating, a very potent effect. In adolescents, increasing leptin acts as a signal of adequate fat stores, to initiate pubertal development and permit fertility. Secretion of leptin is influenced by many factors, such as glucose metabolism (insulin), inflammation (tumour necrosis factor-α (TNF-α)), and steroid pathways (glucocorticoids and oestrogens), indicating leptin's complex interactions with several physiological systems. Leptin deficiency occurs in extremely rare cases, in consanguineous families with leptin gene mutation, and functional leptin deficiency occurs with mutations in the leptin receptor gene. Both result in hyperphagia, decreased energy expenditure, and very severe, disabling obesity presenting in early childhood. Treating leptin deficiency with leptin results in a dramatic reduction in energy intake and massive weight loss, to reach normal BMI, but leptin does not help other forms of obesity.

Adiponectin, a polypeptide secreted exclusively by adipocytes, circulates at high concentrations in the bloodstream. Acting on receptors in liver, muscle, and blood vessels, adiponectin exerts a wide range of broadly beneficial effects that include insulin sensitization, anti-inflammatory effects by antagonizing pro-inflammatory cytokines such as TNF-α, and anti-atherogenic actions by decreasing monocyte adhesion and smooth muscle proliferation in vessel walls. Adiponectin is *inversely* associated with adipose tissue mass so that with obesity, levels of adiponectin are reduced. Thus, weight loss and improved insulin sensitivity are associated with increasing levels of adiponectin. Conversely, inflammatory and insulin-resistant states and coronary heart disease are associated with decreased adiponectin levels.

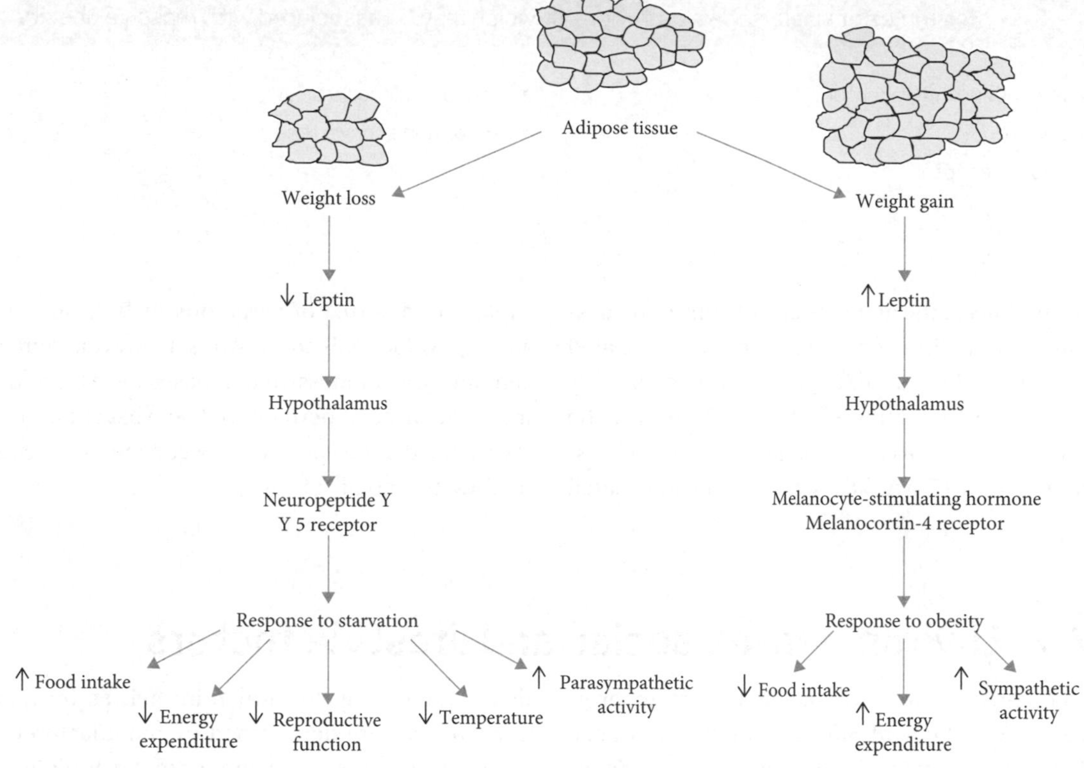


Fig. 17.3 Hormonal and other regulatory responses to alterations in adipose tissue mass.

# 17.6 Genetics of obesity

Although the rapid increase in obesity prevalence cannot be explained by genetic factors, genes and/or epigenetic modulations play important roles in determining variations in weight gain and body fat accumulation amongst individuals. It is a commonplace observation that overweight runs in whole families, and a variety of adoption, twin, and family studies all indicate that adiposity is highly heritable. The weights of adults adopted as children are related more to the weight of their biological parents than to the weight of their adopting family. Overfeeding studies among twins find a strong genetic component in the amount of weight gained by individuals with the same amount of overfeeding. However, the high figures (60–80%) published for 'heritability' of BMI include influences from shared environments as well as genes.

The exceedingly rare single-gene defects such as leptin deficiency prove that genetics can have profound effects. These monogenic causes of obesity are typically associated with massive obesity in early childhood, and usually involve mechanisms associated with appetite regulation, rather than metabolism or physical activity (Box 17.3). They are mainly in offspring of cousin marriages, where homozygosity for very rare mutations is possible. Obesity was well known to be common with Down's syndrome and Turner's syndrome, making genetic factors highly likely. More common pleiotropic syndromes, such as Prader–Willi syndrome (PWS), where obesity occurs together with intellectual disability, organ abnormalities, and dysmorphism, were also known to have a genetic basis. PWS commonly occurs with a deletion in chromosome 15, usually from the father, or by having two copies of the maternal chromosome 15, which are suppressed by epigenetic genomic imprinting. The molecular mechanisms by which this acts to produce severe obesity, and other clinical features, have not been elucidated, but the mechanism involves loss of some non-coding small nucleolar RNAs.

**Box 17.3 Examples of single genes, mutations of which may be associated with massive obesity**

- Melanocortin-4 receptor
- Leptin
- Leptin receptor
- Pro-opiomelanocortin
- Prohormone convertase 1

Estimations of the genetic basis of common obesity have varied. The gene most strongly associated with greater BMI, the *FTO* gene, can explain only 1–2% of variance. Adding effects on BMI from all the genes identified by a conventional genome-wide association study (GWAS) as 'significantly' associated explain under 10% of variations in BMI, an effect size equivalent only to 1–2 kilos. However, more recent analyses of massive databases have computed that influences of perhaps 10,000 weaker and rarer BMI-related gene variants may conspire to account for BMI heritability of 20–40%.

## 17.7 Environmental, social, and lifestyle factors

The potential for environmental factors to overpower genetically determined regulation has been well illustrated within ethnic groups who experience changes in lifestyles over a brief period. Among genetically similar Pima Indians, one group living on social support in reservations in the USA do little physical activity, consume energy-dense 'added-value' convenience foods and sugary drinks, and have exceptionally high rates of obesity and comorbidities. Their distant relatives, still following their traditional subsistence lifestyles in mountainous areas of Mexico, have low rates of obesity and associated diseases. Transition from traditional to more Western ways of life (increased food energy density, effective social marketing, and a sedentary lifestyle) explain the rapid increase in obesity rates and its younger onset in the global pandemic.

### 17.7.1 Food intake

Overweight or obese individuals tend to underestimate total energy consumption, sometimes by as much as 50%, or report what they regard as socially acceptable. Food waste has increased recently, but this relates largely to vegetables and fruit. Manufactured and packaged foods, often energy dense, are usually consumed in their entirety, and portion sizes are determined by manufacturers to optimize sales, not by consumers to satisfy biological need or appetite. Some studies have suggested that energy intakes have declined over the past 40 or 50 years, despite rising BMI levels. However, these studies used self-reported data, and often excluded foods eaten outside the home, over a period when fast foods and convenience foods have increased dramatically. These studies are likely to have seriously underestimated energy intake. Their conclusion that more overweight populations are eating less is implausible.

Foods, food groups, and nutrients are more easily measured than total energy intake. The WHO/FAO Expert Consultation on Diet, Nutrition, and the Prevention of Chronic Diseases (TR 916) identified a number of lifestyle-related factors that were considered to promote or protect against excessive weight gain (Table 17.3). Unsurprisingly, a high intake of energy-dense foods and heavy marketing of them were regarded as important promotive factors. Dietary fibre (non-starch polysaccharide) was considered to be a convincing protective factor and sugar-sweetened beverages and fruit juices were the only group of foods singled out as important contributing factors.

**Table 17.3 Summary of strength of evidence of factors that might promote or protect against weight gain and obesity**

| Evidence | Decreased risk | No relationship | Increased risk |
|---|---|---|---|
| Convincing | Regular physical activity<br>High dietary intake of non-starch polysaccharide | | Sedentary lifestyles<br>High intake of energy-dense, micronutrient-poor foods |
| Probable | Breastfeeding<br>Home and school environments that support healthy food choices for children | | Heavy marketing of energy-dense foods and fast-food outlets<br>High intake of sugar-sweetened soft drinks and fruit juices<br>Adverse socioeconomic conditions[a] |
| Possible | Low glycaemic index foods | Protein content of the diet | Large portions<br>High proportion of food prepared outside the home (developed countries)<br>'Rigid restraint/periodic disinhibition' eating patterns |

[a]Especially for women in developed countries.

*Adapted from:* WHO/FAO (2003) *Diet, nutrition and the prevention of chronic diseases.* Report of a Joint WHO/FAO Expert Consultation, Technical Report Series 916. Geneva: World Health Organization.

## 17.7.2 Physical inactivity

As well as the protective inverse relationship with physical activity, obesity is separately and positively associated with time spent totally inactive. At population level, urbanization, affluence, technology, and modernization of lifestyle have reduced the conventional components of physical activity—occupational, transport, and household. The remaining category, recreational physical activity, has increased, but mainly for those with already-higher exercise levels. Inactivity may result from changes in lifestyle, ageing, injury, or disease (such as arthritis, which restricts mobility, or respiratory diseases, which reduce exercise capacity). The prevalence of obesity in children can be related directly to hours of screen-time as a measureable proxy, their inactivity compounded by inappropriate food selection resulting from television marketing. Very old observational studies by Mayer, in humans and supported by animal research, showed that while appetite and food intake increase to match the need from greater levels of physical exertion, there is a paradoxical rise in appetite and food consumption with complete inactivity—a state only rarely and briefly met during evolution. Studies are underway where school classrooms and workplaces are built around standing and locomotion instead of sedentary behaviour.

An interesting, often overlooked component of energy expenditure is non-exercise activity thermogenesis (NEAT)—the small but still measurable energy expenditure from non-planned movements, such as shifting weight and moving arms, head, and trunk when standing, talking, and moving around a room. Fidgeting and other physical expressions of anxiety can increase energy expenditure. Even laughter has been found to increase energy expenditure in a measurable way.

## 17.7.3 Other factors

Cessation of smoking commonly causes a 1–4 kg weight gain. Treatment with some medications causes weight gain, particularly corticosteroids and some antidiabetic, antiepileptic, and antipsychotic medications (Box 17.4).

Hormonal alterations are often blamed as causes of obesity, but seldom cause major weight gain (Box 17.5). Cushing's syndrome, including high-dose

**Box 17.4** Medications associated with increased appetite and weight gain

- Diabetes management:
  - o insulin
  - o sulphonylureas
  - o thiazolidinediones.
- Steroids
- Antipsychotic medications (including newer antipsychotics)
- Antidepressants, especially tricyclics
- Lithium
- Antiepileptics: valproate, gabapentin
- Beta-blockers
- Antihistamines

**Box 17.5** Hormonal alterations associated with and often blamed as causes of obesity, but seldom causes of major weight gain

- Hypothyroidism
- Acromegaly
- Cushing's syndrome
- Polycystic ovarian syndrome
- Hyperprolactinaemia
- Insulin resistance

corticosteroid use, does increase appetite, body weight, and fat content, and also promotes a central fat distribution and reduces muscle mass. Hypothyroidism does not provoke much weight gain but makes weight loss difficult. Hypothalamic injury and adjacent pituitary tumours are rare causes of major weight gain. Polycystic ovarian syndrome is associated with obesity, but more as a consequence of obesity than a cause. Although it cannot be disputed that the laws of thermodynamics apply in body weight regulation, some additional potential factors have been proposed, summarized in Box 17.6, including the roles of breastfeeding, the microbiome, and reduced sleep. While evidence for causality is weak, and these factors can make only small contributions to positive energy balance, over life-long periods they may become relevant contributors.

**Box 17.6** Environmental factors related to greater prevalence of obesity and proposed as contributory

| |
|---|
| More thermoneutral homes |
| Increasing age of first pregnancy |
| Endocrine disrupting pollutants |
| Viruses |
| Intestinal bacteria (microbiome) |
| Living at lower altitude |
| Reduced non-exercise activity thermogenesis (NEAT) |
| Cannabis use |
| Reduced sleep |
| Intrauterine epigenetic effects |
| Non-breastfeeding |

## 17.8 Clinical consequences of overweight and obesity

The multiple adverse health consequences of obesity are listed in Box 17.7. Some are considered 'minor', but many major disabling and life-threatening diseases include the disease-process of obesity in their aetiology. The clinical consequences fall into several broad categories. For some there is a ready explanation. Fatty liver and type 2 diabetes depend almost entirely on the metabolic and low-grade

**Box 17.7 Medical consequences of obesity**

***Metabolic and low-grade inflammation-related:***

Ectopic fat accumulation, including non-alcoholic fatty liver disease (NAFLD), non-alcoholic steatohepatitis (NASH), and cirrhosis

Insulin resistance, impaired glucose tolerance, type 2 diabetes, metabolic syndrome

Dyslipidaemia:

- increased VLDL, triglyceride, increased LDL cholesterol
- reduced HDL cholesterol
- increased apo B lipoprotein (small dense molecules)

Gallstones (can be aggravated by weight loss)

Polycystic ovarian syndrome/infertility

Erectile dysfunction and impaired sperm function

Gout

***Cardiovascular:***

Hypertension

Coronary heart disease

Heart failure

Peripheral vascular disease, varicose veins

Lymphoedema

Asthma

Vascular and other dementias

***Cancers (include endocrine and pro-inflammatory mechanisms)***

Breast

Endometrium

Prostate

Kidney

Pancreas

Colon

Skin—melanoma

Brain

***Mechanical:***

Osteoarthritis

Back pain

Spinal stenosis

Plantar fasciitis

Obstructive sleep apnoea

Hernias

Urinary stress incontinence in women

Increased falls and road traffic accident risks

***Psycho-social:***

Stigmatization and discrimination

Low self-esteem, depression

***Other:***

Aggravated viral and bacterial infections (but protection against pulmonary tuberculosis)

Increased anaesthetic risk

inflammatory response associated with excess body fat accumulation. For others the mechanism is unknown. Some include contributions from more than one obesity-dependent mechanism. The risks of developing an obesity-mediated disease depend on age, duration lived overweight, and weight or BMI. Commonly, several obesity-mediated diseases coexist.

People of European origin with a BMI >30 kg/m$^2$ have total mortality rates that are approximately double those with a BMI 18.5–25, from a variety of causes, but dominated by accelerated cardiovascular disease and type 2 diabetes. Early weight reduction relieves most symptoms and risks of obesity-mediated conditions. Weight loss of 10–15 kg or more can reverse the underlying disease process of ectopic fat accumulation and type 2 diabetes. Evidence that life can actually be prolonged by intentional sustained weight loss of >10 kg has accumulated from various studies, in which weight loss maintenance has been achieved by dietary measures, bariatric surgery, and some anti-obesity drugs. The improved survival mainly reflects fewer cardiovascular events, but obesity-related cancers may also be reduced.

## 17.8.1 Cardiometabolic risks and metabolic syndrome

The elevated cardiovascular risks with overweight or obesity mainly result from a combination of linked metabolic abnormalities and the low-grade pro-inflammatory state associated with ectopic fat accumulation in intra-abdominal sites and a fatty liver. This can develop in some susceptible individuals even with BMI within the conventionally 'normal' range. A large waist circumference is often present, as a useful signal of elevated risk. Additionally, heart failure, especially with preserved ejection fraction, cor pulmonale, and lymphoedema characteristically occur with severe obesity, and there is evidence linking atrial fibrillation and aortic stenosis to obesity.

Metabolic syndrome (Chapter 21) is a feared phenotype of the disease-process of obesity, doubling mortality rates, that affects some 40% of all adults by retirement age, and more in people of Asian origin. It develops with a variable degree of weight gain in genetically susceptible individuals, manifesting as a progression through ectopic intra-abdominal fat accumulation (with large waist), hepatic steatosis, insulin resistance, and clinically important metabolic disorders including the sequence of impaired glucose tolerance (IGT and prediabetes) to type 2 diabetes and its complications, hypertension, and dyslipidaemia. The criteria of individual components required for diagnosing metabolic syndrome are set at levels below their individual treatment thresholds, but each can progress to reach treatment thresholds for secondary diseases. While clearly familial, the genetic or epigenetic basis for metabolic syndrome is incompletely understood.

Overweight and obesity are potent causes of hypertension. About half of all hypertensions in the US can be attributed to having a BMI >25, and dietary weight loss is a more potent treatment for hypertension than any known medication.

Without excess alcohol consumption, elevated liver transaminases in overweight people usually mark non-alcoholic fatty liver disease (NAFLD) as part of the metabolic syndrome, later leading to non-alcoholic steatohepatitis (NASH) and cirrhosis. There is a moderately raised level of very-low-density lipoproteins (VLDL), raised triglycerides, and low high-density lipoprotein (HDL) cholesterol. The dyslipidaemia profile of metabolic syndrome is commonly accompanied by high insulin levels and insulin resistance, and sooner or later with hyperglycaemia, which can progress.

### 17.8.2 Non-cardiac metabolic complications

The prevalence of gallstones increases with increasing age and weight, and clinical problems can be provoked by very low-fat dieting.

Many women undergoing *in vitro* fertilization are overweight or obese: polycystic ovarian syndrome contributes. The success of such programmes increases substantially with weight loss, as does the likelihood of a natural pregnancy.

### 17.8.3 Obesity-associated cancers

Several major cancers are strongly associated with obesity (Chapter 20). It is possible to understand this association with hormone-dependent tumours (breast, endometrium, and prostate), as obesity is associated with low circulating levels of sex hormone binding globulin (SHBG), and so elevated free hormone concentrations. Adipose tissue contains the enzyme aromatase, which converts oestrogens into androgens, also accounting for the infertility and hirsutism of obesity with polycystic ovarian (PCO) disease. The strong association between obesity and colon cancer and the clear though less striking relationship with other cancers are less well understood, but the chronic low-grade inflammatory milieu of obesity, with elevated secretion of pro-inflammatory cytokines from adipose tissue, may contribute.

### 17.8.4 Mechanical consequences

Osteoarthritis of the weight-bearing joints in obesity is understandable and common. Arthritis of non-weight-bearing joints (e.g. in the hands) is also more frequent. Obstructive sleep apnoea is common among obese individuals, especially in men. They often report symptoms such as snoring, stopping breathing during sleep (apnoea), morning headache, daytime sleepiness, and difficulty in mental concentration. They can be successfully treated with weight loss and/or continuous positive airways pressure (CPAP) administered by a nasal mask during sleep.

### 17.8.5 Psycho-social consequences

People who are overweight often have low self-esteem, especially if they were overweight as children. Obese children are often teased at school and may feel socially isolated. Those who have made many unsuccessful attempts to lose weight may carry the added burden of personal failure. Obesity or its medical consequences impede enjoyable activities, resulting in impairment of quality of life. In some societies, there is a poor perception of obesity by the community at large and obese individuals may experience discrimination in various forms, including reduced employment opportunities.

## 17.9 Management of overweight and obesity

Evidence-based clinical guidelines for obesity treatment, or more properly 'weight management', have only existed for the last 25 years. The first was the Scottish SIGN Guideline in 1996, followed shortly by a very similar guideline from the US Institute of Medicine, and subsequently by largely derivative guidelines in other countries, including the NICE guidance in England. The principles of that first SIGN guideline, retained in the others, including the revised SIGN guideline in 2010, introduced a four-component model for weight management, recognizing that the target is a life-long disease-process that becomes manifest at different ages and stages for different individuals in different settings:

1. Primary prevention of excess weight gain
2. Induction of weight loss for those already overweight
3. Weight loss maintenance: prevention of regain
4. Optimization of elevated risk factors, irrespective of success in weight loss or maintenance

Given the escalating rates of obesity worldwide, there has been a great deal of discussion regarding possible public health measures aimed at the primary prevention of excess weight gain. These have included taxation of sugar-sweetened beverages, reduction or elimination of advertising of energy-dense snack and convenience foods to children, and programmes aimed at ensuring a healthy food environment in schools. While some benefits of such measures have been observed, it seems likely that in order for population-based approaches to have a major impact, wide-ranging measures aimed at reducing the obesogenic environment will be necessary. Changes in the food systems of some countries may be required in order to ensure that appropriate food choices are available and affordable.

As major changes in the food environment seem unlikely to occur in the foreseeable future, every person with the genetic-physiological predisposition to weight gain and obesity may potentially require treatment for all four components of weight management. Each may involve different dietary and lifestyle strategies. Different adjunctive measures, such as psychological support for individuals with particular issues, evidence-based methods to support behaviour change, and evidence-based medications, can also be considered to help individuals at every stage of weight management.

Conventionally, the clinical measures that can be offered for weight management are presented as an escalating hierarchy of care: (1) diet and lifestyle, (2) pharmaceutical intervention, (3) bariatric surgery. This is a familiar categorization, used for managing many diseases, when each step up in intensity, and risk of unwanted effects from treatment, accompanied by greater costs and resource demands, is determined by degree of success and then a balance of health risks from escalated care. In most medical care, the steps are usually viewed as separate and independent. For example, successful coronary artery bypass surgery abolishes the symptoms of angina without the need to continue medication. However, for the treatment of obesity, the three modes of treatment are not independent. If diet and lifestyle advice and support is unsuccessful, a simple switch to medication is usually unsuccessful: there are some effective modern drugs, but no drug is very effective without good dietary advice and support for dietary change. Medication may make it easier for people to adhere to the necessary dietary changes, but weight loss is still determined entirely by the degree and duration of negative energy balance, at approximately 1 kg loss per 7000 kcal net deficit.

Similarly, weight loss after bariatric surgery also depends on the net energy deficit from negative energy balance. So, good dietary advice and support remains vital, to optimize weight loss and prevent regain: late weight regain after bariatric surgery is sadly not rare, and may need addition of medication. Individual management of obesity demands an understanding of the complicated physiology and socio-environmental factors that an individual has to deal with. It can be daunting and mysterious, aggravated by large amounts of readily available misinformation on the internet, in popular books, and from ill-informed or unsympathetic advisors.

The management of a life-long disease-process such as obesity inevitably has several slightly separate phases and its variable time course means that different actions or interventions are appropriate at different life-stages. In every individual case, management may need to address both the biological factors driving appetite and energy balance, and also the plethora of varying environmental factors that tend to facilitate weight gain and its maintenance. A complicating factor is that compensatory physiological changes always tend to preserve current body weight against weight loss. This was a valuable survival mechanism, retained and enhanced for millions of years of our evolution in a world in which food was often scarce. Metabolic rate varies between individuals of the same weight, with variance of about 8–10%, so the strategy to achieve weight loss by dietary restriction will result in slightly different weight changes between individuals. Metabolic rate varies in proportion to body weight, but it is much more affected by, and suppressed by, acute negative energy balance. Normal functions of a well-fed person in energy balance shut down temporarily over days when energy intake is below expenditure. They feel cold, and have cold extremities, bowel function reduces, and voluntary physical activity tends to fall: they feel tired. All this saves stored energy and slows weight loss.

Humans have different physiologies, and as a survival mechanism, most people tend to have appetites that, if unconstrained, tend to push their energy consumptions a little higher than needed for energy balance. The same appetite may not generate unwanted weight gain in young people who are physically active, but minor reductions in day-to-day activities, through social changes in later years (e.g. marriage or cohabiting, new demands from occupation or children, injury), can allow weight gain to occur. It is generally difficult to see the effect until there is substantial weight gain, and it is always difficult to make conscious changes to activities, beliefs, attitudes, and behaviours that are all part of that individual's normality.

Very large numbers of magazines, books, online programmes, coaches, and personal trainers offer diets and physical activity programmes that are claimed to be effective for weight management—or more commonly just for weight loss. As a general rule, physical activity or exercise programmes have rather little effect on the body weight of people with obesity. They tend not to enjoy greater physical activity, partly for physiological reasons (people with metabolic syndrome have more glycolytic type 2 muscle fibres and lack the oxidative type 1 muscle fibres needed for endurance-type activity to control body weight). Increasing physical activity can also do damage to weight-bearing joints of people who are overweight, so exercise plans are best left until after weight loss is achieved by dietary changes. They are principally of value in maintaining rather than achieving weight loss. The emphasis for effective weight management, or for preventing unwanted weight gain, must therefore be on diet, and limiting energy intake. In clinical trials amongst individuals who are able to increase physical activity there is a small added effect on weight, and a more valuable retention of muscle mass during weight loss, but for most people with obesity, trying to increase physical activity at the same time can tend to distract from the all-important diet control.

### 17.9.1 Dietary modification

Weight control always requires control of energy balance, principally by dietary modification and a lesser contribution from physical activity. A plethora of named diets and books are available. Many are completely implausible, but most would incur negative energy balance and lead to weight loss if followed exactly as designed. The most helpful approaches are those that involve the support of appropriately qualified health professionals and take into account two important considerations:

- Modest weight loss (~5 kg) improves multiple weight-related health risks for people with overweight or obesity, and can be achieved by a wide range of conventional dietary methods that achieve negative energy balance. Evidence suggests that for many who are appreciably overweight or obese, this often does not generate sufficient improvement in quality of life to make it sustainable by the individual. Many revert to their life-long 'normal' eating habit, so their weight regains.
- A more substantial weight loss of 10–20 kg (or more) is usually sufficient to reverse some major medical consequences of obesity—i.e. remove the diagnosis—such as type 2 diabetes, sleep apnoea, and infertility. It also usually brings a big improvement in quality of life—people look and feel better—so longer-term weight loss maintenance is more worthwhile and more frequent.

The approach most likely to be effective is to start with a weight loss induction phase with substantial reduction in calorie intake, aiming to lose 10–15 kg or more, over 10–12 weeks. This is best achieved using an initial minimal-choice diet, which can be modified, meal by meal, to achieve weight-loss maintenance in the long term. A daily intake of about 850 kcal has proved acceptable for most people, without provoking hunger. With severe obesity (e.g. weight above 120–130 kg, BMI above 40), 1200 kcal/day is a sensible starting point, reducing to 850 kcal as tolerated. After the weight loss phase, a clear plan is needed for long-term weight maintenance at a lower level of body weight. That demands an energy intake less than the previous level when overweight, so never returning to the previous life-long 'normal' eating habits. People generally find maintaining weight loss much harder than inducing the weight loss: most will regain some, and some will regain all. Guidance and support from friends, family, and primary care providers is most needed for this transition period, to avoid slipping back to the previously 'normal' eating habits. Two approaches may be used for those requiring appreciable weight loss: a well designed self-help weight management programme with some health professional support, or a professionally supported structured formula diet programme.

The very simple Scottish 'No Doubts Diet' (Fig. 17.4) is an example of the former. It provides three basic, popular, small food-based meals per day (porridge, and two types of lentil soup, plus fruit with each meal) with vitamins and minerals needed during weight loss. Weight losses of 10–20 kg are quite frequent. Patients who are severely obese would be advised to increase the portion sizes, perhaps by 30–50% at the start, to give a daily energy consumption of 1000–1200 kcal. An alternative approach involves the use of one of the many commercially available total meal replacements to achieve initial weight loss. To establish long-term weight loss maintenance, after weight loss, patients need to be guided to vary one meal at a time, to gradually increase variety and calorie intake. They often find this phase, with the reintroduction of choices and exposure to varieties of foods, the most challenging. It remains possible to return to the initial weight loss induction diet, such as No Doubts Diet, at any time in the future, in order to lose more weight or to correct regain.

Many formula diets are commercially available. The products themselves are all very similar in content, either as very low energy diets (VLED) at 400–600 kcal day or as low energy formula diets at 800–1000 kcal/day. All the current commercial diets are now required to be nutritionally complete for micronutrients under European law, so they are

| The No Doubts Diet |
|---|
| **Breakfast** |
| 40 g (half cup) porridge oats, made with water (1 cup).<br>1 portion fruit (e.g. small banana, chopped).<br>**Preparation:** add everything into a microwavable dish, and heat for 2 minutes or eat fruit separately. Eat with a splash of cold milk. |
| **Lunch** |
| Bowl of lentil soup, and 1 slice of wholemeal/wholegrain bread<br>**NB.** bread should be of average size, 100 calories or less, per slice<br>**Preparation:** see soup recipe in link below |
| **Dinner** |
| Bowl of lentil soup, and 1 slice of wholemeal/wholegrain bread<br>**NB.** bread should be of average size, 100 calories or less, per slice<br>**Preparation:** see soup recipes |
| **Evening Snack** |
| 1 portion fruit |
| **Fluids** |
| Aim for 3 litres (zero calorie) fluids per day. There is an allowance of 200 mls semi-skimmed milk for teas/coffees, or as a drink. If you prefer not to use milk, have a 150–175 g pot of low fat yoghurt instead.<br>The remaining 2.8 litres as water, teas, herbal infusions, coffee. Try to avoid sweetened 'diet' drinks.<br>**Tip:** Flavour water (including sparkling water) with lemon, cucumber, or orange and keep a jug in the fridge to keep nice and cold. |
| **Nutritional information** |
| Energy: 830 calories, Carbohydrate: 154 g, Fat: 11 g Saturated fat: 3.5 g, Protein: 39 g, Fibre: 19 g (70% Carbohydrate, 11% Fat, 19% Protein).<br>Vitamin B1: 1.4 mg, Vitamin B6: 1.5 mg, Vitamin C: 94 mg, Calcium: 538 mg, Iron: 10 mg. |

**Fig. 17.4** The No Doubts Diet principles. This example of a food-based 850 kcal/day weight loss induction diet can be modified meal-by-meal to reach a more varied, individualized weight-maintenance diet.

*Source:* https://www.directclinicaltrial.org.uk/Documents/The%20Lean%20Team%20No%20Doubt%20Diet%20plan.pdfs

nutritionally safe and medical supervision is not usually necessary, except where medication doses may need to be reduced with weight loss (e.g. drugs for diabetes or hypertension). Weight losses are almost identical between 450 and 850 kcal/day diets, presumably because the VLEDs are less well tolerated than a more liberal diet. There is therefore no good reason to use VLEDs if a formula diet of 800–900 kcal is available. While these products can be purchased in most countries, without support they seldom lead to worthwhile long-term weight control. Several programmes providing the necessary support have been developed. Currently, the best-evidenced example of a formula diet for weight management is Counterweight-Plus, notably used to achieve mean weight losses of over 10 kg, and remissions of type 2 diabetes, in the UK Diabetes Remission Clinical Trial (DiRECT). It is now provided entirely online, with remote support using advanced technologies such as email, WhatsApp, Zoom, and even the telephone. An audit has shown the average weight loss at 12 months is similar to its face-to-face use, at about 15 kg. If the formula diet induction phase is followed completely for 3 months, weight loss of over 20 kg is likely. https://www.counterweight.org/collections/all (accessed 22 April 2022).

### 17.9.2 Medication added to the best available diet

Anti-obesity medication is not a replacement for, or an alternative to, diet and lifestyle changes. Operating with slightly different pharmacological mechanisms, they all only work by interrupting the normal physiological regulation of appetite and eating to generate negative energy intake for a period. Effectiveness varies considerably between individuals, because the mix of endogenous and external or environmental factors contributing to their obesity varies. Use of an approved drug will always be a therapeutic trial, for that individual patient, judged on the result for body weight, and that can only be observed over a period of months. The clinical trials that have provided evidence for regulatory approvals of anti-obesity drugs have all included regular dietitian contact, to optimize weight losses. The pharmaceutical treatment of obesity is undergoing seismic changes because of the recent availability of GLP-1 agonist medications, already widely used in lower doses to treat type 2 diabetes, which reliably generate sustained satiety and 'double-digit' weight losses (>10 kg). They have largely replaced the older anti-obesity classes based on sympathomimetic and serotoninergic mechanisms.

### 17.9.3 Bariatric surgery

Three procedures are in current use: gastric banding to reduce the gastric pouch beyond the oesophagus to a very small size, slowing the speed at which food can be consumed; sleeve gastrectomy to reduce the size of the entire stomach; and gastric bypass, which leaves the stomach as a blind loop, and redirects food directly from the oesophagus into the duodenum just beyond the pancreatic duct. Highly publicized claims that gastric bypass surgery has specific effects on endocrine functions via alteration of gut hormone secretion have led to plausible but poorly evidenced beliefs: ultimately, all these procedures mainly work by limiting food consumption, at least for a substantial period.

All the surgical procedures usually lead to substantial weight loss (20–40 kg), which brings multiple metabolic and physical improvements, potentially life-saving or life-extending for some, but life-long medical and dietetic supervision is needed. Some late weight regain is common. Side effects like dumping syndromes affect most, and most patients require additional treatments for side effects.

## 17.10 Service allocation: clinical grading of obesity

Given the high rates of obesity worldwide and the multiple demands on health services, it is inevitable that even evidence-based treatments described in section 17.9 are likely to be rationed to some extent. Often only those with high grades of obesity as determined by BMI or considered to be at high risk of cardiometabolic consequences of obesity qualify for intensive nutrition management, costly drugs, or surgery. More sophisticated and person-centred schemes for clinical grading and treatment allocation have been devised. The best known is the Edmonton Obesity Staging System (EOSS), a five-point ordinal classification system that considers conditions mediated by obesity and functional status of people with BMI >30 kg/m$^2$, introduced in 2008 (Fig. 17.5). The King's Obesity Staging Criteria (KOSC) classifies patients with BMI >30 kg/m$^2$ as stage 0 to 3 for nine functional domains mediated by obesity (Fig. 17.6).

These schemes are not perfect for every purpose, but they illustrate the growing understanding of

**EOSS:** EDMONTON OBESITY STAGING SYSTEM - *Staging tool*

**Stage 0**

- NO sign of obesity-related risk factors
- NO physical symptoms
- NO psychological symptoms
- NO functional limitations

Case Example:
Physically active female with a BMI of 32 kg/m$^2$, no risk factors, no physical symptoms, no self-esteem issues, and no functional limitations.

*Class I, Stage 0 Obesity*

EOSS Score

WHO Obesity Classification

**Stage 1**

- Patient has obesity-related **SUBCLINICAL** risk factors (borderline hypertension, impaired fasting glucose, elevated liver enzymes, etc.) - *OR* -
- **MILD** physical symptoms - patient currently not requiring medical treatment for comorbidities (dyspnea on moderate exertion, occasional aches/pains, fatigue, etc.) - *OR* -
- **MILD** obesity-related psychological symptoms and/or mild impairment of well-being (quality of life not impacted)

Case Example:
38 year old female with a BMI of 59.2 kg/m$^2$, borderline hypertension, mild lower back pain, and knee pain. Patient does not require any medical intervention.

*Class III, Stage 1 Obesity*

WHO Classification of weight status (BMI kg/m$^2$)

Obes Class I .........30–34.9
Obes Class II ........35–39.9
Obes Class III .......≥40

→ **Stage 0/Stage 1 Obesity**
Patient *does not meet clinical criteria for admission* at this time. Please refer to primary care for further preventative treatment options.

**Stage 2**

- Patient has **ESTABLISHED** obesity-related comorbidities requiring medical intervention (HTN, Type 2 Diabetes, sleep apnea, PCOS, osteoarthritis, reflux disease) - *OR* -
- **MODERATE** obesity-related psychological symptoms (depression, eating disorders, anxiety disorder) - *OR* -
- **MODERATE** functional limitations in daily activities (quality of life is beginning to be impacted)

Case Example:
32 year old male with a BMI of 36 kg/m$^2$ who has primary hypertension and obstructive sleep apnea.

*Class II, Stage 2 Obesity*

**Stage 3**

- Patient has **SIGNIFICANT** obesity-related end-organ damage (myocardial infarction, heart failure, diabetic complications, incapacitating osteoarthritis) - *OR* -
- **SIGNIFICANT** obesity-related psychological symptoms (major depression, suicide ideation) - *OR* -
- **SIGNIFICANT** functional limitations (eg: unable to work or complete routine activities, reduced mobility)
- **SIGNIFICANT** impairment of well-being (quality of life is significantly impacted)

Case Example:
49 year old female with a BMI of 67 kg/m$^2$ diagnosed with sleep apnea, CV disease, GERD, and suffered from stroke. Patient's mobility is significantly limited due to osteoarthrities and gout.

*Class III, Stage 3 Obesity*

**Stage 4**

- **SEVERE** (potential end stage) from obesity-related comorbidities - *OR* -
- **SEVERELY** disabling psychological symptoms - *OR* -
- **SEVERE** functional limitations

Case Example:
45 year old female with a BMI of 54 kg/m$^2$ who is in a wheel chair because of disabling arthritis, severe hyperpnea, and anxiety disorder.

*Class III, Stage 4 Obesity*

Fig. 17.5 The Edmonton Obesity Staging System, which incorporates conditions mediated by obesity and functional status of people with BMI >30 kg/m$^2$.

*Adapted from:* Sharma, A.M. and Kushner, F.F. (2009) *Int J Obes*, **33**(3), 289–95. doi: 10.1038/ijo.2009.2.

| Criteria | Stage 0<br>Normal health | Stage 1<br>At risk of disease | Stage 2<br>Established disease | Stage 3<br>Advanced disease |
|---|---|---|---|---|
| A Airways | Normal<br>Neck < 43 cm | Mild OSA<br>Neck ≥ 43 cm<br>Asthma/COPD | Requires CPAP | – |
| B BMI | NA[a] | 35–39.9 kg/m$^2$ | 40–50 kg/m$^2$ | >50 kg/m$^2$ |
| C CV-risk | <10% | 10–19% | ≥20%<br>Stable CAD | |
| D Diabetes | FPG < 5,6<br>HbAl < 5,7 | IFG<br>HbAlc 5.7–6.4% | DM2<br>HbAlc < 9% | DM2<br>HbAlc ≥ 9% |
| E Economic complications | None | None | Workplace disadvantage | Disabled |
| F Functional limitation | ≥3 h moderate physical activity/week | 1–2 h moderate physical activity/week | <1 h moderate physical activity/week | – |
| G Gonadal dysfunction[b] | Normal | Hyperandrogenemia[c] | PCOS[d] | – |
| H Perceived Health status, I body Image | Normal | Anxiety/depression without medication | Psychoactive drugs<br>Eating disorder | – |

*OSA*, obstructive sleep apnea, *COPD*, chronic obstructive pulmonary disease, *CPAP*, continuous positive airway pressure, *BMI*, body mass index, CV-risk, ten years risk of cardiovascular disease (Framingham risk assessment), *CAD*, coronary artery disease, *FPG*, fasting plasma glucose, *IFG*, impaired fasting glucose, *DM2*, diabetes type 2, *PCOS*, polycystic ovarian syndrome

[a]Patients with BMI < 35 kg/m$^2$ ($n = 42$) were excluded from the analysis since they did not fulfil the criteria for bariatric surgery
[b]Female participants
[c]Hyperandrogenemia denotes a free testosterone index (FTI) above the normal range (FTI > 0.6). FTI was calculated by the formula 100 x serum testosterone (nmol/L)/sex hormone binding globulin (SHBG, nmol/L)
[d]women with known PCOS and those with an FTI > 0.6 or hirsutism combined with oligo-/anovulation were classified as having PCOS

Fig. 17.6 The King's Obesity Staging Criteria (KOSC) classifies patients with BMI >30 kg/m$^2$ as stage 0 to 3 for nine functional domains mediated by obesity.

*Source:* Valderhaug, T.G., Aasheim, E.T., Sandbu, R., et al. (2016) The association between severity of King's Obesity Staging Criteria scores and treatment choice in patients with morbid obesity: a retrospective cohort study. *BMC Obesity*, **3**, 51. DOI 10.1186/s40608-016-0133-1

obesity as a complex disease process with multiple clinical implications. This understanding is essential for service planning, and introduces the concept that, as with many other chronic and progressive diseases, for some patients it becomes inappropriate to pursue treatments directed at weight loss. Good anticipatory care and palliative management is required for all patients with obesity, and it must be recognized that some either will not lose weight, or that weight loss or pursuit of weight loss will not confer significant improvements in health and well-being. Despite these caveats, recent research has shown enormous benefits from modern weight management, and the future looks more positive for many people living with obesity and diseases mediated by obesity.

## Further reading


1. **Al-Gindan, Y.Y., Hankey, C., Govan, L., Gallagher, D., Heymsfield, S.B., and Lean, M.E.J.** (2014) Derivation and validation of simple equations to predict total muscle mass from simple anthropometric and demographic data. *Am J Clin Nutr*, **100**, 1041–51. https://doi.org/10.3945/ajcn.113.070466

2. **Al-Gindan, Y., Hankey, C., Govan, L., Gallagher, D., Heymsfield, S., and Lean, M.** (2015) Derivation and validation of simple anthropometric equations to predict adipose tissue mass and total fat mass with MRI as the reference method. *British Journal of Nutrition*, **114**(11), 1852–67. doi:10.1017/S0007114515003670

3. **Bray, G.A., Katzmarzyk, P.T., Kirwan, J.P., Redman, L.M., Bouchard, C. and Schauer, P.R.** (eds) (2023) *Handbook of Obesity*, 5th edn. Taylor & Francis Ltd. ISBN: 9781032551081

4. **Bray, G.A. and Ryan, D.H.** (2021) Evidence-based weight loss interventions: individualized treatment options to maximize patient outcomes. *Diabetes Obes Metab*, **23**(Suppl 1), 50–62. doi: 10.1111/dom.14200. Epub: 24 November 2020. PMID: 32969147.

5. **Han, T.S., Al-Gindan, Y.Y., Govan, L., et al.** (2019) Associations of BMI, waist circumference, body fat, and skeletal muscle with type 2 diabetes in adults. *Acta Diabetol*, **56**, 947–54. https://doi.org/10.1007/s00592-019-01328-3

6. **Han, T.S., Al-Gindan, Y.Y., Govan, L., Hankey, C.R., and Lean, M.E.J.** (2019) Associations of body fat and skeletal muscle with hypertension. *J Clin Hypertens*, **21**(2), 230–8.

7. **Lean, M.E.J., Han, T.S., and Deurenberg, P.** (1996) Predicting body composition by densitometry from simple anthropometric measurements. *Am J Clin Nutr*, **63**(1), 4–14. doi: 10.1093/ajcn/63.1.4. PMID: 8604668.

8. **Lean, M.E. and Malkova, D.** (2016) Altered gut and adipose tissue hormones in overweight and obese individuals: cause or consequence? *Int J Obes (Lond)*, **40**(4), 622–32. doi: 10.1038/ijo.2015.220. Epub: 26 October 2015. PMID: 26499438; PMCID: PMC4827002

9. **NCF Risk Factor Collaboration (NCD-RisC)** (2017) Worldwide trends in body mass index, underweight, overweight and obesity from 1975–2016. *Lancet*, **390**, 2627–42.

10. **Srivastava, G. and Apovian, C.** (2018) Future pharmacotherapy for obesity: new anti-obesity drugs on the horizon. *Curr Obes Rep*, 7(2), 147–61. doi: 10.1007/s13679-018-0300-4. PMID: 29504049.

11. **WHO** (2000) *Obesity: preventing and managing the global epidemic*. Report of a WHO consultation, Technical Report Series 894. Geneva: World Health Organization.

12. **WHO/FAO** (2003) *Diet, nutrition and the prevention of chronic diseases*. Report of a Joint WHO/FAO Expert Consultation, Technical Report Series 916. Geneva: World Health Organization.

13. **WOF/IDF (2022)** *A policy brief. Obesity and type 2 diabetes: a joint approach to halt the rise*. World Obesity Federation (WOF) & International Diabetes Federation (IDF). Available at: https://www.idf.org/news/261:idf-and-wof-release-new-policy-brief-to-address-obesity-and-type-2-diabetes.html (accessed 27 April 2022).

Global prevalence data are continuously updated by World Obesity Federation and are available on their website: http://www.worldobesity.org/ (accessed 21 April 2022).

# 18 Protein-Energy Malnutrition

A. Stewart Truswell



## 18.1 Biochemistry of negative energy balance

People, young or old, who eat less food than they usually eat and need, lose body weight; the deficit of energy (or calories) in the diet is made up by drawing on the body's energy reserves: first fat, later muscle. The weight loss is carbon dioxide (breathed out) and water (excreted) from oxidation of fat:

$$2(C_{55}H_{106}O_6) + 157O_2 \rightarrow 110CO_2 + 106H_2O + \text{heat}$$

(This representative fat molecule is a triglyceride with oleic (18:1), linoleic (18:2), and palmitic (16:1) acids.) This is undernutrition.

Undernutrition can be mild or severe, beneficial (in someone who was obese) or dangerous. The loss of weight is a manifestation of energy depletion. Essential nutrients, protein, and micronutrients are likely to be depleted at the same time, but some micronutrients have large stores in the body, and requirements of some others are lower when energy intake is reduced. In children, who have higher protein requirements than adults, important depletion of protein is likely to accompany serious undernutrition.

## 18.2 Definitions of malnutrition

- Undernutrition is depletion of energy (calories).
- Malnutrition is serious depletion of any of the essential nutrients (other than energy).
- Fasting is voluntary abstention from food.
- Starvation is involuntary lack of food.
- Famine is severe food shortage of a whole community.
- Wasting is loss of substance, especially muscle (from insufficient food, disuse, or disease).

Protein depletion can affect the body in two different ways:

1. *In somatic protein depletion,* the loss of tissue shows as general wasting of muscles, which together contain the largest amount of the body's protein.

2 *In visceral protein depletion*, the brunt of the protein loss is borne by the liver, pancreas, and gut. This is the less common type of protein malnutrition and nutritional scientists still do not fully agree why it occurs.

Protein-energy malnutrition (PEM) occurs in three situations:

1 In young children in poor communities, usually in developing countries.

2 In adults, even in affluent countries, due to severe illness (hospital malnutrition).

3 In people of all ages in a famine.

In old people, wasting of muscles (sarcopenia) and loss of height are more often part of the ageing process than due to lack of food. They are looked after in different settings by different professionals, and causative factors are different (see Chapter 35).

## 18.3 Protein-energy malnutrition in young children

There are two forms of severe PEM: marasmus and kwashiorkor.

### 18.3.1 Nutritional marasmus

Nutritional marasmus is the common form; it is starvation in an infant or young child. (The word is from the Greek, 'marasmos,' meaning 'wasting.') The child is very thin. Weight is less than 60% of the median reference weight-for-age and there is marked wasting (see Fig. 18.1). There is no oedema.

There is loss of almost all the adipose tissue and (to a smaller extent) wasting of the voluntary muscles. Growth has stopped and it has taken weeks of inadequate feeding for a child to become very wasted like this. The cause is a diet very low in total energy; that is, not enough food, for example, early weaning from the breast on to dilute food, because of poverty, or ignorance. Poor food hygiene leads to gastroenteritis, diarrhoea, and vomiting. This leads to poor appetite, so more dilute feeds are given. Further depletion in turn leads to intestinal atrophy and more susceptibility to diarrhoea.

Not enough food implies not enough protein, because most foods contain some protein. It is most unlikely that a child not getting enough food would still be eating a protein-rich food, since such foods are expensive. With negative energy balance the major fuel to maintain life is free fatty acids, drawn from the adipose tissue. Blood glucose needed for tissues that can only metabolize glucose (brain, red blood cells) is maintained by gluconeogenesis of glucogenic amino acids (e.g. alanine) drawn from the body's proteins—usually the muscles, sometimes the viscera. Although energy depletion predominates in

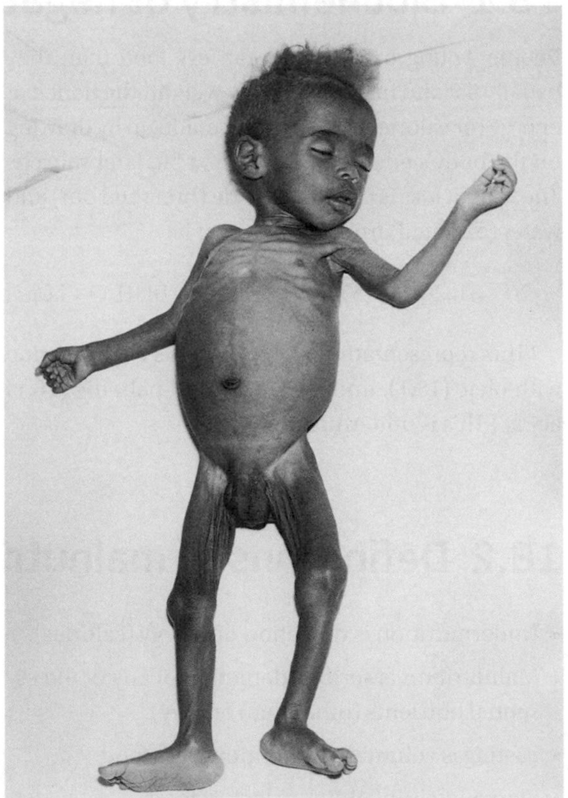

Fig. 18.1 Marasmus.

*Source:* Reproduced with kind permission from Professor J.D.L. Hansen of Cape Town.

**Table 18.1 Classification of malnutrition**

| | Moderate | Severe |
|---|---|---|
| Symmetrical oedema | No | Yes (i.e. kwashiorkor) |
| Weight-for-height (length) | −2 to −3 SD of reference (70–79%), wasting[a] | <3 SD below reference (−70%), severe wasting (i.e. marasmus) |
| Height-for-age | −2 to −3 SD of reference (85–89%), stunting[a] | <3 SD below reference (<85%), severe stunting |

[a]−2 to −3 SD below reference also called Z score −2 to −3.

'Reference' is WHO reference weight-for-height (length) and height-for-age.

*Adapted from:* WHO (1999) *Management of severe malnutrition: a manual for physicians and other senior health workers.* Geneva: World Health Organization.

marasmus, there is insufficient protein intake and inevitably loss of protein inside the body.

Inside the body, the heart, brain, liver, and kidneys are least wasted, but in advanced cases the heart becomes atrophied (wasted) and brain weight is reduced. There is increased mobilization of free fatty acids from adipose tissue, with ketosis (increased plasma concentration of 3(OH) butyrate and acetoacetate). The blood glucose may be subnormal. The basal metabolic rate goes down; an increased proportion of triiodothyronine is in the inactive rT3 form. Plasma insulin is low and leptin is low.

Infections that are only a temporary nuisance in well-nourished children become life-threatening in children with severe PEM. Their bodies are not capable of producing the usual responses to common bacterial infections: pyrexia, and increased white blood cells (leucocytosis). Cell-mediated immunity, the main defence against viruses and tuberculosis, is impaired. Pathogenic bacteria in the intestines can more easily gain access to the blood circulation (Table18.1).

## 18.3.2 Kwashiorkor

Marasmus has been known for centuries, but the other type of severe PEM (Table 18.1), kwashiorkor, was not generally recognized until the WHO report 'Kwashiorkor in Africa' by Brock and Autret in 1952. The first description was by Cecily Williams in the *Lancet* in 1935. She wrote from Accra, Ghana, and gave the syndrome the name that the people used there, in the Ga language.

Typically, a child with kwashiorkor (Fig. 18.2) develops oedema, which is generalized. The child is miserable, withdrawn, obviously ill, and will not eat. Changes can be seen in the skin: there are areas of pigmentation that are symmetrical in distribution, most commonly in the nappy area (Fig. 18.2).

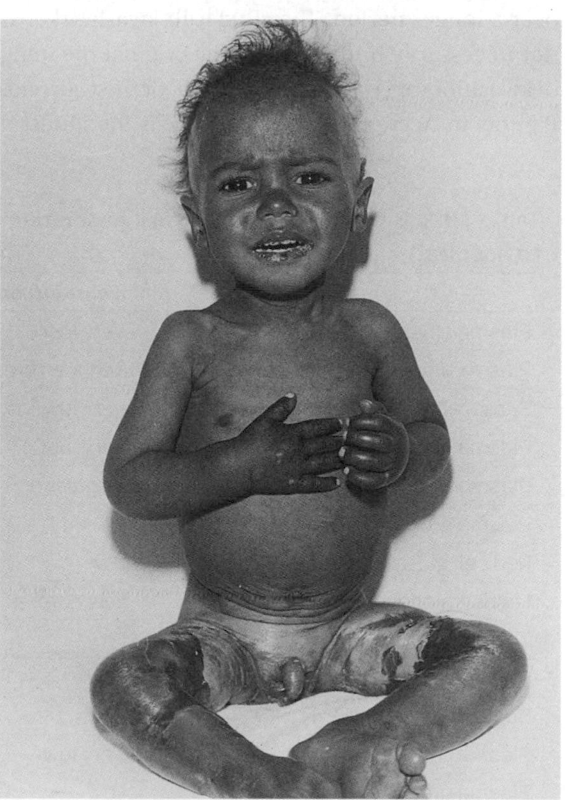

**Fig. 18.2** Kwashiorkor.

*Source:* Reproduced with kind permission from Professor J.D.L. Hansen of Cape Town.

The skin later shows cracks and the superficial layer peels off. The hair is thinned and discoloured, blonde or red or grey, instead of black. There is diarrhoea. Inside the body, the liver is enlarged and its parenchymal cells contain numerous fat droplets. The protein in the liver is reduced and two of the main features of kwashiorkor can be explained by failure of the liver to make two important (export) plasma proteins. Failure to synthesize albumin and the consequent very low plasma albumin may, because of low plasma osmotic pressure, at least partly explain the oedema. Failure to synthesize very-low-density lipoproteins, and inability to transport fat out of the liver to the periphery, explains the accumulation of fat in the liver. There is an abnormal and characteristic pattern of amino acids in plasma (Table 18.2).

The full picture of kwashiorkor develops more quickly than marasmus. One day the oedema appears and the carer seeks medical help—though changes in skin and hair must have been developing over a longer period. The child with kwashiorkor is not necessarily underweight. The original meaning of kwashiorkor is 'the deposed child' or 'first second'. Parents in Accra thought that this was the illness a child can get when a second baby follows and displaces the first one from the breast.

There are two schools of thought on the cause of kwashiorkor. The question is why is there an acute depletion of protein from the liver and other viscera rather than from the muscles in these cases of PEM?

The original theory is that the child who develops kwashiorkor has been fed on a diet moderately adequate in carbohydrate but very low in protein, so that there is a relative deficiency of protein to energy (i.e. protein malnutrition), whereas the diet that leads to marasmus is low in energy with protein. Researchers who disagree with this classical theory argue that in their experience, dietary histories are indistinguishable between children with kwashiorkor and children with marasmus. Something else must explain the visceral protein depletion—'dysadaptation', mycotoxins, or free radical damage have been suggested. A practical trial of antioxidants in over 2000 preschool children in Malawi failed to prevent kwashiorkor (Ciliberto et al., 2005).

Individual dietary histories are not likely to be scientifically reliable from the carer(s) of a child who

**Table 18.2 Biochemical findings in kwashiorkor compared with marasmus (on admission to hospital)**

| | Kwashiorkor | Marasmus |
|---|---|---|
| Plasma albumin | Very low | Usually normal range |
| Plasma amino acids | Reduced branch chain and tryosine | More normal |
| Serum amylase | Very low | Normal/low normal |
| Plasma (total) cholesterol | Very low | Normal/low normal |
| Plasma free fatty acids | Increased | Increased |
| Plasma growth hormone | Raised | Not as high |
| Red cell glutathione | Low | Normal |
| Fasting blood glucose | Low normal | Low |
| T lymphocytes | Low | Low |
| Plasma retinol | Low | Low |
| Somatomedin-C (IGF-1) | Low | Not as low |
| Plasma transferrin | Very low | Low normal |
| Plasma urea | Low | Not as low |
| Plasma urate | Low | Raised |
| Plasma zinc | Low | Not as low |

has become severely malnourished. Kwashiorkor children are not necessarily underweight (energy-deficient). Their very low blood and urinary urea levels indicate low protein intakes. Cure of kwashiorkor has been initiated with a diet consisting only of casein, dextrose, and salts. Kwashiorkor occurs in countries where the staple diets for weaned children have very low protein/energy ratios (e.g. cassava, plantains, sweet potato, or refined maize). Kwashiorkor is rare in developed countries. These infants have been fed a protein-deficient diet by well-intending parents as a result of perceived 'intolerance' to milk or formula. The skin lesions are red in a white baby (Tierney et al., 2010).

Whitehead et al. (1997) made observations comparing children in The Gambia, where the usual form of severe PEM is marasmus, with children in Uganda, where kwashiorkor occurs. Ugandan children grew more in weight and height and had more subcutaneous fat but lower plasma albumin concentrations. They had higher plasma insulin levels and lower plasma cortisols. Protein/energy ratios of their food were lower. Whitehead suggested that on a very low protein diet, but with adequate carbohydrate, the carbohydrate stimulates insulin, which is known to favour deposition of amino acids in muscles. On a very low protein diet, amino acids are in short supply, so muscle proteins can only be maintained at the expense of the liver (and other viscera). The liver is stimulated by infection to put much of its protein synthetic effort into making 'acute phase' plasma proteins that should help to fight the infection. Syndromes resembling kwashiorkor can be produced in monkeys on a diet of cassava with added sugar, and in young rats on a 5% protein ration.

### 18.3.3 The spectrum of severe protein-energy malnutrition

Kwashiorkor and marasmus are distinct diseases, but in communities where both occur, cases of severe PEM often have some features of both (e.g. they are very underweight and also have skin or hair changes). This is marasmic kwashiorkor.

The Spanish name for PEM, *syndrome policarencial infantile*, means the polynutritional deficiency of infants. In severely malnourished cases there are likely to be deficiencies of micronutrients, notably vitamin A, with risk of xerophthalmia, and potassium from diarrhoea, which contributes to the oedema. Niacin deficiency (where maize is the staple) may contribute to the skin lesions, and zinc deficiency further weakens immune response to infections.

Malnourished children have diarrhoea. An infection has probably precipitated the severe illness. These children stand infections poorly; measles is especially lethal. In parts of Africa, HIV infection underlies a proportion of malnourished children.

### 18.3.4 Management of severe malnutrition

Children who are severely wasted or have generalized oedema should be admitted to hospital, where they can be treated and fed day and night. Complications need medical treatment. The nutritional management is (surprisingly) similar for marasmus and kwashiorkor. 'Take it slowly' is the established principle for refeeding severely malnourished children and adults. Children with kwashiorkor cannot cope with high-protein diets for the first days of treatment. Management is in these three stages (WHO, 1999):

1. *Treatment of acute complications*: correction of dehydration and/or electrolyte disturbance, and/or very low blood glucose, and/or low body temperature (hypothermia), and start of treatment for infections. Nearly all severely malnourished children have bacterial infections, though they may not show fever or leucocytosis. All should receive broad-spectrum antibacterial treatment. The fluid given to rehydrate malnourished children should contain less sodium and more potassium than the standard UNICEF oral rehydration solution. ReSoMal contains only 45 mmol sodium per litre of water, but 40 mmol/L potassium, more glucose, and some magnesium and zinc.
2. *Initiation of cure*: refeeding, gradually working up the energy and protein intake, and giving multivitamin drops and potassium, magnesium, and zinc supplements. Children with kwashiorkor have poor appetites. They have to be handfed, with

frequent feeds, preferably in the lap of their carer or a nurse they know. To start refeeding, the standard 'formula' is F75, recommended by WHO, which provides 75 kcal and 1 g protein/100 mL. This is given for the first few days, in small amounts, which add up to around 100 mL/kg/day. Although the child is malnourished, energy and protein must be limited to avoid metabolic stress.

When the child is starting to recover, F100 is introduced, which contains 100 kcal and 2.9 g protein/100 mL (12% protein and 53% fat). This can be given to satiety. F75 and F100 are available as powders that can be reconstituted with water. They contain dried skimmed milk, sugar, vegetable oil, mineral mix, and vitamin mix. Details of the formula can vary in different centres.

3 *Nutritional rehabilitation*: after about 3 weeks the child should be obviously better, with oedema cleared, and mentally bright with good appetite, yet still below the reference weight-for-height. At this stage, catch-up growth should occur if the child is well looked after and given nutritious combinations of local, familiar foods. If the child has been in hospital they may be able to go on to a nutrition rehabilitation unit if one is available in the area. Sooner or later they will go home. It is unlikely that the home of a malnourished child is well resourced with nourishing food. At this stage, locally produced ready-to-use (therapeutic) food (RUTF) is being increasingly useful for rehabilitation of children recovering from severe malnutrition and also to manage children with moderately

**Box 18.1** Ten steps for management of severe PEM

| | Stabilization | | Rehabilitation |
|---|---|---|---|
| | **Days 1–2** | **Days 3–7** | **Weeks 2–6** |
| 1. Hypoglycaemia | → | | |
| 2. Hypothermia | → | | |
| 3. Dehydration | → | | |
| 4. Electrolytes | → | → | → |
| 5. Infection | → | | |
| 6. Micronutrients | → | no iron → | with iron → |
| 7. Cautious feeding | → | → | |
| 8. Catch-up growth | | | → |
| 9. Sensory examination | → | → | → |
| 10. Prepare for follow-up | | | → |

**Box 18.2** Prognosis of children with PEM

Even in well-equipped hospitals, the death rate of children with severe PEM is around 20%. Prognosis is worse for cases with kwashiorkor or HIV infection. Are there lasting effects in those who survive? Follow-up biopsies after kwashiorkor have shown that the liver returns to normal; the fatty change does not progress to cirrhosis (unlike that in alcoholics). In marasmic children who have become severely wasted in the first 2 years of life, growth of the head (circumference easily measured) is retarded so the brain must be smaller than normal, and such children may subsequently have impaired intelligence, unless they are fortunate thereafter and brought up in an excellent environment.

severe malnutrition, so that they do not require expensive (and often distant) hospital treatment.

A typical example of RUTF is 'Plumpy Nut'™, made of peanut butter, milk powder, vegetable oil, sugar, vitamins, and minerals, and packed in air-tight sachets. They are pastes that the child can eat as solid food. They have very low water activity, resist bacterial contamination so can be stored, and do not have to be cooked. Their nutrient content approximates to the water-based WHO's F100. The main difference is the peanut butter in RUTF. Allergic reactions appear to be very rare in malnourished children.

There are several different names for RUTFs. Some are commercial products, some made locally. They are provided by NGOs, charities, food aid, and/or the countries' health services. They should be treated as medicines and not shared with other family members.

## 18.4 Mild to moderate protein-energy malnutrition

For every florid case of marasmus or kwashiorkor, there are likely to be 7–10 children in the community with mild to moderate PEM. Like an iceberg, there is more malnutrition below the surface and not easily recognized. Carers often do not realize that their child is malnourished because he or she is similar in size and vitality to many of the same age in an impoverished neighbourhood. Most children with mild to moderate PEM can be detected, however, by their weight-for-age, which is less than 85% of the international standard (Table 18.1). Such children are either *wasted*, with subnormal weight-for-height/length, or *stunted* (nutritional dwarfism), with subnormal height-for-age (but not wasted), or both. Wasted children have used up body fat, and some muscle, to maintain their fuel supply. Stunted children have adapted in a different way, by stopping or slowing their growth. Reference tables (or graphs) are available from the World Health Organization (WHO) for weight-for-age, height/length-for-age, and weight-for-height for pre-pubertal children (see Chapter 34).

In many developing countries, around 2% of preschool children have severe PEM and 20% (in some places more) have mild to moderate PEM. The importance of this mild to moderate PEM is that affected children are growing up smaller than their genetic potential and have increased susceptibility to severe gastroenteritis and respiratory infections. Mild to moderate PEM is a major underlying reason why the 1–5-year mortality in poorer developing countries is 30–60 times higher than in Europe, North America, or Australasia. Nine of the ten countries with highest 1–5-year mortality are in Africa and one is Afghanistan. Deaths are recorded as due to pneumonia, diarrhoea, and malaria, but micro- and macro-analysis indicate that poor nutrition is the underlying cause of 35–50% of these deaths (Bejan et al., 2008; Black et al., 2013).

**Table 18.3 Child Malnutrition Estimates (2018) UNICEF/WHO/World Bank Group**

| Area | Stunted | Wasted | Overweight |
|---|---|---|---|
| World<br>% all children | 151 million<br>22.2% | 51 million<br>7.5% | 38 million<br>5.6% |
| Asia | 83.6 million | 35.0 million | 17.5 million |
| Africa | 58.7 million | 13.8 million | 9.7 million |
| Latin America and Caribbean | 5.1 million | 0.7 million | 0.7 million |
| Oceania | 0.5 million | 0.1 million | 0.1 million |
| Northern America | 0.5 million | 0.1 million | 1.7 million |

*Adapted from:* UNICEF/WHO/World Bank Group Joint Child Malnutrition Estimates. Key Findings of the 2018 edition.

**Box 18.3 UNICEF's inexpensive measures to prevent PEM**

- *G for growth monitoring*: the carer keeps the simple weight-for-age chart in a cellophane envelope and brings the child to a maternal and child health clinic regularly for weighing and advice.
- *O for oral rehydration*: the UNICEF ORS formula (NaCl 3.5 g, $NaHCO_3$ 2.5 g, KCl 1.5 g, glucose 20 g in clean water to 1 L) is saving many lives from gastroenteritis.
- *B for breastfeeding*: this has overwhelming advantages for a baby in a poor community with no facilities for hygiene. It should be continued as long as possible while solid foods are added. Additional foods, which should be prepared from locally available foods, are not usually needed before 6 months of age.
- *I for immunization*: for a few dollars, a child can be protected against measles, diphtheria, pertussis, tetanus, tuberculosis, poliomyelitis, etc., infections that predispose to and aggravate malnutrition.

## 18.5 Prevention of protein-energy malnutrition

Kwashiorkor most often occurs in the second year of life; marasmus mostly in the first year. Kwashiorkor is more amenable to the medical model of education, for example, education of carers about the need for protein foods for weaned children and encouraging their provision at the political level. Marasmus is a more intractable problem, bound up with poverty, the status and education of women, lack of contraceptive resources, and poor sanitation. UNICEF has achieved reductions in rates of PEM with four simple measures, represented by GOBI (Box 18.3). Rates of immunization are now 80% or more across the world. Measles and tuberculosis immunizations are particularly valuable. In many developing countries, there are extensive programmes for (large-dose) vitamin A supplements to be given at the same time (see Chapter 11).

## 18.6 Famine

The worst famines in recent times have been in areas torn by civil war. The hostilities greatly hamper communication of early warning and confirmation of the severity of the food shortage and transport of relief food into the area. Drought is the major natural cause of famine.

When there is not enough food for an entire community, children stop growing and children and adults lose weight. Starving people feel cold and weak and crave for food. Subcutaneous fat disappears and muscles waste. Pulse is slow and blood pressure low. The abdomen is distended; diarrhoea is common. Infections are to be expected, especially gastrointestinal infections, pneumonia, tuberculosis, and typhus.

The problem in a famine is not so much loss of food as loss of ability to obtain it. People have to sell all their assets in the attempt to buy food. The community's social and economic structures break down.

Aid professionals in relief operations should expect to have a mainly administrative and organizational role. It is impossible to give most time to treatment of a few very sick individuals. Therapeutic feeding is not an effective use of resources. Field workers have three options for distributions of food where supplies are insufficient to provide the minimum requirements of 1900 kcal (7.9 mJ)/day:

- where community and family structure are still intact and community representatives can be identified, let the community decide how the limited food is to be distributed

- where community structures have been disrupted, distribute food selectively to those assessed to be at the highest risk of mortality
- the third alternative is equitable distribution of the same basic ration to all members of the affected population, with selection of particularly vulnerable members.

The standard food aid rations usually consist of cereals, legumes, and some oil. If the cereal is whole grain, milling equipment is necessary. Milk powder is used for malnourished children. Provision of clean water is a priority. Care must be taken that the population gets the critical micronutrients, which are not the same in different areas and situations, e.g. vitamin C and potassium.

In emergency feeding situations, micronutrient powders are distributed for people and children to add to their food when they eat it. 'Sprinkles' and 'Mix Me' contain in one dose the recommended nutrient intakes of vitamins and the minerals iron, zinc, iodine, and selenium.

To assess the degree of undernutrition in individuals, two measures are commonly used: mid-upper arm circumference (MUAC) and weight-for-height in children or body mass index (BMI, kg/m$^2$) in adults. MUAC is obviously quicker and tape-measures can be given to several workers. Weight and height are slower to measure. In children, low MUAC tends to select younger children as malnourished and miss older children with low weight-for-height (or BMI). In adults, MUAC and BMI appear to correlate fairly well. A MUAC of 220 mm in men or 210 mm in women corresponds approximately to a critical BMI of 16 kg/m$^2$. As a general rule, moderate starvation = weight-for-height 80–71% of reference (in children), BMI of 18–16 kg/m$^2$ (in adults); severe starvation = weight-for-height ≤70% of reference (in children) or BMI ≤15.7 kg/m$^2$ (in adults).

## Further reading

Five major books have been written (in English) on PEM of children, the first in 1954. The two most recent are by Suskind and Lewinter-Suskind (1990) and by Waterlow (1992) (see below).


1. **Ashworth, A., Chopra, M., McCoy, D., et al.** (2004) WHO guidelines for management of severe malnutrition in rural South African hospitals: effect on case fatality and the influence of operational factors. *Lancet*, **363**, 1110–15.
2. **Bejan, P., Mohammed, S., Mwangi, I., et al.** (2008) Fraction of all hospital admissions and deaths attributable to malnutrition among children in rural Kenya. *Am J Clin Nutr*, **88**, 1626–31.
3. **Bhutta, Z.A.** (2009) Addressing severe acute malnutrition where it matters. *Lancet*, **374**, 94–6.
4. **Bhutta, Z.A., Das, J.K., Rizvi, A., et al.** (2013) Evidence-based interventions for improvement of maternal and child nutrition: what can be done and at what cost? *Lancet*, **382**, 452–77.
5. **Black, R.E., Victoria, C.G., Walker, S.P., et al.** (2013) Maternal and child undernutrition and overweight in low-income and middle-income countries. *Lancet*, **382**, 427–51.
6. **Ciliberto, J.H., Ciliberto, M., Briend, A., Ashorn, P., Bier, D., and Manary, M.** (2005) Antioxidant supplementation for the prevention of kwashiorkor in Malawian children: randomized, doubled blind, placebo controlled trial. *BMJ*, **330**, 1109–11.
7. **Collins, S., Dent, N., Binns, P., Bahwere, P., Sadler, K., and Hallam, A.** (2006) Management of acute severe malnutrition in children. *Lancet*, **368**, 1992–2000.
8. **Diop, E.H.I., Dossou, N.I., Ndour, M.M., Briend, A., and Wade, S.** (2003) Comparison of the efficacy of a solid ready-to-use food and a liquid, milk-based diet for the rehabilitation of severely malnourished children: a randomized trial. *Am J Clin Nutr*, **78**, 302–7.

9. **Suskind, R.M. and Lewinter-Suskind, L.** (1990) *The malnourished child*. New York, NY: Raven Press.

10. **Tierney, E.P., Sage, R.J., and Shwayder, T.** (2010) Kwashiorkor from a severe dietary restriction in an 8-month infant in suburban Detroit, Michigan: case report and review of the literature. *Int J Dermatol*, **49**, 500–6.

11. **Truswell, A.S. and Brand Miller, J.C.** (1993) Pathogenesis of the fatty liver in protein-energy malnutrition. *Am J Clin Nutr*, **57**, 695–6.

12. **Waterlow, J.C.** (1992) *Protein-energy malnutrition*. London: Edward Arnold.

13. **Whitehead, R.G., Coward, W.A., Lunn, P.G., and Rutishauser, I.** (1997) A comparison of the pathogenesis of protein energy malnutrition in Uganda and the Gambia. *Trans Roy Soc Trop Med Hyg*, **71**, 189–95.

14. **WHO** (1999) *Management of severe malnutrition: a manual for physicians and other senior health workers*. Geneva: World Health Organization.

# 19 Cardiovascular Diseases

Jim Mann and Rachael McLean



Cardiovascular disease (CVD) includes coronary heart disease (CHD), also referred to as coronary artery disease or ischaemic heart disease (IHD), cerebrovascular disease or stroke, and peripheral arterial disease. A similar pathological process underlies each of these three groups of conditions, which affect the heart, the brain, and peripheral arteries. Inappropriate nutrition has most consistently been linked with CHD, and this chapter deals primarily with CHD, those risk factors for the disease that are influenced by diet, and the evidence indicating that dietary modification has the potential to reduce clinical CHD. While there are fewer data directly linking nutritional factors to cerebrovascular disease, there are several shared risk factors, and raised blood pressure is a particularly important causal factor for most types of stroke. Thus, dietary advice is a pivotal component in reducing the risk of CHD and cerebrovascular disease and in the treatment of those with established diseases.

Globally, cardiovascular disease is a leading cause of death and disability. The number of deaths from cardiovascular disease has been estimated by the Global Burden of Disease Study to have been 18.6 million in 2019. Although the age-standardized rates of cardiovascular disease mortality have decreased in many countries, absolute numbers continue to increase globally due to population increase and ageing. There is a striking geographic variation in the frequency of cardiovascular disease. Age-standardized mortality for cardiovascular disease was highest in Tajikistan, the Solomon Islands, and Uzbekistan in 2019, around 6 times higher than in countries with the lowest age-standardized mortality rates (France, Peru, and Japan).

## 19.1 Coronary heart disease

CHD is a common condition in most high-income and many low- and middle-income societies. In most industrialized countries, it is the most common cause of death, often accounting for around one-third of all deaths. In addition, each year there are about as many non-fatal cases as there are deaths.

A high proportion of the healthcare budget in many countries is spent treating CHD and its consequences. Genetic as well as nutritional and other lifestyle-related factors contribute to the aetiology of the condition.

## 19.2 Pathology

The basic pathological lesion underlying CHD is the atheromatous plaque, which narrows one or more of the coronary arteries that supply blood (containing oxygen and other nutrients) to the heart muscle (myocardium) (Fig. 19.1). Post-mortem studies suggest that the pathological process may begin in childhood. In addition, a superimposed thrombus or clot may further occlude the artery. A variety of cells and lipids are involved in the pathogenesis of the atherosclerotic plaque and the arterial thrombus, including lipoproteins, cholesterol, triglycerides, platelets, monocytes, endothelial cells, fibroblasts, and smooth muscle cells. Nutrition can influence the development of CHD by modifying either atherogenesis or thrombogenesis or both these processes.

## 19.3 Clinical conditions

Two readily identifiable clinical conditions result from these pathological processes:

1 *Angina pectoris* is typically characterized by pain in the centre of the chest, which is brought on by exertion or stress, and which may radiate down the left arm or to the neck. It results from a reduction or temporary block to the blood flow through a coronary artery to the heart muscle. The pain

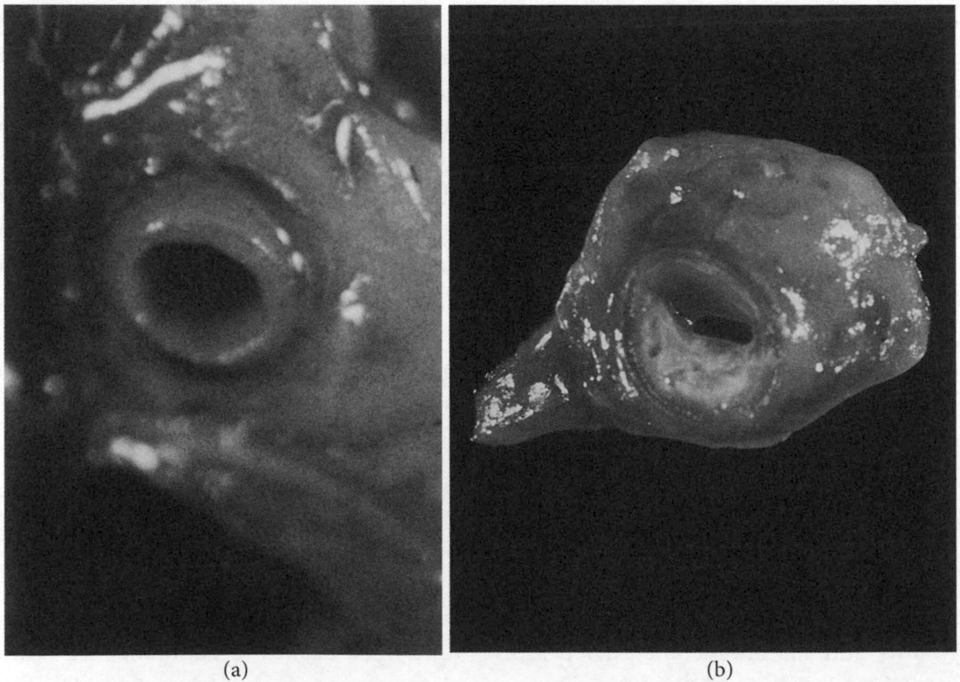


Fig. 19.1 A normal coronary artery (a) is contrasted with an artery showing atheromatous deposits (b).

usually passes with rest and seldom lasts for more than 15 minutes.

2. *Coronary thrombosis* or *myocardial infarction* results from total occlusion of an artery, which causes infarction or death of some of the heart muscle cells and is associated with prolonged and typically severe central chest pain. Atypical presentations of these conditions are relatively common, especially in women. For example, the pain is not always severe and may be mistaken for 'indigestion'. The terms coronary thrombosis and myocardial infarction are used to describe the same clinical condition, although they really describe its two distinct pathological processes.

## 19.4 Epidemiological aspects

Coronary heart disease has often been thought of as a relatively modern disease; however, evidence of calcification, which is suggestive of atheroma in relatively young people, has been found in Egyptian Mummies dating from the tenth century BC, and descriptions of angina are to be found in historical medical texts. Much of the epidemiological data relating to CHD rates has focused on mortality, but differences and trends in rates for non-fatal disease and the burden of disease associated with CHD tend to be closely related to mortality rates.

CHD rates and the associated disability vary widely amongst countries and population groups within countries (see Fig. 19.2). Overall age-standardized rates of CHD are higher in men than women, although as women age, CHD contributes a greater proportion of total mortality. CHD incidence increases with age in both sexes and is largely a disease of middle and older ages. CHD rates have also changed over time. In many high-income countries, age standardized CHD mortality rates increased steadily from the middle of the twentieth century, peaking in the 1960s and 1970s, after which they started to decline. In contrast, in many countries in Eastern Europe, the increase started much later and the decline,

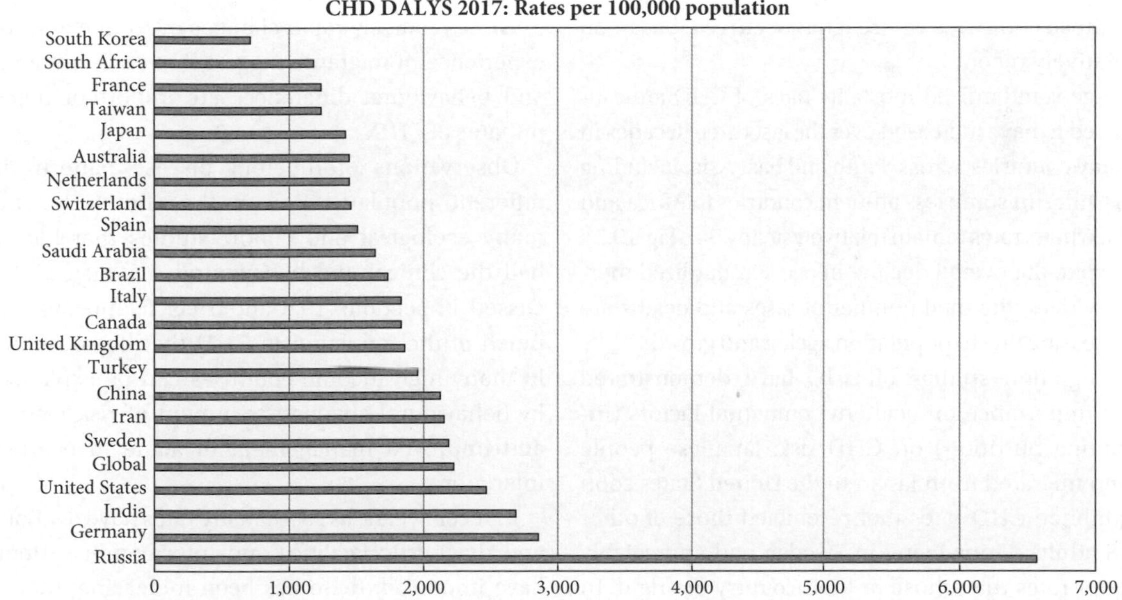


Fig. 19.2 International differences in disability-adjusted life years (DALYS) per 100,000 population from coronary heart disease (CHD) in 2017.

*Adapted from:* Data from Khan M, Hashim M, Mustafa H, et al. (July 23, 2020) Global Epidemiology of Ischemic Heart Disease: Results from the Global Burden of Disease Study. Cureus 12(7): e9349. doi:10.7759/cureus.9349

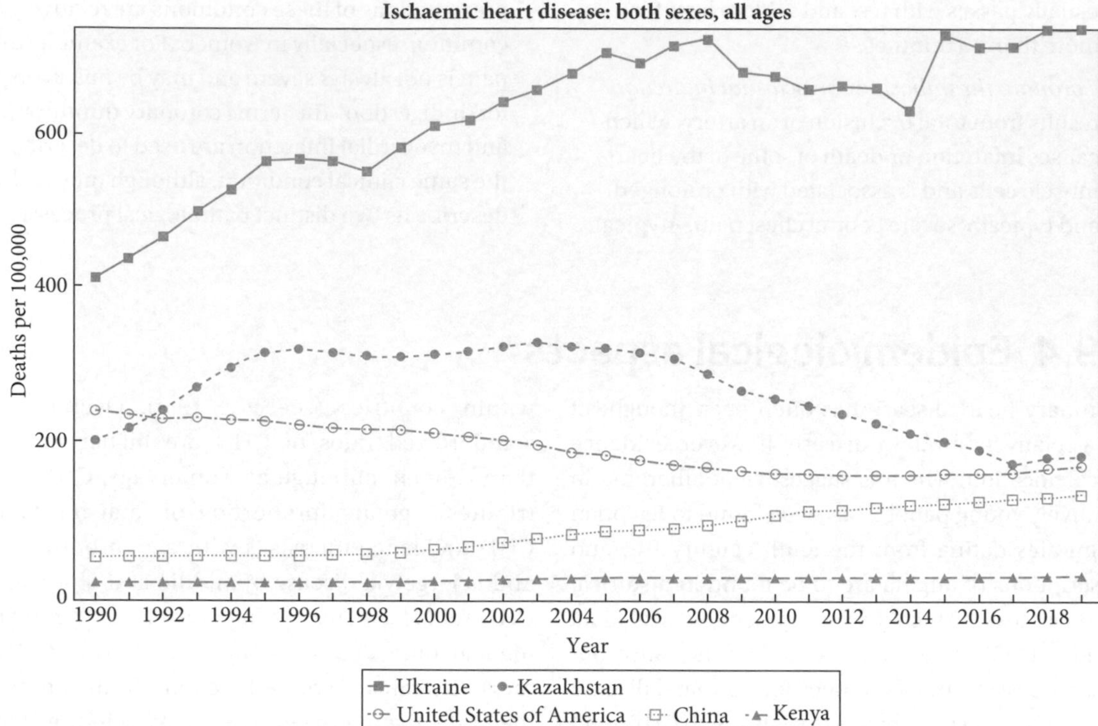


**Fig. 19.3** Trends in age-standardized death rates (per 100,000) from CHD/IHD in men and women in five countries.

*Source:* Institute for Health Metrics and Evaluation (IHME). GBD Compare Data Visualization. Seattle, WA: IHME, University of Washington, 2020. Created from data available at http://vizhub.healthdata.org/gbd-compare. (Accessed 1 November 2021)

in those countries where it has occurred, has been relatively recent.

Age-standardized mortality rates of CHD are estimated to have increased over the last three decades in many countries across South and East Asia, including in China. In some less affluent countries in Africa and elsewhere, rates remain relatively steady (see Fig. 19.3). Despite the overall decline in age-standardized mortality rates, the total number of cases and deaths are increasing due to population ageing and growth.

Migration studies of CHD have demonstrated the importance of local environmental factors (including nutrition) on CHD risk. Japanese people who migrated from Japan to the United States soon exhibited CHD rates that resembled those of other US adults. Finns living in Sweden had appreciably lower rates than those in their country of origin. In the UK, where CHD rates were appreciably higher in Scotland and Northern Ireland than in England, CHD rates depended upon country of residence at the time of death, rather than country of birth.

These relatively rapid changes over time and the experience of migrants suggest that environmental and behavioural differences are important determinants of CHD.

Observations of different disease patterns in different populations were the stimuli for the many ecological and cohort studies that identified the clinical and behavioural risk factors discussed in sections 19.5 and 19.6. It appears that much of the reduction in CHD that has occurred in many high-income countries can be explained by behavioural changes, treatment of risk factors, and improved management of acute myocardial infarction.

In recent years, as case fatality rates have declined and diagnostic facilities and information systems have improved, there has been increasing interest in the overall burden of disease. This is reported in *disability-adjusted life years* (DALYs), a measure of years of life lost due to ill health, disability, or death. In 2019 CHD was the second leading cause of DALYs in the world (second only to neonatal deaths), and the leading cause of DALYs in those over 50 years of age. This is largely a result of increasing incidence and mortality from CHD in low- and middle-income countries. In these countries, CHD events and deaths have increased due to a number of factors, including population growth, ageing populations, and changing patterns of risk factors, which may differ amongst countries. In many low- and middle-income countries, the average age of onset of CHD is lower than in high-income countries, resulting in an increasing burden of disease as CHD increases. The situation is particularly concerning in countries where affluence and poverty co-exist and which are said to be in a state of nutrition transition. In such countries (e.g. India and South Africa), CHD rates are high amongst the relatively affluent and those accumulating wealth, whereas diseases of undernutrition remain prevalent amongst the poor and underprivileged. The situation is the reverse of what is observed in most high-income countries, where the socioeconomically disadvantaged have higher CHD rates than better educated, more affluent groups in the community. The Global Burden of Disease Study has estimated that in 2019, age-standardized DALY rates were highest in Russia as well as other Eastern European countries but also very high in parts of Oceania, Central Asia, North Africa, and the Middle East.

The observation that changes in CHD rates have occurred over relatively short time periods encourages the belief that CHD is potentially preventable, especially in those who are in the prime of life. The challenge is to achieve the behavioural changes required to reduce the risk factors described in the sections that follow, and ensure appropriate medical management for those in whom lifestyle changes alone are insufficient.

## 19.5 Risk factors for coronary heart disease

Attempts to explain the pathological process underlying CHD and to identify individuals at risk show that there is no single cause of the disease. An understanding of the characteristics that put individuals at particular risk of developing CHD provides a useful background against which to examine in more detail the role of diet in the aetiology. The term 'risk factor' is used to describe physical and biochemical attributes, as well as features of lifestyle and behaviour, which predict an increased likelihood of developing CHD. Putative risk factors are identified in ecological studies and when comparisons are made between people who have developed CHD and healthy controls (case-control studies). Such attributes are confirmed by cohort (prospective) studies in which these factors are measured in large groups of apparently healthy people who are then followed to see if they develop the disease or not at some future date. The presence, absence, or degree of each factor can then be related to the risk of developing CHD. Experimental studies in humans and animals and controlled trials contribute to the totality of evidence, which determines whether 'risk factors' are likely to be causally related to the disease. Table 19.1 lists most of the important risk factors for CHD that have been identified in this way. When more than one risk factor is present, the effect is synergistic; that is, the overall increase in risk of CHD is greater than might be expected from simply adding together the risk associated with each. The irreversible, psychosocial, and geographic factors, as well as cigarette smoking and physical activity, are described in textbooks of medicine and epidemiology. This chapter concentrates on potentially modifiable factors, especially those that have been shown to be influenced by diet in humans. It is noteworthy that there is close interaction amongst many of the risk factors, notably that those related to diet have an appreciable influence on biological factors that also have genetic determinants. Furthermore, a number of the diet-related risk factors are also favourably influenced by regular physical activity, so lifestyle-related advice to reduce cardiovascular risk generally involves recommendations relating to increasing exercise in addition to diet.

**Table 19.1 Risk factors for coronary heart disease**

| | |
|---|---|
| Irreversible | • Male sex<br>• Increasing age<br>• Genetic traits, including monogenic and polygenic disorders of lipid metabolism |
| Potentially modifiable | • Diet<br>• Elevated blood pressure and hypertension<br>• Cigarette smoking<br>• Dyslipidaemia: increased levels of cholesterol, triglyceride, low-density and very-low-density lipoprotein, and apolipoprotein B; low levels of high-density lipoprotein; atypical lipoproteins<br>• Oxidizability of low-density lipoprotein<br>• Obesity, especially when centrally distributed (high waist circumference)<br>• Physical inactivity<br>• Diabetes, hyperglycaemia, and insulin resistance<br>• Increased thrombosis: increased haemostatic factors and enhanced platelet aggregation<br>• High levels of inflammatory markers (e.g. CRP, IL-6, TNFα)<br>• Impaired foetal nutrition<br>• High levels of homocysteine |
| Psychosocial | • Stressful situations<br>• Coronary-prone behaviour patterns: type A behaviour |
| Geographic | • Climate and season: cold weather<br>• Air pollution (indoor and outdoor) |

CRP, C-reactive protein; IL, interleukin; TNF, tumour necrosis factor.

## 19.5.1 Dyslipidaemia

Altered levels of blood lipids and lipoproteins (see Chapter 3) place individuals at increased risk of CHD in three main ways. First, a relatively small proportion of people have an exceptionally high risk because of a clearly inherited increase of plasma lipids and lipoproteins. The most common of these is *familial hypercholesterolaemia,* a dominantly inherited condition resulting from mutations of the LDL receptor gene and characterized by markedly raised levels of total and low-density lipoprotein cholesterol. Second, a large number of people (perhaps as many as half the adult population in high-risk countries) have a slight to moderately increased risk because their blood lipids are higher than desirable as a result of an interaction between polygenic (many genes involved) and lifestyle-related factors, principally dietary factors (see section 19.6). They are described as having 'polygenic' or 'common' hyperlipidaemia. The association between total or LDL cholesterol and CHD is characterized by a gradient of risk (see Fig. 19.4), so there are no clear cut-offs beyond which levels

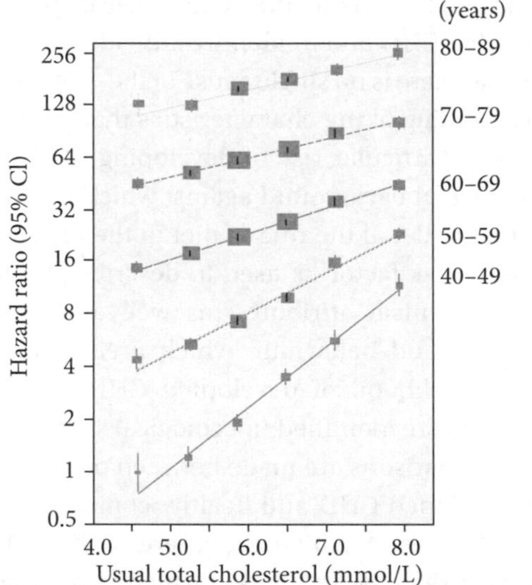


Fig. 19.4 Age-specific associations between usual total cholesterol and IHD mortality; meta-analysis of individual data from 61 prospective studies.

*Adapted from:* Prospective Studies Collaboration (PSC) (2007). Blood cholesterol and vascular mortality by age, sex, and blood pressure: a meta-analysis of individual data from 61 prospective studies with 55 000 vascular deaths. *Lancet*, **370**, 1829–39, with permission from Elsevier.

are regarded as unacceptable. Nevertheless, from a practical perspective, levels below 4 mmol/L for total cholesterol and 2 mmol/L for LDL cholesterol have been suggested as targets for healthy individuals. Lower targets are generally recommended for those who have diagnosed cardiovascular disease. Third, some people are also at increased CHD risk because of low levels of high-density lipoprotein (HDL) often associated with high levels of triglyceride. This lipid profile is most commonly seen in people with the metabolic syndrome, prediabetes, and type 2 diabetes (see Chapter 21) and those who are overweight or obese. Given the synergistic effect of risk factors, the overall risk is based upon assessment of several major risk factors (see Box 19.1). Some people may have altered lipid levels secondary to other medical conditions (e.g. hypothyroidism, renal disease).

## 19.5.2 Elevated blood pressure and hypertension

Increasing levels of both systolic and diastolic blood pressure are associated with increased rates of CHD, strokes (cerebrovascular disease), peripheral vascular disease, cardiomyopathy, and kidney disease. The association between elevated blood pressure and increased risk of CHD and stroke has been clearly established by observational studies and randomized controlled trials of blood pressure-lowering interventions. In the past, blood pressure was categorized into 'hypertension' (indicating increased risk of cardiovascular disease, usually defined as ≥140/90 mmHg) and 'normal blood pressure'. Pooled data from 61 prospective cohort studies have shown a linear relationship between blood pressure and risk of mortality from CHD and stroke above around 115/70 mmHg (see Fig. 19.5). Similarly, recent trials have shown the benefits of blood pressure lowering by intensive medication therapy to levels around 120 mmHg systolic. It is also clear from studies examining multiple risk factors that there are interactive effects (see Box 19.1). For these reasons, many guidelines recommend blood pressure lowering through lifestyle and medical interventions for those determined to be at high risk of cardiovascular disease (CVD), even though they may not have 'hypertension' as defined above.

Worldwide, elevated blood pressure has been ranked as the leading risk factor for burden of

### Box 19.1 Assessment of cardiovascular risk

Because there is a gradient of risk associated with increasing levels of several risk factors, and because cardiovascular risk factors interact, overall cardiovascular risk is determined from algorithms that estimate a person's absolute risk of experiencing a cardiovascular disease event (CHD or stroke) over the next 5 or 10 years. Most of the algorithms currently used are based on knowledge of a fairly small number of risk factors, including total cholesterol (or total: HDL cholesterol ratio), systolic blood pressure, and whether or not they have diabetes or are smokers, their age, and sex. Absolute risk calculators are available electronically (see for example *HEARTS technical package for cardiovascular disease management in primary health care: risk based CVD management. Geneva: World Health Organization; 2020. Licence: CC BY-NC-SA 3.0 IGO*), and are widely used in clinical practice. Treatment decisions are based on an individual's absolute risk of a CVD event over a particular time period, in recognition that moderate reductions in several risk factors may have a greater benefit than substantial reduction of a single risk factor. Lifestyle advice and support, including dietary advice, is usually the first approach to treatment and should be included in treatment recommendations regardless of the level of absolute cardiovascular risk and whether or not drug treatment is required. A number of limitations to most of the currently available risk assessment tools have been identified. As populations age and life expectancy increases, assessment of risk over longer periods becomes more relevant, especially for younger people, than the typical 5 or 10 year risk estimate. Ethnicity and genetic predisposition are not included in most of the current risk equations. Population risk factor patterns have been changing over time (e.g. in many countries smoking rates have declined while obesity prevalence has increased). New biomarkers and societal determinants of risk are emerging. New risk equations are being developed to take these factors into account.

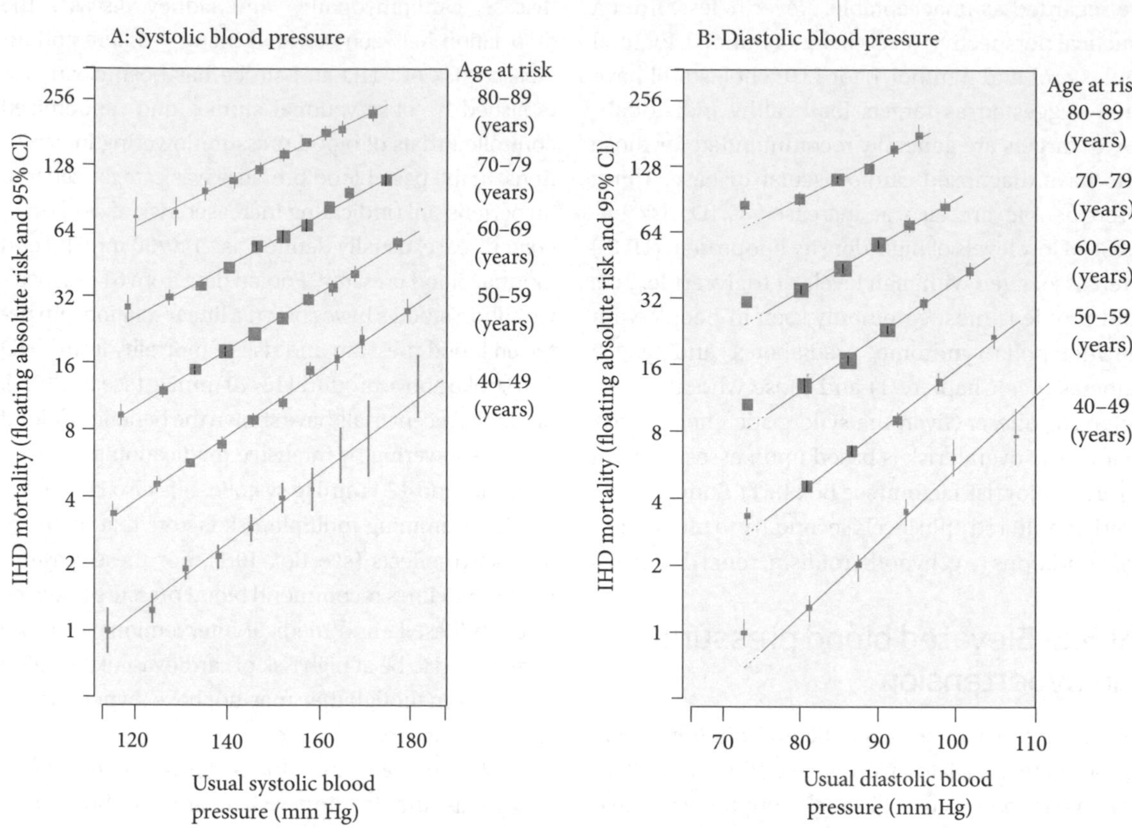


**Fig. 19.5** Ischaemic heart disease (IHD) mortality rate in each decade of age versus usual blood pressure at the start of that decade.

*Reprinted from:* Prospective Studies Collaboration (PSC) (2002). Age-specific relevance of usual blood pressure to vascular mortality: a meta-analysis of individual data for one million adults in 61 prospective studies. *Lancet*, **360**, 1903–13, with permission from Elsevier.

disease, accounting for 9% of DALYs in 2019. It is an important risk factor in high-, middle- and low-income countries, contributing to around 10.8 million deaths worldwide in 2019.

A number of factors contribute to elevated blood pressure and hypertension, including (in most societies) increasing age, genetic factors, and presence of diabetes or the metabolic syndrome and obesity. Some ethnic or racial groups, including African Americans, have an increased risk of developing elevated blood pressure. Lifestyle factors such as smoking and physical inactivity are associated with increased blood pressure. The individual nutritional determinants of blood pressure, notably sodium (salt) and alcohol, which are promotive, and potassium and several foods, which are protective, are discussed in section 19.6.

## 19.5.3 Diabetes and the metabolic syndrome

Many studies have shown that people with diabetes, prediabetes, and the metabolic syndrome have an increased risk of CHD (see Chapter 21). This seems to be largely due to the fact that these conditions are associated with other risk factors for CHD, notably dyslipidaemia, elevated blood pressure, and raised levels of insulin and several inflammatory markers.

## 19.5.4 Obesity

Obesity is an important risk factor for CHD. While increasing body mass index shows a modest graded association with CHD, increasing waist circumference shows a more striking relationship. The effect is not surprising given the association between centrally distributed excess body fat and several other risk factors, notably dyslipidaemia, elevated blood pressure, raised levels of inflammatory markers, insulin resistance, and type 2 diabetes. Weight loss corrects most of the clinical and metabolic derangements seen in overweight and obese individuals (see Chapter 17).

### 19.5.5 Inflammatory markers

Inflammation is acknowledged as a process that can appreciably increase cardiovascular risk. High-sensitivity C-reactive protein (CRP) is regarded as a useful risk indicator, but it is uncertain whether the association reflects a causal relationship. Raised levels of other inflammatory markers (e.g. interleukin 6 and tumour necrosis factor-α) may also be associated with increased risk. Information regarding the extent to which nutritional factors can influence the inflammatory response is limited, but strict adherence to a largely plant-based, low-fat diet has been shown to reduce levels of some inflammatory markers.

### 19.5.6 Thrombogenesis

Factors that increase the tendency to thrombosis (as a result of either increased platelet aggregation or a high level of coagulability of blood) have received less attention than other important risk factors. Platelet aggregation is largely controlled by a balance between the proaggregatory compound thromboxane $A_2$ (synthesized from arachidonic acid released from the platelet membrane after injury to the blood vessel wall) and the antiaggregatory prostacyclins $PGI_2$ and $PGI_3$ and the series 1 prostanoid $PGE_1$. The type of dietary fat has an effect on these compounds and hence platelet aggregation (see section 19.6).

Dietary factors may also influence thrombogenesis via an effect on the *coagulation system*. The physiological function of coagulation is to secure haemostasis after an injury. Thrombin is produced, which enables the conversion of soluble fibrinogen to insoluble fibrin. A high level of coagulability might predispose to thrombosis. Several prospective studies suggest that factors involved in the coagulation system (notably factor VII and fibrinogen) are important predictors of CHD, although this effect may be small compared to other risk factors. Factor VII is associated with dietary factors. High levels of fibrinogen are associated with obesity and cigarette smoking.

### 19.5.7 Raised plasma homocysteine levels

Patients with inborn errors of homocysteine metabolism have very high levels of plasma homocysteine, homocysteinuria, and a high risk of cardiovascular disease. It has also been found that in the general population there is a gradient of CHD risk associated with increasing levels of plasma homocysteine, suggesting that homocysteine may be an independent risk factor for CHD and stroke. Experimental evidence suggests that homocysteine damages the endothelium of coronary arteries by interfering with nitric oxide production. Homocysteine may also increase oxidative stress, and disturb lipoprotein metabolism. Although homocysteine levels can be influenced by nutritional measures, there is no evidence from randomized controlled trials that reducing homocysteine levels can reduce clinical CHD (see section 19.6.9).

### 19.5.8 Impaired foetal nutrition

Barker and colleagues observed some time ago that low birthweight babies, especially those who tended to gain weight rapidly in early life, were more prone than those of normal weight to a range of clinical and metabolic abnormalities (including obesity, hypertension, dyslipidaemia, insulin resistance) that predispose to the increased risk of CHD and diabetes in later life. The foetal origins (Barker) hypothesis suggests that maternal malnutrition at critical stages of foetal development leads to intrauterine growth retardation, including decreased pancreatic islet β cells, decreased number of nephrons, insulin resistance, and a range of other abnormalities that may not be associated with later chronic diseases if the child remains in a relatively deprived nutritional environment. However, problems are proposed to occur if the malnourished foetus is born into conditions of adequate nutrition or over-nutrition and rapid catch-up growth occurs. The main message to be taken from this research at this stage is the importance of adequate and appropriate nutrition for women of childbearing age.

## 19.6 Nutritional determinants of coronary heart disease

Nutritional determinants of CHD and cardiovascular risk factors have been identified in ecological, observational, and biomarker studies, randomized controlled trials, and carefully controlled dietary intervention studies in humans and experimental animals. Associations between CHD and risk factors and a substantial number of foods and nutrients have been observed, so it is essential to examine the totality of evidence and its quality to determine whether the association is likely to be causal. The causes of CHD are complex, involving an interaction between many different genes conferring susceptibility and a range of lifestyle-related and other environmental exposures. However, the association with some foods and nutrients is sufficiently convincing to be regarded as causal and justify strong recommendations for individuals with or at high risk of CHD and for populations with high CHD rates (see Table 19.2).

### 19.6.1 Dietary fats

**Observational studies** It was the pioneering Seven Countries study, started in 1958 and coordinated by Ancel Keys and colleagues, which gave prominence to the suggestion that nutrition plays an important role in determining risk of CHD. Food consumption of people in 16 defined cohorts in seven countries, selected because of their widely varying rates of CHD, was related to subsequent CHD incidence. The strongest associations observed were between percentage of energy derived from saturated fat, serum cholesterol, and CHD (see Fig. 19.6). Weak inverse associations (suggesting protective effects) were found with percentages of energy from mono- and polyunsaturated fat and CHD. Total fat intake was not associated with CHD. These observations led to the suggestion that the diet–heart disease link was principally mediated via an effect of saturated fat on plasma cholesterol, which in turn increased the risk of CHD. Of the other well-known risk factors measured in this study, only blood pressure appeared to explain some of the geographic variation in the frequency of CHD.

Saturated fat intake has also been shown to be positively associated with increased risk of CHD in more recent cohort studies, with the magnitude of effect influenced by the level of intake of other macronutrients. Systematic reviews and meta-analyses show that partially substituting saturated fat with PUFA, plant MUFA, or slowly digested carbohydrates (foods high in dietary fibre and whole grain cereals) is associated with a significantly decreased incidence of CHD. However, substituting saturated fat with rapidly digested carbohydrate is not associated with change in CHD risk. Substitution with animal protein is associated with increased incidence of CHD (see Fig. 19.7). Total mortality is significantly lower in those whose saturated fat intake is less than 10% TE, compared with those whose intake of saturated fat is above this level.

Similar analyses show that higher intakes of *trans*-unsaturated fat are associated with increased incidence of CHD with a dose-response relationship. Substitution of *trans*-fatty acids with *cis*-monounsaturated fat reduces CHD incidence by 20% (Fig. 19.8).

**Dietary intervention studies** Studies examining the effects of dietary fat manipulation on major cardiovascular risk factors make an important contribution to the totality of evidence. This is particularly the case for determinants of total and LDL cholesterol and the ratio of total (or LDL) to HDL cholesterol. The results of such studies have been consistent, regardless of whether they were conducted under carefully controlled

**Table 19.2 Summary of the type of evidence available for linking various nutrients and foods with coronary heart disease. The dietary intervention studies refer to the effects on risk factors; the randomized trials refer to long-term interventions with clinical endpoints**

| Nutrient or food | Increased risk/ protective (INCR/PROT) | Observational studies (cohort) | Dietary intervention trials | Randomized controlled trials | Overall assessment of evidence |
|---|---|---|---|---|---|
| Total fat | INCR | 0 | + | 0 | 0 |
| SFA[a] | INCR | ++ | +++ | +++ | +++ |
| Monounsaturated (*cis*) | PROT | ++ | ++ | + | ++ |
| Monounsaturated (*trans*) | INCR | +++ | +++ | 0 | +++ |
| ω6 PUFA[a] | PROT | + | +++ | +++ | +++ |
| ω3 PUFA | PROT | ++ | ++ | ++ | ++ |
| Dietary cholesterol | INCR | ++ | ++ | 0 | + |
| Total CHO | INCR | 0 | + | 0 | 0 |
| Sugars | INCR | + | ++ | 0 | + |
| Dietary fibre | PROT | +++ | ++ | ++ | +++ |
| Sodium | INCR | +++ | +++ | +++ | +++ |
| Potassium | PROT | ++ | +++ | +++ | +++ |
| Antioxidant nutrients | PROT | ++ | + | 0 | + (in foods, not supplements) |
| B vitamins | PROT | + | + | 0 | 0 |
| Fruit and vegetables | PROT | ++ | ++ | 0 | + |
| Fish/fish oil | PROT | ++ | ++ | ++ | ++ |
| Nuts | PROT | ++ | ++ | 0 | ++ |
| Whole grains | PROT | +++ | + | 0 | +++ |
| Coffee (unfiltered, boiled) | INCR | + | +++ | 0 | + |
| Moderate alcohol | PROT | + | + | 0 | 0 |

0 = no or minimal evidence or the results probably due to confounding.

+, ++, +++ = suggestive, moderate, convincing evidence.

[a]The evidence is especially strong when ω6 PUFA replaces SFA, rather than when SFA or PUFA are considered in isolation.

experimental conditions with all foods provided or amongst free-living individuals responsible for their food preparation. Substitution of saturated and *trans*-unsaturated fatty acids with unsaturated fatty acids from unhydrogenated plant-derived oils (with a *cis*-configuration) results in a reduction in total and LDL cholesterol and an improvement in the ratio of total (or LDL) to HDL cholesterol. Substitution of saturated and *trans*-fats with carbohydrate will also result in a reduction in total and LDL cholesterol (Chapter 3), but the effect on the total (or LDL) to HDL cholesterol ratio depends upon the type of carbohydrate used as replacement (see section 19.6.3).

The effects of individual saturated fatty acids on the lipid profile also differ. Stearic acid has little effect on lipids and lipoproteins, with lauric, myristic, and palmitic acids having an LDL (and total) cholesterol raising effect (myristic and palmitic greater than lauric) when compared with unhydrogenated plant oils or carbohydrate. However, lauric acid has a marked HDL raising effect when compared with carbohydrate so that it appears to be associated with a more favourable ratio of HDL to total or LDL cholesterol than the other saturated fatty acids.

Omega 3 polyunsaturated oils, principally obtained from marine sources, have a somewhat different effect from plant oils and are associated with triglyceride and VLDL lowering and a variable effect on total and LDL cholesterol when replacing other macronutrients.

There are limited data relating to the effects of dietary fat on other cardiovascular risk factors. Replacing saturated fat with mono- and polyunsaturated fatty acids increases insulin sensitivity. There is no clear effect of the individual saturated and unsaturated fatty acids. Longer-chain saturated fatty acids (C14:0, C16:0, C18:0) have been shown to accelerate thrombosis in experimental animals via inhibition of anti-aggregatory prostacyclin and C18:0 may also

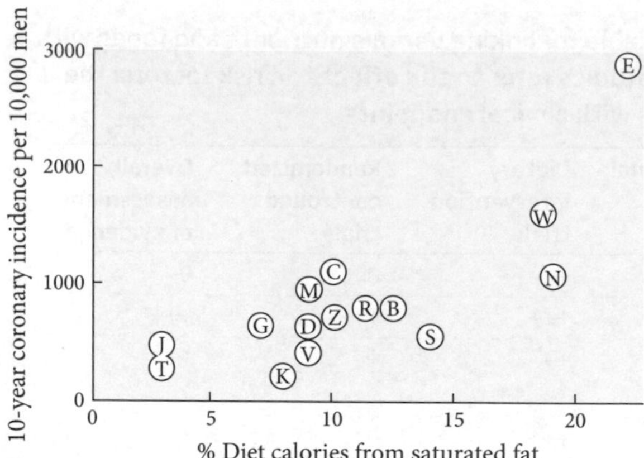


Fig. 19.6 Association between CHD and percentage energy derived from saturated fatty acids in the Seven Countries study. Letters on the graph indicate the location of the cohorts in the seven countries: B, Belgrade; C, Crevalcore; D, Dalmatia; E, East Finland; G, Corfu; J, Ushibuka; K, Crete; M, Montegiorgio; N, Zutphen; R, Rome railroad; S, Slavonia; T, Tanushimaru; V, Velika Krsna; W, West Finland; Z, Zrenjanin.

Reprinted by permission of the publisher from *Seven countries: a multivariate analysis of death and coronary heart disease* by Ancel Keys, p. 253, Cambridge, MA: Harvard University Press, © 1980 by the President and Fellows of Harvard College.

| Replacement | N | Cases | People | RR (95% CI) |
|---|---|---|---|---|
| 5% PUFA or linoleic acid | 17 | 22320 | 448921 | 0.89 (0.81, 0.98) |
| 5% MUFA | 4 | 10133 | 167855 | 1.00 (0.82, 1.21) |
| 5% Plant MUFA | 2 | 4419 | 93384 | 0.83 (0.69, 1.00) |
| 5% Animal MUFA | 2 | 4419 | 93385 | 1.06 (0.80, 1.41) |
| 5% Protein | 2 | 2466 | 40319 | 1.26 (1.06, 1.50) |
| 5% Plant protein | 2 | 2466 | 40319 | 0.83 (0.61, 1.12) |
| 5% Animal protein | 2 | 2466 | 40319 | 1.31 (1.14, 1.50) |
| 5% Carbohydrate | 6 | 10458 | 313066 | 0.98 (0.88, 1.09) |
| 5% Slowly digested CHO | 7 | 12641 | 225278 | 0.94 (0.89, 0.99) |
| 5% Moderately digested CHO | 3 | 4409 | 93963 | 1.03 (0.79, 1.34) |
| 5% Rapidly digested CHO | 7 | 12641 | 225278 | 1.08 (0.99, 1.17) |
| 2% TFA | 2 | 7667 | 127536 | 1.06 (0.89, 1.26) |

0.6 1 1.6

Fig. 19.7 Relative risk of coronary heart disease when dietary saturated fat intake is substituted by other macronutrients.

*Source:* Reynolds, A.N., Hodson, L., de Souza, R., Diep Pham, H.T., Vlietstra ,L., Mann, J., (2022). Saturated fat and trans-fat intakes and their replacement with other macronutrients: A systematic review and meta-analyses of prospective observational studies. Geneva: World Health Organisation; 2022. Licence: CCBY-NC-SA3.0IGO

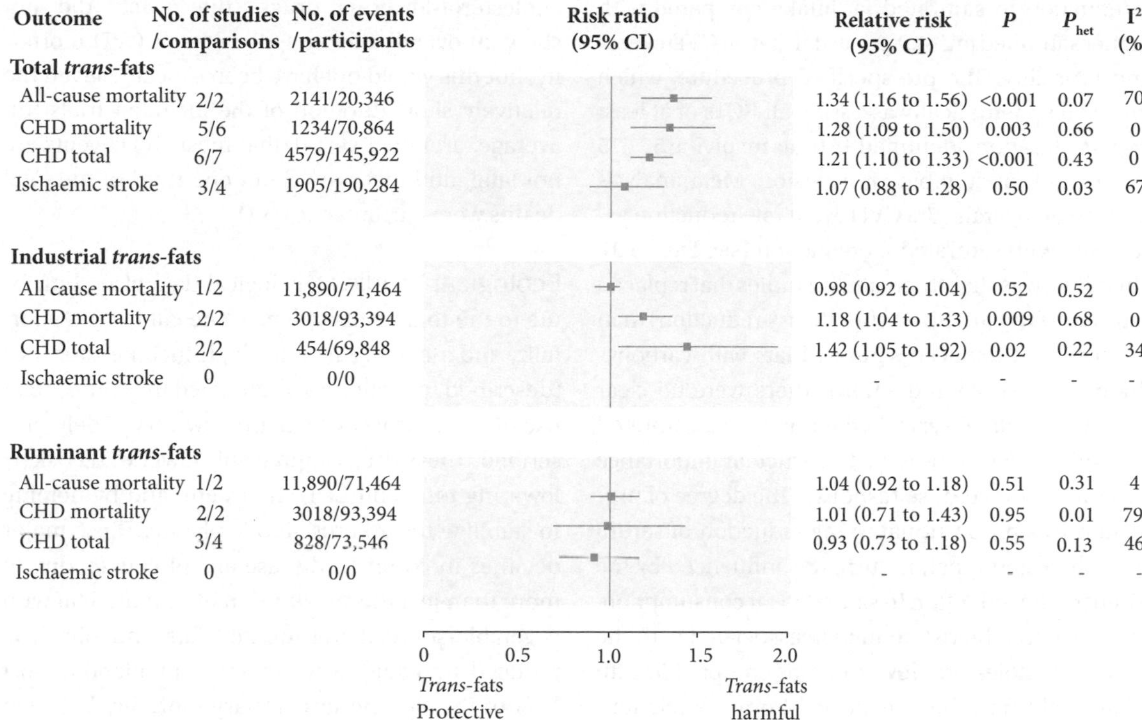

| Outcome | No. of studies /comparisons | No. of events /participants | Relative risk (95% CI) | *P* | $P_{het}$ | $I^2$ (%) |
|---|---|---|---|---|---|---|
| **Total *trans*-fats** | | | | | | |
| All-cause mortality | 2/2 | 2141/20,346 | 1.34 (1.16 to 1.56) | <0.001 | 0.07 | 70 |
| CHD mortality | 5/6 | 1234/70,864 | 1.28 (1.09 to 1.50) | 0.003 | 0.66 | 0 |
| CHD total | 6/7 | 4579/145,922 | 1.21 (1.10 to 1.33) | <0.001 | 0.43 | 0 |
| Ischaemic stroke | 3/4 | 1905/190,284 | 1.07 (0.88 to 1.28) | 0.50 | 0.03 | 67 |
| **Industrial *trans*-fats** | | | | | | |
| All cause mortality | 1/2 | 11,890/71,464 | 0.98 (0.92 to 1.04) | 0.52 | 0.52 | 0 |
| CHD mortality | 2/2 | 3018/93,394 | 1.18 (1.04 to 1.33) | 0.009 | 0.68 | 0 |
| CHD total | 2/2 | 454/69,848 | 1.42 (1.05 to 1.92) | 0.02 | 0.22 | 34 |
| Ischaemic stroke | 0 | 0/0 | - | - | - | - |
| **Ruminant *trans*-fats** | | | | | | |
| All-cause mortality | 1/2 | 11,890/71,464 | 1.04 (0.92 to 1.18) | 0.51 | 0.31 | 4 |
| CHD mortality | 2/2 | 3018/93,394 | 1.01 (0.71 to 1.43) | 0.95 | 0.01 | 79 |
| CHD total | 3/4 | 828/73,546 | 0.93 (0.73 to 1.18) | 0.55 | 0.13 | 46 |
| Ischaemic stroke | 0 | 0/0 | - | - | - | - |



Fig. 19.8 Fat, carbohydrates, and heart disease: estimated percentage of changes in the risk of coronary heart disease associated with isocaloric substitution of one dietary component for another.

*Source:* Li, Y., Hruby, A., Bernstein, A.M., et al. (2015) Saturated fats compared with unsaturated fats and sources of carbohydrates in relation to risk of coronary heart disease. A prospective cohort study. *J Am Coll Cardiol*, **66**(14), 1538–48, page 1545, with permission from Elsevier.

promote thrombogenesis by increasing fibrinogen. Data in humans are limited. Polyunsaturated fatty acids, especially ω3 fatty acids, are potentially important anti-thrombotic nutrients by virtue of their ability to reduce the tendency for platelets to aggregate. The clue to this effect was the observation made in the Inuit people of Greenland, who have low rates of CHD and reduced platelet aggregation compared with people in Western nations, despite high intakes of total fat. However, this fat comes largely from marine foods rich in ω3 fatty acids (C20:5 and C22:6), which form the antiaggregatory prostanoid $PG1_3$; C20:5 and C22:6 inhibit conversion of arachidonic acid to thromboxane $A_2$, as well as facilitate the production of the additional antiaggregatory substance $PGI_3$. Polyunsaturated fatty acids of the ω6 series may reduce platelet aggregation by providing the series 1 prostanoid $PGE_1$, which is also antiaggregatory. Oleic acid may act as an inhibitor of platelet aggregation, though the effect is less than for polyunsaturated fatty acids. Diets high in total fat, notably saturated fat, are associated with increased levels of coagulation Factor VII.

**Randomized controlled trials** Intervention trials, which are of sufficient size and duration to examine the effects of altering diets on mortality or clinical endpoints, are fraught with difficulties, mainly in relation to achieving long-term dietary compliance. Nevertheless, systematic reviews and meta-analyses provide powerful means of testing the associations suggested in observational studies and studies examining the effects of dietary manipulation on risk factors.

The 2020 updated Cochrane Review examining the effect of reducing saturated fat intake on cardiovascular disease risk reported on the randomized controlled trials (RCTs) that assessed the effect of a reduction in saturated fat intake compared with higher saturated fat intake or usual diet on CVD events and mortality. The pre-specified procedure, which involved systematically assessing all RCTs of at least 2 years' duration, identified 15 trials involving 56,675 participants as suitable for inclusion. Meta-analysis showed an overall 17% CVD event rate reduction associated with saturated fat reduction (see Fig. 19.9). This effect was slightly greater in studies that replaced saturated fats with PUFAs (a 21% risk reduction) than in those that replaced saturated fats with carbohydrates (a 16% risk reduction). There were no clear health benefits observed when replacing saturated fats with MUFAs or protein. Of particular importance was the observed dose response. The degree of protection was proportional to the reduction of serum total cholesterol that, in turn, was influenced by the extent of the reduction in saturated fat consumption. Furthermore, the risk reduction associated with the observed cholesterol lowering was compatible with what might have been expected from the relationship between cholesterol and CVD risk observed in prospective observational studies and RCTs of cholesterol-lowering drugs. The review did not show an overall effect on all-cause or CVD mortality, but this would not have been expected given the relatively short duration of the included trials (on average, around 4 years), that most CVD events are not fatal, and that only about one-third of the total deaths were attributed to CVD.

**Ecological studies** Ecological data also contribute to the totality of evidence. The fall in CVD mortality and incidence in many high-income countries (described in section 19.4) predated the widespread use of statin drugs, which are now very widely prescribed. These drugs appreciably lower CHD risk by lowering total and LDL cholesterol and by helping to stabilize the atherosclerotic plaque. These major declines in coronary disease are, of course, due to more than just the substitution of saturated fat with vegetable-sourced unsaturated fats and oils (e.g. reduced smoking). Nevertheless, in Finland almost 40% of the decline in coronary mortality between 1982 and 1997 was attributed to a decline in blood cholesterol, of which about 60% was attributed to

| Study or Subgroup | lower SFA Events | lower SFA Total | higher SFA Events | higher SFA Total | Weight | Risk Ratio M-H, Random, 95% CI |
|---|---|---|---|---|---|---|
| Black 1994 | 0 | 66 | 2 | 67 | 0.3% | 0.20 [0.01, 4.15] |
| DART 1989 | 136 | 1018 | 147 | 1015 | 13.6% | 0.92 [0.74, 1.15] |
| Houtsmuller 1979 | 8 | 51 | 30 | 51 | 4.7% | 0.27 [0.14, 0.52] |
| Ley 2004 | 11 | 88 | 16 | 88 | 4.4% | 0.69 [0.34, 1.40] |
| MRC 1968 | 62 | 199 | 74 | 194 | 12.0% | 0.82 [0.62, 1.07] |
| Moy 2001 | 5 | 117 | 3 | 118 | 1.4% | 1.68 [0.41, 6.87] |
| Oslo Diet-Heart 1966 | 64 | 206 | 90 | 206 | 12.5% | 0.71 [0.55, 0.92] |
| Rose corn oil 1965 | 15 | 28 | 6 | 13 | 4.7% | 1.16 [0.59, 2.29] |
| Rose olive 1965 | 11 | 26 | 5 | 13 | 3.5% | 1.10 [0.48, 2.50] |
| STARS 1992 | 8 | 27 | 20 | 28 | 5.3% | 0.41 [0.22, 0.78] |
| Sydney Diet-Heart 1978 | 37 | 221 | 25 | 237 | 7.5% | 1.59 [0.99, 2.55] |
| Veterans Admin 1969 | 97 | 424 | 122 | 422 | 13.2% | 0.79 [0.63, 1.00] |
| WHI 2006 (1) | 1399 | 19541 | 2145 | 29294 | 16.8% | 0.98 [0.92, 1.04] |
| **Total (95% CI)** | | **22012** | | **31746** | **100.0%** | **0.83 [0.70, 0.98]** |
| Total events: | 1853 | | 2685 | | | |

Heterogeneity: $Tau^2 = 0.04$; $Chi^2 = 36.65$, df = 12 (P = 0.0003); $I^2 = 67\%$
Test for overall effect: Z = 2.17 (P = 0.03)
Test for subgroup differences: Not applicable



**Footnotes**
(1) Total CVD during study period, Prentice 2017

Fig. 19.9 Risk ratios for all cardiovascular events (the primary outcome) when comparing usual diet with diets involving saturated fatty acid reduction.

*Source:* Hooper L, Martin N, Jimoh OF, Kirk C, Foster E, Abdelhamid AS. Reduction in saturated fat intake for cardiovascular disease. Cochrane Database of Systematic Reviews 2020, Issue 8. Art. No.: CD011737. DOI: 10.1002/14651858.CD011737.pub3.

changes in the composition of dietary fats. Of note, trends in total blood cholesterol levels will underestimate the impact of changing dietary saturated fat intake on blood lipid levels, because while LDL cholesterol levels have been steadily falling, HDL cholesterol levels have been rising.

### 19.6.2 Dietary cholesterol

Cholesterol in the blood is derived from endogenous synthesis and dietary intake, principally from dairy fat and meat, which are also important sources of saturated fatty acids, and eggs, which are not. Dietary cholesterol raises plasma LDL and total cholesterol, especially when consumed in substantial amounts and when intake of saturated fatty acids is also high. There is some, though not entirely consistent, evidence from observational studies that increasing intakes are associated with increasing CHD risk. However, no clinical trial has examined the effect of reducing dietary cholesterol without also substantially reducing saturated fatty acids. From a practical point of view, restriction of saturated fatty acids will be associated with a reduction in the dietary cholesterol except in individuals with an unusually high intake of egg yolk.

### 19.6.3 Carbohydrate

The observation that populations consuming high-carbohydrate/low-fat diets (e.g. in many Asian countries where rice is the staple, and in African countries where maize is the staple) were at low risk of CHD, and that high-carbohydrate, low-fat diets were associated with low levels of total cholesterol, led to the suggestion that high-carbohydrate diets were cardioprotective. However, given that sugars and rapidly digested starches now contribute a relatively high proportion of carbohydrate intake in many countries, the various sources of carbohydrate must be considered individually. High intakes of rapidly digested carbohydrates (free sugars, especially when consumed as sugar-sweetened beverages and rapidly digested starches) may be associated with an atherogenic lipid profile (increased ratios of total:HDL cholesterol, increased triglyceride) when compared under isocaloric conditions with fats or proteins. On the other hand, dietary fibre and whole grain cereals have been shown to be protective against CHD (see Fig. 19.10), especially if they are consumed as part of a diet that is low in saturated fat or substituted for saturated fat intake. Intake of slowly digested carbohydrates high in dietary fibre is also associated with decreased incidence of diabetes in observational studies, and foods high in several different soluble forms of dietary fibre (notably pectins, gums, mucilages) have the potential to lower total and LDL cholesterol without unfavourably influencing HDL cholesterol and consequently the ratio to total or LDL cholesterol (see Box 19.2). Given that the proportion of total energy provided by carbohydrate is determined 'by difference' (i.e. the energy required after protein and fat requirements are met), selection

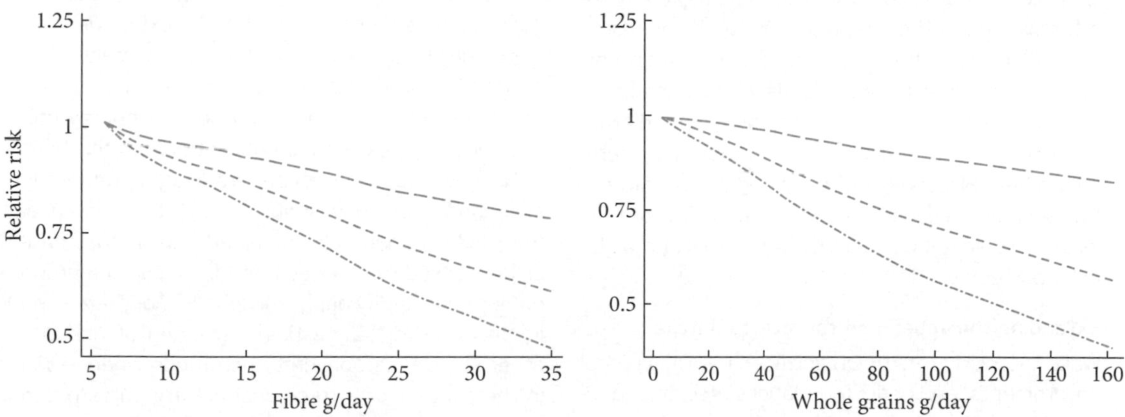


Fig. 19.10 Dose-response relationships between dietary fibre (L) or whole grain (R) intake and coronary heart disease incidence from prospective studies.

*Adapted from:* Reynolds, A., Mann, J., Cummings, J., et al. (2019) Carbohydrate quality and human health: a series of systematic reviews and meta-analyses. Lancet, 393.

### Box 19.2 Confusion about fats, carbohydrates, and coronary heart disease

Misinterpretation of the science relevant to dietary fats and carbohydrates has caused some confusion amongst health professionals and the public at large.

#### Very low carbohydrate diets

A small but vocal group, including protagonists of the so-called Paleo and Ketogenic diets, has claimed that there is no need for either saturated or total fat restriction. Instead, they recommend radical restriction of all carbohydrate and elimination as far as possible of all cereal-based foods, starchy vegetables, and sugars, except for those in intact fruits. The justification provided is that such high-fat, high-protein diets are less likely to promote obesity, facilitate weight loss in those already overweight and obese, and reduce insulin resistance and risk of diabetes and other non-communicable diseases. It is claimed that recommending restriction of saturated and/or total fat will lead to the increased consumption of sugars and rapidly digested starches, which, in addition to promoting body fatness, will result in lowered HDL cholesterol and increased triglyceride levels. Substitution of total and/or saturated fat with free sugars and rapidly digested starches may have undesirable metabolic effects, but dietary recommendations made by governments and health-related non-governmental organizations (e.g. heart foundations) recommend fibre-rich, minimally processed, carbohydrate-containing foods as substitutes for those rich in saturated fats, not sugars and rapidly digested starches. Although very low carbohydrate diets may facilitate weight loss in the short term, there is no evidence that they are more effective than other dietary patterns in long-term weight control. The exclusion of whole grain cereals potentially eliminates foods with wide-ranging health benefits. This very low carbohydrate dietary approach has never been shown to have long-term health benefits and is not endorsed by any authoritative bodies. Widespread adoption of high protein intakes would not be sustainable in terms of the global food supply.

#### Extending the upper limit for total fat intake

While the World Health Organization (WHO), national governments, and heart foundations recommend restriction of SFAs to below 10% total energy, there is some difference of opinion as to the optimum upper limit of total fat intake, or indeed whether there needs to be an upper limit, provided the fat is derived largely from unhydrogenated vegetable (other than coconut or palm) or marine oils. Current recommendations regarding the upper limit of total fat range from 30–40% total energy and some experts would have it even higher. This apparent 'confusion' is readily resolved. In considering only coronary heart disease, the science suggests that what matters is principally the nature of dietary fat, so total fat intake is unimportant. However, in the context of global public health, obesity and its comorbidities are of prime importance. A high fat intake will result in an energy-dense diet, which in many individuals and populations is likely to promote obesity. Thus, in the twenty-first century, for most of the world, it is appropriate to recommend some degree of total fat restriction.

#### Is restriction of rice and other starchy carbohydrate-rich food always necessary?

Traditional diets throughout much of Asia are rice-based. Yet in many of these countries, CHD rates are relatively low and in Japan, life expectancy is amongst the longest in the world. There is concern in China regarding the reduction in rice and total carbohydrate consumption, which is occurring in association with an increase in fat intake and increasing rates of obesity, diabetes, and CHD. It is conceivable that eating practices explain different effects amongst different ethnic groups. Rice consumed at cooler temperatures or following reheating (as may occur to a greater extent amongst traditional high rice consumers than in the context of Western diets) may be associated with increased gelatinization and slowed starch digestion. Whether or not foods are consumed in the context of energy balance or energy excess may also influence metabolic outcomes. Japanese people are taught from early childhood that they should complete their meal feeling less than fully satisfied. Nutritional principles derived from a single ethnic or cultural group do not necessarily apply universally. So, there is no justification for suggesting restriction of rice and other starchy carbohydrate-rich foods when eaten in the context of a traditional dietary pattern and energy balance.

of the most appropriate carbohydrate-containing foods is particularly important. Encouraging the consumption of 'minimally processed' carbohydrate-containing foods, which are high in dietary fibre and have a low glycaemic index, has been recommended as a means of encouraging the most suitable carbohydrate food sources.

### 19.6.4 Sodium

Reduction in dietary sodium in populations has been identified by WHO as one of the most important ways to reduce non-communicable diseases globally. The association between a high sodium intake and elevated blood pressure is well established and has been demonstrated in a range of studies, including animal studies, ecological and observational studies, and randomized controlled trials. There is increasing evidence from high quality cohort studies and RCTs that a high sodium intake is also associated with increased risk of CHD and stroke and that reduction is beneficial.

The best available method for assessing sodium intake is 24-hour urinary sodium excretion. This method and standardized blood pressure measurements were used in the INTERSALT study, which collected cross-sectional data on 10,000 people in 32 countries. The results showed an association between blood pressure and sodium excretion in individuals and populations. Importantly, it also showed that in populations with low salt intakes, there was very little rise in blood pressure with age. The positive association between sodium intake and blood pressure has been confirmed in randomized controlled trials of dietary interventions. A Cochrane review published in 2020, which pooled data from 131 trials lasting at least 7 days, showed that a reduction in sodium excretion resulted in a mean change in systolic blood pressure of between –1.1 mmHg and –4.0 mmHg among white and black participants, respectively, described as 'normotensive', and between –5.7 mmHg and –6.6 mmHg among white and black participants described as 'hypertensive'. Similar decreases were seen in diastolic blood pressure levels.

In recent years there has been some controversy about the association between sodium intake and cardiovascular disease outcomes. This has followed publication of a few cohort studies that demonstrate a J-shaped association, leading to a suggestion of increased risk of cardiovascular disease at lower levels of sodium intake. These studies have been shown to be methodologically flawed, as they include only a single spot urine at baseline as measure of usual sodium intake. Spot urine has been widely demonstrated as being an inaccurate and inappropriate measure of sodium intake in individuals, although it is sometimes used to estimate sodium intake in populations. A recently published cluster randomized controlled trial of use of salt substitutes (containing decreased sodium and increased potassium) among nearly 21,000 Chinese adults aged 60 years and above has demonstrated significant decreases in risk of stroke, major cardiovascular events, and death over a 5-year period. Further evidence comes from long-term follow-up of participants who participated in the Trials of Hypertension Prevention randomized controlled trial of dietary sodium reduction. This has shown that 15 years after the completion of the trial period, there was a reduction in CVD events of 25–30% amongst the people initially randomized into the low sodium diet groups in the original trials, and 20 years after the trial period there was a lower rate of total mortality amongst those in the low sodium group. A 2011 meta-analysis of sodium reduction trials showed that dietary sodium reduction was associated with a 20% reduction in cardiovascular events over a 5-year period.

Several mechanisms have been suggested to explain the association between salt intake and blood pressure, including reduced urinary sodium excretion and fluid retention by some individuals, as well as direct effects on the vascular endothelium. The heterogeneity in the response of individuals to sodium restriction suggests the existence of a group of particularly 'salt-sensitive' individuals, but there is as yet no simple test by which such individuals might be identified.

The WHO has set a global target of a 30% reduction in population sodium intake by 2025 as one of nine strategies in its Global Action Plan to reduce non-communicable disease worldwide. The WHO currently recommends that adults consume no more than 2000 mg (87 mmol) of sodium per day (equivalent to 5 g salt/day) with lower recommended sodium intakes for children.

## 19.6.5 Potassium and calcium

**Potassium** In the INTERSALT study, urinary potassium excretion—an assumed indicator of intake—was negatively related to blood pressure. A pooled analysis of 21 randomized controlled trials demonstrated that increased potassium intake reduced systolic blood pressure by 3.49 mmHg (95% CI: 1.82–5.15) and diastolic blood pressure by 1.96 mmHg (0.86–3.06) in adults, with further benefits seen when a higher potassium intake up to 90–120 mmol/day (3510–4700 mg/day) was consumed. Other studies have indicated an interaction between sodium and potassium, which has led the WHO to recommend a sodium/potassium molar ratio of 1.0.

**Calcium** Intracellular calcium is an important determinant of arteriolar tone, and some claims have been made that increased calcium intake can reduce blood pressure. However, two meta-analyses summarizing the results of more than 20 trials showed that intakes of 1000 mg or more of calcium per day have only a trivial effect on blood pressure, and some clinical trials of calcium supplementation have shown an association with an increased risk of myocardial infarction. On the other hand, the DASH diet, which is high in low-fat dairy products (and consequently also calcium) and promotes consumption of a range of cardioprotective foods, is associated with appreciable blood pressure lowering, especially when recommended in conjunction with reduction in sodium intake (see Box 19.3).

It is generally believed, therefore, that while low-fat dairy products may be a feature of cardioprotective diets, calcium supplements are not recommended for cardiovascular health.

## 19.6.6 Alcohol

Ecological, case-control, and cohort studies all suggest that people who drink moderately have a lower rate of myocardial infarction compared with non-drinkers and heavy drinkers. The effect is

**Box 19.3 The DASH diet**

The DASH diet is widely recommended for blood pressure lowering and prevention of cardiovascular disease. This diet is high in fresh fruit and vegetables and low-fat dairy products, so is relatively low in saturated fat, and high in potassium, calcium, and fibre. The DASH diet also includes nuts, whole grains, fish, and poultry, and limits red meat and sugary foods. Carefully controlled intervention trials, where subjects with moderately elevated blood pressure (120–159 mmHg systolic) were provided with all their food, compared the DASH diet with a typical Western diet, both with sodium content typical of the US diet, and demonstrated that the DASH diet resulted in lower blood pressure. The effect was even more striking when the DASH diet was combined with a low sodium intake (see Fig. 19.11). There is some debate about which of the individual components of the DASH diet might be responsible for its blood pressure lowering effect; however, it is generally regarded that a DASH-style dietary pattern is cardioprotective.

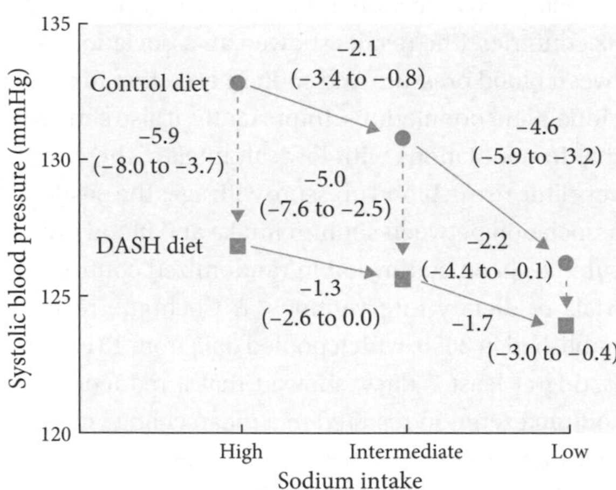


**Fig. 19.11** The effect on systolic blood pressure with a control diet when sodium is reduced (●) and on the DASH diet with similar reductions in sodium (■). Differences between DASH and control diets are significant at each level of sodium intake but smallest when sodium intake is low (1.5 g/day).

*Source:* Sacks, F.M., Svetkey, L.P., Vollmer, W.M., et al. (2001) Effects on blood pressure of reduced dietary sodium and the dietary approaches to stop hypertension (DASH) diet. *N Engl J Med*, **344**(1), 3–10. With permission from Massachusetts Medical Society.

apparent for all alcoholic drinks and has been attributed to the HDL-raising effect of alcohol, or to the antioxidant content of some alcoholic beverages. However, alcohol consumption is associated with increased risk of other important cardiovascular and health risks, including stroke, heart failure, and fatal hypertensive disease. Alcohol intake is associated with elevated blood pressure above around one standard drink per day for women or two standard drinks per day for men. Furthermore, it is possible, indeed likely, that the apparent benefits of a moderate intake of alcohol are attributable to other healthy lifestyle attributes of those choosing to drink alcohol in moderation, and so the degree to which confounding may explain the apparent protective effect demonstrated in observational studies is uncertain.

### 19.6.7 Coffee

Boiled unfiltered coffee raises total and LDL cholesterol because of the cafestol content of coffee beans. Such unfiltered coffee is still widely consumed in Greece, the Middle East, and Turkey. A shift from coffee prepared in this way to filtered coffee is believed to have contributed appreciably to the decline in plasma cholesterol in Finland. There is no evidence that moderate consumption of coffee as usually consumed in most other countries is related to CHD.

### 19.6.8 Antioxidant nutrients

Experimental evidence has clearly demonstrated the potential for antioxidant nutrients (e.g. vitamin E, vitamin C, β-carotene, flavonoids) to reduce the oxidizability of LDL *in vitro*, and prospective studies have shown a decreasing risk of CHD with increasing intakes of these nutrients and flavonoids, which occur in a variety of foods of vegetable origin, such as onions, berries, apples, and tea. However, several large, clinical trials in which vitamins E and C and β-carotene have been given as supplements have shown no consistent benefit. Indeed, vitamin E supplementation has been studied in a sufficient number of studies for meta-analyses to confirm absence of benefit in terms of reducing cardiovascular risk. Thus, it is conceivable that benefit only accrues when these nutrients are provided in combinations present in foods and there is no justification to support the use of these nutrients as dietary supplements with the expectation of reducing cardiovascular risk.

### 19.6.9 B vitamins

High homocysteine levels have been associated with increased risk of CHD and stroke in case-control and cohort studies. Dietary folate and folic acid as a supplement or fortificant result in lowering of homocysteine by facilitating the methylation of homocysteine to methionine (see Chapter 12) and vitamins $B_6$ and $B_{12}$ may further reduce homocysteine. However, randomized controlled trials have not shown consistent benefits of homocysteine lowering on CVD events despite demonstrating lowering of homocysteine levels, and meta-analyses of trials have shown no benefit with respect to CHD or stroke. It has been suggested that homocysteine may be a marker of elevated risk, rather than a cause.

## 19.7 Foods and food groups

**Fruits, vegetables, and berries** Ecological and prospective studies have reported reduced risk of CHD and stroke associated with relatively high intakes of fruits, vegetables, and berries. The DASH trials have shown convincing benefits in terms of reducing blood pressure levels, especially when fruits and vegetables are consumed together with relatively high intakes of low-fat dairy products and a reduced intake of sodium. Antioxidant nutrients, flavonoids, and dietary fibre may further explain the protective effect of these foods.

**Fish** Many prospective studies have shown a reduced risk of CHD in association with fish consumption, especially fatty fish, which are rich sources of long-chain ω3 polyunsaturated fatty acids (C20:5, C22:6), with one recent systematic review suggesting that high-risk populations might halve CHD deaths if fish intake was increased to 40–60 g/day. All-cause mortality may also be reduced by an increase in fish consumption. The DART and GISSI-Prevenzione trials provide some support for the clinical benefits of regular consumption of fish and fish oils (C20:5, C22:6), especially in high-risk individuals with established cardiovascular disease.

Nuts Several large epidemiological studies have reported decreased CHD risk in association with frequent consumption of nuts. Nuts are high in unsaturated fats, and regular consumption results in a favourable alteration in the fatty acid profile of plasma lipids, as well as some reduction in atherogenic lipoproteins. A reduction in clinical events has been reported in one trial, in which daily consumption of nuts has been recommended in the context of a Mediterranean diet. Any recommendations to include nuts in the diet must be tempered with a reminder of their relatively high-energy content and that many commercially available nut products have added salt and may be roasted with added fats or oils.

Soy Soy protein has a favourable effect on several cardiovascular risk factors. An overview of 38 clinical studies suggests that a consumption of 47 g of soy protein daily leads to a 9% decline in total cholesterol and a 13% fall in LDL cholesterol. Soy isoflavones have been shown to lower blood pressure, and there is some evidence of beneficial effects on vascular and endothelial function, platelet aggregation, smooth muscle cell proliferation, and LDL oxidation. Soy protein has been shown to inhibit atherosclerosis in animals.

## 19.8 Nutritional strategies for high-risk populations

It is essential in countries with high CHD rates to have in place a nutritional strategy that is aimed at the entire population. Individuals with extreme levels of risk factors or with several different risk factors are at the highest risk of CHD and will benefit most, as individuals, from dietary modification. However, the majority of cases of CHD will occur in people at moderate risk, i.e. those with one or two risk factors that may be modestly elevated. This is simply because there are far more such individuals in the population than people at very high risk. Thus, in order to reduce (or in those countries where rates are already coming down, to further reduce) the epidemic proportions of CHD, population change is essential. The disadvantage of the 'population approach' is that many individuals are being asked to make changes that are likely to produce a relatively small reduction in their personal CHD risk (the 'prevention paradox') and those who are at high risk will still need to be individually identified because they are likely to require more radical and individually designed lifestyle changes and perhaps medication.

Most high-risk countries have in place nutrient targets as well as dietary guidelines designed to help implement the nutrient targets. Although these are intended to promote health, prevent nutrient deficiencies, and reduce the risk of all nutrition-related non-communicable diseases, they are strongly influenced by the requirement to minimize cardiovascular risk. Table 19.3 summarizes recent targets recommended in the United States and the Nordic countries and by the WHO for those nutrients of particular relevance to CHD. There has been a trend towards liberalizing the total fat intake provided the amount of saturated fatty acids is not increased.

## 19.9 Dietary patterns

More recently, there has been interest in certain dietary patterns that are regarded as protective against CHD (see also Chapter 30). The traditional Mediterranean diets of Italy, Greece, and Spain have been singled out as being of particular benefit, principally because populations consuming

**Table 19.3 Ranges of intake goals (for adults) for macronutrients considered to be of particular relevance to coronary heart disease as recommended for the Nordic countries, the United States, and by the WHO/FAO**

| | Nordic countries[a] (2012) | United States[b] (2020) | WHO[c] (2003, 2007, 2012) AMDR | U-AMDR |
|---|---|---|---|---|
| Total fat | 25–40% | 20–35% | 30% | |
| Saturated fatty acids (SFA) | <10% | <10% | | 10% |
| Total *cis*-polyunsaturated fatty acids (PUFA) | 5–10% (at least 1% from ω3 PUFA) | | | |
| ω6 PUFA<br>ω3 PUFA | | | 2.5–9%<br>0.5–2% | |
| Linoleic acid g/day | | 12–17* | | |
| *cis*-Monounsaturated fatty acids (MUFA) | 10–20% | | By diff. | |
| Dietary cholesterol (mg/day) | | | | 300 |
| Dietary fibre (g/day) | 23–35 | 28–34* | | |
| Fruit and vegetables (g/day) | | | >400 | |
| Sodium (mg/day) | <2400 | <2300 | | <2000 |

*Depending upon age and sex.
Goals are expressed as percentage of total energy, unless otherwise stated.
Total fat = SFA + PUFA + *cis*-MUFA.
AMDR, accepted macronutrient distribution range for adults.
U-AMDR, upper AMDR.
These targets may be readily translated into food-based dietary guidelines appropriate to food preferences of different population groups.
*Adapted from:*
[a]Nordic Council of Ministers (2013) *Nordic nutrition recommendations 2012: summary, principle and use.* Copenhagen: Nordic Council of Ministers.
[b]US Department of Agriculture and US Department of Health and Human Services (2020) *Dietary guidelines for Americans, 2020–2025*, 9th edn. Available at **DietaryGuidelines.gov**.
[c]WHO/FAO (2003) *Diet, nutrition and the prevention of chronic diseases*, WHO Technical Report Series 916. Geneva: World Health Organization. WHO/FAO (2003) updated following WHO/FAO (2007) and WHO Guideline: sodium intake for adults and children (2012).

the traditional diets of countries surrounding the Mediterranean Sea had remarkably low rates of CHD and relatively low levels of several cardiovascular risk factors. There is some evidence from clinical trials that Mediterranean-type diets protect against cardiovascular events. The traditional Japanese diet and the dietary pattern of the Nordic countries, which are characterized by foods very different from those consumed in the Mediterranean countries, are also cardioprotective. Japanese people have low rates of CHD and the greatest life expectancy worldwide.

Vegetarian diets have been promoted for their apparent cardioprotective effect. Vegetarians are indeed at lower risk of CHD than meat eaters, but it has not been established which attributes of the vegetarian diet might be protective, since there are many aspects other than the avoidance of meat that characterize these diets. One study suggests that the lower rates might be due to the relatively low intake of saturated fatty acids, rather than meat avoidance. Although prospective studies have attempted to control for confounding factors, it remains conceivable that the healthy lifestyle of many vegetarians may at least to some extent account for this cardioprotective effect.

While Mediterranean, Japanese, and vegetarian diets may indeed be associated with reduced cardiovascular risk, there are other equally cardioprotective dietary patterns. The attributes of cardioprotective diets listed in Table 19.4 are seen in other traditional diets. Furthermore, the typical Western diet can be modified to follow these dietary principles, as has been the case with the DASH diet (see Box 19.3). It is important to note that adopting individual attributes of potentially appropriate cardioprotective diets may not confer benefit. For example, consuming substantial quantities of olive oil, a well-recognized feature of Mediterranean diets, in the context of an otherwise inappropriate diet may confer little or no advantage. Unfortunately, some populations that previously consumed cardioprotective diets and had low rates of CHD have introduced foods not traditionally consumed. For instance, many modern Mediterranean or Asian-style diets may be high in saturated fatty acids and, thus, at least to some extent, have lost their health benefits.

**Table 19.4 Attributes of cardioprotective dietary patterns**

| |
|---|
| Low intakes of saturated fatty acids |
| High intakes of raw or appropriately prepared fruits and vegetables |
| Whole grains and lightly processed cereal foods are preferred, though white rice not excluded in the context of the traditional Japanese diet |
| Fat intakes are predominantly derived from unmodified vegetable oils[a] |
| Fish, nuts, seeds, and vegetable protein sources are important dietary components |
| Meat, when consumed, is lean and eaten in small quantities |
| Energy balance reduces rates of obesity |

[a]Coconut oil and palm oil are not encouraged because they tend to elevate cholesterol and LDL.

Failure to ensure appropriate energy balance may negate the benefits of adhering to many of the other attributes listed in Table 19.4 because obesity-mediated cardiovascular risk (see Chapter 17) may outweigh favourable trends in other risk factors.

## 19.10 Gene–nutrient interactions

It has long been appreciated that not everyone exposed to a risk factor for CHD will develop the disease and that exposure to foods or nutrients known to influence risk factors will produce a range of responses within a group of individuals all consuming the same diet. For example, while modifying dietary fatty acid composition may alter total and LDL cholesterol to an extent that may, in a population or group of individuals, be predicted by a formula (Box 19.4), some individuals may show little or no change, whereas others will have striking changes in their levels when the nature of dietary fat changes (Fig. 19.12). In the clinical context, failure to respond to dietary modification has often been attributed to failure to adhere to dietary advice. While non-adherence is indeed associated with absence of response or reduced response, variation has been confirmed in carefully controlled dietary experiments in which all participants were fed identical diets. Genetic factors undoubtedly account for this variability, and several genetic polymorphisms that predict plasma lipid responses to change in dietary fat have now been identified. For example, individuals carrying the *APOE4* allele of the *APOE* gene will show a more marked reduction in total and LDL cholesterol than those who do not carry it, when changed to a diet that is reduced in both fat and cholesterol. A common genetic polymorphism in the promoter region of the *APOAI* gene determines the extent to which HDL responds to substantial increases in polyunsaturated fatty acids.

Such observations have led to speculation that this study of nutrigenetics will soon lead to the development of 'designer diets', which will enable individualized dietary advice to be based on genetic characteristics. However, while this type of nutrigenetic research will no doubt continue to flourish,

## Box 19.4

A number of predictive equations for estimating changes in plasma cholesterol and lipoprotein concentrations in response to changes in dietary fatty acids and cholesterol have been developed using data from controlled feeding studies. The first of these was developed by Keys et al. (1965). More sophisticated equations were developed when the effects of different saturated and unsaturated fatty acids were identified. TC and LDL-C are given in mg/dL. These equations are useful for estimating the effects of dietary change at a population level.

| Authors | Equations |
|---|---|
| Keys et al. (1965) | $\Delta TC = 1.35(2\Delta S - \Delta P) + 1.52\Delta Z$ |
| Mensink and Katan (1992) | $\Delta TC = 1.51\Delta S - 0.12\Delta M - 0.60\Delta P$ |
| | $\Delta LDL\text{-}C = 1.28\Delta S - 0.24\Delta M - 0.55\Delta P$ |

Abbreviations: LDL-C, LDL cholesterol; M, percentage of monounsaturated fatty acids, all as percentage of total energy intake; P, percentage of polyunsaturated fatty acids; S, percentage of saturated fatty acids; Δ, change; TC, total cholesterol; Z, square root of daily dietary cholesterol. For mmol/L, multiply by 0.02586.

*Adapted from:* Kris-Etherton, P.M., Yu-Poth, S., Sabaté, J., et al. (1999) Nuts and their bioactive constituents: effects on serum lipids and other factors that affect disease risk. *Am J Clin Nutr*, **70**, 504S–11S.

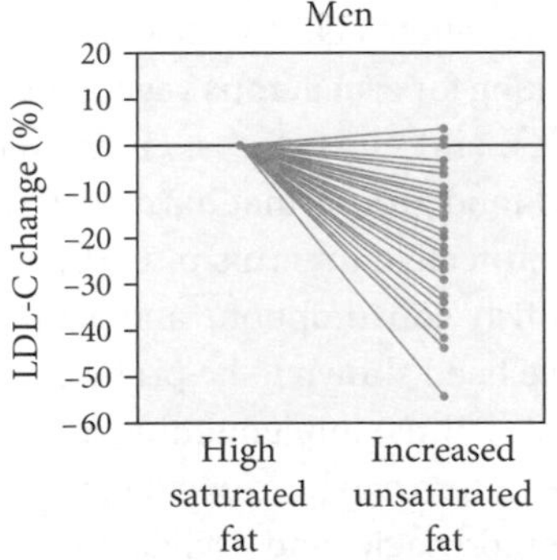


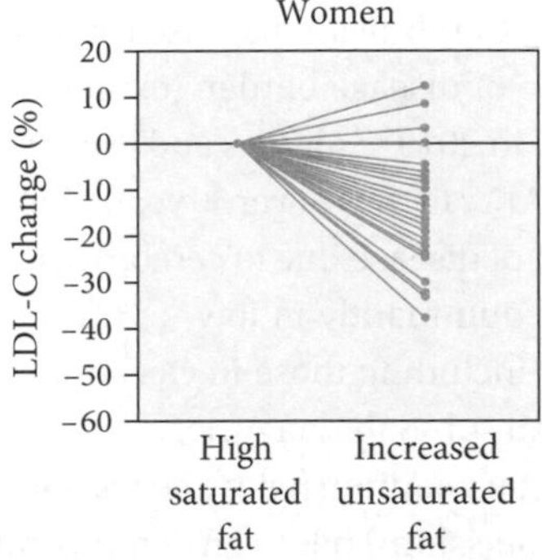


Fig. 19.12 Individual changes in low-density lipoprotein cholesterol (LDL-C) when *cis*-unsaturated fatty acids replace some of the saturated fatty acids in the diet.

*Source:* Schaefer, E.J. (2002) Lipoproteins, nutrition, and heart disease. *Am J Clin Nutr*, **75**, 191–212.

such expectations are premature for several reasons: polygenic, rather than monogenic, factors are almost certainly responsible for most of the heterogeneity in risk factor response to diet; no studies carried out thus far have involved sufficient participants to study the interaction of many genes; most research to date has centred on the variability in plasma lipid and lipoprotein response to changes in nature of dietary fat, and many other risk factors and nutrients are involved; and finally—and perhaps of greatest importance—there is convincing evidence that compliance with cardioprotective dietary patterns can profoundly reduce cardiovascular risk in high-risk populations and the majority of individuals who have raised risk factor levels.

Nutrigenomic research explores how nutrients influence gene expression, rather than the inter-individual differences in relation to the effects of nutrients. Of particular relevance to cardiovascular disease has been the research that has shown the profound effect of ω3 fatty acids on the activation of nuclear receptors, which are associated with inflammation and lipid metabolism, thus helping to explain their cardioprotective effect.

## 19.11 Cerebrovascular disease

Cerebrovascular disease presents clinically as stroke. The clinical features of stroke (typically loss or slurring of speech and weakness of one side of the body) result from the loss of blood supply to a section of the brain. There are two major types of stroke. Ischaemic strokes are the more common type and result from thrombosis and atheroma, the process being similar to that which results in myocardial infarction. The other cause of stroke is a bleed from one of the cerebral arteries (haemorrhagic stroke); it may be associated with a congenital abnormality (aneurysm). Like the clinical manifestations of CHD, stroke also has a major impact on public health because of its high frequency in most countries. Cerebrovascular disease mortality and DALYs have increased globally over the past three decades. Cerebrovascular disease was the third leading cause of disease burden (expressed as DALYs) worldwide in 2019 at all ages and the second leading cause of DALYs in those over 50 years (behind CHD). The burden of disease due to cerebrovascular disease is now predominantly in low- and middle-income countries, including those in Oceania, Asia, Eastern Europe, and sub-Saharan Africa.

Reduction in dietary sodium (with and without increase in dietary potassium) has been demonstrated in observational studies and RCTs to reduce stroke incidence. Other nutritional determinants have been far less studied than is the case for CHD, the risk factors for the two sets of conditions are generally similar, and therefore nutritional factors that predispose to CHD should also apply. Elevated blood pressure plays a uniquely important role in both types of stroke so that nutritional measures aimed at reducing blood pressure (most importantly restriction of dietary sodium but also reduced alcohol intake and increased intake of dietary potassium) is especially relevant in terms of reducing the risk of stroke. Raised levels of total and LDL cholesterol and obesity are clearly described risk factors for ischaemic stroke. Several large prospective cohort studies have suggested that increasing intakes of fruit and vegetables have a protective effect against stroke; the effect is particularly striking for cruciferous vegetables, green leafy vegetables, and citrus fruits. Of the remaining nutrients and food groups that have been identified as causal or protective in terms of CHD, ω3 fatty acids, regular fish consumption, and intake of whole grains have been shown to be protective against ischaemic stroke. Thus, implementing a cardioprotective dietary pattern can be expected to reduce the risk of both haemorrhagic and ischaemic stroke, especially if there is emphasis on reducing intake of sodium.

## Further reading

1. **Chrysant, S.G. and Chrysant, G.S.** (2018) The current status of homocysteine as a risk factor for cardiovascular disease: a mini review. *Expert Review of Cardiovascular Therapy*, **16**(8), 559–65.
2. **Coates, A.M., Hill, A.M., and Tan, S.Y.** (2018) Nuts and cardiovascular disease prevention. *Current Atherosclerosis Reports*, **20**(10), 48.
3. **Cook, N.R., Appel, L.J., and Whelton, P.K.** (2016) Sodium intake and all-cause mortality over 20 years in the trials of hypertension prevention. *J Am Coll Cardiol*, **68**(15), 1609–17.
4. **GBD 2019 Diseases and Injuries Collaborators** (2020) Global burden of 369 diseases and injuries in 204 countries and territories, 1990–2019: a systematic analysis for the Global Burden of Disease Study 2019. *The Lancet*, **396**(10258), 1204–22.

5. **GBD 2019 Risk Factors Collaborators** (2020) Global burden of 87 risk factors in 204 countries and territories, 1990–2019: a systematic analysis for the Global Burden of Disease Study 2019. *The Lancet*, **396**(10258): 1223–49.

6. **Hooper, L., Martin, N., Jimoh, O.F., Kirk, C., Foster, E., and Abdelhamid, A.S.** (2020) Reduction in saturated fat intake for cardiovascular disease. *Cochrane Database of Systematic Reviews* (8).

7. **Joseph, P., Leong, D., McKee, M., et al.** (2017) Reducing the global burden of cardiovascular disease, Part 1. *Circ Res*, **121**(6), 677–94.

8. **Keys, A., Anderson, J.T., and Grande, F.** (1965) Serum cholesterol response to changes in the diet: IV. Particular saturated fatty acids in the diet. *Metab*, **14**(7), 776–87.

9. **Keys, A., Menotti, A., Karvonen, M.J., et al.** (1986) The diet and 15-year death rate in the Seven Countries Study. *Am J Epidemiol*, **124**(6), 903–15.

10. **Li, Y., Hruby, H., Bernstein, A.M., et al.** (2015) Saturated fats compared with unsaturated fats and sources of carbohydrates in relation to risk of coronary heart disease. *J Am Coll Cardiol*, **66**(14), 1538–48.

11. **Mensink, R.P. and Katan, M.B.** (1992) Effect of dietary fatty acids on serum lipids and lipoproteins. A meta-analysis of 27 trials. *Arterioscler Thromb*, **12**(8), 911–19. doi: 10.1161/01.atv.12.8.911

12. **Mozaffarian, D.** (2016) Dietary and policy priorities for cardiovascular disease, diabetes, and obesity. *Circ*, **133**(2), 187–225.

13. **Neal, B., Wu, Y., Feng, X., et al.** (2021) Effect of salt substitution on cardiovascular events and death. *NEJM*, **385**(12), 1067–77.

14. **Reynolds, A., Mann, J., Cummings, J., et al.** (2019) Carbohydrate quality and human health: a series of systematic reviews and meta-analyses. *The Lancet*, **393**(10170), 434–45.

15. **Roth, G.A., Mensah, S.Y., Johnson, C.O., et al.** (2020) Global burden of cardiovascular diseases and risk factors, 1990–2019. *J Am Coll Cardiol*, 76(25), 2982–3021.

16. **Schaefer, E.J.** (2002) Lipoproteins, nutrition, and heart disease. *Am J Clin Nutr*, **75**, 191–212.

17. **WHO/FAO** (2003) *Diet, nutrition and the prevention of chronic diseases*, WHO Technical Report Series 916. Geneva: World Health Organization.

18. **Zanetti, D., Tikkanen, E., Gustafsson, S., et al.** (2018) Birthweight, type 2 diabetes mellitus, and cardiovascular disease. *Circ-Genom Precis Me*, **11**(6), e002054.

# Nutrition and Cancer

Kathryn E. Bradbury and Timothy J. Key



## 20.1 Introduction

Cancer is one of the leading causes of death worldwide. All cancers are characterized by loss of control over normal cell division and replication. Cancers can arise from the cells of different tissues and organs in the body, therefore there are many different types of cancers, and the causes vary between sites. Epidemiological studies showing large variations in cancer rates between populations indicate that the majority of cases of cancer are, at least in theory, preventable, and that environmental and lifestyle factors, including diet, can play important roles in the development of cancer.

Several major causes of cancer have been identified; the most important is tobacco smoke, which causes cancer of the lung and many other types of cancer. There are also known factors that afford protection against cancer, for example childbirth reduces the risk of cancers of the breast, endometrium, and ovary. However, for many cancers the factors that influence risk are not well established. Diet is thought to influence the risk of many cancers, but research to determine which dietary factors are related to risk is challenging and few dietary factors have been firmly established as causally related to cancer risk.

## 20.2 Pathophysiology of cancer

Cancer is defined as a disease in which the normal control of cell division is lost, so that an individual cell multiplies inappropriately to form a tumour. The tumour may eventually spread through the body and overwhelm it, causing death. Most cancers develop from a single cell that grows and divides more than it should, resulting in the formation of a tumour (growth). Cancers developing in most tissues take the form of a lump that grows, invades local non-cancerous tissue or adjacent tissues/organs, and may spread to other distant parts of the body through the bloodstream or lymphatic system. The majority of human tumours are of epithelial origin. Cancers arising in the cells of the blood, such as leukaemia, do not form a lump because the cells are floating freely throughout the bloodstream.

The change from a normal cell into a cancer, termed carcinogenesis, is a multistage process. Cancers represent a form of dedifferentiation that is associated with the loss of growth control and disturbances in the regulation of the cell cycle. Hanahan and Weinberg (2011) suggested that most cancers acquire the capability to do the following: sustain proliferative signalling, evade growth suppressors, resist cell death, enable replicative immortality, induce angiogenesis, activate invasion and metastasis, reprogramme energy metabolism, and evade immune destruction. Underlying these hallmarks is instability of the genome, and inflammation. The fundamental changes that determine carcinogenesis are mostly mutations in the DNA; therefore, cancer can be viewed as a genetic disease at the level of somatic cells. Typically the change from a normal cell to a cancer requires mutations in several different genes. Mutations in many different genes can result in cancer, but certain genes are frequently involved; in particular, cancer usually involves changes in the function of genes that control cell division (mitosis) and cell death (apoptosis). The key genes in carcinogenesis can be considered in two classes: oncogenes, which are genes that when over-activated lead to over-stimulation of cell growth and cell division, such as *Ras* and *Myc*; and tumour suppressor genes such as *p53*, which normally limit the rate of cell division but, if inactivated by a mutation, allow uncontrolled cell division. Recent research has shown that epigenetic changes are also important in tumour development. Epigenetic modifications, for example DNA methylation, can regulate gene expression without changing the DNA sequence, and epigenetic changes can be influenced by environmental factors, including diet (Wild et al., 2020).

The genetic mutations that lead to cancer development may be inherited (germline; see further discussion below) and/or arise during the lifetime of an individual (somatic) due to replication errors or the effects of ionizing radiation, chemical carcinogens, viruses, and endogenous damage (e.g. as caused by oxidants). The development of cells into a new cancer is also strongly influenced by various endogenous and exogenous growth-promoting agents, especially hormones.

Chance plays an important role in determining the occurrence of cancer in an individual. In simplified terms, if mutations in several key genes occur in several different cells then the behaviour of all these cells could remain normal, but if the same mutations all occur together within one cell, then this cell could give rise to a cancer. Chance, however, has little net effect on the incidence of cancer in populations (Swerdlow and Peto, 2020).

Dietary factors may be associated with cancer development in various ways. At the cellular level, these may be via DNA damage and repair, cell proliferation, carcinogen metabolism, apoptosis, and cell differentiation. Hormonal regulation, inflammation, and immune responses can also all be affected by dietary constituents. Experimental data in animals or cultivated human cells can provide the basis for hypotheses of effects of diet in humans and plausible mechanisms to explain observed associations; we do not discuss these data in this chapter, but focus instead on the data from dietary studies in humans.

## 20.3 Descriptive epidemiology of cancer

Worldwide, there were an estimated 19 million new cancer cases and 10 million deaths from cancer in the year 2020 (GLOBOCAN, 2020). The most common cancers (see Fig. 20.1) in terms of new cases were breast (2.3 million), lung (2.2 million), colorectal (1.9 million), prostate (1.4 million), and stomach (1.1 million).

Much of the evidence for the potential role of lifestyle and environmental factors in the determination of cancer risk comes from descriptive epidemiology, which looks at the variations of cancer rates with place and time (Swerdlow and Peto, 2020). These studies have shown that, for all the common cancers, rates vary widely (for many cancers by at least five-fold) between populations in different parts of the world; for example, Fig. 20.2 shows worldwide variation in the incidence of colorectal cancer. These variations are mostly not due to genetic make-up, because rates change with time within populations, and can also change markedly when people migrate from one country to another.

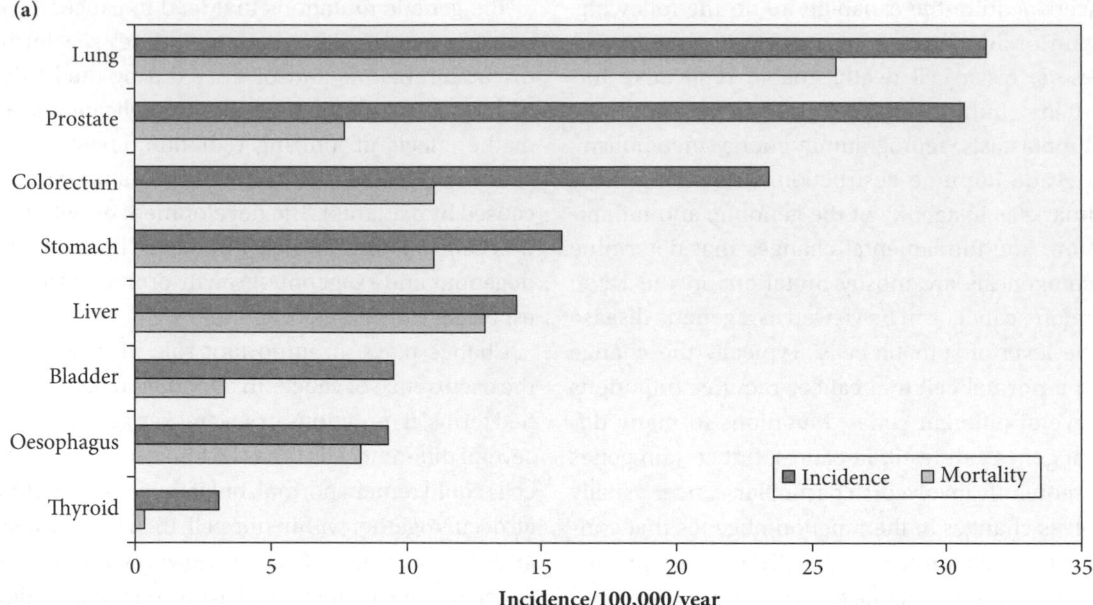


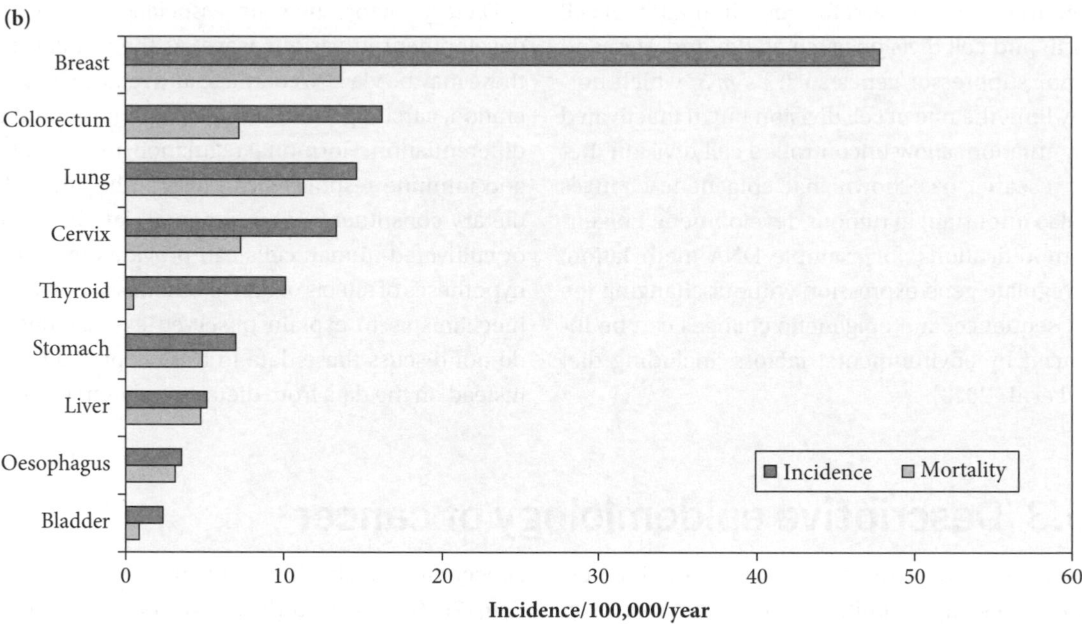


**Fig. 20.1** The most common cancers worldwide: (a) Data for men; (b) Data for women.

*Adapted from:* Ferlay J, Ervik M, Lam F, Colombet M, Mery L, Pineros M., et al. Global Cancer Observatory: Cancer Today. Lyon, France: International Agency for Research on Cancer, 2020. Available at: https://gco.iarc.fr/today (accessed 11 June 2020).

Table 20.1 shows the age-standardized incidence rates for the six most common cancers in eight countries, selected to be representative of different parts of the world with high-quality data on cancer incidence. Stomach cancer rates are high in Japan and China. Colorectal cancer rates are very high in Japan, the UK, and New Zealand. Liver cancer rates are very high in China and high in Japan, especially among men. Lung cancer rates are high in men in all selected countries except India, and relatively high in women in the USA, the UK, New Zealand, and China. Breast cancer rates are low in India. Prostate

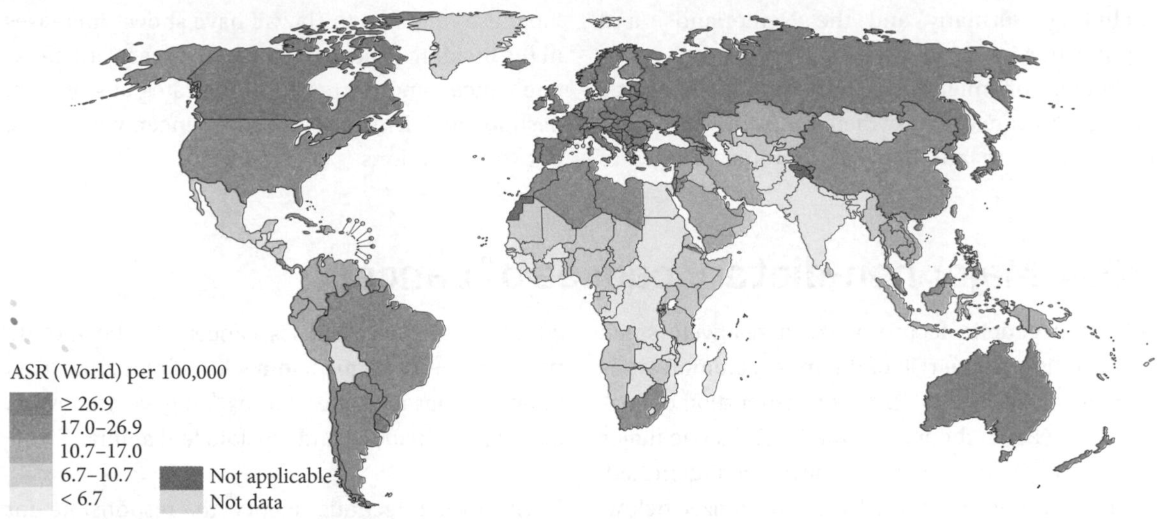


Fig. 20.2 Worldwide variation in the incidence of colorectal cancer.

*Source:* Reproduced with permission from: Ferlay J, Ervik M, Lam F, Colombet M, Mery L, Pineros M, et al. Global Cancer Observatory: Cancer Today. Lyon, France: International Agency – SEE CLEARED PERMISSION.

**Table 20.1 Age-standardized incidence rates, per 100,000 per year, for selected countries with high-quality data, 2020**

| Country | Stomach | | Colorectal | | Liver | | Lung | | Breast | Prostate |
|---|---|---|---|---|---|---|---|---|---|---|
| | Men | Women | Men | Women | Men | Women | Men | Women | Women | Men |
| Argentina | 9.3 | 4.0 | 31.0 | 20.6 | 5.3 | 2.5 | 28.1 | 12.3 | 73.1 | 42.0 |
| China | 29.5 | 12.3 | 28.6 | 19.5 | 27.6 | 9.0 | 47.8 | 22.8 | 39.1 | 10.2 |
| India | 6.1 | 2.9 | 6.0 | 3.7 | 3.6 | 1.6 | 7.8 | 3.1 | 25.8 | 5.5 |
| Italy | 10.5 | 5.5 | 34.2 | 25.2 | 12.0 | 3.7 | 36.0 | 16.4 | 87.0 | 59.9 |
| Japan | 48.1 | 17.3 | 47.3 | 30.5 | 16.1 | 5.3 | 47.0 | 19.5 | 76.3 | 51.8 |
| New Zealand | 6.6 | 3.2 | 38.3 | 29.7 | 7.0 | 2.6 | 24.7 | 25.0 | 93.0 | 92.9 |
| United Kingdom | 5.4 | 2.7 | 40.0 | 29.0 | 7.1 | 3.7 | 35.2 | 29.9 | 87.7 | 77.9 |
| USA | 5.3 | 3.1 | 28.7 | 22.9 | 10.4 | 3.7 | 36.3 | 30.4 | 90.3 | 72.0 |

Data from GLOBOCAN (2020). Available at: https://gco.iarc.fr/ (accessed 11 June 2021)

cancer rates are very high in New Zealand, and high in the UK and the USA.

The rates of some cancers have changed rapidly within a few decades, for example stomach cancer incidence rates have fallen dramatically in most high-income countries. In the UK, age-standardized incidence rates of stomach cancer decreased by over two-thirds within 50 years, from 48.9 per 100,000 in men and 24.6 per 100,000 in women in 1971 to 14.4 per 100,000 in men and 6.2 per 100,000 in women 2017 (Office for National Statistics, 2019). Incidence rates of lung cancer have also changed rapidly within populations where smoking prevalence has been decreasing for several decades, for example among men in the US and the UK, where rates of lung cancer have been declining. In other countries where smoking prevalence has not yet started decreasing, or has only recently begun to decline, lung cancer rates are still increasing, for example among women in many European countries, including Germany and the Netherlands, and among men in China.

Studies of migrants have also shown large changes in cancer rates when people migrate from one country to another. For example, studies of Japanese migrants to Hawaii have shown increases in the incidence rates of colorectal, breast, and prostate cancer, and decreases in the rates of stomach, oesophageal, liver, and cervical cancer, within one or two generations.

## 20.4 Major non-dietary causes of cancer

Inherited genetic factors make a relatively small contribution to the risk of the most common types of cancer (see Box 20.1). A large proportion of cancer is, at least in theory, avoidable, and some major non-dietary causes of cancer have been identified, the most important of which are summarized below.

**Tobacco** Worldwide, the most important preventable cause of cancer is tobacco, which causes approximately 30% of cancers in many high-income countries. Tobacco causes cancers of the mouth, pharynx, oesophagus, stomach, colorectum, liver, pancreas, nasal cavities, larynx, lung, cervix, ovary, kidney, and bladder, and myeloid leukaemia.

**Infections** Infectious agents are responsible for about 13% of cancers worldwide, with the proportion being higher in low-income countries and lower in most high-income countries. The most important numerically are cancer of the liver (hepatitis B virus

### Box 20.1 Inherited genetic factors and cancer risk

For the common types of cancer, inherited highly penetrant genetic factors contribute to around 5–10% of the cases of cancer in a population (Wild et al., 2020). Genetic factors are involved in the determination of whether individuals develop cancer, but are generally less important in determining the variation in cancer risk between populations. Inherited genetic factors (as opposed to mutations in genes that can occur during a person's lifetime) can be considered in two classes: high-risk mutations, and low-risk genetic polymorphisms.

#### High-risk mutations

Inherited high-risk mutations confer a high risk for developing cancer, perhaps 10–50 times higher than the risk in individuals who do not have the mutation. The prevalence of these mutations, however, is low, generally around 1 in 1000 or less. Well-known examples of genes that, when mutated, confer a high risk for common cancers are the mismatch repair genes *MLH1* and *MSH2*, which are associated with hereditary non-polyposis colorectal cancer; and *BRCA1* and *BRCA2* that increase the risk of breast, ovarian, and prostate cancer. At present, there is no clear evidence that dietary factors can modulate the effects of genes such as these on cancer risk.

#### Low-risk polymorphisms

Low-risk polymorphisms are genetic variants that are termed polymorphisms (rather than mutations) because they occur at a prevalence of more than 1% in a population. Such polymorphisms usually confer a risk of cancer only moderately higher or lower (usually around 5–50%) than the risk in individuals with the 'wild type' allele (the most common genotype in the population). In the last two decades, genome-wide association studies have identified numerous new susceptibility alleles in the human genome, many of which are specific to particular types of cancer. Some of these polymorphisms may modify the impact of environmental factors such as diet on cancer risk, such as by affecting the rate of detoxification or activation of mutagenic chemicals present in some foods. For example, a genetic polymorphism appears to modify the effect of alcohol on the risk of oesophageal cancer through the carcinogenic metabolite acetaldehyde; at the same level of alcohol intake, individuals with a polymorphism that reduces the activity of acetaldehyde dehydrogenase 2 have higher circulating acetaldehyde and a higher risk of oesophageal cancer than those without this polymorphism (Wild et al., 2020).

and hepatitis C virus), cancer of the cervix (human papillomavirus (HPV)), and cancer of the stomach (*Helicobacter pylori*). In some parts of the world parasites are important causes of cancer.

Reproductive and hormonal factors Reproductive and hormonal factors are important determinants of three types of cancer in women: cancers of the breast, ovary, and endometrium (lining of the womb). Childbirth reduces the risk for all three of these cancers, probably through inducing terminal differentiation of the epithelial cells in the breasts, and stopping, for the duration of pregnancies, ovulation in the ovaries and cell division in the susceptible cells in the endometrium. Hormonal factors are also important in cancer of the prostate and may be important in the aetiology of cancer of the testis. Hormones are thought to affect cancer risk largely by controlling the rate of cell division, rather than by directly causing mutations.

Other factors Other factors, including ionizing radiation, ultraviolet light, medical drugs, occupational exposures, and pollution, are each estimated to account for around 5% or less of cancers (Swerdlow and Peto, 2020).

## 20.5 Diet and cancer

Research into the effects of diet on cancer presents several challenges. Whereas some of the established non-dietary causes of cancer have large effects on cancer risk (e.g. heavy smoking increases the risk for lung cancer by about 40-fold, and persistent HPV infection increases the risk for cervical cancer by about 100-fold), for dietary factors the size of the effect on risk may be small or moderate (usually less than two-fold). Nevertheless, diet and diet-related factors are important contributors to the global burden of cancer. For example, in the UK nearly 15% of all cancers are estimated to be attributable to dietary factors, alcohol, and obesity (Brown et al., 2018). Clues regarding potential dietary factors may come from animal and human experiments, and ecological studies (comparing rates between populations, and time trends). However, confirmation from a range of epidemiological approaches is required before being confident that a nutrient, food, or dietary pattern promotes or protects against a particular cancer. The relatively small increase in risk typically associated with diet-related factors means that large sample sizes are needed to detect these more modest associations. Case-control and cohort studies provide much of the evidence base for the links between diet-related factors and cancer; the results of case-control studies are affected by biases in recall of diet and in participation; therefore, evidence from cohort studies is preferred. A small number of randomized controlled trials (RCTs) have been conducted in an attempt to confirm the observations made in analytical observational studies. The difficulties associated with assessing dietary intake and the pros, cons, and challenges associated with case-control, cohort studies, and RCTs are discussed in Chapter 1. More recently, biomarkers of nutritional intake (e.g. blood concentrations of micronutrients and fatty acids) have been used as surrogates of dietary intake in 'nested case-control studies' (where the blood samples are collected before the cancer is diagnosed) and, where genetic data are available, the Mendelian randomization approach can sometimes be used to investigate the relationship of nutrients and other dietary factors with cancer risk (Schatzkin et al., 2009; see Chapter 1). Expert bodies have reviewed the evidence for the carcinogenicity of specific dietary factors (see Box 20.2).

### 20.5.1 Diet and the most common cancers

#### 20.5.1.1 Cancers of the oral cavity, pharynx, and oesophagus

Cancers of the oesophagus, lip, oral cavity, and pharynx were estimated to account for 1,298,000 cases and 889,000 deaths worldwide in 2020. There

**Box 20.2 Evaluation of dietary-related risk factors by expert groups**

The World Cancer Research Fund (WCRF) and the International Agency for Research on Cancer (IARC) have systematically reviewed the carcinogenic potential of foods and evaluated the evidence using expert panels. These expert groups have identified that there is convincing (WCRF) or sufficient (IARC) evidence that overweight/obesity and alcohol consumption cause cancer at several sites. The IARC considers that there is sufficient evidence that overweight and obesity cause cancers of the oesophagus (adenocarcinoma), stomach (gastric cardia), colorectum, liver, gallbladder, pancreas, postmenopausal breast, endometrium (corpus uteri), ovary, kidney, meningioma, thyroid, and multiple myeloma. The WCRF and the IARC consider that there is convincing/sufficient evidence that alcohol consumption causes cancers of the oral cavity, pharynx, larynx, oesophagus (squamous cell carcinoma), colorectum, and breast. Both groups have also concluded that processed meat causes colorectal cancer, that aflatoxins cause liver cancer, and that arsenic in drinking water causes lung cancer. The WCRF and IARC have also identified several dietary factors that they judge to be probable causes of cancer; for example, both groups conclude that red meat probably increases the risk for colorectal cancer.

are pockets of extremely high risk of oesophageal cancer in some areas within countries, such as in Linxian in China and Golestan in Iran; in Linxian the incidence rates are more than 100-fold higher than in North America and Europe. In high-income countries the main risk factors for these cancers are alcohol and tobacco, and up to 75% of these cancers are attributable to these two lifestyle factors; alcohol is a risk factor for squamous cell carcinoma of the oesophagus, but not for adenocarcinoma. The mechanism of the effect of alcohol on these cancers is not fully understood, but may involve direct effects of salivary acetaldehyde derived from alcohol acting as a local carcinogen, for example on the epithelium of the oropharynx (Salaspuro, 2020). There is also consistent evidence that consuming drinks and foods at a very high temperature increases the risk for these cancers. Overweight/obesity is an established risk factor specifically for adenocarcinoma (but not squamous cell carcinoma) of the oesophagus; the increased risk for adenocarcinoma of the oesophagus may be due to reflux of gastric contents.

For oesophageal cancer, in the very-high-risk areas of the world, alcohol and tobacco use are not prevalent and do not explain the high risk. It has been proposed that a substantial proportion of cancers of the oral cavity, pharynx, and oesophagus in some developing countries is due to micronutrient deficiencies resulting from a restricted diet that is low in fruit and vegetables and animal products (see Box 20.3 for a discussion of fruit and vegetable intake and risk of cancer). The relative roles of various micronutrients are not yet clear, but deficiencies of riboflavin, folate, vitamin C, and zinc may be involved. However, the results of trials in Linxian in China, aimed at reducing oesophageal cancer rates with micronutrient supplements, have shown no significant protective effects of a range of micronutrient supplements. The Golestan Cohort study found inverse associations between dietary intakes of fruit, vegetables, calcium, and zinc and squamous cell carcinoma of the oesophagus, but it is unclear whether these dietary factors explain the very high risk in this area.

One type of cancer within this group, nasopharyngeal cancer, is particularly common in Southeast Asia, and has been consistently associated with a high intake of Chinese-style salted fish, especially during early childhood, as well as with infection with the Epstein–Barr virus (Swerdlow and Peto, 2020; Wild et al., 2020). Chinese-style salted fish is a special product that is usually softened by partial decomposition before or during salting and can contain high levels of nitrosamines; some other types of preserved foods may also be associated with an increase in the risk for developing nasopharyngeal cancer.

**Box 20.3 Fruit and vegetables and cancer risk**

Fruit and vegetables are an important source of micronutrients and phytochemicals. Early case-control studies suggested a protective role of fruit and vegetables for some types of cancer, and the first WCRF expert report, released in 1997 and based largely on the results from case-control studies, concluded that there was convincing evidence that fruit and/or vegetables reduced the risk for several cancers, such as those of the gastrointestinal tract and lung. However, the evidence that subsequently accumulated from prospective studies is weaker, probably because of a 'healthy control' bias in the case-control studies, and the weak associations still observed in some prospective studies might be due to residual confounding, especially due to smoking and alcohol. In the second WCRF expert report, released in 2007, the judgement for fruit and vegetables was downgraded; the panel concluded that fruit and non-starchy vegetables probably reduced the risk of cancers of the mouth, pharynx, larynx, oesophagus, and stomach. The panel also concluded that fruit probably reduced the risk of lung cancer. In the third expert report, released in 2018, fruit and vegetables were not judged to be convincing or probably protective against any individual cancer type but the panel concluded that high total consumption of non-starchy fruit and vegetables probably reduces the risk of aerodigestive cancers (as an aggregated group of cancers). We think it is more likely that very low intakes might increase the risk, rather than that high intakes are protective, but the current evidence is not conclusive and therefore the role of fruit and vegetables in relation to cancer risk remains uncertain.

### 20.5.1.2 Stomach cancer

Stomach cancer was estimated to account for 1,089,000 cases and 769,000 deaths worldwide in 2020. In 1975, stomach cancer was the most common cancer in the world, but mortality rates have been falling, particularly in most high-income countries, and stomach cancer is now much more common in Asia than in Europe or North America (see Fig. 20.3). Obesity is a risk for gastric cardia cancer. Infection with the bacterium *Helicobacter pylori* is an established risk factor for the development of stomach cancer, and a decline in the prevalence of infection with *Helicobacter pylori* in recent generations in many populations is an important factor in the decline in the incidence of stomach cancer. Dietary changes are also implicated in the recent

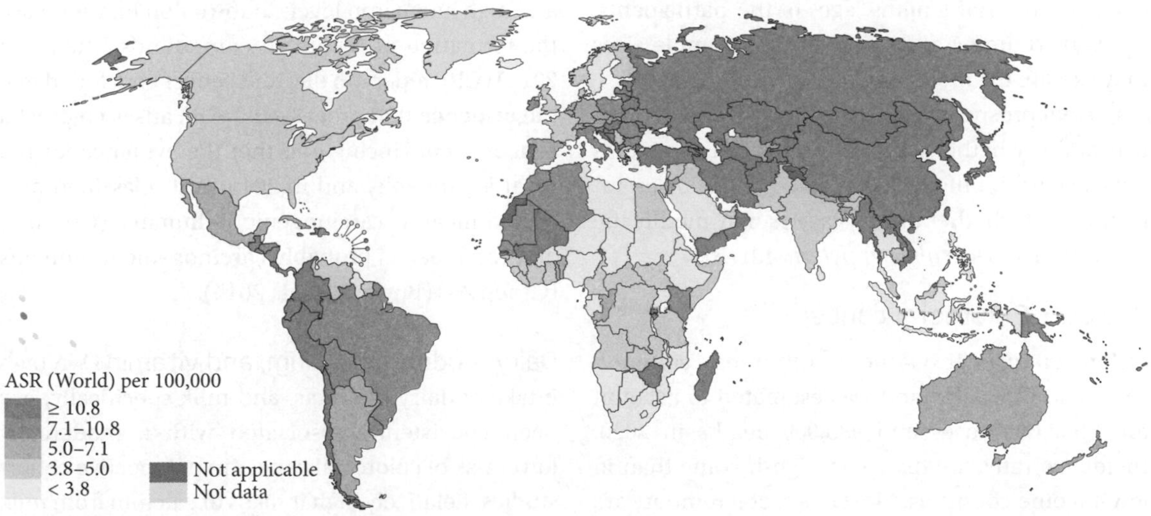


Fig. 20.3 Worldwide variation in the incidence of stomach cancer.

*Source:* Reproduced with permission from: Ferlay J, Ervik M, Lam F, Colombet M, Mery L, Pineros M, et al. Global Cancer Observatory: Cancer Today. Lyon, France: International Agency – SEE CLEARED PERMISSION.

decline in stomach cancer incidence. The introduction of refrigeration has probably led to a reduction in intakes of pickled and salt-preserved foods such as meat, fish, and vegetables, and has also facilitated year-round fruit and vegetable availability.

Substantial evidence suggests that a high total salt intake or a high intake of salted foods, such as salted processed meat and salt-preserved fish, increases the risk of stomach cancer. There are also prospective data showing an increased risk of stomach cancer with high intakes of pickled vegetables. These associations are thought to be causal, and may be due to salt itself or chemical carcinogens derived from nitrites; in some types of pickled vegetables, mould or fungi can develop during preservation and may also contribute to an increased risk of stomach cancer.

There is some evidence from observational studies that higher dietary vitamin C intakes or higher plasma levels of vitamin C are associated with a decreased risk of stomach cancer. In Linxian, China, combined supplementation with β-carotene, selenium, and alpha-tocopherol resulted in a significant reduction in stomach cancer mortality, but no significant benefit was obtained from vitamin C (which was given together with molybdenum). In the few trials in patients with pre-cancerous gastric lesions, some have shown greater regression with the use of supplements of vitamin C and/or β-carotene; differences in the trial lengths, ages of the participants, doses of vitamins and minerals, and sample sizes may explain the inconsistencies between trials.

Further prospective data are needed, in particular to examine whether some of the dietary associations may be partly confounded by *Helicobacter pylori* infection and whether dietary factors may modify the association of *Helicobacter pylori* with risk.

#### 20.5.1.3 Colorectal cancer

Colorectal cancer is the fourth most common cancer in the world and was estimated to account for 1,932,000 cases and 935,000 deaths in 2020. Incidence rates are higher in high-income than in low-income countries. Most colorectal tumours are thought to arise from a precursor lesion, the adenomatous polyp, and prevention or surgical removal of colorectal adenomas can decrease the occurrence of colorectal cancer. Diet-related factors may account for up to 80% of the between-country differences in rates of colorectal cancer. Overweight/obesity increases the risk of colorectal cancer; this has been observed in prospective cohort studies and recently confirmed in Mendelian randomization studies. Adult height, which is partly determined by the adequacy of nutrition in childhood and adolescence, is weakly positively associated with increased risk, and physical activity has been consistently associated with a reduced risk. Alcohol causes a moderate increase in risk. These factors together, however, do not explain the large variation between populations, and there is almost universal agreement that some aspects of a Western diet are a major determinant of risk.

**Meat** International studies show a strong positive association between per capita consumption of meat and colorectal cancer mortality, and several biological mechanisms have been proposed through which meat may increase cancer risk. Mutagenic heterocyclic amines and polycyclic aromatic hydrocarbons can be formed during the cooking of meat at high temperatures, and nitrites and related compounds found in some processed meats may be converted to carcinogenic *N*-nitroso compounds in the colon. In addition, red meats are rich in haem iron, and high haem iron levels in the colon may increase the formation of *N*-nitroso compounds. The latest 2018 WCRF report on diet and cancer concluded that the evidence that processed meat causes colorectal cancer is convincing and that the evidence for red meat is probable, and in 2015 IARC classified processed meat as carcinogenic to humans (Group 1) and red meat as probably carcinogenic to humans (Group 2A) (Bouvard et al., 2015).

**Dairy products, calcium, and vitamin D** A high intake of dairy products, and milk specifically, has been consistently associated with a moderately lower risk of colorectal cancer in prospective cohort studies. Relatively high intakes of calcium from milk or other foods may reduce the risk for colorectal cancer, perhaps by forming complexes with secondary bile acids and haem in the intestinal lumen and thus inhibiting their damaging effects on the epithelium. Several observational studies have supported the hypothesis that calcium may have a protective role (although it is difficult to separate the roles of calcium and milk), and some, but not all, trials have suggested that supplemental calcium may have a modest protective effect on the recurrence of colorectal adenomas; overall, the evidence is not strong enough to draw a firm conclusion, and it is possible that the protective association for milk is due to a component other than calcium. There is also observational evidence that relatively high blood levels of vitamin D are associated with a relatively low risk for colorectal cancer, but this association might be confounded by other aspects of lifestyle such as time spent exercising outdoors, and might also be affected by reverse causation. Evidence from Mendelian randomization studies does not support a causal relationship with vitamin D (Dimitrakopoulou et al., 2017).

**Fibre, fruit, and vegetables** Burkitt suggested in the 1970s that the low rates of colorectal cancer in Africa were due to the high consumption of dietary fibre, and there are several plausible mechanisms for a protective effect. High intakes of fibre increase stool bulk, reduce transit time through the colon, and may thus minimize the absorption of carcinogens by the colonic mucosa. Dietary fibre may also reduce exposure to carcinogens through dilution of the gut contents and/or by binding the carcinogens for faecal excretion. Finally, fermentation of fibre in the large bowel produces short-chain fatty acids, such as butyrate, which may protect against colorectal cancer through the ability to promote differentiation, induce apoptosis, and inhibit the production of carcinogenic secondary bile acids by lowering the intraluminal pH. Since Burkitt's hypothesis, evidence on the association of fibre with risk of colorectal cancer has become available from many prospective epidemiological studies conducted in a range of countries. Meta-analyses of these studies have shown that, on average, a 10 g per day higher intake of total dietary fibre is associated with a small (approximately 7%) reduction in the risk for colorectal cancer (WCRF, 2018); analyses of different sources of dietary fibre have shown that this protection is linked to cereal fibre and whole grains, but is not clearly linked to fibre from fruit or vegetables. Randomized controlled trials of fibre supplementation have not, however, shown any effect on colorectal adenomas or colorectal cancer, but the relatively short duration and small sample sizes of these trials, as well as the specific fibre supplements used and the high-risk background of the participants, may explain the apparent discrepancy between the trials and the prospective studies.

**Fat** As with meat, international correlation studies showed a strong association between per capita consumption of fat and colorectal cancer mortality. One possible mechanism is that a high fat intake may increase the level of potentially mutagenic secondary bile acids in the lumen of the large intestine. The Women's Health Initiative Dietary Modification Trial evaluated the effects of a low-fat eating pattern on risk of colorectal cancer in postmenopausal women. Although the intervention group achieved a moderate reduction in their fat intake (and increased their grain, fruit, and vegetable intake), there was no evidence that the intervention reduced the risk of invasive colorectal cancer during the 8-year follow-up period. Taken together with the results of prospective observational studies, there is overall little evidence that high intakes of fat increase the risk for colorectal cancer, at least in the middle-aged people studied in high-income countries.

**Folate** In several observational studies, high dietary folate intake has been associated with reduced risk of colorectal cancer. Diminished folate status might contribute to carcinogenesis by altering gene expression or by increased DNA damage. It has also been suggested that, while a higher folate intake might be protective against initiation of colorectal neoplasia, it might promote the growth of existing tumours; however, a pooled analysis of randomized controlled trials including 49,621 participants among whom 429 cases of colorectal cancer occurred found no effect of folic acid supplementation.

#### 20.5.1.4 Liver cancer

Liver cancer was estimated to account for 906,000 cases and 830,000 deaths in 2020, worldwide. Approximately 80% of cases of liver cancer occur in low-income countries, and liver cancer rates vary about ten-fold between countries, being much higher in Eastern and Southeast Asia than in Europe and North America. The major risk factor for hepatocellular carcinoma, the main type of liver cancer, is chronic infection with hepatitis B, and to a lesser extent, hepatitis C virus. Aflatoxin is a food contaminant produced by the fungus *Aspergillus*; high levels of aflatoxin can occur in foods such as grains, oilseeds, nuts, and dried fruit stored in hot and humid conditions. Ingestion of foods contaminated with aflatoxin is an important risk factor for liver cancer among people with active hepatitis virus infection in some low-income regions, such as in sub-Saharan Africa and southern China. Chronic, high alcohol consumption is established as the main diet-related risk factor for liver cancer in most high-income countries, probably via the development of cirrhosis and alcoholic hepatitis (Swerdlow and Peto, 2020). Overweight/obesity also increases the risk of liver cancer. Several prospective studies have suggested that high coffee consumption is associated with a lower risk of liver cancer, but this needs further research and might be affected by reverse causation (i.e. individuals with some physiological characteristics or early pathological changes that are associated with the development of liver cancer may experience symptoms that lead them to reduce their coffee intake).

#### 20.5.1.5 Lung cancer

Lung cancer is the most common cancer in the world and was estimated to account for 2,207,000 cases and 1,796,000 deaths in 2020. Heavy smoking increases the risk by around 40-fold, and smoking causes over 80% of lung cancers in high-income countries. The possibility that diet might also affect lung cancer risk was raised in the 1970s and a number of observational studies have found that lung cancer patients often report a lower intake of fruits, vegetables, and related nutrients (such as β-carotene) than controls. However, this association has been weak in prospective studies, and the apparent weak relationship is likely to be due to residual confounding by smoking, since smokers generally consume less fruit and vegetables than non-smokers. Several trials have tested the effects of supplements of β-carotene on lung cancer incidence. The hypothesis was that increased β-carotene intake would reduce the risk for lung cancer, but none of the trials found any evidence of benefit, and in fact two trials with a large number of cases of lung cancer, conducted among persistent smokers, found that the men who took the β-carotene supplements had a higher incidence of lung cancer than the men who did not. In conclusion, the overall evidence to date suggests that diet has little or no effect on the risk of lung cancer, and in public health terms the overriding priority for reducing lung cancer rates is to reduce the prevalence of smoking.

#### 20.5.1.6 Breast cancer

Breast cancer is the second most common cancer in the world and the most common cancer among women. Breast cancer was estimated to account for 2,261,000 cases and 685,000 deaths in women in 2020. Incidence rates are about three times higher in high-income countries than in low-income countries. Much of this international variation is due to differences in established reproductive risk factors such as younger age at menarche, lower parity and older age at childbirth, and less breastfeeding, but differences in dietary habits and physical activity may also contribute. In fact, age at menarche can be partly determined by dietary factors, in that restricted dietary intake during childhood and adolescence can lead to delayed menarche. Adult height is positively associated with risk, and can be partly determined by dietary factors during childhood and adolescence, in that restriction in food supply during growth can reduce adult height. Oestradiol and perhaps other hormones play a key role in the aetiology of breast cancer. Recently, there has been a lot of interest in examining risk factors for different subtypes of breast cancer—primarily oestrogen-receptor (ER)-positive (about 80% of breast cancers) and ER-negative breast cancers, with the recognition that these different subtypes may have distinct risk factors. It is possible that any further dietary effects on risk for ER-positive breast cancer are mediated by hormonal mechanisms, whereas diet might affect the risk for ER-negative breast cancer by other, non-hormonal mechanisms.

**Overweight/obesity** Obesity increases breast cancer risk in postmenopausal women by around 50%, probably by increasing serum concentrations of free oestradiol. Postmenopausal women with obesity have circulating free oestradiol concentrations more than twice as high as those in lean postmenopausal women, because after the menopause the main source of circulating oestrogens is the conversion from androgens by the enzyme aromatase, particularly in the adipose tissue, and higher circulating concentrations of free oestradiol increase risk for cancers of the breast because oestrogens stimulate growth and division of the cells in this tissue. Obesity does not increase risk among premenopausal women, in fact there is convincing evidence of an inverse association between body mass index (BMI) and premenopausal breast cancer, although the underlying mechanisms that explain this association are not well understood (Premenopausal Breast Cancer Collaborative Group, 2018). However, obesity in premenopausal women is likely to lead to obesity throughout life and therefore to an eventual increase in breast cancer risk.

**Alcohol** Apart from obesity, the only other well-established dietary risk factor for breast cancer is alcohol. There are now large amounts of data from well-designed studies that consistently show an increase in risk with higher consumption, with about a 10% increase in risk for an average long-term intake of one alcoholic drink every day. The mechanism for this association is not known, but may involve increases in oestrogen levels.

**Fat** Much research and controversy has surrounded the hypothesis that a high fat intake increases breast cancer risk, and the results of observational studies have varied. In the Women's Health Initiative Dietary Modification Trial, the effect of a low-fat eating pattern on risk of breast cancer in postmenopausal women was tested (see Box 20.4).

### Box 20.4 Dietary fat and breast cancer

The hypothesis that a high dietary fat intake increases the risk of breast cancer originated from ecological correlations between per capita dietary fat consumption of different countries and breast cancer mortality that was first observed by Lea in 1966 and further described by Armstrong and Doll in their seminal 1975 paper, which examined ecological correlations between dietary and environmental factors and cancer incidence and mortality.

Case-control studies conducted largely in the 1970s and 1980s supported the dietary fat and breast cancer hypothesis, with the results of a pooled analysis of case-control studies published in 1990 showing an approximately 50% increase in risk of breast cancer for the highest vs the lowest quintile of total fat intake. Although it was acknowledged that case-control studies were susceptible to bias and may therefore overestimate the effect size, it was widely believed that the magnitude of the association was too strong to be entirely explained by bias inherent in the design of case-control studies. However, subsequent results from prospective cohort studies did not generally support the positive findings from case-control studies.

The Women's Health Initiative Randomized Controlled Dietary Modification Trial was an ambitious study designed to investigate whether a diet low in fat (and high in fruit and vegetables and grains) was effective for the primary prevention of breast cancer. It was hoped that this would provide a definitive answer to the dietary fat and breast cancer hypothesis. In total, 48,835 postmenopausal women were recruited from 40 US clinical centres over a 6-year period (from 1992 to 1998). The intervention group aimed to reduce total fat to 20% of energy, increase consumption of fruit and vegetables to five or more servings a day, and increase the consumption of grains to at least six servings per day, with an intensive behavioural support programme. Weight loss was not a goal of the intervention group.

The main results were published in 2006. Dietary assessment indicated sustained dietary changes in the intervention group, although the dietary goals were not fully met (percentage of energy from fat decreased from 37.8% at baseline to 24.3% after 1 year in the intervention group, and a difference of 8 percentage points between the groups was maintained after 6 years of follow-up). There was also a modest weight loss in the intervention group (about 2 kg after 1 year and a difference of about 1 kg between the groups was maintained after 6 years of follow-up). After an average of 8.1 years of follow-up, 1727 cases of invasive breast cancer occurred. The hazard ratio of breast cancer incidence in the intervention group was 0.91 (95% CI: 0.83–1.01). The trial results were disappointing and highlight the difficulties in designing, executing, and interpreting the findings of dietary RCTs. The 9% decrease in breast cancer risk in the intervention group was not statistically significant, but leaves open the possibilities that if the dietary goals had been fully met, the trial was larger, the follow-up period longer, or if the intervention had started earlier in life, then the reduction in risk may have reached statistical significance. However, even if there had been a significant reduction in the risk of breast cancer, it still would have been difficult to disentangle the effects of the dietary composition and the unintended weight loss of the intervention group. This is because a higher BMI is associated with an increased risk of postmenopausal breast cancer, and therefore weight loss in postmenopausal women would be expected to reduce breast cancer risk. Nevertheless, interest in the dietary fat and breast cancer hypothesis began to wane; many considered that the results of the expensive Women's Health Initiative Randomized Controlled Dietary Modification Trial ruled out any important effect of dietary fat on breast cancer risk. This was supported by a subsequent smaller randomized controlled trial of 4690 women that found no significant effect of a low-fat diet on the risk of invasive breast cancer after 10 years of follow-up (HR 1.19, 95% CI: 0.91–1.55).

**Other dietary factors** The results of studies of other dietary factors including meat, dairy products, fruit and vegetables, fibre, and phyto-oestrogens with overall breast cancer are inconsistent and no convincing associations with these factors have been established. Some recent prospective studies have shown inverse associations between fruit, vegetables, or fibre intakes and ER-negative breast cancer, but further research is needed to confirm these associations and the possible underlying mechanisms. Higher circulating concentrations of insulin-like growth factor-1 (IGF-1) are associated with increased risk of breast cancer. IGF-1 is a growth hormone that stimulates cell proliferation and inhibits programmed cell death (apoptosis). There are some data, mainly cross-sectional, that show that higher intakes of animal or dairy protein increase circulating levels of IGF-1, but more research is needed on these relationships.

### 20.5.1.7 Cervix cancer

Cancer of the cervix was estimated to account for 604,000 cases and 342,000 deaths in women in 2020. The highest rates are in sub-Saharan Africa, Melanesia, and Central and South America. The major cause of cervical cancer is infection with certain subtypes of the human papillomavirus. There is little evidence that dietary factors influence the risk of this cancer.

### 20.5.1.8 Prostate cancer

Prostate cancer was estimated to account for 1,414,000 cases and 375,000 deaths in 2020. Prostate cancer incidence rates are strongly affected by diagnostic practices and therefore difficult to interpret, but mortality rates show that death from prostate cancer is about two to three times more common in North America and Europe than in Asia; mortality is also high in some other parts of the world, including parts of Africa, and the Caribbean. Little is known about the aetiology of prostate cancer, and the only well-established risk factors are increasing age, family history, black ethnicity, and genetic factors. There is also evidence that obesity is associated with a higher risk of aggressive forms of the disease, and height is positively associated with risk.

Hormones control the growth of the prostate, and interventions that lower androgen levels are moderately effective in treating prostate cancer. Observational studies have shown that both free testosterone and IGF-1 are associated with a higher risk of prostate cancer. Diet might affect prostate cancer risk by influencing IGF-1 levels, and data suggest that high intakes of protein, particularly dairy protein, may increase circulating levels of IGF-1; an important focus of current research is better understanding of these relationships. Several observational studies have suggested that prostate cancer risk may be lower in men with relatively high selenium status (estimated by measuring selenium in blood or in toenails); however, a large randomized controlled trial, as well as a Mendelian randomization study, have provided no evidence for a causal link between selenium and prostate cancer. Randomized trials have also found that supplements of β-carotene and vitamin E do not reduce the risk for prostate cancer and a Mendelian randomization study has not supported a protective effect of vitamin D. Lycopene, primarily from tomatoes, has been associated with a reduced risk in some observational studies, but the data are not conclusive. Prospective studies in Asian men have reported that isoflavones, largely from soya foods, are associated with a lower risk of prostate cancer, but further research on this topic is needed.

#### 20.5.1.9 Bladder cancer

Cancer of the urinary bladder was estimated to account for 573,000 cases and 213,000 deaths in 2020. The geographic variation in incidence between countries is about five-fold, with relatively high rates in high-income countries. Tobacco smoking increases the risk for bladder cancer, accounting for between one-third and two-thirds of all bladder cancers. Occupational risk factors, such as exposure to aromatic amines and polyaromatic hydrocarbons, also play a significant role. Aristolochic acid from herbal medicines or crop contamination can cause higher risks of bladder cancer, as can arsenic in contaminated water. There are no other established dietary-related risk factors for bladder cancer.

#### 20.5.1.10 Thyroid cancer

Cancer of the thyroid was estimated to account for 586,000 cases and 44,000 deaths in 2020.

Thyroid cancer incidence has increased rapidly in the past few decades in the US, the UK, and many other high- and middle-income countries. It is more commonly diagnosed in women, compared to men. The increase in rates appears to be due, at least in part, to increased detection of small tumours (primarily papillary tumours) via medical imaging. Incidence rates are highest in South Korea, and high in North America, parts of South America, Europe, and Australia, and very low in sub-Saharan Africa. Mortality rates are low worldwide, but relatively higher in Melanesia, West Asia, and Africa. Exposure to ionizing radiation in childhood increases the risk for thyroid cancer. Overweight and obesity increase the risk, and greater adult height is also associated with an increased risk. There are no established dietary risk factors for thyroid cancer, except perhaps for iodine deficiency.

#### 20.5.1.11 Other cancers

There are over 200 different types of cancer and for many of these, particularly the rare cancers, there has been little research on the possible role of diet. With continued follow-up in the many large existing cohort studies, more data will become available and may show new associations of diet with cancer risk.

## 20.6 Summary

Diet-related factors, including obesity and alcohol, may account for about 20% of cancers in high-income countries. The effects of dietary factors on cancer risk are summarized in Table 20.2. Obesity increases the risk for cancers of the oesophagus (squamous cell carcinoma), stomach (gastric cardia), colorectum, liver, gallbladder, pancreas, breast (postmenopausal), endometrium, ovary, kidney, meningioma, thyroid, and multiple myeloma. Alcohol causes cancers of the oral cavity, pharynx, larynx, oesophagus, colorectum, liver, and breast. High intakes of processed meat increase the risk for colorectal cancer, and high intakes of red meat probably increase the risk for this disease. Milk, calcium, and fibre probably reduce the risk for

**Table 20.2 Dietary risk factors, dietary protective factors, and other major risk factors for common cancers***

| Cancer | Dietary and diet-related risk factors | Dietary protective factors | Major non-dietary risk factors |
|---|---|---|---|
| Oral cavity, pharynx, and larynx | Alcohol | | Smoking<br>Human papillomavirus |
| Nasopharynx | Chinese-style salted fish | | Smoking<br>Epstein–Barr virus |
| Oesophagus | Alcohol (squamous cell carcinoma)<br>Obesity (adenocarcinoma)<br>Probably very hot drinks | | Smoking |
| Stomach | Probably high intake of salt and salt-preserved foods<br>Obesity (cancer of the gastric cardia) | | Infection by *Helicobacter pylori*<br>Smoking |
| Colorectum | Alcohol<br>Obesity<br>Processed meat<br>Probably red meat | Probably milk and calcium<br>Probably dietary fibre | Sedentary lifestyle<br>Smoking |
| Liver | Alcohol<br>Obesity<br>Foods contaminated with aflatoxin | | Hepatitis viruses<br>Smoking |
| Lung | | | Smoking |
| Breast | Alcohol<br>Obesity (postmenopausal) | Obesity (premenopausal) | Reproductive and hormonal factors |
| Cervix | | | Human papillomavirus<br>Smoking |
| Endometrium | Obesity | | Reproductive and hormonal factors |
| Ovary | Obesity | | Reproductive and hormonal factors |
| Prostate | Probably obesity for aggressive disease | | |
| Bladder | | | Smoking<br>Occupational exposures |
| Thyroid | Obesity | | Ionizing radiation in childhood |

*Only factors for which there is strong evidence are presented.

colorectal cancer. Micronutrient deficiencies might contribute to an increase in cancer risk in malnourished populations. In high-income regions, dietary intervention trials with micronutrient supplements have not shown benefit in reducing cancer risk, and high-dose micronutrient supplements can be harmful (vitamin supplements are, however, useful in certain situations in relation to other aspects of health, for example folic acid supplements for women of childbearing age prior to conception).

The most important lifestyle recommendation to reduce cancer risk is to refrain from smoking. For many cancers the importance of dietary factors is not clear. Nevertheless, based on the systematic reviews of evidence, WCRF has ten cancer prevention recommendations, which are to:

- Be a healthy weight
- Be physically active
- Eat a diet rich in whole grains, vegetables, fruit, and beans
- Limit consumption of 'fast foods' and other processed foods high in fat, starches, or sugars
- Limit consumption of red and processed meat
- Limit consumption of sugar-sweetened drinks
- Limit alcohol consumption
- Do not use supplements for cancer prevention
- For mothers: breastfeed your baby, if you can
- After a cancer diagnosis: follow the WCRF recommendations, if you can.

## Further reading

1. **Adami, H., Hunter, D.J., Lagiou, P., Mucci, L.** (eds) (2018) *Textbook of cancer epidemiology*, 3rd edn. Oxford: Oxford University Press.
2. **Bouvard, V., Loomis, D., Guyton, K.Z., et al.** (2015). Carcinogenicity of consumption of red and processed meat. *Lancet*, **16**, 1599–1600.
3. **Brown, K.F., Rumgay, H., Dunlop, C., et al**. (2018) The fraction of cancer attributable to modifiable risk factors in England, Wales, Scotland, Northern Ireland, and the United Kingdom in 2015. *Br J Cancer*, **118**, 1130–41.
4. **Dimitrakopoulou, V., Tsilidis, K.K., Haycock, P.C., et al.** (2017) Circulating vitamin D concentration and risk of seven cancers: Mendelian randomisation study. *BMJ*, **359**, j4761.
5. **Ferlay, J., Ervik, M., Lam, F., Colombet, M., Mery, L., Pineros, M., et al.** (2020) *Global Cancer Observatory: Cancer today*. Lyon, France: International Agency for Research on Cancer. Available at: https://gco.iarc.fr/today (accessed 11 June 2020).
6. **Hanahan, D. and Weinberg, R.A.** (2011) Hallmarks of cancer: the next generation. *Cell*, **144**, 646–74.
7. **Key, T.J., Bradbury, K.E., Perez-Cornago, A., et al.** (2020) Diet, nutrition, and cancer risk: what do we know and what is the way forward? *BMJ*, **368**, m511.
8. **Lauby-Secretan, B., Scoccianti, C., Loomis, D., Grosse, Y., Bianchini, F., and Straif, K.; International Agency for Research on Cancer Handbook Working Group** (2016) Body fatness and cancer—viewpoint of the IARC Working Group. *N Engl J Med*, **374**, 794–8.
9. **Loomis, D., Guyton, K.Z., Grosse, Y., et al.** (2016) Carcinogenicity of drinking coffee, mate and very hot beverages. *Lancet Oncol*, **17**, 877–8.
10. **Mei, H. and Hongbin, T.** (2018) Vitamin C and *Helicobacter pylori* infection: current knowledge and future prospects. *Front Physiol*, **9**, 1103.
11. **Murphy, N., Jenab, M., and Gunter, M.J.** (2018) Adiposity and gastrointestinal cancers: epidemiology, mechanisms and future directions. *Nat Rev Gastroenterol Hepatol*, **15**, 65970.
12. **Office for National Statistics** (2019) Cancer statistics: registrations series MB1. Available at: http://www.statistics.gov.uk/StatBase/Product.asp?vlnk=8843 (accessed 22 June 2021).
13. **Premenopausal Breast Cancer Collaborative Group** (2018) Association of Body Mass Index and age with subsequent breast cancer risk in premenopausal women. *JAMA Oncol*, **4**, e181771.
14. **Salaspuro, M.** (2020) Local acetaldehyde: its key role in alcohol-related oropharyngeal cancer. *Visc Med*, **36**, 167–73.
15. **Schatzkin, A., Abnet, C.C., Cross, A.J., et al.** (2009) Mendelian randomization: how it can—and cannot—help confirm causal relations between nutrition and cancer. *Cancer Prev Res*, **2**, 104–13.

16. **Swerdlow, A. and Peto, R.** (2020) Epidemiology of cancer. In: Firth, J.D., Conlon, C., and Cox, T. (eds) *Oxford textbook of medicine*, 6th edn, pp. 1–86. Oxford: Oxford University Press.
17. **Wild, C.P., Weiderpass, E., Stewart, B.W.** (eds) (2020) *World Cancer Report: cancer research for cancer prevention*. Lyon: International Agency for Research on Cancer. Available at: http://publications.iarc.fr/586 (accessed 18 June 2021).
18. **World Cancer Research Fund/American Institute for Cancer Research (WCRF/AICR)** (2018) *Diet, nutrition, physical activity and cancer: a global perspective—the third Expert Report*. London, UK: World Cancer Research Fund International. Available at: https://www.wcrf.org/dietandcancer. (accessed 18 June 2021).
19. **Yao, Y., Suo, T., Andersson, R., et al**. (2017) Dietary fibre for the prevention of recurrent colorectal adenomas and carcinomas. *Cochrane Database Syst Rev*, CD003430.

# Diabetes and the Metabolic Syndrome

Jim Mann



Diabetes rates have increased in recent years and in many countries worldwide are considered to have reached epidemic proportions. Type 2 diabetes (T2DM) accounts for the majority of cases. While genetic factors determine susceptibility, the rapidly escalating rates are largely explained by the almost worldwide increase in overweight and obesity. Reducing excess adiposity while consuming diets in which carbohydrate is predominantly derived from minimally processed grains, intact fruit, vegetables, legumes and pulses, and fat principally from sources rich in polyunsaturated or *cis*-monounsaturated fatty acids improves blood glucose and reduces cardiovascular risk in those with T2DM. Adherence to dietary advice similar to that recommended for treatment of diabetes substantially reduces the chances of those with prediabetes progressing to diabetes. Appreciable weight loss has the potential to enable the withdrawal of all drug treatment and induce remission of the condition in some with established disease. The reasons for the increase in rates of type 1 diabetes (T1DM) in some countries is uncertain but nutrition therapy is necessary to ensure an appropriate balance between insulin and dietary carbohydrate and a nutrient profile expected to reduce risk of diabetes complications. A wide variety of dietary patterns is compatible with the recommended macronutrient distribution.

## 21.1 Diabetes

Diabetes, a condition associated with loss of sugar in the urine, has been diagnosed by physicians for at least three millennia. Until the discovery of insulin in the 1920s, dietary treatment was all that could be offered to people with diabetes. Although dietary advice has altered over the years, reduction and sometimes near elimination of sugars and other carbohydrates formed the cornerstone of management for much of the time until the 1970s, when the merits of a diet low in carbohydrate, and consequently high in fat, were questioned. The most appropriate macronutrient composition of the dietary prescription for people with diabetes remains a much debated topic.

The term '*diabetes*' is used to describe a group of conditions characterized by raised blood glucose levels (hyperglycaemia) resulting from an absolute or relative deficiency of insulin. In T1DM, previously known as insulin-dependent diabetes (IDDM), there is destruction of the insulin-producing pancreatic islet β-cells, usually resulting from an autoimmune process. Insulin treatment, administered by subcutaneous injection, is essential to maintain life.

In T2DM, previously non-insulin-dependent diabetes (NIDDM), a key abnormality is resistance to the action of insulin, and in the early stages of the disease, insulin levels may actually be raised as the β-cells of the pancreas produce more insulin in an attempt to overcome insulin resistance. In many patients with T2DM, the insulin-producing β-cells of the islets in the pancreas may show a degree of failure at some stage during the course of the disease process. Patients with T2DM are initially treated with 'lifestyle modification' therapy. Oral hypoglycaemic (blood glucose-lowering) agents may be added. Insulin may be required later. Glucose levels that are higher than normal after an overnight fast (impaired fasting glucose, IFG) or after a glucose load (impaired glucose tolerance, IGT) and slightly raised levels of haemoglobin $A_{1c}$ may represent the earliest stages of T2DM and are referred to as 'prediabetes'. Diabetes developing during the second and third trimesters of pregnancy is described as gestational diabetes.

Hyperglycaemia usually leads to glycosuria (glucose in the urine) when the renal threshold for glucose (level up to which glucose is reabsorbed in the renal tubules) is exceeded. Hyperglycaemia and glycosuria result in the classical symptoms of diabetes: polyuria (passing large quantities of urine), polydipsia (thirst), blurring of vision, increased susceptibility to infections, and unintentional weight loss.

## 21.2 Clinical features of diabetes

T1DM is invariably associated with marked symptoms of hyperglycaemia and often striking unintentional weight loss. If untreated, the absence of insulin leads to severe metabolic disturbances (ketoacidosis), during which the patient may become unconscious and which may be fatal without insulin treatment and correction of fluid and electrolyte imbalance. The diagnosis of T1DM is usually straightforward because blood glucose levels are markedly raised.

In T2DM symptoms are invariably less severe and ketoacidosis does not occur. People with IFG, IGT, gestational (pregnancy) diabetes, and about half of all cases of T2DM are without symptoms. Diabetes is diagnosed if random venous plasma glucose is greater than 11.0 mmol/L or the fasting value is greater than 7 (two measurements are required if the person is asymptomatic). A glucose tolerance test (measurement of blood glucose fasting and 2 hours after a 75 g glucose load) may be required to diagnose the prediabetic states and those with borderline blood glucose levels (Table 21.1). Haemoglobin $A_{1c}$ (glycated haemoglobin, $HbA_{1c}$) has typically been used in people with diabetes as a measure of blood glucose control during the preceding weeks or months. It is now increasingly also used as a test for diagnosing diabetes (if levels are 48 mmol/mol or greater) or prediabetes (if levels are between 42 and 47 mmol/mol).

**Table 21.1 Venous plasma glucose levels for the diagnosis of diabetes, impaired glucose tolerance, and impaired fasting glucose**

| | Fasting glucose (mmol/L) | 2 hours after 75 g glucose (mmol/L) |
|---|---|---|
| Diabetes | >7.0 | >11.0 |
| Impaired fasting glucose | 6.0–7.0 | <7.8 |
| Impaired glucose tolerance | <6.0 | 7.8–11.0 |
| 'Normal' glucose tolerance | <6.0 | <7.8 |

Both types of diabetes may be associated with hypertension and a range of metabolic disturbances, which together with the hyperglycaemia, help to explain the wide-ranging complications of diabetes that account for much of the ill health and premature death associated with diabetes. The complications result chiefly from the effects on the arterial and nervous systems. They include diabetic retinopathy, which may lead to blindness, diabetic nephropathy, potentially resulting in kidney failure, foot ulceration, which may lead to gangrene, several different neurological conditions, and cardiovascular disease (coronary heart disease and stroke), the most common causes of early death. Interestingly, age-standardized coronary heart disease rates are similar in men and women with diabetes, whereas in the general population, women have appreciably lower rates than men.

## 21.3 Epidemiology and aetiology of type 2 diabetes

### 21.3.1 Epidemiology of type 2 diabetes

Worldwide, T2DM is overwhelmingly more common than T1DM. Thus, global statistics relating to 'diabetes' principally reflect frequencies of T2DM. Data relating to trends over time are not particularly reliable because diagnostic criteria have changed, as have diagnostic facilities and rates of screening. These are clearly important determinants of prevalence rates, especially as so many cases are asymptomatic and may not be diagnosed until complications occur. Despite these reservations, there is no longer any doubt about the true worldwide increase in prevalence, resulting both from a tendency towards ageing populations (T2DM has been clearly shown to increase in frequency with age) and increased incidence at all ages. In the developed world, in populations principally of European descent, the increase has been steady. However, in some other population groups, there have been dramatic increases in prevalence associated either with migration or with a rapid change from a traditional (non-Western) lifestyle to increased consumption of energy-dense foods, high in fats and sugars, and reduced levels of physical activity. The American Pima Indians, Polynesians and Melanesians in the South Pacific, Australian Aboriginals, and Asian Indian migrants to the UK and other countries are examples of populations in which the process of rapid acculturation has led to diabetes prevalence rates far higher than those observed in European populations. Diabetes is believed to affect 9.3% of the world's adult population, with 463 million having the condition (IDF Diabetes Atlas, 9th edn (2019) https://www.diabetesatlas.org/en/ (accessed 3 May 2021)).

India and China have relatively high rates of diabetes and are two of the most populous countries in the world. Thus, China is the country with the most people with diabetes, with a current figure of 116.4 million, followed by India with 77 million. If the current rates of growth continue unchecked, it has been estimated that by 2045, the total number of people with diabetes worldwide will exceed 700 million. In addition to those with diabetes, there are many with prediabetes. For example, amongst adult Maori, the indigenous people of New Zealand, almost half the adult population has been found to have diabetes or prediabetes. In some of these groups, T2DM, previously regarded as a disease of the middle-aged and elderly, has recently been diagnosed in teenagers and children. Thus, T2DM may be regarded as one of the major epidemic diseases of the twenty-first century. Worldwide prevalence estimates of diabetes are shown in Fig. 21.1.

### 21.3.2 Aetiology of type 2 diabetes and its complications

From the earliest times until recently, the sugary urine and raised levels of blood glucose have led to the assumption that an excessive intake of sucrose (table sugar) must be an important cause of the condition. However, the most consistent observation in

Fig. 21.1 Diabetes prevalence (%) estimates of diabetes (20–79 years) 2019.

*Source:* International Diabetes Federation. IDF Diabetes Atlas, 9th edn. Brussels, Belgium: International Diabetes Federation. 2019. Available at: https://www.diabetesatlas.org (accessed 27 June 2022).

prospective and cross-sectional studies is the striking association between risk of T2DM and increasing obesity, particularly when the excess body fat is centrally distributed, i.e. in association with a high waist circumference. In the United States the increase in rates of diabetes has closely followed the increasing prevalence of obesity.

It is also clear that there is a strong genetic component to this condition. The risk of developing T2DM is greatly increased when one or more close family members have the condition, although the precise mode of inheritance has not yet been resolved. It appears therefore, that in predisposed populations or families, genetic and lifestyle factors combine to result in the development of insulin resistance and, consequently, diabetes. The relative importance of genetic and lifestyle factors is well illustrated by diabetes rates in Pima Indians. Although rates amongst the American Pima Indians are amongst the highest in the world, a genetically similar group living in the mountainous regions of Mexico have relatively low rates. The Mexican Pima Indians still follow a traditional way of life unexposed to Western ways and diet and have a high level of physical activity.

There have been several attempts to implicate individual foods or nutrients as causal or protective factors in the aetiology of T2DM independent of any association they have with overweight or obesity. It is conceivable that more than one of the attributes that characterize the Western lifestyle contribute to the aetiology. A study including data from 184 countries suggest that 70% of cases of T2DM are associated with dietary factors. High intakes of fibre rich whole grains, vegetables, fruit, nuts, seeds and yoghurt are associated with a reduced risk whereas refined grains (e.g. white bread, white rice, white pasta), red and processed meat and sugary beverages, including fruit juices, increase the risk. Many of these foods are also protective or promotive of coronary heart disease and several commonly occurring cancers. These findings are compatible with those of the large prospective study of US nurses, which suggested that diets with a high glycaemic load (reflecting the glycaemic index and total amount of carbohydrate) increase the risk of T2DM.

A high intake of saturated fatty acids increases resistance to the action of insulin and is therefore likely to increase the risk of developing T2DM. Intrauterine growth retardation leading to babies born small for dates and prematurity have been suggested as risk factors for the development of T2DM in later life. This appears to be especially the case when rapid catch-up growth occurs in infancy and childhood (foetal programming or 'Barker' hypothesis).

Several complications of diabetes (e.g. retinopathy, nephropathy, and neuropathy) appear to be a result of hyperglycaemia, other metabolic consequences of diabetes, and hypertension. Diet plays a role in their prevention and treatment principally by helping to improve blood glucose control and lower blood pressure (see section 21.4). Cardiovascular disease, which is responsible for the greatest number of deaths and much non-fatal illness in people with T2DM, seems to be even more closely linked to dietary factors. Risk factors for coronary heart disease (CHD) in the general population apply also to those with diabetes, though some appear to be especially important (Table 21.2). This may help to explain the great excess of CHD in this condition. Many are diet-related (see Chapter 19). Thus, a major focus of dietary recommendations for people with diabetes relates to the importance of reducing cardiovascular risk. In countries with low CHD rates in the general population, CHD is also relatively infrequent in people with diabetes.

## 21.3.3 Reducing the risk of type 2 diabetes

While the precise mechanisms by which genes and lifestyle interact to result in T2DM remain elusive, the geographic variation, rapid changes over time, and dietary patterns related to risk suggest that lifestyle modification might help to prevent, or at least delay, the onset of T2DM in predisposed individuals. Randomized controlled trials involving people with IGT, first reported from Finland, the USA, and China, and more recently from other countries, confirm that this is indeed the case. The target interventions (Box 21.1) have generally resulted in an approximately 60% reduction in rates of progression from IGT to T2DM over an initial approximately 4-year follow-up period. Of particular interest is the fact that in the Finnish study, remarkably few of those individuals who complied with most of the five target interventions progressed from IGT to T2DM (Fig. 21.2). Follow-up data suggest that even after withdrawing the intensive intervention provided in the Finnish and US studies, the risk reduction has been maintained for prolonged periods. Similar lifestyle interventions have been shown to increase

**Table 21.2 Risk factors for coronary heart disease (CHD) in people with type 2 diabetes**

| Risk factors that may be of particular relevance | General CHD risk factors that are also relevant in type 2 diabetes |
|---|---|
| ↑ Triglycerides and VLDL triglycerides | ↑ Total cholesterol and low-density lipoprotein |
| ↑ Small, dense, low-density lipoprotein particles | Hypertension |
| ↓ High-density lipoproteins | Cigarette smoking |
| ↑ Oxidation of low-density lipoprotein | Obesity, especially when centrally distributed |
| ↑ Platelet aggregation | Physical inactivity |
| ↑ Plasminogen activator inhibitor | |
| ↑ Pro-insulin-like molecules | |
| Microalbuminuria | |

↑, increased; ↓, decreased.

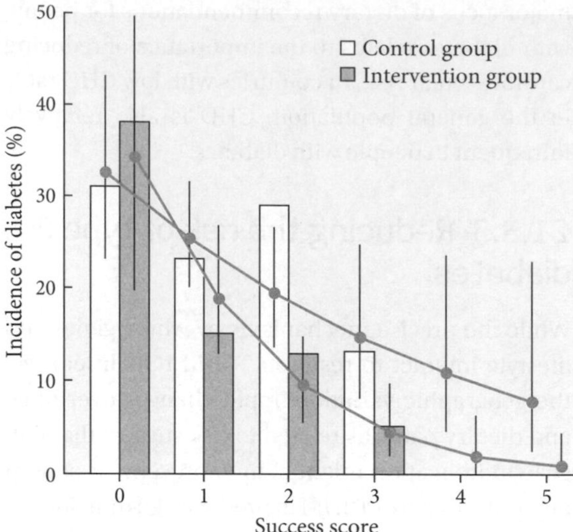


No. with diabetes/total no.

| | | | | | | |
|---|---|---|---|---|---|---|
| Intervention group | 5/13 | 10/66 | 9/69 | 2/38 | 0/25 | 0/24 |
| Control group | 15/48 | 25/107 | 14/48 | 2/15 | 0/11 | 0/4 |

Fig. 21.2 Incidence of diabetes during follow-up, according to the success score. At the 1-year visit, each subject received a grade of 0 for each intervention goal that had not been achieved and a grade of 1 for each goal that had been achieved; the success score was computed as the sum of the grades. The association between the success score and the risk of diabetes, with 95% confidence intervals, was estimated by means of logistic regression analysis of the observed data. The curves show the model-based incidence of diabetes according to the success score as a continuous variable; the curve in which data points align with the open bars represents the model-based incidence for the control group, and the curve in which data points align with the shaded bars represents the model-based incidence for the intervention group.

*Source:* Tuomilehto, M.D., Lindström, M.S., Eriksson, J.G., et al. (2001) Prevention of type 2 diabetes mellitus by changes in lifestyle among subjects with impaired glucose tolerance. *N Engl J Med*, **344**, 1343–50. Reprinted with permission from Massachusetts Medical Society.

insulin sensitivity in insulin-resistant individuals prior to the development of IGT or diabetes.

These observations suggest that in groups or populations with high rates of diabetes, it is appropriate to screen high-risk individuals (in particular those with central adiposity, a family history of diabetes, and those with other cardiovascular risk factors) so that preventative measures may be started in those with prediabetes and treatment initiated in those who have already developed the disease. Current recommendations from organizations such as the American Diabetes Association and Diabetes UK strongly support the weight loss targets and physical activity recommendations employed in the Finnish Diabetes Prevention Study (DPS) and the US Diabetes Prevention Programme (DPP) to reduce the risk of progression from prediabetes to T2DM. However, rather than providing specific ranges of macronutrients, they suggest that weight loss and weight maintenance may be achieved by a number of different dietary patterns (e.g. Mediterranean, Nordic, vegetarian, DASH) in which carbohydrate is predominantly derived from minimally processed grains, intact fruits and vegetables, and often also legumes and pulses, and fat sources are those that contain largely polyunsaturated or *cis*-monounsaturated fatty acids. A relatively low intake of carbohydrate is considered acceptable but it should be noted that the evidence quoted as supporting such an

### Box 21.1 Lifestyle modifications suggested to the intervention group in the Finnish Diabetes Prevention Study

- Weight loss of 5–7% initial body weight or a weight loss of 5–10 kg, depending upon degree of obesity.
- Reduction of total fat to less than 30% energy by encouraging low-fat dairy and meat products.
- Reduction of saturated fat to less than 10% energy by using unsaturated soft margarines and vegetables oils rich in monounsaturated fatty acids.
- Increase in fibre to at least 15 g per 1000 kcal by increasing whole grains, vegetables, and fruit.
- Physical activity, of at least moderate intensity, for a minimum of 30 minutes daily.

*Adapted from:* Lindström, J., Ilanne-Parikka, P., Peltonen, M., et al. (2006) Sustained reduction in the incidence of type 2 diabetes by lifestyle intervention: follow-up of the Finnish Diabetes Prevention Study. *Lancet*, 368, 1673–9.

approach relates to intakes of around 40% total energy, which in many countries would fall within the usual range. In addition, population-based advice to adopt dietary measures and increase physical activity to reduce overweight and obesity will be essential for primary prevention and reducing the epidemic proportions of the disease.

## 21.4 Epidemiology and aetiology of type 1 diabetes

In many countries, there have been considerable increases in incidence in recent years and, as with T2DM, there is marked geographic variation. Rates in Finland in children under 15 years (60/100,000/year) are amongst the highest in the world. The lowest incidence is seen across East and Southeast Asia, where rates as low as 3/100,000/year are reported. Although the total number of cases in all countries is appreciably lower than the number of cases of T2DM, the proportional increase has been comparable or greater in many countries. The frequency does not parallel that of T2DM and there is no clear explanation for the variation from one country to another or the change over time. Genetic factors are important in T1DM, although only about 10% of people with the condition have a clear family history. It is not clear what triggers the autoimmune process that leads to the destruction of the pancreatic islet β-cells. Various nutritional factors have been suggested, but the evidence is far from conclusive. Several epidemiological studies suggest that the early introduction of cows' milk into the diet of infants is associated with an increased risk of developing T1DM later in life, but the extent to which breast milk may be protective or cows' milk detrimental remains to be confirmed. The mechanisms by which infant nutrition might operate as a risk factor are far from clear. It is possible that cows' milk protein might be immunogenic in susceptible individuals. Several immunosuppressive drugs have been suggested as potentially useful means of reducing the risk of T1DM, but these drugs have side effects and none has been demonstrated to be of benefit in randomized controlled trials. Clinical trials involving various infant dietary regimens have also been suggested. Rapid growth in early childhood increases the risk of diabetes, possibly by increasing the workload on β-cells.

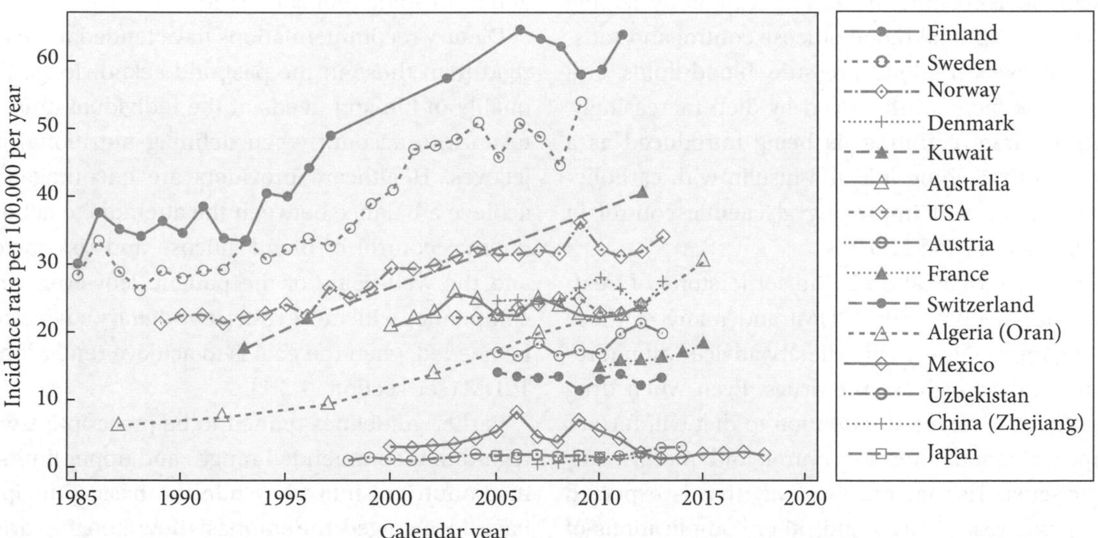


Fig. 21.3 Time-trends in incidence of type 1 diabetes in children from selected populations around the world.
*Source:* Tuomilehto, J., Ogle, G.D., Lund-Blix, N.A., Stene, L.C. (2020) Update on Worldwide Trends in Occurrence of Childhood Type 1 Diabetes in 2020. Ped. Endocrinol.Rev, 17, (Suppl1), 198–209.

Early growth velocity and obesity are much more common now than previously, but there is no direct evidence that this is a determinant of the marked increase in incidence in the young. A possible protective role for vitamin D has also been suggested, but is not clearly established. As with T2DM, many of the complications of T1DM are associated with unsatisfactory metabolic control, and cardiovascular disease in people with T1DM is associated with risk factors similar to those in the general population. In the aetiology of kidney damage (nephropathy), a high intake of protein, especially animal protein, may be associated, but the evidence is not conclusive.

## 21.5 Treatments for diabetes

All people with T1DM require insulin replacement treatment. For them, the goals of lifestyle and dietary advice are to minimize short-term fluctuations in blood glucose and especially to reduce the risk of hypoglycaemia by balancing injected insulin with carbohydrate-containing food and physical activity. The use of insulin pumps to provide continuous subcutaneous insulin infusions that can be adjusted according to amount of carbohydrate eaten and blood glucose levels helps to reproduce as closely as possible normal insulin secretion and has improved quality of life for many with T1DM. However, given the cost of insulin pumps, many people with T1DM worldwide continue to require insulin injections several times a day. Dietary modification also reduces the risk of long-term complications by helping to achieve optimal blood glucose control and satisfactory levels of blood pressure, blood lipids, and other risk factors influenced by diet. Increasingly, 'carbohydrate counting' is being introduced as a means of matching injected insulin with carbohydrate intake and improving glycaemic control in T1DM (see section 21.5.4).

Dietary modification is the cornerstone of treatment for people with T2DM, and many of those who comply closely with dietary advice will not require oral hypoglycaemic drugs. Even when drug treatment is required, attention to diet will further improve blood glucose control and modify cardiovascular risk factors in a way that is expected to reduce risk of CHD and other complications of diabetes. Intensive dietary modification aimed to achieve appreciable weight loss can for some achieve remission of T2DM even when the condition has been present for some time and been treated by drugs (see section 21.5.1 and Box 21.3).

The principles of dietary advice for people with T1DM and T2DM are similar to those recommended for entire populations at high risk of the major non-communicable diseases, and this means that there is no need for people with diabetes to have meals that differ from those of the rest of the family. Evidence-based dietary recommendations for people with diabetes have been issued in many countries. The evidence for the benefit of implementing the recommendations derives principally from systematic reviews and meta-analyses of trials in which dietary manipulations have been shown to improve glycaemic control or level of risk factors, and from epidemiological studies.

Dietary recommendations have tended to be less rigid than those in the past and acknowledge that quality of life and needs of the individual must be taken into account when defining nutritional objectives. Healthcare providers are encouraged to achieve a balance between the attempts to achieve optimal control of blood glucose and risk factors and the well-being of the patient. However, strict compliance with more restrictive dietary advice may be needed when the goal is to achieve remission of T2DM (see section 21.5.1).

Earlier guidelines tended to be prescriptive with regard to recommended ranges and upper limits of macronutrient intakes. While the basic principles have not changed, the emphasis now is on the variety of dietary patterns and the range of foods that have the potential to contribute to optimum glycaemic control and reduced risk of complications. The traditional Mediterranean and Nordic diets and the DASH diet are frequently quoted examples of acceptable dietary patterns, but provided the principles summarized in the text below and Table 21.3 are taken into account, the food preferences of most individuals can be incorporated into an appropriate eating pattern for people with T1DM and T2DM. Very low carbohydrate, high fat, and Paleo diets are

**Table 21.3 Key aspects of the current recommendations for diabetic diet and lifestyle**

| | |
|---|---|
| **Energy and body weight** | Achieve and/or maintain BMI of 18.5–25<br>To achieve remission of T2DM, aim for initial 15 kg weight loss<br>Reducing energy intake is essential to achieving weight loss; increasing physical activity may contribute |
| **Fat** | Saturated and *trans*-unsaturated fat should comprise <10% and <1% total energy, respectively |
| | When reducing saturated and *trans*-fat, replacement should be with polyunsaturated (n-6 and n-3) or *cis*-monounsaturated fats or high fibre minimally processed CHO foods |
| | Reduction of total fat to <30% TE may facilitate weight loss |
| | Plant-based foods high in both mono- and polyunsaturated fats, such as seeds, nuts, and non-hydrogenated non-tropical vegetable oils, are preferred sources of dietary fat<br>Supplements not recommended |
| **Carbohydrate, dietary fibre, and glycaemic index** | A wide range of total carbohydrate is acceptable provided recommendations relating to fibre, fats, and proteins are met |
| | Dietary fibre intake should be at least 35 g/day (or 20 g/1000 kcal/day)<br>Foods naturally high in dietary fibre should be encouraged Minimally processed whole grains, vegetables, legumes, seeds, nuts, and whole fruits are recommended sources (fibre-enriched foods and supplements may be considered when sufficient intakes cannot be obtained from diet alone) |
| | Low glycaemic index may be a useful marker of suitable carbohydrate-rich foods, provided other attributes are appropriate<br>Total free sugars should not exceed 10% total energy (less for those who are overweight)<br>Sugar-sweetened beverages should be avoided<br>Carbohydrate counting is a useful technique to determine meal insulin dose<br>Very low carbohydrate diets (e.g. ketogenic diets) are not recommended |
| **Protein and renal disease** | For most people, protein should provide 10–20% total energy, corresponding to 0.8–1.2 g/kg body weight in women and 0.9–1.3 g/kg in men<br>Total protein intake at lower end of normal range (0.8 g/kg/day) for patients with established nephropathy<br>Higher protein intakes (up to 30% TE) may be consumed in the short term (up to 12 months) to facilitate weight loss in those with normal renal function |
| **Vitamins, antioxidant nutrients, minerals, and trace elements** | Increase foods rich in tocopherols, carotenoids, vitamin C and flavonoids, trace elements, and other vitamins |
| | Fruits, vegetables, and whole grains rather than supplements recommended |
| | Restrict salt to less than 6 g/day (less than 2.3 g sodium)<br>Further reduction if hypertensive |
| **Alcohol** | Up to 10 g for women and 20 g for men per day is acceptable for most people with diabetes who choose to drink alcohol |

not recommended for long-term use as there is no evidence that benefit observed in the short term is sustained, nor that they are safe when used for prolonged periods.

### 21.5.1 Energy balance and body weight

The key recommendation for those who are overweight (BMI >25 kg/m$^2$) is that energy intake should be reduced so that the BMI moves towards the recommended range (18.5–25 kg/m$^2$). Prevention of weight regain is an important aim once weight loss has been achieved. For those who are overweight or obese, reducing energy-dense foods (those high in fats and free sugars) may be sufficient to achieve weight loss. Prescription of precise energy requirements is necessary for those unable to achieve the desired weight reduction. Increased energy expenditure may also facilitate weight loss but the majority of people with T2DM do not achieve the level of physical activity required to make a meaningful contribution. Even modest weight reduction (a loss of less than 10% body weight) in the overweight or obese improves insulin sensitivity, glycaemic control, blood lipids, blood pressure, and other cardiovascular risk factors. Weight loss may reduce or even eliminate the need for hypoglycaemic drug therapy in T2DM and lead to a reduction of insulin dose and improved glycaemic control in T1DM. The reduced life expectancy of overweight people with diabetes is improved in those who lose weight.

Appreciable weight loss can achieve remission of T2DM, defined as HbA$_{1c}$ less than 48 mmol/mol after at least two months off all antidiabetic medication. This has been demonstrated after bariatric surgery in people who are markedly obese. More recently the DIRECT study has shown that remission may also occur with weight loss of 15 kg or more achieved by initial use of liquid very-low-calorie diets followed by the reintroduction of solid food under the supervision of dietitians. The remission achieved by bariatric surgery and intensive medical nutrition therapy may continue in the long term but should not be regarded as a 'cure', as recurrence may occur if weight loss is not sustained.

### 21.5.2 Protein

Protein intake in most Western populations ranges between 10% and 20% TE, corresponding approximately to 0.8–1.3 g/kg body weight, and this range is appropriate for most people with diabetes. Intakes at the higher end of the range are acceptable for people with diabetes who do not have evidence of diabetic renal disease and where preference is to have lower intakes of carbohydrate. Intakes ranging between 15–20% TE may also help to maintain muscle mass and avoid sarcopenia (muscle mass loss) in older people. Higher intakes (up to 30% TE) are acceptable in the short term (up to 12 months) in those with normal renal function, if desired, to facilitate weight loss. In patients with T1DM and evidence of established nephropathy, protein intakes should be at the lower end of this range (0.8 g/kg body weight/day). Such restriction has been shown to reduce the risk of end-stage renal failure or death when compared in randomized controlled trials with more usual intakes (1.2 g/kg/day). The evidence for appreciably reducing protein intake is less convincing for T2DM patients with established nephropathy or for T1DM or T2DM patients with microalbuminuria (incipient nephropathy).

### 21.5.3 Dietary fat

Dietary fat should mostly come from the consumption of foods containing plant-based mono- and polyunsaturated fats rather than saturated or *trans*-fats. This may be achieved by using non-tropical vegetable oils (e.g. olive, canola/rapeseed, soybean, sunflower, linseed) and through the consumption of seeds, nuts, avocado, and fish while limiting fats from meats, processed meats, butter, coconut products, and palm oil.

The striking relationships between saturated and *trans*-unsaturated fatty acids, total and low-density lipoprotein (LDL) cholesterol, and CHD justify the recommendation to restrict intakes (see Chapter 19).

Replacement of SFA with *cis*-monounsaturated and polyunsaturated fatty acids and high fibre, minimally processed, carbohydrate-containing foods is associated with a reduction in LDL cholesterol levels, an improvement in several other cardiometabolic risk factors, and reduction in CHD risk. Studies in people with diabetes confirm that these benefits initially observed in healthy populations also apply to those with the condition. Most guidelines do not specify an upper limit of intake for total fat. This permits a flexible approach to the selection of food choices and dietary patterns. However, high intakes of polyunsaturated fatty acids may be associated with increased lipid oxidation or reduction in HDL, so intakes should not exceed 10% total energy. Furthermore, given the energy density of fat, some restriction of total fat (to no more than 30% TE) may help to reduce body fatness in those who are overweight and obese.

There is no justification for the suggestion by some individuals and groups that restriction of saturated fat is unnecessary for people with diabetes.

### 21.5.4 Carbohydrate, dietary fibre, and glycaemic index

Most sets of dietary guidelines do not include quantitative recommendations regarding intake of total carbohydrate, but emphasize the importance of carbohydrate quality, especially for those who choose relatively high carbohydrate intakes. Vegetables, legumes, intact fruits, and whole grains are the preferred carbohydrate sources because they are rich sources of dietary fibre.

There is convincing evidence from meta-analyses that high fibre diets have the potential to improve measures of glycaemic control, blood lipids, body weight, and inflammation and reduce premature mortality regardless of fibre type and type of diabetes. Benefits are apparent across the range of intakes but the data suggest that increasing fibre by 15 g or to 35 g/day would be a reasonable target and contribute to the reduction of premature mortality in adults with diabetes. Five or more servings per day of fibre-rich vegetables and fruit and four or more servings of legumes per week help to provide minimum requirements for fibre intake, provided most cereal-based foods are whole grain-fibre-rich.

High intakes of starchy or highly processed foods with a high glycaemic index and which are rapidly digested and absorbed (see Table 2.2) may be associated with poor glycaemic control, increased triglycerides, and low levels of HDL. For people with these attributes it is particularly important that carbohydrate-containing foods should be rich in dietary fibre and have a low glycaemic index. Furthermore, those with dyslipidaemia (high triglycerides, low HDL) and poor glycaemic control may be more satisfactorily controlled in terms of their metabolic derangement on relatively low carbohydrate intakes. However, radical restriction of carbohydrate to levels specified for ketogenic diets is inappropriate due to lack of convincing evidence of long-term benefit, difficulty for many in achieving long-term compliance, and potential adverse consequences of an associated increase in saturated fat. The discussion around the pros and cons of 'very low', 'low', or 'lower' carbohydrate diets is confused because there is no agreement regarding terminology (see Box 21.2). This further endorses the need to emphasize the importance of quality rather than quantity of dietary carbohydrate.

Sugars in intact fruit and vegetables and in dairy foods are not generally restricted. When incorporated in modest amounts (<50 g/day) into diets of appropriate macronutrient composition and energy content, sucrose appears not to be associated with any measurable untoward clinical or metabolic effect, hence the recommendation that modest amounts of free sugars might be included in a diabetic dietary prescription. As for the general population, it is advised that total free sugars do not exceed 10% TE, though more restrictive advice concerning free sugars is appropriate for those needing to lose weight, those with poor glycaemic control, and those with dyslipidaemia (high triglycerides, low HDL). Sugar-sweetened beverages should be avoided.

The glycaemic index may provide a useful means of identifying suitable carbohydrate-containing foods. However, there are some important limitations to the use of this index. Foods high in fats and sugars generally have a low glycaemic index. Yet they may be energy dense and include inappropriate fat sources and are therefore poor food sources for people with diabetes. The glycaemic index concept is only meaningful when used to classify predominantly carbohydrate-containing foods and when

**Box 21.2 'Very low', 'lower', and 'low' carbohydrate diets**

The suggestion that reduced carbohydrate intakes may be suitable or even preferable to higher carbohydrate intakes for people with type 2 diabetes is based on several observations. In the United States, rates of obesity and diabetes have increased in parallel with an increase in dietary carbohydrate. Weight loss has been greater and $HbA_{1c}$ and lipid profile appear to improve in people with diabetes when comparing lower with higher carbohydrate intakes. Remission rates for T2DM are reported to be greater.

While the arguments might appear to be persuasive, there are several reasons why they do not necessarily translate into recommendations. Benefits reported for lower compared with higher intakes apply only in the short term, typically for up to 12 months. Furthermore, many of the trials in which the effects of diets relatively high or low in carbohydrates have been compared do not report details regarding the source of carbohydrate. Without this information the results are difficult to interpret given the importance of carbohydrate quality on key outcome measures.

Formal systematic reviews and meta-analyses that attempt to synthesize existing knowledge and typically underpin recommendations are not always helpful given that there is no universal agreement regarding the definitions of 'very low', 'lower', and 'low' carbohydrate intakes. A recent report from the authoritative UK Scientific Advisory Committee on Nutrition (SACN) illustrates this. SACN has suggested that 'lower' carbohydrate intakes may confer benefits for people with T2DM compared with 'higher' intakes over a period of 6 months. The recommendation was based on evidence where 'lower' referred to intakes that ranged from 13% to 47% total energy, an intake well within the usual range in many countries.

Given that there is no evidence of long-term safety when using a very low carbohydrate diet, it seems more appropriate to conclude that a wide range of carbohydrate intakes is acceptable, with carbohydrate quality being a more important consideration than quantity. There is no justification for the routine use of very low carbohydrate ketogenic diets.

comparing foods within a comparable food group (e.g. breads, fruits, pasta, and rice). The glycaemic index of foods must therefore be interpreted in relation to energy content and content of other macronutrients.

For those on tablets to lower blood glucose or insulin, timing and dosage of medication should match quantity, nature, and timing of carbohydrate intake. Failure to do so may result in symptoms of hypoglycaemia and reduce the potential to achieve good blood glucose control. Measurement of glucose on a finger-prick blood sample or by means of newer non-invasive technology and increased availability of insulin pumps and use of carbohydrate counting have facilitated this process.

### 21.5.5 Antioxidant nutrients, vitamins, minerals, and trace elements

Foods naturally rich in dietary antioxidants (tocopherols, carotenoids, vitamin C, flavonoids), trace elements, and other vitamins are encouraged. Daily consumption of a range of vegetables, fruits, and whole grain breads and cereals should provide adequate intakes of vitamins and antioxidant nutrients and there is currently no evidence that dietary supplements confer benefit in those who are not deficient. Given the tendency towards high levels of blood pressure in people with diabetes, restriction of salt intake to less than 6 g/day (2300 mg sodium) is advised, with further restriction considered appropriate for those with elevated blood pressure levels.

### 21.5.6 Alcohol

Alcohol may be a relevant source of energy in those who are overweight and may be associated with raised levels of blood pressure, increased triglycerides, and an increased risk of hypoglycaemia, especially in insulin-treated individuals or those on some oral hypoglycaemic agents. While it is recommended that moderate use of alcohol (up to 10 g/day for women and 20 g/day for men) is acceptable for those with diabetes who choose to drink alcohol, restrictions are recommended for some. When alcohol is taken by those on insulin, it is essential that it be taken with carbohydrate-containing food in order to avoid the risk of potentially profound and prolonged hypoglycaemia. Alcohol should be limited by those who are overweight, hypertensive, or hypertryglyceridaemic. Abstention is advised for women who are pregnant and those with a history of alcohol abuse or pancreatitis, appreciable hypertriglyceridaemia, and advanced neuropathy.

## 21.5.7 Diabetic foods, functional foods, and supplements

Foods advertised as being of particular benefit for people with diabetes ('diabetic' foods) are generally sucrose-free, but may nevertheless be high in fructose or other nutritive sweeteners, and sometimes also fat. These have no substantial advantages over sucrose-containingfoodsforpeoplewithdiabetesand should not be encouraged. Although non-nutritive sweeteners (e.g. aspartame) are often considered to be useful, especially in drinks, comprehensive systematic reviews and meta-analyses have revealed no compelling evidence of benefit when considering a wide range of health outcomes and their authors have concluded that potential harm from the consumption of non-nutritive sweeteners cannot be excluded. Many functional foods and supplements are currently also being promoted for diabetes management or for reducing the risk of diabetes or its complications. These include fibre-enriched products and margarines containing plant sterols or stanols, and supplements containing various dietary fibres, ω3 fatty acids, minerals, trace elements, and some herbs. Some research suggests benefit of dietary fibre regardless of source but most of these products have not been tested in long-term clinical trials and are therefore not encouraged at present.

## 21.5.8 Translating nutritional principles into practice

Most medical practitioners do not have the training or the time to help people with diabetes translate the nutrition principles described here into practice. Dietitians or appropriately trained nutritionists play a key role in translating these general principles into specific advice for individuals. The high intakes of total and saturated fat by people with diabetes in Europe (Fig. 21.4) provide an indication of the extent of dietary change required by many, as discussed earlier (section 21.5.3). The wide ranges of acceptable intake for unsaturated fatty acids and carbohydrates enable the nutrient recommendations to be translated into a variety of dietary patterns (see Chapter 29). Behaviour modification techniques and various special lifestyle programmes can be of considerable value. Ongoing encouragement and reinforcement of lifestyle changes are essential, and one-off advice is rarely adequate.

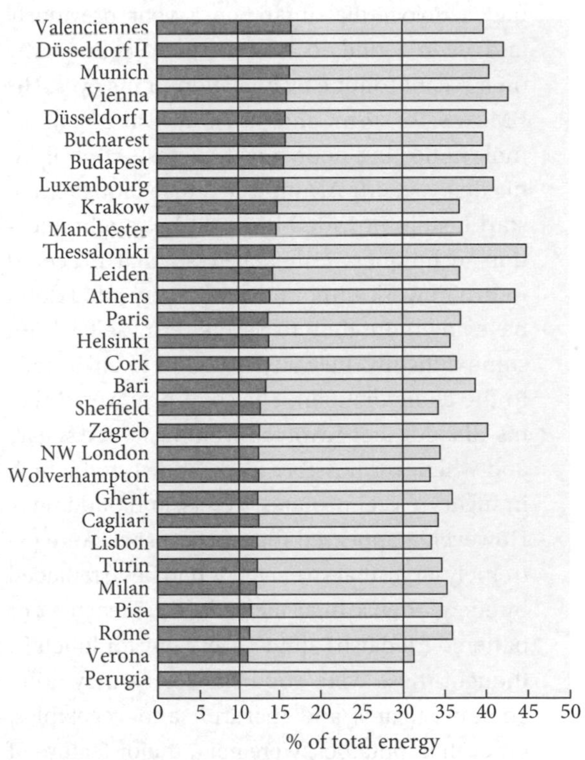


Fig. 21.4 Intake of total (□) and saturated (□) fatty acids as proportion of total dietary intake, calculated from 4-day diet records collected from people with T1DM. Data from European centres participating in the EURODIAB study.

*Source:* Toeller, M., Klischan, A., Heitkamp, G., et al. (1996) Nutritional intake of 2868 IDDM patients from 30 centres in Europe. EURODIAB IDDM Complications Study Group. *Diabetologia*, **39**, 929–39.

New drugs have revolutionised the treatment of T2DM but the case study in Box 21.3 provides an indication of what can be achieved in practice by closely following dietary advice.

**Box 21.3 Case study**

A 63-year-old school principal, Mr AD, with T2DM was referred to our hospital diabetes clinic. He was being treated with maximum doses of oral hypoglycaemic agents available 10 years ago, gliclazide, metformin, and pioglitazone, and was in theory following a 'diabetic diet'. His general practitioner had referred him for conversion to insulin treatment, believing that this was the only option for improving diabetes control, which was considered to be essential and of some urgency, given that he had had a myocardial infarction a year previously and been found to have diabetic retinopathy on a recent routine examination of his eyes. His BMI was 32 kg/m$^2$ and his HbA$_{1c}$ was 81 mmol/mol, indicating poor blood glucose control. At his first visit, Mr AD indicated his willingness to start insulin provided he could be taught injection techniques immediately so that he could return to work without delay ('In my job, I don't have any more time to devote to myself'). With some difficulty, he was persuaded to talk briefly to the clinic dietitian. She soon discovered that his 'diabetic diet' involved avoiding added sugars and whenever possible foods known to be high in sugars (e.g. chocolates, sweets, and puddings). However, it appeared that portion sizes were extremely large, that sweet foods had been replaced by energy-dense high-fat foods (e.g. meat pies, or battered fried fish) almost every day for lunch ('I thought these were good choices as they don't contain sugar'), and that fruit and vegetables, other than potatoes, were not a major feature of his dietary pattern. As a result he was persuaded that dietary modification with some regular daily exercise should precede the initiation of insulin treatment. The possibility that insulin treatment might promote further weight gain provided the impetus required for Mr AD to agree to devote some time to think about his lifestyle in relation to his health. The dietary advice was relatively simple: a reduction in portion size, radical reduction in high-fat, energy-dense foods, an increase in fruit and vegetables other than potatoes, and the use of dense whole grain, low-glycaemic-index bread instead of his usual white bread. Being an intelligent man, having been persuaded to make the changes, he was meticulous regarding implementation. Within days and before his weight had changed, his blood glucose levels were down to between 5 and 9 mmol/L throughout the day. After 3 weeks he had lost nearly 5 kg in weight and was experiencing some hypoglycaemic episodes. Over the subsequent months his weight continued to fall and the pioglitazone dose was reduced and then stopped. As HbA$_{1c}$ continued to improve, the gliclazide dose was also reduced and then stopped, and by the end of the year after first referral, his BMI was 26 kg/m$^2$, his HbA$_{1c}$ was 50 mmol/mol, and metformin was his only hypoglycaemic medication. At the time of writing, some 10 years after the referral, he had lost a total of 14 kg, weight loss had been maintained, and metformin had been stopped as HbA$_{1c}$ was in the 'normal' range. In addition to the improvement in glycaemic control, blood pressure and lipid profile have also improved with the dietary modification and weight loss. However, he continues on his statin drug and other cardiac medications prescribed following his myocardial infarction. In his retirement, Mr AD has taken over much of the meal preparation from his wife, and faces this phase of his life course with a much improved prognosis. Interestingly, he does not miss his earlier dietary pattern and enjoys being adventurous in his culinary skills.

**Comment**

While almost everyone will benefit from improved diet and exercise habits, whether or not tablet treatment or insulin treatment can be reduced or stopped will vary from patient to patient, given that type 2 diabetes is a heterogeneous condition. Dosage of tablets may be reduced if blood sugar levels are consistently low or patients experience hypoglycaemia, as happened in the case of Mr AD. Typically, metformin is continued since this drug tends to enhance insulin sensitivity. Many patients with T2DM do require insulin treatment after having had the condition for many years and this requirement may continue even with meticulous attention to diet. The β-cells become 'exhausted' after a prolonged period of producing larger than usual amounts of insulin in an attempt to overcome insulin resistance, the underlying abnormality in most people with T2DM. While it is never too late to recommend enhanced lifestyle advice, the most striking results are seen when it is implemented relatively early in the disease process. Mr AD has had a dramatic response in that he has shown prolonged remission of his T2DM typically seen with intensive nutrition therapy when weight loss has been induced by very low calorie liquid formula diets, as described in section 21.5.1.

### 21.5.9 Families and communities

Individual compliance with dietary advice is improved if the general aspects of advice are understood by the family and are of potential benefit to them. In view of the strong genetic component to T2DM, the evidence that lifestyle changes reduce the risk of developing the condition, and that the dietary principles are similar to those recommended for the population at large in countries with high rates of non-communicable disease, it seems reasonable to suggest that foods and meals that are suitable for people with diabetes are appropriate for their families. Adoption of public health measures aimed at facilitating healthy food choices, such as ensuring the availability of appropriate foods at reasonable cost and taxing sugary drinks, will further enhance compliance.

## 21.6 Metabolic syndrome

Resistance to the action of insulin, an important underlying abnormality in T2DM, is also associated with a range of additional clinical and metabolic abnormalities (Box 21.4) that are often seen in association with T2DM, IGT, or IFG, but may also occur with normal blood glucose levels. Several national and international organizations have suggested sets of diagnostic criteria for what has become known as the metabolic syndrome (Box 21.5). The validity of describing this constellation of abnormalities as a 'syndrome' has been questioned on the grounds that the criteria are clearly arbitrary and its causes ill understood. Nevertheless, its existence does help to define a group of individuals who are at high risk of cardiovascular disease and of developing T2DM if they have not already done so. As many as one-quarter (and in some instances appreciably more) of all adults in many affluent and some developing

**Box 21.4 Clinical and metabolic features other than hyperglycaemia associated with insulin resistance**

- Central obesity
- Raised blood pressure
- *Dyslipidaemia*:
    - Increased triglyceride
    - Low HDL
    - Predominance of small, dense LDL particles
    - Increased uric acid and gout
- Increased plasminogen activator inhibitor (PAI-1)
- NAFLD (non-alcoholic fatty liver disease)
- Endothelial dysfunction
- Increased pro-inflammatory cytokines
- Increased homocysteine

**Box 21.5** Diagnostic criteria suggested for the metabolic syndrome by different international organizations

| WHO (1999) | NCEP, ATP111 (2001) | IDF (2005) |
|---|---|---|
| Hyperglycaemia or insulin resistance plus two or more of: | Three or more of: | Central obesity:<br>Waist (ethnic specific)<br>Europeans ≥94 cm (M), ≥80 cm (F)<br>South Asians/Chinese ≥90 cm (M), ≥80 cm (F) |
| Obesity:<br>W/H >0.9 (M), 0.85 (F), or BMI >30 kg/m$^2$ | Central obesity:<br>Waist >102 cm (M), 88 cm (F) | Plus two of:<br>Fasting glucose ≥5.6 mmol/L |
| Dyslipidaemia:<br>Triglyceride ≥1.7 mmol/L or HDL <0.9 mmol/L (M), 1.0 mmol/L (F) | Fasting glucose ≥6.1 mmol/L<br>Hypertriglyceridaemia:<br>Triglyceride ≥1.7 mmol/L<br>Low HDL <1.0 mmol/L (M), 1.3 mmol/L (F) | Treated dyslipidaemia: or raised triglyceride ≥1.7 mmol/L<br>Reduced HDL <1.03 mmol/L (M), <1.29 mmol/L (F) |
| Hypertension:<br>Blood pressure ≥140/90 mmHg | Hypertension:<br>Blood pressure ≥135/85 mmHg or treatment | Treated hypertension or:<br>Raised blood pressure:<br>>130 systolic or >85 diastolic |
| Microalbuminuria | | |

ATP, adult treatment panel; BMI, body mass index; F, female; HDL, high-density lipoprotein; IDF, International Diabetes Federation; M, male; NCEP, National Cholesterol Education Programme; W/H, waist/hip ratio; WHO, World Health Organization.
*Adapted from:* WHO (1999) *Definition, diagnosis and classification of diabetes mellitus and its complications*, Report of a WHO consultation. Geneva, Switzerland: World Health Organization; NCEP, ATP 111 (2001) Executive summary of the Third Report of The National Cholesterol Education Programme (NCEP) Expert Panel on Detection, Evaluation, and Treatment of High Blood Cholesterol in Adults (Adult Treatment Panel III). *JAMA*, **285**, 2486–97; Alberti, K.G., Zimmet, P., and Shaw, J. (2005) The metabolic syndrome—a new worldwide definition. *Lancet*, **366**, 1059–62.

countries will fit the criteria set by the WHO and the US National Cholesterol Education Programme (NCEP). However, this figure is appreciably greater if the criteria suggested by the International Diabetes Federation are implemented. Lifestyle-related risk factors are similar to those described for T2DM, as are the measures for prevention and treatment. Lifestyle modification is the pivotal component of management, since weight loss associated with the appropriate dietary measures offers the only means of favourably influencing the broad range of abnormalities associated with the syndrome. Drug treatments may be required for specific metabolic abnormalities.

## Further reading

1. **Åberg, S., Mann, J., Neumann, S., Ross, A.B., Reynolds, A.** (2020) Whole-grain processing and glycemic control in type 2 diabetes: a randomized crossover trial. *Diabetes Care*, **43**, 1717–23.
2. **American Diabetes Association (ADA)** (2021) Classification and diagnosis of diabetes. *Diabetes Care*, **44**, S15–S33.

3. **American Diabetes Association (ADA)** (2021) Prevention or delay of type 2 diabetes. *Diabetes Care*, **44**, S34–S39.
4. **American Diabetes Association (ADA)** (2021) Obesity management for the treatment of type 2 diabetes. *Diabetes Care*, **44**, S100–S110.
5. **Becerra-Tomás, N., Blanco, M.S., Viguiliouk, E., et al.** (2020) Mediterranean diet, cardiovascular disease and mortality in diabetes: a systematic review and meta-analysis of prospective cohort studies and randomized clinical trials. *Crit Rev Food Sci Nutr*, **60**(7), 1207–1227. doi: 10.1080/10408398.2019.156528
6. **Chiavaroli, L., Lee, D., Ahmed, A., Cheung, A., Khan, T.A., et al.** (2021) The effect of low glycemic index/load dietary patterns on glycemic control and cardiometabolic risk factors in diabetes: a systematic review and meta-analysis of randomized controlled trials. *BMJ*, **374**, n1651 doi: https://doi.org/10.1136/bmj.n1651
7. **Chiavaroli, L., Viguiliouk, E., Nishi, S.K., et al.** (2019) DASH dietary pattern and cardiometabolic outcomes: an umbrella review of systematic reviews and meta-analyses. *Nutrients*, **511**(2), 338. doi: 10.3390/nu11020338.
8. **Churuangsuk, C.H.J., Reynolds, A., Griffin, S., et al.** (2021) Diets for weight management in adults with type 2 diabetes: an umbrella review of published meta-analyses and systematic review of trials of diets for diabetes remission. *Diabetologia* **65**(1), 14–36. doi: 10.1007/s00125-021-05577-2.
9. **Evert, A.B., Dennison, M., Gardner, C.D., et al.** (2019) Nutrition therapy for adults with diabetes or prediabetes: a Consensus Report. *Diabetes Care*, **42**, 731–55.
10. **Glenn, A.J., Viguiliouk, E., Seider, M., et al.** (2019) Relation of vegetarian dietary patterns with major cardiovascular outcomes: a systematic review and meta-analysis of prospective cohort studies. *Front Nutr*, **13**(6), 80. doi: 10.3389/fnut.2019.00080.
11. **IDF** (2019) *IDF diabetes atlas*, 9th edn. Brussels, Belgium: International Diabetes Federation (IDF). Available at: https://www.diabetesatlas.org (accessed 24 August 2021).
12. **Kahleova, H., Salas-Salvadó, J., Rahelić, D., et al.** (2019) Dietary patterns and cardiometabolic outcomes in diabetes: a summary of systematic review and meta-analyses. *Nutrients*, **13**(11), 9.
13. **Korsmo-Haugen, H.K., Brurberg, K.G., Mann, J., Aas, A.M.** (2019) Carbohydrate quantity in the dietary management of type 2 diabetes: a systematic review and meta-analysis. *Diabetes, Obesity and Metabolism*, **21**, 15–27.
14. **Lean, M.E., Leslie, W.S., Barnes, A.C., et al.** (2018) Primary care-led weight management for remission of type 2 (DiRECT); an open-label cluster-randomised trial. *Lancet*, **391**, 541–51.
15. **Pfeiffer, A.E.H., Pedersen, E., Schwab, U., et al.** (2020) The effects of different quantities and qualities of protein intake in people with diabetes. Communication. *Nutrients*, **12**, 365.
16. **Reynolds, A.N., Akerman, A.P., and Mann, J.** (2020) Dietary fibre and whole grains in diabetes management: systematic review and meta-analysis. *PLoS Med*, **17**(3). doi: 10.1371/journal.pmed.1003053
17. **Reynolds, A., Mann, J., Cummings, J. et al.** (2019) Carbohydrate quality and human health: a series of systematic reviews and meta-analyses. *Lancet*, **393**, 434–45.
18. **SACN** (2021) *The Scientific Advisory Committee on Nutrition.* Report: Lower carbohydrate diets for type 2 diabetes. Public Health England. Available at: https://www.gov.uk/government/publications/sacn-report-lower-carbohydrate-diets-for-type-2-diabetes (accessed 25 August 2021).
19. **Sainsbury, E., Kizirian, N.V., Partridge, S.R., et al.** (2018) Effect of dietary carbohydrate restriction on glycemic control in adults with diabetes: a systematic review and meta-analysis. *Diabetes Res Clin Pract*, **139**, 239–52.
20. **Schwab, U., Reynolds, A.N., Sallinen, T., Rivellese, A.A., and Risérus, U.** (2021) Dietary fat intakes and cardiovascular disease risk in adults with type 2 diabetes: a systematic review and meta-analysis. *Eur J Nutr*. doi: 10.1007/s00394-021-02507-1.
21. **Tuomilehto, J., Ogle, G.D., Lund-Blix, N.A., and Stene, L.C.** (2020) Update on worldwide trends in occurrence of childhood type 1 diabetes in 2020. *Ped Endocrinol Rev*, **17**(Supp1), 198–209.

22. **Uusitipua, M., Khan, T.A., Viguiliouk, K. et al.** (2019) Prevention of type 2 diabetes by lifestyle changes: a systematic review and meta analysis. *Nutrients*, **11**, 261. doi: 10.3390/nu11112611.
23. **Viguiliouk, E., Glenn, A.J., Nishi, S.K., et al.** (2019) Associations between dietary pulses alone or with other legumes and cardiometabolic disease outcomes: an umbrella review and updated systematic review and meta-analysis of prospective cohort studies. *Adv Nutr*, **10**(Suppl4), S308–S319. doi: 10.1093/advances/nmz113
24. **Wylie-Rosett, J. and Hu, F.B.** (2019) Nutritional strategies for prevention and management of diabetes: consensus and uncertainties. *Diabetes Care*, **42**, 727–30.

# 22 The Eating Disorders: Anorexia Nervosa, Bulimia Nervosa, Binge-Eating Disorder, and OSFED

Hannah Turner and Robert Peveler



The term 'eating disorders' includes three principal conditions, anorexia nervosa (AN), bulimia nervosa (BN), and binge-eating disorder (BED). However, it is also recognized that a significant proportion of patients seen in routine clinical practice present with a syndrome that falls short of a 'full-blown' diagnosis and the term 'other specified feeding or eating disorder' (OSFED) is currently used to describe this clinical group. It is now recognized that OSFED is as serious a condition as any other eating disorder diagnosis. Despite OSFED replacing 'eating disorder not otherwise specified' (EDNOS) as the term used to describe atypical presentations in the fifth edition of *Diagnostic and statistical manual of mental disorders* (*DSM-V*; APA, 2013), the debate as to what constitutes 'clinical' eating disorders continues. Descriptions of AN date back to 1873, but BN was not described as a clinical disorder until 1979. The term EDNOS has only been in use since 1980, recently being replaced by OSFED in 2013. Whilst BED was initially included in *DSM-IV* as a provisional diagnosis within EDNOS, it was formally recognized as a condition in 2013. Eating disorders are a significant cause of physical and psychosocial morbidity, and whilst historically associated primarily with female adolescent and young women, it is now widely accepted that eating disorders can affect people of all ages, genders, sexual orientations, ethnicities, and socioeconomic backgrounds.

## 22.1 Definitions

### 22.1.1 Anorexia nervosa

The diagnostic criteria for AN (as given in *DSM-V*) are shown in Table 22.1. The first is the persistent restriction of energy intake leading to significantly low body weight (in the context of what is minimally expected for age, sex, developmental trajectory, and physical health). Weight loss is commonly achieved through

**Table 22.1 Diagnostic criteria for anorexia nervosa, bulimia nervosa, binge-eating disorder, and other specified feeding or eating disorder**

| Anorexia nervosa |
|---|
| Restriction of energy intake relative to requirements, leading to a significantly low body weight in the context of age, sex, developmental trajectory, and physical health. *Significantly low weight* is defined as a weight that is less than minimally normal, or for children and adolescents, less than that minimally expected. |
| Intense fear of gaining weight or of becoming fat, or persistent behaviour that interferes with weight gain, even though at a significantly low weight. |
| Disturbance in the way in which one's body weight or shape is experienced, undue influence of body shape and weight on self-evaluation, or persistent lack of recognition of the seriousness of the current low body weight. |
| **Bulimia nervosa** |
| Recurrent episodes of binge eating. An episode of binge eating is characterized by both of the following: (1) eating, in a discrete period of time (e.g. within any 2-hour period), an amount of food that is definitely larger than most people would eat during a similar period of time and under similar circumstances; and (2) a sense of lack of control over eating during the episode (e.g. a feeling that one cannot stop eating or control what or how much one is eating). |
| Recurrent inappropriate compensatory behaviours to prevent weight gain, such as self-induced vomiting, misuse of laxatives, diuretics, or other medications, fasting, or excessive exercise. |
| The binge eating and inappropriate compensatory behaviours both occur, on average, at least once a week for 3 months. |
| Self-evaluation is unduly influenced by body shape and weight. |
| The disturbance does not occur exclusively during episodes of AN. |
| **Binge-eating disorder** |
| Recurrent episodes of binge eating. An episode of binge eating is characterized by both of the following: (1) eating, in a discrete period of time (e.g. within any 2-hour period), an amount of food that is definitely larger than most people would eat during a similar period of time and under similar circumstances; and (2) a sense of lack of control over eating during the episode (e.g. a feeling that one cannot stop eating or control what or how much one is eating). |
| The binge-eating episodes are associated with three (or more) of the following:<br>• eating much more rapidly than normal<br>• eating until feeling uncomfortably full<br>• eating large amounts of food when not feeling physically hungry<br>• eating alone because of feeling embarrassed by how much one is eating<br>• feeling disgusted with oneself, depressed, or very guilty afterward. |
| Marked distress regarding binge eating is present. |
| Binge eating occurs, on average, at least once a week for 3 months. |
| The binge eating is not associated with the recurrent use of inappropriate compensatory behaviours as in bulimia nervosa and does not occur exclusively during the course of bulimia nervosa or anorexia nervosa. |
| **Other specified feeding or eating disorder (OSFED)** |
| *Atypical anorexia nervosa*: all criteria are met, except that despite significant weight loss, the individual's weight is within or above the normal range. |
| *Binge-eating disorder (of low frequency and/or limited duration)*: all of the criteria for BED are met, except that the binge eating occurs, on average, less than once a week and/or for less than 3 months. |
| *Purging disorder*: recurrent purging behaviour to influence weight or shape in the absence of binge eating |
| *Night eating syndrome:* Recurrent episodes of night eating. Eating after awakening from sleep, or by excessive food consumption after the evening meal. There is awareness of recall of the eating. The behaviour is not better explained by environmental influences or social norms. The behaviour causes significant distress/ impairment. The behaviour is not better explained by another mental health disorder (e.g. BED). |

extreme dietary restriction, although a subgroup of patients will also engage in other weight-loss behaviours such as compulsive exercise, self-induced vomiting, and laxative misuse. The remaining diagnostic features are concerned with a characteristic set of attitudes and values concerning shape and weight. Patients often experience an intense fear of gaining weight or becoming fat, even though they are significantly underweight. They also express a level of dissatisfaction with their body weight and shape that far exceeds that typically seen in the general population, and they tend to judge their self-worth almost solely in terms of their weight, shape, and ability to control their food intake. Those with AN are often unaware of the seriousness of their current low body weight.

### 22.1.2 Bulimia nervosa

The *DSM-V* diagnostic criteria for BN are also shown in Table 22.1. The first is recurrent episodes of binge eating during which an objectively large amount of food is consumed, with associated loss of perceived control of eating. The second feature is the use of compensatory behaviours designed to prevent weight gain. The third feature is the same set of attitudes and values seen in AN, with self-worth being determined almost exclusively on the basis of weight, shape, and ability to control food intake. Although most people with BN are within the normal weight range, a proportion will have a history of AN. Where BN occurs in the context of AN, the latter diagnosis takes precedence.

### 22.1.3 Binge-eating disorder

The *DSM-V* diagnostic criteria for BED are also shown in Table 22.1. Patients with BED engage in recurrent episodes of binge eating. Such episodes are very distressing and are typically associated with eating when not physically hungry, eating more rapidly than normal, or eating until feeling uncomfortably full. Patients with BED often feel disgusted, depressed, or very guilty after an episode of binge eating, but unlike those with BN, they do not engage in the recurrent use of compensatory behaviours.

### 22.1.4 Other specified feeding or eating disorder

A diagnosis of OSFED may be given to a person who has a feeding or eating disorder that causes clinically significant distress and impairment in areas of functioning, but does not meet the full criteria for any of the other feeding and eating disorders. Examples of OSFED are given in Table 22.1 and include patients who present with all the key features of BN or BED, but fail to fulfil the diagnostic criteria relating to frequency of occurrence of behavioural symptoms. Others may present with all the diagnostic features of AN, but despite significant weight loss, body weight remains within or above the normal range.

## 22.2 Epidemiology

AN is most likely to develop during adolescence and young adulthood. The most common age of onset is 15 years (range 9–24 years). During their lifetime, 0.9–2.2% of females and 0.2–0.3% of males are diagnosed with AN. However, the true rate is likely to be higher given that a significant percentage of those who develop the condition never seek treatment. The incidence of AN has not changed significantly over the past five decades; however, there is evidence to suggest a slight increase in incidence among younger children. BN affects 1.5–2.0% of females and 0.5% of males. Incidence studies suggest an increase in diagnoses in the 1980s and mid-1990s, followed by a slight decrease in incidence in the late 1990s, with stability since that time. Compared with AN, the age of onset is later, typically developing during late teenage years or early adulthood. The lifetime prevalence of BED is estimated at 3.5% among adult females and 2.0% in males, and a significant percentage also present with obesity. The prevalence of atypical eating disorders remains unclear, although reports suggest that it is the most common presentation seen in clinical practice.

## 22.3 Anorexia nervosa

### 22.3.1 Development of the disorder

The onset of AN is usually in adolescence, although cases of pre-pubertal and adult onset have been reported. The disorder typically starts with an episode of dieting, although the path into this behaviour can vary. For some, natural bodily changes that accompany puberty, or a negative weight-related or shape-related comment from another person, may lead to a conscious decision to diet. For others, an episode of physical illness with associated weight loss, such as glandular fever, may lead to more intentional dietary restriction. Commonly occurring in the context of low self-worth, positive feedback in the form of attention from others initially serves to reinforce dieting behaviour and further weight loss. Patients with AN often report a sense of euphoria at being in control of their weight. For others, it brings feelings of success and a fleeting sense of superiority at achieving something few in the general population can accomplish. An example of a food diary for a patient with anorexia nervosa is given in Fig. 22.1.

### 22.3.2 Clinical features

As dieting intensifies, so weight falls, and the physiological and psychological effects of starvation develop. Behaviours around food become increasingly rigid and deceptive, and the range of acceptable foods slowly diminishes. Foods that are viewed as fattening are typically avoided and behaviours such as calorie counting, cutting food into small pieces, hiding food, and avoiding eating with others are all common. The average energy intake is in the region of 600–900 kcal/day, with the proportion of energy derived from fat being particularly low. Mineral intake is also low, although mineral deficiencies are rare. It is possible that zinc deficiency may contribute to the maintenance of the disorder through an effect on appetite and taste. However, true 'anorexia'

**Food Diary**

| | | Place | * | V/L | Comments |
|---|---|---|---|---|---|
| | | Kitchen | | | Feel OK |
| 12 pm | Salad of lettuce, cucumber and tomato<br>3 crab sticks<br>1 cracker<br>Black coffee | Lounge | | | Worried that I've eaten too much |
| 2 pm | Apple<br>Glass of water | Kitchen | | | |
| 6 pm | Steamed vegetables<br>1/2 piece of chicken | Lounge | | | Feel full + bloated<br>Worried that my weight will go up tomorrow |

* = binge, V = vomiting, L = laxatives

Fig. 22.1 Example food diary for a patient with anorexia nervosa.

is rare insofar as most patients report feeling persistently hungry. A proportion will feel driven to engage in excessive and/or compulsive exercise, leading to further weight loss. Some will also engage in other forms of weight-control behaviour, such as self-induced vomiting, laxative misuse, or taking appetite suppressants, such as slimming or dieting pills. A minority of patients will intermittently lose control over their eating and binge, although the amounts consumed tend not to be large.

Reduction in food intake is accompanied by several cognitive changes, including an increase in compulsive traits and cognitive rigidity. Many will view either their body, or particular parts, as being bigger than their true size, and this is typically accompanied by an intense dislike or loathing of the body or body part. Cognitive rumination about weight and shape becomes all-consuming and an increasing amount of time is given to thinking about food. Over time, general functioning becomes increasingly impaired, interest in other areas of life diminishes, and day-to-day routines become characterized by social withdrawal and isolation. Depression, irritability, and anxiety are common, as are obsessional features. Typically, all features get worse with further weight loss. Chronic presentations may also be accompanied by thoughts of hopelessness and suicide.

**Physical health** AN is associated with a range of physical abnormalities, most of which are now believed to be secondary to disturbed patterns of eating and low weight. Although patients often present with few physical complaints, further enquiry often reveals heightened sensitivity to cold and a variety of gastrointestinal symptoms, including constipation, fullness after eating, bloatedness, and vague abdominal pains. Other symptoms include restlessness, lack of energy, low sexual appetite, early morning wakening, and dizziness on standing. In post-menarchal females who are not taking an oral contraceptive, amenorrhoea is often present, with infertility posing a concern for many women. On examination, patients are typically emaciated and underweight. Those with a pre-pubertal onset may be short in stature and show failure of breast development. Often there is fine downy hair (lanugo) on the back, arms, and side of the face. The skin tends to be dry and the hands and feet cold. Blood pressure and pulse are low and there may be dependent oedema. The findings on investigation are shown in Table 22.2.

**Table 22.2 Anorexia nervosa and bulimia nervosa: common abnormalities on investigation**

| Anorexia nervosa |
|---|
| *Endocrine* |
| Low levels of female sex hormones (luteinizing hormone, follicle stimulating hormone, and oestradiol) |
| Low triiodothyronine ($T_3$) level but normal levels of thyroxine ($T_4$) and thyroid-stimulating hormone |
| Raised growth hormone and cortisol levels |
| *Cardiovascular* |
| Low blood pressure (especially postural) |
| Bradycardia |
| Other arrhythmias |
| *Haematological* |
| Slightly lowered white cell count |
| Anaemia (normocytic normochromic) |
| Low erythrocyte sedimentation rate |
| *Other metabolic abnormalities* |
| Raised blood cholesterol |

(Continued)

**Table 22.2 Anorexia nervosa and bulimia nervosa: common abnormalities on investigation (*Continued*)**

| |
|---|
| Increased serum carotene |
| Low blood sugar |
| Dehydration |
| Electrolyte disturbance (in those who vomit frequently or misuse large quantities of laxatives or diuretics), especially hypokalaemia |
| *Other findings* |
| Skeletal abnormalities (raised rate of osteopenia with risk of fractures) |
| Delayed gastric emptying and prolonged gastrointestinal transit time |
| Hypertrophy of salivary glands (especially parotids) |
| Enlarged cerebral ventricles and external cerebrovascular fluid spaces ('pseudoatrophy') |
| **Bulimia nervosa** |
| *Endocrine* |
| Electrolyte disturbance, especially hypokalaemia, in those who vomit frequently or misuse large quantities of laxatives or diuretics |
| *Gastrointestinal* |
| Prolonged digestion |
| Oesophageal damage and/or irritation of the oesophagus and/or pharynx due to contact with gastric acids |
| Perforation of upper digestive tract, oesophagus, or stomach |
| Abdominal pain and distention |
| Hypertrophy of salivary glands |
| *Cardiac* |
| Cardiac arrhythmias |
| *Other* |
| Dental erosion—gastric acids may cause deterioration of tooth enamel (perimolysis) |
| Hand callouses |
| Blood in vomit |
| Sore throat |
| Fatigue |
| Nausea |
| Weight gain |

## 22.3.3 Aetiology

**Predisposing factors** AN is a complex illness that develops over time, often as a result of multiple influences. Cultural and environmental factors, such as the idealization of thinness, weight/shape-related teasing, and hobbies or careers where weight is salient, have all been identified as potential triggers. Dieting appears to be a general risk factor, although only a small percentage of those who diet go on to develop a clinical eating disorder. This suggests the importance of other aetiological factors. Other risk factors identified in the literature include a genetic heritability, and premorbid characteristics such as long-standing low self-esteem, an anxious temperament, obsessive-compulsive traits, and perfectionism. Patients with AN are often high achievers who have an intense need to be accepted by others. Whilst early theories of AN attributed a causal role to the family, it is now widely accepted that this is not the case. AN does, however, have a significant impact on the entire family, as meal times become tense, relationships strained, and family life disrupted. A history of eating disorders and depression are also commonly seen within patients' families.

Maintaining factors As weight loss continues, so the physical and psychological sequelae of starvation become more prominent. Some perpetuate the disorder. For example, delayed gastric emptying results in fullness even after eating small amounts of food, a situation that can fuel concern about uncontrollable weight gain. Weight loss may initially be accompanied by positive feedback from others, but over the course of time, social withdrawal can lead to isolation from peers. Controlling food intake can lead to a powerful sense of self-control and often serves to enhance self-esteem. Those with AN can also hold a powerful position within the family.

## 22.3.4 Assessment

A large proportion of people with AN are ambivalent about seeking treatment and many will have been persuaded to seek help by concerned relatives or friends. Thus, assessment forms a crucial part of the treatment process. Assessment should cover psychological, social, and physical needs, as well as assessment of risk to self and others. Where possible, the diagnosis should be made using a standardized diagnostic instrument, such as the Structured Clinical Interview for DSM-IV (SCID-I). No physical tests are required to make the diagnosis, and unless there are positive reasons to suspect the presence of another physical condition, no tests are required to exclude other medical disorders, when it is apparent that weight loss is self-induced.

A proportion of patients will present in a general medical setting and will report physical symptoms such as gastrointestinal symptoms, amenorrhoea, or infertility. However, once it has been established that weight loss has been self-induced, a diagnosis of AN can be explored. All patients with AN should have a thorough physical examination. Those who vomit frequently or misuse significant quantities of laxatives or diuretics should have their electrolytes checked. Given the increased risk of osteoporosis, it can be useful to conduct a dual energy X-ray absorptiometry (DEXA) scan to assess bone density in patients who have been underweight for an extended period of time. It may also be helpful to conduct a pelvic ultrasound to assess ovarian and uterine maturity in those with persisting amenorrhoea.

## 22.3.5 Management

Management and treatment of AN is best delivered by a multidisciplinary team, which is likely to include medical, dietetic, and psychological input. There are three aspects to the management of AN. The first is to help patients to recognize that they have an illness. Acceptance and motivation to change are crucial given the recalcitrant nature of the illness. The second goal is the normalization of eating habits and weight restoration. Due to the risk of refeeding syndrome, calorie intake should be increased gradually and it may be necessary to monitor phosphate, magnesium, and potassium levels during this initial phase. Weight gain may be achieved through outpatient, day-patient, or inpatient treatment, and is typically facilitated through providing a combination of nutritional advice and psychological support. Drugs have almost no role, although dietary supplements can be of value in assisting weight restoration. The third aspect involves addressing patients' over-evaluation of weight and shape, their extreme control over eating patterns, and underlying psychosocial functioning. These aspects of treatment are typically addressed in specialized treatments such as family-based therapy, eating-disorder-focused cognitive behavioural therapy (CBT-ED), MANTRA (Maudsley Model of Anorexia Nervosa Treatment for Adults), or SSCM (Specialist Supportive Clinical Management). It is also now widely recognized that the family often have an important role to play in creating an environment that promotes and supports recovery.

### 22.3.6 Course and outcome

Response to treatment varies widely. For those who are willing and able to change, the illness duration can be relatively short. However, for others it can become a chronic problem commonly characterized by a resistance to change, despite a wish to recover. Although there are few consistent predictors of outcome, a long history and late onset have both been associated with a poor prognosis. Low weight and a history of premorbid psychosocial problems also tend to be associated with a poor outcome. More recently, illness duration has been identified as a key predictor of treatment outcome, with improved outcomes seen in those with a duration of illness that is <3 years. This has led to the development of service pathways that promote rapid access to treatment for those aged 16–25 years with early onset eating disorders (e.g. FREED; First Episode Rapid Early Intervention for Eating Disorders).

Long-term follow-up studies indicate that at 6-year follow-up, 55% had no eating disorder, 27% continued with AN, 10% had BN, 2% were classified with OSFED, and 6% were deceased. AN is associated with increased mortality, the standardized mortality ratio over the first 10 years after presentation showing a 10-fold increased risk. Most deaths are either a direct result of medical complications or due to suicide. The mortality rate appears to be higher for those presenting with lower weight during their illness and for those who require multiple episodes of inpatient treatment.

## 22.4 Bulimia nervosa

### 22.4.1 Clinical features

As in AN, those with BN tend to judge their self-worth almost exclusively on the basis of their weight, shape, and their ability to control their food intake. They also use extreme forms of weight control, such as fasting and self-induced vomiting. However, people with BN tend to lie within the normal weight range, and they regularly engage in episodes of 'binge eating'. Binges will vary in size; they typically involve the consumption of 2000 kcal or more and tend to consist of foods the person is attempting to avoid. Patients typically become stuck in a vicious cycle of dietary restriction, bingeing, and purging, and are plagued by a constant fear of weight gain. This cycle invariably has a detrimental impact on other areas of functioning, such as work and social relationships, and it can have significant financial implications, leading some to steal money or food from others. People with BN tend to 'value' their symptoms less compared with those with AN and often binge and purge in secret.

Evidence suggests that a significant proportion of people with BN have difficulty regulating their emotions, and for many bingeing may serve as a form of emotional regulation: a means of reducing the intensity of emotions when they become intolerable. Many also have impulse control problems and a history of interpersonal difficulties. A subgroup will also present with comorbid depression and/or emotionally unstable personality disorder, and many of these will engage in a range of self-destructive behaviours such as cutting, overdosing, and substance misuse, with the dominant behaviour changing over time.

**Physical health** Physical complications most commonly associated with BN include irregular or absent menstruation, weakness and lethargy, vague abdominal pains, and toothache. On examination, appearance is usually unremarkable. Parotid gland enlargement may be present and there may be significant erosion of the dental enamel, particularly on the lingual surface of the upper front teeth. The most important abnormality on investigation is the electrolyte disturbance that is encountered in those who vomit frequently and in those who take large quantities of laxatives or diuretics. Clinically serious electrolyte disturbance may require treatment with potassium supplements until the eating disorder has been

**Food Diary**

| Time | Food and drink consumed | Place | * | V/L | Comments |
|---|---|---|---|---|---|
| 9 am | Black coffee | Kitchen | | | Feel OK – today is going to be a good day |
| 1 pm | Ham and Salad sandwich | Office desk | | | So far so good |
| 6 pm | 2 cheese rolls | Kitchen | | | Ate whilst preparing dinner—wish I hadn't but know I'll bring it up later |
| 6:30 pm | Lasagne and garlic bread | Kitchen | * | V | Shouldn't have had seconds |
| 7 pm | Bowl of cereal<br>2 bowls of ice-cream<br>Chocolate bar<br>2 slices of toast<br>Bottle of coke (small) | Lounge<br>"<br>"<br>"<br>" | * | V | Feel horrible – can't believe I lost control after being so good at work |

* = binge, V = vomiting, L = laxatives

**Fig. 22.2** Example food diary for a patient with bulimia nervosa.

resolved. An example of a food diary for a patient with bulimia nervosa is given in Fig. 22.2.

## 26.4.2 Aetiology

Although many patients with BN will report a history of AN, a number of factors have been identified that preferentially increase the risk of BN. These include childhood and parental obesity, early menarche, and parental alcoholism. A history of trauma, including childhood sexual abuse (CSA) has also been associated with BN. However, while the rate of CSA in patients with BN is higher than that amongst matched subjects in the general population, it is not higher than that found amongst young women with other psychiatric disorders. This suggests it may serve as a general risk factor for psychiatric disorder, rather than for BN per se.

## 22.4.3 Assessment

Many patients with BN feel too guilty and ashamed of their illness to ask for help, and often live with their disorder for years before seeking treatment. As with AN, they may present complaining of physical symptoms such as gastrointestinal or gynaecological symptoms. The lack of clear markers can make diagnosis difficult and highlights the importance of conducting a thorough assessment.

Those with BN typically present with a loss of control over their eating that is characterized by frequent episodes of binge eating and compensatory behaviours. Assessment and diagnosis is often relatively straightforward, although chaotic eating patterns can sometimes make it difficult to establish the presence of discrete episodes of binge eating. As with AN, no physical tests are needed to establish the diagnosis. However, the electrolytes should be checked of all those who vomit frequently or misuse large quantities of laxatives or diuretics.

## 22.4.4 Management

Most patients with BN can be treated on an outpatient basis. Eating-disorder-focused cognitive behavioural therapy (CBT-ED) represents the most extensively researched and validated psychological therapy for BN. CBT-ED is a time-limited intervention that consists of up to 20 outpatient sessions. The first phase focuses on engagement and education, supporting patients to establish a regular pattern of eating whilst also addressing eating and weight change-related anxiety. The second phase involves working with concerns about weight and shape, as well as the tendency to binge in response to difficult thoughts and feelings. Towards the end of treatment, the focus moves to maintaining gains and minimizing the risk of relapse. A subset of patients will respond well to less intensive behavioural interventions, such as bulimia-nervosa-focused guided self-help programmes, although this is unlikely to be sufficient for the majority.

### 22.4.5 Course and outcome

Relatively little is known about the long-term course and outcome of BN. Outcome studies conducted to date suggest that at best only 50% of those who receive treatment are likely to achieve full recovery, 20% can expect to continue with a full diagnosis, while the remaining 30% will experience episodes of relapse or will continue with a subclinical form. In relation to identifying who is likely to respond well to psychological therapy, early symptom change has been identified as a consistent predictor of a positive treatment outcome.

### 22.4.6 Binge-eating disorder

BED is commonly associated with obesity and it affects up to 30% of obese patients seeking weight-loss treatment. Treatment for BED is primarily psychological and will include either a binge-eating-disorder-focused guided self-help programme or CBT-ED delivered in either a group or individual format. It is important that patients are made aware that treatment will have a limited effect on body weight and that weight loss is not a target for therapy in itself. Studies examining the natural course of BED suggest recovery rates of up to 85% at 5 years.

## 22.5 Eating disorders and comorbidity

A number of other high-risk groups have also been identified. These include those presenting with an eating disorder and type 1 diabetes (T1DM), and athletes with eating disorders. In relation to the former, it has been found that full and subclinical eating disorders are more common among those with T1DM compared with age-matched peers. Patients with T1DM may adopt the underuse or omission of insulin as a means of weight control, and it has been shown that even relatively short periods of impaired metabolic control can lead to increased risk of the physical complications associated with diabetes, such as retinopathy, nephropathy, or neuropathy, as well as increased mortality. It is widely accepted that the treatment and management of patients with type 1 diabetes and an eating disorder requires a multidisciplinary approach that includes diabetes teams and mental health teams, each having an awareness of the complexities of both disorders. It is important that physical and psychological care is integrated from the outset, and psychological interventions should draw upon evidence-based treatments for the eating disorder, as well as addressing difficulties associated with living with diabetes, including diabetes distress and fear of hypoglycaemia. Psychological sessions should complement the other interventions offered to address nutrition and diabetes management, with close multidisciplinary working remaining central to care delivery.

Athletes have also been identified as a group at increased risk of developing disturbed eating and associated problems. Pressure to conform to a sport-specific 'ideal' in disciplines such as distance running, gymnastics, and figure skating may lead a proportion to diet in order to improve performance. In order to capture the risks associated with eating disorders in athletes, terms such as 'anorexia athletica' and the 'female athlete triad' have been developed, the latter referring to the following three conditions: disordered eating, amenorrhoea, and osteoporosis. However, male athletes can also be at risk, especially those competing in sports that tend to place an emphasis on appearance, size, or weight, such as cycling or boxing. Identifying disordered eating among athletes must include the identification and assessment of a wide range of weight-control behaviours.

## 22.6 Summary

Whilst a considerable number of those with an eating disorder will present with AN, BN, or BED, atypical presentations are not uncommon in routine clinical settings. Eating disorders are associated with significant psychological and physical complications. While many of these features will remit with weight restoration and/or cessation of weight-control behaviours, those presenting with a chronic course may experience long-term complications such as osteoporosis and infertility. The co-occurrence of subclinical disordered eating and conditions such as T1DM is also associated with significant physical morbidity and mortality. Patients with eating disorders present in a range of clinical settings and thus careful assessment is crucial for accurate identification and diagnosis. Comprehensive treatment requires the management of both physical and psychological aspects of the illness. While a proportion of patients will recover, a minority will present with an unremitting pattern of symptoms that may require longer-term management.

### Further reading

1. **Birmingham, C.L. and Treasure, J.** (2019) *Medical management of eating disorders*, 3rd edn. Cambridge: Cambridge University Press.
2. **Brownell, K.D. and Walsh, B.T.** (eds) (2017) *Eating disorders and obesity: a comprehensive handbook.* New York, NY: Guilford Publications.
3. **Fairburn, C.G.** (2008) *Cognitive behavioural therapy and eating disorders.* New York, NY: Guilford Publications.
4. **National Institute for Health and Care Excellence** (2017) *Eating disorders: recognition and treatment.* London, UK: National Institute for Health and Care Excellence.
5. **Ozier, A.D. and Henry, B.W.** (2011) Position of the American Dietetic Association: nutrition intervention in the treatment of eating disorders. Example food diary for a patient with bulimia nervosa. *J Am Diet Assoc*, **8**, 1236–41.
6. https://www.rcpsych.ac.uk/docs/default-source/improving-care/better-mh-policy/college-reports/college-report-cr233-medical-emergencies-in-eating-disorders-(meed)-guidance.pdf?sfvrsn=2d327483_55
7. **Waller, G., Turner, H., Tatham, M., Mountford, V.A., and Wade, T.D.** (2019) *Brief cognitive behavioural therapy for non-underweight patients: CBT-T for eating disorders.* London: Routledge.
8. **Walsh, B.T., Attia, E., Glasofer, D.R., and Sysko, R.** (eds) (2015) *Handbook of assessment and treatment of eating disorders.* London: American Psychiatric Publications.

# Part 5

# Foods

# 23 Food Groups



## 23.1 Grains

Andrew Reynolds

Grains have been a staple food since *Homo sapiens* learned to cook. Considered the 'seeds of civilization,' it was only after nomadic hunter-gatherers learned to farm crops such as grains that permanent human settlements could be maintained. Grains are cheap, filling, and portable staples that form the basis of food patterns around the world. While their appearance in the food supply and diet may have changed in recent decades, grain from cereals and pseudocereals remains the most consumed food group in the modern world.

Grains are the dried fruiting body of grasses (cereals) or certain seeds (pseudocereals) that have the appearance and culinary use of cereal grains. Their purposeful cultivation can be traced to 10,000 years ago in the Euphrates valley (wheat), the Fertile Crescent (barley), Mexico (maize), and China (rice). Wheat and barley were staple foods of ancient Egypt, Greece, and Rome; rice in India and Southern China; maize in the Americas; oats and rye in Northern Europe; and millet was important in the development of Africa and parts of Asia. Analyses of the stomach contents of Ötzi, a mummified European believed to have lived 5200–5400 years ago, indicated he consumed wheat, most likely in the form of a crude bread.

Wheat, rice, and maize are the predominant grains worldwide both in terms of the land devoted to them and their consumption. Wheat covers more of the Earth's surface than any other crop, largely having replaced rye, barley, and oats in Northern Europe, and increasingly replacing sorghum and millet in Africa. About 50% of the food protein available on the globe is derived from grain, with consumption greatest in the developing countries (providing two-thirds of energy and protein). In general, grain intake decreases with increasing income, being replaced in the diet by animal products.

The structure of most grains includes an endosperm, germ, bran, and hull. The endosperm comprises the bulk of the grain, and is primarily starch. The germ is a small component of the total grain, and is rich in polyunsaturated fats. Bran forms the outer layer of the edible grain, and is fibre- and micronutrient-dense. The hull, an inedible husk, is removed before

**Table 23.1 Nutrient content of grains**

| | Nutrients per 100 g (dry grains) | | | | | |
|---|---|---|---|---|---|---|
| | Whole grain wheat flour | Refined wheat flour | Refined rice | Maize | Oats | Quinoa |
| Energy (kJ) | 1420 | 1520 | 1510 | 1530 | 1590 | 1540 |
| Water (g) | 10.7 | 11.9 | 12.9 | 10.4 | 10.8 | 13.3 |
| Protein (g) | 13.2 | 10.3 | 6.6 | 9.4 | 13.2 | 14.1 |
| Fat (g) | 2.5 | 1.0 | 0.6 | 4.7 | 6.5 | 6.1 |
| Total carbohydrate (g) | 72.0 | 76.3 | 79.3 | 74.3 | 67.7 | 64.2 |
| Fibre (g) | 10.7 | 2.7 | 2.8 | 7.3 | 10.1 | 7.0 |
| Thiamin (mg) | 0.5 | 0.1 | 0.1 | 0.4 | 0.5 | 0.4 |
| Niacin (mg) | 5.0 | 1.3 | 1.6 | 1.6 | 1.1 | 1.5 |
| Calcium (mg) | 34 | 15 | 9 | 12 | 52 | 47 |
| Iron (mg) | 3.6 | 1.2 | 0.8 | 2.7 | 4.3 | 4.6 |

Values adapted from the USDA FoodData Central Database. https://fdc.nal.usda.gov/ (accessed 21 March 2022)

consumption. Grain foods can be categorized as either whole grain (endosperm, bran, and germ) or refined (endosperm). Broadly speaking, the nutritional value of different grains is similar (see Table 23.1), with variation due to food processing, genetic factors, and environmental conditions. Grains are high in carbohydrate, particularly starch, while being low in sugar. Gluten is the major protein in wheat and rye, and oryzenin is the major protein in rice. In most cereal grains the limiting amino acid is lysine, with maize additionally low in tryptophan. Pseudocereals are not limited in lysine or tryptophan, providing consumers with a complete source of all essential amino acids. All cereals and pseudocereals are low in fat, with most of the fat polyunsaturated.

## 23.1.1 Whole grains

Whole grain foods use every edible part of the grain. Definitions of whole grains permit food processing techniques such as milling, with some definitions also permitting the reconstitution of separate grain fractions in the same relative proportions of an intact grain. Definitions also allow for small losses of components during grain processing, such as up to 2% of the grain weight.

Whole grains are a good source of dietary fibres, thiamin, and vitamin E when the germ is retained, and contain significant amounts of minerals, especially potassium, phosphorus, magnesium, iron, and zinc plus selenium, copper, and manganese. They provide more fibre and micronutrients than refined grains, as these nutrients are concentrated in the bran layer. The bran layer may also contain phytates, which bind some minerals to inhibit their uptake; however, mineral uptake remains higher for whole grains than refined grains. Whole grains are also rich in 'bioactive' components, including antioxidants (especially phenolics), phyto-oestrogens (lignans), and phytosterols.

Whole grains are an important part of a healthy diet. Epidemiological studies indicate health benefits with whole grain intakes, such as the reduced risk of heart disease, type 2 diabetes, certain cancers (notably colorectal cancer), and overall premature mortality. Health trials of increasing whole grain intake have observed body weight reductions, despite no advice to reduce energy intake or lose weight, in part due to their greater satiety. Due to their fibre content, whole grain foods are digested more slowly than refined grains, which is beneficial in blood glucose management. Current national or international dietary guidelines include the phrasing 'make most of your grains whole grains', or 'choose mainly whole grains' to reflect these associations between higher whole grain intakes and improved health.

### 23.1.2 Refined grains

Refined grains are the endosperm, with the bran and germ removed by food processing techniques such as polishing. Refined grains provide considerably less dietary fibre than whole grains, as well as a reduced micronutrient content. The majority of grains consumed are refined grains, such as white rice and foods made with refined wheat flour. Refined grains have been viewed culturally as symbols of purity or wealth, as they required additional steps to produce and were not visibly contaminated. Refined grains traditionally demanded higher prices than less refined or whole grains due to the increased processing and reduced yield. In the current food supply, however, refined grains are cheaper due to greater current consumption, and the health benefits associated with whole grains driving prices higher. Dietary guidelines tend to minimize the intake of refined grains in favour of higher whole grain intakes, and often combine them with sugars using such phrasing as 'limit sugars and refined starches'.

### 23.1.3 Common types of grain

Wheat was the traditional staple of many cultures, and is now ubiquitous in the global food supply. Wheat undergoes a wide range of food processing techniques, most commonly milling, and is the basis of breakfast cereals, breads, noodles, cakes and pastries, etc. Wheat is consumed as both a whole grain and a refined grain. Different wheat cultivars and environmental conditions produce variation in the protein content and 'hardness' of wheat. Hard, high-protein wheat is most suitable for bread; soft, low-protein wheats are most suitable for biscuits; extra-hard durum wheat is used to make pasta.

Rice is the staple grain of over half the world's population, especially in Asia. After harvesting, rice is cleaned and de-hulled, with the bran layer retained in the whole grain brown rice. Brown rice can be further milled and polished to produce refined grain white rice, which is the more commonly consumed form. Brown and white rice can be cooked and consumed whole, or further processed into a versatile range of snack foods such as rice cakes, rice bran, rice noodles, and rice crackers. Compared with other grains, rice is often lower in dietary fibre and protein. White rice intake is of concern in many Asian nations, given its popularity as a dietary staple and the loss of valuable nutrients like protein, thiamin, and iron when the bran is removed.

Maize is somewhat unique in that fresh corn is considered by most to be a vegetable, while dried corn, or maize, is deemed a grain. Maize is a common staple of Mexico, being the primary ingredient of masa harina, which is used to make tortillas. Maize is also traditionally consumed in most African countries, such as in a stiff porridge (ugali) or thin gruel (uji), which can also be used as the basis for alcoholic or non-alcoholic beverages. Dry milling maize produces grits, corn meal, flour, and hominy meal. Maize is also flaked to produce breakfast cereals (corn flakes). Maize contains little niacin or its amino acid precursor, tryptophan, which has led to widespread pellagra in populations reliant on maize, where additional food processing has not increased niacin bioavailability or content.

Oats are able to be grown in cooler climates such as Northern Europe and Scotland, leading to traditionally higher intakes in these areas. Oats are generally eaten as a breakfast cereal (porridge or muesli), and are increasingly used as the basis of a non-dairy milk alternative. Oats are processed by steaming or kiln-drying before de-hulling. The resultant 'groats' can be cut to produce a coarse meal, which is steamed, then rolled to make oat flakes, or granulated to produce a fine oatmeal.

Barley was a staple grain of Egypt, Greece, and the Roman empire, and remains a staple grain of Tibet. Barley is consumed in the form of pearled grains for soups, flour for breads or noodles, and ground grain for porridge. Malted barley is used in brewing and baking, and for making vinegar and flavouring breakfast cereals.

Rye was traditionally grown in cooler regions of Europe. Cracked rye is used for porridge and other breakfast cereals while rye flour can be baked into bread and crispbreads. Rye is high in dietary fibre (15 g per 100 g of dry grain) when compared with other grains.

Millet is a name given to a group of cereal grains that includes sorghum. Millet is consumed in Africa, parts of India, Pakistan, and China. The grains are pounded into flour and mixed with water to make porridge. In Ethiopia, finely ground millet grains (teff) are left to ferment slightly then cooked to produce injera bread.

Amaranth is a pseudocereal that was a staple food of the Aztecs 500–700 years ago. Amaranth can be milled into flour, puffed, made into breakfast cereal flakes, or cooked and eaten like rice. Like all pseudocereals, amaranth does not contain gluten, so may be a useful alternative grain for those with coeliac disease.

Buckwheat is not to be confused with cereal wheat, and is a pseudocereal consumed in Asia and Eastern Europe for around 8000 years. Buckwheat now is used as flour, or made into foods such as soba noodles in Japanese cuisine. Buckwheat groats, i.e. de-hulled kernels, can be made into a porridge.

Quinoa is a pseudocereal similar in size to a sesame seed. Quinoa can be white, red, purple, or black in colour. Quinoa was a staple food in Peru and Bolivia around 3000–4000 years ago. Quinoa seeds are naturally rich in saponin, a toxic glycoside, so are prewashed before sale. Any bitterness in quinoa is from residual saponins. Quinoa can be milled to flour, eaten as a whole grain after cooking, or made into other cereal-based products, such as breakfast cereals and bread.

### 23.1.4 Grains and food processing

All grains are processed to some extent, such as de-hulled or cooked before consumption. Current trends in the global food supply indicate grain foods are becoming more processed and refined, which has repercussions on their digestibility, nutrient profile, and the health consequences of their intake.

Milling is a series of grinding and sifting steps to produce flour. Both whole and refined grains can be milled. Flour is then used in cooking, or often undergoes further food processing techniques before consumption. Common milling methods are stone ground or, more often in commercial production, roller ground. Milling degrades the polysaccharide and non-polysaccharide structures within grains, making them available to enzymatic digestion, increasing the speed of their uptake as glucose into the bloodstream. Milling of whole grains is associated with poorer blood glucose management in type 2 diabetes than when whole grains are left more intact.

Polishing is undertaken on whole grains such as brown rice, to remove the bran layer and produce refined grains, such as white rice. Because fibre and micronutrients are more concentrated in the bran layer, polishing reduces the nutrient contribution of grain foods to the diet.

Extrusion is where whole or refined grains are milled to some degree and then forced under pressure through a narrow opening. This pressure binds flour particles together to create a new food matrix. A common example of extrusion is dried pasta or noodles.

Nixtamilization is the soaking or cooking of grain, typically corn, in an alkaline solution. This process is important to remove aflatoxins that may be present in contaminated grain, and to boost the bioavailability of niacin. Nixtamilization of corn was important in Mexico to prevent pellagra (niacin deficiency state). Conversely, diets reliant on untreated corn in Northern Italy in the late 1800s led to over 100,000 cases.

Puffing can be done with whole or refined grains such as rice and corn, such as by explosion puffing, to produce a new grain food with a dry, crunchy texture. A common example of explosion puffing is puffed rice or rice cakes.

Flaking or rolling can be done with both whole and refined grains, often with steam, to disrupt the native grain structure. Common examples of flaking and rolling are rolled oats or corn flakes.

Malting is the partial germination of grains. Soaking whole grains over time leads to enzymatic conversion of starch into sugars, increasing the digestion or fermentability of grains. A grain commonly malted is barley, to then be used in baking or alcoholic and non-alcoholic beverage production.

Fermentation is an important stage in beer brewing and leavened bread production, where whole or refined grains are degraded by yeast or microbial cultures. During fermentation the enzyme phytase is produced, which breaks down phytate, further increasing grain digestibility and mineral uptake.

Fortification is where whole or refined grain foods are fortified with micronutrients to meet dietary targets for the population or subgroups of the population. Given the regularity of their intake, low cost, and availability, grains are considered a suitable vehicle for such fortification. Common micronutrients added to grain foods are folic acid and iodine.

Parboiling is a partial cooking of a grain, most commonly rice. Parboiling whole grain rice before polishing to produce white rice results in an inward migration of water-soluble vitamins to the endosperm.

### Further reading


1. **Åberg, S., Mann, J., Neumann, S., et al.** (2020) Wholegrain processing and glycaemic control in type 2 diabetes: a randomised crossover trial. *Diabetes Care*, **43**, doi.org/10.2337/dc20-02631.1
2. **Baker, P., Machado, P., Santos, T., et al.** (2020) Ultra-processed foods and the nutrition transition: global, regional and national trends, food systems transformations and political economy drivers. *Obes Rev*, **21**(12), e13126. doi: 10.1111/obr.13126
3. **McKeown, N.M., Troy, L.M., Jacques, P.F., et al.** (2010) Whole- and refined-grain intakes are differentially associated with abdominal visceral and subcutaneous adiposity in healthy adults: the Framingham Heart Study. *Am J Clin Nutr*, **92**(5), 1165–71.
4. **Piperno, D.R., Weiss, E., Holst, I., et al.** (2004) Processing of wild cereal grains in the Upper Palaeolithic revealed by starch grain analysis. *Nature*, **430**(7000), 670–3.
5. **Reynolds, A.N., Mann, J., Cummings, J., et al.** (2019) Carbohydrate quality and human health: a series of systematic reviews and meta analyses. *The Lancet*, **393**(10170), 434–45. doi.org/10.1016/S0140-6736(18)31809-9.
6. **Sanders, L.M., Zhu, Y., Wilcox, M.L., et al.** (2021) Effects of whole grain intake, compared with refined grain, on appetite and energy intake: a systematic review and meta-analysis. *Adv Nutr*, **12**(4), 1177–95.
7. **Slavin, J.** (2003) Why whole grains are protective: biological mechanisms. *Proc Nutr Soc (UK)*, **62**, 129–34.
8. **Slavin, J.L., Jacobs, D., Marquart, L.** (2000) Grain processing and nutrition. *Critical Reviews in Food Science and Nutrition*, **40**(4), 309–26.


## 23.2 Legumes

Andrew Reynolds and A. Stewart Truswell

Legumes are the edible seed from the Leguminosae family (Fabaceae) and include dried or split peas, beans, soy, peanuts, and lentils. Cultivated since the earliest civilizations, leguminous plants grow in a wide range of climates. Legumes were once considered an inferior food eaten by peasants, described as being 'poor man's meat'. Nowadays, legumes are valued as a cheap, nutritious, plant-based food that can be incorporated into many meals. There are environmental benefits to legume consumption as well, as they require only 1% of the environmental inputs of red meat, and are often used as a crop in sustainable farming practices to introduce nitrogen into the soil.

Legumes are a good source of the essential amino acid lysine, which is low in cereals. Conversely, legumes are generally limited in methionine, which is abundant in cereals. Consuming these two food groups together has played an important synergistic role in meeting nutritional requirements in many cuisines. Beans and tortillas are eaten in Mexico, rice and dhal in South Asia, soy products with rice in East and South East Asia, lentils and rice in the Middle East, split peas and teff in Ethiopia. The ratio of quantities of cereal and legumes consumed is similar among cuisines, with cereals providing the main source of energy and legumes used as accompaniments.

Legumes have been called a nutrition powerhouse. They are high in dietary fibre, low in fat, and supply adequate protein while being low in energy density and a good source of minerals and some vitamins. Cooked legumes contain about 6–9% protein, which is about twice as much protein as in most cereal foods. The exceptions are soybeans and peanuts, which contain 18 and 24%, respectively, when cooked. Legumes provide around 10–13 g carbohydrate that is the slowly digested type, and 6–9 g of fibre per 100 g of edible portion. Legumes also contain oligosaccharides, which escape digestion in the small intestine, to be fermented by bacteria in the large bowel. This is responsible for the abdominal discomfort and flatulence that can be experienced and is perhaps a factor limiting consumption. However, these compounds may have beneficial effects for gastrointestinal health. Most legumes are low in fat (<3%), but soybeans and peanuts are much higher. This fat is mostly monounsaturated or polyunsaturated. Soybeans contain the ω3 fatty acid linolenic acid.

Legumes supply vitamins and minerals, including thiamin, niacin, iron, zinc, calcium, and magnesium (see Table 23.2). Like almost all plant foods, legumes do not contain $B_{12}$. Legumes have several qualities associated with the prevention of chronic diseases, such as being high in fibre, low in energy, and highly satiating. Higher intakes of legumes have been associated with a decreased risk of coronary heart disease, stroke, and type 2 diabetes over time when compared with lower intakes. Increasing intakes of

**Table 23.2 Nutrient content of legumes**

| | Nutrients per 100 g (edible portion) | | | | | |
|---|---|---|---|---|---|---|
| | **Soybean** | **Chickpea** | **Kidney bean** | **Split pea** | **Lentil** | **Peanut (dry roasted)** |
| Energy (kJ) | 721 | 610 | 564 | 493 | 485 | 2460 |
| Water (g) | 63 | 64 | 63 | 70 | 70 | 2 |
| Protein (g) | 18.2 | 8.2 | 8.7 | 8.3 | 9.0 | 24.4 |
| Fat (g) | 9.0 | 3.2 | 0.6 | 0.4 | 0.4 | 49.7 |
| Total carbohydrate (g) | 8.4 | 22.5 | 24.7 | 21.1 | 20.1 | 21.3 |
| Fibre (g) | 6.0 | 7.3 | 8.0 | 8.3 | 7.9 | 8.4 |
| Thiamin (mg) | 0.2 | 0.1 | 0.2 | 0.2 | 0.2 | 0.2 |
| Niacin (mg) | 0.4 | 0.2 | 0.8 | 0.9 | 1.1 | 14.4 |
| Calcium (mg) | 102 | 58 | 48 | 14 | 19 | 58 |
| Iron (mg) | 5.1 | 2.1 | 2.1 | 1.3 | 3.3 | 1.6 |

Values adapted from the USDA FoodData Central (online searchable database): https://fdc.nal.usda.gov/index.html (accessed 1 November 2021).

legumes is associated with reduced total and LDL cholesterol concentrations.

Although largely considered beneficial and widely promoted in dietary guidelines across the globe, legumes may contain antinutrients. The majority of these substances are destroyed or inactivated by cooking before consumption. Trypsin inhibitors may reduce the effectiveness of the digestive process; haemagglutinins appear to reduce the efficiency of absorption of digestive products; phytate binds metals, like zinc and iron, decreasing their absorption; goitrogens interrupt the absorption of iodine. Substances in beans can cause specific health conditions. *Lathyrus sativus* (khesari dal) can precipitate lathyrism (a neurological disorder) when consumed in large amounts. Broad beans (*Vicia faba*) may result in favism, a haemolytic anaemia, in the genetically susceptible. Peanuts are common food allergens and can cause anaphylaxis (see Chapter 34). Care must be taken when storing legumes. Peanuts are susceptible to *Aspergillus* mould, which produces aflatoxin, a potent liver carcinogen. Bacteria capable of causing food poisoning may be present on bean sprouts, even in developed countries. For this reason, the US Food and Drug Administration advise that children, the elderly, and people with a compromised immune system should not eat raw or lightly cooked bean sprouts.

## 23.2.1 Legume processing and consumption

Uncooked dried legumes are virtually indigestible, tasteless, and may contain toxins or antinutrients. Cooking in some form is essential to make legumes safe, edible, and to improve nutrient availability. For dried legumes, pre-soaking and rinsing before cooking is recommended to remove antinutrients and minimize gastrointestinal discomfort and flatulence that can occur. Germination and fermentation may improve the nutritional quality of legumes further, resulting in increases in vitamin C, niacin, riboflavin, thiamin, and Vitamin E content.

In India and some parts of Africa, raw legumes are milled to remove fibrous seed coats. Cotyledons are split along natural cleavage lines to form split peas. The peas or beans are then boiled, roasted, fermented, germinated, or ground into flour or paste. In Western countries, South America, and much of Africa, legumes are soaked and cooked for long periods of time to inactivate toxic substances and improve digestibility. Legumes are then either eaten whole, puréed and fried, or made into cakes. In Asia, immature beans are eaten whole in salads, vegetable dishes, or soups; mature soybeans are not usually eaten as whole beans, but rather processed into curd, cheese, sauces, or pastes, fermented, or sprouted as a vegetable.

Today, peanuts and soybeans account for most of the legume products. Soybeans are made into a range of products, including soy proteins and concentrates or isolates, extrusion-textured products, soy flour, soy milk, and tofu. Many soy products are used in manufactured food products, including bread and baked goods, processed meat products, soy ice cream, sauces, and low-fat spreads. Peanuts are made into nut butter, and used in confectionery or baked goods.

### Further reading

1. **Afshin, A., Micha, R., Khatibzadeh, S., et al.** (2014) Consumption of nuts and legumes and risk of incident ischemic heart disease, stroke, and diabetes: a systematic review and meta-analysis. *AJCN*, **100**(1), 278–88.
2. **Darmadi-Blackberry, I., Wahlqvist, M.L., Kouris-Blazos, A., et al.** (2004) Legumes: the most important dietary predictor of survival in older people of different ethnicities. *Asia Pacific J Clin Nutr*, **13**, 217–20.
3. **Marventano, S., Pulido, M.I., Sánchez-González, C., et al.** (2017) Legume consumption and CVD risk: a systematic review and meta-analysis. *PHN*, **20**(2), 245–54.

4. **McGee, H.** (2004) Legumes: beans and peas. In: McGee, H. (ed) *On food and cooking. The science and lore of the kitchen*, 2nd edn, pp. 483–501. New York, NY: Scribner.
5. **Messina, M.** (2019) Legumes and soybeans: overview of their nutritional profiles and health effects. *AJCN*, **70**(3), 439s–450s.
6. **Ritchie, H. and Roser, M.** (2020) *Environmental impacts of food production. Our World in Data.* Available at: https://ourworldindata.org/environmental-impacts-of-food (accessed 1 November 2021).

## 23.3 Nuts and seeds

Margaret Allman-Farinelli

Nuts and seeds have been valued for their oils as much as for food, and have been an important source of nutrients and energy since the earliest civilizations. Traditional cuisines have utilized locally grown nuts and seeds for both savoury and sweet dishes. In recent years, there has been much interest in the nutritional properties of nuts that promote health. Today, nuts and seeds are processed for their oil, ground into pastes, used as ingredients in baked goods, eaten raw, or roasted as snack foods.

Common types of nuts include almonds, walnuts, pecans, cashews, Brazil nuts, macadamias, hazelnuts, pine nuts, and pistachios. Sunflower, sesame, and pumpkin seeds are the most common seeds eaten as foods. Caraway and poppy seeds are used as seasonings.

Nuts and seeds have similar nutritional qualities; their low water content and high content of energy, protein, vitamins, and minerals makes them nutritious foods (Table 23.3). The energy content of nuts is mostly due to their high fat content. The fat in nuts varies in both quantity and type. Chestnuts are low in fat, but other nuts contain from 45% to 75% fat. The majority of nuts contain unsaturated fatty acids, either monounsaturated (e.g. macadamia) or polyunsaturated. Walnuts are rich in ω6 polyunsaturated fatty acids, but also a good source of ω3 linolenic acid. The protein content of nuts ranges from 2% to 25%; lysine is the limiting amino acid. They are good sources of dietary fibre, B vitamins (thiamin, riboflavin, and niacin), vitamin E, and iron, zinc, magnesium, potassium, and calcium. Nuts and seeds contain no vitamin C.

**Table 23.3 Nutrient content of raw nuts and seeds**

| Nutrient | Nutrients per 100 g |
|---|---|
| Protein (g) | 2.0 (chestnuts) to 24.4 (pumpkin seeds) |
| Fat (g) | 2.7 (chestnuts) to 77.6 (macadamias) |
| Carbohydrate (g) | 3.1 (Brazil nuts) to 36.6 (chestnuts) |
| Dietary fibre[a] (g) | 1.9 (pine nuts) to 7.9 (sesame seeds) |
| Thiamin (mg) | 0.18 (roasted peanuts) to 0.93 (sesame seeds) |
| Riboflavin (mg) | 0.06 (macadamias) to 0.75 (almonds) |
| Niacin equivalents (mg) | 0.9 (chestnuts) to 9.1 (sunflower seeds) |
| Calcium (mg) | 11.0 (pine nuts) to 670.0 (sesame seeds) |
| Iron (mg) | 1.6 (macadamias) to 10.4 (sesame seeds) |
| Zinc (mg) | 0.5 (chestnuts) to 6.6 (soybeans) |

[a]Englyst method (see section 36.2.4).

*Adapted from:* Holland, B., Unwin, I.D., and Buss, D.H. (1992) *Fruit and nuts*. First supplement to: *McCance and Widdowson's the composition of foods*, 5th edn. London: Royal Society of Chemistry, MAFF.

Regular consumption of nuts is associated with lower all-cause mortality, cardiovascular mortality, and even cancer mortality. However, there is almost no evidence of an association with stroke. A recent meta-analysis indicated that there is no association of tree nut or peanut consumption with type 2 diabetes. The constituents likely to confer cardiovascular benefits are the amino acid arginine, vitamin E, and/or unsaturated fat. The components that are hypothesized to prevent cancer are the phytochemicals, which are antioxidants, and other bioactive constituents that are anti-inflammatory, such as alpha-linolenic acid and resveratrol. It must be acknowledged that there is some likely confounding in these associations. Studies have also indicated that people regularly consuming nuts have either the same or a lower body weight than those not eating nuts. Nuts make up only a small percentage of the total diet but are a useful addition to a healthy eating pattern for most people. Fewer than 2% of people are allergic to tree nuts and around 2% are estimated to have an allergy to peanuts.

## Further reading

1. **Afshin, A., Micha, R., Khatibzadeh, S., and Mozaffarian, D.** (2014) Consumption of nuts and legumes and risk of incident ischemic heart disease, stroke, and diabetes: a systematic review and meta-analysis. *Am J Clin Nutr*, **100**, 278–88.
2. **Becerra-Tomás, N., Paz-Graniel, I., Hernández-Alonso, P., et al.** (2021) Nut consumption and type 2 diabetes risk: a systematic review and meta-analysis of observational studies. *Am J Clin Nutr*, **113**(4), 960–71.
3. **Falasca, M., Casari, I., and Maffucci, T.** (2014) Cancer chemoprevention with nuts. *J Natl Cancer Inst*, **106**, ii.
4. **Grosso, G., Yang, J., Marventano, S., et al.** (2015) Nut consumption on all-cause, cardiovascular, and cancer mortality risk: a systematic review and meta-analysis of epidemiologic studies. *Am J Clin Nutr*, **101**, 83–93.
5. **McWilliam, V., Koplin, J., Lodge, C., et al.** (2015) The prevalence of tree nut allergy: a systematic review. *Curr Allergy Asthma Rep*, **54**. doi: 10.1007/s11882-015-0555-8.

# 23.4 Fruit

A. Stewart Truswell

In its strict botanical sense, a 'fruit' is the fleshy or dry ripened ovary of a plant enclosing the seed and so includes corn grains, bean pods, tomatoes, olives, cucumbers, and almonds and pecans (in their shells). We usually restrict the term 'fruit' to mean the ripened ovaries that are sweet and either succulent or pulpy and usually eaten as an appetizer or a dessert. Fresh fruits were formerly only available seasonally, but with technology and imports, many are available most of the year. Most of us eat more species in the fruit food group than in any other food group, for example, 30 or more species—some daily (e.g. apples, oranges, bananas) and others rarely (e.g. elderberries, mulberries, loganberries, passion fruit).

Our ancestors originally ate fruits for their sweetness. They learned that those fruits that were not bitter were unlikely to be toxic. The sugar that makes them sweet provides energy, and may be fructose, glucose, and/or sucrose and others (such as sorbitol). Fruits generally provide useful amounts of potassium, vitamin C, carotenoids, and folate, and very little sodium (see Table 23.4). Avocados are unusual because they have a high content of fat (mostly monounsaturated) and contain a 7-carbon sugar.

Fruits are preserved in jams by boiling with sugar; the pectin (soluble dietary fibre) they contain makes a gel. Some fruits are dried, especially grapes, plums, dates, bananas, and apricots. Drying fruit however reduces Vitamin C content.

Quite large amounts of some fruits, especially citrus and apples, are consumed in the form of fruit juices, which contain most of the nutrients, but less of the fibre, and give less satiety.

Grapes are high in sugar (glucose), which makes crushed grapes a good substrate for fermentation into wine; apples are fermented to make cider.

Most fruits taste somewhat sour, because they contain organic acids such as citric, malic, tartaric, ascorbic, and in some cases, benzoic or sorbic acids. Only citric and malic acids provide small amounts of energy in the mammalian body. Because of the sourness, sugar is added to some fruits to make them palatable.

Fruits contain other substances, not classical nutrients, which can be biologically active, for example, flavonoids, salicylates, and limonoids. Tiny amounts of many different natural esters, aldehydes, and ketones contribute the distinctive volatile flavours of fruits (e.g. in an apple, 103 flavour compounds have been identified).

People have taken these other substances for granted. It was a surprise when Canadian pharmacologists discovered that grapefruit juice increases the blood concentrations of a number of commonly used potent drugs, for example, most statins and felodipine, by downregulation of cytochrome P450 3A4, which is involved in their first-pass metabolism. The active substance in grapefruit is probably 6′,7′-dihydrobergamottin, a furanocoumarin.

People cannot live for long on fruits alone unless they include nuts. Fruits are inadequate in protein, sodium, calcium, iron, and zinc. Nutritionists, however, look very favourably on fruit as part of a mixed diet and urge people to eat plenty of fruits each day. This is because they are low in energy, fat, and sodium, and make valuable contributions to the intakes of vitamin C, carotenoids, folate, and dietary fibre. Numerous epidemiological studies have shown that people who eat above-average amounts of fruit and vegetables (intakes of these two food groups are usually recorded together) have below-average rates of heart disease, stroke, and probably cancer. This may be partly because more fruit is eaten in privileged sections of society, and partly because of the good things in fruits, such as antioxidant nutrients (vitamin C and carotenoids) or other substances yet to be fully elucidated. Dietary guidelines consistently recommend people should eat some fruit every day, right up to Dietary Guidelines for Americans 2020.

There is little to be said about fruits on the negative side, except that in the orchard they are often sprayed with pesticide. These pesticides should have decomposed before sale, but it is safer to wash fruits before eating. The seeds or stones of some fruits contain cyanogenic glycosides, which are potentially toxic.

**Table 23.4 Nutrients of interest in fruits per 100 g raw edible portion**

| Nutrient | Lowest concentration | Typical fruit (orange) | Highest concentration |
|---|---|---|---|
| Protein (g) | 0.3 (apples) | 1.1 | 2.6 (passion fruit) |
| Fat (g) | 0.1 (most) | 0.1 | 19.5 (avocado) |
| Sugars (g) | trace (olive) | 8.5 | 20.9 (bananas) |
| Dietary fibre (g) | 0.1 (watermelon) | 1.7 | 3.7 (guava) |
| β-Carotene (mg) | 7.0 (lemon) | 47 | 700 (mangoes) |
| Vitamin C (mg) | 3.0 (grapes) | 54 | 230 (guava)* |
| Folate (mg) | trace (nectarines) | 31 | 34 (blackberries) |
| Potassium (mg) | 88.0 (watermelon) | 150 | 400 (banana) |
| Sodium (mg) | 1 (peaches) | 5 | 32 (honeydew melon) |

*see Table 13.1 for uncommon fruits with even more vitamin C.

*Adapted from:* Food Standards Agency (2002) *McCance and Widdowson's The composition of foods*, 6th summary edn. Cambridge, Royal Society of Chemistry.

## Further reading

1. **Bailey, D.G., Malcolm, J., Arnold, O., and Spence, J.D.** (1998) Grapefruit juice—drug interactions. *Br J Clin Pharmacol*, **46**, 101–10.
2. **Joshipura, K.J., Ascherio, A., and Manson, J.E.** (1999) Fruit and vegetable intake in relation to risk of ischemic stroke. *JAMA*, **282**, 1233–9.
3. **Paine, M.F., Widmer, W.W., Hart, H.L., et al.** (2006) A furanocoumarin-free grapefruit juice establishes furanocoumarins as the mediators of the grapefruit juice–felodipine interaction. *Am J Clin Nutr*, **83**, 1097–105.
4. **Steffen, L.M.** (2006) Eat your fruit and vegetables. *Lancet*, **267**, 278–9.
5. **Gomes, F.S. and Reynolds, A.N.** (2021) Effects of fruit and vegetable intakes on direct and indirect health outcomes – background paper for FAO/WHO International Workshop Background paper for the FAO/WHO International Workshop on fruits and vegetables 2020. Rome. FAO and PAHO. https://doi.org/10.4060/cb5727en.

# 23.5 Vegetables

Meika Foster

Vegetables comprise any plant part, other than fruit and seeds, which is used as food. They include roots and tubers, such as potatoes, taro, turnips, parsnips, carrots, cassava, and yams; bulbs, such as onions, leeks, and garlic; stems, like celery; leaves, such as lettuce, cabbage, and parsley; and flowers, such as broccoli and cauliflower. Some fungi (e.g. mushrooms) are also consumed as vegetables. Zucchini, squash, cucumber, and tomatoes, although strictly fruits, are usually treated as vegetables by the consumer. Peas and beans are legumes, but when immature and green are treated as vegetables. The nutritional composition and usage pattern of the roots and tubers is somewhat different to that of the stems, leaves, and flowers.

Potatoes have their origins in South America, being a staple of the diet of the Incas; other tubers, such as yams and taro, have been staple foods in the Pacific Islands for many years. In New Zealand, Māori traditionally consumed parts of native plants, such as fern roots and shoots, sowthistle (pūhā), and an acidic form of spinach (kōkihi), as vegetables. Australian native vegetables include warrigal greens and a native variety of carrot. Leafy vegetables were grown in monastery gardens during the Middle Ages. Today, most nations have cereal staples like rice or wheat, but some countries have vegetable staples. Cassava is a staple energy source for about 200 million people in tropical countries. Potatoes, whether boiled or baked, or as chips, remain a much-consumed commodity in developed nations. While leafy vegetables are commonly eaten in developed countries, they are less popular in developing nations. It is estimated that in India there were 40 species of leaves grown 70 years ago, but that number has now diminished.

Potatoes supply moderate amounts of protein in many people's diets. The biological value is good, with the limiting amino acid being methionine. Stem, bulb, leaf, and flower vegetables usually provide smaller quantities of protein to the diet, although their content may be similar. All vegetables contain negligible fat. Starch predominates in tubers, mainly amylopectin; the other vegetables contain sugars. Vegetables are a good source of dietary fibre. The fibre includes the soluble type (e.g. pectin), but also insoluble fibre, like cellulose.

Green leafy vegetables have a very high water content and are exceptionally low in energy, while being relatively high in micronutrients (Darmon et al., 2005). Some vegetables are rich in specific micronutrients—potatoes are a major source of vitamin C (because of the amount consumed), carrots are exceptionally high in β-carotene, and spinach is rich in folic acid. Dark-green leafy vegetables like spinach are a good source of lutein, and orange capsicums are a good source of zeaxanthin. These two carotenoids function in the macula lutea in the centre of the retina. Broccoli is relatively rich in calcium and spinach in iron, although neither is necessarily consumed in sufficient quantities to make a large contribution to mineral intake. The other factor to consider is the micronutrient bioavailability. Studies have shown that β-carotene from vegetables is more poorly absorbed than the pharmaceutical preparation. This is unfortunate because leaves are rich in β-carotene, and blindness from vitamin A deficiency remains a major public health risk in many developing countries. It is well established that the absorption of non-haem iron from vegetables is not as good as the haem form found in meat, but the presence of vitamin C will enhance non-haem iron absorption. Green leafy vegetables are the main dietary source of vitamin $K_1$ (see Chapter 14).

Cooking reduces the vitamin C and folate content of vegetables, often by a considerable amount. Vegetables should be cooked, therefore, for the shortest possible time in a small amount of water.

Apart from the well-recognized nutrients (Table 23.5), vegetables contain a variety of other substances that may be beneficial for health. Epidemiological studies indicate that vegetable consumption is associated with a lower prevalence of certain types of cancer, like bowel (van Duijuhoven et al., 2009), lung, and stomach. Initially this was attributed to their high content of antioxidant vitamins (β-carotene) but randomized controlled trials with β-carotene were disappointing (ATBC Cancer Prevention Study Group, 1994), and other constituents (such as fibre) may be more important. Flavonoids like quercetin and kaempferol, found in onions and broccoli, and glucosinolates, found in cruciferous vegetables like broccoli and Brussels sprouts, may protect against cancer. The flavonoids may also be cardioprotective because they function as antioxidants and may decrease platelet aggregation. In the Nurses' Health Study and the Health Professionals Follow-Up (combined cohort of 109,635 participants), higher vegetable consumption was associated with lower risk of cardiovascular disease (Hung et al., 2004); and in the Nurses' Health Study and Nurses' Health Study II (combined cohort of 182,145 participants), a higher intake of fruits and vegetables, specifically cruciferous and yellow/orange vegetables, was associated with a reduced risk of breast cancer (Farvid et al., 2019).

**Table 23.5 Nutrient content of typical leafy vegetables and roots and tubers**

| Nutrient | Nutrients per 100 g edible portion (leafy vegetables), raw | | | Nutrients per 100 g edible portion (roots and tubers), raw | | |
|---|---|---|---|---|---|---|
| | Cabbage, Chinese (pak-choi) | Lettuce, iceberg | Spinach | Carrots | Cassava | Potato, white, with skin |
| Energy (kJ) | 55 | 58 | 97 | 173 | 667 | 288 |
| Protein (g) | 1.50 | 1.13 | 2.86 | 0.93 | 1.36 | 1.68 |
| Fat (g) | 0.20 | 0.14 | 0.39 | 0.24 | 0.28 | 0.10 |
| Carbohydrate (g) | 2.18 | 2.92 | 3.63 | 9.58 | 38.1 | 15.7 |
| Fibre (g) | 1.0 | 1.2 | 2.2 | 2.8 | 1.8 | 2.4 |
| Vitamin C (mg) | 45.0 | 6.0 | 28.1 | 5.9 | 20.6 | 10.0 |
| β-Carotene (μg) | 2681 | 2371 | 5626 | 8285 | 8 | 5 |
| Folate (μg) | 66 | 34 | 194 | 19 | 27 | 18 |
| Calcium (mg) | 105 | 27 | 99 | 33 | 16 | 9 |
| Iron (mg) | 0.80 | 0.64 | 2.71 | 0.30 | 0.27 | 0.52 |
| Potassium (mg) | 252 | 168 | 558 | 320 | 271 | 407 |
| Zinc (mg) | 0.19 | 0.16 | 0.53 | 0.24 | 0.34 | 0.29 |

*Source:* US Department of Agriculture, Agricultural Research Service. FoodData Central, 2019. fdc.nal.usda.gov (accessed 16 August 2022)

Seasoning meals with a generous amount of herbs and spices to enhance flavour and increase the appeal and consumption of nutritious foods is a useful means of promoting health, protecting against chronic disease, and reducing our reliance on salt as a flavour enhancer.

## Further reading


1. **Carlsen, M.H., Halvorsen, B.L., Holte, K., et al.** (2010) The total antioxidant content of more than 3100 foods, beverages, spices, herbs and supplements used worldwide. *Nutr J*, **9**, 3.
2. **Craig, W.J.** (1999) Health-promoting properties of common herbs. *Am J Clin Nutr*, **70**, 491S–9S.
3. **Greenberg, S. and Ortiz, E.L.** (1983) *The spice of life*. London: Mermaid Books/Channel 4 TV.


# 23.12 Food processing and additives

Andrew Reynolds and A. Stewart Truswell

All the foods we eat are living matter, made of cells that contain enzymes, and many foods, especially those of animal origin, are inhabited by micro-organisms. Hence, food processing is necessary to prevent food decaying and to keep it safe for consumption. Food processing destroys the growth of micro-organisms, including pathogens like *Salmonella* and *Listeria*, and will inactivate autolytic enzymes and some natural toxins (e.g. trypsin inhibitors). The storage life of the food is increased, which means that it can be grown some distance from the point of consumption. This enables us to benefit from economies of scale by growing large quantities of food on the most suitable land.

Food processing is also undertaken to enhance the appearance and flavour of foods or make them easier to prepare before consumption. Given the convenience and increased palatability of these foods, there is a current global transition towards higher intakes of highly processed foods. Several methods to categorize the level of food processing now exist, while the association between food processing categories or techniques and human health is a growing area of research. There is emerging observational evidence that indicates that the regular consumption of ultra-processed foods is associated with adverse health outcomes. The NOVA classification system is a proposed method to categorize foods according to processing rather than in terms of nutrients. This classification system is used widely; however, it has drawn criticism due to ambiguity in classifying leading to high inter-individual variation, and for not taking into account existing scientific evidence on nutrition and food processing effects. A summary of the NOVA food classification system is shown in Box 23.1.

## 23.12.1 Methods of food processing

More specific food processing methods are discussed as they relate to certain foods in other parts of this chapter (e.g. section 23.1.2).

### Drying

One of the earliest methods of food preservation used was *drying* to reduce water content and inhibit microbial growth. The ancient Greeks sun-dried grapes to produce raisins, which lasted longer than the fresh fruits. Salting foods, or the inclusion of sugars such as in jams, reduces available water content due to their osmotic effects. Smoking foods serves a similar function. Modern technology includes tunnel drying, spray drying, and freeze drying of food to make milk powders, egg powders, and coffee powders.

### Cooling

Low temperatures such as by refrigeration decrease both bacterial growth and enzyme activity, while freezing makes the water in foods unavailable to

**Box 23.1** Summarized NOVA food classification system

| Food classification | Description | Examples |
|---|---|---|
| Unprocessed or minimally processed foods | Plant or animal food with no alteration or submitted to: cleaning, removal of inedible or unwanted parts, fractioning, grinding, drying, fermentation, pasteurization, cooling, or freezing. No oils, fats, sugar, salt, or other substances added | Frozen vegetables, whole grains, pasta, eggs, legumes, unsalted nuts and seeds, pasteurized milk |
| Processed culinary ingredients | Products extracted from natural foods by processes such as: pressing, grinding, crushing, pulverizing, and refining. Used in moderation in homes and restaurants to season and cook food | Vegetable oils, butter, salt, and honey or sugar |
| Processed foods | Food products manufactured by industry with processed culinary ingredients added to natural or minimally processed foods to preserve or to make them more palatable. They are recognizable as versions of the original foods. Most processed foods have two or three ingredients | Canned legumes or vegetables, canned fish, dried or salted meats, cheese, tomato paste, salted or sugared nuts |
| Ultra-processed foods | Industrial formulations made entirely or mostly from substances extracted from foods (oils, fats, sugar, starch, and proteins), derived from food constituents (hydrogenated fats and modified starch), or synthesized in laboratories from food substrates or other organic sources. Manufacturing techniques include extrusion, moulding, and pre-processing by frying | Breakfast cereals and bars, pastries, cakes, biscuits, chips, soft drinks, energy drinks, processed meats, ice cream, frozen desserts, chocolate, and confectionery |

*Adapted from:* Monteiro CA, Cannon G, Levy RB et al. (2016). NOVA. The star shines bright. [*Food classification. Public health*] *World Nutrition*, **7**: 1–3, 28–38

most enzymic function. Temperature control is used as one preservation method when shipping foods with a short shelf-life, such as vegetables and fruit.

### Heating

*Heat* is used in several ways to prevent food spoilage by microbial growth. Pasteurization of milk by heating to 72°C for 15 seconds destroys pathogenic organisms. Blanching of food (75–95°C for 1–8 minutes) before freezing and canning inactivates autolytic enzymes. Canned and sealed foods are sterilized by the application of heat.

### Pressing or juicing

Pressing or juicing is used to produce edible oils from oilseeds like canola and sunflower seeds, or wine and juices. Juicing mechanically shears the cell walls of vegetables or fruit, reducing their fibre content.

### Packaging

Packaging provides a physical barrier that minimizes contaminants being introduced into foods, or environmental microbes landing on foods. Sealing sterilized foods in cans or vacuum packs prevents microbial growth because oxygen is unavailable. Modified atmosphere packaging replaces the air within packaging with a gas that inhibits microbial spoilage to extend shelf-life.

### Food irradiation

Food irradiation can be used to inhibit sprouting of potatoes, delay the ripening of fruits, kill insect pests in fruit, grains, or spices, reduce or eliminate food spoilage organisms, and reduce micro-organisms on meats and seafood without altering the food structure. The practice is restricted by legislation in many countries at present.

## 23.12.2 Effects of food processing on nutrient content

Food processing can diminish or sometimes enhance the nutrient content of foods. *Cooking*, especially in water, can result in appreciable losses of vitamins and leaching of minerals. Most labile of the vitamins are vitamin C, folate, and thiamin, which are unstable with heat, although low pH will protect vitamin C. *Salting* can introduce high levels of sodium, just as use of sugar in preserving can appreciably increase added sugar intakes. *Juicing* and *milling* can reduce the fibre content of food. Both *drying* and *juicing* can lead to a greatly increased consumption compared with whole fresh fruit. More advanced food techniques are applied to *ultra-processed foods*. They can contain added salt, sugar, oils, and fats, as well as additives used to imitate sensorial qualities. Ultra-processed foods can contain residual components of foods; however, they are unlikely to provide the health benefits associated with their intake. Ultra-processing can also turn inedible by-products, such as chicken fat and epithelial tissue, into edible foods, such as chicken nuggets. The health effects of ultra-processed food intake is a growing area of research.

There are some nutritional benefits beyond enhanced food safety with food processing:

- freezing can stabilize vitamins that would otherwise degrade in transit
- heating can inactivate inhibitors in foods to increase bioavailability.

## 23.12.3 Food additives

During food processing, small quantities of food additives are often added for a variety of reasons. *Preservatives* may be added to specific foods to prevent bacterial growth. Examples include benzoic acid, propionic acid, sorbic acid, and sodium metabisulphite. *Antioxidants* may be added to slow the oxidation of oils and fats, preventing rancidity. *Emulsifiers* keep the oil and aqueous phases together in mayonnaises and sauces. *Humectants* prevent food from drying out (e.g. glycerol in cake frostings). *Thickeners* may be added to sauces or jams to improve texture. *Anticaking agents* may be added to ensure that powdered foods do not become lumpy (e.g. flavoured coffee mixes). *Food acids* may be used for flavour or to adjust the pH. *Colours* enhance the appearance of foods, with about half of those used from natural sources (e.g. β-carotene). *Artificial sweeteners* are used to replace sugars and reduce the energy content of foods. Examples are sucralose, saccharine, allulose, aspartame, and steviol glycosides. Saccharine and cyclamate break down with heat so they cannot be added to foods before cooking. All these sweeteners have been researched and monitored for safety. Aspartame is a dipeptide of phenylalanine and aspartic acid. It should be avoided by people with phenylketonuria. *Flavours* are usually a 'trade secret' so that the names of individual flavours are not declared on the label. However, in most developed nations, food legislation only permits those that have been shown to be safe. In the USA they are known as GRAS, meaning 'generally regarded as safe'. See Table 23.12 for a list of common food additives.

Other types of food additive are *vitamins* or *minerals*. Some foods may have their micronutrient

**Table 23.12 Some common food additives, the code number used in the European Union and other countries, and one of the foods or drinks to which it is added**

| Food additive | Code no. | Foods |
|---|---|---|
| Preservatives | | |
| Sorbic acid | 200 | Cheesecake |
| Benzoic acid | 210 | Fruit juices |

(Continued)

Table 23.12 **Some common food additives, the code number used in the European Union and other countries, and one of the foods or drinks to which it is added (*Continued*)**

| Food additive | Code no. | Foods |
|---|---|---|
| Sodium metabisulphite | 223 | Wine |
| Lactic acid | 270 | White bread |
| Propionic acid | 280 | Bread |
| Fumaric acid | 297 | Confectionery |
| **Antioxidants** | | |
| Ascorbic acid | 300 | Stock cubes |
| α-Tocopherols | 307 | Oils |
| Propyl gallate | 310 | Gelatin desserts |
| Butylated hydroxyanisole | 320 | Ice cream |
| **Emulsifiers** | | |
| Lecithins | 322 | Chocolate |
| Mono- and diglycerides of fatty acids | 471 | Potato crisps |
| **Humectants** | | |
| Sorbitol | 420 | Chewing gum |
| Glycerol | 422 | Pastilles |
| **Thickeners** | | |
| Alginic acid | 400 | Ice cream |
| Guar gum | 412 | Salad dressings |
| Xanthan gum | 415 | Bottled sauces |
| Pectin | 440 | Jams |
| Methyl cellulose | 461 | Jelly |
| **Anticaking agents** | | |
| Magnesium carbonate | 504 | Icing sugar |
| **Food acids** | | |
| Acetic acid | 260 | Tomato ketchup |
| Malic acid | 296 | Canned tomatoes |
| Citric acid | 330 | Marmalades |
| **Colours** | | |
| Curcumin | 100 | Curry powder |
| Erythrosine | 127 | Glacé cherries |
| β-Carotene | 160a | Margarines |
| Canthaxanthin | 161g | Biscuits |
| **Artificial sweeteners** | | |
| Acesulphame K | 950 | Baked goods |
| Aspartame | 951 | Soft drinks |
| Cyclamate | 952 | Tea and coffee |
| Saccharine | 954 | Tea and coffee |

content *restored* (e.g. the thiamin removed with the bran during milling of flour is replaced), while other food may be enriched (increasing the content of a nutrient already present in the food) or *fortified* (adding a nutrient to a food that doesn't naturally contain that nutrient). Vitamins may be added to a food that becomes a *replacement* for a food in which the vitamin naturally occurs; such an example is margarine, to which vitamin A and D are added, as both occur in butter.

### 23.12.4 Safety aspects of food additives

A series of strict tests must be conducted before a food additive is permitted. The Food and Agriculture Organization/World Health Organization (FAO/WHO) have a Joint Expert Committee on Food Additives (JECFA) and most countries have an expert body that prepares food legislation, for example, the US Food and Drug Administration (FDA), the European Scientific Committee for Food, the UK Food Standards Agency, and Foods Standards Australia New Zealand (FSANZ). The acute toxicity of the additive must be tested in both male and female animals in a minimum of three species and distribution of the compound in the body is assayed. Short-term feeding trials are conducted in at least two species of animal (only one can be a rodent) and reproduction is studied over two generations. After this, both mutagenicity and carcinogenicity are tested for in bacteria and tissue culture. Effects of food additives in humans are continually reviewed. It is not sufficiently realized that the most deleterious substances in our foods are naturally occurring ones rather than food additives. Additives are the most thoroughly monitored and tested of all chemicals in the food supply chain (see also Chapter 25).

### 23.12.5 Summary

Food processing reduces the occurrence of foodborne illness through pathogen control, and can enhance the appearance and flavour of foods. Virtually all foods undergo some processing, such as washing, cooking, or preserving, which enables a year-round food supply. Food processing can diminish or sometimes enhance the nutrient content of foods. Food processing also facilitates the addition of salt, sugars, saturated fats, and food additives to the diet. Food additives are used for a variety of reasons. They are rigorously tested and must be regarded as safe for consumption before being added to the food supply. Heavily processed foods that are manufactured to be highly appealing and no longer resemble their native ingredients are often termed ultra-processed foods. There is a current global transition towards higher intakes of these ultra-processed foods, with the health effects of their consumption a growing area of research.

## Further reading


1. **Baker, P., Machado, P., Santos, T., et al.** (2020) Ultra-processed foods and the nutrition transition: global, regional and national trends, food systems transformations and political economy drivers. *Obesity Reviews*, **21**, e13126.
2. **Coultate, T.P.** (2002) *Food: the chemistry of its components*, 4th edn. York: Royal Society of Chemistry.
3. **FAO/IAEA/WHO** (1999) *High dose irradiation: wholesomeness of food irradiated with doses above 10 kGy*, Report of a Joint Study Group, WHO Technical Report Series 890. Geneva: World Health Organization.
4. **Food Standards Agency** (2002) *McCance and Widdowson's the composition of foods*, 6th summary edn. Cambridge: Royal Society of Chemistry.
5. **International Food Information Service** (2005) *Dictionary of food science and technology*. Oxford: Blackwell.
6. **Monteiro, C.A., Cannon, G., Lawrence, M., Costa Louzada, M.L., and Pereira Machado, P.** (2019) Ultra-processed foods, diet quality, and health using the NOVA classification system. Rome: FAO.

7. **Petrus, R., Sobral, P., Tadini, C., and Gonçalves, C.M.** (2021) The NOVA classification system: a critical perspective in food science. *Trends in Food Science & Technology*, **16**, 603–08. https://doi.org/10.1016/j.tifs.2021.08.010 (accessed 28 August 2022).
8. **Wang, X., Ouyang, Y., Liu, J., et al.** (2014) Fruit and vegetable consumption and mortality from all causes, cardiovascular disease, and cancer: systematic review and dose-response meta-analysis of prospective cohort studies. *BMJ*, **349**, g4490.
9. **Williams, C.** (1995) Healthy eating: clarifying advice about fruit and vegetables. *BMJ*, **310**, 1453–5.

## 23.13 Food fortification

A. Stewart Truswell and Andrew Reynolds

Not everyone manages to obtain sufficient dietary intakes of all nutrients. Over the last 100 years, governments have acted to increase key micronutrient intake through regulation of the food industry using mandatory or voluntary food fortification legislation. A summary of major micronutrient fortification is listed here in approximate historical order. Additional nutrients added to foods in some countries include iron, calcium, and riboflavin.

### 23.13.1 Iodine

Endemic goitre was a serious health problem in Switzerland, where food and waters are low in iodine. Goitre was associated with impaired cognition, hearing impairment, and other disorders. Iodization of salt (NaCl) was started in the canton of Appenzell Ausser-Rhoden in 1922 and over the next 10 years was adopted by the other 25 cantons. The iodine content was cautiously increased and the number of goitres in army recruits declined 100-fold. An unexpected extra benefit was disappearance of congenital iodine-deficiency syndrome (historically referred to as endemic cretinism), which in Bern canton alone affected 700 people in 1923. Salt iodization started in the USA in Michigan in 1924 and has spread to many countries since Basil Hetzel (Australia) initiated the ICCIDD (International Consultative Council for Iodine Deficiency Disorders) in 1985 with support of WHO and UNICEF. It has been an effective campaign. As of 2015, iodine status is reported as adequate in 116 countries and there no longer appears to be any country with severe iodine deficiency, although it does still occur in regions. In many countries it is now mandatory that all salt is iodized, including Canada. However, in the USA, the UK, Australasia, and European countries, consumers can choose between iodized and non-iodized salt. The addition in Australian iodized salt is 4.0 mg iodine/100 g salt (in the form of potassium iodide).

### 23.13.2 Niacin

For the first 40 years of the twentieth century there was an epidemic of pellagra affecting many thousands in the southern states of the USA. Mortality was about 40%, with many more institutionalized in psychiatric hospitals. Following Goldbergen's research and the discovery of nicotinamide's potential to cure and prevent pellagra in 1937, voluntary bread enrichment started in 1938, and was followed by federal government action to fortify all cereal flours in the USA from 1941. The American Medical Association and the Food and Nutrition Board of the National Research Council pressed for this to be mandatory, which was enacted in 1943. Thiamin and riboflavin were included with niacin in the B vitamin addition. By 1953, mortality from pellagra in the USA had fallen dramatically from 3200 per 10,000 to very few, as it is now. Cases now occur alongside chronic alcoholism, due to the associated malnutrition and impaired metabolism. Pellagra is present in Africa in those subsisting largely on refined maize. It was only in 2003 that South Africa made enrichment of maize meal mandatory.

### 23.13.3 Vitamin A

An outbreak of xerophthalmia in young children appeared in Denmark in 1916–1920. Denmark produced plenty of butter, a good source of vitamin A, but it was too expensive because it commanded a high price in Germany. Poor people were eating margarine, then lacking in vitamin A. In 1918, the Danish government introduced a butter ration and weekly family allowance that alleviated the problem. In the 1930s, manufacturers started to add vitamin A to margarines (some added vitamin D as well). In Britain at the start of World War II in 1939, the Ministry of Food made it mandatory to add vitamin A to margarine, to bring the level up to that in butter in a good summer. This became the standard for margarines in all countries.

Another place for fortification with vitamin A is in dried milk powder distributed by the World Food Program for children where there are food shortages. This is another important addition of vitamin A to a staple food and must have prevented many cases of xerophthalmia.

### 23.13.4 Vitamin D

In the first quarter of the twentieth century, rickets affected probably the majority of infants and young children in New York, London, and dark northern cities, although it was not always severe. Sunlight and butter gave protection to rural children. Cod liver oil was available, but many doctors were sceptical of its value. Fortification of liquid milk with vitamin D started in the USA in the 1930s, and Britain followed during World War II. The original form of vitamin D used was ergocalciferol, $D_2$, which was discovered before $D_3$. Today, bottle-fed infants are protected from rickets by the vitamin D in all infant formulas, and in high-latitude countries, children and adults in the USA, Canada, and Scandinavia get vitamin D in enriched milk and some other foods. In Britain and Australasia vitamin D is added to margarine, but intake is low. The case for allowing greater fortification of vitamin D to milk and cereals has been made in the UK and Australia (Shrapnel and Truswell, 2006; Lanham-New et al., 2011). Meanwhile, advice to receive a safe amount of sun exposure is valuable, as vitamin D can be synthesized in the skin on exposure to sunlight (290–300 nm); otherwise people at risk of osteoporosis are advised to take a supplement.

### 23.13.5 Fluoride

Many people have filled or missing teeth or are edentulous as a result of dental caries. This has reduced in developed countries due to the preventive fortification of the drinking water supply with fluoride. Ainsworth (1933) showed that in Maldon, Essex, children had dental mottling, the drinking water had (high) 5 ppm fluoride, and the prevalence of caries was unusually low. H.T. Dean (1942) in the USA carried out extensive surveys of water and caries in thousands of children (aged 12–14 years) in several states. The result was a classic graph (see Fig. 10.7) showing that the caries rate falls as water fluoride increases to 1 part per million, and then stays about the same prevalence to 3 ppm.

The first controlled experiment of water fluoridation occurred in Michigan, USA, when caries rates in Grand Rapid (1 ppm fluoride added to the water) were compared with Muskegon (no fluoride added to water). In total, 24,000 children were surveyed, and after 6.5 years (in 1951) the caries rate was halved in Grand Rapids. Other trials in the USA and Britain confirmed these results. No toxic effects occur with these low levels of fluoride, except for some cases of dental mottling.

Today, the USA, Australia, New Zealand, Colombia, Ireland, Israel, and Malaysia have 80–100% of the population supplied with water at 1 ppm of fluoride. However, in Britain and across Europe only a minority of communities have fluoridated water. The WHO and all leading medical, health, and dental authorities recommend water fluoridation. The US Centers for Disease Control consider it one of the ten great public health achievements of the twentieth century. Water fluoridation is an economical public health initiative given the substantial reductions in healthcare system access, and one shown to reduce health inequities. Use of fluoride-containing toothpaste in fluoridate areas is seen as providing an additive benefit.

### 23.13.6 Thiamin

Many countries have their bread flour fortified with thiamin, following the lead of the USA and Britain, where fortification started in 1940 in World War II. However, this was not applied in Australia. In the 1970s and 1980s, the frequency of Wernicke's encephalopathy and beriberi in most hospitals in Australia was noticeable enough to cause concern. These thiamin-deficiency diseases were associated with alcoholism. If Wernicke's encephalopathy is not promptly and adequately treated with thiamin, the patient is left with the memory disorder Korsakoff's psychosis, which can be permanent. In 1987, the National Health and Medical Research Council (NHMRC) recommended mandatory enrichment of beer and cask and flagon wines with thiamin but the proposal met with vigorous opposition lobbying from different parts of the alcohol industry. Then, at a symposium of medical colleges, psychiatrists, nutritionists, and food scientists in October 1988, an alternative proposal, to enrich flour and bread with thiamin, as was already done in other developed countries, was agreed to. The federal government enacted the mandatory addition of thiamin to bread-making flour at 6.4 mg/kg (the level used in the USA) to start on 1 January 1991. Since then, the incidence of Wernicke–Korsakoff syndrome has moved from uncommon to rare in Australian hospitals.

### 23.13.7 Folate

Folate fortification followed after evidence indicated that an increased intake of folic acid in early pregnancy can reduce the risk of neural tube defect. First were observational studies, then randomized controlled trials by the British Medical Research Council Vitamin Study Research Group (1991) and by Czeizel et al. (2004) in Hungary. In Australia an Expert Panel in 1994 cautiously recommended voluntary fortification with folate to be permitted in cereal foods (that naturally contain some folate). The food standards authority modified regulations to permit this and food companies started to add folic acid to breakfast cereals.

In 1996 the US Food and Drug Administration authorized the addition of folic acid to grain products, and it became mandatory in 1998. Canada shared in this nutritional change. The amount of folic acid added was 140 μg folic acid per 100 g flour. South American countries and others have followed, and this fortification now happens in about 57 countries. The UK is still hesitant over mandatory fortification. Australia made folic acid mandatory in bread in 2005.

In countries where folic fortification has been implemented, up to 50% fewer neural tube defects have followed, with spina bifida usually more reduced than anencephaly. Fortification has significantly reduced, but not abolished, neural tube defect. The additional need for greater folate intakes to further reduce the risk of neural tube defect is now met with supplementation in the 12 weeks before and after conception. It is unknown whether non-pregnant people who chose to take folate supplements could have any adverse effects from excessive intakes. Possibilities of cognitive decline in older age and accelerated growth of pre-existing cancers do not appear to have happened.

### 23.13.8 Guidelines

Ideally, *enrichment* should mean adding a nutrient to a food that already contains some of it. *Fortification* then would be adding a nutrient to a food that naturally doesn't contain it. *Restoration* is addition of nutrient(s) to replace those lost in processing (e.g. flour milling); see Box 23.2.

**Box 23.2** Rules for adding a nutrient to a food

1. There is a demonstrated inadequacy (public health problem).
2. The food will be eaten by those in need.
3. The added nutrient will be stable.
4. It won't spoil the taste or appearance of the food.
5. The nutrient will be bioavailable.
6. It is not toxic.
7. It won't cause dietary imbalance.
8. It shouldn't deceive consumers about unintentional merit.
9. The additional cost is reasonable.
10. Methods are available to measure and control based on Codex Alimentarium, 1994.

## Further reading

1. **Czeizel, A.E., Dobó, M., and Vargha, P.** (2004) Hungarian cohort-controlled trial of periconceptional multivitamin supplementation shows a reduction in certain congenital abnormalities. *Birth Defects Res A Clin Mol Teratol*, **70**(11), 853–61. doi: 10.1002/bdra.20086
2. **Lanham-New, S.A., Buttriss, J.L., Miles, L.M., et al.** (2011) Proceeding of the Rank Forum on Vitamin D. *Br J Nutr*, **105**, 144–56.
3. **Lazarus, J.H.** (2015) The importance of iodine in public health. *Environ Geochem Health*, **37**, 605–18.
4. **Medical Research Council Vitamin Study Research Group** (1991) Prevention of neural tube defects: results of the Medical Research Council Vitamin Study. *Lancet*, **338**(8760), 131–7.
5. **Park, Y.K., Sempos, C.T., Barton, C.N., Vanderveen, J.E., and Yetley, E.** (2000) Effectiveness of food fortification in the United States: the case of pellagra. *Am J Publ Health*, **90**, 727–38.
6. **Rolland, S. and Truswell, A.S.** (1998) Wernicke–Korsakoff syndrome in Sydney hospitals after 6 years of thiamin enrichment of bread. *Publ Health Nutr*, **1**, 117–22.
7. **Shrapnel, W. and Truswell, A.S.** (2006) Vitamin D deficiency in Australia and New Zealand: what are the dietary options? *Nutr Dietet*, **63**, 206–13.
8. **Whelton, H.P., Spencer, A.J., Do, L.J, and Rugg-Gunn, A.J.** (2019) Fluoride revolution and dental caries: evolution of policies for global use. *J Dent Res* **98**(8), 837–46.

## Useful website

**Food Fortification Initiative**
https://www.ffinetwork.org/

# Functional Foods and Health Claims

Martijn B. Katan



This textbook teaches you who needs to eat which nutrients and in what amounts. However, translating this knowledge into foods and meals is no simple task. Most people do not have access to a dietitian or other nutrition professional, and even if they did, many would still want to decide for themselves which foods are good for them. This has created a market for so-called functional foods that promise to increase the well-being or health of the consumer. Health claims are integral to such foods.

## 24.1 What is a functional food?

One definition of functional foods is 'foods that provide health benefits beyond basic nutrition'. However, that ignores the commercial character of functional foods. Tap water provides benefits beyond basic nutrition; a liberal intake reduces the risk of cystitis, and kidney and bladder stones, but no one would call tap water a functional food. A functional food always comes in a bottle or package, and carries a brand name, logo, and a suggestion that it promotes health. Therefore, functional food can be defined as: a branded food that claims explicitly or implicitly to improve health or well-being.

An example may clarify the role of branding and health claims. Polyunsaturated oils such as sunflower or soybean oil reduce plasma cholesterol and the risk of coronary heart disease (CHD), but few people would call a generic bottle of sunflower oil a functional food. However, if a manufacturer developed a proprietary brand of sunflower oil and marketed it with a cholesterol claim, then that oil could well be called a functional food. Box 24.1 illustrates the role of marketing in creating a functional food.

Functional foods and dietary supplements are related. The active ingredients of functional foods, such as vitamins, plant sterols, lactic acid bacteria, or herbal extracts, can also be packaged into a capsule or tablet to create a supplement. The terms 'nutraceutical' and 'nutriceutical' have been used both for foods and supplements; there is no consensus on what these words mean. Finally, some products are halfway between foods and supplements, e.g. candies or sweets with added vitamins.

**Box 24.1 Functional foods have four layers**

1. The active ingredient (e.g. plant sterols)
2. The food matrix (e.g. orange juice, yoghurt, a cookie, chocolate, margarine)
3. The package with the health claim or health suggestions
4. Other marketing efforts, including flyers, TV commercials, and sponsored media coverage

## 24.2 Typical ingredients of functional foods

### 24.2.1 Established nutrients

Many functional foods employ ingredients that are also available from regular foods. You can get lycopene from special drinks and supplements, but the same lycopene is found in tomatoes or tomato ketchup. The newness of such functional foods is in the way in which known ingredients are incorporated into a palatable and attractive food that can be patented and marketed. That is also the potential benefit of functional foods—they may provide nutrients in a form that is more attractive or convenient for the consumer than regular foods.

Table 24.1 lists some nutrients typically found in functional foods, and the quality of the evidence for their efficacy. Whether the food itself is efficacious depends on the amount and the processing of the ingredient; thus, 'whole wheat cookies' may contain

**Table 24.1 Examples of established nutrients that are used as functional food ingredients, and the evidence for their efficacy**

| Ingredient | Examples of products | Health claim | Strength of evidence in humans |
|---|---|---|---|
| Folic acid | Cereals | Protects against neural tube defects | ++ |
| Dietary fibre | Drinks | Relieves constipation | ++ |
| Low in sodium and/or high in potassium | Drinks, soups, margarine | Reduces blood pressure | ++ |
| Unsaturated fatty acids | Spreads, cookies | Reduces risk of heart disease | ++ |
| Sugar alcohols | Chewing gum | Reduce caries risk | ++ |
| Soluble fibre from whole oats or *Psyllium* husk | Cereals, cookies | Reduces cholesterol and risk of heart disease | ++ For cholesterol reduction |
| Calcium | Cereals, fruit juices, milk products, spreads | Protects against osteoporosis, helps maintain bone density | + For consumers with a low calcium intake |
| Zinc | Sweets, lozenges | Prevention/cure of common cold | + – |
| Vitamin C | Drinks, sweets | Protects against cardiovascular disease | + – In observational studies<br>– In clinical trials |

++, Proven efficacy, consistent effect seen in multiple high-quality studies; +, reasonable evidence for efficacy, effect seen in a limited number of studies, or some inconsistency between studies; –, absence of an effect evident from a limited number of studies.

Evidence was graded according to the Australia New Zealand Food Authority criteria for levels and kinds of evidence for public health nutrition. The evidence consisted of randomized trials in humans, unless indicated otherwise.

too little wheat bran to affect defecation, or the bran may have been ground to a powder, which is less active than coarse bran.

The health claims for some of these ingredients are well substantiated. For other ingredients, the evidence is weaker. Intakes of vitamins and minerals from functional foods, and especially from supplements, may be much higher than from regular foods, and the adverse effects of such high intakes are a cause for concern. For example, there is evidence that excessive intake of calcium promotes prostate cancer.

### 24.2.2 Novel ingredients

Table 24.2 lists more novel or exotic ingredients of functional foods. Most of the claims for benefits of novel or 'exotic' ingredients have not been substantiated in clinical trials. However, a few have been well investigated and show some promise, such as the following examples.

**Sterols and stanols** Plant sterols and stanols block cholesterol absorption. Margarines and other foods enriched with stanols or sterols lower low-density lipoprotein (LDL) cholesterol. That effect is well documented, but the effect on heart disease has not been evaluated and remains uncertain.

**Pre- and probiotics** Probiotics are viable bacteria that survive passage through the gastrointestinal tract and exert beneficial effects on the consumer. Probiotics can be patented because each bacterial strain is unique. That makes probiotics attractive to industry, but documented beneficial effects are scarce.

**Table 24.2 Newer functional food ingredients and their efficacy**

| Ingredient | Product examples | Health effect or claim | Evidence in humans |
|---|---|---|---|
| Plant stanols and sterols | Margarine, yoghurt, cereal bars | Lower cholesterol and risk of CHD | ++ For LDL cholesterol lowering<br>+ − For CHD |
| *Lactobacillus* GG bacteria | Yoghurt | Reduce diarrhoea | + For rotavirus-induced diarrhoea in infants<br>+ For antibiotic-induced infections |
| *Lactobacillus* GG bacteria | Yoghurt | Reduce risk of early atopic disease | + − Results of trials contradictory |
| Other 'probiotic' live bacteria, plus fermentable sugars ('prebiotics') | Yoghurt | Enhance immunity | + − Some effects on biomarkers but none on disease |
| Isoflavones (phyto-oestrogens) | Soy products | Reduce menopausal symptoms, osteoporosis, cardiovascular disease | − For hot flushes<br>+ − For osteoporosis and heart disease |
| Catechins and flavanols | Tea, cocoa | Reduce cardiovascular risk | + − Some epidemiological evidence<br>No trial data |
| Conjugated linoleic acid (CLA) | Supplements | Reduces body weight, protects against heart disease and cancer | − Minimal effects on body weight in humans<br>− CLA lowers HDL and raises LDL |

++, Proven efficacy, consistent effect seen in multiple high-quality studies; +, reasonable evidence for efficacy, effect seen in a limited number of studies, or some inconsistency between studies; + −, evidence for no effect, absence of an effect evident from a limited number of studies; −, proven not to work, absence of an effect evident in multiple high-quality studies; LDL, low-density lipoprotein.

Prebiotics are non-digestible carbohydrates that selectively stimulate growth of beneficial bacteria. Inulin and fructo-oligosaccharides are examples. Health effects of prebiotics appear to be limited to improved bowel function.

Polyphenols High intakes of tea and chocolate rich in catechins and other flavonoid polyphenols have been associated with a reduced risk of CHD, but the association has not been tested in clinical trials.

### 24.2.3 Herbs and herbal extracts

Herbal ingredients are used both in supplements and in foods, but amounts in foods are much lower. Safety is a concern, as exemplified by herbal teas with *Aristolochia*, which causes renal cancer, and products with *ephedra*, which causes hypertension, strokes, and seizures. Most herbal remedies have not been shown to be safe and effective.

## 24.3 How to prove health effects of foods

Every type of nutrition research has its limitations and judgements on efficacy and safety of foods must rely on a synthesis of the totality of the scientific evidence. The various classes of evidence are reviewed below and summarized in Table 24.3.

### 24.3.1 Cell studies

A true understanding of the effect of a nutrient on health requires insights at the molecular level. Such comprehension remains the ultimate goal of nutrition science. However, as phrased by Willett (1998), 'our understanding of biological mechanisms remains far too incomplete to predict confidently the ultimate consequences of eating a particular food or nutrient'. That remains true today; cell and molecular studies cannot by themselves establish efficacy and safety of a food ingredient.

### 24.3.2 Animal research

When a food is known to affect human health, that effect can sometimes be recreated in a laboratory animal. The animal is then called a model for the

**Table 24.3 Types of research used to investigate the relation between diet and disease, and their strengths and weaknesses**

| Type | Strengths | Limitations |
|---|---|---|
| (Sub)cellular studies | Mechanistic insights | Extrapolation to humans uncertain |
| Animal feeding trials | Long term<br>Hard endpoints<br>Show cause and effect | Extrapolation to humans uncertain<br>Conditions often extreme and unphysiological |
| Epidemiological observations | Long term<br>Hard endpoints<br>Applicable to humans | Confounding<br>Associations do not prove causality |
| Mendelian randomization | Long term<br>Hard endpoints<br>Less confounding | Mutation may act through other paths than the one of interest (pleiotropy)<br>Populations may be heterogeneous and stratified |
| Randomized trials with surrogate endpoints | Human<br>Prove cause and effect | Short duration<br>Validity of surrogate endpoints uncertain |
| Randomized clinical trials | Hard endpoints<br>Prove cause and effect | Duration sometimes too short<br>Selected groups |

human disease. A model is like a map; it helps scientists to navigate human metabolism just like a map helps to navigate a country. Animal models are indispensable; they have provided insight into nutrition and health ever since Eijkman and Grijns used thiamin-deficient chickens to discover the cause of beriberi (see Chapter 12). However, the mere existence of a nutritional effect in animals does not prove that this effect also exists in humans. Consider the medieval maps that depicted a huge island called Atlantis in the middle of the Atlantic Ocean; such maps do not prove that Atlantis existed. Similarly, an animal 'model' is of little use until the relevant effect of diet on health has been demonstrated in humans. This holds all the more now that animal 'models' can be expressly constructed to reflect a hypothetical effect of a nutrient on a disease. The nutrient will be effective in this model because that is what it was built to do, but extrapolation to humans is unwarranted.

### 24.3.3 Observational epidemiology

Epidemiology is a valuable source of information on the effects of foods on disease, but associations do not by themselves prove causality unless the association is very strong. Unfortunately, relative risks in nutritional epidemiology are mostly weak. Confounding then becomes a problem; someone who eats lots of vegetables may also exercise more, smoke less, and do other healthy things, and even the most sophisticated statistical techniques cannot completely separate these factors. 'Data-dredging' is also a problem in epidemiology; hundreds of associations may be tested and only the 'statistically significant' ones may get published.

When epidemiological findings are consistent with other forms of evidence a causal link becomes probable. That is why a causal role of *trans*-fat in heart disease is likely; the association between *trans*-fatty acids intake and CHD in epidemiological studies is corroborated by the adverse effects of *trans*-fatty acids on blood lipids in metabolic trials.

### 24.3.4 Mendelian randomization

Genetic polymorphisms that affect the metabolism of a nutrient or blood levels of a metabolite can help to identify the role of diet in disease causation. For example, a genetic ability to tolerate lactose and to drink large amounts of milk is associated with prostate cancer risk. That strengthens earlier suspicions about calcium and prostate cancer (Torniainen et al., 2007). Mendelian randomization has also refuted the idea that vitamin D reduces obesity; causality is the other way around, and obesity lowers blood vitamin D (Vimaleswaran et al., 2013).

The number of such 'Mendelian randomization' studies has increased exponentially. They do not suffer from conventional confounding because carrying a particular genetic variant does not cause people to smoke, exercise, or eat differently (Davey Smith et al., 2005). However, confounding does occur in incompletely mixed populations that contain people of different ethnicity. Randomness is then incomplete; carriers of a particular genotype may differ from non-carriers in other genes, history, and lifestyle. Another weakness of Mendelian randomization is that the mutation may be pleiotropic. It may affect disease through a pathway different from the one relevant for nutrition.

### 24.3.5 Trials with surrogate endpoints

Trials that measure the effect of diet on an intermediate disease marker, such as blood pressure or insulin sensitivity, can show cause-and-effect in humans. However, even established markers can lead us astray. For instance, drugs that normalized blood glucose in type 2 diabetics failed to reduce risk of cardiovascular disease. Thus, changes in risk markers cannot be automatically equated with changes in the disease that they predict. Also, there are few validated markers outside the cardiovascular field. One can measure the effect of diet on hundreds of variables involved in immune response, but the relevance of each of them to prevention of infection is uncertain.

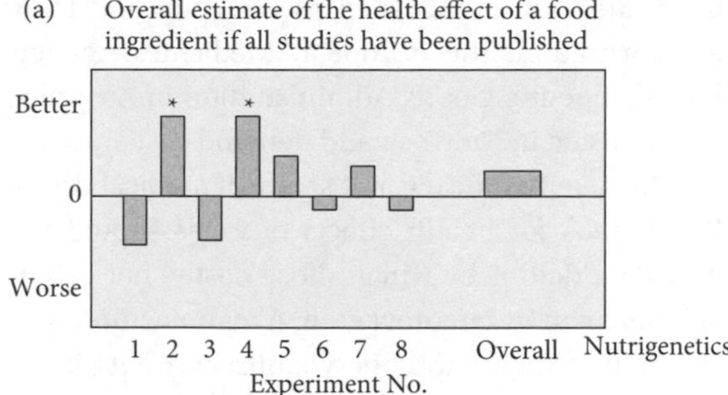


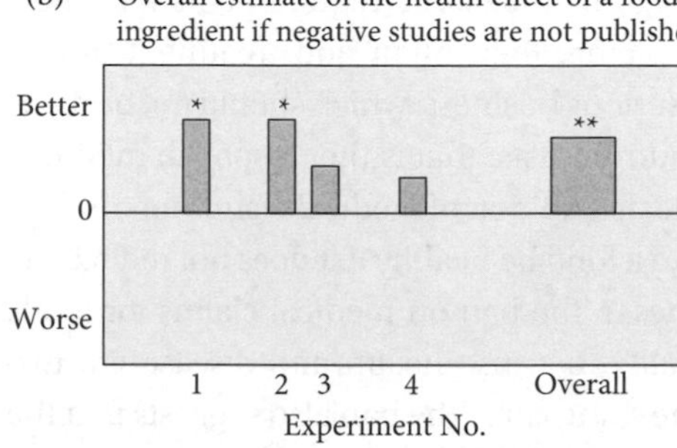


Fig. 24.1 Hypothetical effect of publication bias on the perceived health effect of a food ingredient. The vertical axis plots the effect of a food ingredient on a risk indicator, e.g. blood pressure. Each bar represents one experiment. All experiments studied the same effect; the differences are due to chance and biological variability. (a) The overall effect in eight trials is small and not significant. (b) If experiments with nil or negative outcomes remain unpublished, the overall effect in the four published studies becomes significant.

* = $P < 0.05$; ** = $P < 0.01$.

Finally, selective publication is a general issue both for epidemiology and for animal and human trials. A study that fails to support a health effect may remain unpublished (see Fig. 24.1). Alternatively, emphasis may be put on a secondary outcome when the primary outcome shows no benefit. As a result, the scientific literature may offer a biased view of what a food or ingredient really does. This explains why meta-analysis of small studies may show beneficial effects, while large studies do not; large studies usually get published, even if the results are negative.

### 24.3.6 Randomized clinical trials

Randomization eliminates confounding. Therefore, randomized clinical trials with hard disease and death endpoints are the gold standard in biomedical research, and they offer a level of confidence that no other type of research can match.

Randomized clinical trials also have their weaknesses. Chief of these is that they last too short a time. Many benefits of diet may not be reaped within the 3–5 years of a randomized clinical trial, and a negative outcome is thus less than definitive, especially if contradicted by the outcomes of observational epidemiological studies.

### 24.3.7 Summary

The costs of establishing properly that a functional food promotes health is huge and, understandably, there is pressure from industry to adopt *in vitro* and 'functional' tests as substitutes for more expensive and lengthy clinical trials. However, the history of β-carotene—which seemed to prevent cancer *in vitro* but promoted cancer in clinical trials—shows the risks of relying on soft evidence.

## 24.4 Health claims

The health effect of a food cannot be perceived directly by the consumer, which is why it is communicated in the form of a health claim. Health claims need to be regulated, otherwise market forces will produce a proliferation of unsubstantiated health claims.

### 24.4.1 Categories of claims

Most jurisdictions recognize multiple categories of health claims. The lowest category is called structure/function, nutrition, or nutrition content claims, such as 'low fat', 'no added sugar', and 'high in fibre'. Such claims may be used for every food that meets the criteria; manufacturers do not need to submit special documentation. An intermediate category, called health claims or general-level health claims, allows hinting at diseases without actually naming them, e.g. 'Iodine is necessary for normal production of thyroid hormones'. The highest category contains claims for disease or disease risk reduction; demands for evidence are more stringent there. Examples of such high-level health claims are: 'docosahexaenoic acid (DHA) intake contributes to the normal visual development of infants' (European Union) and 'a high intake of fruit and vegetables reduces risk of coronary heart disease' (Australia/New Zealand).

The existence of multiple categories provides producers with a choice. They can save effort and money by using a generic, low-level claim, or they can go for a more ambitious disease-reduction claim that may require large investments in research and marketing. However, the categories also create confusion because the distinctions between categories are mostly lost on consumers. 'This food contains no cholesterol' is a claim of the lowest category, 'reduces blood cholesterol' is intermediate, and 'reduces risk of heart disease' is the highest level, but to consumers they all mean the same: 'this food prevents heart disease' (Williams, 2005). However, that is a medical claim that paradoxically is forbidden in most countries.

### 24.4.2 The ban on medical claims

A ban on medical claims for foods does not make sense scientifically. Foods can be used to prevent, treat, and cure diseases; that is what nutrition science is about. Lemons cure scurvy, folic acid prevents neural tube defects, and polyunsaturated oils reduce heart attacks. The ban on medical claims for foods is a legal fiction, but it is a useful fiction because it protects patients from treatments that are ineffective or unsafe. Authorities such as the Food and Drug Administration, the European Medicines Agency, the Therapeutic Goods Administration in Australia, and Medsafe in New Zealand demand vast amounts of evidence for efficacy and safety of medical drugs. Regulations for health effects of foods should not provide a detour by which those costly, but essential, rules can be circumvented. A manufacturer who claims that their probiotic yoghurt prevents influenza must follow the pharmaceutical road and show that the yoghurt meets the requirements for a new drug or vaccine. Prevention and treatment of diseases is a serious business, which should not be regulated by nutritionists; that is the unspoken rationale behind the prohibition of medical claims for foods.

How can a food be healthy if it does not reduce the risk of illness? The ban on medical claims for foods denies reality, because health and disease are two sides of the same coin. The ban also suggests that the impact of foods on health is not in the same class as that of medicines. There is some truth to that. In developed countries, overt nutrient deficiency diseases have become rare. Nutrition does play an important role in diabetes, cardiovascular disease, cancer, and other non-communicable diseases, but these are caused more by an excessive intake of calories, alcohol, salt, sugar, and saturated fat than by an inadequate intake of functional foods. Still, specific foods can reduce the risk of non-communicable diseases. Examples are sugar-free chewing gum against caries, zinc-plus-antioxidants supplements against macular degeneration, potassium against hypertension and its sequelae, and iron and iodine to improve foetal brain development (Katan et al., 2009). Evidence for beneficial bacteria is still thin, but in the coming decades the boom in microbiome research might yield bacteria that do promote health. Food producers and authorities need to walk a fine line to communicate such benefits to consumers without trespassing into medical territory (Katan, 2012).

## 24.5 Nutrient profiling

Claims are allowed only if claimed beneficial components are present in a certain minimum amount. Claims for low levels of adverse components similarly require a certain minimal reduction compared with regular foods. Also, claims may not be used to make unhealthy foods look healthier. In the USA, this legislation is known as the 'jelly bean rule'; foods high in sugar, salt, fat and saturated fat, sodium, or cholesterol cannot be given a healthy appearance by adding vitamins or minerals. Worldwide, nutrient profiling is a bitterly contested area, with shifting rules.

## 24.5.1 Regulation of health claims around the world

The trustworthiness of health claims is largely dependent on government regulation. Consumers assume that claims have been approved by government authorities, but many countries do not require that the health benefits claimed or suggested for functional foods be supported by proper scientific evidence. In the USA, effects of foods on health may be claimed on the package even if the Food and Drug Administration (FDA) has concluded that there is very little scientific evidence for this claim, and food supplements can be marketed without proof that they work or that they are safe. Legislation in Australia and New Zealand seems similar to that of the European Union (EU), but the law contains a loophole that allows manufacturers to construct their own claims and decide whether the claim is valid, based on a review that they have written or commissioned. That gives rise to claims such as 'Elderberry fruit extract (*Sambucus nigra*) contributes to protection from certain virus strains including influenza' and '15 mL of apple cider vinegar daily contributes to weight loss.' Box 24.2 lists other examples.

**Box 24.2 Examples of self-substantiated food-health relationship permitted in Australia and New Zealand**

- Psyllium contributes to heart health.
- Green tea catechins (GTCs) with caffeine contribute to modest weight loss in overweight and obese adults.
- Probiotics promote intestinal health.
- Elderberry fruit extract (*Sambucus nigra*) contributes to normal immune system function.

Self-substantiated claims may be used in Australia and New Zealand without having been evaluated by the authorities. Similar claims were earlier evaluated—and rejected—by the EU.

In essence, manufacturers can claim what they wish as long as they notify their 'self-substantiated food-health relationship' to the authorities.

In contrast, the EU puts strict demands on nutrition and health claims. A list of 'nutrition claims' can be used to state that a food is high in beneficial or low in harmful components. Any manufacturer may use these; Box 24.3 gives examples. The next level is health claims, and food producers have filed applications for several thousands of claims for specific foods, nutrients, and ingredients. After careful evaluation of the evidence by the European Food Safety Authority (EFSA), 90% were rejected, including all claims for antioxidants, pre- and probiotics, and for improvement of immune function.

The EU approved about 250 health claims; Box 24.4 shows examples. The wheat bran claim cannot mention constipation, but most consumers will make that link themselves. The claim for magnesium and protein synthesis asserts a basic fact of cell physiology that will interest few; many approved claims are of this type. The claim for vitamin A and vision illustrates a dilemma that regulators face. People need vitamin A in order to maintain eyesight. However, to Europeans with poor eyesight, the claim may suggest that cod liver oil and carrots may help them. That will rarely if ever be true; Europeans with vision problems are more likely to need spectacles or surgery. Thus, people may still be misled in spite of all the efforts to the contrary.

Health claims cannot mention a disease, but most readers will assume that 'Guar gum contributes

**Box 24.3 Examples of generic nutrition claims permitted in the European Union**

- Low energy
- Low saturated fat
- Salt-free
- With no added sugars
- High calcium
- Naturally high [in] riboflavin
- Source of protein

*Adapted from:* European Commission—Nutrition Claims. http://ec.europa.eu/food/safety/labelling_nutrition/claims/nutrition_claims/index_en.htm

**Box 24.4 Examples of health claims approved by the European Union**

- Wheat bran fibre contributes to an increase in faecal bulk.
- Substituting two daily meals of an energy-restricted diet with meal replacements contributes to weight loss.
- Magnesium contributes to normal protein synthesis.
- Vitamin A contributes to the maintenance of normal vision.
- Guar gum contributes to the maintenance of normal blood cholesterol levels.

The EU has approved ('authorized') 256 health claims.

*Source:* European Commission—EU Register of nutrition and health claims made on foods. https://food.ec.europa.eu/safety/labelling-and-nutrition/nutrition-and-health-claims_en

to the maintenance of normal blood cholesterol levels' means it prevents heart disease. That may explain why so few food producers went to the trouble and expense of filing for a disease risk reduction claim, which does allow naming a disease. Fourteen such claims were approved. Typical examples are glucans, unsaturated fatty acids, and plant sterols that lower blood cholesterol and, by implication, the risk of CHD; vitamin D to prevent muscle weakness and falls, which are risk factors for bone fractures; and sugar-free chewing gum to reduce risk of dental caries.

European regulations for health claims are not perfect, but they have done a lot to eliminate disinformation from supermarket shelves. The 2000 rejected health claims in the EU Register provide a fascinating, thoroughly reviewed collection of flawed claims for health effects of diets, foods, and ingredients. The 256 approved claims were also thoroughly reviewed and each review published in the EFSA journal. Together, these 'authorized' claims provide an overview of what modern nutrition can do for the health of consumers in affluent countries. Given proper regulation, health claims can play a valuable role in making healthy foods available and communicating their benefits to consumers.

## Further reading


1. **Davey Smith, G., Ebrahim, S., Lewis, S., Hansell, A.L., Palmer, L.J., and Burton, P.R.** (2005) Genetic epidemiology and public health: hope, hype, and future prospects. *Lancet*, **366**, 1484–98.
2. **Katan, M.B.** (2004) Editorial: health claims for functional foods. *BMJ*, **328**, 180–1.
3. **Katan, M.B.** (2012) Why the European Food Safety Authority was right to reject health claims for probiotics. *Benef Microbes*, **3**, 85–9.
4. **Katan, M.B., Boekschoten, M.V., Connor, W.E., et al.** (2009) Which are the greatest recent discoveries and the greatest future challenges in nutrition? *Eur J Clin Nutr*, **63**, 2–10.
5. **Katan, M.B. and De Roos, N.M.** (2004) Promises and problems of functional foods. *Crit Rev Food Sci Nutr*, **44**, 369–77.
6. **McClements, D.J., Decker, E.A., Park, Y., and Weiss, J.** (2009) Structural design principles for delivery of bioactive components in nutraceuticals and functional foods. *Crit Rev Food Sci Nutr*, **49**, 577.
7. **Nestle, M.** (2013) *Food politics: how the food industry influences nutrition and health.* London: University of California Press, Ltd.
8. **Torniainen, S., Hedelin, M., Autio, V., et al.** (2007) Lactase persistence, dietary intake of milk, and the risk for prostate cancer in Sweden and Finland. *Cancer Epidemiol Biomarkers Prev*, **16**, 956–61.
9. **Verhagen, H., Vos, E., Francl, S., Heinonen, M., and van Loveren, H.** (2010) Status of nutrition and health claims in Europe. *Arch Biochem Biophys*, **501**, 6–15.
10. **Vimaleswaran, K.S., Berry, D.J., Chen, L., et al.** (2013) Causal relationship between obesity and vitamin D status: bi-directional Mendelian randomization analysis of multiple cohorts. *PLoS Med*, **10**, e1001383.
11. **Willett, W.C.** (1998) *Nutritional epidemiology*, 2nd edn. New York, NY: Oxford University Press.
12. **Williams, P.** (2005) Consumer understanding and use of health claims for foods. *Nutr Rev*, **63**, 256–64.

# Food Toxicity and Safety

Peter Williams and Paul Brent



Until recently, eating food in modern industrialized countries has usually been regarded as a low-risk activity, but several highly publicized food safety scares have raised consumer concerns about the safety of our food supplies (Table 25.1).

Very few of the foods that we commonly eat have been subject to any toxicological testing and yet they are generally accepted as being safe to eat. However, all chemicals, including those naturally found in foods, are toxic at some dose. Laboratory animals can be killed by feeding them glucose or salt at very high doses, and some nutrients such as vitamin A and selenium are hazardous at

**Table 25.1 A chronology of recent food scares**

| | |
|---|---|
| 1986 | First cases of mad cow disease in Britain |
| 1990 | Benzene in Perrier mineral water in France |
| 1996 | *Salmonella* in peanut butter in Australia |
| 1997 | Contagious swine fever in The Netherlands |
| 1999 | Contaminated Coca-Cola in Belgium |
| 1999 | Pollen from genetically modified maize reported to kill Monarch butterflies in the USA |
| 2001 | Foot-and-mouth disease all over Europe |
| 2002 | Acrylamide found in starchy foods cooked at high temperature in Europe |
| 2003 | Outbreak of bird flu in Asian poultry |
| 2004 | Warnings about mercury in shark, mackerel, and swordfish in the USA |
| 2008 | Melamine contamination of baby milk formula in China |
| 2009 | Hepatitis A in semi-dried tomatoes from Turkey in Australia |
| 2009–2010 | *Listeria* in ready-to-eat (RTE) foods in Canada |
| 2011 | *E. coli* H:104 in sprouts in Germany |
| 2011 | Fukushima nuclear reactor accident in Japan |
| 2015 | Hepatitis A in frozen berries from China in Australia |
| 2016 | *Salmonella* in bean sprouts in Australia |
| 2018 | *Listeria monocytogenes* in rock melons in Australia |
| 2018 | Needles in strawberries in Australia |
| 2019 | *Salmonella enteritidis* in eggs in the USA and Australia |
| 2019 | *Salmonella welterevden* in frozen meals in the USA |

intakes only a few times greater than normal human requirements. Even very common foods, such as pepper, have demonstrated carcinogenic activity. Toxicity testing of a food or ingredient can tell us what the likely adverse effects are and at what level of consumption they may occur, but by itself this does not tell us whether it is safe to eat in normally consumed amounts. 'Risk' is the probability that the substance will produce injury under defined conditions of exposure. The concept of risk takes into account the dose and length of exposure, as well as the toxicity of a particular chemical, and is a better guide to the safety of a food. Consequently, any attempt to examine the safety of the food supply should not be based on the question 'Is this food or ingredient toxic?' (the answer is always 'yes'), but rather by finding out if eating this substance in normal amounts is likely to increase the risk of illness significantly, i.e. 'Is it safe?'.

## 25.1 Hazardous substances in food

Three general classes of hazards are found in foods:

1. Microbial or environmental *contaminants*
2. Naturally occurring toxic *constituents*
3. Those resulting from intentional *food additives* or *novel foods or ingredients*

The most dangerous contaminants are those produced by infestations of bacteria or moulds in food and, more recently, viruses transmitted in foods (hepatitis A), which can produce toxins that remain in the food even after the biological source has been destroyed. Other contaminants, such as pesticide residues or heavy metals, are usually well controlled in modern food supplies, but can be significant hazards in particular localities. Naturally occurring toxic constituents are usually present in doses that are too small to produce harmful effects when foods are eaten normally, except in the cases of atypical consumers who may be sensitive to individual ingredients, or where food containing natural toxins is not prepared properly (e.g. cassava must be properly prepared to avoid cyanide poisoning). However, the chemical acrylamide, which is the reaction product of cooking starchy foods, such as potatoes, at high temperatures in the presence of the amino acid asparagine (called the Maillard reaction), is considered to be a genotoxic carcinogen in animals and thus a human health risk by the WHO Joint Expert Committee on Food Additives (JECFA) and the European Food Safety Authority (EFSA). Efforts are being made worldwide to reduce our exposure to this cooking by-product by avoiding high-temperature frying and limiting these types of starchy foods in the diet. Food additives or novel foods are generally the least dangerous hazards because their toxicology is well studied and the conditions of use are tightly controlled. Table 25.2 summarizes the types of hazardous substances that may be present in food. It should be noted that with all of the chemicals listed in Table 25.2, even for the very toxic chemicals, the general rule of toxicology is that 'the dose makes the poison'. Every chemical presents an intrinsic hazard, but the risk is always the hazard × exposure.

The US Food and Drug Administration (FDA) has ranked the relative importance of health hazards associated with food in the following descending order of seriousness:

1. Microbiological contamination
2. Inappropriate eating habits
3. Environmental contamination
4. Natural toxic constituents
5. Pesticide residues
6. Food additives

This list is very different from that found in public opinion polls, which show that most people rate pesticide residues and food additives as some of their major concerns about the safety of the food supply.

Table 25.2 **Potential hazards in foods**

| Hazards | Examples |
|---|---|
| **Microbial contamination** | |
| Pathogenic bacteria | Toxins from *Clostridium botulinum* |
| Mycotoxins | Aflatoxin from mould on peanuts |
| Viruses | Hepatitis A in various frozen foods |
| **Environmental contamination** | |
| Heavy metals and minerals | Arsenic and mercury in fish |
| Criminal adulteration | Aniline in olive oil; melamine in milk |
| Packaging migration | BPA and phthalates from plastics |
| Industrial pollution | PCB and dioxins, radioactive fallout |
| Changes during cooking or processing | Carcinogens produced in burnt meat<br>Acrylamide in high-temperature cooked starchy foods |
| **Natural toxins** | |
| Inherent toxins | Cyanide in cassava |
| Produced by abnormal conditions | Ciguatera poisoning from fish |
| Enzyme inhibitors | Protease inhibitors in legumes |
| Antivitamins | Avidin in raw egg white |
| Mineral-binding agents | Goitrogens in *Brassica* vegetables |
| **Agricultural residues** | |
| Pesticides, herbicides, fungicides | DDT, dieldrin, carbendazim |
| Hormones | Bovine somatotrophin |
| **Intentional food additives** | |
| Artificial sweeteners | Cyclamate, neotame, advantame |
| Preservatives | Sodium nitrite |

BPA, bisphenol A; DDT, dichlorodiphenyltrichloroethane; PCB, polychlorinated biphenyl.

# 25.2 Microbial contamination

## 25.2.1 Pathogenic bacteria

Outbreaks of acute gastroenteritis caused by microbial pathogens are usually called food poisoning. They can be caused by food-borne *intoxication* (where microbes in food produce a toxin that produces the symptom) or food-borne *infection* (where the symptoms are caused by the activity of live bacterial cells multiplying in the gastrointestinal system). Table 25.3 lists the most common bacterial causes of food poisoning, in order of the rapidity of onset of symptoms. In general, the intoxications have a more rapid onset.

The most important pathogens are *Clostridium botulinum*, *Staphylococcus aureus*, *Salmonella* species, and *Clostridium perfringens*. The last three organisms account for about 70–80% of all reported outbreaks of food-borne illness, but there are also many others, as well as some viral and protozoan agents. For example, recent food-borne viral infec-

**Table 25.3 Common bacterial and virological food poisoning organisms**

| Organism | Symptoms | Time after food | Typical food sources |
|---|---|---|---|
| **Toxins** | | | |
| *Staphylococcus aureus* | Vomiting, diarrhoea, abdominal pain | 1–6 hours (mean 2–3 h) | Custard and cream-filled baked goods, cold meats |
| *Clostridium perfringens* | Diarrhoea and severe pain, nausea | 8–24 hours (mean 8–15 h) | Meat products incompletely cooked or reheated |
| *Bacillus cereus* | Nausea, vomiting<br>Abdominal pain, watery diarrhoea | 1–5 hours<br>6–16 hours (mean 10–12 h) | Rice dishes, vegetables, sauces, puddings |
| *Clostridium botulinum* | Dry mouth, difficulty swallowing and speaking, double vision, difficulty breathing. Often fatal | 2 h–8 days (mean 12–36 h) | Home-canned foods (usually meat and vegetables) and inadequately processed smoked meats |
| **Infection** | | | |
| *Vibrio parahaemolyticus* | Diarrhoea, abdominal cramp, nausea, headache, vomiting | 4–96 hours (mean 12 h) | Fish, crustaceans |
| *Salmonella* spp. | Diarrhoea, fever, nausea, vomiting | 8–72 hours (mean 12–36 h) | Undercooked poultry, reheated food, cream-filled pastries |
| *Yersinia enterocolytica* | Fever, abdominal pain, diarrhoea | 24–36 hours | Raw and cooked pork and beef |
| *Escherichia coli* | Fever, cramps, nausea, diarrhoea | 8–44 hours (mean 26 h) | Faecal contamination of food or water |
| *Shigella* spp. | Diarrhoea, bloody stools with mucus, fever | 1–7 days (mean 1–3 d) | Faecal contamination of food |
| *Campylobacter jejuni* | Fever, abdominal pain, diarrhoea | 1–10 days (mean 2–5 d) | Raw milk, poultry, eggs, meat |
| *Listeria monocytogenes* | Septic abortion, septicaemia, meningitis, encephalitis. Often fatal | 1–7 weeks | Milk and dairy products, raw meat, poultry and eggs, vegetables and salads, seafood |
| *Hepatitis A* | Liver toxicity | 30–60 days | Frozen foods, processed foods |

tions from hepatitis A, and serious, life-threatening bacterial infections from *E. coli* 0157 and 0:104:H4 are increasing. The four most frequently identified factors contributing to food poisoning incidents are: improper cooling of food, lapses of 12 hours or more between cooking and eating, contamination by food handlers, and contaminated raw foods or ingredients.

The reported incidence and cost of food-borne illness in most countries is increasing, although it is difficult to measure this exactly. It is estimated that less than 1% of cases are captured in existing notification schemes. Some of the reasons for increasing rates of food-borne illness are: new and emerging pathogens, changes in the food supply (including more intensive animal husbandry and longer shelf-life fresh-chilled products), ageing populations, and a greater proportion of food eaten away from home. Around 60–80% of food-borne illness arises from the food service industry.

## 25.2.2 Control of food poisoning

The trend in all countries today is to require more formal training of all food handlers and the development of food safety plans wherever food is prepared and served to the public, based on the principles of Hazard Analysis of Critical Control Points (HACCP).

HACCP is a preventative approach to quality control, used worldwide in all segments of food production, from primary production to food manufacture and food service settings. It is based on seven principles:

- identify all potential *hazards* at each step in the food chain and possible preventative actions
- determine the *critical* points in the operation where the hazards must be controlled
- establish *limits* at each critical control point: examples of control procedures are washing hands, sanitizing food preparation surfaces and tools, cooking food to a specific temperature, and maximum food storage times
- set up procedures to *monitor* each critical control point
- plan the *corrective actions* to be taken if a critical limit is exceeded
- establish a *recording system* to document performance of the process
- *verify* that the HACCP process is working.

Table 25.4 outlines an example of some parts of a HACCP plan for a commercial food product sold as ready-to-eat.

### 25.2.3 Mycotoxins

Moulds, or fungi, are capable of producing a wide variety of chemicals that are biologically active. Humans have used some of these as effective antibiotics, but there are also a number of diseases resulting from accidental exposure to fungal products that contaminate food. Some examples are as follows.

**Aflatoxins** These are a group of highly toxic and carcinogenic compounds from the common

**Table 25.4 An example of six steps from a HAACP plan for chilled chicken salad**

| Step | 1. Growing and harvesting | 2. Raw material processing | 3. Supply storage temperatures | 4. Ingredient assembly | 5. Bagging | 6. Labelling |
|---|---|---|---|---|---|---|
| Hazard | Chemicals, antibiotics | Chemical, microbiological, virological | Microbiological, virological | Microbiological, virological | Microbiological, virological | Incorrect dates, traceability |
| Control | Raw material specifications | Certified supplier | Raw material specifications | Temperature control specs | Correct seal settings | Legible, correct dates and codes |
| Limit | Regulatory approved residues | Free of pathogens and foreign material | Chicken < −12°C, vegetables < 4°C | Food < 4°C | Upper tolerance limit on sealer | Use proper labels |
| Monitoring | Certificate of compliance | Monitor supplier HACCP programme | Check cool room records daily | Check temperature once per shift | Check setting every 15 min | Each batch at changeover |
| Action if limit exceeded | Reject lot | Reject as supplier | Investigate time/ temperature abuse | Report to supervisor | Examine all packages | Destroy incorrect labels |
| Responsibility | Receiving operator | Purchaser | Store person | Cook | Seal inspector | Packer |

*Source:* Adapted from Microbiology and Food Safety Committee of the National Food Processors Association (1993) HACCP implementation: a generic model for chilled foods. *J Food Prot*, **56**, 1077.

*Aspergillus* fungus species. They are heat stable and survive most forms of food processing. Aflatoxin contamination can occur whenever environmental conditions are suitable for mould growth, but the problem is more common in tropical and semitropical regions. Aflatoxins were first recognized in the 1960s in peanuts. On a worldwide basis, maize is the most important food contaminated with aflatoxin.

Patulin Patulin is an antibiotic that is produced by the mould *Penicillium caviforme*. It has been implicated as a possible carcinogen from one study in rats, although other studies have not confirmed this. Patulin is primarily associated with the apple-rotting fungus and so apple juices and some baked goods with fruit can contain patulin.

Fumonisins Fumonisins (e.g. deoxynivalenol (DON)) are carcinogenic mycotoxins from the *Fusarium* fungus associated with maize. These were first characterized in 1988 and are known to be potent inhibitors of sphingolipid synthesis. Ingestion of fumonisin-affected corn has been shown to be carcinogenic in rats. In 1990 it was reported that use of mouldy corn with high levels of fumonisins to make beer in the Transkei of South Africa was associated with a very high incidence of oesophageal cancer.

## 25.2.4 New food-borne diseases

Three of the most serious food pathogens today (*Campylobacter*, *Listeria*, and enterohaemorrhagic *Escherichia coli*) were unrecognized as causes of illness 50 years ago. Some of the more important new organisms are described here.

Campylobacter jejuni *Campylobacter jejuni* was a well-known bacterium in veterinary medicine before it was identified as a human pathogen in 1973. It is now recognized as one of the most important causes of gastroenteritis in humans, of similar importance to *Salmonella*. It is present in the flesh of cattle, sheep, pigs, and poultry, and can be introduced wherever raw meat is handled.

Listeria monocytogenes *Listeria monocytogenes* is a bacterium widely distributed in nature, but is unusual in that it grows at refrigeration temperatures (down to 0°C). Listeriosis can cause abortions, as well as death in the elderly and those with compromised immune systems, such as people with AIDS. Listeriosis outbreaks have been attributed to RTE food products such as pâté, unacidified jellied pork tongue (i.e. in aspic), rillettes, frankfurters, certain deli-meats, chicken wraps, cheese made from either raw or pasteurized milk, pasteurized milk (including chocolate milk), butter, frozen ice cream cake, whipping cream, coleslaw, fruit salad, RTE fish products such as smoked mussels, gravlax (aka. gravad) and cold-smoked trout, imitation crab meat, shrimp, pre-packaged sandwiches, as well as rice and corn salads. In Canada, there were two large outbreaks involving 57 and 40 confirmed cases, respectively (Public Health Agency of Canada, 2009; Public Health Agency of Canada, 2010). France, the United Kingdom, and several other European countries have also reported increases in the incidence of listeriosis over the last several years. In these countries, the increase has been predominantly driven by an increased incidence in patients >60 years of age.

Escherichia coli *Escherichia coli* 0157:H7 and H:104 are bacteria that can damage the cells of the colon, leading to bloody diarrhoea and abdominal cramps. Raw or undercooked hamburger meat was a major vehicle of transmission in a number of well-publicized outbreaks in the USA in 1993, contaminated metwurst was responsible for a major outbreak of illness in Australia in 1995, and a type of sprouts caused an outbreak in Germany in 2011.

Salmonella typhimurium *Salmonella typhimurium* is a multidrug-resistant strain that became a major pathogen in the UK in the 1990s. As well as being highly virulent, it can survive at low pH and be infectious in very low numbers.

Norwalk virus Norwalk virus is found in the faeces of humans, and illness is caused by poor personal hygiene among infected food handlers. Symptoms include nausea, vomiting, diarrhoea, abdominal pain, and fever. Because it is a virus, it does not reproduce in food, but remains active until the food is eaten.

Hepatitis A and E Hepatitis A has been associated with several food-borne disease outbreaks around the world, the cause being attributed to bad hygiene practices by food industry workers. There have been several serious outbreaks in Australia from consumption of infected frozen berries purchased in supermarkets (see www.foodstandards.gov.au). Reports have emerged regarding hepatitis E infection in humans from eating raw or undercooked pig meat or pig liver products, but the evidence is limited. A 2017 report by EFSA concluded that raw and undercooked pork meat and pork liver were the most common cause of hepatitis E virus in the EU. The report estimates that less than 10% of pigs carry infectious virus particles at the time of slaughter.

'Mad cow disease' Mad cow disease (or bovine spongiform encephalopathy (BSE)) is a slowly progressive and ultimately fatal neurological disorder of adult cattle that results from infection by a unique transmission agent called a prion. Prions seem to be modified forms of normal cell surface proteins. BSE was first confirmed in Britain in 1986, but has now spread to cattle in other countries of Europe, Japan, and North America. The same infective agent is also responsible for variant Creutzfeldt–Jakob disease (vCJD), a fatal disease of humans, mostly affecting young adults. By October 2009, it had killed 166 people in Britain and 44 elsewhere, and the number was expected to rise because of the disease's long incubation period. However, it is worth noting here that, thankfully, by 2015, this increase in numbers of vCJD cases did not occur. Three principal controls have been put in place to keep infected meat out of the food chain: banning slaughter of beef aged over 30 months (before the age at which BSE typically develops), removal of parts of the body with the highest levels of infection (e.g. nervous and bone tissue), and a ban on feeding meat and bonemeal to any farmed livestock. Milk and gelatin products from beef do not appear to be affected.

COVID-19 (SARS-CoV-2) When the Covid-19 pandemic began in late 2020, not much was known about SARS-CoV-2 (the coronavirus) and its survival in food, on various materials, and on surfaces. Since then, several food safety agencies have assessed the risk of potentially acquiring the virus from contaminated food or food packaging. The consensus is that, currently, there is no evidence that it's a food safety risk.

## 25.3 Environmental contamination

### 25.3.1 Heavy metals and minerals

Selenium Selenium is one of the toxic essential trace elements. The level of selenium in foods usually reflects the levels in the soil and in a few high-selenium areas, such as North Dakota and parts of China, excessive selenium intake has been associated with gastrointestinal disturbances and skin discolouration.

Mercury Fish and shellfish concentrate mercury in their bodies, often in the form of methylmercury, a highly toxic organomercury compound. Fish products have been shown to contain varying amounts of heavy metals, particularly mercury and fat-soluble pollutants from water pollution. Species of fish that are long-lived and high in the food chain, such as marlin, tuna, shark, swordfish, king mackerel, and tilefish, contain higher concentrations of mercury than others (*Mercury levels in commercial fish and shellfish (1990–2010)*; archived 2015-05-03 at the Wayback Machine US Food and Drug Administration).

Mercury is known to bioaccumulate in humans, so bioaccumulation in seafood carries over into human populations, where it can result in mercury poisoning. Mercury is dangerous to both natural ecosystems and humans because it is a metal known to be highly toxic, especially due to its ability to damage the central nervous system (Park et al., 2008).

In human-controlled ecosystems of fish, usually done for market production of wanted seafood species, mercury clearly rises through the food chain via fish consuming small plankton, as well as through non-food sources such as underwater sediment (Cheng, 2011).

Fish can contain 10–1500 mg/kg of organic mercury, and even higher levels when mercury wastes are released into lake waters. Serious poisonings from mercury in fish have occurred in Japan, the most famous being that in Minamata Bay (from 1953 to 1960). Another example of widespread mercury intoxication occurred in Iraq in 1971–72 as a result of bread made from wheat treated with mercury-based pesticides. Most countries have now established maximum permitted levels on mercury in fish in the range of 0.4–1.0 mg/kg. There has been much debate in relation to getting the right balance between exposure to mercury levels from eating large fish and the health benefits of a diet high in fish consumption. To this end, most food regulators around the world have specific advice on their websites in relation to the benefits and risks of fish consumption, especially with respect to species high in mercury, and also warnings about consumption for pregnant women and young children.

Cadmium Cadmium is a toxic element that accumulates in biological systems. Chronic exposure at excessive levels can lead to irreversible kidney failure. Plants readily take up cadmium from the soil, and there has been a slow increase in the cadmium levels in soils due to the use of phosphate fertilizers and the effect of air and water pollution. The average food-based cadmium intake is now approximately 10–50 μg per day, which is approaching the provisional tolerable weekly intake. Measures to control cadmium contamination include controls on waste disposal and developing new crops that accumulate less cadmium, and avoiding types of soils and crops that accumulate cadmium (National Cadmium Management Strategies).

Other heavy metals such as arsenic (in seafood and rice) and lead (in seafood) are also often contaminants of common and similar concern due to their presence in foods and serious human health concerns at low doses, especially for lead, for which the provisional tolerable daily intake (PTDI) was recently withdrawn by JECFA (https://apps.who.int/food-additives-contaminants-jecfa-database/chemical.aspx?chemID=3511; accessed 16 August 2021).

### 25.3.2 Criminal adulteration

Modern food regulations began in the nineteenth century when there were widespread examples of adulteration of foods to increase profits. Milk was diluted with water, cocoa with sawdust, and butter with borax. Today, standards in the food industry are much higher and risks from illegal adulteration are rare. However, some recent trends, known as economically motivated adulteration (EMA), or commonly known as 'food fraud', are a cause of concern for both consumers and governments alike, such as the addition of often banned or dangerous medicines to foods to achieve weight loss (the banned blood pressure drug, sibutramine, in various foods) or improved sexual function (Viagra in coffee), as well as adulteration of milk, oils, herbs, fruit juices, and meat (e.g. European horse-meat substitution). Although this last example (horsemeat substitution) was not a health concern, it does indicate the extent of this potentially serious human health problem.

There are some notorious instances. In Spain in 1981, there was an outbreak of an apparently new disease characterized by fever, rashes, and respiratory problems. Many thousands were hospitalized and over 100 people died. The agent responsible was identified as cooking oil that had been fraudulently sold as pure olive oil, but in fact was mostly rapeseed oil intended for industrial uses, which was contaminated with aniline. In China in 2008, at least six children died of acute kidney failure and nearly 300,000 more fell ill after consuming tainted infant formula. Melamine, a synthetic nitrogenous product found in many industrial goods, was found to have been illegally added to milk-based foods to make them appear higher in protein than they really were.

### 25.3.3 Packaging migration

The materials used to package food can sometimes contaminate the food itself. At one time, the lead used in the solder of metal cans was a significant source of contamination of infant formulas, but this problem has been eliminated by the introduction of non-soldered cans. Bisphenol A (BPA) is an industrial chemical used as the starting material for the production of polycarbonate plastics and synthetic resins. BPA is found in containers that come into contact with foodstuffs such as drinking vessels, baby bottles, and the internal coating on cans for tinned food. BPA belongs to a group of substances that can act in a similar way to some hormones, and studies in laboratory animals originally suggested that low levels may have an effect on the reproductive system. In 2010, the FDA released a report on the safety of BPA, which raised concerns about its potential effects on the brain, behaviour, and prostate gland in foetuses, infants, and young children. Recent exhaustive studies from the US FDA and the EFSA, however, have clearly shown that BPA is a safe chemical at the low levels (parts per billion; parts per trillion) that humans, including small infants and children, are exposed to in the diet. Despite this, subsequently manufacturers of baby bottles around the world have agreed to move to BPA-free bottles as soon as possible.

### 25.3.4 Industrial pollution

Throughout the industrial era, many potentially hazardous substances have been released into the environment and are now widely distributed in the food chain. Among the most important are the polychlorinated biphenyls (PCBs). PCB is a generic term for a wide range of highly stable derivatives of biphenyl that have been used in a vast number of products, including plastics, paints, and lubricants. Although manufacture has now ceased, their stability and lipid solubility has meant that they accumulate in fatty tissue and they have become widespread, particularly in seafood. They can be found at low levels now even in human milk. The health effects of PCBs are not well established, although they are thought to be mild carcinogens. In one incident in Japan in 1978, when rice oil was contaminated with 2000–3000 ppm PCB, growth retardation occurred in young children and the foetuses of exposed mothers.

### 25.3.5 Radioactive fallout

The most important dangerous radioisotopes in fallout are strontium-90 and caesium-137, with half-lives of 28 and 30 years, respectively. Strontium is absorbed and metabolized like calcium and stored in bones. Because it is concentrated in milk, it is particularly dangerous for infants and children. Since the Nuclear Test Ban Treaty of 1963, the level of radioactive contamination from atmospheric dust has markedly declined, but accidental exposure can still occur, such as that after the Chernobyl and Fukishima disasters, and lead to dangerous food contamination over widespread areas.

### 25.3.6 Changes during cooking or processing

Food is frequently exposed to high temperatures during cooking. In roasting and frying, localized areas of food may be subjected to temperatures that lead to carbonization and under these circumstances any organic substance is likely to give rise to carcinogens. The major compounds are polycyclic aromatic hydrocarbons (PAH), produced mainly by burning of fats, and heterocyclic amines (HCA), produced from amino acids. Char-broiling or barbecuing is particularly likely to lead to carcinogen formation.

Acrylamide In 2002, the Swedish National Food Authority announced that acrylamide could be found in starch-containing foods cooked at high temperatures, such as fried or roasted potato products, and cereal-based products, including sweet biscuits and toasted bread. In 2010, a World Health Organization (WHO) expert committee (JECFA) and more recently EFSA (2015) determined that there is evidence that acrylamide can cause cancer in laboratory animals and, while there is currently no scientific evidence that links acrylamide with cancer risk in humans, all food regulatory agencies around the world are encouraging new technological strategies aimed at reducing its formation.

Irradiation Irradiation can be used to sterilize foods, control microbial spoilage, eradicate insect infestations, and inhibit undesired sprouting. Despite the great potential of the technology, there has been substantial opposition from consumer groups concerned about the process producing toxic chemicals in foods. Extensive studies have shown that the products formed are no different from those produced in normal cooking, and over 1300 studies have consistently found no adverse effects from feeding irradiated food to animals or humans. Food irradiation is approved by the WHO and currently more than 30 countries allow some form of use.

## 25.4 Natural toxins

Many plant species contain hazardous levels of toxic constituents. Intoxications from poisonous plants usually result from the misidentification of plants by individuals harvesting their own foods, or inadequate cooking processes, but many ordinary foods also contain potential toxicants at less harmful levels.

### 25.4.1 Inherent natural toxins

There are many examples of potentially dangerous toxins in natural food products: cyanogenic glycosides in plants such as almond kernels, cassava, and sorghum, alkaloids in herbal teas and comfrey, and lathyrus toxin in chickpeas. In Japan the puffer fish, which contains a potentially fatal neurotoxin, is considered a delicacy and is consumed to produce a tingling sensation. However, natural toxicants are a generally accepted hazard because the foods that contain them have been eaten in traditional diets for many generations. We are protected from their harmful effects in three ways: avoidance, removal, and detoxification.

First, traditional knowledge has been passed down about which foods are safe and which are not. Thus, we know it is safe to eat certain mushrooms and not others. Second, traditional preparation methods have evolved to reduce harmful effects. Specialist chefs prepare puffer fish to remove the parts with the highest toxin concentration. People in South America and Africa use complex chopping and washing procedures in their preparation of cassava, which removes much of the cyanide naturally found in the raw product. Third, the body has numerous detoxification systems, mainly enzymes in the liver, to deal with toxins we do ingest. So we can still happily eat nutmeg and sassafras, even though both contain the naturally occurring carcinogen safrole.

### 25.4.2 Abnormal conditions of the animal or plant used for food

Some foods only become hazardous during particular conditions of growth or storage.

**Ciguatera poisoning** This is serious human intoxication, caused by eating contaminated fish, causing gastrointestinal disorders, neurological problems, and, in severe cases, death. There are over 400 species of fish that may become ciguatoxic, but almost all of the fatal cases are attributable to barracuda. The poisoning is particularly insidious because it occurs in tropical and subtropical fish that are normally safe to eat, but not so when they have been feeding on certain dinoflagellates that produce toxins that accumulate in the flesh.

**Paralytic shellfish poisoning** It has been known for many centuries that shellfish can occasionally become toxic. Symptoms include numbness of the lips and fingertips and ascending paralysis, which can lead to death within 24 hours. The poisoning, which primarily affects mussels and clams, occurs when dinoflagellates undergo periods of rapid growth ('blooms' or 'red tides') in areas where the shellfish grow. The toxin cannot be removed by washing or destroyed by heat.

**Glycoalkaloids** In potatoes, solanine is one of a range of heat-stable glycoalkaloid compounds found in the green parts of the potato plant that are toxic above concentrations of 20 mg/100 g. In normal peeled potatoes there is about 7 mg solanine/100 g. Solanine synthesis can be induced by exposing the tubers to light and also by simple mechanical injury. In very green potatoes, the levels can reach up to 100 mg/100 g. These glycoalkaloids possess anticholinesterase activity, which can produce gastrointestinal and neurological disorders, and deaths have occasionally been reported from consumption of excessive amounts of green potatoes.

### 25.4.3 Enzyme inhibitors

**Protease inhibitors** Substances that inhibit digestive enzymes are widespread in many legume species, and trypsin inhibitors are found in oats and maize, as well as Brussels sprouts, onion, and beetroot. These inhibitors are proteins and therefore are denatured and inactivated by cooking. Thus, for humans, these substances are not a problem, although feeding raw legumes to animals can result in pancreatic enlargement.

### 25.4.4 Antivitamins

One of the best known antivitamins is the biotin-binding protein, avidin, in raw egg white. Biotin deficiency induced by eating raw egg white is rare because biotin is well provided in most human diets. The few cases that have been reported involved abnormally large amounts of raw egg white, so the occasional raw egg is perfectly safe. Avidin is inactivated when heated. Other antivitamins, such as the pyridoxine antagonist amino-D-proline in flax seeds and a tocopherol oxidase in raw soybeans, are only of importance in animal feeding.

### 25.4.5 Mineral-binding agents

**Goitrogens** There are a number of glucosinolate and thiocyanate compounds in foods that interfere with normal utilization of iodine by the thyroid gland and can result in goitres. Goitrogens are widely distributed in cruciferous vegetables such as cabbage, Brussels sprouts, and broccoli. The average intake of glucosinolates from vegetables in Great Britain is 76 mg per day and clinical studies have found that intakes of 100–400 mg per day may reduce the uptake of iodine by the thyroid. There is no evidence that normal consumption of these foods by humans is harmful, but it is possible that eating large amounts of *Brassica* plants might contribute to a higher incidence of goitre in areas where dietary iodine intake is low.

**Phytate** In wholemeal cereals, phytate can bind minerals and make them less available for absorption. In leavened bread, phytases in the yeast break down the phytate, but in some parts of the Middle East, where unleavened bread is a dietary staple, phytate has been reported to be the cause of zinc deficiency.

**Oxalate** Certain plants, including rhubarb, spinach, beetroot, and tea, contain relatively high levels of oxalate. Oxalate can combine with calcium to form an insoluble complex in the gut that is poorly absorbed, and high intakes can lower plasma calcium levels. Kidney damage and convulsions can accompany oxalate poisoning. However, the average diet supplies only 70–150 mg oxalate per day, which could theoretically bind 30–70 mg calcium. Since calcium intakes are usually 10 times this amount, food oxalates do not normally have any detrimental effect on mineral balance.

**Tannins (polyphenols)** These are present in tea, coffee, and cocoa, as well as broad beans. Tannins inhibit the absorption of iron, and in Egypt, in children with low iron intakes, regular consumption of stewed beans has been associated with anaemia. High levels of tea consumption may contribute to low iron status in people with marginal iron intakes.

## 25.5 Agricultural residues

### 25.5.1 Pesticides

The most common agricultural chemicals found in foods are pesticides, albeit at very low levels. The chlorinated organic pesticides (such as dichlorodiphenyltrichloroethane (DDT) and chlordane) were among the first modern pesticides to be used. In general, they have low toxicity to mammals and are highly toxic to insects. However, they are very stable compounds, which persist in soils, and they are stored in the fat tissue of animals. Because of concern about their effect on the reproduction of certain birds and possible carcinogenic activity, use of these compounds has been restricted. Surveys of foods show that the levels of organochlorine compounds have been declining in recent years. Alternative insecticides now in use, such as organophosphates, do not accumulate in the environment. No food poisonings have ever been attributed to the proper use of insecticides on foods, but in 1997 there were 60 cases of food poisoning in India attributed to indiscriminate organophosphate spraying in a kitchen.

### 25.5.2 Fungicides and herbicides

Most fungicides and herbicides show very selective toxicity to their target plants and therefore present very little hazard to humans. In addition, most do not accumulate in the environment.

### 25.5.3 Hormones and growth promotors

The use of hormones, such as bovine somatotrophin (BST), to improve yields of meat and milk has been controversial in many countries. Although low levels of BST can be detected in the milk of treated cows, the hormones in humans are digested and inactivated in the stomach when consumed in food. The US FDA approved the commercial use of BST in 1993 and later reviews by Canadian authorities, the Codex Alimentarius, and FAO/WHO JECFA have agreed that there are no health risks to humans. However, in several countries, BST use is not permitted on animal welfare grounds.

Technologies that increase the efficiency and sustainability of food animal production to provide meat for a growing population are necessary and must be used in a manner consistent with good veterinary practices, approved labelled use, and environmental stewardship. Compounds that bind to beta-adrenergic receptors (β-AR), termed beta-adrenergic receptor ligands (β-ligands), are one such technology, and have been in use globally for many years. Though all β-ligands share some similarities in structure and function, the significance of their structural and pharmacological differences is sometimes overlooked. Structural variations in these molecules can affect absorption, distribution, metabolism, and excretion as well as cause substantial differences in biological and metabolic effects. Specifically for use in cattle production, several β-ligands are available for use. Ractopamine and zilpaterol are beta-adrenergic agonists (β-AA) approved to increase weight gain, feed efficiency, and carcass leanness in cattle. They both bind to and activate β1 and β2-AR. Lubabegron is a new selective beta-adrenergic modulator (SβM) with unique structural and functional features. Lubabegron displays antagonistic behaviour at the β1 and β2-AR, but agonistic behaviour at the β3-AR. Lubabegron is approved for use in cattle to reduce ammonia emissions per unit of live or carcass weight. Additionally, lubabegron can withstand prolonged use, as the β3-AR lacks structural features needed for desensitization. Due to these unique features of lubabegron, this new β-ligand provides an additional option in cattle production and the maintenance of a sustainable, safe food supply (Dilger et al., 2021).

## 25.6 Intentional food additives

### 25.6.1 Approval process for food additives

Each country has its own legislation to control the approval of additives in foods, but most follow the same general principles that are used by the two main international bodies of experts organized by WHO and the Food and Agriculture Organization (FAO): the Joint Expert Committee on Food Additives (JECFA), and the Codex Alimentarius Committee on Food Additives and Contaminants (CCFA). The aim of the evaluation of a food additive is to establish an acceptable daily intake (ADI). The ADI is usually expressed in mg/kg of body weight and is defined as the amount of a chemical that might be ingested daily, even over a lifetime, without appreciable risk to the consumer. The evaluation process consists of a number of steps:

1. Toxicity testing is carried out in experimental animals—usually mice and rats, but other species may also be employed. Three types of testing are performed: (a) acute toxicity studies at high doses to determine the range of possible toxic effects of the chemical, (b) short-term feeding trials at various doses, and (c) long-term studies of 2 years or more to examine the effect of exposures over several generations.
2. From the feeding trials, the level of additive at which observed health effects do not appear in the animals is determined. This is called the 'no observed effect level' (NOEL).
3. The lowest NOEL is divided by a safety factor to derive an exposure level that is regarded as acceptable for humans, the ADI. Most commonly, a safety factor of 100 times is used, but for some substances factors of up to 1000 have been used. This safety factor allows for possible differences in susceptibility between experimental animals and humans and also the differences in sensitivity of individual people.

Not all additives have been evaluated for safety using modern testing procedures. Some have been used for many years without apparent harm, and in the USA, ingredients not evaluated by prescribed testing procedures can be classified as generally recognized as safe (GRAS). This list includes common ingredients such as salt, sugar, seasonings, and many food flavourings.

While the 100-fold safety factor is accepted for most additives, in the USA the Delaney Clause prohibits the use in *any* amount of substances known to cause cancer in animals or humans. When the bill was introduced in 1958, chemicals could be detected down to 100 parts per billion; anything less was considered zero. Improved analytical techniques can now detect substances at parts per trillion and there might be only trivial risks from such minute quantities. The FDA has now changed the interpretation of the clause so that if a food additive increases the chance of developing cancer over a lifetime by less than one case per million of cancer, the threat is considered too small to be of concern.

Ames et al. (1987) ranked the level of carcinogenic risk associated with a variety of chemicals we may be commonly exposed to. The Human Exposure/Rodent Potency Index (HERP) expresses the typical human intakes as a percentage of the dose required to produce tumours in 50% of rodents. The values in Table 25.5 show that the risk from the alcohol in a glass of wine is almost 100 times higher than that from the saccharin in a can of diet cola, and more than 10,000 times the hazard from the residues of the pesticide ethylene dibromide. That the risks from wine appear more acceptable to most consumers seems to relate to the fact that benefit is easily perceived, that wine is seen as 'natural', and because the risk is voluntary. Although the risks from other additives and contaminants may be far smaller, they arouse suspicion because they are risks that people generally cannot control (see Table 25.5).

### 25.6.2 Artificial sweeteners

**Saccharin** Saccharin is one of the oldest artificial sweeteners, having been used in foods since the twentieth century. Studies in rats have linked very high doses (7.5% of the diet by weight) of saccharin with bladder cancer and because of this there have been attempts to ban its use in human foods. However, these initial studies in rats were later shown to have been performed in a species that is prone to these types of cancers, leading to lifting of bans on its use, and at lower doses, such as 1%, no adverse effects are found and large epidemiological studies of people with diabetes who have had lifetime exposure to saccharin have found no increased incidence of cancer in humans.

**Cyclamate** Dietary cyclamate appears to promote bladder cancer and induce testicular atrophy in rats, although carcinogenicity testing in mice, dogs, and

**Table 25.5 Rankings of possible carcinogenic hazards**

| Daily human exposure | Carcinogen and dose per 70 kg person | Index of possible hazard (HERP, %) |
|---|---|---|
| **Natural dietary toxins** | | |
| Wine (250 mL) | Ethyl alcohol, 30 mL | 4.7 |
| Basil (1 g of dried leaf) | Estragole, 3.8 mg | 0.1 |
| Peanut butter (32 g, one sandwich) | Aflatoxin, 64 ng | 0.03 |
| Cooked bacon (100 g) | Dimethylnitrosamine, 0.3 μg | 0.003 |
| **Food additives** | | |
| Diet cola (1 can) | Saccharin, 95 mg | 0.06 |
| **Pesticides** | | |
| DDE/DDT (daily diet intake) | DDE, 2.2 μg | 0.0003 |
| EDB (daily diet intake) | EDB, 0.42 μg | 0.0004 |

DDE, dichlorodiphenyldichloroethylene; DDT, dichlorodiphenyltrichloroethane; EDB, ethylene dibromide.

*Adapted from:* Ames, B.N., Magaw, R., and Gold, L.S. (1987) Ranking of possible carcinogenic hazards. *Science*, **236**, 271–80.

primates have all been negative. The US FDA banned the food use of cyclamate in 1969, but in over 50 other countries it is still a permitted sweetener, and there is no good evidence from mutagenicity testing or epidemiological studies that it is a health risk to humans.

**Aspartame** Aspartame is a dipeptide of two amino acids, phenylalanine and aspartic acid. Aspartame is metabolized to phenylalanine and therefore carries a risk for people with phenylketonuria, but for the normal population it is an extremely safe sweetener that is digested like any other protein.

**Stevia** More recently, over the last 10 years, a series of newer, intense sweeteners have been approved worldwide, including neotame, advantame, alitame, and stevia, which are much sweeter than cyclamate and aspartame, resulting in less being needed, and which exhibit fewer adverse effects in safety studies in animals and humans. The latter sweetener, stevia, is derived from a plant from South America, and is considered more attractive to consumers because it is seen as being a 'natural' product distinct from the artificial sweeteners.

### 25.6.3 Preservatives

Preservatives are used in foods as antioxidants and to prevent the growth of bacteria and fungi. Most pose no toxicological problems, but a few have generated some concerns.

**Sodium nitrite** Sodium nitrite is used as an antimicrobial preservative that is very effective in preventing the growth of *Clostridium botulinum*, as well as acting as a colour-fixing agent (to preserve the red colour) in cured meat products such as bacon and ham. Nitrite reacts with primary amides in foods to produce *N*-nitroso derivatives, many of which are carcinogenic. However, the risk to human health from dietary nitrite is difficult to assess. While food additive nitrites are significant, a substantial amount is also produced by bacterial reduction from naturally occurring nitrate in vegetables. In recent years, manufacturers have worked to reduce the levels of nitrite used in cured meats and have added agents such as ascorbic acid, which help to prevent the formation of nitrosamines in the stomach.

**Sulphur dioxide and its salts (sulphites)** These are commonly used as inhibitors of enzymic browning, dough conditioners, antimicrobials, and antioxidants. Although sulphites have been used for many centuries, with no adverse effect for most consumers, 1–2% of asthmatics are sensitive to sulphites and in those individuals the reaction can be fatal. Many food additives, such as sulphites, are known to cause food intolerances in some sensitive individuals, and these adverse effects can result in varying degrees of incapacitation depending on the individual (e.g. rashes, migraine headache).

### 25.6.4 Colours and flavours

All colours and flavours approved for use in foods are rigorously evaluated before being approved for use.

Red No. 2 (amaranth) In the early 1970s, data from Russian studies raised questions about Red No. 2's safety. The FDA Toxicology Advisory Committee evaluated numerous reports and decided that there was no evidence of a hazard, but concluded that feeding it at a high dosage results in a statistically significant increase in malignant tumours in female rats. The FDA ultimately decided to ban the colour, but it is still found in foods in Canada and Europe.

Tartrazine (E102) Food sensitivity (intolerance) to tartrazine can be experienced by a small number of individuals, but claims related to clinical problems such as asthma and hyperactivity are not well supported by scientific studies. Tartrazine is still a permitted additive, but its presence has to be declared in ingredient lists so sensitive individuals can avoid it.

Monosodium glutamate (MSG) The flavour enhancer MSG is a sodium salt of glutamic acid, one of the most common amino acids. It is present in virtually all foods and found in high levels in tomatoes, mushrooms, broccoli, peas, cheese, and soy sauce. 'Chinese restaurant syndrome' has been claimed to be caused by foods with a lot of added MSG, but most controlled studies have not demonstrated this effect.

Traditional methods for evaluating the safety of colours have not usually considered their potential behavioural effects. New research published in 2007 in the *Lancet*, using a mixture of six permitted colours (sunset yellow, tartrazine, carmoisine, ponceau, quinoline yellow, and allura red) at relatively high doses, concluded that there was limited evidence that these colours could affect the activity and attention of children in the general population. However, the European Food Safety Authority, FSANZ, and the US FDA, amongst others, have all concluded that uncertainties in the study meant that there was insufficient evidence to change current permissions for use of these colours.

## 25.7 Novel foods

Technology now allows the development of many new ingredients or whole foods that do not have a history of traditional use in the human food supply. Many of these novel foods have been developed to have improved nutritional quality. Recent examples include genetically modified foods, foods produced using new plant breeding techniques (such as new mutagenic techniques—oligo-directed mutagenesis, zinc finger nucleases, GM rootstock grafting techniques, gene silencing, CRISPR-Cas9), artificial fat substitutes for energy-reduced foods, new algal sources of omega-3 fatty acids, and phytosterols to reduce cholesterol. In addition, the advent of cultured meat (meat grown in culture in a laboratory or factory facility) and dairy-free milk and eggs are considered novel foods that may require some regulatory oversight.

### 25.7.1 Approval process for novel foods

There are significant practical difficulties in assessing the long-term safety of modified whole foods or ingredients. Unlike additives, which can be fed at very high doses to assess their toxic effects, it is not possible to feed large amounts of one single food to animals without making their diet nutritionally unbalanced. Animals also prefer a mixture of foods and are likely to refuse to eat if offered a single food in large amounts. These difficulties, and welfare concerns about the use of animal studies that were unlikely to result in meaningful information, led to the development of the concept of 'substantial equivalence', particularly for the assessment of genetically modified (GM) foods. This type of assessment does not quantify the safety or risk of a food, but aims to determine whether novel foods are as safe as traditional counterparts. For GM foods, the process involves assessment and comparison of a wide range of factors, including:

- source and nature of any new protein
- stability of any genetic changes
- potential toxicity of the new protein
- levels of naturally occurring and newly introduced allergens
- nutritional composition
- levels of antinutrients
- ability of the food to support normal growth and well-being
- potential unintended environmental consequences.

### 25.7.2 Genetically modified foods

Modern biotechnology now allows specific individual genes to be identified, copied, and transferred into other organisms in a much more direct and controlled way. For example, genes for the enzyme chymosin from beef have been inserted into yeast, and the GM chymosin from these organisms has now widely replaced natural rennet from animals in cheese making. Genetic modification can also allow individual genes to be switched on or off: the gene that controls fruit softening can be repressed to maintain a higher solids content in tomatoes designed for use in tomato paste.

In Australia, for example, all the cotton grown is GM, and this has resulted in at least a 90% decrease in the use of pesticides in the cotton industry. Uptake of GM technology has proceeded rapidly. The annual report on the worldwide commercial use of genetically modified plants is published by the agro-biotechnology agency ISAAA (International Service for the Acquisition of Agri-Biotech Applications). According to the report, 18 million farmers use GM plants worldwide. The United States had the largest area of genetically modified crops worldwide in 2019, at 71.5 million hectares, followed by Brazil with a little over 52.8 million hectares.

In 2009, 134 million hectares of GM crops were planted worldwide and 90% of all soy is now grown from GM varieties. However, by 2019 this figure for worldwide planting of GM crops had increased to 190.4 million hectares (see https://www.statista.com/statistics/271897/leading-countries-by-acreage-of-genetically-modified-crops/) (accessed 30 June 2021).

In terms of acreage, the most commonly genetically modified crops are soybeans, corn, cotton, and canola, as of 2019. In that year, 13.5% of all the cotton grown worldwide was genetically modified. The same was true for 48.2% of all soybean plants. There are many future uses planned that will bring more direct consumer benefits: oils with improved fatty acid profiles, rice with higher levels of vitamins (golden rice), nuts with lower levels of allergens, and potatoes that absorb less fat during frying.

Most countries have now established stringent approval processes for GM foods, including mandatory labelling to inform consumers when foods include GM-modified ingredients. Assessments to date have usually found GM foods to be as safe as their normal counterparts and there are likely to be increasing numbers of GM foods in the marketplace in the future. By 2015, there had been hundreds of approvals for GM foods across the world by respected food safety regulators, such as US FDA, Health Canada, FSANZ, and EFSA, and the overwhelming weight of scientific opinion is that GM foods are safe to consume beyond any reasonable doubt.

Newer types of genetic modification are now being used by the life sciences industry (use of small inhibitory RNA technology and others), which is presenting new benefits to the consumer, such as changes to the nutrient profiles of foods (increased amounts of healthy fatty acids in foods; omega-3 fatty acids expressed in terrestrial plants such as canola). It is also proving challenging for food regulatory agencies to keep abreast of this fast-moving technology.

It is worthy of specific mention because of its potential utility in safe food production. CRISPR-Cas9 system is a plant breeding innovation that uses site-directed nucleases to target and modify DNA with great accuracy (Harvard University (2015) CRISPR: A game-changing genetic engineeringtechnique).Seehttp://sitn.hms.harvard.edu/flash/2014/crispr-a-game-changing-genetic-engineering-technique/ (accessed 30 June 2021).

Developed in 2012 by scientists from the University of California, Berkeley, CRISPR-Cas9 has received a lot of attention in recent years due to its range of applications, including biological research, breeding and development of agricultural crops and animals, and human health applications. These include gene silencing, DNA-free CRISPR-Cas9 gene editing, homology-directed repair (HDR), and transient gene silencing or transcriptional repression (CRISPRi).

CRISPR, or Clustered Regularly Interspaced Short Palindromic Repeats, is an integral part of a bacterial defence system. It is also the basis of the CRISPR-Cas9 system. The CRISPR molecule is made up of short palindromic DNA sequences that are repeated along the molecule and are regularly spaced. Between these sequences are 'spacers', foreign DNA sequences from organisms that have previously attacked the bacteria. The CRISPR molecule also includes CRISPR-associated genes, or Cas genes. These encode proteins that unwind DNA, and cut DNA, called helicases and nucleases, respectively.

The CRISPR-Cas9 system can be applied to nearly every organism. Early studies using CRISPR-Cas9 for gene editing have focused on crops important for agriculture. It was realized early on that the system could be used in crops to improve traits, such as yield, plant architecture, plant aesthetics, and disease tolerance.

### 25.7.3 Fat substitutes

There are a number of fat substitutes now in use, including *Simplesse* (microcapsules of milk proteins or egg white), *Splendid* (derived from pectin), and *N-oil* (derived from tapioca). In the USA, *Olestra*, a mixture of heat-stable sugar polyesters that are not digested and yield no energy, has been controversial because it can reduce the absorption of fat-soluble vitamins. The FDA approved use of Olestra in a limited range of foods in 1996, but required addition of vitamins A, D, and K, as well as further monitoring of the health impacts and warning labelling that it may cause abdominal cramping and loose stools. In 2003, after a scientific review of several post-market studies, the FDA concluded that the warning statement was no longer warranted. Olestra is not yet approved in the UK, Europe, or Australasia.

### 25.7.4 Phytosterols

In many countries, plant sterols are now approved to be added to a range of foods to help lower blood cholesterol. They reduce the absorption of cholesterol from the gut, but have a side effect of also lowering absorption of carotenoids. A typical daily dose of 2–3 g per day can reduce serum β-carotene levels by 20–25%. Safety reviews have concluded that since there is no evidence of reduction in serum retinol levels, this effect is not a significant health concern and that advice to maintain adequate fruit and vegetable intakes can ensure adequate carotene intakes.

### 25.7.5 Nanotechnology

Nanotechnologies comprise a range of technologies, processes, and materials that involve manipulation of substances at sizes in the nanoscale range (from 1 nm to 100 nm). Food and drinking water naturally comprise particles in the nanometre scale. Humans ingest many millions of organic and inorganic nanoscale particles every day in their food and it is estimated that people inhale around 10 million nanometre-scale particles in every breath. Generally, proteins in foods are globular structures 1–10 nm in size and the majority of polysaccharides and lipids are linear polymers with thicknesses in the nanometre range. Milk is an example of an emulsion of fine fat droplets of nanoscale proportions.

It has been claimed that some of the nanomaterials now being used in foods and agricultural products introduce new risks to human health because they may be absorbed more easily. Examples include nanoparticles of silver, titanium dioxide, zinc, and zinc oxide materials now used in nutritional supplements and food packaging. However, reviews have concluded that safety cannot be determined from the size alone, and it is novelty and not size that raises concern and has to be considered in undertaking risk assessments (EFSA, 2020).

## 25.8 Regulatory agencies

Although all regulators use similar processes to evaluate scientific evidence and assess the safety of foods, the management of food safety legislation varies between countries.

The Codex Alimentarius Commission (Codex) was created in 1963 by the FAO and WHO to develop international standards, guidelines, and codes of practice related to food composition and safety, with the aim of harmonizing food regulations between countries. Over 165 countries are members of Codex. While not legally binding on individual countries, Codex standards are very influential and form benchmarks for key World Trade Organization agreements, such as those on the Application of Sanitary and Phytosanitary Measures and Technical Barriers to Trade, which make it increasingly difficult for countries to adopt food standards that are significantly different from Codex.

In the USA, the FDA develops standards for food composition, quality, and safety, as well as being responsible for approval of therapeutic drugs and cosmetics. It also has food inspection and monitoring responsibilities nationally. The work of the FDA, such as the GRAS listings, is influential internationally because of the high quality and resourcing of the many expert scientific staff.

In Australasia, the binational authority, Foods Standards Australia New Zealand, sets standards for all manufactured foods for both countries, including standards for additives and contaminants, and assesses the safety of novel foods. Primary food production and food service safety standards are set separately in each country. In Australia compliance is the responsibility of individual state governments, not the national standard setting agency.

In Britain, an independent food safety watchdog, the Food Standards Agency, was established in 2000 to protect the public's health and consumer interests after concerns raised by the BSE outbreak. The FSA provides advice and information to the public and government on food safety from farm to fork, nutrition, and diet. It also protects consumers through effective food enforcement and monitoring.

In Europe, the European Food Safety Authority (EFSA) was created in 2002 to provide independent scientific advice on all matters linked to food and feed safety. The EFSA principally deals with requests for risk assessments from the European Commission, Parliament, and Council and has clearly established itself as one of the leading and most influential food safety agencies in the world today.

One of the key roles of all regulatory agencies is risk assessment and management.

**Risk assessment** This is a scientific process consisting of four steps:

1. Hazard identification (biological, chemical, or physical agents capable of causing adverse health effects)
2. Hazard characterization (qualitative and quantitative evaluation of the hazards, including dose–response effects)
3. Exposure assessment (the likely intake of the risk factor from food, taking into account typical dietary patterns)
4. Risk characterization (estimating the probability and severity of potential adverse effects)

Risk management This is the process of weighing policy options in the light of the risk assessment results and selecting appropriate control measures. Control options can include prohibiting certain substances in foods entirely (some carcinogenic herbs, for example), setting maximum permitted levels in foods (e.g. additives or agricultural residues), the development of codes of good manufacturing practice, labelling requirements (e.g. warnings about allergens), and public education about safe use of foods (e.g. in relation to mercury in fish).

Risk communication This is the process of making the risk management information comprehensible to food producers, policy makers, and consumers.

## 25.8.1 Key points

- Despite the many potential health risks associated with foods, in practice the degree of risk associated with the modern food supply is extremely low.
- By far the most important hazards of significance are those from biological agents: pathogenic bacteria, viruses, fungi, and a few toxic seafoods.
- Trends to larger-scale production, longer distribution chains in the food supply, increased eating away from the home, and the emergence of new pathogens mean food-borne illness continues to be a significant public health issue.
- Assessment of the safety of food additives is led internationally by JECFA, but each individual country still develops and determines its own local regulations and food standards.
- The ADI is defined as the amount of a chemical that might be ingested daily, even over a lifetime, without appreciable risk to the consumer.
- Genetically modified foods, novel foods, and nanomaterials pose new challenges for traditional safety assessment processes, but as the food supply becomes increasingly global, food regulations about food safety are becoming more harmonized internationally.



## Further reading


1. **Ames, B.N., Magaw, R., and Gold, L.S.** (1987) Ranking of possible carcinogenic hazards. *Science*, **236**, 271–80.
2. **Barlow, S.** (2009) Scientific opinion of the scientific committee on a request from the European Commission on the potential risks arising from nanoscience and nanotechnologies on food and feed safety. *EFSA J*, **958**, 1–39.
3. **Barlow, S. and Schlatter, J.** (2010) Risk assessment of carcinogens in food. *Toxicol Appl Pharmacol*, **243**, 180–90.
4. **Blackburn, C.D. and McClure, P.J.** (eds) (2010) *Food-borne pathogens: hazards, risk analysis and control*, 2nd edn. Cambridge: Woodhead Publishing.
5. **Branen, A.L., Davidson, P.M., Salminen, S., and Thorngate, J.H.** (eds) (2002) *Food additives*, 2nd edn. New York, NY: Marcel Dekker.
6. **Cheng, Z.** (2011) Mercury biomagnification in the aquaculture pond ecosystem in the Pearl River Delta. *Archives of Environmental Contamination and Toxicology*, **61**(3), 491–9. skab094, https://doi.org/10.1093/jas/skab094
7. **Dilger, A.C., Johnson, B.J., Brent, P., and Ellis, R.L.** (2021) Comparison of beta-ligands used in cattle production: structures, safety, and biological effects. *Journal of Animal Science*, **99**(8), 1–16.
8. **EFSA** (2013) Scientific opinion on the re-evaluation of aspartame (E 951) as a food additive. *EFSA J*, **11**(12), 3496.
9. **EFSA** (2015) Scientific opinion on acrylamide in food. *EFSA J*, **13**(6), 4104.
10. **EFSA** (2015) Scientific opinion on the risks to public health related to the presence of bisphenol A (BPA) in foodstuffs. *EFSA J*, **13**(1), 3978.
11. **EFSA** (2020) Environmental Risk Assessment (ERA) of the application of nanoscience and nanotechnology in the food and feed chain. *EFSA J*, **17**(11). https://doi.org/10.2903/sp.efsa.2020.EN-1948

12. **FAO/WHO** (2002) *Health implications of acrylamide in food.* Report of a Joint FAO/WHO Consultation. Geneva, Switzerland: WHO/FAO.
13. **Fischer, A.R., De Jong, A.E., Van Asselt, E.D., De Jonge, R., Frewer, L.J., and Nauta, M.J.** (2007) Food safety in the domestic environment: an interdisciplinary investigation of microbial hazards during food preparation. *Risk Analysis*, **27**, 1065–82.
14. **Food Standards Australia New Zealand (FSANZ)** (2005) *Safety assessment of genetically modified foods.* Canberra: FSANZ.
15. **FSANZ** (2021) FSANZ advice on hepatitis A and imported ready-to-eat berries. www.foodstandards.gov.au (accessed 1 July 2021).
16. **Gillespie, I.A., McLauchlin, J., Grant, K.A., Little, C.L., Mithani, V., Penman, C., et al.** (2006) Changing pattern of human listeriosis, England and Wales, 2001–2004. *Emerg Infect Dis*, **12**, 1361–6.
17. **Jackson, L.S.** (2009) Chemical food safety issues in the United States: past, present, and future. *J Agric Food Chem*, **57**, 8161–70.
18. **McCann, D., Barrett, A., Cooper, A., et al.** (2007) Food additives and hyperactive behaviour in 3-year-old and 8/9-year-old children in the community: a randomised, double-blinder, placebo-controlled trial. *Lancet*, **370**, 1560–7.
19. **Microbiology and Food Safety Committee of the National Food Processors Association** (1993) HACCP implementation: a generic model for chilled foods. *J Food Prot*, **56**, 1077.
20. **National Cancer Institute** (2017) Acrylamide in food and cancer risk. http://www.cancer.gov/cancertopics/causes-prevention/risk-factors/diet/acrylamide-fact-sheet (accessed 1 July 2021).
21. **Nyachuba, D.G.** (2010) Food-borne illness: is it on the rise? *Nutr Rev*, **68**, 257–69.
22. **Omaye, S.T.** (2004) *Food and nutritional toxicology*. Boca Raton, FL: CRC Press.
23. **Park, K.S., Seo, Y.-C., Lee, S.J., and Lee, J.H.** (2008) Emission and speciation of mercury from various combustion sources. *Powder Technology*, **180**(1–2), 151–6.
24. **PHAC (Public Health Agency of Canada)** (2009) Listeria monocytogenes outbreak. https://www.canada.ca/en/public-health.html (accessed 23 March 2023).
25. PHAC (Public Health Agency of Canada) (2010) Update to 2008 Listeria monocytogenes case numbers. https://www.canada.ca/en/public-health.html (accessed 23 March 2023).
26. **Shibamoto, T. and Bjeldanes, L.F.** (eds) (2009) *Introduction to food toxicology*, 2nd edn. Burlington, MA: Academic Press.
27. **Stevenson, K.E.** (2006) *HACCP: a systematic approach to food safety*, 4th edn. Washington, DC: Food Producers Association
28. **US FDA** (n.d.) Bisphenol A (BPA). https://www.fda.gov (accessed 1 July 2021).
29. **US FDA** (n.d.) Bovine somatotropin (bST). https://www.fda.gov (accessed 1 July 2021).

## Some key food safety websites

1. World Health Organization
http://www.who.int/foodsafety/en/
2. US Food and Drug Administration
http://www.cfsan.fda.gov/list.html
3. Food Standards Australia New Zealand
http://www.foodstandards.gov.au
4. European Food Safety Authority
http://www.efsa.europa.eu
5. Health Canada
http://www.health.canada.gov.ca

# Part 6

# Changing Food Habits

# Food Habits

A. Stewart Truswell and Helen Leach



All health professionals who give advice for dietary change, based on other chapters in this book, come up against the strength of food habits: why people choose to eat all the foods they do, even if they have been advised that doing so may be unhealthy. As Jean Ritchie put it 'There is never a vacuum surrounding food, waiting for a nutritionist to come along and fill it by providing information about what foods are good for people. Everyone already has definite ideas on this subject' (Ritchie, 1979).

## 26.1 Studying food habits

A number of social scientists also locate their subject matter in this broad field of food habits: the American school of nutritional anthropology (e.g. Bryant, Fitzgerald, Jerome, Robson); a group of social anthropologists and sociologists interested in symbolism and structural order in food habits (e.g. Lévi-Strauss, Douglas, Nicod); anthropologists committed to more materialist explanations of food habits (e.g. Harris, Mintz); sociologists, economists, and social historians concerned with explaining change in food habits (e.g. Mennell, Burnett, Charsley); and social psychologists who have an interest in the interface between cultural beliefs and individual behaviour, as they affect food preference (e.g. Rozin).

Historically, the science of human nutrition developed in tandem with other medical sciences. At a clinical level, therefore, it has concentrated on the individual, and at a policy level, on the population, using epidemiological studies as a starting point. It has made by far the greatest progress with the biological aspects of human nutrition and with the development of effective treatment of nutritional disorders in individuals. However, in those aspects of human nutrition where individuals cease to behave as biological organisms or as members of biological populations, nutrition as an explanatory and applied science has had much less success. Of course, humans have to eat to survive, the environment constantly sets limits on food production, and populations will continue to show change in genetic factors affecting metabolism, but where humans exhibit any choice at all in what they eat, what they select is more likely to be socially influenced than the result of a biological craving, environmental determinism, or idiosyncratic whim.

In discussing food habits, therefore, a definition must be used that stresses the socially influenced food-related behaviour of humans as members of groups. Murcott's usage of the phrase 'food habits' as 'a provisional, convenient, and inclusive shorthand to cover the widest possible range of food choice, preferences, and meal patterns and cuisines' (Murcott, 1988) will be followed here.

In families, the parents have the responsibility of guiding their children to eat enough of the main foods to provide the nutrients needed for growth and health. They also have a role in advising on portion sizes of foods. Establishing appropriate portion sizes of foods low in energy and dense in nutrients is important in prevention of overweight and obesity.

## 26.2 Who chooses?

There can be no denying that in the course of a human lifetime, the decision whether to ingest a particular food item lies ultimately with the individual (except in cases of forced or tube feeding). Infants are particularly adept at exercising the right to reject food. However, the power to decide what food items are made available for the selection process, and in what form, frequently lies beyond the individual. For the baby, the parents or carers usually decides what should be offered, starting with the decision of either breast milk or formula. Her choice on behalf of the baby is usually influenced by advice from relatives, or health professionals. In the case of the former, this permits a family tradition of infant feeding to be passed on, compatible with the beliefs of the family's ethnic group and with their religious affiliation. A nurse's or doctor's advice will normally reflect the society's prevailing scientific paradigm concerning infant nutrition, for example that solids should not be introduced until a certain age or that certain foods should be avoided.

The growing child has little control over the household menu, though at certain times (such as illness or the celebration of milestones in development), the person responsible for food acquisition and preparation will deliberately produce an item known to be the child's favourite. In some cultures, gender, and position in the family, may affect what is offered to each child in both quantity and variety (see Box 26.1). Similarly, the menu selected for the

### Box 26.1 Fruit and vegetables at dinner: French different from American

Nutrition research indicates that for health, many people should eat more fruits and vegetables (see Chapters 23 and 30). Public health initiatives to increase fruits and vegetables (F&V) have mostly been in school curricula and have had little success.

The French eat more F&V than Americans. Kremer-Sadlik et al. (2015) thought that the ecology of eating and parents' behaviours might have an important influence. They set up video recordings of family dinners with eight middle-class dual-earner families in Los Angeles and eight in Paris. Each had two or three children, with at least one between 7 and 11 years. During weekday dinners for each family, the videographers set up the cameras on tripods and left the room.

The recordings showed that the French children were exposed to a greater amount and variety of fruit and it was an integral part of the dinner. In the American families, most of the children were not observed to eat any fruit. The French dinner had multiple courses, and regular-size vegetable dishes were central in the meal. In the American dinners, often single course, meat and starchy foods were more prominent. Half the American children did not touch their vegetable dishes, but only 10% of the French children. The French children were treated as equal diners and expected to eat the same food as the adults. American parents treated vegetable foods as optional and of lesser value. French parents both gently pressed their children to taste a vegetable; American parents let them opt out.

household may be strongly influenced by the preferences of a senior adult member. In societies where it is traditional for women to cook for their families, the desire to please their husbands may dictate the dishes they prepare for the whole household. Finally, in old age, any former control over the menu may be lost through institutionalization or displacement from the kitchen by a younger household member. Thus, for many humans, selection of food is subject to significant constraints for much of their lifetime, despite the apparent freedom of the individual to eat as they choose.

## 26.3 Social and cultural influences on food choice

It is not surprising then that the cook, food purchaser, housewife, or househusband, indeed any member of a group who makes food choices on behalf of that group, is sometimes referred to in research literature as the 'key kitchen person' (KKP), focal person, or gatekeeper, all terms that recognize a pivotal role in the diet and food habits of their group. What social factors influence their choice of food to prepare? The broad answer is that they select according to the unwritten rules or norms of the culture to which they belong. Even when they respond to the food preferences of a particular member of their group, they are choosing items, composing them into dishes, and combining the dishes into menus within particular culinary traditions.

Culinary traditions operate at many levels, from small, kin-based group traditions to the nearly global culinary styles of Western cultures. In some isolated situations in low-income countries, only one culinary tradition may be relevant, but for most groups in high-income countries, food providers and KKPs choose to work within the tradition that they see as most appropriate for a particular eating situation. For many important social occasions, the family culinary tradition of the organizers and chief participants will be followed. The significance of the occasion (e.g. wedding breakfast, Christmas dinner, religious festival) and the desire for a successful outcome usually constrain the menu within the traditional family or community pattern.

Even at this lowest organizational level, the family pattern of eating may be distinctive from that of its neighbours, though both may belong to the same ethnic and religious group and occupy the same socioeconomic position. Family food habits may be comparatively resistant to change in places where culinary knowledge and skills are learned primarily within the household and passed down between generations. In societies where people can afford to choose food and drink, there are often differences in food habits and attitudes between women and men and in different age groups.

Urbanized communities also feature a wide range of household arrangements, from two-parent and two-generational families to groups of unrelated young adults. Food habits here are influenced by the background and social network of the group member who becomes the KKP for each main meal. If that person learned culinary skills in his or her own family setting, many of these will be transferred to the new household and applied according to the type of meal. However, in urbanized Western societies, such knowledge is frequently acquired outside the home (e.g. from the formal education system). Cooking training in schools has been motivated by the goals of 'good nutrition' and has been responsible for reinterpreting Western culinary traditions within a scientific paradigm. After schooling is complete, recipe repertoire is influenced by interaction with peers, within social networks, and by the media.

Magazines, newspapers, and television reflect an amalgam of culinary traditions depending on the contributing sources. A locally based food writer for a monthly newspaper column may work within a regional food tradition, combining new variants with well-established and familiar dishes. A national or internationally distributed magazine may offer more cosmopolitan fare, strongly influenced by international food fashion trends. However, there is little research on whether these fashions have a lasting impact on household food habits. For example, will the currently fashionable grain couscous become an important cereal in cosmopolitan cuisine, or will it be as short-lived as the fondue party?

The relatively recent and continuing trend of a cosmopolitan culinary repertoire has probably been influenced by the range of dishes cooked in the commercial kitchens of restaurants and fast-food outlets and sampled as part of the phenomenon of 'eating out'. For the household member who is not actively involved in food preparation, eating out means more choice is possible from an extended menu. For the KKP, the responsibility for selecting food for others is removed, along with control over the dishes on the menu and their composition. However, the characterization of restaurants and other food outlets into well-defined categories, such as vegetarian, wholefood, seafood, Italian, Thai, Cantonese, or other specified ethnic type, suggests that consumers still prefer to choose their food from an identifiable culinary tradition, even if it is not that of their own birth culture.

Thus, for the purpose of food and menu selection, many Western households operate within multiple layers of culinary traditions, not simultaneously, but moving between them from meal to meal, through the weekly cycle and the annual calendar of festive occasions. Perhaps the unwritten rules for deciding which culinary tradition is appropriate are part of evolving culinary traditions in their own right.

## 26.4 Food habits in nutrition practice

From the viewpoint of the nutritionist offering clinical advice to an individual, it is clearly important to know how the patient relates to the KKP of the household where he or she normally resides. Without the informed support of the KKP, long-term dietary modification is unlikely to occur, and even with that support, pressure from other members of the household or a high-status member might frustrate any change.

It is also valuable for the nutritionist to have some knowledge of the culinary tradition from which most of the patient's meals are derived. Any modification of intake must be compatible with that tradition. We tend to categorize other traditions by their starch staple and typical flavouring substances. Culinary traditions are more than just the combinations of dishes/recipes that characterize and distinguish the eating patterns of particular human groups. They also include the rules for selecting and preparing food for consumption (such as butchering practices and preservation techniques), rules for composing the menu according to an acceptable structure of courses, and the rules or norms concerning eating behaviour (e.g. time, location, participants). As with all human social behaviour, these rules or norms are culturally transmitted. They are not immutable, but variation and innovation tend to take place more frequently within the structure than radical alteration of the overall framework. Culinary traditions also include the non-nutrient-related meanings of food, including those that concern ethnic identity, values, and religious beliefs.

Within certain traditions, there is strong social pressure to prepare and serve more food on occasions than might be considered nutritionally desirable. In such instances, the food is satisfying more than biological needs, and the participants are well aware that it is carrying a coded message concerning social relationships and value systems. For example, the Oglala Sioux of Pine Ridge Reservation in South Dakota require participants at communal high feasts to eat beyond satiety, and to bring containers for the removal of leftovers.

Particular foods may encode metaphysical meanings that completely transcend any nutritional value. For the Christian taking communion, the bread and the wine are clearly not food, but have been spiritually transformed.

For many societies, food choice is constrained by religious proscription. Many of these food taboos are of such antiquity that they are written into the formal teachings of the religion, such as Jewish and

Muslim prohibition of pig meat, Hindu prohibition of beef, and Muslim avoidance of alcohol. However, more recent dietary regimens based on particular philosophical or ethical principles have evolved food taboos of equal force for their adherents. As a secular movement, vegetarianism has been important for nearly two centuries, throughout this period drawing for justification on the accumulating data from the science of nutrition. Not surprisingly, nutritionists faced with evidence of health benefits from vegetarian diets for adults have treated the movement with respect. However, the twentieth and twenty-first centuries have seen the development of far more restrictive dietary regimens allied to the 'health food' movement and the Western preoccupation with weight reduction. Cultural and family traditions and personal taste can resist health professional advice, but the other competition is that the public is bombarded almost daily by new threats to health and safety. We are in a Tower of Babel of nutritional breakthroughs. From newspapers, magazines, TV, radio, advertisements, supermarket shelves, and the internet there is information from competing interests in the food industry, from journalists, and from some apparent experts. It is unfortunate that to be newsworthy, the research is usually preliminary and contrary to received opinion. Nutrition research that is accepted and established is not news. The woman or man in the street seems to be adapting to the threats and concludes that all the news and advertisements cancel one another out: 'It can't matter much what you eat, so I'll eat what I enjoy and am used to.'

Food symbolism is highly developed in these regimens, with the polarization of 'natural' (= healthy) versus synthetic (= dangerous) constantly reworked and elaborated by those who benefit commercially from sales of diet books and the approved foods. For the clinical nutritionist, disentangling the pseudo-scientific justifications and powerful symbolism of these restrictive and potentially damaging diets is a major challenge. Furthermore, the social networks into which adherents of 'fad' or extreme diets are drawn for mutual support are likely to counteract what is perceived as criticism by the health professionals.

## 26.5 Changing food habits in the modern world

So far, this introduction to food habits has stressed the inbuilt tendency to conservatism within a culinary tradition, the individual's lack of control over food choice at various periods in his or her life, and the symbolic loading on certain foods, which may exert powerful pressure against any change in usage or avoidance. Taken together, these factors explain why nutritional advice aimed at individual or group level may be listened to politely (because of the status of the health professional) but not subsequently acted upon.

At the same time, the success of multinational food companies provides ample evidence that throughout the world, even within highly conservative and symbol-rich culinary traditions, some substantial changes have occurred in diet during the twentieth century. These have included the introduction of many new food items, and even new menus borrowed from other traditions. Food manufacturers have invested heavily in research on food acceptance and have realized that restrictive rules affecting food habits apply to meals rather than snacks. This observation, also made by Michael Nicod from his structuralist study of working-class food habits in Britain (Douglas and Nicod, 1974), led to his definition of the meal as a structured food event and the snack as an unstructured event. Although meals have courses that must be served in a prescribed order and contain certain food categories, for snacks consumers are free to eat the items in any order or combination, at any time, and with or without company. It is not surprising then that snack-food manufacturers have had a global impact, because their promotional advertising does not have to confront long-established culinary norms. The lack of structure in snacking behaviour has allowed it to become a global arena where multinationals compete with new products, or old ones, newly packaged.

Nutritionists have had some success in promoting snacks with lower fat and sodium content, but realize that because snacking is usually supplementary feeding, their long-term objectives can only be met by dietary reform of meals, the area in which the multinationals have made much less headway. It has taken several decades to achieve public acceptance of pizza- or burger-based main meals, when these items were originally conceived of more as snacks. Fast-food outlets have had to supply additional items such as potatoes, salads, and even desserts to meet the public's norms of what constitute 'proper' meals.

The key to effecting change in meal composition is to provide substitutions for existing elements without threatening the overall structure (see Box 26.2). Studies of dietary change in indigenous societies following contact with Europeans have shown that the way in which the new foods are slotted into existing native classification systems is a useful guide to their acceptability. The most rapidly accepted foodstuffs are those that are judged similar to existing foods, in attributes such as taste, appearance, style of preparation, or growth habit. In most Pacific cultures, more than one starch staple was available, although they were seldom served at the same meal. This choice may have increased their readiness to accept the new staples introduced by Europeans. Monostaple cultures seem much more resistant to change. The depth of their attachment to their single staple may often be seen in the extension of the word used for the staple to mean food in general.

The effects of globalization on developing countries constitute a special case of culture contact. The accompanying economic changes are of concern to nutritionists because they may lead to deleterious substitutions in traditional diets. These often involve replacement of higher-nutrient crops (such as millets in central Africa) with crops of lower nutritional value (such as cassava, which is more tolerant of drought and degraded soils). Studies have shown that, given the local economic context, such diet shifts are based on rational choices designed to increase food security, even at the expense of optimal nutrition. Labour migration, pressure to produce cash crops, and increasing poverty produce adaptations in traditional food systems, which can transform adequate nutrition into chronic undernutrition or, in the most vulnerable groups, even malnutrition. In these cases, nutritional reform is unlikely to succeed unless poverty can first be reduced.

The message for nutritionists from such studies of culture contact is that the structural elements of culinary traditions are highly resistant to change. It is precisely these elements that, in giving stability and continuity, define the tradition. Where dietary change has occurred, it has involved substitution of new foods for old within the indigenous food system, using the classification process as a guide to how the new food is to be used.

Judging from historical studies, actual structural changes involving things like course order and essential elements may take a century or longer. If thought nutritionally desirable, reform at such a fundamental level might be expected to add a new dimension to the notion of long-term planning! Ultimately, nutritionists must translate their findings into recommendations that will work within the culinary tradition, not against it.

### Box 26.2 Easy changes without disturbing main food/drink

- Fortification(s), e.g. folate, iodine, thiamin
- Artificial sweeteners for sugar
- Polyunsaturated margarine for butter
- Gluten-free bread
- Adding protein powder to a dish
- Tasteless fluoride in the water supply

## Further reading


1. **Cox, D.N. and Anderson, A.S.** (2004) Food choice. In: Gibney, M.J. (ed) *Public health nutrition*, pp. 144–66. Oxford: Blackwell Science.
2. **Craig, P.L. and Truswell, A.S.** (1988) Changes in food habits when people get married: analysis of food frequencies. In: Truswell, S. and Wahlquist, M.L. (eds) *Food habits in Australia*, pp. 94–111. Richmond, VA: Heinemann.
3. **Deutsch, R.M.** (1977) *The new nuts among the berries*. Palo Alto, CA: Bull Publishing Co.
4. **Douglas, M.** (1970) *Purity and danger: an analysis of concepts of pollution and taboo*. Harmondsworth, England: Penguin Books.
5. **Douglas, M. and Nicod, M.** (1974) Taking the biscuit: the structure of British meals. *New Soc*, **30**, 744–7.
6. **Fisher, J.O., Goran, M., Hetherington, M., and Rowe, S.** (2014) Forefronts in portion size. An overview and synthesis of a roundtable discussion. *Appetite*, **88**, 1–4.
7. **Harris, M.** (1986) *Good to eat: riddles of food and culture*. London: Allen and Unwin.
8. **Kremer-Sadlik, T., Morgenstern, A., Peters, C., et al.** (2015) Eating fruits and vegetables. An ethnographic study of American and French family dinners. *Appetite*, **89**, 84–92.
9. **Larsen, M.H.** (2015) Nutritional advice from George Orwell. Exploring the social mechanisms behind the overconsumption of unhealthy foods by people with low socio-economic status. *Appetite*, **91**, 150–6.
10. **Leach, H.M.** (1993) Changing diets—a cultural perspective. *Proc Nutr Soc NZ*, **18**, 1–8.
11. **Lentz, C.** (ed) (1999) *Changing food habits: case studies from Africa, South America and Europe*. Amsterdam: Harwood.
12. **Messer, E.** (1984) Anthropological perspectives on diet. *Annu Rev Anthropol*, **13**, 205–49.
13. **Murcott, A.** (1984) *The sociology of food and eating*. Aldershot: Gower.
14. **Murcott, A.** (1988) Sociological and social anthropological approaches to food and eating. *World Rev Nutr Diet*, **55**, 1–40.
15. **Ritchie, J.A.S.** (1979) *Learning better nutrition*. FAO Nutritional Studies number 20. Rome: FAO.
16. **Sharman, A., Theophano, J., Curtis, K., and Messer, E.** (1991) *Diet and domestic life in society*. Philadelphia, PA: Temple University.
17. **Sleddens, E.F.C, Kroeze, W., Kohl, L.F.M., et al.** (2015) Correlates of dietary behaviour in adults: an umbrella review. *Nutr Rev*, **73**, 477–99.
18. **Truswell, A.S. and Darnton-Hill, I.** (1981) Food habits of adolescents. *Nutrition Reviews*, **39**, 73–88.
19. **Truswell, A.S. and Wahlqvist, M.L.** (1988) *Food habits in Australia*. Richmond: Heinemann.
20. **Visser, M.** (1988) *Much depends on dinner*. Toronto: McClelland and Stewart.

# Nutritional Recommendations for the General Population

A. Stewart Truswell, Leanne Hodson, and Jim Mann



Nutrition research generates results that may be translated by the media, or sometimes researchers themselves, into potentially confusing and conflicting messages. It is therefore important for governments, which develop food and nutrition policies, for those involved in health and nutrition education, and for consumers to have authoritative nutrition recommendations that represent consensus opinions of expert nutrition scientists. There are three major sets of recommendations: nutrient intake recommendations, dietary goals, and dietary guidelines (Box 27.1).

*Nutrient intake recommendations* were the earliest types of recommendations. They are based on authoritative quantitative estimates of human requirements for essential nutrients. The first set, which included only a few micronutrients, was issued by the nutrition committee of the old League of Nations in 1937. The best-known series of nutrient intake recommendations are the US recommended dietary allowances (RDAs), which were first published in 1943 and have been revised and augmented regularly since. Other countries or groups of countries (e.g Australasia, Europe, UK) and the FAO/WHO have their own sets of recommended nutrient intakes.

More recently there has been the recognition that in most relatively affluent countries, and in affluent groups in some developing countries, inappropriate intakes of macronutrients are adversely affecting health status through increasing the prevalence of chronic diseases, e.g. coronary artery disease (Chapter 19), diabetes (Chapter 21), and some cancers (Chapter 20). One problem that arises in trying to set quantitative nutrient recommendations in relation to chronic disease prevention is the level of certainty surrounding the amount of a specific nutrient required for prevention of what is often a multicausal condition. However, since the late 1960s, many countries have developed *dietary goals,* which principally provide targets for macronutrient intakes, aimed to achieve reductions of chronic diseases. They may include some mention of the foods or food groups that provide sources of important nutrients (e.g. specific types of fats).

*Dietary guidelines* are the tools that help consumers to select a diet that will give them a better chance of long-term health. They assume knowledge

**Box 27.1 Nutrient intake recommendations, dietary goals, and dietary guidelines: an explanation of the terms**

***Nutrient intake recommendations*** deal only with nutrients, usually essential nutrients, and are expressed in weight units (μg, mg, or g) per day. The primary function of nutrient intake recommendations is to prevent nutrient deficiency. The terms recommended dietary allowance (RDA), recommended dietary intake (RDI), and recommended (or reference) nutrient intake (RNI) all refer to the amount of nutrients required to ensure adequacy for the majority of an apparently healthy population.

***Dietary goals*** include macronutrient targets for the whole population, useful for bureaucrats to direct national policy. They can be expressed quantitatively, e.g. 'average free sugars intake should be less than 10% of energy'.

***Dietary guidelines*** are written for individuals in the population at large, for shoppers, and cooks. They advise on foods and food components. They often say 'eat more' or 'eat less' compared with an estimated usual intake.

Nutrient intake recommendations are well established scientifically, based on physiological depletion experiments and observations of treating malnutrition. They have been fairly fixed since the 1980s, except when new functions have been discovered for a particular nutrient, e.g. folate.

Dietary goals and guidelines are more provisional, based on mostly indirect evidence about the complex role of food components in multifactorial diseases with long incubation periods. They rely largely on epidemiological evidence.

about nutrient intake requirements and generally address the balance of food groups required to attain a diet that will provide all the essential nutrients without the excesses that may increase chronic disease prevalence. The guidelines generally apply to the total diet and not to the 'healthiness' of particular foods. They are meant to be used as a framework for food choice at a time when the food supply in many affluent countries can be bewilderingly large and in some developing countries can be changing rapidly. Worldwide, over 100 countries have developed food-based dietary guidelines that are adjusted to fit food availability, culinary cultures, eating habits, and nutrition situation. They are considered an essential component of countries' public health strategies (see Chapter 28). To ensure dietary food-based guidelines and food guides are in line with current scientific evidence, the FAO assists member countries in developing, revising, and implementing them. Dietary guidelines in some countries include goals for intakes of macronutrients and recommendations for individual nutrients.

## 27.1 Nutrient intake recommendations

The first US National Research Council's RDAs in 1943 were defined as 'tentative goals towards which to aim in planning dietaries.' They were later decribed in 1964 as 'designed to afford a margin of sufficiency above average physiological requirements to cover variations among practically all individuals in the general population.' Essentially the same definition continued in the 1989 edition: 'RDAs are the levels of intake of essential nutrients that, on the basis of scientific knowledge, are judged by the Food and Nutrition Board to be adequate to meet the known nutrition needs of practically all healthy persons.' The term 'RDA' continues to be used in North America. Elsewhere the terms recommended nutrient or dietary intake (RNI, RDI) or reference nutrient intake (RNI) are used to describe the same concept. There are several other groups (or categories) of reference values that serve different purposes, different terms being used in different countries and by different organizations (Table 27.1).

**Table 27.1 Different terms used for requirements of essential nutrients**

| Terminology | Definition | Used by |
|---|---|---|
| Estimated average requirement (EAR) | Daily intake value estimated to meet the requirement—as defined by the specified indicator of adequacy—in half of the healthy individuals in a population group | USA/Canada<br>Australia/New Zealand<br>UK<br>FAO/WHO |
| Recommended dietary allowance (RDA)<br>Recommended dietary intake (RDI)<br>Recommended nutrient intake (RNI)<br>Reference nutrient intake (RNI) | Amount to cover the majority (97–98%) of an apparently healthy population group | USA/Canada: RDA<br>Austria/Germany/Switzerland: RDI<br>Australia/New Zealand: RDI<br>FAO/WHO: RNI<br>UK: RNI |
| Adequate intake (AI)<br>Safe intake (SI)<br>Estimated values and Guiding values (certain nutrients) | Amount to cover most members of the apparently healthy population group but where data are limited and certainty much less | USA/Canada, Australia/New Zealand: AI<br>UK: SI<br>Germany/Austria/Switzerland |
| Acceptable macronutrient distribution ranges (AMDR)<br>Suggested dietary targets and acceptable macronutrient distribution ranges<br>Protective nutrient intake<br>Guiding values (certain nutrients) | Suggested optimal range of intake. Intakes that give 'high-level' wellness or reduced risk of one, or more, degenerative diseases | USA/Canada: AMDR<br>Australia/New Zealand: suggested dietary targets and acceptable macronutrient distribution ranges<br>FAO/WHO: protective nutrient intake<br>Germany/Austria/Switzerland: guiding values |
| Tolerable upper intake level (UIL)<br>Upper level (UL)<br>Safe upper levels for vitamins and minerals | The highest level of daily nutrient intake that is likely to pose no risk of adverse health effects in almost all individuals in a specific population group | USA/Canada, EU: UIL<br>Australia/New Zealand: UL<br>UK: safe upper levels for vitamins and minerals |

Setting an RDA, RNI, or RDI for a nutrient is dependent on being able to set an *estimated average requirement* (EAR), which is the daily intake value of a nutrient estimated to meet the requirement in half (50%) of the healthy individuals in a specified population group. Determining the EAR of each nutrient requires review of the available scientific data to establish a value that relates to physiological function and/or disease prevention. The RDA is then set at the EAR plus twice the standard deviation (SD) if known. If there are insufficient data to calculate the SD, then a coefficient of variation for the EAR of 10% is assumed (i.e. RDA = 1.2 × EAR).

When insufficient or inadequate scientific data are available to calculate an EAR, population requirements are expressed in terms of *adequate intake* (AI). The AI may be based on data from a single experiment, on estimated dietary intakes in apparently healthy population groups, or on a review of data from different approaches that considered alone do not permit a reasonably confident estimate of EAR. As there may only be limited data available, the variation in the requirement for the nutrient across life stages may be less certain. Nutrients in this group include vitamins (biotin and pantothenic acid) and the trace elements (copper, chromium, fluoride, manganese, and molybdenum).

The current recommended intakes for selected nutrients for adult men and women are set out for America, Australia/NZ, Germany/Austria/Switzerland, the UK, and for FAO/WHO in Table 27.2. Although they are all produced by separate committees

**Table 27.2 Recommended nutrient intakes (per day) from four national sets and FAO/WHO Codex, for selected nutrients for men and women**

| Nutrient | USA 2020–2025 RDA (31–50 y) | Australia/New Zealand 2017 RDI or AI (31–50 y) | Germany/Austria/Switzerland 2017-2020 RDI (25–51 y) | UK 2021 RNI (19–50 y) | FAO/WHO Codex NRV 2019 (19–51 y) |
|---|---|---|---|---|---|
| Protein (g) | 56M, 46F | 64M, 46F | 57M, 48F | 56M, 45F | 50 |
| Vitamin A (mg) | 0.9M, 0.7F | 0.9M, 0.7F | 0.85M, 0.70F | 0.7M, 0.6F | 0.8 |
| Vitamin D (μg) | 15 (or 600 IU) | 5.0 | 20 | 10 | 5.0–15 |
| Vitamin E (mg) | 15 | 10.0M, 7.0F | 14M, 12F | — | 9 |
| Thiamin (mg) | 1.2M, 1.1F | 1.2M, 1.1F | 1.2M, 1.0F | 1.0M, 0.8F | 1.2 |
| Riboflavin (mg) | 1.3M, 1.1F | 1.3M, 1.1F | 1.4M, 1.1F | 1.3M, 1.1F | 1.3 |
| Niacin (mg) | 16M, 14F | 16M, 14F | 15M, 12F | 17M, 13F | 15 |
| Vitamin $B_6$ (mg) | 1.3 | 1.3 | 1.6M, 1.4F | 1.4M, 1.2F | 1.3 |
| Folate (μg) | 400 | 400 | 300 | 200[a] | 400 |
| Vitamin $B_{12}$ (μg) | 2.4 | 2.4 | 4.0 | 1.5 | 2.4 |
| Vitamin C (mg) | 90M, 75F | 45 | 110M, 95F | 40 | 100 |
| Calcium (mg) | 1000 | 1000 | 1000 | 700 | 1000 |
| Iron (mg) | 8M, 18F | 8M, 18F | 10M, 15F | 8.7M, 14.8F | 14–22 |
| Zinc (mg) | 11M, 8F | 14M, 8F | 14*M, 8*F | 9.5M, 7.0F | 14 |
| Iodine (μg) | — | 150 | 200; S150 | 140 | 150 |
| Selenium (μg) | — | 70M, 60F | 70M, 60F | 75M, 60F | 60 |

NRV, nutrient reference values; M, males; F, females; IU, international units; S, Switzerland
[a]More in preganancy; *average phytate intake

of independent nutrition scientists, the difference between most of the numbers are so small, it has been suggested they could almost be harmonized. But for some nutrients (e.g. vitamin C), the measures of adequacy have differed between committees. While recommended intakes are of great importance from a public health and health promotion perspective, it is important to appreciate that there is considerable individual variation in human requirements (Table 27.3).

For some nutrients, there may be evidence for a health benefit above the recommended intake. The optimal intake will be between the recommended intake and *upper level* beyond which undesirable effects may occur. Publication of upper levels of intake in some sets of recommendations has been considered useful because of the widespread use of nutrient supplements with tablet dosage ranging well above daily intake requirement, which increases the potential risk of adverse effects. For example, vitamin C is available in tablets up to 1000 mg strength, over 10 times the recommended intake, which may have untoward effects on some individuals (e.g. those with impaired renal function). Fat-soluble vitamins A and D and water-soluble vitamin $B_6$ can be toxic at moderately high intakes. Examples of upper intake levels for three groups of countries are shown in Table 27.4.

In recent years the nutrient intake recommendations from some countries (including North America and Australasia) have added lower levels for diagnosis of nutritional adequacy to their recommended intakes, as well as upper levels for the approach of undesirable effects. Different approaches have been

**Table 27.3 Range of requirements for selected essential nutrients**

| Essential nutrients | Adult daily requirements (rounded) |
|---|---|
| Vitamin $B_{12}$ | 2–5 µg |
| Chromium, molybdenum, biotin | 30–40 µg |
| Selenium, vitamin K | 70–100 µg |
| Iodine | 140–200 µg |
| Folate | 200–400 µg |
| Vitamin A, thiamin, riboflavin, vitamin $B_6$, fluoride, copper | 1–3 mg |
| Pantothenate, manganese, vitamin E, niacin (equivalents), zinc, iron | 5–15 mg |
| Vitamin C | 40–110 mg |
| Magnesium, sodium | 400–500 mg |
| Calcium, phosphorus, essential fatty acids, potassium | 1–2 g |
| Protein | 45–64 g |

**Table 27.4 Proposed upper levels for 12 nutrients from three different countries for non-pregnant adults**

| | USA (31–50 y) | Germany/Austria/ Switzerland (19+ y) | Australia/New Zealand (31–50 y) |
|---|---|---|---|
| Vitamin A (mg) | 3 | 3 | 3 |
| Vitamin D (µg) | 100 | 100 | 80 |
| Vitamin E (mg) | 1000 | 800 | 300 |
| Niacin (mg) | 35 | 10* 900~ | 35 |
| Vitamin $B_6$ (mg) | 100 | 25 | 50 |
| Folate (µg) | 1000 | 1000 | 1000 |
| Vitamin $B_{12}$ (µg) | Not possible to set | 5000 | Not possible to set |
| Vitamin C (mg) | 2000 | 1000 | Not possible to set |
| Calcium (mg) | 2500 | 2500 | 2500 |
| Iron (mg) | 45 | — | 45 |
| Iodine (µg) | 1100 | 600 | 1100 |
| Selenium (µg) | 400 | 300 | 400 |

* = Nicotinic acid; ~ = Nicotinamid

used to define a diagnostic level for a 'deficiency', with some simply using two-thirds of the recommended intake. The concepts of deficiency, recommended daily intake, optimal intake, and upper level (high intake) are illustrated diagrammatically in Fig. 27.1.

No matter what nomenclature is used, the purpose of the recommended figures are to set a standard for an adequate intake of each essential nutrient for individuals or groups in the population. They advise people how much, on average, they should aim to eat each day of these nutrients. They have a prescriptive or health promotion role. They also serve as the reference unit for each essential nutrient (Table 27.3). We need these because adult human requirements for individual nutrients range 1 billion-fold from just over 1 µg for vitamin $B_{12}$, through 1 mg for thiamin, to 1 g for calcium and greater than 1 kg for water.

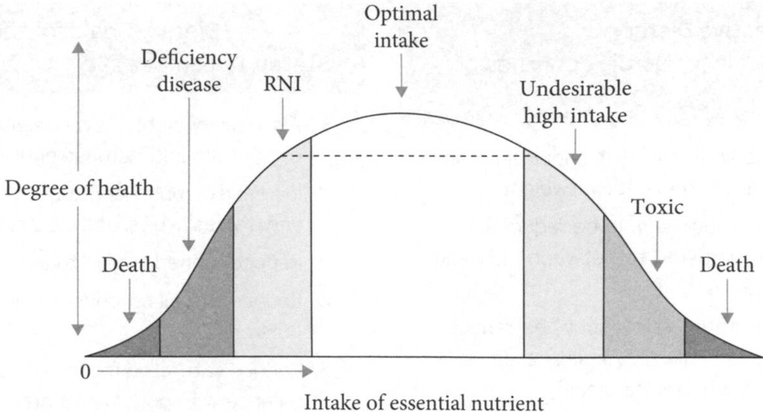


Fig. 27.1 The concept of optimal intake above the population recommended nutrient intake (i.e. RNI/RDA/RDI). Ingestion of the RNI/RDA/RDI should guarantee no deficiency disease, but beyond this there may still be additional health benefits (e.g. partial protection from a degenerative disease). The top of the dome beyond the RNI/RDA/RDI is then the optimal intake range.

Why do some countries and organizations have different sets of these numbers when undertaking a major review of these figures is both time-consuming and expensive? Many countries do use figures derived by another equivalent country (e.g. with a population with similar body size/culture/environment). However, others choose to vary their figures in the light of newer research data, a different philosophical approach, country-specific data, concerns about interpretation of the research data, or a perception that some key information was not considered. There may also be differences in the age groups used by different countries in relation to their own particular policy and assessment needs.

## 27.2 Proportions of macronutrients and dietary goals

The *acceptable macronutrient distribution ranges* (AMDR) are often given by countries as part of their *dietary goals*, which are aimed at achieving reductions in chronic diseases. For a given energy intake, increases in the proportion of one macronutrient (as percentage of total energy) necessarily involves a decrease in the proportion of one, or more, of the other macronutrients—a high-fat diet is usually relatively low in carbohydrate, and vice versa. Imbalance in the proportions of macronutrients can increase risk of chronic disease and may adversely affect micronutrient intake. In addition, the form of fat (e.g. saturated, polyunsaturated, monounsaturated, or specific fatty acids) or carbohydrate (e.g. starches or sugars, high or low glycaemic index) and fibre has to be considered.

It became clear that there was more to public health nutrition advice than providing recommendations regarding essential nutrients in the 1960s, when the world was settling down after the food shortages in World War II. Different claims were being made and disputed about 'unhealthy' and 'healthy' foods.

The professors of nutrition in Nordic countries (Sweden, Denmark, Finland, and Norway) proposed the first set of authoritative dietary goals in 1968 (Box 27.2). They were translated and published by Ancel Keys in the USA the same year. The first consideration was that, with mechanization, some people were eating more calories than they spent. So, fats and sugar ('empty calories') should be reduced. This could also help protect against coronary heart disease (less saturated fat, more polyunsaturated) and dental caries. More milk and meat would reduce calcium and iron deficiency. Here too was an early statement of the value of habitual regular exercise.

**Box 27.2 Collective Dietary Recommendations for Nordic Countries (1968)**

- The calorie supply in the diet should in many cases be reduced to prevent overweight.
- Total fat consumption should be reduced from the present around 40% to between 25% and 30% of total calories.
- The use of saturated fats should be reduced and consumption of polyunsaturated fats should be increased simultaneously.
- Consumption of sugar and sugar-containing products should be reduced.
- Consumption of vegetables, fruits, potatoes, skimmed milk, fish, lean meat, and cereal products should be increased.
- From the medical and nutritional standpoint it is essential to emphasize the importance of regular exercise habits from childhood for all individuals with mainly sedentary work.

**Box 27.3 Dietary Goals for the United States, December 1977**

1. Avoid overweight. If overweight, decrease energy intake and increase energy expenditure.
2. Increase complex carbohydrates (includes vegetables, fruits, and dietary fibre).
3. Reduce refined sugars to about 10% energy.
4. Reduce overall fat consumption, from 40% to 30% energy.
5. Reduce saturated fat to 10% energy and balance with polyunsaturated and monounsaturated fats, each 10% energy.
6. Reduce cholesterol consumption to about 300 mg/day.
7. Limit sodium, by reducing salt to about 5 g NaCl/day (85 mmol Na).

Nine years later, Dietary Goals for the United States first appeared (1977) and it was revised later the same year (Box 27.3). It was published by a committee of the US Senate, not by the US government. It stimulated much controversy by affected food industries, but was welcomed by a *Lancet* editorial. The components of the December 1977 revision were fairly similar to the Nordic guidelines, with two differences: advice to reduce cholesterol intake and to limit sodium intake by reducing salt intake to below 5 g NaCl/day (85 mmol Na) was included.

There appears to be quite a wide range of relative intakes of proteins, carbohydrates, and fats that are acceptable in terms of chronic disease risk. The risk of chronic disease (and of inadequate micronutrient intake) may increase outside these ranges, but data in free-living populations are limited at these extremes of intake. Much of the evidence is based on epidemiological studies with clinical endpoints, but they generally show associations rather than causality and are often confounded by other factors that can affect chronic disease outcomes. The acceptable distribution ranges for macronutrients from the USA, Australia/New Zealand, and Nordic countries are shown in Table 27.5.

**Table 27.5 Acceptable distribution ranges for macronutrients to reduce chronic disease risk while still ensuring adequate micronutrient status**

| Nutrient | Australia/New Zealand 2017 | USA 2022 | Nordic 2012 |
|---|---|---|---|
| Protein | 15–25% of energy | 10–35% of energy | 10–20% of energy |
| Carbohydrate | 45–65% of energy (predominantly from low-energy-density and/or low-glycaemic index foods) | 45–65% of energy | 45–60% of energy |
| Fat | 20–35% of energy | 20–35% of energy | 25–40% of energy |
| Linoleic acid (ω6 fat) | From 4–5% to 10% of energy | 12–17 grams | 5–10% of energy |
| α-Linolenic acid (ω3 fat) | From 0.4–0.5% to 1% energy | 1.1–1.6 grams | ≥1% of energy |

When recommendations are made about nutrient intake for chronic disease prevention, it is thus important to be clear whether the cut-off numbers refer to populations or individuals. If the goal is set for the population, not every individual needs to conform for the population goal to be met (see Chapters 19, 20, and 21).

## 27.3 Dietary guidelines

Dietary guidelines initially focused on encouraging foods that were considered necessary to achieve a healthy diet and to ensure that people consumed sufficient vitamins and minerals to prevent nutrient deficiency. Over time, guidelines have evolved, in light of available scientific evidence, from a focus on nutrient adequacy to include recommendations relating to foods that may impact on chronic disease risk. For example, dietary guidelines now include advice on what to eat and drink to maintain a healthy body weight, along with guidance on limiting dietary components such as sugar, fat, saturated fat, cholesterol, sodium, and alcohol.

Most countries base their guidelines on food groups and, although the number of food groups may vary, food guides in most industrialized countries contain between four and six food groups. Groupings commonly used include:

- meat/fish/poultry and alternatives (including legumes)
- milk and milk products
- fruits, vegetables
- cereals
- fats and oils.

Because of the degree of commonality in their nutrient profile, fruits and vegetables often form a single group. Sometimes the 'starchy' foods, such as potato, rice, and cereals, are grouped together. Some countries also include an additional food group category of 'less healthy' extra or indulgence foods—generally, these are energy-dense foods or drinks, high in fat, salt, sugar, or alcohol. They are addressed in recognition that these foods and drinks do form part of the food supply in many countries and can form part of a healthy diet if consumed in small amounts.

In most countries, the information relating to food groups is presented as a food choice guide, which can be summarized in a graphical form with some additional explanatory text. The graphic display of food group recommendations varies from country to country and has evolved over time (Fig. 27.2). In some countries, a pyramid form has been used. In 2011, the US changed from using a Pyramid Guide to a simpler Plate Guide (MyPlate) (Chapter 28.2.1). Some Nordic countries, together with health organizations in China, Australia, and New Zealand, have also used pyramid or triangle designs. In the UK, Australia, many European countries, and the Caribbean, the national food guide is depicted as a plate or in a circular or semicircular design, often divided up in proportion into the recommended amounts for the various food groups. Sometimes a food square is used in developing countries, with each group equally represented (e.g. Iran). Other designs include a quarter-rainbow, which has been changed to an Eat Well Plate (Canada), steps (New Zealand), a wine glass (Israel), a spinning top (Japan), a bicycle (Korea), and a traffic light (UK Health Education Council), which is also often used in diabetic education materials. Nutrition education programmes and wide-ranging public health nutrition approaches are essential for the dissemination and adoption of nutrition guidelines. These are discussed further in Chapter 28.

Traditionally, in most countries a range of expert groups, appointed by national governments, international agencies, and health related non-government organizations have determined recommended nutrient intakes and developed dietary goals (including acceptable macronutrient distribution ranges), which have then been translated by other groups into dietary guidelines. The 9th and

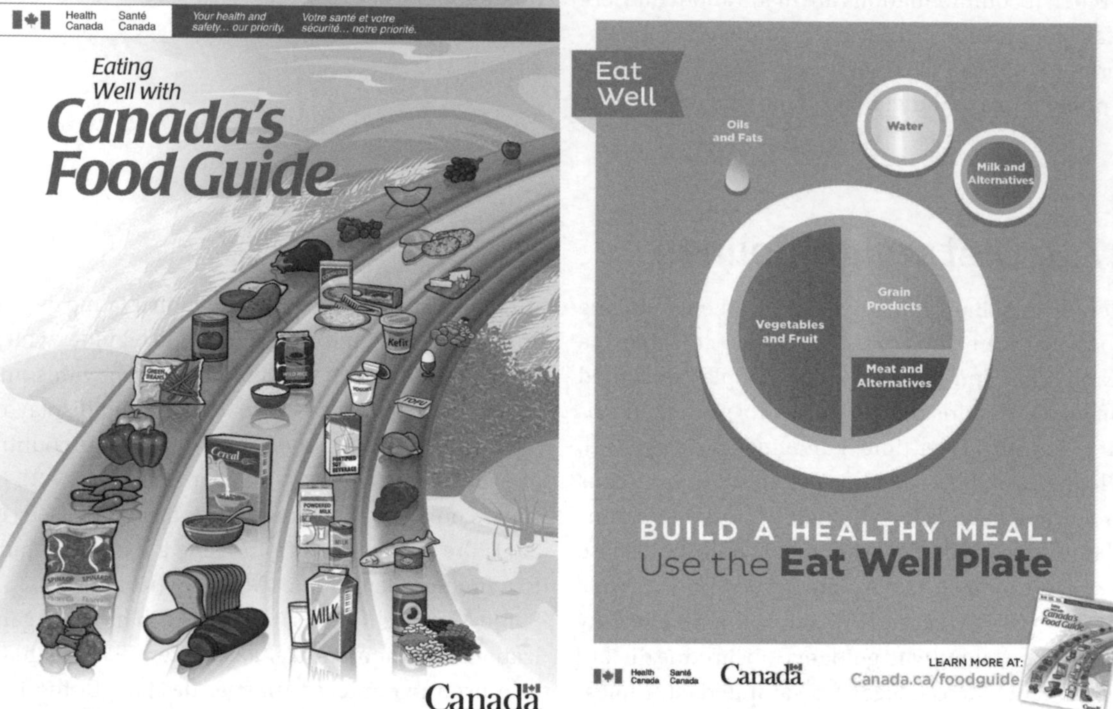


Fig. 27.2 Example of the evolution of a national food guide for Canada, which moved from a Food Rainbow to an Eat Well Plate.

(History of Canada's Food Guides from 1942 to 2007 – Canada.ca and Government Of Canada Launches The Eat Well Plate! – Half Your Plate. Other examples for Australia and UK can be found here: Australian guide to healthy eating | Eat For Health and The Eatwell Guide – GOV.UK (www.gov.uk) (accessed 8 November 2022)

most recent edition of the Dietary Guidelines for Americans 2020–2025, has for the first time made recommendations by life stage, from birth to older adulthood, which are grounded in robust scienitific reviews of current evidence on key nutiriton and health topics. It was noted that although many recommendations have remained relatively consistent over time, this edition of the guidelines has built upon previous editions and evolved as scientific knowledge has grown in three important ways: i) recognition that diet-related chronic diseases are very prevalent and pose a major public health problem, ii) benefits of focusing on dietary patterns rather than on individual nutrients, foods, or food groups, and iii) importance of healthy eating across the lifespan. A summary of the overarching guidelines and four key recommendations are shown in Box 27.4. In addition to the key recommendations, the guidelines

## Box 27.4 Dietary Guidelines for Americans, 2020–2025

**Make every bite count with the Dietary Guidelines for Americans**

There are four overarching guidelines that encourage healthy eating patterns at each stage of life and recognize individuals will need to make shifts in their food and beverage choices to achieve a healthy pattern:

1. *Follow a healthy dietary pattern at every life stage:* infancy, toddlerhood, childhood, adolescence, adulthood, pregnancy, lactation, and older adulthood—it is never too early or too late to eat healthfully.

2. *Customize and enjoy nutrient-dense food and beverage choices to reflect personal preferences,*

*cultural traditions, and budgetary considerations:* a healthy dietary pattern can benefit all individuals regardless of age, race, or ethnicity, or current health status.

3. *Focus on meeting food group needs with nutrient-dense foods and beverages, and stay within calorie limits:* an underlying premise is that nutritional needs should be met primarily from specifically nutrient-dense foods and beverages. The core elements that make up a healthy dietary pattern include:
   - vegetables, of all types
   - fruits, especially whole fruit
   - grains, at least half of which are whole grain
   - dairy, including fat-free or low-fat milk, yogurt, and cheese, and/or lactose-free versions and fortified soy beverages and yoghurt as alternatives
   - protein foods, including lean meats, poultry, and eggs; seafood; beans, peas, and lentils; and nuts, seeds, and soy products
   - oils, including vegetable oils and oils in food, such as seafood and nuts
4. *Limit foods and beverages higher in added sugars, saturated fat, and sodium, and limit alcoholic beverages:* a healthy dietary pattern doesn't have much room for extra added sugars, saturated fat, or sodium—or for alcoholic beverages. Limits are:
   - added sugars—less than 10% of calories per day starting at age 2
   - saturated fat—less than 10% of calories per day starting at age 2
   - sodium—less than 2300 milligrams per day—and even less for children younger than age 14
   - alcoholic beverages—drink in moderation by limiting intake to two drinks or less in a day for men and 1 drink or less in a day for women, when alcohol is consumed.

The Guidelines explicitly emphasize that a healthy dietary pattern is not a rigid prescription. Rather, they are a customizable framework of core elements within which individuals make tailored and affordable choices that meet their personal, cultural, and traditional preferences. Several examples of healthy dietary patterns that translate and integrate the recommendations in overall healthy ways to eat are provided, along with recipes to achieve these within meals.

The Guidelines are supported by Key Recommendations that provide further guidance on healthy eating across the lifespan. The comprehensive 150-page document can be downloaded online.

include comprehensive information regarding goals for macronutrients and recommended intakes for vitamins and minerals for healthy individuals in specified population groups. There is also clear messaging to 'start simple' and 'make every bite count', with examples of food choices, recipes, menus, and how to use the MyPlate as a guide to aid in achieving nutritional goals and a healthy diet.

## 27.4 Dietary goals and guidelines in developing countries

In developing countries, one of the major nutritional problems is that large sections of the population cannot afford or cannot grow enough food to meet all their family's requirements for essential nutrients. This is further exacerbated by ongoing wars, famine, and climate change, which also negatively impact the nutritional status of these populations. There are, however, reasons why dietary goals and guidelines in developing countries also need to include advice aimed at reducing nutrition-related non-communicable diseases:

- diet-related, non-communicable diseases in developing countries account for an increasing share of national mortality
- low-income countries cannot afford to add the burden of medical care of premature degenerative diseases to their already overstretched health budgets

- preparation and production of dietary guidelines is a low-cost measure
- the affluent middle-class in a country such as India or Indonesia may only be 5% of the population, but this means many millions of people (perhaps as many as 70 million in India—more than many whole nations). Many of these more affluent people in the community play a key role in the nation's development.

For India, Gopalan (1989) proposed two sets of recommendations. For the 'relatively poor' majority, diets should be least expensive and conform to tradition and cultural practice as far as possible. Some legumes (pulses) should be eaten along with the high-cereal diet, with some milk and leafy vegetables eaten each day. For affluent Indians, he recommended restriction of energy and of fat (especially ghee), sugar, and salt, with emphasis on unrefined cereals and green leafy vegetables in the diet.

As is the case in more affluent societies, nutrition education programmes and public health nutrition measures are essential for the implementation of goals and guidelines. One method is Community-based Management of Acute Malnutrition (CMAM) programmes, which are available in areas where at least 10% of children are considered to be moderately malnourished. These CMAM programmes provide community health workers with educational tools to help them identify and treat acute malnutrition in children. Moreover, these programmes enable health workers to then provide this education to parents and individuals in the community to empower them to provide both themselves and their children with the proper nutrition they need to reach an acceptable nutritional status.

Furthermore, efforts have been made to produce policies that encourage individuals to consume more vitamins and minerals to tackle the abundance of deficiencies. A large proportion of the Indian population suffers from vitamin A deficiency (VAD), which can cause xerophthalmia, night blindness, delayed growth, reproductive problems, and a higher risk of infections. The high prevalence of VAD is possibly due to the expensive nature of high quality sources of vitamin A (e.g. liver and dairy products), poor absorption rates, and sociocultural constraints. In order to address this issue, the Food Safety and Standards Authority of India (FSSAI) set a policy in place that mandates vitamin A fortification of all cooking oils to be sold. Not only are cooking oils relatively cheap, but they are also widely used among a large variety of the population, making them a good resource for fortification. The FSSAI has stated that fortified oil can help an individual reach 25–30% of their recommended amount of vitamin A daily, thus providing the country with a cheap solution to combat the high prevalence of VAD and VAD-related issues.

## 27.5 Nutrition labelling on foods

Nutrition labelling is part of a public health strategy to help people adopt a healthier diet (section 28.2.1.2). In many countries, labels on food packaging carry information regarding energy and nutrient content in a set amount (often 100 g) of the product or in a typical serving and relating the latter to the daily requirement of the nutrients listed. Because of the limitations of size on food packs, a single reference figure for the daily requirement of a given nutrient is derived from the range available across gender and age bands. In the USA, the reference figure is called a daily value (DV), which is expressed in grams, milligrams, or micrograms of nutrients to consume or not to exceed each day, and the nutrient content of a food may be expressed as % DV per serving or per unit weight.

The wordings and the reference values used for food for labelling purposes are generally laid down by the Codex Alimentarius Commission (Codex, or CAC), which has established intenational food safety standards, or statutory bodies such as the Food and Drug Administration (FDA) in the USA, or Food

Standards Australia New Zealand (FSANZ) in Australia and New Zealand, who are active participants in the Codex process. The same reference value may be used in controlling permitted additions of micronutrients in food fortification (e.g. up to 50% daily values may be added per reference quantity of food 'X').

Nutritional labels are typically displayed on the back of food and drink packaging. However, some countries and organizations have developed simplified labelling systems such as the traffic light system (UK) and the health star rating (Australia and New Zealand) for use on the front of packaged foods (Fig. 27.3). This system can be helpful when consumers want to compare the calorie, fat, sugar, and salt content of different food products at a glance and can help make informed choices. The traffic light system, which was created by the UK Food Standards Agency (FSA) in 2006, is currently only used within the UK. This system relates to four main nutrients (fat, saturates, sugar, and salt), which are placed into one of three colours. Red indicates that the product is high in that nutrient and that intake should be limited, amber indicates a medium level and that foods in that category can be eaten in moderation, and green indicates a low level and that with regard to that particular nutrient, the product represents a good food choice. The traffic light system was set up to encourage the UK population to make healthier choices, by advising them to choose products with amber or green labels.

## 27.6 Summary

Nutritional recommendations are developed by groups of experts usually appointed by governments or authoritative national or international health-related organizations. They are based on systematic reviews and where appropriate meta-analyses of existing research data. There are three types of recommendations.

Advice relating to recommended daily intakes of essential nutrients is typically based on experimental evidence examining amounts of the nutrients necessary for relevant physiological processes or to avoid overt clinical deficiencies. Requirements may differ in women and men, at different stages of the lifecycle, and in many pathological situations, so tables of recommended nutrient intakes are presented by age and sex groups. Requirements may also differ in some disease states and occasionally by geographic location, such as with vitamin D, where higher intakes are required by those with limited exposure to sunlight.

Recommendations relating to optimal proportions and types of dietary fat and carbohydrate are key components of dietary goals that have become increasingly important with the recognition that nature as well as amount of these macronutrients are determinants of risk of many non-communicable diseases, which contribute a high proportion of the globabl burden of disease in most countries.

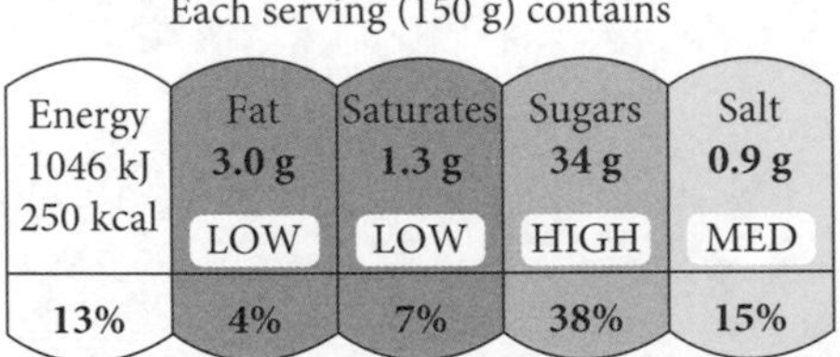


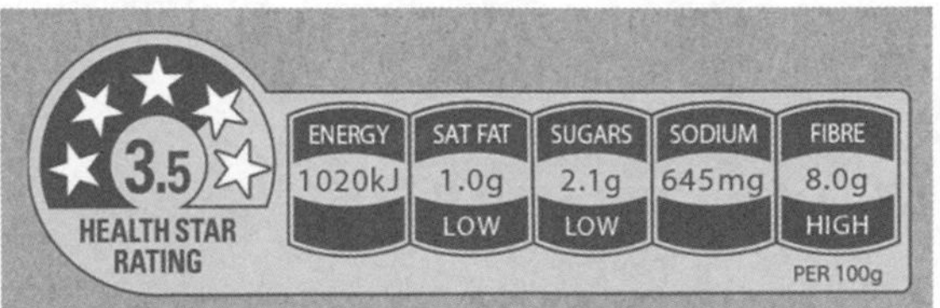


Fig. 27.3 Examples of the traffic light label found on some food labels in the UK and the health star rating found on some food labels in Australia and New Zealand.

Advice relating to acceptable macronutrient distribution ranges is derived principally from prospective epidemiological studies and sometimes randomized clinical trials. Dietary goals may also include advice relating to energy balance and food sources of recommended and restricted macronutrients, as well as essential nutrients.

Recommended nutrient intakes and dietary goals are intended for governments, other authorities charged with the responsibility of ensuring an adequate and appropriate food supply, and health professionals. Dietary guidelines are necessary to translate this information into advice that can be implemented by individuals wishing to make healthy food choices, by those responsible for food preparation, and by food manufacturers. Guidelines relate to food groups and individual foods and how they are prepared. They are the tools by which dietitians and other health professionals advise those with individual health needs and form the basis of public health nutrition approaches aimed at reducing the worldwide consequences of inadequate or inappropriate nutrition.

## Further reading


1. **Food Standards Agency** (2020) How to use nutritional labels on pre-packed foods to find calorie, fat, saturates, sugars and salt content information. https://www.food.gov.uk/safety-hygiene/check-the-label (accessed 27 October 2022).
2. **Gopalan, C.** (1989) Dietary guidelines from the perspective of developing countries. In: Latham, M.C. and van Veen, M.S. (eds) *Dietary guidelines: Proceedings of an International Conference. International Monograph 21*. Ithaca, NY: Cornell.
3. **Government Dietary Recommendations UK** (2016) Government recommendations for energy and nutrients for males and females aged 1–18 years and 19+ years. https://assets.publishing.service.gov.uk/government/uploads/system/uploads/attachment_data/file/618167/government_dietary_recommendations.pdf (accessed 2 November 2022).
4. **National Health and Medical Research Council, Australian Government Department of Health and Ageing, New Zealand Ministry of Health** (2006; updated 2017) *Nutrient Reference Values for Australia and New Zealand*. Canberra: National Health and Medical Research Council. https://www.nhmrc.gov.au/sites/default/files/images/Nutrient-reference-aus-nz-executive-summary.pdf (accessed 27 October 2022).
5. **The German Nutrition Society (DGE), Austrian Nutrition Society (ÖGE), Swiss Society for Nutrition Research (SGE)** (2002) *Reference values for nutrient intake*. Frankfurt am Main: Umschau/Braus.
6. **US Department of Agriculture and US Department of Health and Human Services** (2020) *Dietary Guidelines for Americans, 2020–2025*, 9th edn. www.dietaryguidelines.gov (accessed 13 October 2021).
7. **World Health Organization (WHO)** (2020) Healthy diet. https://www.who.int/initiatives/behealthy/healthy-diet (accessed 2 November 2022).

# Public Health Approaches to Implement Dietary Recommendations

Susan A. Jebb



## 28.1 Introduction

Most countries have a process to develop dietary recommendations for their population. They may be produced by national expert committees or based on the work of regional groups such as the European Food Safety Agency (EFSA) or international bodies, e.g. World Health Organization (WHO). Dietary recommendations provide the prescription for a healthy diet, but in order to improve what people eat, these recommendations need to be implemented. Public health nutrition policy is aimed at supporting this goal.

Traditionally, most efforts to change dietary intake have focused on educational initiatives, including both individual-level counselling, usually delivered in a clinical context by healthcare professionals, or through large-scale public campaigns. While adults make decisions about what, where, when, and how much to eat, making healthy food and beverage choices can be challenging because what may appear to be a conscious choice is more often than not an automatic process, born from past experience and habits or the availability of food in the immediate environment. Moreover, there is growing recognition that individual choices are shaped by marketing activities, price promotions, and social and cultural norms. Accordingly, while the final step in the process of eating is an individual action, implementing recommendations that require a change in dietary behaviours needs action across society to create a culture in which healthy lifestyle choices at home, school, work, and in the community are easy, accessible, affordable, and normative. Public health interventions to implement dietary recommendations need to encourage and empower individuals with practical strategies suited to the prevailing environment, whilst also making the healthier choices easier by changing the environmental determinants to favour diets that meet recommendations for good health.

## 28.2 Interventions to improve dietary intake

A range of approaches have been used to encourage the adoption of healthier diets. These can be subdivided into actions in four broad areas: people, products, promotions, and place, though there are many interrelationships between them (Fig. 28.1).

### 28.2.1 People

#### 28.2.1.1 Nutrition education

Dietary recommendations are complex and frequently expressed in the form of nutrients such as the proportion of energy from saturated fat or sugars, or grams of a particular micronutrient such as sodium. To improve consumer understanding, many countries have translated nutrient recommendations into food-based guidelines that can be communicated in a range of formats. Best known is the 'MyPlate' option, which has long been used to provide guidance for Americans. This shows the approximate proportion of each of the major food groups to constitute a healthy diet. These tools form the basis of much of the consumer education, whether in one-to-one consultations by health professionals or in written or digital formats.

In recent years, public health campaigns have adopted a more behaviourally informed approach, often described as social marketing. These campaigns seek to provide a mix of information and support, including goal setting, incentives, monitoring, and feedback to more effectively engage with the target audience and to stimulate behaviour change. Evaluation of these campaigns tends to show improvements in awareness with short-term changes in reported and some objectively measured behaviours. For example, the Change4Life campaign in England conducted a campaign to encourage reductions in sugar intake, which was associated with a significant decrease in some high-sugar foods and drinks, at least in the short term (Wrieden and Levy, 2016). For example, in week 3, 32% of the intervention group versus 19% in the comparison groups ($p = 0.01$) had purchased a lower-sugar drink, and 24% versus 12% purchased a lower-sugar cereal ($p = 0.009$).

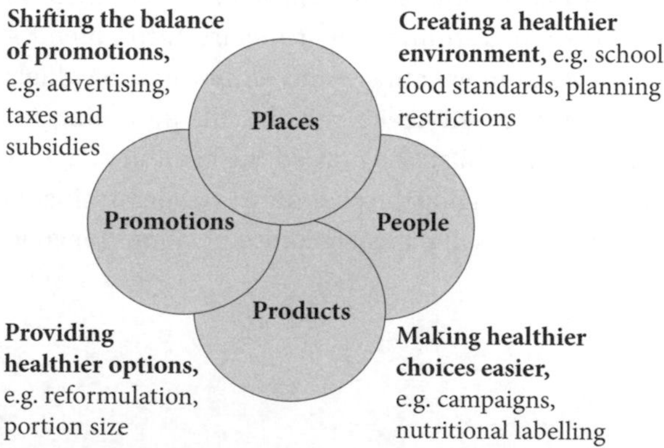


Fig. 28.1 A public health approach to dietary change.

These initiatives may also help to build public support for other policy actions by raising awareness of the importance of the health risks and the steps needed to change diets.

#### 28.2.1.2 Nutritional labelling

Providing accurate and accessible nutrition information is vital if people are to be able to select healthier products. In addition to the mandatory back-of-pack nutrient declarations, a growing number of manufacturers now provide front-of-pack information in a variety of formats, including numerical information and colour-coded guidance. Nutrition information is currently less available for products sold for immediate consumption in out-of-home settings. However, New York State has led the way in calorie labelling on menus and this is gradually being adopted across the USA and internationally. A Cochrane review has shown that calorie labelling on menus leads to a modest reduction in the energy content of foods purchased (Crockett et al., 2018). The impact on foods sold in grocery stores is less clear, perhaps because customers are less focused on individual food items or may be purchasing food for a number of people with diverse needs. For people who look at labels, this information may change the selection or purchase of foods, but more work is needed to motivate others to seek out this information. However, there are consistent reports from the food industry that front-of-pack labelling acts as a stimulus during product development to enhance the nutritional profile of foods (reformulation).

## 28.2.2 Products

### 28.2.2.1 Reformulation

Monitoring of the nutritional composition of food and drinks within countries, or at an international level, shows that apparently similar products may differ markedly in their nutritional composition. Encouraging companies to strive to achieve the healthiest composition for their product can help to reset the nutritional composition of a range of food and drinks. Such action may result from consumer demand or a planned strategy developed by industry or policy makers.

Reformulation of processed foods to reduce the amount of saturated fat, sugar, and salt is a strategy that has been proven to help improve diet quality at a population level, since it improves the nutrient composition of mainstream products, effectively shifting the default, without requiring any conscious change on the part of consumers. This is sometimes described as 'health by stealth.' The most well documented example is the salt reduction programme in the UK, which has seen salt intake decrease by approximately 15% in a decade.

This success of voluntary agreements with industry is dependent on the willingness of companies to participate. It has proved hard to identify incentives to encourage participation or to introduce disincentives for inaction, and progress is often patchy. To avoid this problem, some countries have introduced mandatory targets for reformulation on a limited range of products to create a 'level playing field' whereby the more progressive companies, who invest in reformulation, are not unduly penalized. For example, in the case of *trans*-fats, some countries have introduced mandatory bans, which prohibit the addition of any industrially hydrogenated *trans*-fats. Whether or not mandation is more effective than voluntary measures hinges on the breadth and ambition of the targets set and the strength of monitoring and enforcement efforts.

### 28.2.2.2 Portion size

In recent decades there has been an increase in the portion size of a wide range of food and drinks. A Cochrane systematic review has shown that people consistently consume more food and drink when offered larger-sized portions or packages of food than when offered smaller-sized versions (Hollands et al., 2015). This suggests that policies and practices that successfully reduce the size, availability, and appeal of larger-sized portions can contribute to meaningful reductions in a range of nutrients of concern, including energy, saturated fat, sugar, and salt. Voluntary agreements with industry have led to some notable reductions in portion size, such as a voluntary cap on the portion size of individual servings of confectionery to no more than 250 kcal across three of the leading manufacturers in the UK. But despite this agreement, sales of confectionery have increased as a consequence of other compensatory actions, such as the introduction of 'sharing bags' of confectionery. Moreover, some attempts to introduce mandatory caps on portion size have met with consumer resistance, especially because larger portions are often sold at disproportionately low prices. In New York, an attempt to limit the portion size of soft drinks to no more than 16 oz (approximately 0.5 L) was overturned following a campaign funded by the soft drink industry, highlighting the importance of broad support from citizens for mandatory action to be effectively implemented.

## 28.2.3 Promotions

### 28.2.3.1 Advertising

Available research evidence shows that all forms of marketing consistently influence food preference, choice, and purchasing, at both a brand and category level. Since the nature of food and drink that is advertised is less healthy than that which is recommended, advertising is likely to be an important contributor to poor dietary habits. Several studies have assessed the effects of food and non-alcoholic beverage advertising on food intake, especially among children, with results that strongly support public health policy action to reduce children's exposure to unhealthy food advertising. Despite mounting evidence of the impact of advertising on children's food choices, there is a lag in policy actions to reduce exposure in most countries.

Many food companies have signed up to voluntary codes of practice, but there are numerous examples of breaches of the codes. In some countries there are legal restrictions, but given the broad and complex nature of advertising, these only cover a small fraction of the overall marketing activity. In the UK, where a partial ban that prohibits the advertising of foods high in fat, sugar, and salt during children's television programming is in place, there has been a reduction in exposure during the protected television viewing time, but much of children's viewing occurs during peak hours when no such restrictions exist. This has now been recognized by government, and legislation to introduce a '9pm watershed' for TV advertising and, crucially, a complete ban online will come into force in 2023. Controlling online advertising is important, since expenditure on internet advertising already exceeds that of television advertising.

#### 28.2.3.2 In-store promotions

A wide variety of promotional strategies are used by manufacturers and retailers to encourage the purchase of particular foods and drinks in-store. These strategies are designed to change dietary habits with the purpose of boosting profits. Many operate through changing the price of a product, such as multi-buy offers or temporary price reduction. These strategies may be used in conjunction with other forms of promotion, such as prominent product positioning, for example on an eye-level shelf or at the end of a shopping aisle.

The net effect of these promotions is that people buy more, fostering overconsumption, often of food and drinks high in saturated fat, sugar, and salt. In the UK the use of in-store promotions of food is particularly high, accounting for more than a third of all food purchases. Reducing the impact of such promotions is increasingly seen as an important strategy to improve diet quality. Recent research has shown that removing less healthy foods from prominent locations such as end-of-aisle can lead to a marked decrease in overall purchasing (Ejlerskov et al., 2018). In England, legislation to limit the use of prominent location or volume-based promotions of foods high in saturated fat, free sugars, and salt was introduced in 2022.

In contrast, promotions to encourage healthier food choices have been less successful. Reductions in price stimulate sales but once the promotion is removed there is no evidence of a lasting change in food habits. Placing healthier foods in more prominent locations, such as eye-level shelves, has had limited impact on sales, for example, of breakfast cereals with higher fibre or lower sugar content, perhaps because this is not an impulse purchase and people are seeking specific brands.

There is a clear need for more detailed information on 'what works' to shift purchasing behaviour to meet health rather than commercial goals. Alongside, with the growing importance of sales of food for consumption out-of-home, more research is needed to consider how these canteen, café, or restaurant environments shape food choices.

#### 28.2.3.3 Tax

There is considerable interest in the use of taxation to shift dietary habits. Price increases have been an important component in public health strategies to decrease alcohol consumption and reduce the prevalence of smoking. Extensive modelling work suggests a similarly positive effect on diet quality, though the magnitude of changes in intake is difficult to forecast. Several countries have now introduced a tax on sugar-sweetened beverages. In Mexico, an excise tax of approximately 10% was associated with a decrease of 6% in purchases of sugar-sweetened drinks and corresponding increases in purchases of untaxed beverages (Colchero et al., 2016). The decrease was greatest among low-income groups where intake was initially the highest, suggesting proportional benefits. The UK soft drink industry levy, which was intentionally constructed to drive reformulation, led to a decrease in the mean sales weighted sugar content of these drinks by 34% from 2015 to 2018, and a concomitant decrease of 30% in the per capita sales of sugar, from 15.5 g per person per day in 2015 to 10.8 g in 2018 (Bandy et al., 2020). Despite all the public health campaigns to raise awareness of the high sugar content of these products, there was an overall increase in consumption of soft drinks, reiterating the public health benefits of pressure on the food supply chain, rather than relying on changes in consumer demand to feedback to changes in supply.

### 28.2.4 Place

#### 28.2.4.1 Local communities

The availability of food and drink increases the likelihood of consumption. Traditional meal eating at home is in decline and more food is consumed 'on-the-go', increasing the number of eating occasions each day. Food and drink, especially snacks high in fat, sugar, and salt, are often available in non-food stores, often at locations intended to prompt impulse purchases. A range of strategies has been proposed that target the placing of food and drink in the micro-environment within stores, for example restrictions on foods and drinks sold at checkouts, especially in supermarkets, though these usually rely on voluntary action by stores themselves.

There are also opportunities to take action at a community level. Research shows some associations between the number of food outlets within a community and the dietary habits or risk of obesity in an area (Liu et al., 2020). This has prompted some local governments to take action, for example to limit food outlets near schools or to use planning laws to control the overall number of food outlets. In other cases, efforts have been made to boost the availability of healthier options, especially fresh fruit and vegetables. This includes encouraging local stores to stock fruit and vegetables, as well as community gardening or allotment schemes, often set up in areas of deprivation as a means of developing social cohesion and boosting food skills, as well as encouraging consumption of fresh food.

In places where food is provided, there are opportunities to provide healthier choices through changes in the procurement, provision, and sales of food and drinks. Schools have been the focus for a large number of place-based initiatives around the world, including mandatory standards. Food consumed in schools makes an important contribution to children's diets, but also provides an opportunity to educate children about healthy eating and to use role modelling and peer support to encourage children to try new foods, increasing dietary diversity. There is potential to adopt health-promoting food policies in other public environments, such as hospitals or community leisure facilities.

#### 28.2.4.2 Food delivery

In recent years there has been a marked rise in home delivery food services, accelerated in many countries by Covid-19 restrictions. A small number of online aggregators control a large proportion of a highly fragmented out-of-home market, in effect exerting similar power to grocery stores within the in-home environment. This offers new opportunities to consider enhanced regulation of this sector, from point-of-choice nutrition labelling, through to controls on promotions and advertising.

## 28.3 A framework for public health interventions

Implementing interventions to change dietary habits will require coordinated action by a wide range of stakeholders and requires strong political leadership. There is a lively debate about how far public policies can and should go to change what is considered by many people to be a personal decision about what to eat.

The Nuffield Council on Bioethics set out a stewardship model to describe the broad policy options to achieve improvements in public health in the form of a ladder of interventions (Box 28.1). This recognizes that health is a key asset and that better health is associated with greater economic productivity and well-being, justifying societal action. Their model, based on the principles of libertarian paternalism, proposes that action should be confined to the lower levels of the intervention ladder until it proves ineffective. At this point, further action should

**Box 28.1 Nuffield ladder of intervention**

Possible government actions ranging from doing nothing or simply monitoring to the elimination of choice.

*Eliminate choice.* Regulate in such a way as to entirely eliminate choice, for example through banning confectionary in schools.

*Restrict choice.* Regulate in such a way as to restrict the options available to people with the aim of protecting them, for example removing unhealthy ingredients from foods, or unhealthy foods from shops or restaurants.

*Guide choice through disincentives.* Fiscal and other disincentives can be put in place to influence people not to pursue certain activities.

*Guide choices through incentives.* Regulations can be offered that guide choices by fiscal and other incentives, for example by introducing subsidies or vouchers for fruit and vegetables.

*Guide choices through changing the default policy.* For example by positioning products in grocery stores so healthier options are more prominent.

*Enable choice.* Enable individuals to change their behaviours, for example by offering participation in a NHS 'stop smoking' programme, building cycle lanes, or providing free fruit in schools.

*Provide information.* Inform and educate the public, for example as part of campaigns to encourage people to walk more or eat five portions of fruit and vegetables per day.

*Do nothing or simply monitor the current situation.* For example through introducing taxes on soft drinks.

https://www.nuffieldbioethics.org/wp-content/uploads/2014/07/Public-health-ethical-issues.pdf (accessed 21 March 2022).

be proportional to the public health risks and used only where there is good evidence that the intervention will produce the intended effect. This approach seeks to balance the infringement of personal autonomy with the societal gains to be achieved.

At the lower levels of the ladder, action is mostly confined to knowledge-based interventions, equipping individuals with the education and skills to make informed choices for themselves or those they care for. This might include actions such as social marketing campaigns or nutritional labelling. The middle steps describe changes in the choice architecture, sometimes known as 'nudge' theories. These interventions are predicated on the fact that many 'choices' about what to purchase or to eat are not the product of conscious decision making but represent the easiest or default 'choice'. Interventions here seek to create an environment in which choices are not restricted but designed such that the healthier option becomes the more common choice. This might include the reformulation of food products or menu options where the default is the healthier choice, leaving customers to make a conscious decision if they wish to take alternative action. This could include serving dressings for salads on the side, not adding salt in cooking but leaving customers to add salt at the table if they wish, making smaller glasses of wine the standard measure, or more substantive changes such as serving meals with non-fried potatoes in place of fried options, or offering meat-free meals with meat dishes as the alternative. Changing the choice architecture in retail outlets to favour the healthier option has profound implications for the promotion of foods and drinks. For example, there is a marked sales uplift for products positioned at the end of the shopping aisle. If this space was reserved for healthier options, with less healthy variants only available in their regular position within the aisles, this would not reduce choice available in the store, but would give greater prominence to the healthier item and likely shift the pattern of purchases.

Next are interventions that attempt to guide choice through the use of incentives. Here there are opportunities across all forms of marketing to promote healthier choices using a raft of established techniques, including advertising and promotions. Using disincentives to guide choice, most prominently by introducing a tax, is a more intrusive policy since it introduces tangible penalties into the system to discourage some food and drink choices.

The strongest policy action actively restricts the choices available either by controlling who can purchase certain items, such as age restrictions on alcohol, or the settings in which foods are available for purchase or consumption, such as the restrictions on the sale of certain foods and drinks in schools. Total bans on specific food or drink items are rare in a modern democracy, though it is conceivable that food consumption could be limited in some settings to reduce availability.

## 28.4 Developing a public health nutrition strategy

In practice the complexity of the diet and the scale of the challenge are likely to need a portfolio of actions. Many of the policy options can be used alongside each other to create a complex behavioural intervention. For example, public health campaigns can motivate people to make healthier choices; this is enabled by nutritional labelling, which in turn creates a commercial advantage to reformulate, and hence incentivizes change across the industry.

Over the years there have been a number of models proposed to guide policy development. The Analysis Grid for Environments Linked to Obesity (ANGELO) was developed to guide interventions to prevent obesity, but can be applied more broadly. It includes a consideration of physical environments that govern the food that is available, economic aspects, the political economy, and sociocultural context. More recently the NOURISHING framework was developed through reviews of theoretical models such as ANGELO and national policy interventions that have been proposed or introduced (Hawkes et al., 2013). The NOURISHING framework describes a comprehensive package of policies to improve the availability, affordability, and acceptability of healthier diets (Table 28.1). It sets out actions across three domains: food environment, food system, and

**Table 28.1 The NOURISHING policy framework to promote healthy diets**

| Domain | | Policy area | Examples of potential policy actions |
|---|---|---|---|
| Food environment | N | Nutrition label standards and regulations on the use of claims and implied claims on foods | e.g. Nutrient lists on food packages; clearly visible 'interpretive' and calorie labels; menu, shelf labels; rules on nutrient and health claims |
| | O | Offer healthy foods and set standards in public institutions and other specific settings | e.g. Fruit and vegetable programmes; standards in education, work, health facilities; award schemes; choice architecture |
| | U | Use economic tools to address food affordability and purchase incentives | e.g. Targeted subsidies; price promotions at point of sale; unit pricing; health-related food taxes |
| | R | Restrict food advertising and other forms of commercial promotion | e.g. Restrict advertising to children that promotes unhealthy diets in all forms of media; sales promotions; packaging; sponsorship |
| | I | Improve the nutritional quality of the whole food supply | e.g. Reformulation to reduce salt and fats; elimination of *trans*-fats; reduce energy density of processed foods; portion-size limits |
| | S | Set incentives and rules to create a healthy retail and food service environment | e.g. Incentives for shops to locate in underserved areas; planning restrictions on food outlets; in-store promotions |

*(Continued)*

Table 28.1 **The NOURISHING policy framework to promote healthy diets (*Continued*)**

| Domain | | Policy area | Examples of potential policy actions |
|---|---|---|---|
| Food system | H | Harness the food supply chain and actions across sectors to ensure coherence with health | e.g. Supply-chain incentives for production; public procurement through 'short' chains; health-in-all policies; governance structures for multi-sectoral engagement |
| Behaviour change communication | I | Inform people about food and nutrition through public awareness | e.g. Education about food-based dietary guidelines, mass media, social marketing; community and public information campaigns |
| | N | Nutrition advice and counselling in healthcare settings | e.g. Nutrition advice for at-risk individuals; telephone advice and support; clinical guidelines for health professionals on effective interventions for nutrition |
| | G | Give nutrition education and skills | e.g. Nutrition, cooking/food production skills on education curricula; workplace health schemes; health literacy programmes |

*Adapted from:* the WCRF NOURISHING Framework Domains—last updated 24 November 2015. https://www.wcrf.org/policy/policy-databases/nourishing-framework/ (accessed 21 March 2022).

behaviour change communication; within which policy makers have the flexibility to select specific policy options suitable for their national/local contexts and target populations. This recognizes the need to tailor the policy response to local conditions.

A mapping exercise to relate existing international policies to the NOURISHING framework shows clearly that the majority of current policy actions relate to the provision of information and other educational components of the model, with relatively few measures directed at changes in the food environment and a paucity of interventions that run through the whole food supply chain. Using this framework to monitor the actions taken by individual countries to help achieve dietary recommendations shows that no country has yet implemented a comprehensive strategy across all of these domains.

## 28.5 Summary

Dietary recommendations do not in themselves improve public health unless they drive change in the food system that leads to a sustained change in dietary habits. An individual's food habits represent the final step in a complex system of food production, processing, and promotion, shaped by personal preferences, experience, and circumstances. Efforts to implement dietary recommendations need to recognize this complex system and engage the full range of stakeholders in a coordinated set of responses. Some interventions will be dependent on other interventions to yield their full benefits. For example, social marketing campaigns may be a necessary prerequisite to change public attitudes to facilitate the introduction of other, more effective interventions.

One of the lessons from tobacco control policies is that a portfolio of coordinated interventions is necessary to shift behaviour. This broad approach is likely to be even more important for dietary change, where there is clear potential for compensatory responses, with improvements in one nutrient or food group counterbalanced by a worsening in other components of the diet, or improvements in dietary intake at home offset by greater consumption out of home.

Given that a poor diet is gradually overtaking tobacco as the leading modifiable risk factor for disease, it is vital that far greater attention is paid to translating the scientific evidence in relation to diet and health into practical strategies to implement recommendations. The hope is that, over time, as interventions start to shift the social and cultural norms related to food, the pace of change will accelerate and bring much needed improvement in public health. As individual countries take new actions to enhance public health nutrition, it is vital that there is independent evaluation of the impact of these policies that can be shared as a case study to inform action elsewhere.

## Further reading


1. **Bandy, L.K., Scarborough, P., Harrington, R.A., et al.** (2020) Reductions in sugar sales from soft drinks in the UK from 2015 to 2018. *BMC Med* **18**, 20. https://doi.org/10.1186/s12916-019-1477-4 (accessed 29 June 2022).

2. **Testing availability, positioning, promotions, and signage of healthier food options and purchasing behaviour within major UK supermarkets: Evaluation of 6 nonrandomised controlled intervention studies** Carmen Piernas 1, Georgina Harmer 1, Susan A Jebb 1 DOI: 10.1371/journal.pmed.1003952

3. **Colchero, M.A., Popkin, B.M., Rivera, J.A., Ng, S.W. et al.** (2016) Beverage purchases from stores in Mexico under the excise tax on sugar sweetened beverages: observational study. *BMJ*, **352**, h6704. doi: 10.1136/bmj.h6704

4. **Crockett, R.A., King, S.E., Marteau, T.M., et al.** (2018) Nutritional labelling for healthier food or non-alcoholic drink purchasing and consumption. *Cochrane Database Syst Rev*, **2**(2), CD009315. doi: 10.1002/14651858.CD009315.pub2

5. **Ejlerskov, K.T., Sharp, S.J., Stead, M., et al.** (2018) Supermarket policies on less-healthy food at checkouts: Natural experimental evaluation using interrupted time series analyses of purchases. *PLoS Med*, **15**(12), e1002712. doi: 10.1371/journal.pmed.1002712

6. **Gressiere, M., Swinburn, B., Frost, G., Segal, A.B., and Sassi, F.** (2021) What is the impact of food reformulation on individuals' behaviour, nutrient intakes and health status? A systematic review of empirical evidence. *Obes Rev*, **22**(2), e13139.

7. **Hawkes, C., Jewell, J., and Allen, K.** (2013) Food policy package for healthy diets and the prevention of obesity and diet-related non-communicable diseases: the NOURISHING framework. *Obes Rev*, **2**, 159–68.

8. **Hollands, G.J., Shemilt, I., Marteau, T.M., et al.** (2015) Portion, package or tableware size for changing selection and consumption of food, alcohol and tobacco. *Cochrane Database Syst Rev*, **14**(9), CD011045. doi: 10.1002/14651858.CD011045.pub2

9. **Liu, B., Widener, M., Burgoine, T., and Hammond, D.** (2020) Association between time-weighted activity space-based exposures to fast food outlets and fast food consumption among young adults in urban Canada. *Int J Behav Nutr Phys Act*, **17**(1), 62.

10. **Mah, C.L., Luongo, G., Hasdell, R., Taylor, N.G.A., and Lo, B.K.** (2019) A systematic review of the effect of retail food environment interventions on diet and health with a focus on the enabling role of public policies. *Curr Nutr Rep*, **8**(4), 411–28.

11. **Mahesh, R., Vandevijvere, S., Dominick, C., and Swinburn, B.** (2018) Relative contributions of recommended food environment policies to improve population nutrition: results from a Delphi study with international food policy experts. *Public Health Nutr*, **21**(11), 2142–8.

12. **Micha, R., Karageorgou, D., Bakogianni, I., et al.** (2018) Effectiveness of school food environment policies on children's dietary behaviors: a systematic review and meta-analysis. *PLoS One*, **13**(3), e0194555.

13. **Theis, D.R.Z. and White, M.** (2021) Is obesity policy in England fit for purpose? Analysis of government strategies and policies, 1992–2020. *Milbank Q*, **99**(1), 126–70.

14. **Wrieden, W.L. and Levy, L.B.** (2016) 'Change4Life Smart Swaps': quasi-experimental evaluation of a natural experiment. *Public Health Nutr*, **19**(13), 2388–92. doi: 10.1017/S1368980016000513

# Dietary Patterns

C. Murray Skeaff



## 29.1 Defining dietary patterns

There is tremendous variation in the food habits of individuals and in the food choices, preferences, meal patterns, and cuisines that define their diets. Yet, despite this variation, there also exists amongst communities, both small and large, many similarities in the mix or 'pattern' of foods and beverages that make up their diets. These similarities are often referred to or labelled as 'dietary patterns'. Dietary patterns describe or classify the quantity, combinations, and frequency of typical foods and beverages consumed amongst groups of people, and are often used as examples of diets to be copied, imitated, or, in some cases, to be avoided.

### 29.1.1 Patterns defined by geographical region

The first and most common types of dietary patterns are those that describe the diets in a particular geographical region; for example, the Mediterranean diet, the Nordic diet, the Asian diet, or the Western diet. In the case of the Mediterranean or Asian diet, interest in these dietary patterns was part of the early efforts to understand, using epidemiological methods, why people in those regions had particularly good health, lower rates of cardiovascular disease, and longer life expectancy. In the case of the Western diet, it was to understand the opposite; the dietary pattern associated with a higher burden of several non-communicable diseases, such as cardiovascular disease, seen in high-income Westernized countries.

Dietary patterns based on geographical region are generalizations and often overlook differences between countries in the same region; moreover, the patterns in the present day may have changed from those of decades past, and usually differ between age groups. The Mediterranean diet has some common features including an abundance of fruit and vegetables, legumes, whole grains, olive oil, some nuts, wine, dairy, and lean meat; but this pattern is more common to Spain, Greece, and Italy than to other countries that border the Mediterranean Sea, such as Morocco, Egypt, or Turkey, for example. Even across Spain, Greece, and Italy, differences exist in the Mediterranean diet. For example, results from the European Prospective Investigation into Cancer and Nutrition, a prospective study of 500,000 people in ten European countries, found much lower consumption of legumes and higher consumption of cereals in Italy. Spain had very high consumption of eggs and fish, whereas Greece was very high in olive oil and legumes. This begs the question, exactly where is the Mediterranean diet to be found?

### 29.1.2 Patterns defined by dietary guidelines

The second category of dietary patterns are often constructed to conform to a set of food or nutrient-based guidelines; well-known examples include the low-fat high-carbohydrate diet, the Dietary Approaches to Stop Hypertension (DASH) diet, and the low-carbohydrate high-fat diet. Some dietary patterns, such as vegetarian or vegan, are defined more by avoidance than by consumption of particular foods.

### 29.1.3 Other dietary patterns

Scores of other dietary patterns, sometimes referred to as 'populist' or 'fad' diets, abound in the weight-loss industry, or have been promoted commercially or philanthropically along with bold claims to achieve particular health outcomes; for example, the Atkins diet, Dukan diet, South Beach diet, Cambridge diet, detox diet, Paleo diet, Scarsdale diet, raw food diet, intermittent fasting diet (e.g. the 5:2 diet), and the glycaemic index diet, to name a few. Unfortunately, some of the unorthodox 'dietary patterns' require extreme dietary behaviours that belie the history of human eating patterns. Furthermore, some of the diets involve practices that pose risk to health both in the short and long term (see Box 29.1).

### 29.1.4 Patterns defined mathematically

A third, more complex and recent approach to defining dietary patterns is to take a dataset from a dietary survey of a large population and apply statistical methods—such as factor analysis, principle component analysis, or cluster analysis, reduced rank regression, or discriminant analysis—to identify foods that cluster into particular patterns. These dietary patterns are often given names, somewhat arbitrarily, that signal the predominant foods or food groups that cluster together; 'healthy', 'prudent', 'unhealthy', 'Western', 'meat and potatoes', 'fats and processed meat', 'vegetables and fish', 'fruit, vegetables, and dairy', 'animal', 'customary', 'convenience', 'ultra-processed', or 'fast-food' are but a few

#### Box 29.1 Common types of dietary patterns

- *Mediterranean diet*: plenty of fruit, vegetables, legumes, and cereals. Fish frequently. Meat and dairy, mainly cheese, infrequently; regular use of extra virgin olive oil, and moderate alcohol consumption.
- *Western diet*: high in red and processed meats, eggs and egg-based meals, refined grains, sugary and salty desserts and snack foods, high-fat dairy foods, potatoes.
- *Vegetarian diet*: defined by absence of animal meat or fish from the diet. Tends to be higher in fruits and vegetables, breads, cereals, whole grains, and other plant products.
- *Lacto-ovo vegetarian diets*: include eggs and dairy products, whereas vegan diets exclude all animal products and include only food from the plant kingdom. Some consider that fish can be included in a 'semi-vegetarian diet', called a pescetarian diet.

**Weight-control diets**

- *Low-fat diets*: focus on avoidance of high-fat and saturated fat-rich foods, consumption of lean meats and low-fat dairy products, plenty of fruit and vegetables, high consumption of carbohydrate-rich foods, preferably rich in unrefined whole grains, and minor amounts of polyunsaturated or monounsaturated fat-rich plant oils.
- *Low-carbohydrate (high-fat) diets*: focus on avoidance of foods that contain carbohydrate—whether sugars or starch—emphasis on protein-rich foods such as lean meats, also eggs, vegetables, nuts and seeds, and fats and oils; usually no specific advice to avoid saturated fat-rich foods.

examples. However, in this approach the dietary patterns are determined by the dataset and this uniqueness makes it difficult to compare dietary patterns from one population to the next.

## 29.2 A widening focus from nutrients to dietary patterns

Interest in dietary patterns has surged in recent times, partly fuelled by the comprehensive failure of randomized controlled trials to demonstrate the efficacy of single or multiple vitamin and mineral supplements (e.g. antioxidants) to prevent cancer or cardiovascular disease. This failure has caused a serious re-evaluation of the assumptions on which the nutrient supplementation trials were justified and initiated. These assumptions arose from large prospective studies in which people with high intakes of antioxidants from food, or in some cases supplements, had lower risk of

**Box 29.2** Case study: the PREDIMED Trial (Estruch et al., 2018)

**Description of study**

- Randomized controlled trial to prevent cardiovascular disease.
- Conducted in Spain and included 7477 participants (55–80 years) with type 2 diabetes, or at least three major risk factors for cardiovascular disease.
- Participants were randomized to one of three groups:
  - a Mediterranean diet group supplemented with extra virgin olive oil
  - a Mediterranean diet group supplemented with mixed nuts
  - a control group following a reduced-fat version of the local Mediterranean diet.
- The trial was stopped prematurely after 5 years because there was certainty that the intervention was beneficial.
- The rate of major cardiovascular events was 31% lower in the Mediterranean diet group supplemented with extra virgin olive oil, and 28% lower in the Mediterranean diet group supplemented with nuts compared with the control group.

**Interpretation of study**

- The results of this trial have been widely interpreted as proving that adherence to a Mediterranean diet prevents cardiovascular disease.
- One potentially important bias in the study design was that during the first 3 years of the study, participants in the control group were met only once per year and were given no motivational support, whereas participants in the Mediterranean diet groups were given individual and group support with meetings once every 3 months.
- The participants in the Mediterranean diet groups made very few changes to their diets other than consuming the extra virgin olive oil or nuts that were provided to them at no cost by the study organizers.
- The only other statistically significant differences between the Mediterranean diet group and the control group were a higher consumption of legumes (3–4 g/day) and fish (6–7 g/day), changes that were both very small.
- Thus, based on what participants reported eating, it was the extra virgin olive oil and nuts, within the context of the participants' usual Mediterranean diet, which lowered rates of cardiovascular disease.
- This raises several questions about the independent effects of extra virgin olive oil or nuts. Will they occur in people who do not follow a Mediterranean diet? How much needs to be consumed?
- The trial also shows how difficult it is for participants in dietary intervention trials to change their dietary patterns, even when supported by the best dietitians and study organizers.

cardiovascular disease and some types of cancer. In light of the lack of proven effect, in clinical trials, of vitamin or mineral supplementation on cancer and cardiovascular disease, the evidence from observational studies of association between antioxidants and cancer and cardiovascular disease was re-interpreted to recognize that high intake of specific nutrients act as markers for consumption of particular foods, food groups, or dietary patterns. The protective effects associated with specific nutrient intakes may have arisen from the interaction of a complex mix of nutrients and phytochemicals delivered in a particular balance by a healthy pattern of food intake high in the indicator nutrients. As the results of clinical trial after clinical trial showed virtually no effect of nutrient supplements on major disease outcomes, the positive results of a small trial, conducted in Lyon, of diet and cardiovascular disease prevention amongst heart attack survivors drew attention to the potential importance of dietary patterns in the management and prevention of non-communicable diseases.

The Lyon Diet and Heart Study (de Lorgeril et al., 1999) was a 5-year randomized controlled trial of a Mediterranean-type diet for the prevention of cardiovascular disease in 605 heart attack survivors. The results showed that a Mediterranean diet pattern reduced the risk of fatal cardiovascular disease by 76%. Unfortunately, the actual changes participants made in the consumption of foods associated with a Mediterranean diet were very small compared with the remarkably large protective effect of the dietary intervention. Despite this mismatch and the natural scepticism it created about the size and veracity of the effect, the results of the Lyon Diet and Heart Study encouraged further efforts to test the effectiveness of the Mediterranean diet and other dietary patterns on health (see Box 29.2).

## 29.3 Dietary patterns and health

The main interest, insofar as the science of health and nutrition is concerned, lies not in classifying and describing patterns of food and beverage consumption, but rather in identifying the dietary patterns, and their common features that confer better health on those who follow them. The intense focus on learning about healthy dietary patterns has evolved for several reasons, not least of which is the long-held nutrition adage that 'people eat food not nutrients'. On one hand, this simple truism states the obvious, that food is the vehicle that delivers to the body the nutrients it needs for life. However, on the other, it reflects a universal acceptance that nutrition for living longer, healthier lives—healthy longevity—requires more than a fixed number of essential macro- and micronutrients that can be delivered in capsules. Rather, with respect to diet, healthy longevity requires and is influenced by a wide range of non-essential or complementary nutrients, particularly phytochemicals (i.e. chemical constituents of plants). The range is so diverse, and the balance of amounts so difficult to prove by clinical trials of supplementary mixtures of purified phytochemicals, that it seems sensible to evaluate food-based approaches—dietary patterns—while bringing to bear evolving understanding about the metabolic effects of the mix of nutrients and non-nutrients associated with particular dietary patterns.

Current knowledge and understanding about the different types of dietary patterns, and more specifically their effects on health, is dominated heavily by research generated in Europe and North America and published in English-language journals. Yet the majority of the world's population resides in Asia (60%) and Africa (17%), not in Europe (10%) or North America (5%). The citizens of Japan have the longest life expectancy and considerably lower age-standardized rates of non-communicable diseases, yet awareness, at least in the West, about the dietary patterns that may contribute to this longevity is sparse.

### 29.3.1 Scoring adherence to dietary patterns

One research approach to learning about healthy dietary patterns has been to follow-up participants prospectively over many years and examine whether high adherence to a given dietary pattern is associated with lower rates of disease. Adherence to dietary patterns in these types of studies is assessed by a scoring system that usually has 7–10 dietary components, with points awarded by comparing a person's food consumption, and in some cases nutrient intake (e.g. sodium), to cut-offs of each component. The criteria that define low or high consumption within each component are often set according to the median intakes in the study population, or may be set by absolute targets as defined in the dietary guidelines (see Box 29.3). The body of this evidence shows that good adherence to dietary patterns based on dietary guidelines, such as the Health Eating Index or Dietary Guidelines for Americans, are associated with lower risk of cardiovascular disease

**Box 29.3** Adherence score for a Mediterranean dietary pattern (PREDIMED)

| Question | Foods and frequency of consumption | Criteria for one point[a] |
|---|---|---|
| 1 | Do you use olive oils as the main culinary fat? | Yes |
| 2 | How much olive oil do you consume in a given day (including oil used for frying, salads, out-of-house, meals, etc.)? | 4 or more tablespoons |
| 3 | How many vegetable servings do you consume per day? (1 serving = 200 g, consider side dishes as 1/2 serving) | 2 or more (at least 1 portion raw or as salad) |
| 4 | How many fruit units (including natural fruit juices) do you consume per day? | 3 or more |
| 5 | How many servings of red meat, hamburger, or meat products (ham, sausage, etc.) do you consume per day? (1 serving = 100–150 g) | Less than 1 |
| 6 | How many servings of butter, margarine, or cream do you consume per day? (1 serving = 12 g) | Less than 1 |
| 7 | How many sweet/carbonated beverages do you drink per day? | Less than 1 |
| 8 | How much wine do you drink per week? | 7 or more glasses |
| 9 | How many servings of legumes do you consume per week? (1 serving = 150 g) | 3 or more |
| 10 | How many servings of fish or shellfish do you consume per week? (1 serving: 100–150 g fish, or 4–5 units or 200 g shellfish) | 3 or more |
| 11 | How many times per week do you consume commercial sweets or pastries (not homemade), such as cakes, cookies, biscuits, or custard? | Less than 3 |
| 12 | How many servings of nuts (including peanuts) do you consume per week? | 3 or more |
| 13 | Do you prefer to consume chicken, turkey, or rabbit meat instead of veal, pork, hamburger, or sausage? | Yes |
| 14 | How many times per week do you consume vegetables, pasta, rice, or other dishes seasoned with *sofrito* (sauce made with tomato and onion, leek, or garlic, simmered with olive oil)? | 2 or more |

[a]Zero points if criteria are not met.

*Adapted from:* Estruch, R., Ros, E., Salas-Salvadó, J., et al. (2013) Primary prevention of cardiovascular disease with a Mediterranean diet. *N Engl J Med*, **368**, 1279–90.

and other important health outcomes. This is, to some extent, a self-fulfilling prophecy because the dietary guidelines, by definition, are developed to include and combine components of the diet, whether food- or nutrient-based, for which there is convincing and reliable scientific evidence of health benefit. Therefore, one would expect that adherence to a dietary pattern that conforms to the guidelines is associated with better health outcomes. Evidence from prospective studies for the health benefits of a Mediterranean diet is also strong, but whether it is superior compared with dietary patterns based on dietary guidelines remains uncertain.

One of the limitations of using prospective epidemiological approaches to evaluate the health effects of dietary patterns is the difficulty of controlling for confounding. Without exception, people in Europe and North America who score high on adherence to a 'healthy' dietary pattern also have other demographic and lifestyle characteristics that confer better health expectancy. In general, they tend to have lower rates of smoking, exercise more, have more years of formal education, and are of higher socioeconomic status. A healthy dietary pattern, therefore, is also a marker for a healthy lifestyle. To examine the independent effects of dietary patterns on health outcomes, statistical methods must be used to try and disentangle diet from the confounding effects of other demographic and lifestyle factors. This is an imperfect process and there is always the possibility that residual confounding remains and the beneficial effects of adherence to dietary patterns is overestimated.

Many randomized controlled trials have been conducted to examine the effects of dietary patterns on risk factors for cardiovascular disease. For example, the DASH diet, which includes an emphasis on fruit and vegetables, low-fat dairy products, and fewer sweets and snacks, is one of the more well known, and has been shown in both highly controlled or community settings to lower blood pressure substantially. Randomized controlled trials, however, have a different set of limitations than observational studies; these include selection bias, short duration, difficulties in changing food-related behaviours, and lack of adherence. Refer to the Further reading section for further information about clinical trials.

## 29.4 Key food components of a healthy dietary pattern

The diverse range of dietary patterns that has been associated with positive health outcomes leads naturally to an examination of features that are common to most of them. These features include regular consumption of fruit and vegetables, cereals and whole grains, low-fat dairy products, nuts, and legumes. Fish is common, as is moderate consumption of alcohol, but consumption of red and processed meats and sugar-sweetened foods and beverages is low. Dietary patterns that are more plant-based, nutrient-dense, and minimally processed seem to offer particular benefits. These common features across healthy dietary patterns suggest that specific foods or food groups, such as fruit and vegetables or grains and cereals, can have a positive influence on health across a range of dietary patterns. They need not, for example, be combined with olive oil, as in the Mediterranean diet, to be effective, but can be consumed in a range of dietary patterns, for example in a low-fat diet with plenty of rice, as in the Japanese diet. Of course, it is relevant to know the features of healthy dietary patterns followed by people living in North American and Europe, but the direct relevance and applicability of this knowledge to the larger proportion of the world's population living in Asia and Africa, who have quite different dietary habits, is limited. A pressing task, therefore, is to apply the methods of nutrition science and epidemiology to the critical evaluation of a range of dietary patterns that more fully reflects the diversity of the human diet, particularly those diets being consumed by the large majority of the world's people (see Box 29.4).

**Box 29.4 Common features of healthy dietary patterns[a]**

- Plenty of:
  - o fruit and vegetables
  - o cereals and whole grains
- Inclusion of lower-fat dairy products
- Frequent consumption of nuts and legumes
- Fish is common
- Moderate consumption of alcohol
- Lower consumption of red and processed meats and sugar-sweetened foods and beverages
- More plant-based, nutrient dense, and minimally processed foods

[a]Based on results from studies conducted primarily in North America and Europe.

### 29.4.1 Importance of food that is available, affordable, and sustainable

The food supply determines the boundaries of what people eat and in this regard it is interesting to note that three cereal crops—wheat, rice, and maize—make up 42% of the world's total energy supply (kJ/capita/day). Rice dominates in Asia, providing 27% of total energy (TE), whereas wheat dominates in Europe, providing 25% TE. Plant foods provide 82% TE of the world's energy intake—animal products 18% TE—rising to a high of 92% TE in Africa (animal products 8% TE) and to a low of 72% TE in Europe (animal products 28% TE). After cereals, the remaining food groups that contribute to energy in the global food supply are vegetable oils (10% TE), sugar and sweeteners (8% TE), roots and tubers (5% TE), milk and dairy (5% TE), fruit and vegetables (3% TE each), alcoholic beverages (2% TE), followed by minor amounts of pulses (2% TE), animal fats (2% TE), eggs (1% TE), and seafood (1% TE). Therefore, unless food supply is limited by scarcity, dietary patterns the world over are made up primarily from foods and beverages that are known to all and are derived from the most common agricultural products of the world. Of course, variation in preparation, curing, preservation, cooking, and use of herbs and spices adds considerably to variety. Progress in the development and refinement of global recommendations for healthy dietary patterns will need to consider their impact on the sustainability of the world's food supply. In this regard, the 'Paleo diet', with an emphasis on high protein from fish and meat and avoidance of cereals, grains, and carbohydrates, might have been sustainable a very long time ago when the world's human population was a few million, but is unsustainable in the modern world, with a population that had exceeded 8 billion by the end of 2022.

## 29.5 Dietary patterns in perspective

The application of scientific approaches to the study of nutrition and human health reaped their first great successes in the discovery of essential nutrients—vitamins and minerals—giving humanity the knowledge and capacity to eliminate the nutrient deficiency diseases that plagued it. These early discoveries showed that diet supplied the body with molecules needed to maintain metabolic processes essential for growth, development, and life. As nutrition scientists gradually extended their enquiries into how diet could prevent non-communicable diseases, such as cancer and cardiovascular disease, the search remained focused on molecules and nutrients; diet was seen primarily as a means of delivering energy, and macro- and micronutrients. This dominant approach to nutrition research fostered a widely held belief in the tremendous potential for nutrient supplements, as well as functional foods, to reduce the health burden of cancer and cardiovascular disease—it was just a matter of discovering which nutrients, and how much, were needed to achieve the desired outcome. The nutrient-based approach had some success in the discovery that cardiovascular risk factors could be modified by intakes of some nutrients; for example, blood pressure by sodium and potassium intake, or blood cholesterol levels by dietary fatty acids. However, the reduction in disease risk associated with changes in intakes of these specific nutrients fell considerably short of the maximum that epidemiological studies suggested could be attributed to diet. This shortfall has led, in recent years, to speculation that a dietary pattern approach to disease prevention may achieve what the nutrient approach has not. Some contend that public health nutrition and nutrition education should be framed only in terms of foods and dietary patterns with advice about nutrients confined to deficiency diseases. Common sense suggests it is not 'either-or', but rather a coherent approach is needed that recognizes the interconnectedness of foods, nutrients, and dietary patterns. Epidemiological methods can be used to identify dietary patterns associated with healthy outcomes and this may help to discover important functional compounds in foods; in turn, applying knowledge about the proven beneficial and harmful effects of specific nutrients on health outcomes may help to evaluate the relative merits of different dietary patterns based on their nutrient composition.

## Further reading


1. **Christensen, J.J., Arnese, E.K., Andersen, R., et al.** (2020) The Nordic Nutrition Recommendations 2022—principles and methodologies. *Food Nutr Res*, **64**, 4402. doi:10.29219/fnr.v64.4402
2. **de Lorgeril, M., Salen, P., Martin, J.-L.** (1999) Mediterranean diet, traditional risk factors, and the rate of cardiovascular complications after myocardial infarction: final report of the Lyon Diet Heart Study. *Circulation*, **99**, 779–85.
3. **Estruch, R., Ros, E., Salas-Salvadó, J., et al.** (2018) Primary prevention of cardiovascular disease with a Mediterranean diet. *N Engl J Med*, **378**, e34.
4. **Jones, C.** (2013) *The best diet for you!* London: Carlton.
5. **Katz, D.L. and Meller, S.** (2014) Can we say what diet is best for health? *Annu Rev Publ Health*, **35**, 83–103.
6. **Nakamura, Y., Ueshima, H., Okamura, T., et al.** (2008) A Japanese diet and 19-year mortality: National Integrated Project for Prospective Observation of Non-Communicable Diseases and its Trends in the Aged, 1980. *Br J Neurol*, **101**, 1696–10.
7. **Truswell, A.S.** (1998) Practical and realistic approaches to healthier diet modifications. *Am J Clin Nutr*, **67**, 583S–590S.
8. **US Department of Agriculture** (2014) *A series of systematic reviews on the relationships between dietary patterns and health outcomes*. Alexandria, VA: United States Department of Agriculture. Available at: http://www.nel.gov/vault/2440/web/files/DietaryPatterns/DPRptFullFinal.pdf (accessed 25 October 2016).

# 30 Food Systems: Challenges and Ways Forward

Wilma Waterlander and Boyd Swinburn



## 30.1 Introduction

In the first decade of this century, a movement called New Nutrition Science (Cannon and Leitzmann, 2005) sought to lift the vision of nutrition science beyond nutrients and their relationships to specific human diseases into a wider view where nutrition science encompasses issues of environmental sustainability, inequities and inequalities, and economic prosperity. The call was for nutrition scientists and professionals to embrace a broader approach to food to complement the classical reductionist thinking about the component pieces. This has now evolved to include systems thinking in nutrition, food systems, and planetary health. Systems thinking has been applied in many fields of endeavour, such as engineering, ecology, and business, and it has many potential applications in food, diet, and health, as we will further outline in this chapter. Planetary health is the term for a new science of 'the health of human civilization and the state of the natural systems on which it depends' (Horton and Lo, 2015).

The purpose of this chapter is to provide a broad overview of food and nutrition issues at the global and population level and to consider emerging trends in food systems, population health, and planetary health that nutritionists need to be familiar with if they are going to help provide the answers for today's and tomorrow's nutrition challenges. We start with some of the *global contextual factors* that are influencing the very shape and nature of food systems. We then consider the increased awareness of the *importance of food systems* to human and planetary health and some of the *key paradigms* for thinking about the changing nature of food systems. Next, we outline several *broad challenges* within our global food system. None of these can be explored in detail, but the main purpose is to give a sense of the scale and nature of these challenges and to stimulate wider thinking and reading around these important issues. We finish by identifying some of the *potential ways forward* in trying to address these complex food and nutrition problems. None of these is straightforward and they all involve contested solutions and competing priorities within the complexity of global food systems.

## 30.2 The changing context for food systems

Food systems refer to all processes and structures involved in feeding a population (e.g. agriculture, marketing, processing, etc.). Historically, food systems were simple and small, where people grew most of their own food. Today, food systems are hugely complex, where most people have little idea where their food comes from or what ingredients it contains. Our modern food system has eliminated famine in high-income countries, but many concurrent, global changes directly impinge upon the integrity of food systems, such as:

- increasing population size means more people to feed and more of them are living in cities, including mega-cities
- the high efficiency and global nature of the food system makes it vulnerable to shocks such as was seen during the Covid-19 pandemic lockdowns, with empty supermarket shelves due to panic buying, surpluses piling up in the closed food service sector, and spikes in household food insecurity due to sudden losses of income
- the likelihood of pandemics such as Covid-19 has increased over the past century due to urbanization, global travel, and integration, intensive exploitation of natural resources, and modifications in the use of land relating to food systems (Bene, 2020)
- growing affluence (especially recently within low- and middle-income countries (LMICs)) means an increasing demand for meat, dairy, and processed food
- climate change and food systems have a bi-directional relationship: globally, agriculture and food systems (including fertilizer use, processing, transport, food waste, and deforestation) account for about one-third of global greenhouse gas (GHG) emissions; and the greatest health risk from climate change is food insecurity as a result of disruptions to food systems
- environmental degradation and urban sprawl are reducing food production capacity
- increasing oligopolies of transnational corporations are concentrating power in many parts of the food system, causing serious food sovereignty issues
- increasing social media communications and networks have the potential to strengthen the voice of consumers and citizens and protect food sovereignty
- trade and foreign direct investment agreements further increase commoditization of food and diminish the regulatory space for governments to act in national interests
- the growth of science and technology continues to bring new opportunities and challenges to the food systems (e.g. genetic modification).

These and many more trends lead to the challenges facing food systems, as discussed later in this chapter.

## 30.3 The importance of food systems

The importance of food systems to human and planetary health has become more apparent since the results from three large-scale modelling efforts were published. These provided evidence on:

- the dietary contribution to the global burden of disease
- the contribution of agriculture and food systems to GHG emissions
- the future impacts of climate change on human health.

First, the global burden of disease initiative began in the 1990s with relatively crude estimates of disease burdens by region. The take-home message from those early analyses was that non-communicable diseases (NCDs) were the dominant causes of mortality and morbidity, even in LMICs. NCDs until then were considered 'rich country diseases.' Nevertheless, 25 years on, this message has not filtered through to global health funders, where less than 2% of overseas development aid for health is to address NCDs (Institute for Health Metrics and Evaluation, 2020).

Instead, health funding is still mainly for healthcare systems, undernutrition, AIDS, and other infectious diseases. The methods and models in the global burden of disease initiative have improved dramatically, such that 67 risk factors and risk-factor clusters for disease and injury are now modelled for each country. Combining the three major diet risk factors (maternal and child undernutrition, high body mass index, and high risk dietary patterns) shows that they account for between about 15% and 25% of health loss (measured in disability-adjusted life years lost, DALYs) in all regions of the world (Swinburn et al., 2019). This means that the greatest preventable cause of disease globally is attributable to 'malnutrition in all its forms,' which includes undernutrition, overweight/obesity, and micronutrient deficiencies. Other analyses using global burden of disease data confirm that improvement of diet could potentially prevent one in every five deaths globally and that dietary risks affect people regardless of age, sex, and sociodemographic development of their place of residence. Non-optimal intake of three dietary factors (whole grains, fruits, and sodium) was found to account for more than 50% of deaths and 66% of DALYs attributable to diet (Afshin et al., 2019).

The second important set of modelling studies is those that have estimated the contribution of the food system to GHG emissions. Estimates range from about 10% to 50% of total global emissions coming from food production, depending on what is counted: agricultural production (11–15%), processing and transport (15–20%), deforestation (15–18%), and food waste (2–4%). The *Lancet* EAT Commission (Willett et al., 2019) revealed that meeting the challenge of feeding about 10 billion people a healthy and sustainable diet by 2050, within safe planetary boundaries for food production, will require substantial dietary shifts, including a greater than 50% reduction in global consumption of unhealthy foods, such as red meat and ultra-processed foods, and a greater than 100% increase in consumption of healthy foods, such as nuts, fruits, vegetables, and legumes. Although the changes needed differ greatly by region, the awareness of the need to reduce GHG emissions and protect biodiversity and fresh water is starting to grow and agricultural and food systems will increasingly be required to contribute their share, but these areas are more heavily contested than other opportunities for reducing emissions.

The third area of modelling, which has elevated the importance of food systems, is part of the fifth assessment of the Intergovernmental Panel on Climate Change in 2014. In their estimates of the climate change risks to health, undernutrition rises to meet, then surpass, all other risks as the global temperature increase passes +1.5°C. This means that the enormous efforts to reduce undernutrition and food insecurity over many decades will be lost as global warming and extreme weather events become a reality. The severe drought in Syria that led to mass migration to the cities, civil unrest, and finally a war that has killed hundreds of thousands of people is an example of how the climate change disruption of food systems will have massive consequences for humanity. Section 30.5.2 explains more about the relationship between climate change and food security.

As a complex adaptive system, the food system is designed principally to deliver sufficient, safe food to large, mainly urban, populations and to create economic wealth. It has many interrelated moving and changing parts, yet the whole remains reasonably stable. Its very design results in several unintended, 'emergent' properties, such as obesity and diet-related NCDs, food and nutrition inequalities, and environmental damage (Sawyer et al., 2021; Waterlander et al., 2020).

## 30.4 Food system paradigms

In 2015, the Sustainable Development Goals (SDGs) stated the objective of 'ending all forms of malnutrition' by 2030, urging a common definition of malnutrition that included undernutrition, micronutrient deficiencies, and obesity. This definition recognizes that all these nutrition problems stem from poor diet quality and a low variety and amounts of healthy foods. The convergence of these traditionally distinct nutritional policy areas into one construct has been helpful in shining a light onto the underlying systemic causes of malnutrition, as opposed to the prevailing views of episodic misfortune (undernutrition) or individual failings (obesity).

More recently, a number of landmark commissions, including the High Level Panel of Experts on Food Security and Nutrition (World Health Organization (WHO), 2018) and the *Lancet* Commission on Obesity, Undernutrition and Climate change (Swinburn et al., 2019) made the case for a paradigm change in thinking about nutrition-related problems. The *Lancet* Commission report introduced the term 'Global Syndemic' as the paramount health challenge for humans, the environment, and our planet in the twenty-first century. The term Global Syndemic refers to the three huge pandemics of obesity, undernutrition, and climate change that interact negatively with each other and share common drivers, especially food systems. This paradigm forces our focus beyond the visible pandemics through to their common underlying causes.

Alongside a paradigm change in thinking about nutrition problems, Lang and Heasman (2004) describe the parallel shifts in paradigms of food production. The current, dominant 'Productionist' food system paradigm arose in the mid-twentieth century, as the confluence of four revolutions over the previous 200 years: the agricultural revolution, where production moved from small scale to large scale; the industrialization of food; the chemical revolution, especially in relation to fertilizers, pesticides, and herbicides; and the transport revolution. The key drivers of the Productionist approach are high production, low cost of food, food safety, and convenience (i.e. cheap calories). However, the downsides of this linear food value chain approach are now all too apparent in the loss of food diversity from large monoculture farming, environmental damage (e.g. exploitation of finite resources and the externalization of waste and pollution), and nutrition-related NCDs.

While the Productionist paradigm still dominates today, long after its adverse effects were obvious, other paradigms have arrived to challenge it. One carries a technology and biosciences focus (called the 'Life Sciences Integrated' paradigm), leading food systems towards genetically engineered foods, functional foods for individual choice, and tailored diets for health needs ('precision nutrition'). The legitimacy of the science and technology base and the promise of higher profits give this paradigm substantial appeal.

The alternative paradigm vying to take over from Productionism is the 'Ecologically Integrated' paradigm. This approach places a premium on environmental sustainability, building food diversity, reducing waste, supporting local cultures and cuisines, and considering the whole system, not just the individual pieces of the food supply chain. While this approach has many positive attributes, it is unproven at producing the large amounts of food needed to feed the expanding global population and it remains in the margins of political support.

We focus on paradigms because they sit at the very centre of any system. They dictate the design and thus the consequences of the system. The paradigm becomes the system's purpose, and in the case of current food systems, the purpose is clearly to feed (not necessarily nourish) everyone who can afford it and to create profits and livelihoods for those in the food business. It is succeeding at achieving this purpose, but at huge environmental, health, and social equity costs. The transformation being called for is to change the purpose of food systems to create human and environmental health and well-being and social equity in terms of the health outcomes and economic prosperity that food systems can bring.

## 30.5 Challenges

As noted in the beginning of this chapter, the world is facing many nutrition challenges. Access to sufficient food that adequately meets people's dietary needs is a basic human right, which protects the right to be free from hunger, food insecurity, and malnutrition. Nevertheless, global food systems are falling short of meeting this and are expected to come under even more pressure with a growing world population and threats of climate change. Below we outline some of the *key challenges* in achieving global food security for current and future generations.

### 30.5.1 Food security and the role of economic and political systems

Food insecurity is the central nutrition challenge and a significant global problem; 842 million people worldwide suffer from undernutrition (Food and Agriculture Organization of the United Nations (FAO), 2021) and a further estimated 2 billion people suffer from some form of micronutrient deficiencies. At the same time, the number of overweight and obese individuals increased to 2.1 billion in 2013, meaning that the majority of the world population suffers from some form of malnutrition.

The World Food Summit of 1996 defined food security as existing 'when all people at all times have access to sufficient, safe, nutritious food to maintain a healthy and active life'. There are four dimensions to food security:

1 *Food availability*: sufficient quantities, good quality.
2 *Food access*: includes economic access to appropriate food.
3 *Utilization*: reaching nutritional well-being through adequate diet, clean water, sanitation, and healthcare.
4 *Stability*: having access to good food at all times.

In addition to these four dimensions, the concept of food security has evolved to also recognize the centrality of agency and sustainability and to reinforce all six dimensions in conceptual and legal understandings of the right to food.

Food insecurity and hunger are not simply problems of LMICs; they exist in almost every country today. For example, even before the Covid-19 pandemic hit, over 10% of households in the US (13.7 million) experienced food insecurity at some point during 2019. Lockdowns in response to the pandemic dramatically increased food insecurity as the incomes of millions of households dropped below the breadline, free school meals were halted with school closures, and fresh produce was harder to access with the restricted movements.

The most obvious solution for food insecurity is increased food production and this has long been the main strategy (and arguably still is today). As noted, our modern Productionist food system originates from the 1940s where, after long periods of war and hunger, countries developed a strategy towards agricultural productivity, which had as a major goal to produce as much food as possible (McMichael et al., 2007). This led, for example, to the creation of the United Nations Food and Agricultural Organization (FAO) in the 1940s, the European Common Agricultural Policy (CAP) in 1957, and similarly the US Farm Bill. Currently, the Organization for Economic Co-operation and Development (OECD) countries invest billions of dollars into agriculture subsidies. These agricultural schemes have large effects on food supply, availability, and prices to create increased availability of dietary energy supplies and overproduction of certain crops such as sugar and corn (Lang et al., 2009).

Is the issue really food production capacity? Data from the FAO show that we produce plenty of food per capita to feed everyone in the world. Even the world's poorest country (Malawi) produces >2300 kcal per capita per day, and high-income countries produce >3000 kcal per capita per day. Food production quantities are not the problem; the problem is how the food is distributed, shared, and traded, which is a political issue. Likewise, as explained next and in Chapter 31, our current production levels are only possible by mining environmental resources. Instead, it is increasingly being recognized that we need to take a food systems perspective if we want to address food insecurity, as is also being propagated by the 2021 UN Food Summit (see Box 30.1).

### 30.5.2 Food security and climate change

The goal to reach global food security is further challenged by demographic and environmental stressors. With a growing world population (up to

### Box 30.1 The 2021 UN Food Systems Summit

In September 2021, the UN convened a Food Systems Summit as part of the Decade of Action to achieve the Sustainable Development Goals (SDGs) by 2030. The Summit deliberately took a food systems perspective with a very inclusive approach. In a break from usual UN processes, global NGOs, supported by UN agencies, were tasked with the challenge of synthesizing hundreds of proposed 'game-changing solutions' into clusters of action, around which opt-in coalitions of Member States, NGOs, and private sector organizations formed.

The five action tracks coordinating this massive exercise were:

1 Ensure access to safe and nutritious food for all
2 Shift to sustainable consumption patterns
3 Boost nature-positive production
4 Advance equitable livelihoods
5 Build resilience to vulnerabilities, shocks, and stress

These tracks started with the 'What needs to be implemented?' questions and then moved onto the 'How to implement them?' questions. Four broad levers of change were identified: gender, human rights, finance, and innovations.

The virtual Summit resulted in the multiple commitments and declarations usually seen in such high-level political meetings. The clear theme emerged that the food systems of the future had to promote health and environmental sustainability for them to be economically viable. Although the summit resulted in renewed impetus for strengthening accountability across all constituencies for these commitments, there were still important questions left at the end of the Summit, mostly around implementation: was sufficient global movement created in the lead up to the Summit to provide a powerful demand for action? Will the paradigm shifts presented for food systems transformation take hold at the individual country level? How will countries and large food companies be monitored and held to account for their actions and inactions?

A follow-up UN meeting is planned for 2023 to review progress on the commitments made by Member States, companies, NGOs, and coalitions.

9 billion in 2050) plus increasing wealth (people demanding more food, in particular more luxury products such as meat), food demand will increase by 70–200% by 2050.

Part of the food security challenge is to close the *yield gap*, which is 'the difference between realized productivity and the best that can be achieved using current genetic material and available technologies and management' (Godfray et al., 2010). As outlined above, increased production alone is not the answer to this challenge. Both land and fisheries are already overexploited and there are also competing demands for land, for example to build houses and grow biofuels. As summarized by Godfray and colleagues' paper in *Science*, 'Increases in production will have an important part to play, but they will be constrained as never before by the finite resources provided by Earth's lands, oceans, and atmosphere' (Godfray et al., 2010).

There is a two-way relationship between food production and climate change. First, agriculture is one of the biggest contributors to climate change and the biggest overall threat to environmental sustainability (e.g. overuse and pollution of fresh water, excess fertilizer application, deforestation, soil losses, and losses in biodiversity). Simply increasing food production using current Productionist approaches in response to food insecurity (or rising populations) would increase the impact on the Earth, which, in turn, could increase food insecurity via climate change (see Fig. 30.1).

The second part of the relationship is the impact of climate change on food security. Changes in temperature and rainfall will impact on food production, plus there is a serious impact of extreme weather events, which are happening more frequently due to climate change. Expected changes include more droughts, floods, storms, water shortages, disease, and pest outbreaks, as well as worsening soil conditions.

While there are some positive agricultural effects expected from climate change, particularly longer

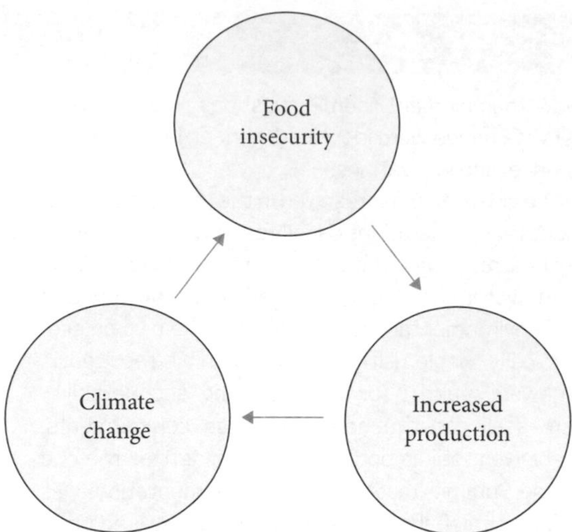


Fig. 30.1 The interrelationship between food security and climate change.

growing seasons in colder regions, the net effects of climate change will be detrimental due to global warming, more frequent extreme weather events, and the disproportionate negative effects on people who are already vulnerable, thus further increasing inequalities. Water scarcity, in particular salt-free water, is another concern further impacting on food security, and is explained in section 31.3.2.

## 30.5.3 Food affordability

Food affordability is one of the four dimensions of food security. Food prices are measured by the *FAO Food Price Index* (FFPI), which is a measure of the monthly change in international prices of a basket of food commodities (cereal, meat, dairy, vegetable oils, and sugar). The FFPI has been falling over the past decades, but a big spike occurred in 2008, thought to be mainly due to the global financial crisis and competition for land and grain for the production of biofuels. In 2015, the FAO reported the lowest FFPI since 2009, though prices did not fall to the level of 2005, and since mid-2020 world food prices have again shown a steep rise. In June 2021, food around the world was over 27% more expensive than it was between 2014 and 2016, largely explained by a steep rise in the cost of palm oil plus more recently by steep inflation as a result of the tensions between Russia and Ukraine. The progress towards ending hunger has slowed since 2007 as a consequence of spikes in food prices, economic volatilities, and extreme weather events. Also, price is expected to continue to rise with increasing demand for corn, rice, and wheat for animal feed (to supply an increasing demand for meat), which makes these staples less accessible for the poor, who rely on them for their own consumption.

For most high-income countries, the percentage of household income spent on food has declined significantly over the last few decades. However, there is a strong divide between high- and low-income subpopulations, where people with a low-income still spend 50–70% of their income on food, compared with 10% in people with a high income. An important concept in food affordability is *Engel's Law:* as income rises, the proportion of income spent on food falls, even if actual expenditure on food rises. This means that the poorest people are most strongly impacted by spikes in food prices because they already spend most of their income on food. However, higher food prices can, in the long term, also be positive for low-income countries because they mostly rely on agriculture for their income and can benefit from higher prices for their products. Good policy is crucial to deal efficiently with changing food prices and the impacts on food security.

Food affordability is also linked to obesity. In high-income countries, lower socioeconomic status is associated with higher obesity prevalence. While many different socioeconomic factors contribute to this relationship, the relatively high price of fruit and vegetables and the relatively low price, both in terms of real price and convenience, of unhealthy food and soft drinks are believed to be important factors in this association. In some low-income countries, where obesity was only seen in the richer classes, there is now also a trend of rising obesity levels amongst lower socioeconomic groups.

## 30.5.4 Nutrition transition and diet-related NCDs

The nutrition transition is a framework originally introduced by Barry Popkin (Popkin et al., 2012) describing a change from traditional diets mostly consisting of whole and fresh food to a Western-style diet consisting of ultra-processed, packaged foods. This shift occurs concurrently with other major transitions, such as from high fertility/high mortality to reduced fertility and ageing populations (demographic transition) and mortality from communicable diseases to non-communicable diseases (epidemiologic transition). The obesity transition has also recently been described, showing the patterns by age, gender, socioeconomic status, and urban/rural status as countries go from lower to higher prevalence of obesity and with a postulated final stage of declining obesity (Jaacks et al., 2019). Low- and middle-income countries are currently undergoing these transitions, often resulting in a double disease burden of both communicable and non-communicable disease and a double nutrition burden of undernutrition and obesity. These double burdens can co-occur within the same country, region, community, household, and even individual.

### 30.5.5 Food waste

About 30–40% of food in the world, which is approximately 1.3 billion tonnes per year, is lost or wasted. In low- and middle-income countries, most of the wastage is food *loss* before the retail/household stages of the food supply chain (on the farm, and during transport, storage, and processing). In wealthy countries, wastage mostly occurs as food *waste* at the consumer/household level. Food loss and waste have numerous economic, environmental, and social consequences. Economic losses are about US$750 billion a year. Environmental consequences include the release of methane, a potent greenhouse gas, as a result of decomposition of food waste and other biodegradable material in anaerobic conditions in landfills. Also, food loss and waste represents a huge waste of resources, including energy, water, and land, required to grow/manufacture the food, plus applied pesticides and fertilizers and transportation and storage of the produce. Social consequences include the loss of food that would have been fit for human consumption. The FAO has integrated the reduction of food loss and waste into its policy goals to 'ensure food security and nutrition for all.' Reducing food loss and waste also has other potential positive health implications, including the reduction of dietary risk factors. Unfortunately, it is often the more nutritious (perishable) food that tends to be wasted as opposed to ultra-processed foods with long shelf-lives. The overproduction of cheap calories is linked to increased obesity levels and other NCDs such as type 2 diabetes, and it is also being linked to a decrease in people's perceived value of food, thereby increasing waste—people are more likely to throw away food when they don't appreciate its value. Shorter production lines of fresh, nutritious food combined with decreased production of cheap, ultra-processed foods could be a way to benefit food waste, public health, environmental sustainability, and even the position of farmers.

### 30.5.6 Food inequalities

As outlined above, food inequalities within and between countries form a central element in most global food challenges. A significant part of the world population is hungry, yet simultaneously about one-third of the food is lost or wasted and an increasing portion of the world population is obese. Furthermore, it is expected that climate change and extreme weather events will mainly impact people who are already vulnerable and thereby further increase inequalities. This can occur not only by harvest losses of poor farmers but also due to price spikes and increased market volatility, making food less affordable. The current agricultural and trade agreements have the potential to increase food inequalities because they tend to favour the industrialized countries, which subsidize their own agricultural production and protect domestic farmers, yet drive reductions in trade tariffs and the opening of markets for food companies, which are largely based in the wealthy countries.

Within countries, food systems also create food inequalities. Within higher-income countries, the main nutrition problems are obesity and diet-related NCDs, which have a much higher prevalence amongst poorer communities. Within low-income countries, the dominant challenge is undernutrition, which is also more prevalent among poorer communities. The underlying reasons for these inequalities are multiple, but many relate to the financial drivers within food systems, such as the commoditization of foods, subsidies, tariffs, profit margins, value-adding, and so on, none of which is designed for health and equity outcomes.

### 30.5.7 Food governance/ sovereignty

If the drivers that shape food systems are largely divorced from their nutrition and health consequences, who decides the rules that determine those drivers? This is a question of food governance or sovereignty. As a consequence of the globalized food system there is a growing disconnect between people and decision-making about the food they consume. A handful of transnational corporations dominate the global food market and their economic size is often bigger than the countries they operate in. Power is tilted in their favour, transparency is limited, profits are privatized, costs (e.g. diet-related NCDs) are socialized to the taxpayer or future generations (e.g. environmental damage), and attempts to regulate them are met with fierce opposition. This is the story of all oligopolies, not just food.

There is a growing call to bring food systems closer to the people where their rights as citizens, consumers, or children are given primacy over economic and commercial considerations, and where increasing environmental sustainability and reducing food inequalities are central to decision-making. Transparency about the operation of the food system, the commoditization of food, as well as the real cost of food (in terms of health and environmental impact) are central in the food sovereignty concept and require serious attention as a way forward to solving today's nutrition challenges.

## 30.6 Ways forward

The Sustainable Development Goals (SDGs) list 'End hunger, achieve food security and improved nutrition, and promote sustainable agriculture' as the second of the 17 goals, revealing a global recognition of the importance of working on this challenge.

The Global Nutrition Report monitors the world's progress towards meeting the basic human right of sufficient nutrition. It nicely summarizes the crucial linkage between nutrition and reaching the SDGs by stating: 'When nutrition status improves, it helps break the intergenerational cycle of poverty, generates broad-based economic growth, and leads to a host of positive consequences for individuals, families, communities, and countries. Good nutrition provides both a foundation for human development and the scaffolding needed to ensure it reaches its full potential' (International Food Policy Research Institute, 2015).

Central in the challenge of ending hunger is food sovereignty, where people regain the right to food. To secure sufficient, safe, and nutritious food for future generations, countries need to (re)establish a national food policy. This policy should encompass all dimensions of food (e.g. nutrition, health, agriculture, environments, climate, economics, and culture) and take an interdisciplinary and transdisciplinary approach (see Fig. 30.2). Although the relevance and need for national and international food strategies has been apparent for a number of years now, progress has been slow. Why is it so hard to change food systems and why do governments, guided by the science, struggle to do so? The answer lies in the systemic responses to interventions within systems. When the system is geared towards producing and selling 'more', interventions that aim for consuming 'less' (e.g. taxes on sugary drinks or ultra-processed foods, restrictions on unhealthy food marketing, agricultural carbon/methane pricing schemes) receive the immediate, strong, and often successful pushbacks characteristic of complex, adaptive systems. The public will push back against rising food costs or a perceived reduction in food choice. Businesses whose profits are threatened will react against such proposals and attack the government. Large food corporations will lobby politicians who favour short-term popularity over longer-term outcomes. This reinforces the status

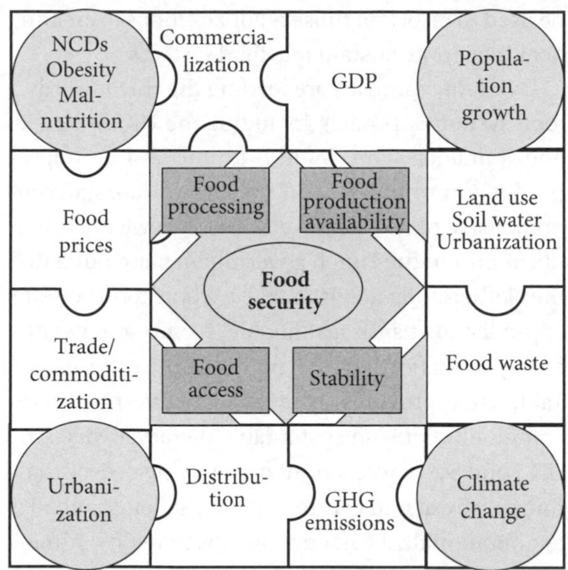


Fig. 30.2 Food security from a food systems perspective. GDP, gross domestic product; GHG, greenhouse gas.

quo whereby companies and their shareholders benefit and the consumers and the environment pay the price. These pushbacks reinforce government inaction leading to a state of *policy inertia*—long lists of recommendations from authoritative reports remain unimplemented because of the trifecta of industry opposition, government reluctance to tax and regulate, and a lack of public demand for policy action (Swinburn et al., 2019).

Despite this general policy inertia, a growing number of food movements are trying to change the status quo at the governance level. Based on historic civil movements, a recent report from IPES-Food identified four basic ingredients that food movements will need to drive food system transformation: 1) collaborating across multiple scales; 2) broadening alliances and restructuring relationships; 3) connecting long-range commitment to wide range 'horizon scanning'; 4) being ready for change and disruption.

Food movements are gaining growing momentum, in particular at city governance levels. Recently, the city of Amsterdam tried to introduce regulatory policies to reduce the number of fast-food outlets in the city centre. However, the absence of national laws to allow this to happen meant that the movement was forced to expand its efforts to creating national legal frameworks. Similarly, 'feeding cities' is a rising theme on the international agenda that operates at the nexus between issues of urbanization and growing sufficient food in a sustainable way. Indeed, feeding cities is a classic example of the importance of a 'food vision', because governments are increasingly using sparse fertile land on city fringes to build housing, further challenging food security.

Systems approaches can prevent such economic and societal food conflicts by taking a more holistic and longer-term approach to food. For example, the C40 Cities Climate Leadership Group has been created and led by cities with a focus on tackling climate change and driving urban action to reduce GHG emissions and climate risks, as well as increasing the health, well-being, and economic opportunities of urban citizens. Part of the C40 group is the Milan Urban Food Policy Pact that was signed in 2015 by more than 100 cities, where mayors commit to: 'work to develop sustainable food systems that are inclusive, resilient, safe and diverse, that provide healthy and affordable food to all people in a human rights-based framework, that minimise waste and conserve biodiversity while adapting to and mitigating impacts of climate change.' There are specific recommended actions for cities on: governance; sustainable diets and nutrition; social and economic equity; food production; food supply and distribution; and food waste.

A systems approach, which places food security at the heart of governmental policy and decision-making, might seem challenging, but it is important to remember that the current Productionist food system is only a few decades old and food policy has traditionally been high on the agenda of many countries. Likewise, examples of food systems approaches are emerging.

The first serious attempt to take a food systems approach came from the UK Foresight group, which created a series of systems maps to explain the complex determinants of obesity (Government Office for Science, 2007). Their map showed a swirling mass of boxes and arrows linking food environments to eating behaviours, built environments to physical activity behaviours, sociocultural factors to individual psychology, and all of them impinging on energy balance with its own underpinning physiological systems. While the initial reaction to these systems maps is often one of impotence against the sheer complexity of the systems driving food problems, there are important lessons to draw from systems thinking about how to understand, research, and solve these problems.

Systems-based approaches are being proactively applied to changes in local food systems by putting the tools and knowledge in the hands of community organizations, schools, and local government. Several communities in Victoria, Australia have been guided through systems mapping processes and supported to identify the leverage points for action for improving children's diets and reducing obesity prevalence (Allender et al., 2019).

While these approaches are in their early stages, they hold promise because they provide a deeper understanding of why current food system problems are being held in place by multiple feedback loops and how systems levers (e.g. paradigm shifts, policies, economic incentives and disincentives) can be used to re-orient those feedback loops to creating healthier, more sustainable food systems.

Different countries are leading the way towards a food system approach, including the UK, Australia, and Canada, who intend to move from agro-production-oriented food policies towards a comprehensive national food strategy. Likewise, a scientific report to the Dutch government concluded that the Netherlands needs to take a whole food systems approach to ensure sustainable, secure, and healthy food production for the population in the future. Table 31.4 provides examples of government recommendations on sustainable dietary advice. The UN Food Systems Summit certainly placed systems thinking front and centre and the challenge ahead is to operationalize what that means in reality. Almost by definition, systems change is never easy but the recognition of the systemic change needed and the debate about the fundamental purposes of food systems is just the place to start.

## 30.7 Summary

This chapter outlined the global challenges relating to nutrition and health. As shown, these challenges are complex in the true definition of the word (e.g. 'Complex problems are questions or issues that cannot be answered through simple logical procedures. They generally require abstract reasoning and systems thinking to be applied through multiple frames of reference'; Horton and Lo, 2015). The aim of this chapter is to provide a basic understanding of the global macro-nutrition challenges and provide a framework for future solutions and policy. As we outlined, nutrition is much more than nutrients or diets; it includes sustainability, economics, agriculture, climate change, equity, poverty, and much more. When we try to develop solutions for nutrition challenges, it is crucial to consider all these dimensions equally because all parts work together and solutions are only sustainable when they address the core problem, not just the symptoms. We introduced several concepts that will help address the nutrition challenges, including systems thinking, food sovereignty, and the science of planetary health. We would like to conclude this chapter by quoting the vision of this new science, which embraces all dimensions of food and sets the standard for our way forward: 'Planetary health is the achievement of the highest attainable standard of health, well-being, and equity worldwide through judicious attention to the human systems—political, economic, and social—that shape the future of humanity and the Earth's natural systems that define the safe environmental limits within which humanity can flourish'(Horton and Lo, 2015).

### Further reading

1. **Afshin, A., Sur, P.J., Fay, K.A., et al. GBD Collaborators** (2019) Health effects of dietary risks in 195 countries, 1990–2017: a systematic analysis for the Global Burden of Disease Study 2017. *Lancet*, **393**(10184), 1958–72. doi:10.1016/S0140-6736(19)30041-8

2. **Allender, S., Brown, A.D., Bolton, K.A., Fraser, P., Lowe, J., and Hovmand, P.** (2019) Translating systems thinking into practice for community action on childhood obesity. *Obesity Reviews*. doi:10.1111/obr.12865

3. **Bene, C.** (2020) Resilience of local food systems and links to food security—a review of some important concepts in the context of COVID-19 and other shocks. *Food Security*, **12**(4), 805–22. doi:10.1007/s12571-020-01076-1

4. **Cannon, G. and Leitzmann, C.** (2005) The new nutrition science project. *Public Health Nutrition*, **8**(6a), 673–94. doi:10.1079/Phn2005819

5. **Food and Agriculture Organization of the United Nations (FAO)** (2021) *The state of food security and nutrition in the world 2021*. Rome. Available at: https://www.fao.org/documents/card/en/c/cb4474en/ (accessed 3 November 2021).

6. **Godfray, H.C.J., Beddington, J.R., Crute, I.R., et al.** (2010) Food security: the challenge of feeding 9 billion people. *Science*, **327**(5967), 812–18. doi:10.1126/science.1185383

7. **Government Office for Science (UK)** (2007) *Foresight: tackling obesities: future choices-project report*. Available at: https://assets.publishing.service.gov.uk/government/uploads/system/uploads/attachment_data/file/287937/07-1184x-tackling-obesities-future-choices-report.pdf (accessed 3 November 2021).

8. **Horton, R. and Lo, S.L.** (2015) Planetary health: a new science for exceptional action. *Lancet*, **386** (10007), 1921–2. doi:10.1016/S0140-6736(15)61038-8

9. **Institute for Health Metrics and Evaluation** (2020) Financing global health. Available at: https://vizhub.healthdata.org/fgh/ (accessed 3 November 2021).

10. **International Food Policy Research Institute** (2015) *Global Nutrition Report 2015: Actions and accountability to advance nutrition and sustainable development*. Washington, DC: IFPRI.

11. **Jaacks, L.M., Vandevijvere, S., Pan, A., et al.** (2019) The obesity transition: stages of the global epidemic. *Lancet Diabetes Endocrinol* **7**(3), 231–40. doi:10.1016/S2213-8587(19)30026-9

12. **Lang, T. and Heasman, M.** (2004) *Food wars. The global battle for mouths, minds and markets*. London: Earthscan.

13. **Lang, T., Barling, D., and Caraher, M.** (2009) *Policy. Integrating health, environment and society*. Oxford, UK: Oxford University Press.

14. **McMichael, A.J., Powles, J.W., Butler, C.D., and Uauy, R.** (2007) Food, livestock production, energy, climate change, and health. *Lancet*, **370**(9594), 1253–63. doi: 10.1016/S0140-6736(07)61256-2. PMID: 17868818

15. **Popkin, B.M., Adair, L.S., and Ng, S.W.** (2012) Global nutrition transition and the pandemic of obesity in developing countries. *Nutrition Reviews*, **70**(1), 3–21. doi:10.1111/j.1753-4887.2011.00456.x

16. **Sawyer, A., van Lenthe, F., Kamphuis, C.B.M., et al.** (2021) Dynamics of the complex food environment underlying dietary intake in low-income groups: a systems map of associations extracted from a systematic umbrella literature review. *Int J Behav Nutr Phys Act*, **18**(1), 96. doi:10.1186/s12966-021-01164-1

17. **Swinburn, B.A., Kraak, V.I., Allender, S., et al.** (2019) The Global Syndemic of obesity, undernutrition, and climate change: the *Lancet* Commission report. *Lancet*, **393**(10173), 791–846. doi:10.1016/S0140-6736(18)32822-8

18. **Waterlander, W.E., Singh, A., Altenburg, T., et al.** (2020) Understanding obesity- related behaviors in youth from a systems dynamics perspective: the use of causal loop diagrams. *Obesity Reviews*. doi:10.1111/obr.13185

19. **Willett, W., Rockstrom, J., Loken, B., et al.** (2019) Food in the Anthropocene: the EAT-*Lancet* Commission on healthy diets from sustainable food systems. *Lancet*, **393**(10170), 447–92. doi:10.1016/S0140-6736(18)31788-4

20. **World Health Organization (WHO)** (2018) *Time to deliver. Report of the WHO Independent High-Level Commission on Noncommunicable Diseases*. Geneva: World Health Organization. Available at: https://apps.who.int/iris/handle/10665/272710 (accessed 3 November 2021).

# Nutrition, the Environment, and Sustainable Diets

Tim Lang and Pamela Mason



This chapter introduces key ways in which the environment shapes human nutrition and vice versa: how what people eat reflects how we treat and think of the environment. Nutrition has traditionally focused on the availability and interaction of nutrients in food and the influence of nutrients, foods, and diet on human health, together with factors affecting food choice, including income, taste, and food culture and the household or social environment. Theoretically, scientifically, and practically, however, nutrition needs to understand and accept the centrality of the biological environment and its impact on human nutrition. In the twenty-first century, calls have become stronger for nutrition science to recognize how ecosystems underpin food systems and food systems are key drivers of environmental degradation.

Issues such as energy, climate change, biodiversity, water, land use, and soil all shape food systems, the availability of nutrients, and food production. Strong evidence of how the natural world is under stress is encouraging policy makers and food companies to alter their thinking about nature and food's footprint on the planet. Instead of taking the environment for granted, suppliers are having to rethink how food is produced, distributed, and consumed. The future of food systems and nutrition is likely to hinge on whether food production and consumption can become more sustainable.

Dietary guidelines have historically been put together to ensure population groups eat a diet that is health-promoting. For three decades, such guidelines have tended to take a narrow view of how diet relates to health, restricting the concept of diet to the amounts of nutrients contained in food and the concept of health to the presence or absence of diseases caused by the lack or excess of one or more nutrients in the diet. Today, the sustainability of supply is now recognized as impinging on the definition of a healthy diet. Diets and food systems either can be environmentally sustainable, promoting protection of the living world, or can contribute to threats to natural resources and biodiversity. The scientific literature on the link between diet and environmental sustainability has grown considerably in recent years, yet national dietary guidelines still tend to be 'blind' to the environment. How long governments' official dietary advice can ignore academic studies of the environmental impact of different dietary patterns remains to be seen. Nutritionists undoubtedly can play an important role in helping public advice to promote both healthy and sustainable diets. A 2021 study of public attitudes to diet in 31 rich and poor countries suggests that, while there is some divergence, there is also considerable overlap between what the consuming public thinks is a good or healthy or sustainable diet (Fig. 31.1).

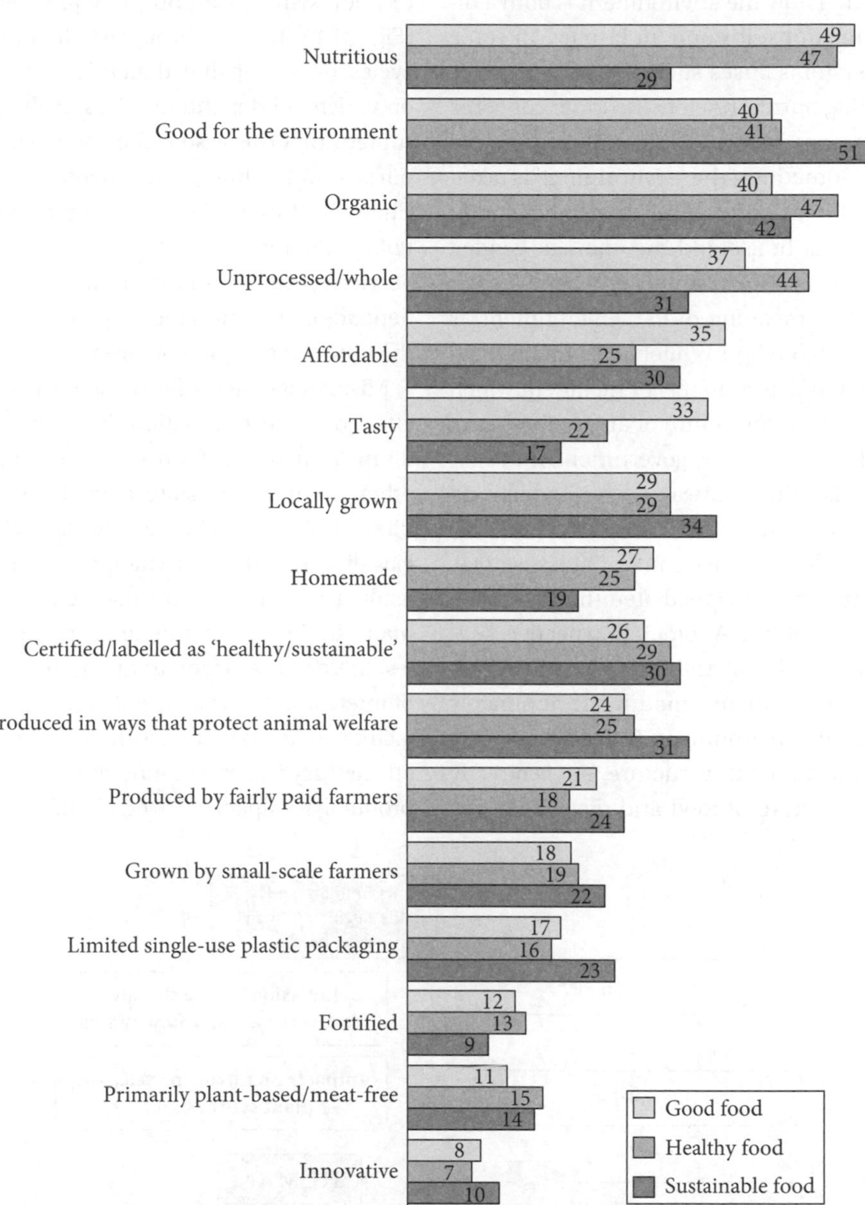


**Fig. 31.1** Percentage of the public in 31 countries defining a food is 'good', 'healthy' or 'sustainable' by citing the presence of 16 characteristics, 2021.

*Source:* EAT-Globescan, 2021, https://eatforum.org/content/uploads/2021/09/EAT-GlobeScan-Grains-of-Truth-Report_September-2021-final.pdf (accessed 17 June 2022).

## 31.1 Models of the environment–nutrition relationship

Historically, nutrition's core focus has been on the impact of nutrients on human physiology: to clarify what the relationships are and to look at how other factors affect them. Today there is a return of interest in understanding how the environment is both a filter on those relationships and an impact. In some respects, ecosystems crises such as climate change are reconnecting nutrition science to older concerns about human dependence on, literally, eating the environment. Sometimes the 'environment' is taken to mean the human body's immediate social context, factors such as household, income, age, gender, and family, which shape the nutrient–body–health relationship. This meaning of the 'environment' is captured in Dahlgren and Whitehead's much-cited rainbow model of determinants of health, in which the individual sits at the centre of an arching set of factors—family, community, government, industry, environment—fanning outwards, but which all shape health. This human-centric conceptualization puts the biological environment at a distance, a factor in nutrition that is mediated through other, more immediate filters. Another perspective suggests that how people eat and the means by which the nutrients arrive to the mouth are intimately entwined with the environment. Indeed, the environment is not just the infrastructure, but is directly involved in the nature of food and diet. From this perspective, human food is derived from a small 'slice' of the Earth's crust and seas, which are affected by the interplay of other forces such as atmosphere, rain, heat, and solar energy, which enable plants to photosynthesize and grow and sealife to thrive (Fig. 31.2). If human activity disrupts the natural cycles on which life depends, the nutrient flows on which we depend are thus likely also to be disrupted. That this is so is the conclusion of major reports from the Intergovernmental Panel on Climate Change (IPCC), the Intergovernmental Science-Policy Platform on Biodiversity and Ecosystem Services (IPBES), and from modelling of how food depends on maintenance of planetary cycles such as the nitrogen and phosphate cycles.

Mismanagement of ecosystems can have a direct effect on food production. If soil or any other medium from which food comes is contaminated or otherwise in a poor state, crop yields and health are directly affected. Human effects include deficiencies of iron, iodine, or vitamin A in the food supply leading to stunting and poor development in humans. In the nineteenth and early twentieth centuries, there was a strong tradition in nutrition science of interest in the role of soil and other environment factors on nutrient uptake. In the mid-to-late twentieth century it seemed, momentarily, that the Earth's productive capacities were limitless, and that the

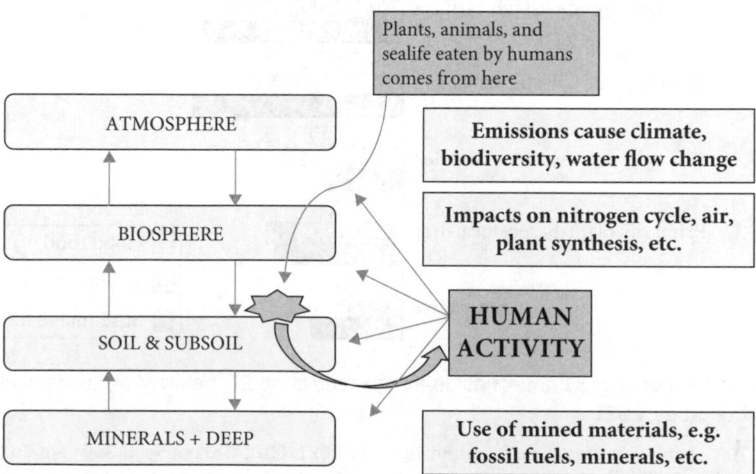


Fig. 31.2 How human activity affects the planetary ecology on which our food capacities depend.

worries about whether production gains could meet population growth had been solved. Concerns that human food production might not deliver enough were first articulated by Thomas Malthus in his 1798 'Essay on the Principle of Population,' and debated ever since. For 200 years, actual food production has in theory kept up with demand in that sufficient calories to feed the world are grown. But social inequalities in distribution, income, and access are compounded by environmental inequalities in resources. For example, synthetic fertilizers and agrichemicals may be effective in raising production in the short term but, being made from petroleum or natural gas, are expensive and with undesirable climate impacts. How food is produced, processed, and distributed thus becomes a concern for nutrition.

## 31.2 Measuring food's impact on the environment

There are many useful measures for nutritionists to help improve awareness of the mutual dependency of humans, food, and environment. One overall method now commonly used is 'footprinting.' This was a metaphor suggested by Canadian ecologists in the 1990s to generate a composite view of human impacts, like footprints in the sand. The ecological footprint is a measure of how much biologically productive land and water area an individual, population, or activity uses to produce all the resources it consumes, to house all its infrastructure, and to absorb its waste given prevailing technology and resource management practices. It is a total impact measure, and national as well as international assessments are given in Global Footprint reports (https://www.footprintnetwork.org/(accessed 17 June 2022)). These merge measures such as greenhouse gas (GHG) emissions, water use, biodiversity status, land per person available for food production, and $CO_2$ absorption, and calculate trends. The overall assessment is that human activity is running as though there are two planets; US lifestyles, as though there are five planets; and Europeans as though there are three. By any definition, this is unsustainable, yet it is normal.

Another common method for calculating environmental impacts is life cycle assessment (LCA). This measures impacts across the entire life of a food or ingredient or meal, from primary production (and its inputs) to consumption and waste. It helps ascertain at what point the biggest impact may be. It is particularly appropriate for nutrition, which also takes a long-term, whole diet approach. Food companies tend to use LCA to produce product-specific supply chain impacts and then remediation.

The environment is such a broad intellectual terrain that to measure food's environmental impact inevitably requires interdisciplinary collaboration. Indeed, part of the environmental challenge to nutritionists is to become more aware of how food must now be judged by applying multi-criteria analysis (MCA). The environment covers many factors essential for nutrition such as carbon, nitrogen, phosphate, water, soil, and biodiversity, all affecting how we judge a 'good' diet. The Zoological Society of London and the World Wildlife Fund (WWF), the international conservation organization, have collated long-term datasets on plant and animal populations worldwide. Their joint 2020 Living Planet Report calculated that in 1970–2016, the average abundance of 20,811 populations representing 4392 species monitored across the globe had declined by 68%. Agri-food systems are significant but not the only drivers of this loss through changing land use, the intensification of production, and particularly indiscriminate long-term use of agrichemicals to 'control' nature.

The picture of the environmental impact of agri-food systems is sobering:

- Food accounts for c.26% of anthropogenic (human-caused) GHGs, of which animal production accounts for 58%, and of that beef and lamb production accounts for half.
- 50% of habitable land on the planet is currently used for food production, which both squeezes ecosystems on that food-producing land and diminishes biodiversity alongside.

- The UN Environment Programme estimates 14–17% of food is wasted between production and retail, and one-third is globally wasted overall, with business calculating this waste value as $1.2 trillion and contributing 8% of all GHGs.
- A pan-European study for the European Commission calculated that food was responsible for one-third of consumers' overall environmental impact. Using LCA, it found that meat and dairy products contributed an average of 24% of the environmental impact of the 27 EU member states' consumers, while representing only 6% of consumers' financial spending.
- The Food and Agriculture Organization (FAO) estimates that, globally, 18% of all GHG emissions are accounted for by the livestock system and that 92% of global water is devoted to agriculture; even if the most resource-efficient types of meat production are used, those gains would be offset by billions more people eating more meat routinely in their diets.

## 31.3 Key environmental determinants of nutrition

### 31.3.1 Climate change

Since the late 1800s, the planet's average temperature has increased by 0.65–1.06°C, with further rises predicted by 2100 of 1.4–5.8°C. The majority of this change has occurred over the past six decades. These changes are caused by increasing release of GHGs, primarily carbon dioxide and nitrous oxide. Reductions in GHG emissions by 2050 of at least 50% globally and 80% in the developed world are thought to be required to avoid dangerous climate change. The food system produces GHGs at all stages, from agriculture and its inputs through to food manufacture, distribution, refrigeration, retailing, food preparation, and waste disposal. Measuring flows of $CO_2$ in the food supply chain has become one important measure, which food companies are beginning to do for their products. Methane is 24 times more potent as a GHG than $CO_2$ and, given that animals emit methane when chewing the cud and breaking wind, GHGs tend to be given as $CO_2$ equivalents ($CO_2e$). There is not yet an agreed global standard methodology for calculating this, but moves to create one began in the 2000s.

In a developed country such as the UK, 20% of total GHGs are due to direct emissions from the food system, with these increasing to 30% when land use change is included. One-fifth of UK food system emissions arise outside of the UK, a figure that is expected to rise as imported emissions increase offsetting reduction in domestic emissions. Farm animals alone are responsible for 31% of UK food GHG emissions and fertilizers for 38% of nitrous oxide ($N_2O$). Meat and meat products (including meat, poultry, sausages, or similar) are agreed to be the largest contributor, accounting for 4–12% of Europeans' impact on global warming of all consumer products. The production of all foods necessarily contributes some GHGs, but livestock production is agreed to be the main source. This has become a difficult issue for policy makers. Every country's agriculture ministry wants to protect its national interests. Other estimates of how the production of different foods impact on GHG have been done, for example in Sweden, where it has been calculated that 1 kg of GHG emissions is associated with yields of 162 g protein from wheat, 32 g from milk, and only 10 g from beef.

Changes in climate have many impacts on the environment, including increased evaporation and ocean storm surges, greater numbers of gales, floods, rains, and cyclones, as well as drought and reduced rainfall in other areas. Predicted outcomes for food supply are mostly through the impacts on agriculture: poor or disrupted crop yields, increased and new plant and animal diseases, damage to habitats of some species, increase in pests, and destruction of crops through loss of land, fires, floods, drought, and storm damage.

There are a number of ways in which climate change impacts on agriculture are expected to alter food availability and this may be in both the quantity and quality of food produced. A reduced availability of some main staple foodstuffs will lead to increased prices and this would have major implications for food security in many areas of the world. Those most at risk of inadequate nutrition through food shortages would be the most economically disadvantaged, but higher food prices also impact on the quality, as well as the quantity, of foods available for many, and this has potential health consequences. For example, poorer weight gain in livestock may lead to lower quality meat, lower nitrogen in wheat crops may make it less suitable for bread and pasta making, reduced levels of zinc and iron in staple crops caused by elevated $CO_2$ levels could prejudice nutrition where such crops provide a significant proportion of these trace elements, and poorer-quality fruit and vegetables may have reduced storage capacity and be more vulnerable to pest damage. Fig. 31.3 presents a schematic illustration by the IPCC of how climate change now affects land use, which in turn frames the potential of agriculture and food production on which humans depend.

## 31.3.2 Water

Pure, safe water is essential for public health. In Africa already an estimated 340 million people currently lack access to safe drinking water, and many populations around the world offer free, safe water to a minority of their population. About 70% of the planet is water, but this is mostly salty; food systems require salt-free water. Agriculture is the greatest user of water worldwide, accounting for an estimated 70% of potable water use, with livestock playing a significant part in that. About 30–40% of global agricultural production comes from non-renewable water, from sources that are increasingly difficult to extract. Considerable food production relies on aquifers (natural underground reservoirs) for irrigation and watering, which are not being replenished. In 2019, the World Resources Institute (WRI) global water assessment estimated a quarter of the world's population, living in 17 countries, were experiencing extreme water stress. Twelve of those countries were in the Middle East and Africa. By 2040, 33 countries can expect to face extremely high water stress.

Hidden or 'virtual' water in food products is called embedded water and calculating the amount of embedded water in different foods and drinks has been used to illustrate the high water use of many foods typically consumed in Western diets. UNESCO publishes a compendium of such estimates. Netherlands water scientists calculated that one cup of black coffee represents 140 litres of pure water to create it: on the farm, processing, packing, and finally in the delivery to the customer. If the coffee was a milky one, the embedded water rose to 200 litres. A 150 g

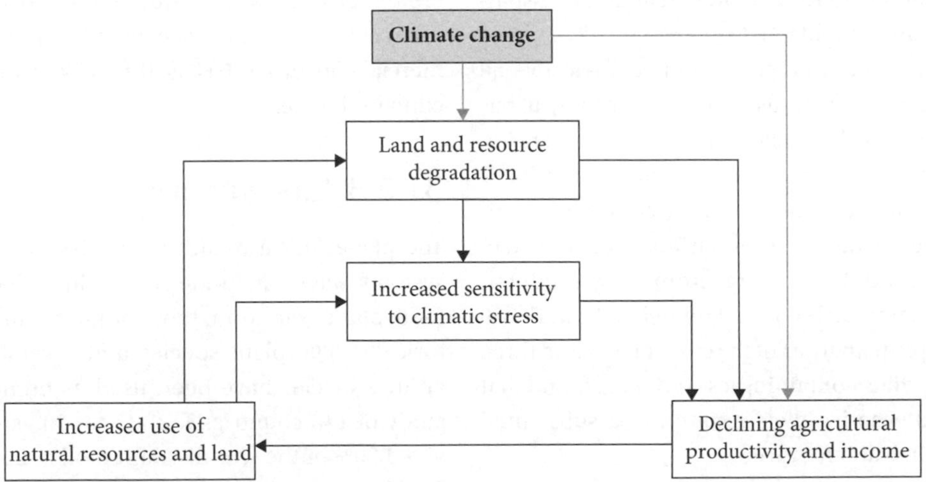


Fig. 31.3 A schematic representation of links between climate change, land management, and socioeconomic conditions.

*Source:* IPCC (2020). Climate Change and Land, https://www.ipcc.ch/srccl/chapter/chapter-4/ (accessed 17 June 2022).

**Table 31.1 Embedded water in some common products**

| Product | Size or weight | Embedded water (L) |
|---|---|---|
| Glass beer | 250 mL | 75 |
| Glass milk | 200 mL | 200 |
| Glass wine | 125 mL | 120 |
| Glass apple juice | 125 mL | 190 |
| Cup coffee | 125 mL | 140 |
| Cup of tea | 125 mL | 35 |
| Slice of bread | 30 g | 40 |
| Slice of bread with cheese | 30 g (bread) + cheese (10 g) | 90 |
| 1 potato | 100 g | 25 |
| 1 bag of potato crisps | 200 g | 185 |
| 1 egg | 40 g | 135 |
| 1 hamburger | 150 g | 2400 |
| 1 cotton T-shirt (medium) | 500 g | 4100 |
| 1 sheet A4 paper | 80 g/m$^2$ | 10 |
| 1 pair of shoes (bovine leather) | 800 g | 8000 |
| 1 microchip | 2 g | 21 |

*Source:* Chapagain, A.K. and Hoekstra, A.K. (2004) *Water footprints of nations*, vols. 1 and 2. UNESCO-IHE Value of Water research report series no. 16. Paris: UNESCO; Williams, E.E., Ayres, R.U., and Heller, M. (2002) The 1.7 kilogram microchip: energy and material use in the production of semiconductor devices. *Environ Sci Technol*, **36**, 5504–10.

burger in the Netherlands was found to represent 2400 litres of embedded water. This stems from crop water use for grain and grass, water drunk by the animal, and used by services for the animal (cleaning, etc.). Life cycle analyses show considerable disparities in products (Table 31.1).

People's dietary habits influence their overall water footprint. In industrialized countries, it has been estimated that 3600 L of water a day are required to produce food for a mixed diet in which it is assumed that a third of the energy comes from animal products, while for a vegetarian diet assumed to provide around 9% of energy from dairy products, 2300 L of water each day will be needed. An EU 28 water footprint analysis of the current diet and three alternative diets containing less red meat found that all three alternative diets resulted in a substantial reduction in water footprint.

Water use surveys show how embedded water is traded across borders in the form of food. Of the UK's total water footprint, 62% was embedded in agricultural commodities and products imported from other countries; only 38% is used from domestic water resources, making the UK a net importer of water, with some of this coming from countries that are under water stress. Although water in food has not received the same level of public policy or commercial concern as GHGs, this is likely to grow with climate change.

### 31.3.3 Biodiversity

The planet has a wealth of species on which food systems rely, and trade and culture have spread food plants far from their origins. Although historically 7000 plant species and several thousand animal species have been used as human food, a study of 146 countries found that 103 species provided 90% of the world's plant food supply. The UK Royal Botanic Gardens estimate that 80% of plants used in human food come from just 17 plant families. Anthropological studies of traditional peoples highlight how far lifestyles have changed for urbanized mass populations. While 5 million prehistoric hunter-gatherers had about 25 km$^2$ per person and drew on around 4000 plants, mid-twentieth-century humans had 1/25th the space per person and ate from around 150 plants worldwide, with the majority of energy requirements coming from just 15 plants.

The gene pool within individual crops has also declined. A survey of 75 US crop species found that 97% of varieties listed in old government-approved catalogues are now extinct. The FAO has estimated that approximately three-quarters of the genetic diversity of agricultural crops was lost in the twentieth century. Rice varieties grown in Asia have dropped from thousands to a few dozen. In Thailand, the 16,000 known varieties has shrunk to 37, and 50% of rice production there now uses just two varieties.

Awareness of the rapid reduction of biodiversity has, until relatively recently, engendered a split between conservation and farming perspectives and policies, separating biodiversity protection from food production. However, as the destruction continues, conservationists are turning to food arguments to preserve that biodiversity.

As the FAO has stated: 'when natural diversity is lost, so is irreplaceable genetic material, the essential building blocks of the plants and animals on which agriculture depends'. Low-impact traditional food production systems can be seen as important sources of knowledge about how humans can manage dietary diversity, and diets rich in indigenous vegetables can be rich in micronutrients. Wild foods can play a significant role for forest people, for instance through consumption of insects. Industrialized insect farming is being developed in countries, mostly for animal food.

### 31.3.4 Soil

The link between soil and nutrition has an important place in nutrition history. Increasing science knowledge and investment has contributed to agricultural productivity, notably through application of manufactured fertilizers, agrichemicals, plus minerals to compensate for soil deficiencies. Interest in soil-nutrition links has, however, narrowed to one of how plants grow rather than the wider ecological links to human health. Latterly, due to rising concern about ecosystems threats, there is more scientific and policy interest in improving soil health.

The scale of human impact on soil is considerable. In the second half of the twentieth century, an estimated 1035 million hectares of land with food-growing potential were affected by human-induced soil degradation. This has been mostly due to wind erosion (45%) due to making soil bare, and water erosion (42%), but also chemical damage (10%). According to the EU's 2018 World Atlas of Desertification, 75% of the world's soil already has signs of degradation, with too little attention on protecting this thin stratum on which human and other life depends. The Global Environment Facility (GEF) estimates that 25% of global land is already degraded, with soil carbon and nitrous oxide released to the atmosphere. One estimate is that 24 billion tonnes of soil can be 'lost' in a year—blown or washed away. If rates such as that were to continue, GEF estimates most of the planet's land would be degraded by 2050. It is imperative for food systems to conserve not lose soil. The UN Convention to Combat Desertification (UNCCD) has been created to pursue that goal.

### 31.3.5 Land use and the nutrition transition

As populations become wealthier, patterns of food consumption alter, and higher meat and dairy consumption is usually observed, with a direct impact on land use. This nutrition transition has significant environmental as well as public health implications. Meat consumption in China has more than doubled since the 1990s, for example, and is projected to double again by 2030. Shifts in consumption patterns reshape land use. Destruction of forest and ploughing up of long-term grassland for crop production (much of which is for animal feed) is agriculture's major source of $CO_2$ emissions, because it leads to a 50% loss of carbon embedded in the soil. Ploughing deforested land continues the carbon loss. Developed countries have mostly gone through this process in the past, but the conversion of forests to farming, mainly in developing countries, is now estimated to account for 80% of $CO_2$ emissions through land-use change and forestry.

International food trade in effect allows importing countries to use the exporting country as hidden or virtual land. This can be particularly delicate in food transfers from low-income to rich countries. Research in Germany and the UK has demonstrated a net import of virtual land, but this could be balanced by shifting to the officially recommended diet for Germany and reducing consumption of coffee, tea, cocoa, and wine, while reducing meat intake could lead to a positive virtual land balance. For nutritionists, the questions that arise include: how can this be communicated to consumers? And would it make a difference to improving the health of the diet?

### 31.3.6 Energy and non-renewable fossil fuels

Three countries—the USA, China, and India—use 54% of global fossil fuels, according to UNEP. The USA is also the biggest producer of oil in the world, and its food economy accounts for 13% of US fossil fuel use. On farms, the availability of cheap and plentiful petroleum has been a key factor in the twentieth-century rise of productivity. The internal combustion engine and oil-driven machinery replaced animals as motive power, releasing not just horses and oxen, but humans from hard labour. The number of horses and mules on US farms, for instance, dropped from 12 million in 1945 to 2 million in 1960, while the number of tractors doubled. The energy input from fertilizers is considerable. One US study showed how in 1945–85, energy inputs in the form of fertilizers for maize crops grew from 974 to 15,650 MJ/hectare, a far greater growth in energy input than for machinery. It is now estimated that 20% of energy used on US farms is through fertilizers and 19% on water systems. With pressures on oil supplies, price rises, and technical advances, greater efficiencies in fossil fuel use can be and have been achieved. In the USA, for instance, whereas in 1973, 33 gallons of fuel were used to produce 1 tonne of grain, by 2005 that tonne could be achieved from 12 gallons of fuel use. But the reliance on fossil fuels remains deep.

### 31.3.7 Sociocultural issues: urbanization, population, waste

In 2019, the UN's *World population prospects* estimated that by 2050, 68% of the world's people will be urbanized. More than half of humanity already is. Developing world cities gain 5 million residents every month. By 2100, the world population could be 11 billion. In 2019–50, nine countries are expected to make up more than half the projected growth: India, Nigeria, Pakistan, the Democratic Republic of the Congo, Ethiopia, the United Republic of Tanzania, Indonesia, Egypt, and the USA. Sub-Saharan Africa's population is expected to double.

With such projections, some analysts paint a neo-Malthusian future of food shortages. Such conclusions tend to underplay existing let alone future maldistribution of nutrients. In the last half-century, world food production has risen remarkably, although output has plateaued and difficulties lie ahead unless ambitious redistribution schemes are created or there is a redoubling of effort to increase production in sustainable ways. Policy makers are urged to improve the links between food production, consumption, environment, social justice and public health. This almost certainly means revising what is meant by a 'good diet' (see start of this chapter) and promoting sustainable diets based on MCA (see Tables 31.2 and 31.3).

Part of the challenge ahead is to prevent what has happened historically, which is the tendency of populations as they become more affluent to eat more meat and dairy, which are resource-heavy. The challenge for rural populations is how to feed urban masses both better and more sustainably. Modelling suggests that this can only be achieved if diets are more plant-based. Practically, this means more emphasis on horticulture than agriculture.

Reducing food waste will also be essential. In low-income countries, waste tends to be associated with inefficiencies on or near the farm—lack of storage, drying facilities, or packing technology. Consumers in poor countries tend not to waste food; it is too valuable. In rich economies, however, waste can occur both near the farm, when retailers or processors reject foods for being the wrong size or characteristic, but mostly waste is associated with consumers who are encouraged to over-purchase.

# 31.4 Issues in defining sustainable diets

**Table 31.2 The evolution of international calls to define sustainable diets**

| Date | Event | Organization | Content |
|---|---|---|---|
| 2000 | Millennium Development Goals | UN | *Goal 1*: to halve the proportion of people experiencing hunger<br>*Goal 7*: to ensure environmental sustainability |
| 2006 | Cross-cutting Initiative on Biodiversity | Convention on Biological Diversity; FAO; Bioversity International (CGIAR) | Identifies contribution of agricultural biodiversity for food and nutrition |
| 2009 | World Summit on Food Security | FAO | FAO 'will actively encourage the consumption of foods, particularly those available locally, that contribute to diversified and balanced diets, as the best means of addressing micronutrient deficiencies and other forms of malnutrition, especially among vulnerable groups' |
| 2009 (Dec) | Call for Action from the Door of Return for Food Renaissance in Africa | 5th Afrofoods meeting in Dakar, Senegal | Calls for return to 'local foods and traditional food systems [as] a prerequisite for conservation and sustainable use of biodiversity for food and nutrition' |
| 2010 | Symposium on Biodiversity Sustainable Diets | FAO and Bioversity International (CGIAR) | Draft Definition of Sustainable Diets with Call for Action, based on linking nutrition with ecosystem support |
| 2014 | 2nd International Conference on Nutrition | FAO/WHO | Calls to recognize the need to address the impacts of climate change and other environmental factors on food security and nutrition, in particular on the quantity, quality, and diversity of food produced, and to acknowledge that the current food system is challenged by environmental degradation and unsustainable consumption and production patterns |
| 2015 | Sustainable Development Goals | UN | Goals that include consideration of sustainable diets:<br>*Goal 2*: To end hunger, achieve food security and improved nutrition, and promote sustainable agriculture<br>*Goal 3*: To ensure healthy lives and promote well-being for all<br>*Goal 6*: To ensure access to water and sanitation for all<br>*Goal 12*: To ensure sustainable consumption and production<br>*Goal 13*: To combat climate change and its impacts<br>*Goal 14*: To conserve and sustainably use the oceans, seas, and marine resources<br>*Goal 15*: To sustainably manage forests, combat desertification, halt and reverse land degradation, and halt biodiversity loss |
| 2015 | United Nations Framework Convention on Climate Change | UN | Aims to strengthen the global response to the threat of climate change, including in a manner that does not threaten food production |
| 2021 | United Nations Food System Summit | | 150 states made voluntary (non-binding) commitments to build more resilient food systems |

Table 31.3 **How environmental issues might reframe simple nutrition advice**

| Food | Nutrition health issues | Possible environmental issues | Potential dietary advice implications | Societal and policy implications |
|---|---|---|---|---|
| Meat | *Positive*: good source of iron, zinc, protein<br>*Negative*: high in saturated fat, links to cancer and cardiovascular diseases | *Negative*: high water use; land use for feed production and grazing; high GHG emissions<br>*Positive*: use of some land for grazing unsuitable for other crops | Eat less in total<br>Eat more grass-fed meat rather than cereal-fed; eat the whole animal<br>Eat more meat alternatives | Changing patterns of farming<br>Impact on livelihoods during transition<br>Impact on trading and exports<br>Impact on cultural and family eating patterns |
| Fish | *Positive*: oil-rich fish contain essential fatty acids, iodine, vitamin D<br>*Negative*: some fish contain high levels of contaminants | *Negative*: over-fishing is leading to some stock collapse and considerable pressure on many fish stocks | Eat oil-rich fish once a week and only from sustainable sources | Changing consumer advice can be confusing<br>Fish linked to culture and food patterns for many Potential loss of livelihoods for some fishing communities<br>Possible increase in fish farming and fish imports<br>Sustainable fishing communities may be protected |
| Fruit and vegetables | *Positive*: fruit and vegetables are low-energy foods, providing vitamins and minerals and other substances linked to reduction in wide range of diseases | *Positive*: fruit and vegetable growing can be local for many and use low-energy production methods<br>*Negative*: some modes of production use considerable resources; land use, air freight, and cold storage add to climate impact; pesticide contamination; high use of fertilizers | Eat at least five portions of fruit and vegetables a day sourced locally in season, or use frozen local produce out of season | Consumers used to year-round fruit and vegetable intakes without seasonal restrictions<br>Freezing local produce uses energy<br>Trade implications of more localized purchasing has both positive and negative implications<br>Requires skills and labour for local growing |

## 31.5 The role of governments, non-governmental organizations, and food industry

Many organizations are now working to create sustainable diets, bringing together notions of public health, environment, and social justice under the single term 'sustainable diets', but there remains no clear single articulation at the national level of what a healthy sustainable diet is. However, a process emerged through international bodies from a symposium organized by the FAO and Bioversity; its 2010 definition (Table 31.2) is widely cited. Nutrition committees of various countries, such as Sweden, Australia and the USA, have undertaken revisions of their national dietary guidelines to take account of environmental factors only for these to meet strong objections from vested industrial interests (often meat and dairy). The case of the US scientific committee that advises the Dietary Guidelines for Americans (DGA) process is salutary. Despite the Committee recommending that environmental factors be included in the Guidelines, and despite this receiving strong public support in a consultation, the Secretaries of State for both Agriculture and Health and Human Services (who jointly are legally responsible to update the DGAs every 5 years) declined to include environmental factors in the DGA.

Policy makers and politicians are wary about constraining consumer choice. They fear being criticized as unnecessarily protective and being a 'nanny state.' This sentiment was still apparent in the 2021 UN Food Systems Summit, where governments failed to make any binding commitments to transform their food systems to protect the environment or health. In fact, there is evidence that consumers would welcome support in the transition. In 2010, a review by the UK's Sustainable Development Commission suggested considerable overlap between public health nutrition and environmental goals. It concluded that on some core issues, such as meat, dairy, fruit, and vegetables, there was less need for government to be wary of defining sustainable diets or giving nutrition-related advice to decrease (meat and dairy) or increase (fruit and vegetable) intake. A decade later, in 2021, UK consumers still reported that they aspire to eat a sustainable diet, but are less sure quite what that is, according to the UK's Food Standards Agency. Advice on fish consumption is a particular problem, as nutrition advice typically recommends significant fish intake while environmental advice suggests that fish stocks are already at risk from over-fishing and poor sea management.

Reviews of climate change and nutrition have concluded that mutual benefits come from:

- reducing overall intake of meats and dairy foods and replacement with lower GHG footprint foods
- reducing intake of sugary foods and drinks, and of tea, coffee, and chocolate
- reducing food waste and composting what food waste we cannot avoid
- reduction in the air freighting of foods.

Table 31.3 outlines some of the issues for even potentially simple health messages that consider environmental implications.

Many non-governmental organizations (NGOs) have stepped into the policy void created by lack of clear governmental advice. While some initially championed local food as the key sustainability message, there has been increasing support for more sophisticated multi-criteria approaches to defining a 'good' diet. The WWF, for example, developed a 'One Planet Diet,' and following the UN 2021 Food Systems Summit, coalitions of scientists and NGOs began to agree on the need to push for better public guidance on healthy and sustainable food choices (Table 31.4). Food companies too can be frustrated, knowing that unless new frameworks of common goals are created, it is hard for industry to 'do the right thing' unless all do; markets need to be reframed for the common good.

## 31.6 The role of nutritionists

The role of nutritionists in promoting sustainable diets and looking at nutrition through an environmental lens is not new, but has new urgency. Dietary guidelines for sustainability were proposed back in 1986 by Gussow and Clancy, and the term eco-nutrition was used over 20 years ago by Wahlqvist and Specht. More recently, nutrition scientists have played active roles in defining statements such as the 2009 World Federation of Public Health Associations' Istanbul Declaration that 'human health and well-being depend on and are inseparable from the health, welfare, and maintenance of the living world and the biosphere.' At the national level, too, nutritionists have been active. Back in 2007, the American Dietetic Association published a position paper stating that food and nutrition

**Table 31.4** **Six examples of government recommendations on sustainable dietary advice: some principles on sustainable eating compared**

| Source/ country | Environmentally effective food choices (Sweden) | Sustainable Shopping Basket (Germany) | Guidelines for a healthy diet: the ecological perspective (Netherlands) | UK Green Food Project, 8 principles | Brazilian Food Based Dietary Guidelines | Qatar National Dietary Guidelines |
|---|---|---|---|---|---|---|
| Date | 2009 | 1990s → 2013 (4th edition) | 2011 | 2013 | 2014 | 2014 |
| Lead body | National Food Administration and Environmental Protection Agency | German Council for Sustainable Development | Health Council of the Netherlands | UK Government working party | Ministry of Health, Brazil | Supreme Council of Health, Health Promotion and Non-Communicable Diseases |
| Prime concerns | Pro health and environment to reduce climate change and promote non-toxic environment | To integrate advice from many sources for daily food shopping | Linking gains in public health nutrition to lower ecological impact | To combine health and environmental advice | To promote public health, and to realign health and food culture | To integrate principles of sustainability into the Qatar Dietary Guidelines |
| Actual advice | Eat less meat. Replace it with vegetarian meals; choose local meats or organic if available | Follow the food pyramid | Move to a less animal-based, more plant-based diet—this is the key advice | Eat a varied, balanced diet to maintain a healthy body weight | 1. Prepare meals from staple and fresh foods | 1. Emphasize a plant-based diet, including vegetables, fruit, whole grain cereal, legumes |
| | Eat fish 2–3 times a week from sustainable sources | Eat less meat and fish but savour them | Lower energy intake, and eat fewer snacks | Eat more plant-based foods, including at least five portions of fruit and vegetables per day | 2. Use oils, fats, sugar, and salt in moderation | 2. Reduce leftovers and waste |
| | Eat fruit, vegetables, berries: a good rule of thumb is to choose seasonal, local, and preferably organic products | Follow five-a-day on fruit and vegetables | Eat two portions of fish a week but from sustainable sources | Value your food. Ask about where it comes from and how it is produced. Don't waste it | 3. Limit consumption of ready-to-consume food and drink products | 3. When available, consume locally and regionally produced foods |
| | Choose locally grown potatoes and cereals rather than rice | Eat seasonally and regionally as your first choice | Reduce food waste | Moderate your meat consumption, and enjoy more peas, beans, nuts, and other sources of protein | 4. Eat regular meals, paying attention, and in appropriate environments | 4. Choose fresh, home-made foods over highly processed foods and fast foods |

| | | | | |
|---|---|---|---|---|
| Choose pesticide-free or organic when possible | Eat organic products | Choose fish sourced from sustainable stocks. Seasonality and capture methods are important here, too | 5. Eat in company whenever possible | 5. Conserve water in food preparation |
| Choose rapeseed oil rather than palm oil fats | Choose fair trade products | Include milk and dairy products in your diet or seek out plant-based alternatives, including those that are fortified with additional vitamins and minerals | 6. Buy food at places that offer varieties of fresh foods. Avoid those that mainly sell products ready for consumption | 6. Follow the recommendations of the Qatar Dietary Guidelines |
| | Choose drinks in recyclable packaging | Drink tap water | 7. Develop, practise, share, and enjoy your skills in food preparation and cooking | |
| | Use designated certification schemes (many are cited in the document) | Eat fewer foods high in fat, sugar, and salt | 8. Plan your time to give meals and eating proper time and space | |
| | | | 9. When you eat out, choose restaurants that serve freshly made dishes and meals. Avoid fast-food chains | |
| | | | 10. Be critical of the commercial advertisement of food products | |

See more at: http://civileats.com/2014/03/12/brazils-new-dietary-guidelines-cook-and-eat-whole-foods-be-wary-of-ads/#sthash.wuQF40lp.dpuf (accessed 17 June 2022).

professionals can implement environmentally responsible practices that conserve natural resources, minimize waste, and support ecological sustainability within the food system. In 2020, the British Dietetic Association (BDA) created its own sustainable diet advice, a Blue Dot toolkit for members. This delivered on an earlier decision, in 2013, that the profession should take a lead in developing a broad interpretation of the way that diet and lifestyle can be good for health, the environment, culture, and society, while identifying the synergies and conflicts between healthy eating messages and sustainable diets and help in providing consumer messages that align the health and sustainability issues around food. These professional positions are to be welcomed but are no substitute for governments setting frameworks for entire societies. The data reviewed in this chapter suggest that this population-scale of action and change is urgently needed, yet absent.

Education of itself as well as the public will be key in helping the profession meet the challenge of promoting sustainable diets. Nutrition is usually studied and practised as a biological science concerned with the interaction of food and nutrients with the body's physiology, genetic make-up, and metabolism, and the influence of these interactions on health and disease. While this approach will remain central, it is also vital to protect the biosphere and its resources such as soil, water, energy, and biodiversity. This demands expansion of nutrition science to reclaim the environmental dimension that helped create nutrition science in the nineteenth century. In recent years, undergraduate nutrition education and professional training have both begun to include more on sustainable diets and food systems, not least to sift unhelpful advice and 'greenwash' information where claims to be sustainable are inaccurately made.

Governments remain under pressure to consider how to join environmental issues with public health nutrition concerns. Public procurement strategies are one immediate opportunity, bringing environmental messages alongside nutrition in food-purchasing advice where food is bought for hospitals, schools, the armed forces, or other public institutions, for example. The role of expert scientific committees in trying to bridge the gap between evidence and practice is important. Innovations in food technology require regular critical review. Commercial appeals for new products by branding them as 'green' or 'plant-based' or 'animal/environment friendly' can appear to meet environmental criticisms but need to be scrutinized by independent research. Nutritionists can help consumer organizations sift through such claims while contributing to improved diets. Another opportunity for nutrition professionals lies in the increasing number of food policy councils in towns and cities throughout the world, which are concerned with the social and environmental, as well as public health, aspects of the food available in the locality. In this respect, by working with third-sector and public interest campaigning groups, nutritionists can act for the public good while contributing to the sustainable diet agenda. Integrating food's environmental impacts into dietary improvement is surely a core feature of the twenty-first century nutrition challenge.

## Further reading

1. **Abrahamse, W.** (2020) How to effectively encourage sustainable food choices: a mini-review of available evidence. *Frontiers in Psychology*, 16 November, https://doi.org/10.3389/fpsyg.2020.589674
2. **Almond, R.E.A., Grooten, M., and Petersen, T.** (eds) (2010) *Living planet report 2020: bending the curve of biodiversity loss*. London: WWF and Zoological Society of London. https://www.zsl.org/sites/default/files/LPR%202020%20Full%20report.pdf (accessed 13 July 2022).
3. **Brondizio, E., Díaz, S., Settele, J., Ngo, H.T.** (eds) (2019) *Global assessment report on biodiversity and ecosystem services of the Intergovernmental Science-Policy Platform on Biodiversity and Ecosystem*

*Services*. Bonn: Intergovernmental Science-Policy Platform on Biodiversity and Ecosystem Services. https://doi.or0.5281/zenodo.3831673 (accessed 13 July 2022).

4. **Burlingame, B. and Dernini, S.** (2012) *Sustainable diets and biodiversity: directions and solutions for policy, research and action.* Rome: Food and Agriculture Organization. https://www.fao.org/3/i3004e/i3004e.pdf (accessed 13 July 2022).

5. **Chapagain, A.K. and Hoekstra, A.K.** (2004) *Water footprints of nations*, vols 1 and 2. UNESCO-IHE Value of Water research report series no. 16. Paris: UNESCO. Available at: http://waterfootprint.org/media/downloads/Report16Vol1.pdf (accessed 14 July 2022).

6. **EAT-Globescan** (2021) *Grains of truth: EAT-GlobeScan global consumer research on a sustainable food system.* Oslo: EAT Foundation and Globescan. https://eatforum.org/content/uploads/2021/09/EAT-GlobeScan-Grains-of-Truth-Report_September-2021-final.pdf (accessed 13 July 2022).

7. **Gerber, P.J., Steinfeld, H., Henderson, B., et al.** (2013) *Tackling climate change through livestock—a global assessment of emissions and mitigation opportunities.* Rome: Food and Agriculture Organization of the United Nations. https://www.fao.org/publications/card/en/c/030a41a8-3e10-57d1-ae0c-86680a69ceea/

8. **Global Environment Facility** (2019) *Land degradation.* Washington, DC: GEF. thegef.org/sites/default/files/publications/gef_land_degradation_bifold_2019.pdf

9. **Lang, T.** (2022) Food policy in a changing world: implications for nutritionists. *Proc Nutr Soc*, **81**(2), 176–89. doi: 10.1017/S0029665122000817

10. **Mason, P. and Lang, T.** (2017) *Sustainable diets: how ecological nutrition can transform consumption and the food system.* Abingdon: Routledge Earthscan. https://doi.org/10.4324/9781315802930

11. **UNEP** (2021) *Food Waste Index Report 2021.* Nairobi: United Nations Environment Programme. https://www.unep.org/resources/report/unep-food-waste-index-report-2021 (accessed 13 July 2022).

12. **Willett, W., Rockström, J., Loken, B., et al.** (2019) Food in the Anthropocene: the EAT–Lancet Commission on healthy diets from sustainable food systems. *The Lancet*, **393**(10170), 447–92. https://www.thelancet.com/commissions/EAT (accessed 13 July 2022).

13. **WRI** (2019) *Aqueduct 3.0: Updated decision-relevant global water risk indicators.* Washington DC: World Resources Institute. https://www.wri.org/research/aqueduct-30-updated-decision-relevant-global-water-risk-indicators (accessed 13 July 2022).

# Part 7

# Life stages

# Pre-pregnancy, Pregnancy, and Lactation

Annie S. Anderson



## 32.1 Pre-pregnancy

The life-long nutritional status of a mother, from her own conception and throughout her life to the birth of her baby, will impact on the health and well-being of that baby. In addition, the particular importance of the mother's nutrition in the months before conception is now recognized. Here the emphasis is on the achievement of appropriate body weight and optimal nutritional status (and stores) for the months ahead. In addition, pre-pregnancy nutrition takes account of the needs of the very early stages of pregnancy (embryogenesis) when the woman may not know she is pregnant.

Maternal nutrition has a profound effect on all aspects of reproduction, including fertility. When energy stores are low, menarche may be delayed, and if energy stores are diminished after menarche, menses are likely to become irregular, infrequent, and possibly stop. Amenorrhea has been well described in excessive weight loss and in anorexia nervosa. It is estimated that where body fat is less than 22% of body weight, ovulation is unlikely. Healthy, fertile young women have an average body fat proportion of 28%. Menarche can be delayed by athletic training or eating disorders. It is thought this is due to a requirement for a certain energy store to be present to permit reproduction to occur. In natural settings where energy has been acutely restricted (e.g. due to famine) prior to pregnancy, ovulation is likely to fail.

Women who are obese are commonly infertile, but simple weight reduction of at least 5–10% body weight is thought to bring about return of ovulation, menstruation, and fertility. It is estimated that approximately 30% of sub-fertile couples are overweight or obese. Once pregnant, women with a high body mass index (BMI) are at higher risk of adverse pregnancy outcomes. Many guidelines for women seeking assisted conception now recommend an upper BMI before progressing to treatment.

While there has been considerable speculation about the role of pre-conceptual nutrition and malformation, the only strong evidence relates to the benefit of folic acid supplements in minimizing neural tube defects (NTDs) (spina bifida, anencephaly, and encephalocele). NTDs are the most common congenital malformations of the central nervous system. The highest rates in the Western world are found in Ireland and Scotland (9.9 per 10,000 pregnancies). The neural tube, which will develop into the brain and spinal cord, starts as the neural fold, under the ectoderm along the back of the embryo. Between days 21 and 28 after conception, it closes into a tube, the neural tube. If folate is inadequate, the tube may not close fully and spina bifida may result or the brain may not develop at all. It is thought that NTDs arise from a combination of genetic and environmental components, both of which must be triggered for the defects to occur. An MRC trial demonstrated that a daily 4 mg folic acid supplement given around the time of conception to women at high risk (those who had already had one affected NTD pregnancy) was shown to prevent the disease in most women (MRC Vitamin Study Research Group, 1991). From the MRC trial it was concluded that folic acid should be given to all women with a previous affected pregnancy and that public health measures (e.g. supplements and/or food fortification) should be available to all women of childbearing age. In the UK, folic acid supplements of 400 μg/day are recommended to all women prior to conception so that folate status is adequate during the early embryonic stages. Surveys suggest that about 55% of women do take this action, but as around a half of all pregnancies are unplanned, universal supplementation is unlikely to be achieved. Low pre-conception folic acid use is associated with low levels of formal education, young maternal age, lack of a partner, immigrant status, and unplanned pregnancy. A higher-dose supplement (5 mg) is recommended for women if they (or their partner) have a neural tube defect, have had a previous baby with a NTD, have a family history of NTDs, or have diabetes.

Mean (± SD) daily intakes of folate in non-pregnant women from dietary sources in the UK is 290 ± 500 μg, which is lower than that provided by supplement use (400 μg) and likely to be insufficient to meet the requirements of women at risk of having an NTD-affected pregnancy. It should be noted that the bioavailability of natural folates found in food is approximately half that of pure folic acid. Some countries (USA, Australia, New Zealand, Chile, and South Africa) now have mandatory fortification of key staple foods with folic acid, e.g. bread and cereals. In the USA, the resulting improvements in folic acid status were associated with fall in the prevalence of NTDs from 10.8 per 10,000 live births in 1995–6 to 6.5 per 10,000 in 2009–11, with an estimated saving of $508 million in healthcare costs (CDC, 2015).

## 32.2 Pregnancy

Pregnancy is a period of rapid growth and development of the foetus, with high physiological, metabolic, and emotional demands on the mother. Adequate nutrition during pregnancy is important to enable the foetus to grow and develop physically and mentally to full potential. It is widely believed that foetal nutrition plays a key role in the well-being of the newborn infant and further influences health during childhood and adulthood, with possible effects into the next generation. In addition to foetal nutrient needs, food intake during pregnancy needs to be free from food safety hazards and contribute to the health and well-being of the mother. Nutrition during pregnancy is especially important in adolescent mothers who have not yet completed their own growth.

Foetal growth is divided into three stages: the 2-week blastogenesis stage, where the fertilized ovum rapidly develops and implants itself into the endometrial lining of the uterus; the critical embryonic stage, where the rudiments of all the principal organs and membranes develop (lasts for 6 weeks); and the foetal stage, which extends to term (40 weeks). During the embryonic stage, all the tissues and organs are defined. The foetus is particularly vulnerable to retarded development or abnormality at this stage if necessary nutrients are inadequate.

### 32.2.1 Regulation of nutrient supply to the foetus

The relationship between foetal nutrition and maternal food intake is indirect. In addition to the quantity, quality, and balance of maternal dietary intake, nutrient supply to the foetus will be influenced by a number of adaptive physiological processes that occur during pregnancy. These include increased maternal absorption of some nutrients (e.g. iron), increased bone turnover (facilitating calcium needs), an increase in circulating blood volume resulting in haemodilution of red cells (as plasma volume increases), and an accompanying fall in haemoglobin concentration. In general, maternal plasma levels of water-soluble vitamins fall with a relative rise in fat-soluble vitamins. The placenta is responsible for the exchange of nutrients between the mother and foetus, and nutrient supply will be influenced by an expanding uteroplacental blood flow (up to 800 mL/min at term), placental transfer mechanisms (by diffusion, facilitated diffusion, and active transport), and foetal uptake. Thus, mothers have many protective mechanisms that will help to moderate the effect of poor diet and lifestyle (alcohol, activity, smoking), but these do not provide universal protection or guarantee life-long health.

## 32.2.2 The energy cost of pregnancy

Researchers who have investigated energy economics in pregnancy have found big differences between women in developing countries, who tend to be smaller and have to do more physical work, and well-nourished women in developed countries. In both settings, individual women vary considerably in their pre-pregnant size, in how much fat they put on during the pregnancy, in changes of basal metabolic rate, and in reduction of their physical activity.

Recommendations for average energy intake are based on energy expenditure for women within the healthy weight category. In pregnancy, this is estimated from:

- the energy value of the foetus and placenta and the extra maternal tissues: uterus, breasts, and adipose tissue
- plus any extra energy expenditure (basal metabolic rate plus physical activity for the heavier body) at the different stages of pregnancy.

The energy cost of pregnancy is not evenly distributed over the gestational period and it is agreed that

**Table 32.1 Extra energy recommendations in pregnancy for the average woman (MJ/day)**

| | North American 2002 and Australasian 2006 | WHO 2004 | UK 2011 |
|---|---|---|---|
| First trimester | 0 | +0.375 | 0 |
| Second trimester | +1.4 | +1.2 | 0 |
| Third trimester | +1.9 | +1.95 | +0.8 |

there is little extra energy needed in the first trimester (current estimates suggest around 105 kJ/day). Most weight is gained in the second and third trimesters at rates of 0.45 kg and 0.40 kg per week. It is estimated that the *average* energy cost of pregnancy for a woman who gains 12 kg in weight works out to 321 MJ. From this, the expert committees in North America and Australasia both recommend an extra 1.5 MJ/day in the second trimester and 1.9 MJ/day in the third trimester (Table 32.1).

However, many measurements of energy intake throughout pregnancy in women in ten different countries (nearly 1000 women in all) found that most of them ate less than one extra MJ/day and the average was +0.3 MJ/day. This finding forms the basis of the 1991 and 2011 UK recommendation of 0.8 MJ extra/day (only) in the third trimester (Table 32.1), but women who are underweight at the beginning of pregnancy and women who do not reduce activity may need more.

Weight gain throughout pregnancy is thought to average 11%, 47%, and 42% of the total in the first, second, and third trimesters. Table 32.2 shows the distribution of maternal weight gain at birth. Increase in maternal tissue (i.e. uterine and breast tissue), blood and other body fluids, and adipose tissue occurs mainly in the second trimester, while growth of the foetus, placenta, and amniotic fluid occurs mainly in the third.

Adequate maternal weight gain during pregnancy is the principal means of ensuring adequate foetal growth and, hence, infant birth weight. Excessive gain is associated with large infants (>4200 g), increased likelihood of caesarean delivery, and post-partum obesity. Because of the wide variation

**Table 32.2 The distribution of maternal weight gain at 40 weeks gestation**

| | Weight gain distribution (g) |
|---|---|
| Foetus | 3300–3500 |
| Placenta | 650 |
| Increase in blood volume | 1300 |
| Increase in uterus and breasts | 1300 |
| Amniotic fluid | 800 |
| Fat stores and additional fluid retention | 4200–6000 |
| *Total* | 11,550–13,550 |

in weight gain among women who give birth to optimally grown infants, a range of weight gains is regarded as acceptable for each BMI. A normal weight gain for most healthy women is between 11 and 15 kg, averaging about 12.5 kg. The US Institute of Medicine guidelines, which are based on observational data, indicate that healthy American women who are a normal weight for their height (BMI 18.5–24.9 kg/m$^2$) should gain 11.5–16 kg (25–35 pounds) during pregnancy. Overweight women (BMI 25–29.9 kg/m$^2$) should gain 7–11.5 kg (15–25 pounds), and obese women (BMI >30 kg/m$^2$) should only put on 5–9 kg (11–20 pounds). However, these figures are not recommended by NICE (UK), who note that there are no evidence-based guidelines from the UK government or professional bodies on what constitutes appropriate weight gain during pregnancy.

Maternal anthropometry differs between ethnic groups and separate guidelines should be used for other ethnic groups, such as Chinese and Polynesian.

### 32.2.3 Birth weight: the effect of maternal age, maternal weight, and energy intake

Birth weight is regarded as one of the best indicators of overall nutritional status of the infant and its well-being. The normal birth weight range is considered to be between 2500 and 4200 g. Low birth weight (LBW), i.e. babies weighing less than 2500 g, is a major cause of infant mortality and has been linked with long-term morbidity, including deficits in growth and cognitive development in childhood and diabetes and heart disease in adult life.

UNICEF estimates that around 19 million newborns each year in the developing world weigh less than 2500 g at birth, and more than half of them are born in South Asia. India has the highest number of LBW babies each year: 7.4 million.

In Australia, babies born to indigenous (Aboriginal) women in 2007 were twice as likely to be LBW (12.5%) as those born to non-indigenous women (5.9%). LBW is caused by pre-term birth, intrauterine growth retardation, or both.

Adolescents who are still growing are one population subgroup at greater risk of having LBW babies. Even when their own weight gain is sufficient to ensure adequate fat stores, they do not appear to mobilize these stores to enhance foetal growth late in pregnancy. Consequently, their nutritional requirements are greater, and this is reflected in higher recommended energy intakes for younger pregnant women in the UK. Low maternal body weight is also associated with LBW. However, in the developed world, current evidence suggests that chronic low maternal energy intake does not significantly contribute to LBW, and attempts to increase birth weight through energy supplements have had negligible effects. One (positive) potential nutritional influence on birth weight is a diet rich in long-chain ω3 polyunsaturated fatty acids (LCP), and randomized controlled trials have shown an increase in gestation (6 days) with LCP supplements, although overall the effects are rather modest. A wide range of nutrients has been examined in an attempt to influence birth outcomes (including iron, folate, zinc, vitamin D), but these have had little effect. Supplementation with magnesium from the 25th week of gestation has been shown to result in fewer pre-term and LBW deliveries, but these results did not differ significantly from placebo groups. It should be noted that protein supplements have a negative effect on birth weight.

### 32.2.4 Obesity in pregnancy

Data from England (2007) showed that maternal obesity doubled from 7.5% to 15.6% over 19 years. In addition, around a quarter of all women in the first trimester are in overweight category and 1 in every 1000 women giving birth in the UK has a BMI of at least 50 kg/m$^2$. Women who are obese when they become pregnant have higher health risks, including impaired glucose tolerance and gestational diabetes, miscarriage, pre-eclampsia, thromboembolism, and maternal death. In the UK enquiry of maternal deaths for the period 2000–2002, 30% of 261 maternal deaths were in obese women. Additionally, it has been observed that even a relatively small gain of 1–2 BMI units (kg/m$^2$) between pregnancies may increase the risk of gestational hypertension and gestational diabetes. Obese women are more likely to have an induced or longer labour, instrumental delivery, caesarean section, or post-partum haemorrhage. Reduced mobility during labour can result in the need for more pain relief and possibly general anaesthesia, with its associated risks. Babies born to obese women also face health risks, including a higher risk of foetal death, stillbirth, congenital abnormality, shoulder dystocia, macrosomia (large baby), and subsequent obesity. However, in the absence of trial data demonstrating effective clinical outcomes of the impact of weight modification during pregnancy (and beyond), obstetricians remain cautious and recommend that dieting for weight loss during gestation should be discouraged, as this may result in LBW infants if there is serious caloric restriction in the third trimester.

In the UK, the National Institute for Health and Care Excellence (NICE, 2010) advises that during pregnancy the key message is to ensure that the myth to 'eat for two' is dispelled. Professionals should be clear that extra calories (above non-pregnant intake) are required, but only during the last trimester, and it is also important to communicate that slimming is not appropriate during pregnancy. For women with a BMI >30 kg/m$^2$, weight loss during pregnancy is not recommended. Personalized guidance on healthy eating (with appropriate assessment) and ways to become more physically active are advised. In terms of starting to increase activity, women should begin with no more than 15 minutes of continuous exercise, three times a week, increasing gradually to daily 30-minute sessions. Encouragement should be given to lose weight after pregnancy.

### 32.2.5 Foetal nutrition, birth outcome, and health in later life

An increasing body of evidence suggests that early nutritional status (as indicated by birth weight and other parameters) modifies the risk of disease (notably cardiovascular) in later life. Birth weight has been used as a proxy measure of foetal nutrient exposure, although it may not be a sensitive enough measure to describe inadequate or unbalanced maternal nutrition. The hypothesis for the relationship between nutrition and early origins of disease is based on the concept that in foetal life, the tissues and organs of the body go through periods of rapid development, termed critical periods. Critical periods may coincide with periods of rapid cell division. Thus, if the foetus is deprived of nutrients or oxygen at such times, it may adapt by slowing the rate of cell division, especially in tissues undergoing critical periods. Even brief periods of undernutrition may permanently reduce the numbers of cells in particular organs. It is postulated that foetal undernutrition may change or programme the body with respect to distribution of cell types, hormonal feedback, metabolic activity, and organ structure.

Extensive research by David Barker and colleagues (Barker, 1998) has related causes of adult mortality and morbidity to foetal and infant life. These observational studies have related LBW to adverse health outcomes in adulthood, including hypertension, type 2 diabetes, and coronary heart disease. Variations in newborn ponderal index (kg/m$^3$) and placenta weight/birth weight ratios have also been related to subsequent hypertension. These observations have led to the foetal origins hypothesis, which states that foetal undernutrition in middle-to-late gestation leads to disproportionate foetal growth and programmes the later development of several diseases that are common in affluent societies and other groups undergoing rapid acculturation.

This hypothesis has been challenged by a number of researchers. Other studies only show a direct association between small size in early life and later adult health outcomes if body size at some intermediate period has been adjusted for. Researchers have suggested that this finding implies that it is probably the change in size across the whole time interval (postnatal centile crossing), rather than foetal biology, that is implicated. Thus, it remains to be resolved whether people who are small in early life and then grow rapidly are more at risk than those who remain small.

### 32.2.6 Nutrient requirements during pregnancy

**Energy** Women do not eat or need to eat for two (see section 32.2.2), but they do need to eat a high-quality nutritious diet to ensure that they obtain their extra requirements for several essential nutrients.

**Protein** A summary of recommended intakes for protein and other nutrients for a number of countries is presented in Table 32.3. Additional protein is required during pregnancy to provide for the synthesis of foetal, placental, and maternal tissue. Maternal and foetal growth accelerates in the second month of pregnancy and continues to increase until just before term. The need for protein follows this growth. However, the extra amount is relatively small (6–10 g/day) and is usually readily provided in a normal pre-pregnant Western diet.

**Folate** Pregnancy is a period characterized by extra cell division and growth. This increases folate requirements more than any other nutrient. Folate requirement is increased in the first 12 weeks (see section 32.1) and supplements are recommended pre-conceptually and in early pregnancy. Increases in dietary intakes of folates are also recommended

**Table 32.3 Recommended daily nutrient intakes during pregnancy**

| | USA[a] | | | Australia/NZ[b] | UK |
|---|---|---|---|---|---|
| | Women *not* pregnant 2005 | Change for pregnancy | Pregnant 2005 | Pregnant 2006 | Pregnant 1991 |
| Protein (g) | 46 | +54% | 71 | 60 | 51 |
| Vitamin A (μg) | 700 | +10% | 770 | 800 | 700 |
| Vitamin D (μg) | 5 | 0 | 5 | 5 | 10 |
| Vitamin E (mg) | 15 | 0 | 15 | 7 | – |
| Vitamin C (mg) | 75 | +13% | 85 | 60 | 50 |
| Thiamin (mg) | 1.1 | +27% | 1.4 | 1.4 | 0.9 |
| Riboflavin (mg) | 1.1 | +27% | 1.4 | 1.4 | 1.4 |
| Niacin (NE) (mg) | 14 | +29% | 18 | 18 | 13 |
| Folate (μg) | 400 | +50% | 600 | 600 | 300[c] |
| Vitamin $B_{12}$ (μg) | 2.4 | +8% | 2.6 | 2.6 | 1.5 |
| Calcium (mg) | 1000 | 0 | 1000 | 1000 | 700 |
| Magnesium (mg) | 310 | +13% | 350 | 350 | 300 |
| Iron (mg) | 18 | +50% | 27 | 27 | 15 |
| Zinc (mg) | 8 | +38% | 11 | 11 | 7 |
| Iodine (μg) | 150 | +47% | 220 | 220 | 140 |

[a]Numbers for USA are recommended daily allowances (RDAs) from dietary reference intake (DRI) recommendations for females 19–50 years.

[b]These recommendations are RDIs 2004, except for vitamin D and vitamin E, which are adequate intakes.

[c]A supplement of 400 μg/day folic acid is now advised.

NE, niacin equivalents.

throughout pregnancy to avoid megaloblastic anaemia in late pregnancy or the puerperium. Differences in recommended dietary intakes by country (Table 32.3) have arisen due to differences in primary indicators of status.

**Calcium** Around two-thirds of the calcium in the foetus is deposited during the last 10 weeks of gestation, mostly in the foetal skeleton. Alterations in maternal calcium metabolism, including a substantial increase in the absorption of dietary calcium, occur early in pregnancy to facilitate this increase in foetal demand. There has been concern that inadequate dietary intake of calcium during pregnancy was compensated by mobilization of skeletal calcium, leading to an increased risk of osteoporosis in later life. However, recent evidence suggests that the adaptation in calcium metabolism that occurs in pregnancy is sufficient to maintain foetal growth, even if dietary calcium is not increased. Current guidance promotes an adequate calcium intake throughout life, rather than increasing dietary intake during the gestation period. A Cochrane review on calcium supplementation during pregnancy for preventing hypertensive disorders and related problems (Hofmeyr et al., 2006) has reported that calcium supplementation appears to be beneficial for women at high risk of gestational hypertension and in communities with low dietary calcium intake. The WHO now recommends 1.5–2 g of oral elemental calcium for women in communities with low dietary calcium intake, especially those at high risk of pre-eclampsia.

**Iron** The demand for iron is not evenly distributed throughout gestation. During the first trimester, requirements are minimal, as iron is no longer lost during menstruation. Iron absorption increases as pregnancy progresses, from an absorption rate of around 7% at 12 weeks' gestation to 66% at 36 weeks. Many women enter pregnancy with low iron stores or even frank iron-deficiency anaemia (IDA). Maternal IDA may increase the risk of pre-term delivery, resultant LBW, and perinatal mortality. As well, evidence is accumulating that maternal IDA reduces infant iron stores post-partum, leading to the possibility of impaired development in the infant. There remains considerable debate on the use of iron supplements during pregnancy with respect to whether this is an attempt to alter a natural physiological process that in some way enhances nutrient supply. Dietary guidance in the UK suggests that no dietary changes to iron intake are required due to physiological changes, although women with poor iron status at the time of obstetric booking may need iron supplements. The WHO recommend a cut-off of 110 g/L for anaemia throughout pregnancy, and in the UK, NICE recommends that iron supplementation should be considered for women with haemoglobin concentrations below 110 g/L in the first trimester and 105 g/L at 28 weeks.

**Zinc** Zinc is necessary for DNA and RNA synthesis. Maternal zinc deficiency may be responsible for growth retardation, pre-term delivery, and abnormality in the foetus, and birth complications in the mother. Factors that limit the absorption of zinc, such as high intakes of dietary phytate, calcium, and iron supplements, may cause secondary zinc deficiency. Recently published dietary recommendations include an increase of zinc intake in pregnancy. The extra zinc should come from foods (see section 10.1.5). It is possible that iron supplements could reduce zinc intake.

**Iodine** Iodine deficiency during pregnancy remains a major public health problem in many areas of the world. Congenital iodine-deficiency syndrome, caused by severe lack of iodine during foetal development, is characterized by both mental and physical impairment. Millions of babies are born each year at risk of mental impairment due to iodine-deficient diets. Iodization of salt is commonly used to prevent deficiency. In areas where there is congenital iodine-deficiency syndrome and extensive goitres, expectant mothers should be given an injection of iodized oil, preferably before conception (see section 10.3.5). There has been increasing concern over iodine adequacy in developed countries among pregnant women, and recent debate has emerged following results of a cross-sectional nationwide survey of adolescent schoolgirls in the UK and some smaller studies of women of childbearing age. The AVON longitudinal cohort study of pregnant women and child pairs reported that women who had mild to moderate iodine deficiency (based on urinary concentrations) had children within the lowest quartile of verbal IQ. However, there is uncertainty about how robust spot urinary iodine measures are as an indication of status given that there is a diurnal variation, and it has been emphasized that a single urine spot is not adequate for assessment of an individual's iodine status. At the population level, a median urinary iodine excretion above 100 μg/day is an indicator of a low prevalence of iodine deficiency (SACN, 2014). In the UK, it is thought that there is insufficient evidence to revise current dietary reference values for pregnancy (no increment during pregnancy), although it is recognized that maternal iodine needs and metabolism during pregnancy have not been well characterized. Good dietary sources are marine fish and shellfish. Levels in milk, meat, chicken, and dairy products reflect iodine content of animal feed. UK recommendations contrast with those of Australia and New Zealand, where pregnant women are advised by the National Health and Medical Research Council to take a daily supplement of 150 μg.

**Vitamin A** Vitamin A deficiency, seen mostly in some developing countries, may be associated with blindness, depressed immune function, and increased morbidity and death from measles and other infectious diseases, as well as increased mother-to-child transmission of AIDS. Encouraging foods rich in this vitamin (see Chapter 11) is the most appropriate approach to ensuring adequacy. However, in pregnancy, intakes above the recommended daily intake (>3000 μg retinol daily) resulting from the use of supplements, excessive intakes of fortified foods, and, occasionally, high intakes of liver can be teratogenic, causing central nervous and heart defects. High levels of vitamin A have been detected in the liver of farmed animals in the UK (due to the composition of animal feedstuffs); thus, all pregnant women are advised to avoid liver, liver products, vitamin supplements, and fish oil supplements that are high in retinol. In developing countries where appropriate foods are not readily available, supplementation may be considered (700 μg/day).

**Vitamin D** Populations at risk for vitamin D deficiency are those whose skin exposure to sunlight is low, particularly dark-skinned women living in northern climates and southern climates, e.g. the South Island of New Zealand, particularly over winter months. Also, women who do not spend time outdoors or wear veils, such as some ethnic minority groups in the UK, may be at risk. The infants born in such populations have low vitamin D stores. Vitamin D insufficiency during pregnancy is associated with lower maternal weight gain, neonatal tetany, and biochemical evidence of disturbed skeletal homeostasis in the infant. In extreme situations, reduced bone mineralization, radiologically evident rickets, and fractures may occur. The WHO states that there is insufficient evidence to identify beneficial and adverse effects of vitamin D supplements taken during pregnancy on maternal or neonatal outcomes. NICE (UK) is one of the few organizations to recommend that pregnant women take vitamin D supplements (10 μg/day) throughout their pregnancy to maintain adequate maternal vitamin D stores. Globally, those deemed at risk should be recommended supplementation.

A review by the Scientific Committee on Nutrition (2015) in the UK found no data to support a case for changing the recommendation from 10 μg/day.

**Vitamin C** It appears that the foetus concentrates vitamin C at the expense of maternal stores and circulating vitamin levels. Accordingly, it is recommended that dietary intake of foods rich in vitamin C are increased during pregnancy.

In summary, while mineral requirements increase during pregnancy, most of the extra nutrients can be attained through physiological adaptation rather than dietary change (in the well-nourished). With respect to vitamins, extra dietary intakes of vitamin C and vitamin A are required, which may mean increasing intakes by rich dietary sources (but not by supplements). Increasing intakes of folate and vitamin D should be met by increasing dietary sources, as well as by supplements.

### 32.2.7 Lifestyle factors that impact on pregnancy outcome

Alcohol Heavy drinkers have a greatly increased risk of inducing the foetal alcohol syndrome, with characteristic underdevelopment of the mid-face, small body size, and impaired cognition. Any effect of alcohol is likely to be greatest in the first few weeks after conception, the embryogenesis stage. There is not a recommended safe level of alcohol intake during pregnancy. Abstinence is recommended as the safest option both during pregnancy and for women who wish to become pregnant. https://www.jogc.com/article/S1701-2163(17)30584-4/abstract (accessed 21 February 2022)

Smoking Smoking causes retardation of foetal growth, thereby increasing the risk of producing an LBW baby. The older the mother and the more cigarettes smoked, the greater the effect. A decrease in birth weight of approximately 200–250 g is usually found in infants of mothers who smoke more than 20 cigarettes a day. Smoking also increases the risk of spontaneous abortion, pre-term delivery, and sudden infant death.

Exercise Most recent studies show that moderate exercise during pregnancy does not harm the foetus and benefits the mother. Benefits can include reduced fat gain, lower risk of gestational diabetes, maintenance of aerobic fitness, shorter labour, quicker delivery, and fewer surgical interventions. However, high-impact exercise or hard physical work can affect foetal development and may result in LBW babies and a higher frequency of obstetric complications.

### 32.2.8 Other diet and health concerns

Environmental carcinogens Diet is an important source of carcinogens, including dioxins, polychlorinated biphenyls (PCBs), polycyclic aromatic hydrocarbons (PArHs), as well as acrylamide formed during cooking processes. The recently reported European New Generis biomarker study (Kleinjans et al., 2015) has demonstrated that dietary carcinogens do pass the placenta barrier and that foetal exposure occurs. The findings suggest that exposure may induce molecular events that indicate increased risk of childhood cancer (notably leukaemia) when there is increased genetic predisposition. High (versus low) exposures of dietary carcinogens were also associated with lower birth weight and gestational age. The authors suggest that it is difficult to reduce exposure in the overall diet given such widespread sources, but pregnant women might consider limiting intake of processed meats and fried and grilled foods, and food manufacturers have a role in reducing the presence of carcinogenic agents through reduction of nitrite and changes to potato crisp production.

Coffee Caffeine freely crosses the placenta. The risk of spontaneous abortion and LBW appears to increase with high maternal caffeine intake during pregnancy. It is appropriate to limit caffeine consumption in pregnancy, and current guidelines suggest pregnant women have no more than 200 mg of caffeine a day (approximately two mugs of instant coffee or four cups of tea). Some women develop aversion to coffee when they are pregnant. The recent dietary guidelines published in the USA note that caffeine beverages are heavily marketed and note women who are pregnant or considering pregnancy should not consume very high levels of caffeine from beverages or supplements (e.g. energy shots, fortified foods), in addition to moderating intake of caffeine-containing drinks.

Oil-rich fish Due to contaminants such as mercury, which may be stored in high concentrations in fatty fish, pregnant women are recommended to restrict intake of oily fish. These recommendations will vary by region. In the UK, pregnant women are advised not to eat marlin, shark, or swordfish, to limit tuna to four cans per week, and not to exceed oily fish consumption above two portions per week. However, it is considered beneficial to consume at least one portion of oil-rich fish, e.g. herring, salmon, and sardines, weekly during pregnancy.

**Food safety** Pregnant women and their newborns are 10–20 times more likely than the general population to be infected with *Listeria monocytogenes*, which can contaminate uncooked foods (see http://www.cdc.gov/ (Centers for Disease Control and Prevention); accessed 21 February 2022). After an incubation period of 2–6 weeks, infection can result in a mild chill or more severe illness, premature birth, or stillbirth. There can also be effects in the newborn, including meningitis. Listeriosis responds to antibiotics when it is diagnosed. In the UK, the incidence is estimated at 1 in 30,000 live and stillbirths. The Department of Health recommends that pregnant women avoid certain ripened soft cheeses, such as brie, camembert, and blue-veined cheese, and any type of pâté.

**Salmonella, toxoplasma, and food poisoning** More generally, these also need to be avoided with greater care during pregnancy than at other times.

**Allergens** In recent years, women were advised to avoid eating peanuts during pregnancy if there was a history of allergy in the family, as it was thought that this would diminish the likelihood of the infant developing peanut allergy. However, recent work has failed to support this hypothesis and studies suggest that pregnant women who consume peanuts (who are not nut allergic) have a *lower* risk of peanut allergy in their offspring.

## 32.2.9 Physiological effects of pregnancy that impact on dietary intake

**Nausea and vomiting** It is hypothesized that nausea and vomiting are part of the maternal system to protect against toxins. In support of this hypothesis, a comprehensive review reported that:

- symptoms peak when embryonic organogenesis is most susceptible to chemical disruption (weeks 6–18)
- women who experience morning sickness are less likely to miscarry than women who do not
- women who vomit suffer fewer miscarriages than those who experience nausea alone
- many pregnant women have aversions to alcoholic and non-alcoholic (mostly caffeinated) beverages and strong-tasting vegetables, especially during the first trimester.

There is no generally effective remedy for morning sickness and, given the vulnerability of the foetus in the early stages of development, it is important to find alternatives to drug therapy.

Changes in eating habits, such as small, frequent meals and avoidance of strong food odours, appear to help some women. A Cochrane review using data from six double-blind randomized controlled trials with a total of 675 participants and a prospective observational cohort study indicated that ginger was effective in relieving the severity of nausea and vomiting episodes, with no significant side effects or adverse effects on pregnancy outcomes. The authors concluded that more observational studies are needed to confirm this encouraging preliminary data on ginger.

**Hyperemesis gravidarum (HG)** This is a condition that causes severe nausea and vomiting in early pregnancy, often resulting in hospital admission. The incidence of HG varies from 0.1% to 1% and appears higher in multiple pregnancies, hydatidiform mole, and other conditions associated with increased pregnancy hormone levels. Both the aetiology and pathogenesis of HG remain unknown, although a range of pregnancy hormones (progesterone, oestrogen, and human chorionic gonadotrophin) and other hormones have been implicated. Infants from HG pregnancies have significantly lower birth weight, younger gestational age, and a greater length of hospital stay. The persistent vomiting can lead to Wernicke's encephalopathy (also see section 12.1.4), so it is important that thiamin is given with replacement intravenous fluids.

**Cravings and aversions** Some change in liked and disliked foods is common in pregnancy. There can be cravings or aversions. Food cravings are popularly believed to be related to the nutritional needs of the mother, to have symbolic value, or to be related to sensory or physiological causes. Food aversions have been defined as 'a definite revulsion against food and drink not previously disliked'. Typical examples are tea, coffee, and alcohol. The explanation for these cravings and aversions is incomplete. They may relate to changes in olfactory and taste sensitivity during pregnancy.

Constipation Around 40% of women report having been constipated at some time during pregnancy. Its aetiology is complex and includes depressed gut mobility in pregnancy, increased fluid absorption from the large intestine, decreased physical activity, and dietary changes. However, an increase in fibre, from an average intake of about 18–27 g/day, has been shown to be effective in treating constipation.

## 32.3 Lactation

The decision to breastfeed is influenced by psychobiological and psychosocial factors that vary between and within cultures. The proportion of babies who were ever breastfed varies widely, as does the duration of exclusive feeding. Across Europe, less than 70% of babies in France and Ireland to around 100% in Denmark, Norway, and Sweden initiate breastfeeding. In Australia in 2005, around 88% of mothers started breastfeeding, but only 17% were exclusively breastfeeding at 6 months. In the UK, around 78% of mothers initiate breastfeeding, but less than 3% are still exclusively breastfeeding at 6 months.

This compares with Rwanda and North Korea, where 88% and 65% are still exclusively breastfeeding at 6 months. In most countries, initiation and duration of breastfeeding are positively associated with maternal education status, maternal income, and marital status.

The physiology of lactation is complex, but may be briefly summarized as follows: the suckling infant stimulates the mother's pituitary gland to release prolactin, a hormone required for the synthesis of breast milk. Milk-producing cells synthesize most of the protein and some of the fats and sugars, which combine with other nutrients derived from the mother's circulation. A second pituitary hormone, oxytocin, is responsible for releasing the milk from the cells into the ducts that carry the milk to the nipples.

### 32.3.1 Composition of breast milk and implications for maternal nutrition

Human breast milk is the optimal source of nutrition for the new infant and should provide adequate nutrition for the first 6 months of life, assuming that maternal diet and stores are adequate and the milk is successfully transferred to the infant. Beyond the nutritional composition, human milk provides significant protection from a range of anti-infective components including macrophages, lymphocytes, immunoglobulins (especially IgA), lactoferrin, lysozyme, complements, interferon, oligosaccharides, and sialic acid. These components decrease risk from GI (and other) infection and thus decrease likelihood of diarrhoea and associated malnutrition. The composition and volume of breast milk progressively changes with the stage of lactation and can be influenced by maternal nutritional factors. Current evidence indicates that infant demand is the major determinant of the quantity of milk produced. The nutritional demands of lactation on the mother are directly proportional to volume and duration of milk production.

The daily milk volume varies over the duration of lactation, but is fairly consistent except in extreme maternal malnutrition or severe dehydration. Breast milk intake among healthy infants averages 750–800 g/day and ranges from 450 to 1200 g/day.

The composition of milk will be influenced by time of day, gestational age (prematurity vs term), stage of lactation, parity, month (i.e. seasonal food intake), nutritional status of mother, and maternal dietary intake. Lipids are the most variable constituent in human milk. In the first week after birth, colostrum is produced, which has a fat content of 2.6 g/100 mL. This is followed by transitional milk between days 7 and 14, and then finally mature milk, which has a fat content of 4.2 g/100 mL.

### 32.3.2 Maternal nutrient requirements to support lactation

During pregnancy, the mother's body prepares for lactation by storing some nutrients and energy. It is difficult to determine precise nutrient requirements for lactation since there is variation in nutritional status before and during pregnancy, and limited knowledge about utilization of maternal nutrient stores and adaptations of maternal metabolism during lactation.

The total energy cost of lactation is derived from the energy content of the milk plus the energy required producing it. The energy value of breast milk is between 2.7 MJ/L and 3.1 MJ/L. The ratio between the energy content of milk and the total energy cost of lactation is the efficiency of milk production. Current estimates for exclusive breastfeeding suggest that the energy cost of lactation is around 2.625 MJ/day (625 kcal) based on a mean milk production of 750 g/day with an energy density of milk of 2.8 kJ/g and energetic efficiency of 0.80. In well-nourished women, this will be partially met by energy mobilization from fat tissues of about 0.65 MJ/day (for weight reduction of 0.5 kg fat/month), resulting in a net increment of around 2.0 MJ/day (480 kcal) over non-pregnant, non-lactating energy requirements. This assumes the ideal situation where the new mother gradually uses up the extra 2–5 kg fat put on during pregnancy over 6 months of breastfeeding. The value will vary when complementary feeding is introduced or the baby becomes only partially breastfed. Women who exclusively breastfeed for 6 months may require as much as 2.4 MJ extra/day.

### 32.3.3 Other nutrients

Table 32.4 shows recommended nutrient intakes for five countries. The North American recommended nutrient intakes were published between 1998 and 2002 and the Australasian in 2005–2006. The British recommendations date from 1991. In developed countries, there is sufficient protein and most other nutrients in usual diets so that the average extra 2.0 MJ of food per day will cover the extra nutrient needs, with two exceptions. In the northern winter, a vitamin D supplement is advisable, and vegans must take a vitamin $B_{12}$ supplement (these are available from microbiological, i.e. non-animal, sources). Although some 260 mg of calcium are secreted per day in the mother's milk, epidemiological studies have found no increase in osteoporosis or fracture in women who have breastfed compared with those who did not. In some countries, including USA, Australia, and New Zealand, 150 μg supplement of iodine is recommended in addition to dietary sources while breastfeeding. Some research suggests that mild iodine deficiency is associated with subtle neurodevelopmental deficits and that iodine supplementation might improve cognitive function in mildly iodine-deficient children (see https://ods.od.nih.gov/factsheets/Iodine-HealthProfessional; this link has a good summary of the research around this (accessed 21 February 2022)).

The National Academy of Medicine (NAM) (https://nam.edu/about-the-nam/ (accessed 7 June 2022)), formerly the US Institute of Medicine, sums it up:

> The loss of calcium from the maternal skeleton that occurs during lactation is not prevented by increased dietary calcium, and the calcium lost appears to be regained following weaning. There is no evidence that calcium intake in lactating women should be increased above that of non-lactating women.

Some nutrients in the breast milk are increased if there is more of them in the mother's diet: water-soluble vitamins, vitamin A, and polyunsaturated fatty acids. Most other constituents in the milk—protein, lactose, total fat, and calcium—do not appear to be influenced by maternal intake.

Table 32.4 **Recommended daily nutrient intakes during lactation**

| Nutrient | USA | USA and Canada, Australia/NZ[a] | USA and Canada | UK 1991 |
|---|---|---|---|---|
| | NPNL women | Lactating | % Increase | Lactating |
| Protein (g) | 46 | 67 | +46 | 56 |
| Vitamin A (μg) | 700 | 1300 (1100) | +86 | 950 |
| Vitamin D (μg) | 5 | 5 | 0 | 10 |
| Vitamin E (mg) | 15 | 19 (11) | +27 | – |
| Vitamin C (mg) | 75 | 120 (80) | +60 | 70 |
| Thiamin (mg) | 1.1 | 1.4 | +27 | 1.0 |
| Riboflavin (mg) | 1.1 | 1.6 | +45 | 1.6 |
| Niacin (NE) (mg) | 14 | 17 | +21 | 15 |
| Folate (μg)[b] | 400 | 500 | +25 | 260 |
| Vitamin $B_{12}$ (μg) | 2.4 | 2.8 | +17 | 2.0 |
| Calcium (mg) | 1000 | 1000 | 0 | 1250 |
| Iron (mg) | 18 | 9[c] | −50 | 15 |
| Zinc (mg) | 8 | 12 | +50 | 13 |
| Iodine (μg) | 150 | 290 (270) | +93 | 140 |

[a]Where the Australasian RDI is different, it is shown in brackets.
[b]These are RDIs for dietary folate equivalents in USA, Australia, and NZ.
[c]For first 6 months of lactation; assumes menstruation not restarted.
NPNL = not pregnant, not lactating.

It is now no longer considered necessary to advise mothers with a family history of allergy to avoid peanuts or peanut butter during pregnancy and lactation. After a cup of coffee or glass of wine, the concentration of caffeine or alcohol in the milk is about the same as in the mother's plasma; the infant gets a lower dose per kg than the mother, but has less metabolizing capacity. A safe level of alcohol intake has not been established. Should a woman choose to consume alcohol, she should delay breastfeeding or expressing milk until the alcohol is completely cleared from her breastmilk—approximately two to three hours for each standard drink consumed.

Adequate fluid intake (especially water) should be advised for breastfeeding mothers. The exact amount required is unknown, but 6–8 glasses is generally sufficient. The nutrient reference values for Australia and NZ (2005) recommend 2.6 L per day during lactation (2.3 L during pregnancy).

### 32.3.4 Lactation and maternal obesity

Pregnancy is a risk factor for obesity. Some women put on more than the standard 2–5 kg of fat. The question arises whether milk production and the baby will suffer if the lactating mother restricts her food intake. Lovelady et al. (2000) tested this out in a randomized controlled trial in overweight (not obese) women. They lost approximately 0.5 kg per week between 4 and 14 weeks post-partum from moderate food restriction and exercise; their infants gained the same weight and length as the controls, but some of the control mothers put on weight. Overall, current evidence (reviewed by NICE, 2010) suggests that weight management interventions (addressing diet and physical activity) during lactation have little or no adverse effects on breastfeeding outcomes, including milk volume, infant intake, and time and frequency of feeding.

Perhaps more importantly, the UK Million Women Study has demonstrated that (at a mean age of 57.5 years) at every parity, women who had breastfed their babies had a mean BMI significantly lower than those who had not, indicating that childbearing patterns have a persistent effect on adiposity in women (Bobrow et al., 2013).

## Further reading


1. **Allan, L.H.** (2005) Multiple micronutrients in pregnancy and lactation: an overview. *Am J Clin Nutr*, **81**, 1206S–12S.
2. **Anderson, A.S. and Wrieden, W.L.** (2010) Teenage pregnancies. In: Symonds, M.E. and Ramsey, M.M. (eds) *Maternal-fetal nutrition during pregnancy and lactation.* Cambridge: Cambridge University Press.
3. **Barker, D.J.P.** (1998) *Mothers, babies and disease in later life.* London: BMJ Books.
4. **Bartley, K.A., Underwood, B.A., and Deckelbaum, R.J.** (2005) A life cycle micronutrient perspective for women's health. *Am J Clin Nutr*, **81**, 1188S–93S.
5. **Bobrow, K.L., Quigley, M.A., Green, J., Reeves, G.K., Beral, V., and Million Women Study Collaborators** (2013) Persistent effects of women's parity and breastfeeding patterns on their body mass index: results from the Million Women Study. *Int J Obes (Lond)*, **37**, 712–7.
6. **Butte, N.F. and King, J.C.** (2005) Energy requirements during pregnancy and lactation. *Publ Health Nutr*, **8**, 1010–27.
7. **Centers for Disease Control and Prevention (CDC)** (2015) Updated estimates of neural tube defects prevented by mandatory folic acid fortification—United States, 1995–2011. *Morb Mort Weekly Rep*, **64**, 1–5.
8. **Hofmeyr, G.J., Atallah, A.N., and Duley, L.** (2006) Calcium supplementation during pregnancy for preventing hypertensive disorders and related problems. *Cochrane Database Syst Rev*, CD000145.
9. **Kleinjans, J., Botsivali, M., Kogevinas, M., and Merlo, D.F.** (2015) Fetal exposure to dietary carcinogens and risk of childhood cancer: what the New Generis project tells us. *BMJ*, **351**, h4501.
10. **Lovelady, C.A., Garner, K.E., Moreno, K.L., and Williams, J.P.** (2000) The effect of weight loss in overweight lactating women on the growth of their infants. *N Engl J Med*, **343**, 449–53.
11. **MRC Vitamin Study Research Group** (1991) Prevention of neural tube defects: results of the Medical Research Council Vitamin Study. *Lancet*, **338**, 131–7.
12. **NICE** (2008) *Guidance for midwives, health visitors, pharmacists and other primary care services to improve the nutrition of pregnant and breastfeeding mothers and children in low income households.* London: National Institute of Health and Clinical Excellence. Available at: http://guidance.nice.org.uk/PH11 (accessed 21 February 2022).
13. **NICE** (2010) *Dietary interventions and physical activity interventions for weight management before, during and after pregnancy.* London: National Institute of Health and Clinical Excellence. Available at: http://guidance.nice.org.uk/PH27 (accessed 21 February 2022).
14. **Public Health England** (2009) *SACN Report to CMO on folic acid and colorectal cancer risk.* Available at: https://www.gov.uk/government/publications/sacn-report-to-cmo-on-folic-acid-and-colorectal-cancer-risk (accessed 21 February 2022).
15. **Ramachenderan, J., Bradford, J., and McLean, M.** (2008) Maternal obesity and pregnancy complications: a review. *Aust NZ J Obstet Gyneacol*, **48**, 228–35.
16. **SACN** (2014) *SACN statement on iodine and health.* Available at: https://www.gov.uk/government/uploads/system/uploads/attachment_data/file/339439/SACN_Iodine_and_Health_2014.pdf(accessed 21 February 2022).
17. **Scholl, T.O.** (2005) Iron status during pregnancy: setting the stage for mother and infant. *Am J Clin Nutr*, **81**, 1218S–22S.

18. **Scientific Advisory Committee on Nutrition** (2006) *Folate and disease prevention.* London: Food Standards Agency. Available at: https://www.gov.uk/government/publications/sacn-folate-and-disease-prevention-report (accessed 21 February 2022).
19. **Scientific Advisory Committee on Nutrition** (2011) *The influence of maternal, fetal and child nutrition on the development of chronic disease in later life*. London: Food Standards Agency. Available at: https://www.gov.uk/government/publications/sacn-early-life-nutrition-report (accessed 21 February 2022).

## Useful website

**First Steps Nutrition Trust: Eating well during pregnancy**
https://www.firststepsnutrition.org/eating-well-in-pregnancy

# Infant Feeding and Eating Well for Toddlers

Anne-Louise Heath, Helen Crawley, and Rachael Taylor



## 33.1 Introduction

Nutrition in the first 1000 days of life (pregnancy and the first 2 years) is fundamentally important to the growth and development of children and to population health. Ensuring all infants and young children receive optimal nutrition is the focus of the *Global strategy for infant and young child feeding* developed by WHO and UNICEF and endorsed in 2002. The key aspects of this strategy are to support exclusive breastfeeding in the first 6 months of life and continued breastfeeding alongside the introduction of a wide variety of appropriate solid foods, for 2 years and beyond. The strategy also reiterates the importance of commitment to continuing action to implement the UNICEF Baby Friendly Initiative, the International Code of Marketing of Breastmilk Substitutes, and the Innocenti Declaration on the Protection, Promotion, and Support of Breastfeeding. Mothers and babies form a biological and social unit and the health and nutrition of one cannot be separated from the other. Good diet during pregnancy prepares the foetus for life outside the womb, and it is important that women are supported pre-pregnancy, and during pregnancy, to eat well (see Chapter 32).

## 33.2 Infant feeding

### 33.2.1 Breast milk and breastfeeding

Breast milk is uniquely suited to a human baby and more than 800,000 young children's and 20,000 women's lives a year could be saved globally if all babies were optimally breastfed. However, despite the importance of exclusive breastfeeding to population health, there are often low rates of breastfeeding in both low- and high-income countries. It is impossible to make a substitute for breast milk as its composition is dynamic—it is a living substance that changes in composition during feeds and as babies grow and develop. It is unique to each mother for her baby and for the environment in which they live. There are many bioactive molecules in human milk that cannot be reproduced, which protect infants from infections and help them develop a strong immune system for the future. These include:

- immunoglobulins and anti-infective agents that protect the infant from infections
- *lactoferrin*: a protein that has strong antibacterial and antiviral properties, and may enhance iron absorption
- special fatty acids, which promote brain growth and development
- antiviral factors, antibacterial substances, and living white blood cells to offer protection against disease.

The milk in the first few days after a baby is born is called colostrum. It is produced in small amounts but is rich in anti-infective agents, which protect the baby when it is most vulnerable. The composition of the milk changes as the baby grows, and after 4–6 weeks becomes the mature breast milk that will support a baby throughout the first year of life and beyond. With the exception of vitamins D and K, breast milk produced by adequately nourished mothers provides all the nutrients needed by a normal, healthy, full-term infant for the first 6 months of life, and makes an important nutritional contribution for as long as the mother wants to continue breastfeeding. Vitamin K is routinely given at birth and therefore the low content in breast milk is not a concern. Sunlight is the major source of vitamin D (see Chapter 14). Water or other fluid supplementation is unnecessary in the otherwise-healthy infant in the first 6 months of life, even in a hot, humid climate, if the infant nurses more often, particularly if their mother is in the same environment, as would usually happen if the infant is being fed directly from the breast. The changing composition of breast milk during a feed, and over time, is one of the unique properties that is thought to be linked to better satiety control in those who have been breastfed.

Breastfeeding has several known advantages for both mother and infant. Although reduced risk of infection in early life is the most accepted benefit, there is now considerable interest in how breastfeeding can protect infants from non-communicable diseases (NCD) such as obesity and diabetes. One potential mechanism, which is receiving increased attention, is the role that breast milk may play in the development of the human gut microbiota, which has been implicated in a number of diseases (e.g. inflammatory bowel disease, and obesity). There is even evidence, including from one randomized controlled trial, to suggest that being breastfed, or being breastfed for longer, is associated with a small increment in intelligence. One of the main benefits for the mother's health is a reduced incidence of breast cancer, with the risk decreasing by more than 4% for every 12 months of breastfeeding. Breastfeeding seems particularly protective against more aggressive breast cancers seen in younger women and a long total duration of breastfeeding appears to be also associated with a substantial reduction in the overall risk of ovarian cancer, although this may vary according to histological subtype. Immediate benefits also include a delayed resumption of ovulation, which can be important in many low-income countries in terms of birth spacing and reduced iron losses as menstruation is delayed. Numerous other benefits have been proposed for both mother and child. Those supported by systematic reviews are summarized in Box 33.1. In addition, it is accepted clinically that breast milk is protective against necrotizing enterocolitis in pre-term infants. There is some evidence that prolonged breastfeeding may increase the risk of tooth decay, so it is important that good dental care is followed (as is discussed below).

Breastfeeding is rarely contraindicated. A woman's capacity to breastfeed is not affected by her age, body shape or size, ethnicity, breast size, the climate, or any experiences her mother or another relative may have had previously. To breastfeed successfully, at least one functioning breast is needed and a willingness to allow nature to do its work by letting a baby suckle and stimulate milk production to establish the supply and demand nature of breast milk production. Women can still breastfeed when they have had breast implants, when they are pregnant with another baby, if they have flat nipples, if they are diabetic, have a disability, or have had a nipple

**Box 33.1** Breastfeeding benefits supported by systematic reviews

| Accepted benefits of breastfeeding *for the infant* | Accepted benefits of breastfeeding *for mothers* |
|---|---|
| Decreased risk of the following diseases:<br>Mortality due to infectious disease<br>Gastrointestinal infection<br>Severe respiratory infection<br>Ear infections<br>Sudden infant death syndrome<br>Type 2 diabetes (probable)<br>Obesity (probable)<br>And:<br>Reduced risk of malocclusion of the teeth<br>Higher intelligence | Decreased risk of the following diseases:<br>Breast cancer<br>Ovarian cancer (possible)<br>Type 2 diabetes (possible)<br>And:<br>Longer time between births |

piercing. Breastfeeding is contraindicated in the rare cases where infants have galactosaemia, or where mothers have untreated active tuberculosis, active herpetic breast lesions, or human T-lymphotropic virus type I infection, or are receiving ongoing radiation therapy or certain chemotherapeutic agents. In low-income countries it is often more dangerous for the infant of an HIV-positive mother to be formula fed than to receive breast milk, and there are WHO guidelines supporting breastfeeding unless an acceptable, feasible, affordable, sustainable, and safe alternative is available. However, because HIV can be transmitted in breast milk, HIV-positive women in high-income countries where the water supply is clean and infant formula is affordable are not recommended to breastfeed, even if their viral load is low. Neither smoking nor environmental contaminants are contraindications to breastfeeding. Moderate, infrequent alcohol ingestion, the use of most prescription and over-the-counter drugs, and many maternal infections, including mastitis, do not preclude breastfeeding.

### 33.2.2 Formula feeding

When a mother chooses not to, or cannot, breastfeed, the only acceptable alternative is a commercial infant formula. Infant formulas are nutritionally adequate for the first year of life, but despite considerable marketing claims made by manufacturers, no component added to formula can increase IQ, offer protection to an infant from infection or allergy development, or reduce the risk of those diseases to which infants who are not breastfed are at increased risk. Most standard infant formulas are based on cows' or goats' milk protein, and include carbohydrate as lactose, maltodextrins, or glucose polymers (in some areas corn syrups and sucrose can be added), a variable blend of vegetable oils (palm oil, sunflower oil, coconut oil, soy oil, and canola oil are commonly used), and an added vitamin and mineral mix. The composition of infant formula is regulated by local, regional, or international standards produced by Codex, and in many countries the composition of all infant formula sold will be very similar. In some countries, such as the USA, infant formulas are available in two versions: low iron and iron fortified. Low-iron formulas, which have a very low iron content, similar to breast milk (<0.1 mg/100 mL), may be attractive to some parents who believe, incorrectly, that it may help with gas and constipation, but are contraindicated as the low bioavailability of iron in these products will not prevent iron deficiency. Many 'low-iron' formulas, however, now have 0.6–0.7 mg iron per 100 mL, which is the standard amount recommended in Europe, where the concept of low iron and iron-fortified formulas is not known. In contrast, 'iron-fortified' formula can have >1 mg iron/100 mL, which exceeds current upper limits in the regulations in Europe (i.e. 0.84 mg iron per 100 mL).

Soy-based formulas made from soy protein, vegetable oils, and glucose polymers are available, but in some areas these are not recommended for use in infants under 6 months of age without medical direction because there are concerns that high levels of phyto-oestrogens in soy-based formulas may be damaging to a developing infant. They are also not recommended for infants with an allergy to cows' milk protein, since an allergy to soy protein is also commonly present.

Manufacturers differentiate formula products on the market through the addition of non-essential ingredients about which they can make claims. However, there is a considerable difference between the claimed benefits of some components added to infant formula by manufacturers and accepted scientific opinion. Isolating an ingredient in breast milk, artificially reproducing it, and adding it to formula milk does not make a product 'close to breast milk' and it is agreed that formula milk cannot mimic human milk with respect to its energy and protein content, for example. Components such as arachidonic acid (AA), eicosapentaenoic acid (EPA), non-digestible oligosaccharides (prebiotics), probiotics or synbiotics, taurine, and nucleotides are all considered unnecessary in infant formula, as are phospholipids as a source of long-chain polyunsaturated fatty acids or triacylglycerols with palmitic acid esterified in the *sn*-2 position. Despite the fact that these factors have been shown to be of no benefit to infant health, many claims are made by manufacturers for benefits of these added components, ranging from prevention of illness, reduced risk of allergy, better digestion, reduced wind or colic, and reduced crying, to better brain development. Many ingredients are added without prior approval by an independent scientific body.

Of even greater concern than the claims relating to possible benefits of prebiotics is the addition of probiotic bacterial strains to formula milk. There are concerns about the use of a daily supply of a 'monoculture' of a single, specific microbial strain in large quantities over a prolonged period of time, particularly for vulnerable infants. Furthermore, infant formulas that contain probiotics must be made up with water at a temperature lower than recommended as safe, putting vulnerable infants at greater risk of infection (see 'making up milks safely' in section 33.2.6).

There remains conflicting support for the addition of docosahexaenoic acid (DHA) to full-term infant formula. DHA accumulates in the brain and eye of the foetus, especially during the last trimester of pregnancy, and some studies suggest that infants may benefit from direct consumption of DHA in infant formula. There is limited evidence of any long-term benefit to full-term infants, but some still-debated evidence that the addition of DHA may be beneficial to visual acuity, with the European Food Safety Association (EFSA) allowing a health claim that 'DHA contributes to the visual development of infants.' The DHA added to infant formula comes from fish oils, and increasingly from microbial synthesis, which provides a vegetarian source and does not impact on the world's fish stocks.

Despite the obvious damage these claims make to breastfeeding choices in populations, aggressive marketing remains in many areas because of weak local and regional regulation around the World Health Organization (WHO) International Code of Marketing of Breast-milk Substitutes and subsequent World Health Assembly (WHA) resolutions. In 2016, WHA resolution 69/7 clarified that all milks marketed specifically to infants and young children aged under 3 years are breast milk substitutes, and so should not be promoted. The inappropriate marketing of both foods and drinks to infants and young children undermines exclusive breastfeeding, length of breastfeeding, and increases the risk of conditions that breastfeeding has been shown to protect against (Box 33.1). Many countries do not put the Code and resolutions into local regulation and a report by WHO in 2020 showed that only 25 countries out of the 194 reporting have passed laws that are substantially aligned with the Code. In fact, from 2000 to 2019, formula consumption in the first 6 months of life actually increased in upper middle-income countries, East Asia, the Pacific, Latin America, the Caribbean, the Middle East, North Africa, Eastern Europe, and central Asia. The formula milk market is dominated by five large multinational companies: Nestlé, Danone, Abbott, Mead Johnson, and Friesland-Campina, and in a recent review by the International Baby Food Action Network, all of these companies fell well short of compliance with the Code. Renewed efforts are underway within the global health community to strengthen commitments to these important Codes and resolutions.

A detailed review of UK infant formula composition and current expert opinion on efficacy of ingredients used can be accessed at the First Steps Nutrition website.

**Follow-on formulas** These products, marketed for infants over 6 months of age, were developed by manufacturers to avoid regulation around marketing of infant formula and have no benefit to infant health. It is agreed by all health bodies globally that there is no need for follow-on formula, which the WHO is clear should be treated as a breast milk substitute and should therefore be regulated as part of the Code. Infants who are consuming infant formula should continue to do so until one year of age rather than changing to a follow-on formula when they reach 6 months.

**Other specialized infant formula** A range of specialized milks are available, often called 'foods for special medical purposes.' Some of these are only available through medical practitioners, but some are sold over the counter, despite no risk assessment being undertaken. For example, lactose-free milks are marketed for babies with 'diarrhoea, tummy ache, and wind'; however, routine use amongst infants is not recommended. A lactose-free formula may not be beneficial in the longer term: lactase is an inducible enzyme and requires the presence of some lactose in the intestine for optimal development. 'Anti-reflux milks' contain a thickener and have to be made up with water at room temperature so that they do not get lumpy. This means that there is a potential microbiological risk (see section 33.2.6). Claims that partially hydrolysed milks can prevent allergies are also not substantiated by expert health bodies. The use of high-energy formula in premature and smaller babies leading to rapid catch-up growth has potential long-term health implications related to chronic disease, and there has been a considerable shift in thinking in many neonatal units as to the importance of establishing breastfeeding to support the infant, not only when they are in hospital but also to ensure they can be exclusively breastfed when they leave hospital care. Some specialized milks are, however, essential, but these should always be used under medical supervision.

**Home-made infant formula** Historically, people made home-made formulas from diluted evaporated milks, but these are nutritionally incomplete and should not be used. Some confusion has been noted in countries where a standard dried milk product has been marketed with a cartoon bear logo that makes it appear similar to a formula product, and local regulation is needed to prevent misuse of non-formula products. More recently, amongst concerns about the ingredients in commercial formulas, some families are making infant formula based on complicated recipes circulating on the internet. These may be nutritionally adequate but may present other difficulties around food safety. In the US there are concerns about genetically modified (GM) ingredients in infant formula that may stimulate the idea of personally controlling what infant formula is made from. With the increasing interest in 'paleo' and 'keto' diets in adult populations in recent years, recipes for 'bone broth' endorsed as a home-made alternative to infant formula have appeared on the internet. This is very concerning because as well as being nutritionally inadequate, some recipes include large amounts of liver (and therefore potentially toxic amounts of vitamin A) or raw egg (a food safety concern in this age group), and many contain added salt (which can be harmful to the developing kidneys). This highlights the importance of health professionals being aware of social media movements that may influence parents, leading to dangerous infant feeding practices.

### 33.2.3 Allergy to cows' milk protein

Cows' milk protein (CMP) allergy in infancy is the most common food allergy, with estimates of prevalence ranging from 1% to 5%. Immunoglobulin E (IgE)-mediated reactions are caused by the release of histamine and other mediators. The reactions are acute and frequently have a rapid onset (up to 2 hours after milk ingestion). Non-IgE-mediated reactions are thought to be caused by T cells. The reactions are non-acute and generally delayed (manifest up to 48 hours or even 1 week after milk ingestion). Although often severe in infancy, most children will grow out of CMP allergy in the first few years of life, but early diagnosis and treatment is needed to prevent faltering growth. Mothers can continue to exclusively breastfeed an infant with CMP allergy if the mother excludes cows' milk protein from her diet, and this should always be considered the first line of treatment, with tailored support given for a mother to do this. A supplement of 1000 mg of calcium and 10 µg of vitamin D to the mother to prevent nutritional deficiencies is frequently recommended in the UK. If the mother is not exclusively breastfeeding or is formula feeding, then an extensively hydrolysed formula is usually recommended. For infants with severe CMP allergy, or multiple food allergies, infants are prescribed formulas based on free amino acids as the protein source. The more hydrolysed or broken down the protein source is, the more expensive and unpalatable the formula will become, and it is important that infants are regularly reviewed so that cows' milk can be safely reintroduced where this is possible. Goats' milk-based formula cannot be used as treatment for CMP allergy as this has some similar antigens to cows' milk, and soy formulas are not recommended for the reasons given previously (section 33.2.2).

### 33.2.4 Lactose intolerance

The majority of adverse reactions to foods do not involve the immune system and are known as food intolerance. The most common food intolerance in humans is lactose intolerance, which is caused by lack of the lactase enzyme that breaks lactose into its component sugars in the intestine. Congenital lactase deficiency in infants is rare, while primary lactase deficiency is very common in the adult population due to a normal developmental decrease in lactase activity with increasing age. It is uncommon for infants under the age of 2 to have primary lactase deficiency and, as a result, other conditions should be excluded prior to treatment. Certain ethnicities are more affected, including Chinese people, people of African descent, Ashkenazi Jews, and North American indigenous people. Secondary lactose intolerance develops in infants as a result of intestinal mucosal damage caused by infectious gastroenteritis, malnutrition, CMP enteropathy, coeliac disease, giardiasis, bacterial overgrowth, inflammatory bowel disease, or drugs. Common symptoms of lactase deficiency include wind, stomach cramps, and diarrhoea. Diagnosis can be obtained by performing a non-invasive breath-hydrogen test in older children or through intestinal biopsies to investigate enzyme activity; however, on a more practical level, diagnosis is usually confirmed clinically by a trial of a lactose-free diet. Infants under the age of 12 months with lactose intolerance can be given a lactose-free cows' milk formula. Infants and children with primary lactase deficiency may be able to tolerate small amounts of lactose-containing foods such as yoghurt and cheese, particularly if taken in small amounts and with time between exposures. Special attention is needed to ensure affected infants have an adequate calcium intake. Following secondary lactase deficiency, reintroduction of small amounts of milk should be tried at regular intervals to return to a balanced diet as soon as possible. It is important that any decision about the presence of lactose intolerance is made with the advice of a GP or paediatrician so that the correct diagnosis is made, so that dairy is not excluded from the diet unnecessarily, and so that adequate calcium intake can be ensured.

### 33.2.5 Responsive bottle feeding

Whichever formula is used, all families should be supported to use responsive feeding techniques to follow their infant's cues around eating. Because breastfeeding requires the active participation of both mother and infant, it is easier for a breastfed infant to stop feeding when they are full. For parents or carers who feed using a bottle, it is suggested that they watch for signs of readiness for a feed so that babies are not crying when the bottle is introduced, that they hold their baby in a semi-upright position and reassure them by looking into their eyes and talking to them during a feed. Gently introducing the teat to the baby's mouth and following their cues to start and stop the feed, and never forcing a baby to finish a bottle, are important. These recommendations also apply when bottle feeding a baby with expressed breast milk. Responsive feeding helps parents to build close and loving relationships with their babies, as well as helping families to respond to their baby's appetite rather than following standard feeding guidance. Over-feeding with formula milk is a potential risk for overweight and obesity in later life.

### 33.2.6 Making up infant milks safely

Powdered infant milks are not sterile and may contain harmful bacteria. *Salmonella enterica* and *Cronobacter* species are the organisms of greatest concern in infant formula. *Salmonella enterica* and *Cronobacter* species do not survive the pasteurization process, but recontamination may occur during handling or from production methods where ingredients are mixed. *Salmonella enterica* and *Cronobacter* species can grow in reconstituted formula if stored above 5°C and can multiply rapidly at room temperature, so it is essential that good hygiene practices are observed during preparation, storage, and feeding in order to avoid recontamination and/or multiplication in the reconstituted product. *Cronobacter* is regarded as an emerging opportunistic human pathogen, with younger infants being more susceptible to infection than older infants, and neonates at greatest risk if they are pre-term or low birthweight, as are those who are immunocompromised. Few countries have a notifying system so it is difficult to be certain how frequently powdered formula may be contaminated and how many infants become infected, but there are potentially serious complications, including bacteraemia, meningitis, and necrotizing enterocolitis. Some cases are fatal. Although contamination is primarily an issue for infant formula users, there have been cases of infection from a breast pump or expressed breast milk, so it is important that care is taken to carefully clean and sanitize breast pumps, and to follow good hygiene practices when collecting, handling, and storing breast milk.

The key recommendation from a wide range of international bodies to reduce risk to infants from bacterial infection has been to encourage the reconstitution of infant formula with water at no less than 70°C. Despite this, many manufacturers sell products that they recommend should be made up at lower temperatures, and there is concern about the increasing use of probiotics in infant formula, which require products to be made up with water below 70°C, despite the lack of any agreed benefit for the probiotic addition. Infant formula should always be made with fresh water, boiled before cooling, bottles and teats should be sterilized carefully, and reconstituted infant formula must be refrigerated if it is not going to be consumed immediately. The recommendation that formula be mixed with water that is at least 70°C is not, however, universally supported. This is because of the practicality of the advice (studies of parents preparing formula show that parents do not know how to determine water temperature), the risk of increased bacterial growth (because parents do not accurately determine water temperature they often mix the formula at lower temperatures that are ideal for bacterial growth), and because of the risk of scalding (infants can be scalded in less than a second by water at 65°C, so mixing multiple bottles of formula a day at temperatures above this poses substantial risk).

Milk powder needs to be carefully measured, as small differences in scoop weights can lead to under- or over-nutrition. This is particularly likely to happen when poverty means infant formula is too expensive to use at full strength, leading to insufficient intake, or if parents are tempted to 'tap' the formula scoop when it is filled so that the formula is more densely packed, leading to excess intake. The difficulties in many areas of having the facilities and equipment to make up and store infant formula safely is one of the reasons for the high number of preventable deaths in infants who are formula fed.

### 33.2.7 Vitamin D

There has been debate about the vitamin D status of populations worldwide, and low vitamin D status is being reported across the lifespan in many countries. Sunlight (UVB) is the main source of vitamin D. Infants depend on their own synthesis and metabolism of vitamin D, as there is little vitamin D in breast milk. Major risk factors for low vitamin D in infants are naturally dark skin, a mother who is deficient or at higher risk of deficiency (for example, as a result of using veils and clothing that cover the whole body), pre-term low birthweight, and exclusive breastfeeding over the winter months in countries far from the equator. Consuming 500 mL of formula a day is protective, but exclusive breastfeeding and consumption of a vitamin D supplement is preferable for overall health. In many countries it is now recommended that all breastfeeding women receive vitamin D supplementation of typically 10 μg/day, and that infants who are breastfed are given a supplement from birth. The conservative recommended dose of vitamin D for infants who are breastfeeding is about 10 μg/day.

## 33.3 Complementary feeding

After about 6 months of age, exclusive breastfeeding no longer meets the growing infant's energy and nutrient needs, and complementary foods must be added alongside breast milk (or infant formula). By introducing solids at this time, infants can develop their chewing and swallowing skills and become familiar with a range of tastes and textures. The additional foods offered as solids are not intended to replace breast milk but are considered complementary to breast milk in the first year.

### 33.3.1 Introduction of solids

There is a tension between the recommendation that infants are exclusively breastfed until 6 months of age and need to start consuming iron-rich solids, and recommendations based on signs of infant 'readiness' for solids. Commonly used signs of readiness are: being able to hold their head upright, and sit up with less help; the ability to take solid foods into the mouth, move them to the back of the mouth, and swallow them (rather than foods being thrust back out on the tongue as is seen in young infants); and showing signs of biting and chewing. These signs are rarely present before about 6 months of age, and families often mistake waking in the night, chewing on fists, or interest in other people eating as indications of readiness for solids. Ideally the signs of readiness will be present at around 6 months. If they appear either before or after 6 months it is important to discuss this with a health professional before acting on the presence or absence of signs of readiness alone. Before about 6 months of age a baby can be made to swallow smooth puréed food, but is not actively swallowing, and purées commonly offered to babies are often less nutritionally dense than the breast milk they replace and sweet in taste, which may impact on their nutrient intake, and their acceptance of a range of other flavours. There is currently no conclusive evidence that introducing solids earlier to breastfed babies can help prevent food allergies, and there is some evidence that earlier introduction of solids may be linked to greater weight gain in the early years. It is recommended that common allergy-causing foods, including well-cooked egg, smooth peanut butter, wheat-based foods, dairy-based foods (e.g. cheese, yoghurt), cooked fish and shellfish, and soy products should be offered during complementary feeding, as this may help reduce the chances of allergies developing. Once they have been offered, they should continue to be offered to maintain baby's tolerance to them.

Historically, there were suggestions that infants should be introduced to smooth foods one at a time and move slowly to more lumpy food, but recommendations about the timing of first solids have changed. It is now recognized that infants at about 6 months can often manage a range of textures very quickly once solids are introduced. This includes food offered on a spoon that may be smooth or mashed at first, rapidly moving to forked and chopped, as well as finger foods offered from about

6 months that encourage infants to be involved in the eating experience.

Scientific evidence in support of traditional recommendations for the order and progression of introducing solids is limited. Infants should be introduced to nutrient-rich solid foods, and require a good source of iron from 6 months of age, when breastfeeding cannot meet their iron needs (see Table 33.1). The most commonly used first food has been iron-fortified infant cereal, but iron-rich meat, fish, eggs, tofu, and pulses are also good choices because of their iron content. While it has been common practice in high-income countries to offer bland foods, often baby rice and fruit purée, babies are able to enjoy spices and strong flavours. It is also now encouraged that infants are involved in the eating process from when solids are introduced, by being offered finger foods that they can hold and use to feed themselves. See section 33.3.2 for advice on ensuring that finger foods are safe.

### 33.3.2 Novel approaches to complementary feeding

Following many years of fairly consistent complementary feeding practices, there have been two recent, potentially far-reaching changes to the infant feeding landscape: 'baby food pouches' and

**Table 33.1 Eating well guidance for infants 6 to 12 months of age**

| Foods and drinks recommended at about 6 months of age | Foods and drinks to avoid in the first year of life |
|---|---|
| **Foods** | **Foods** |
| First foods for babies at about 6 months of age can include a wide range of foods: meat, poultry, fish, shellfish, eggs, tofu, nut butters and seed butters, pulses (peas, beans, and lentils), vegetables, potatoes and other starchy roots such as yam, taro, sweet potato, and cassava, cereal foods such as bread, pasta, rice, oats, polenta, semolina, teff, quinoa, unsweetened whole-fat dairy products such as cheese and yoghurt, fruits.<br><br>Some commercial baby foods such as iron-fortified cereals can be a useful addition to the weaning diet in moderate amounts. Many commercial foods are expensive and are often smoother and sweeter than needed. They do not aid an infant's recognition of flavours and textures associated with family foods. If they are used it is important to follow the manufacturer's instructions carefully. | Never add salt, sugar, or artificial sweeteners to foods for infants.<br><br>Foods that should be avoided include:<br>• ready meals or take-away foods<br>• processed meat or fish dishes (e.g. ones covered in breadcrumbs or batter, canned meats, smoked or cured meat, smoked or cured fish) that may be high in salt<br>• savoury or salty snacks<br>• very high-fibre foods (such as high bran-type cereals)<br>• any foods with special ingredients designed for adults—for example, low-fat or low-sugar products<br>• artificially sweetened foods<br>• foods that are high in salt, either added to them, or intrinsic (e.g. soy sauce, gravy)<br>• honey<br>• whole nuts or chunks of food such as apple that might be a choking risk (this also applies to foods such as whole apples or carrots from which the baby can break off pieces). |
| **Drinks** | **Drinks** |
| No drinks other than breast milk (or infant formula) or water are recommended in the first year. | Drinks that should be avoided:<br>• soft drinks, squashes, fruit juices, cordials<br>• drinks with added caffeine or other stimulants<br>• artificially sweetened drinks<br>• plant 'milks' (e.g. rice milk, coconut milk, almond milk, soy milk)<br>• tea or coffee<br>• alcohol. |

'baby-led weaning'. Neither of them are well understood from a scientific or health point of view, but both approaches are popular with parents.

Pouches are squeezable containers with a plastic spout that are a new way of presenting puréed foods. They have a large and increasing market share with pouches now occupying the majority of supermarket shelf space for commercial infant foods in many countries. Although manufacturers often recommend that the food is squeezed onto a spoon rather than being fed straight from the nozzle, anecdotal reports suggest that at least some infants are consuming the food straight from the pouch, with the potential for increased intake because the food is easy to squeeze into the mouth in large amounts (a standard pouch holds approximately 22 teaspoons of food), and for prolonged and repeated exposure of the teeth to these often sweet and acidic (and therefore presumably cariogenic) foods. There is also concern because the wet nature of the food means that it is not iron-fortified, which could result in a substantial reduction in iron intake if pouch cereals are fed instead of iron-fortified dry cereal. In addition, the convenience of these foods for parents, and their palatability for infants, may lead to them being used well beyond the early weeks of complementary feeding, despite their smooth liquid texture (necessary so that the food can be extruded through the nozzle). This could delay the introduction of a wider range of textures and tastes, with concomitant effects on oral motor development and taste and texture preference. While many concerns have been expressed about the use of baby food pouches, the only research to date has been on the composition of these foods, with no data on how pouches are used by parents and infants, or on the health impacts of their use.

Baby-led weaning is an alternative approach to introducing solids in which infants feed themselves all their food from the start of complementary feeding, so that there is no spoon-feeding by a parent, and solids are only offered as 'finger foods'. This approach differs considerably from the more traditional approach advocated by most agencies internationally in which the infant gradually learns how to eat solid foods by eating foods with progressively increasing textures from puréed to mashed to chopped to whole. It is hypothesized that baby-led weaning may improve infants' relationship with food, reducing food fussiness, and may result in less obesity as infants can better regulate how much they eat. Parents report that it is convenient because baby can join family meals (although it is messy!). The majority of research to date has investigated the characteristics of parents who follow baby-led weaning (they tend to be more highly educated, older, have lower restrained eating scores, and be less anxious), and attitudes to baby-led weaning expressed by health professionals and parents. Very few studies have investigated nutrient intake (iron, zinc, and vitamin $B_{12}$ intakes may be a particular concern). Some studies have reported that infants following baby-led weaning may have a reduced risk of becoming overweight, or fussy about food, and that their choking risk may not be higher than that of infants following traditional spoon-feeding; however, these studies have been cross-sectional so are not conclusive given the differences in parental characteristics that are apparent. Health professionals have raised concerns about potential increased risks with baby-led weaning of iron deficiency (due to the omission of iron-fortified baby rice, which infants cannot pick up and feed themselves at this age), growth faltering (if infants are not able to pick up and feed themselves enough food to meet their energy requirements), and choking (because of the immediate introduction of solid foods without the opportunity to develop skills while eating 'easier' textures). The Baby-Led Introduction to SolidS (BLISS) study demonstrated in a randomized controlled trial that it is possible for a baby-led approach to infant feeding to be followed without increasing the risk of growth faltering, iron deficiency, or choking, but a modified version of baby-led weaning was used that was designed specifically to address these concerns, so the impact of unmodified baby-led weaning is still unclear.

Whatever approach to infant feeding is being used, it is important that finger foods are long enough for the infant to be able to hold them in their fist and eat from the protruding end, care needs to be taken that pieces cannot break off easily and become a choking hazard, and most importantly, foods given to the infant should be tested by a parent first to make sure that they can be squashed on the roof of the mouth with the tongue.

## 33.3.3 Safety issues around feeding

Infants have a lesser degree of immunity to bacteria and viruses in the digestive tract than older children and adults and therefore foods provided to infants must be, as far as possible, free of pathogens. The risk of choking on foods with the potential for aspiration and asphyxia is also highest for infants and toddlers, and therefore care needs to be taken that foods are appropriate in size and texture, and fed safely (see Box 33.2). The greatest risk of choking and aspiration on food occurs in children under the age of 4 years, with a significant peak in the 12–24-month age group. Small, round, or chewy foods such as sausage or hotdog rounds, whole grapes, whole nuts (including peanuts), hard or chewy sweets, stringy foods like celery lengths, and popcorn are the most dangerous, as they can slip into the pharynx before they have been adequately chewed and with a quick gasp for breath be drawn downward and become lodged in the airway. Highly viscous foods, such as peanut butter or chocolate spread, can plug the airway and should not be served by themselves. In addition to the shape and texture of foods, environmental factors such as distractions, inadequate supervision during eating, and crawling or walking while eating increase the risk of food asphyxiation.

## 33.3.4 Food allergies

Adverse food reactions are divided into two general categories: food allergy and food intolerance. A true allergic reaction to a food involves the body's immune system. In the paediatric population, estimates of food allergy range from 1% to 8%, with the highest frequency in the first year of life. There is general agreement that the prevalence is increasing, but the cause of this is not fully understood. Fortunately, food allergies that start in childhood are often outgrown. Food allergies commonly present in infancy with the intake of milk, egg, or peanuts. Along with soy, fish, nuts, and wheat, these foods are responsible for about 95% of food allergies in infants and toddlers. Allergy to milk, egg, soy, or wheat is more likely to be outgrown than allergy to tree nuts or peanut. However, resolution of a food allergy may not occur until as late as the teenage years.

A host of risk factors have been proposed as influencing food allergy and these include gender (male children at greater risk), race/ethnicity, genetics (familial associations and specific genes), dietary fat choices (reduced consumption of omega-3-

**Box 33.2** Guidelines for feeding infants safely

| Reducing the risk of food-borne illness | General safety tips |
|---|---|
| Avoid unpasteurized dairy products as they can introduce pathogens such as *Escherichia coli*, *Salmonella*, *Campylobacter*, or *Cryptosporidium*, which can cause diarrhoea or other more serious infections. | To avoid burns to an infant's palate or face, formula or food warmed in a microwave should be shaken or stirred thoroughly, and the temperature tested before serving. |
| To prevent botulism, infants under 1 year of age should not be fed honey. | To avoid choking, do not offer hard, small and round, stringy, or sticky solid foods, such as whole nuts, whole grapes, cherry tomatoes, popcorn, celery lengths, or sausage rounds. |
| To prevent *Salmonella* poisoning, raw or not fully cooked eggs and foods containing raw eggs should not be fed to infants. | Infant cereal or other solids should not be added to milk in a bottle as it may put the infant at risk for choking and aspiration. |
| | Ensure that infants are always supervised and sitting during feeding. If using a highchair, make sure they are strapped in safely. |

polyunsaturated fatty acids), reduced consumption of antioxidants, obesity (being an inflammatory state), increased hygiene, and the timing and route of exposure to foods.

The issue of preventing allergy is controversial. It no longer appears that elimination diets during pregnancy and lactation play a role in decreasing the risk of infants developing food allergies. Prebiotics and probiotics have also not proven to be effective in allergy prevention. Surprisingly, it is now recommended that even infants with a close family history of allergic disease should be given common allergy-causing foods during complementary feeding (as described in section 33.3.1), with the foods introduced one at a time to help identification of any issues, then regular feeding of the food to maintain tolerance. However, this advice should be discussed with the infant's doctor. Management of diagnosed food allergies involves strict avoidance of the allergenic food and requires careful reading of food labels to detect hidden sources. As sensitivity to many foods disappears within a few years, retesting with the offending food should therefore occur at regular intervals in a controlled and supervised manner.

### 33.3.5 Dietary management of acute diarrhoea

In high-income countries, the typical infant with acute diarrhoea is well nourished, presents with mild to moderate dehydration, and has viral-induced diarrhoea with low stool electrolyte losses. In low-income countries, children with acute diarrhoea are more likely to be malnourished and severely dehydrated with viral- or bacterial-induced diarrhoea with high stool electrolyte losses. Many infants presenting with acute bacterial-induced diarrhoea will have been formula fed. Worldwide, pneumonia and diarrhoeal diseases are the two major killers of children younger than 5 years and each year more than 500,000 children die from diarrhoeal diseases. Oral rehydration therapy has been shown to be an important intervention that appreciably decreases mortality associated with diarrhoeal diseases. Breast milk is well tolerated during diarrhoea and may reduce its severity and duration; therefore, breastfeeding should continue throughout the diarrhoeal illness with additional fluids given as oral electrolyte solutions. Infants with severe dehydration should receive intravenous rehydration, otherwise oral electrolyte solutions containing specified concentrations of carbohydrate, sodium, potassium, and chloride should be used to promote fluid and electrolyte absorption. Fluids such as juices, soft drinks, tea, jelly, or broth do not aid rapid rehydration.

Early and rapid re-feeding should occur as soon as rehydration is achieved and vomiting stops (ideally within 6–12 hours of beginning treatment) as infants treated with oral rehydration therapy and early re-feeding have reduced stool output, shorter duration of diarrhoea, and improved weight gain. Routine change to lactose-free infant milk where the infant is formula fed, or diluted standard formulas, is unnecessary in well-nourished infants with mild to moderate gastroenteritis. Infants and toddlers who were fed solid food before the onset of the diarrhoea should continue to receive their usual diet once rehydration occurs. Although not based on strong science, starchy foods are generally well tolerated as the initial foods for re-feeding. Use of a low-residue diet (commonly called the BRAT diet: bananas, rice, apple sauce or apple juice, and tea or toast) can worsen the clinical state as it supplies less than one-half of an infant's daily energy and protein needs.

In malnourished children, zinc supplements given during an episode of acute diarrhoea reduce the severity and duration of the episode and reduce the recurrence of diarrhoea. The WHO currently recommends 10–20 mg zinc/day for 10–14 days (10 mg for infants under 6 months of age) from the start of an episode of diarrhoea.

### 33.3.6 Constipation

The stooling patterns of infants range from a bowel movement after each feed to one every few days. Stool frequency and consistency are influenced by the volume and type of feeding (e.g. breast milk or formula, how many fruits and vegetables are included in the diet). Bowel frequency decreases with age as a result of the maturing gut's ability to conserve water. After 3–4 years of age, the frequency of bowel movements does not change. Breastfed and non-breastfed infants receiving an adequate diet, and where the formula is made up correctly, are rarely constipated. Although an infant may appear to be straining, it is normal for an infant to grimace or have a red face when having a bowel movement. Educating parents about the wide variation in stooling patterns is important for avoiding overtreatment of normal stooling habits. Hard and painful bowel movements, abdominal distension, or blood in the stool may be signs of true constipation.

Approximately 90–97% of infants and children with constipation have idiopathic non-organic constipation, most often due to a conscious or unconscious decision to delay defecation after experiencing a painful or frightening evacuation (e.g. due to an anal fissure). This is *functional* constipation or withholding constipation. Functional constipation is uncomfortable, but not dangerous, and therefore benign. An infant less than 6 months who is believed to be truly constipated should be referred to a physician for investigation of organic causes. The earlier constipation occurs, the greater the chance of an underlying problem (e.g. Hirschsprung's disease). Common recommendations regarding therapy for constipation in infants are based on theory, not scientific evidence. Increasing free fluid intake by giving additional water, encouraging infants to move, and encouraging a diet of minimally processed foods high in soluble fibre such as fruits, vegetables, and pulses are approaches that can be tried. However, there is no evidence to support the use of dietary factors to alleviate chronic constipation once stool withholding and stool retention have become a problem. In a small number of infants, chronic constipation can be a manifestation of intolerance to cows' milk.

## 33.4 Eating well for toddlers 12–24 months of age

In the second year of life, infants become more involved with their own feeding, which gives them a sense of control and independence. With more teeth and the advancement in fine motor skills and oral coordination, infants can successfully use feeding utensils and enjoy foods with a range of textures. For those who have been bottle-fed, a transition from bottle feeding should be encouraged to promote drinking from a cup from 1 year of age, and no fluid other than milk should ever be given to an infant in a bottle.

As young children gain mobility and are more active, food intake does not always adjust immediately, and alongside more frequent illness when children start attending early childhood education centres, weight can fluctuate, which can cause concern among families. With increasing age, toddlers' likes and dislikes can also be intensified, and this can lead to strong and sometimes limited food preferences. Efforts must be made to continue offering a wide variety of foods, allowing toddlers to handle their food, and making eating an experience the family enjoys together (see also section 33.4.6).

### 33.4.1 Iron deficiency

Iron deficiency is common among infants and young children between the ages of 6 and 24 months and increases the risk of iron deficiency anaemia, which is a risk factor for lower cognitive function that may persist into early childhood if the anaemia is not identified and treated. The major risk factors for iron-deficiency anaemia in infants relate to low socioeconomic status and include early consumption of cows' milk, use of low iron infant formula, and inadequate funds for appropriate foods. Other high-risk groups include toddlers who were low birth weight or premature and may have had delayed introduction of solid food, and those who drink large amounts of cows' milk (>500 mL/day) or sweetened beverages. It is important to prevent, rather than treat, iron deficiency anaemia wherever possible, given that the

**Box 33.3 Strategies to prevent iron-deficiency anaemia**

- Introduction of iron-rich foods such as meat, chicken, fish, seafood, beans, lentils, tofu, or iron-fortified infant cereal from 6 months of age.
- Use of appropriately iron-fortified formula for infants weaned early from the breast or formula fed from birth.
- Delaying introduction of unmodified cows' milk until the second year of life and limiting the amount to less than 500 mL a day alongside continued iron-rich nutritious foods.
- Avoidance of substances with meals that limit iron absorption such as tea, foods high in phytates (e.g. unleavened bread), and encouraging foods high in vitamin C, such as fruits and vegetables, which enhance iron absorption.
- Treatment for helminthic infections, which cause iron deficiency in areas where there is poor sanitation.

only way to diagnose anaemia is by blood test, and the symptoms are so non-specific (see Box 33.3).

### 33.4.2 Obesity

Worldwide, 39 million children under five years of age were overweight or obese in 2020. With so many preschool children overweight and obese, increased attention is being paid to feeding practices. The first protective risk factor is exclusive breastfeeding, which is associated with a reduced risk of overweight and obesity even after correcting for maternal body mass index, and socioeconomic status. Children from low socioeconomic households are at greater risk of obesity, and this may also be linked to lower breastfeeding rates in poorer households, as well as poorer choice of foods from complementary feeding onwards.

Some evidence suggests early introduction of solid foods at <4 months, as well as putting children to bed with a bottle, increase the likelihood of obesity. Unhealthy infant feeding practices may explain the link between socioeconomic status and early childhood obesity. Interventions that combine high levels of parental involvement and interactive school-based learning with targeted physical activity, reducing sedentary behaviour, and promoting dietary change with long-term follow-up, appear most effective in preventing obesity. Policy makers are looking at restrictions on the marketing of high-energy-density processed foods, taxing sugar-sweetened beverages, and improvement of food provision in public places as ways of changing the food environment to help prevent obesity in children. There is, however, little evidence on how obesity should be treated in preschool children.

### 33.4.3 Oral health

Tooth decay in infants is a preventable infectious condition, and as with many common diseases, poor oral health is linked to lower socioeconomic status in many areas. Toothache and dental infections can have a profound effect on infant and toddler eating and sleeping habits, leading to effects on nutrition, growth, and weight gain, as well as causing unnecessary suffering. Acids produced by bacteria (*Streptococcus mutans*) ferment dietary carbohydrate and attack the teeth, eventually causing demineralization and cavitation, particularly during sleep, when salivary secretions are decreased. Severe tooth decay may warrant tooth extraction. The most important modifiable aspect of early childhood caries is decreasing the amount (and frequency) of carbohydrate exposure. Feeding practices such as never putting an infant to bed with a bottle, never offering a sweetened drink in the night, avoiding the use of pacifiers dipped in sugar, syrup, or honey, avoiding bottle-feeding past 12 months of age, avoiding the intake of any fruit-based or carbonated drinks in the first year of life, and reducing intakes of other sugary items can all help in avoiding early caries. Milk or water are the only drinks that should be given in a bottle. Even if drinks are advertised as 'sugar free,' if they are fruit-based or carbonated they will contain acids, which can damage the tooth enamel.

Fluoridation of the water supply has proven to be the most effective, cost-efficient means of preventing dental caries. In areas with low fluoride levels in the water source, fluoride supplements are recommended. The increased availability of fluoride (fluoridated water, foods or drinks made with fluoridated water, toothpaste, mouthwashes, and fluoride supplements) has resulted in an increasing incidence of mild forms of dental fluorosis in both fluoridated and non-fluoridated communities. This sign of excess fluoride intake has led to modifications in fluoride recommendations, including later introduction and lower doses of fluoride supplements, and caution to parents of children to use small amounts and to discourage the swallowing of toothpaste. Mild dental fluorosis has not been shown to pose any health risks and, while there may be mild cosmetic effects, the teeth remain resistant to caries.

Good oral health care in infants and toddlers is often overlooked but needs to start as soon as the first tooth appears. Parents should 'lift the lip' of their infant or toddler once a month to check for any signs of tooth decay, and a dental health practitioner should also check children's teeth regularly for decay, starting when they are toddlers. Teeth should be brushed twice a day (especially before bed time) with a small, soft toothbrush and a small smear of regular strength fluoride toothpaste.

### 33.4.4 Food choices

Toddlers still require nutrient-dense foods, and to ensure all the nutrients required for growth and development are consumed, 'empty calories' should be kept to a minimum. Guidance on eating well for this age group, which has quantified the sorts of foods, and amounts of food, that meet energy and nutrient requirements, can be used to provide guidance to both families and early childhood education settings. Ultra-processed foods should be avoided, but so should very high-fibre foods, as they can be too bulky for small children who require energy and nutrient-dense diets. Many foods marketed specifically as 'children's foods' are less nutritionally dense, as the addition of batter or breadcrumbs to which fat, salt, and sugar are added dilute the original foodstuff. There is no need for special children's foods, and the whole family can eat the same healthy foods so that toddlers can learn food choices from others around them. Many of the food restrictions for infants shown in Table 33.1 should be maintained for toddlers. Particular care needs to be taken with take-away foods and ready meals, as the portion sizes are not appropriate. Toddlers should be encouraged to have a minimum of five different types of fruits and vegetables a day. In many countries, as a nutritional safety net, children from 1–4 years are recommended vitamin drops containing, for example, vitamins A, C, and D.

### 33.4.5 Drinks

Breastfeeding is still encouraged in the second year of life, but where the infant has been weaned from the breast, or where an infant was formula fed in the first year, whole animal milk can be the main milk drink from 1 year of age. In some cases where eating is poor, or there are health problems, infant formula may be encouraged in the second year of life.

A comparison of the nutritional content of a variety of animal milks and plant-based milk alternatives is shown in Table 33.2. A variety of plant-based 'milks' is available for those avoiding animal milks but care is required that the diet remains sufficiently energy- and nutrient-dense, as many are much lower in energy and nutrients than whole animal milk. Soya milk is generally the best alternative because its energy, fat, and protein content is closer to that of whole cows' milk than other plant-based milks. Milk alternatives should be unflavoured, unsweetened, and calcium fortified, with vitamin $B_{12}$ fortification if the toddler is following a vegan diet and is not taking a vitamin $B_{12}$ supplement.

Soft drinks and fruit juices should be restricted as these provide sugar, which is not needed, and can damage teeth. Even if drinks do not contain sugar they will contain acids that can damage teeth (see section 33.4.3).

Herbal teas are of no known benefit to young children and may be harmful. Tea and coffee should be

**Table 33.2 Composition of mature breast milk, cows' milk, sheep's milk, goats' milk, and unsweetened calcium-fortified soy milk and almond milk**

| /100 mL | Mature breast milk | Cows' milk (average summer and winter) | Sheep's milk | Goats' milk | Unsweetened calcium-fortified soya milk | Unsweetened calcium-fortified almond milk |
|---|---|---|---|---|---|---|
| Energy (kcal kJ) | 69 290 | 63 265 | 93 388 | 62 260 | 33 138 | 13 55 |
| Fat (g) | 3.9 | 3.6 | 5.8 | 3.7 | 2.0 | 1.1 |
| Protein (g) | 1.1 | 3.4 | 5.4 | 3.1 | 3.4 | 0.4 |
| Carbohydrate (g) | 7.2 | 4.6 | 5.1 | 4.4 | 1.8 | 0.1 |
| Calcium (mg) | 34[a] | 120 | 170 | 100 | 120 | 120 |
| Phosphorus (mg) | 15 | 96 | 150 | 90 | 49 | 20 |
| Iodine (µg) | 7 | 30 | 310 | 80 | NK | NK |
| Iron (mg) | 0.07[a] | 0.06 | 0.03 | 0.12 | 0.40 | 0.7 |
| Zinc (mg) | 0.3 | 0.5 | 0.7 | 0.5 | 0.23 | 1.5 |
| Retinol equivalents (vitamin A) (µg) | 62 | 38 | 83 | 44 | 3 | NK |
| Vitamin D (µg) | 0.2–3.1 | 0 | 0.2 | 0.1 | 0 | 0 |
| Vitamin E (mg) | 0.34 | 0.06 | 0.11 | 0.03 | 0.01 | 10.0 |
| Riboflavin (mg) | 0.3 | 0.23 | 0.32 | 0.04 | 0.2 (where fortified) | 0.2 (where fortified) |
| Folate (µg) | 5 | 8 | 5 | 1 | 1.5 | NK |
| Vitamin C (mg) | 4 | 2 | 5 | 1 | 0 | 0 |

[a]High bioavailability of these nutrients from breast milk makes comparison with other milks challenging.

NK, not known.

avoided as they contain tannins, which strongly inhibit iron absorption, and caffeine, which should be restricted for young children.

## 33.4.6 Encouraging eating well

One of the most common concerns, and frustrations, for families is their young child's fussy eating or neophobia (fear of new foods). These are considered normal development stages in young children and affect about 10–20% of children under 5. Severe selective eating is rare and generally has its roots in early feeding difficulties or significant health problems. Neophobia typically emerges in the latter half of the second year of life in children and is thought to be an innate predisposition. Studies have shown that fussy eating may lead to a reduced overall food intake, with changes to nutrient intake as a result of a lack of variety in the diet, and lower fruit and vegetable intake, but these findings are not unanimous. A number of factors are suggested as contributing to fussy eating:

- *Giving frequent drinks of milk or juice*: many young children prefer drinking to eating and readily fill themselves up with drinks.
- *Frequent snacking*: some children end up eating most of their food between meals and the snack food often tends to be high in fat, sugar, and salt. There is often little or no incentive for the toddler to eat an appropriate meal if they have already filled up on confectionery, biscuits, and crisps.
- *Snacks or desserts being given when a meal is refused*: children may prefer snack foods and sweet foods and refuse meals in order to be given snacks and desserts instead.

- *Coercing children to eat more and/or extending mealtimes when the child has indicated they have had enough to eat*: this situation can sometimes be exacerbated by parents becoming very anxious at mealtimes.

To promote healthy eating behaviours, toddlers should be offered food at three meals and two nutritious snacks per day. Establishing a daily routine with meals and snacks at regular times, evenly spaced throughout the day (with the last snack at least 90 minutes before a meal), can be helpful. Mealtimes should be a happy, social occasion and, where possible, parents and carers should eat with children. Children should be sitting comfortably and utensils should be appropriate for the child's age. Up to 30 minutes should be allocated for meals, with the food then cleared away without comment. Sweet foods should never be used as bribes or treats.

Mess should be accepted as a normal part of the feeding process and toddlers encouraged to feed themselves. As far as is practical they should be permitted to choose from a variety of nutritious, appropriate foods and left to make their own decisions about quantity. Avoiding distractions like television, toys, or electronic devices at mealtimes can also support better eating habits.

### 33.4.7 Vegetarian, vegan, and other dietary choices

Vegetarian diets, where meat, poultry, and fish are excluded from the diet, have been consumed in some parts of the world for many hundreds of years. In Western countries vegetarian diets may take a number of forms, and are increasingly chosen for reasons related to animal welfare, or environmental or health reasons. Previous concerns about the nutritional adequacy of a vegetarian diet have been allayed by studies demonstrating health benefits related to more plant-based diets, but evidence about vegetarianism in early childhood remains limited. Mothers who are themselves vegetarian may choose to breastfeed for the first year and beyond, and like all breastfeeding mothers are encouraged to consume a good diet. In some countries they may be advised to take a vitamin D supplement, and give supplements to their infant (which could be vitamin drops containing A, C, and D from 6 months of age and vitamin D from birth), but such advice is not considered to be necessary in other countries such as New Zealand. Vegetarian diets that contain full-fat dairy products and eggs, and offer a wide range of other protein sources in the form of beans, peas, lentils, cereals, nuts, and seeds are unlikely to present a nutritional challenge to young children. Vegetarian infants fed adequate amounts of breast milk or infant formula and a balanced, varied diet grow similarly to non-vegetarian infants. There are, however, few infant formulas sold that are suitable for strictly vegetarian families, since many contain fish oil or use rennet in their production.

More rarely, families choose to avoid all foods that have involved the use of, or exploitation of, an animal in their diet, eating only plant-based foods. Vegan mothers who are breastfeeding should be advised to take supplements of vitamin $B_{12}$ (and in some countries will be advised to take vitamin D and iodine supplements) and to select a varied vegan diet. There are no infant formulas suitable for vegan families, since the vitamin D is sourced from sheep wool lanolin, but some families will choose soy-based formula. Soy-based formula is not normally recommended for routine use under 6 months of age. Offering a vegan diet to an infant and toddler requires care and attention, and families should seek expert advice. Very restrictive diets, such as macrobiotic diets, fruitarian, or raw diets, are not suitable for infants and young children.

## 33.5 Assessing nutritional adequacy

In general, it is assumed that an infant or toddler's nutritional status is normal, and nutritional needs met, if he or she has a normal rate of growth, drinks adequate amounts of breast milk (or suitable formula or milk for age), and eats a variety of age-appropriate foods from each of the food groups.

When nutritional status or growth is questionable, in comparison with established standards, the infant or toddler's intake should be evaluated in terms of recommended nutrient intakes or allowances. Developmental milestones are also used to assess a young child's health and well-being.

### 33.5.1 Energy and nutrient requirements

Due to their rapid rate of growth and higher metabolic rate, energy requirements for infants and toddlers are higher than at any other time of life. Recent energy-balance studies in infants and young children found lower energy requirements than previously predicted, starting from 110 kcal/kg/day in the first month of life to 80 kcal/kg/day at 1 year of age. During the second year of life, energy requirements stay stable on a per-kilogram basis, but the total calorie requirement increases as the child grows. Iron requirements are relatively higher in the second and third year of life, and poor diets are associated with lower intakes of vitamin A, iron, zinc, and dietary fibre, and higher intakes of refined sugars and salt. Vitamin D status is also increasingly reported as low, as is fruit and vegetable intake in many areas.

When an infant or toddler's energy requirements are met from a well-balanced, varied diet, the risk of other nutrient deficiencies is minimized.

### 33.5.2 Monitoring growth

Postnatal growth and development of the central nervous system are most rapid during the first year of life. The typical infant doubles birth weight during the first 4–5 months, triples it in the first year, and quadruples it by 2 years of age, by which time a child has grown to half of their adult height. Historically, we have used regular weighing to monitor the sufficiency of infant feeding. In the longer term, growth in height is an index of health and well-being, with slow or stunted growth associated with many disorders of childhood. Conversely, if a child is growing normally they are unlikely to have any major underlying medical problems. After the early days, babies only need to be weighed at the time of routine checks (typically at around 2, 3, 4, and 12 months of age).

International in scope, WHO growth charts reflect the growth of children from six high- and low-income countries around the world who were raised under favourable conditions for supporting optimal growth. Based on the growth pattern of breastfed infants, the charts reflect current international guidelines for optimal feeding of infants and toddlers and, as such, are considered growth *standards*, rather than growth *references*.

In most children, height and weight measurements follow fairly consistently along a channel (i.e. on or between the same centile(s)). Normal growth is indicated by weight-for-length and length-for-age tracking along similar percentiles or growth channels; however, it is common for infants to shift percentiles for both length and weight in the first 2–3 years of life, with the majority settling into a channel towards the 50th percentile (i.e. regression towards the mean), rather than away. When length-for-age and weight-for-length percentiles are disproportional or weight and/or height measurements are on a downward or upward trend, investigation of potential nutritional imbalances is indicated.

## 33.6 Summary

1. Nutrition during infancy and the second year of life (the first 1000 days) is an important determinant of short- and long-term health.
2. Exclusive breastfeeding provides the only nutrition a healthy infant needs for the first 6 months of life, and is recommended as the main milk drink alongside complementary foods throughout the second 6 months, for the first 2 years of life, and beyond.
3. Infant formula is nutritionally adequate, but cannot replicate breast milk, which offers protection from ill health, as well as nutrition. Better regulation to support the WHO Code of Marketing of Breastmilk Substitutes and subsequent WHA

resolutions is needed to protect all families from misleading infant formula claims, which undermine breastfeeding.

4. Complementary foods should be added to the diet alongside breast milk from about 6 months of age, and the best foods are diverse, locally sourced, nutritious foods without added salt or sugar.
5. First foods should include iron-rich foods such as meat, fish, seafood, chicken, beans, peas, lentils, tofu, or iron-fortified infant cereal, as baby will have exhausted their iron stores by 6 months of age.
6. Baby food pouches and baby-led weaning are recent, and increasingly popular, approaches to complementary feeding. Although strong opinions have been expressed about both, there has been little research on their potential health advantages, or disadvantages.
7. After 1 year of age, whole animal milk or a suitable alternative can replace breast milk (or infant formula) as the main milk drink, but continued breastfeeding is encouraged.
8. Adequate nutritional intake can be monitored using infant and toddler growth charts and monitoring of development around developmental milestones.

## Further reading


1. **Ballard, O. and Morrow, A.L.** (2013) Human milk composition: nutrients and bioactive factors. *Pediatr Clin North Am*, **60**, 49–74.
2. **Crawley, H.** (2020) *Eating well: the first year. A practical guide to introducing solids and eating well up to baby's first birthday.* Available at: https://www.firststepsnutrition.org/eating-well-infants-new-mums (accessed 12 November 2021).
3. **European Food Safety Authority** (2013) *Scientific opinion on nutrient requirements and dietary intakes of infants and young children in the European Union.* Available at: http://www.efsa.europa.eu/en/efsajournal/pub/3408.htm (accessed 12 November 2021).
4. **Victora, C.G., Bahl, R., Barros, A.J.D., et al.** (2016) Breastfeeding in the 21st century: epidemiology, mechanisms and lifelong effect. *Lancet*, **387**, 475–90.
5. **World Health Organization** (1981) *The international code of marketing of breastmilk substitutes.* Geneva: WHO.
6. **World Health Organization** (2003) *Global strategy for infant and child feeding.* Geneva: WHO.
7. **World Health Organization** (2007) *Safe preparation, storage and handling of powdered infant formula: guidelines.* Geneva: WHO.


## Useful websites

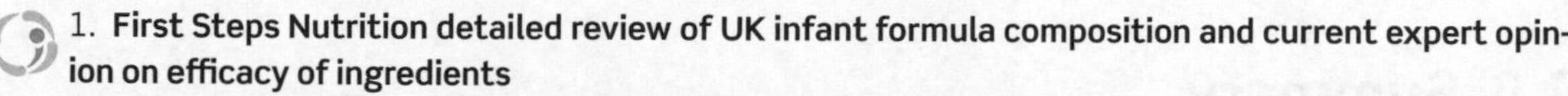

1. **First Steps Nutrition detailed review of UK infant formula composition and current expert opinion on efficacy of ingredients**
https://www.firststepsnutrition.org/composition-and-claims
2. **New Zealand Ministry of Health advice on preventing food related choking**
https://www.health.govt.nz/your-health/healthy-living/food-activity-and-sleep/healthy-eating/food-related-choking-young-children
3. **World Health Organization Child Growth Standards**
https://www.who.int/tools/child-growth-standards

# 34 Childhood and Adolescent Nutrition

Rachael Taylor and Anne-Louise Heath



## 34.1 Introduction

Childhood and adolescence are periods of rapid growth, learning, and development. Nutritional needs are high and differ in many respects from those of adults. Ensuring adequate food intake remains the challenge for many of the world's children. In contrast, for most children in high income countries, making more appropriate food choices and developing and maintaining healthy eating habits is paramount. Most of this chapter refers to both childhood (age 2–12 years), and adolescence (age 13–18 years). The final section (34.9) covers issues that predominantly relate to adolescents.

## 34.2 Dietary recommendations

Dietary recommendations for children and adolescents are expressed in two very different, but complementary, formats: dietary guidelines (providing advice on food intake and related behaviours in a way that is accessible to members of the general public) and nutrient intake values (levels of nutrient intake calculated to minimize the risk of deficiency or disease).

Ideally, each country should develop its own dietary guidelines to ensure that the food- and nutrient-related health concerns of the country are addressed in the context of customary dietary patterns. However, the dietary guidelines for children shown in Table 34.1 are fairly typical (although in the case of these Australian guidelines they are embedded within a set of guidelines for all ages).

**Table 34.1 Dietary guidelines for children and adolescents in Australia**

- Encourage, support, and promote breastfeeding.
- To achieve and maintain a healthy weight, be physically active and choose amounts of nutritious food and drinks to meet your energy needs.
  - Children and adolescents should eat sufficient nutritious foods to grow and develop normally.
  - They should be physically active every day.
  - Their growth should be checked regularly.
- Enjoy a wide variety of nutritious foods from these five groups every day:
  - Plenty of vegetables, including different types and colours, and legumes/beans
  - Fruit
  - Grain (cereal) foods, mostly whole grain and/or high cereal fibre varieties, such as breads, cereals, rice, pasta, noodles, polenta, couscous, oats, quinoa, and barley
  - Lean meats and poultry, fish, eggs, tofu, nuts and seeds, and legumes/beans
  - Milk, yoghurt, cheese, and/or their alternatives, mostly reduced fat (reduced fat milks are not suitable for children under the age of 2 years).
- Drink plenty of water.
- Limit intake of foods containing saturated fat, added salt, added sugars, and alcohol.
  - Limit intake of foods high in saturated fat such as many biscuits, cakes, pastries, pies, processed meats, commercial burgers, pizza, fried foods, potato chips, crisps, and other savoury snacks.
  - Replace high-fat foods that contain predominantly saturated fats, such as butter, cream, cooking margarine, coconut oil, and palm oil, with foods that contain predominantly polyunsaturated and monounsaturated fats, such as oils, spreads, nut butters/pastes, and avocado.
  - Low-fat diets are not suitable for children under the age of 2 years.
  - Limit intake of foods and drinks containing added salt.
  - Read labels to choose lower sodium options among similar foods.
  - Do not add salt to foods in cooking or at the table.
  - Limit intake of foods and drinks containing added sugars, such as confectionery, sugar-sweetened soft drinks and cordials, fruit drinks, vitamin waters, and energy and sports drinks.
- Care for your food; prepare and store it safely.

Nutrient intake values such as dietary reference intakes (in Canada and the USA), dietary reference values (in the UK and Europe), and nutrient reference values (in Australia and New Zealand) provide specific information on nutrient requirements and safe upper limits of intake. Nutrient requirements alter so rapidly during childhood and adolescence that three age bands are usually reported for children (1–3 years, 4–8 years, 9–13 years), and one for adolescents (14–18 years). A typical set of nutrient intake values is shown in Table 34.2.

**Table 34.2 Dietary reference intakes for children and adolescents**

| Nutrient | Age group (years) and gender | | | | | |
|---|---|---|---|---|---|---|
| | Children | | Boys | | Girls | |
| | 1-3 | 4-8 | 9-13 | 14-18 | 9-13 | 14-18 |
| ***Protein (g/day)*** | | | | | | |
| EAR (g/kg/day) | 0.87 | 0.76 | 0.76 | 0.73 | 0.76 | 0.71 |
| RDA (g/day) | 13 | 19 | 34 | 52 | 34 | 46 |

Table 34.2 **Dietary reference intakes for children and adolescents (*Continued*)**

| Nutrient | Age group (years) and gender | | | | | |
|---|---|---|---|---|---|---|
| | Children | | Boys | | Girls | |
| | 1–3 | 4–8 | 9–13 | 14–18 | 9–13 | 14–18 |
| ***Calcium (mg/day)*** | | | | | | |
| EAR | 500 | 800 | 1100 | 1100 | 1100 | 1100 |
| RDA | 700 | 1000 | 1300 | 1300 | 1300 | 1300 |
| ***Iron (mg/day)*** | | | | | | |
| EAR | 3.0 | 4.1 | 5.9 | 7.7 | 5.7 | 7.9 |
| RDA | 7 | 10 | 8 | 11 | 8 | 15 |
| ***Vitamin C (mg/day)*** | | | | | | |
| EAR | 13 | 22 | 39 | 63 | 39 | 56 |
| RDA | 15 | 25 | 45 | 75 | 45 | 65 |
| ***Vitamin A ( μg/day)*** | | | | | | |
| EAR | 210 | 275 | 445 | 630 | 420 | 485 |
| RDA | 300 | 400 | 600 | 900 | 600 | 700 |

EAR, estimated average requirement; RDA, recommended dietary allowance.

*Source:* Dietary reference intake documents prepared for the USA and Canada by the Food and Nutrition Board and Institute of Medicine (2001, 2002, 2005) available at http://www.nap.edu. Calcium values are from the revised 2011 report.

# 34.3 Growth

## 34.3.1 Growth and body composition

The rapid growth period of infancy sees a tripling of birth weight during the first year; growth then slows during childhood, and is relatively consistent and predictable, with annual increments typically ranging from 6 to 7 cm in height and from 2 to 3 kg in weight. Body proportions also change over this time, from that in infancy, where the head and trunk predominate, to a more adult profile resulting from considerable lengthening of the limbs and a reduction in the size of the head relative to the rest of the body. The onset of puberty signals a second accelerated phase in growth, typically lasting 2 or so years, with a younger onset in females. Although the age of onset can vary considerably (see Fig. 34.1), males typically gain around 20 kg of weight and 20 cm in height, and females 16 kg and 16 cm.

Marked age and sex differences in body composition are apparent between birth and the end of adolescence. Children are relatively lean at birth (approximately 15% body fat), but increase steadily in adiposity during the first year of life. Adiposity tends to decline during the preschool years and reach a nadir at approximately 5–6 years of age, before increasing again through later childhood. This phenomenon is known as the adiposity rebound (AR) and has received considerable attention over recent years as a potential 'critical period' in the development of childhood obesity. Children who undergo AR at an early age are more likely to be overweight in later adolescence and early adulthood. However, because AR cannot be determined until it has passed, it has limited diagnostic ability. During adolescence, males gain approximately twice as much lean tissue as females, with a rapid acceleration during the peak growth spurt, and much smaller increments in body

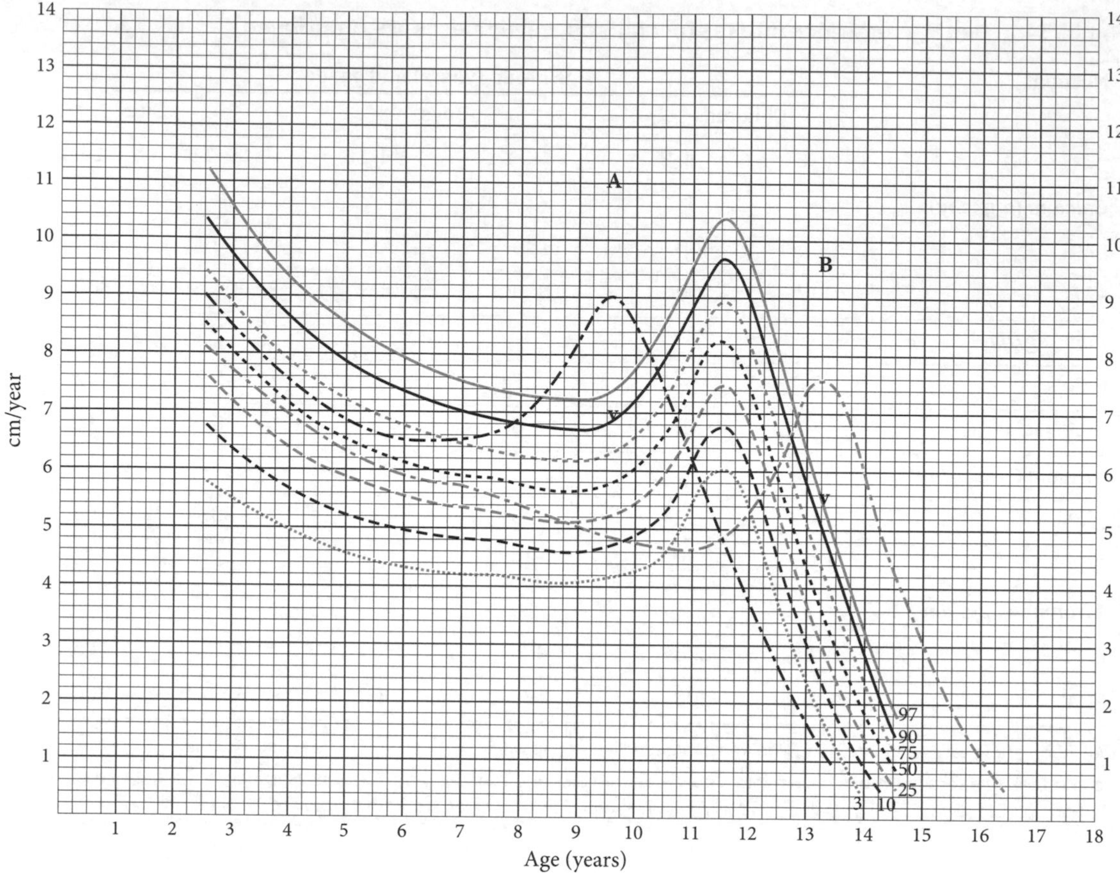


Fig. 34.1 Height velocity for girls. The outlying peaks are an early (A) and a late (B) maturer.

*Source:* Tanner, J. and Davis, P.S.W. (1985) Clinical longitudinal standards for height and weight velocity for North American children. *J Pediatr*, **107**, 317–29. With permission from Elsevier.

fat. By contrast, gains in lean tissue mass appear more consistent during growth in females, and body fat continues to be laid down until late adolescence/early adulthood.

## 34.3.2 Assessment of growth and body composition

An accurate record of growth remains one of the most useful tools for the assessment of both under- and over-nutrition. Serial measurements at multiple time points reveal more information about a child's health than single measurements because of individual variation in growth and development. Anthropometric indices such as height and weight can most usefully be interpreted by plotting them on reference growth charts. These provide a *reference* for an individual child's growth, describing how children grew in a particular place and time. Many countries have their own growth reference data; a widely used example is the Centers for Disease Control and Prevention (CDC) 2000 growth charts (http://www.cdc.gov/growthcharts (accessed 10 August 2022)), which are based on breastfed and formula-fed American children (Fig. 34.2).

Because there are known differences in growth between breastfed and formula-fed infants, the World Health Organization (WHO) published the WHO Child Growth Standards (https://www.who.int/tools/child-growth-standards) in 2006 (accessed 10 August 2022), created from the Multicentre Growth

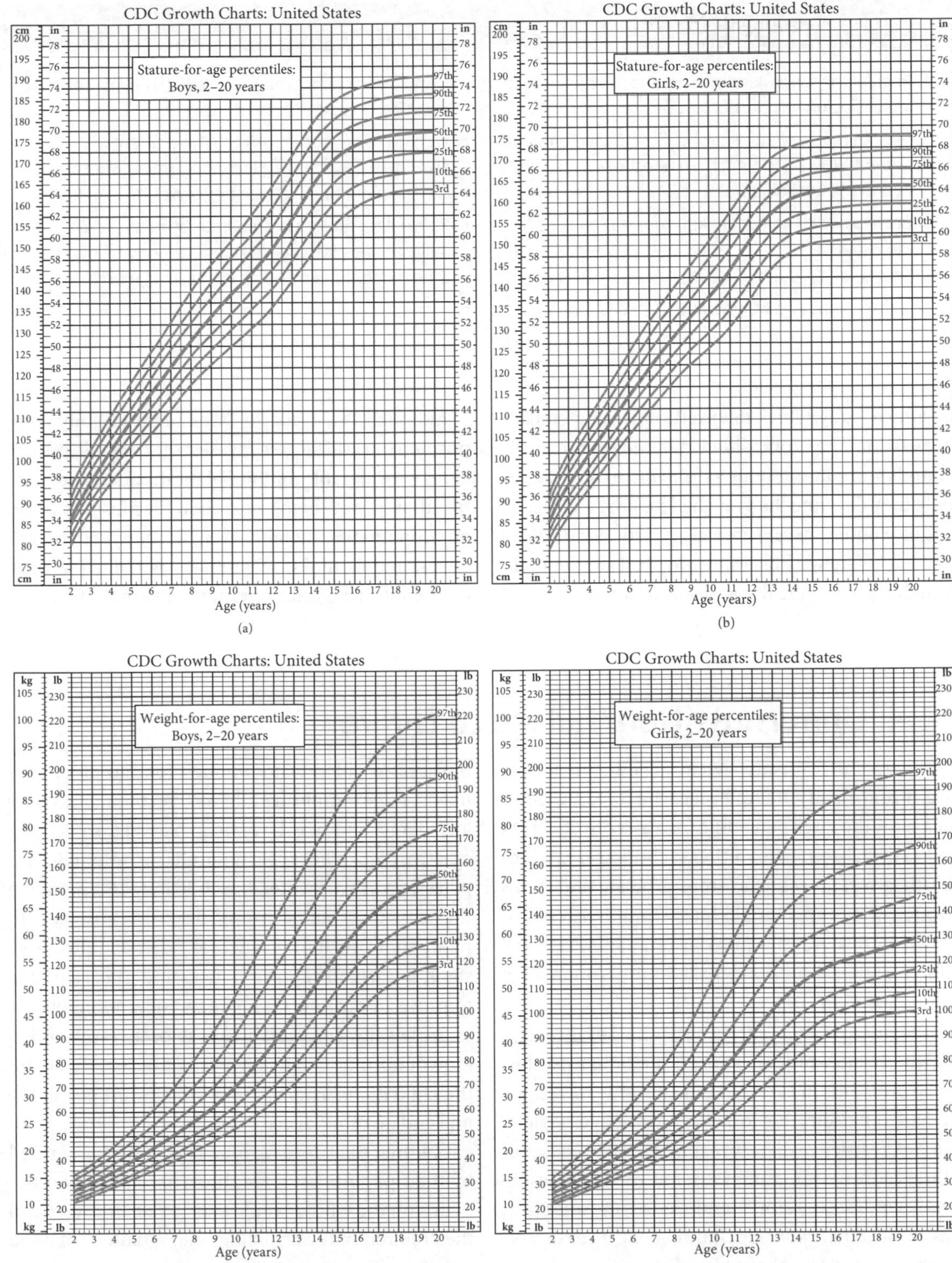


**Fig. 34.2** CDC growth charts for boys and girls aged 2–20 years: (a) boys' stature (height) for age; (b) girls' stature (height) for age; (c) boys' weight for age; (d) girls' weight for age, with percentile lines.

*Source:* National Center for Health Statistics (NCHS)/National Center for Chronic Disease Prevention and Health Promotion (CDC), USA (2000).

**Box 34.1 Measuring the child's growth**

1 Regular measurement of weight is the best way of assessing undernutrition or obesity.
2 Growth reference charts should be based on current WHO charts or those relevant to the child's country (if available).
3 Growth trends give more information than one measurement.
4 A child who is approximately the same percentile for height and weight, and whose height and weight track over time, is unlikely to have a serious nutrition or chronic health problem.
5 A child crossing the 90th or 10th percentiles should be assessed again to determine whether it is a transient or longer-lasting shift.
6 Referral for growth hormone treatment should only be made if the child is consistently below the third percentile and no other cause is found.

Reference Study, which followed approximately 8500 breastfed infants from widely different ethnic backgrounds and cultural settings from birth to 5 years. As such, they describe the growth of healthy children in optimal conditions and are thus growth *standards*. In practice, reference data often are used as standards, even though they are theoretically different.

Although the accurate assessment of body composition (fat and fat-free masses; see Chapter 38) provides considerable additional information about the health and well-being of a child, in most instances anthropometry (height, weight, body circumferences, and skinfold thicknesses) remains the most commonly used way to monitor growth (Box 34.1).

# 34.4 Undernutrition/failure to thrive

Failure to thrive occurs when a child of normal height and head circumference has a weight-for-age below the third percentile (or a $z$-score of −2). As well as this absolute definition, a child who was previously growing well who stops growing and drops through several percentile lines should also be included in this group. In low income countries, the same children are usually called 'undernourished' or 'underweight'.

In high income countries about half of the cases are due to non-organic causes, including psychosocial problems in carers (e.g. depression) or simply an inadequate diet due to lack of knowledge, neglect, or a very restrictive dietary regime. Organic causes of failure to thrive are numerous and include chronic renal, cardiorespiratory, and endocrine diseases. Gastrointestinal causes include coeliac disease, cystic fibrosis, and Hirschsprung's disease.

More than half of the children diagnosed as 'failure to thrive' in primary care have a relatively simple nutrition problem—not enough food to meet their needs. For example, a child may be thought to have an 'allergy' and may have been placed on a very restrictive diet, or because the adults in the families are placed on a low-fat diet, the child may also be eating the same diet. Many chronic diseases cause growth retardation. In the absence of other symptoms or signs, it is usually appropriate to undertake a trial of improved nutrition rather than going immediately to further investigation. Referral for investigation of growth hormone deficiency is not necessary unless the child is consistently below the lowest percentile line and no other disease is present.

Children may develop food fads and be described by their parents as 'difficult eaters'. Prevention depends on the whole family eating a healthy diet and the children naturally falling into a pattern of good nutrition. The use of food, snacks, and sweets in reward and punishment systems for inappropriate behaviour may lead to poor eating habits.

# 34.5 Childhood overweight and obesity

## 34.5.1 Prevalence and health effects

Overweight or obesity in childhood is generally defined as having a body mass index (BMI) value greater than certain age- and sex-specific reference standards (e.g. CDC or WHO growth standards; see section 34.3.2). However, which cut-offs are used to denote overweight and obesity in these different datasets varies. The CDC data have generally used a BMI value of ≥85th but <95th percentile to denote overweight, with ≥95th percentile classified as obese. By contrast, WHO recommends using *z*-score values of above +1 (5–19 years) or +2 (under 5 years) for overweight, and above +2 (5–19 years) or +3 (under 5 years) to indicate obesity. In 2000, Cole et al. published BMI criteria for defining overweight and obesity in children and adolescents created from a dataset of children from six countries (Brazil, Great Britain, Hong Kong, the Netherlands, Singapore, and the USA); the criteria are often referred to as the International Obesity Taskforce (IOTF) cut-offs. These BMI values were created from determining the percentiles corresponding to a BMI of 25 kg/m$^2$ (overweight) and 30 kg/m$^2$ (obesity) at 18 years of age. Use of the IOTF cut-offs (see Table 34.3) is advantageous for international comparisons of the prevalence of overweight and obesity. However, the CDC or WHO growth charts are more useful for tracking weight status in individual children over time, as exact percentiles or *z*-scores can be calculated.

**Table 34.3 Body mass index levels (kg/m$^2$) to diagnose overweight and obesity in children**

| Age (years) | Overweight | | Obesity | |
|---|---|---|---|---|
| | Boys | Girls | Boys | Girls |
| 2 | 18.41 | 18.02 | 20.09 | 19.81 |
| 3 | 17.89 | 17.56 | 19.57 | 19.36 |
| 4 | 17.55 | 17.28 | 19.29 | 19.15 |
| 5 | 17.42 | 17.15 | 19.30 | 19.17 |
| 6 | 17.55 | 17.34 | 19.78 | 19.65 |
| 7 | 17.92 | 17.75 | 20.63 | 20.51 |
| 8 | 18.44 | 18.35 | 21.60 | 21.57 |
| 9 | 19.10 | 19.07 | 22.77 | 22.81 |
| 10 | 19.84 | 19.86 | 24.00 | 24.11 |
| 11 | 20.55 | 20.74 | 25.10 | 25.42 |
| 12 | 21.22 | 21.68 | 26.02 | 26.67 |
| 13 | 21.91 | 22.58 | 26.84 | 27.76 |
| 14 | 22.62 | 23.34 | 27.63 | 28.57 |
| 15 | 23.29 | 23.94 | 28.30 | 29.11 |
| 16 | 23.90 | 24.37 | 28.88 | 29.43 |
| 17 | 24.46 | 24.70 | 29.41 | 29.69 |
| 18 | 25 | 25 | 30 | 30 |

*Adapted from*: Cole, T.J., Bellizzi, M.C., Flegal, K.M., et al. (2000) Establishing a standard definition for child overweight and obesity worldwide: international survey. *BMJ*, **320**, 1240–3.

In most high income countries, childhood overweight and obesity represents a major health issue, with national estimates indicating that 25–40% of children are above a healthy weight. Much of the recent focus has been on the rapid increase in prevalence that has been observed in many countries. Recent data, however, offer promise that these increases in prevalence appear to be abating, at least in some countries, and perhaps more so in younger children. Regardless of the trends over time, the fact remains that too many of our children are carrying excess body weight, with resultant implications for health. In young children, this translates to deleterious effects on psychosocial health in terms of bullying and impaired self-esteem, and the increased risk of transition to an overweight state as an adult, with all the accompanying health effects. In older children and adolescents, other physical health risks become more apparent, including, but not limited to, an increased risk of type 2 diabetes, high blood pressure, high cholesterol, and bone and joint problems, especially fractures of the forearm.

## 34.5.2 Causes

Obesity is a result of excessive food intake in relation to energy requirements, and may arise due to imbalance from either (or both) sides of the energy balance equation. Genetics also plays a major role in determining weight, and susceptibility to weight gain, although very few cases of obesity are a result of an endocrine disorder, chromosomal abnormality, or genetic disorder. Children in affluent countries live in an 'obesogenic' environment, which discourages physical activity and promotes the consumption of nutrient-poor, energy-dense foods.

Dietary factors While there is general acceptance that a high consumption of energy-dense foods is an important cause of obesity, there is little certainty regarding the role of individual foods, drinks, or nutrients. High intakes of sugar-sweetened beverages are associated with increased weight or weight gain in children, but the few randomized controlled trials in which advice to reduce intake of such drinks has been the key intervention have not provided convincing evidence of benefit, perhaps because of poor compliance. 'Fast foods' and 'snack foods' have also been implicated, but these foods are generally high in sugars and fats and therefore energy-dense, which likely explains any contribution to excess weight. More recent research has examined the role that ultra-processed foods may play in children's health. Ultra-processed foods form a large part of the diet of children in Western countries and high intakes have been related to weight gain, although interventions in children to date are rare.

Activity factors The extent to which physical activity has reduced in recent years and thus contributed to the 'epidemic' of childhood obesity is uncertain. However, a number of studies have shown a fairly strong association between television viewing and excess body weight accumulation in young children, either directly from lost opportunities for physical activity during prolonged periods of inactivity or from consumption of energy-dense snack or fast foods while watching television. The use of computers, computer games, and labour-saving devices, and reduction of the use of active transport and school-based physical education, are other examples of the obesity-promoting environment that characterizes Western and some other affluent societies.

Other behaviour factors Insufficient amounts of good quality sleep have been related to excessive weight gain, especially in young children. Although few interventions have been undertaken, early work shows promise that promoting good sleep habits might be an effective approach to obesity prevention during growth. The reduced risk is likely due more to changes in how, what, or when children eat, than changes in activity or hormone levels (e.g. ghrelin and leptin), although more mechanistic research is needed. Family environment, role modelling, and parental support are also considered as potential determinants of excessive weight gain. Some of the behavioural and societal factors that have been implicated in the development of childhood obesity are listed in Table 34.4.

### 34.5.3 Prevention and treatment of overweight and obesity

Changes in the obesogenic environment are essential if the prevalence of overweight and obesity in children and adolescents is to be reduced (see Chapter 17). This will not be achieved without commitment from relevant opinion leaders and policy

**Table 34.4 Behavioural and societal factors implicated in the development of childhood obesity**

| | |
|---|---|
| Larger portion sizes | Television viewing |
| Sweetened beverages | Computer use |
| Ultra-processed foods | Video games |
| Inappropriate snack choices | Labour-saving devices |
| Food advertising | Built environment |
| Skipping breakfast | Community access to resources |
| Reduced sleep (quantity or quality) | Declines in active transport (cycling or walking to school) |
| Family environment | Reduced physical education in schools |
| Parenting styles | Less school recess time |
| Poverty | Parental work commitments |

makers, including governments, local and education authorities, media, food industries, and health professionals. Voluntary as well as legislative measures are required. Abolition or limiting of television advertisements promoting inappropriate foods to children, provision of opportunities for physical activity, and increasing availability of affordable healthy food choices are but three of the many measures that require implementation.

Treating overweight children should in theory be relatively straightforward, since only small changes to eating and/or activity behaviour should be required for most overweight children to successfully reduce their weight over time, given that the energy imbalance is relatively small on a daily basis. Such children need to 'grow into' their weight by restoring the rate of excessive weight gain to a more appropriate level while maintaining linear growth. The fact that so many children are now overweight or obese and that the health consequences are generally not immediately apparent have led to parents often failing to recognize overweight in their children. Furthermore, health professionals find it difficult to initiate appropriate management, which principally involves the promotion of physical activity and appropriate food choices. For children who are obese, the level of intervention will depend upon the severity and duration of the obesity. Some form of behavioural therapy, including expert dietary advice, given within the context of the family environment, is usually regarded as an essential component. Management of the overweight and obese child is made all the more difficult by the current obesity-promoting environment.

## 34.6 Some critical micronutrients

### 34.6.1 Iron

In adults, iron status reflects the balance between iron losses and iron intake. In children, iron intakes also need to be sufficient to support growth. In addition to their increased requirements for growth, adolescent girls need to absorb 0.45 mg of iron a day to cover menstrual losses once they reach menarche. This is equivalent to approximately 2.5 mg a day of additional dietary iron. Some girls will require considerably more than this as losses can be three times higher than the median. The age at which this increased requirement starts varies from child to child.

The iron status of primary school-age children is usually adequate, and just 4% of preschool children 3–5 years of age in the 2007–10 US National Health and Nutrition Examination Survey (NHANES) were iron-deficient. In the UK and the USA, there are higher rates of iron deficiency amongst the poor, immigrants, and non-Caucasian populations. The rates of iron deficiency are much higher for adolescent girls, with NHANES 2003–06 estimating that 9–16% of girls aged 12–19 years were iron-deficient. Approximately 4% of adolescent girls have the severe form of iron deficiency—iron-deficiency anaemia.

Iron-deficiency anaemia is associated with poorer mental, motor, and behavioural development in young children, and fatigue and poorer cognitive function (especially verbal learning and memory) in adolescents. The effects of non-anaemic iron deficiency are less clear, but probably include subtle negative effects on cognitive function and fatigue. It is important that non-anaemic iron deficiency is treated, because it increases the risk of developing iron-deficiency anaemia with rapid growth or onset of menstruation.

Dietary measures may play a role in preventing and treating non-anaemic iron deficiency, although supplementation is required to treat iron-deficiency anaemia. Red meat, a rich source of haem iron, is an effective way to boost intake, noting that processed food such as sausages contain much less iron than lean meat. Beans, peas (including chickpeas), lentils, and iron-fortified foods are useful sources of non-haem iron. Absorption of non-haem iron is enhanced by ascorbic acid found in many fresh fruits and vegetables, and by a yet-to-be identified factor (the MFP factor) in meat, fish, and poultry. Phytates and polyphenols (including tannins) are inhibitors of non-haem iron absorption. Whole grains, nuts, and legumes are sources of phytates but valuable sources of many nutrients. Wholemeal products when leavened with yeast (e.g. wholemeal bread), and legumes when canned, have less phytate. Tea and coffee, rich sources of tannins, should be discouraged. If dietary measures are unable to achieve satisfactory iron status, iron supplements are necessary.

Iron deficiency and its prevention and treatment are covered more fully in Chapter 9.

### 34.6.2 Vitamin D

Although severe vitamin D deficiency is now uncommon amongst children in relatively affluent countries such as the UK, Europe, USA, Australia, and New Zealand, clinicians have recently reported a concerning increase in the number of cases of rickets, particularly amongst children with darker skin and from cultures in which the skin is kept covered. Rickets is described in Chapter 14. It has also become apparent that less severe vitamin D insufficiency is common amongst children and adolescents, even in countries such as the USA that have widespread fortification programmes in place. In the US NHANES 2011–14, suboptimal serum 25-hydroxyvitamin D concentrations (defined as serum <50 nmol/L) were present in 7% of preschoolers but 28% of adolescents. Interestingly, supplementation trials in unselected adolescent girls in Lebanon and Finland both reported increased bone mineral density on supplementation with vitamin D, even in the absence of rickets.

Factors that increase the risk of vitamin D deficiency in children and adolescents include darker skin pigmentation (children with dark skin pigmentation may need 5–10 times longer in the sun to generate the same amount of vitamin D), lack of sunlight exposure (including use of sunscreen or concealing clothing), obesity, clinical conditions resulting in fat malabsorption (e.g. cystic fibrosis), and medication that alters vitamin D metabolism (e.g. anticonvulsants).

A number of foods including milk, margarine, and juice are fortified with vitamin D in the USA, but children and adolescents would need to consume 1.5 L of milk a day to get the 600 IU currently recommended by the American Academy of Pediatrics. Children in countries such as New Zealand, where milk is not fortified with vitamin D, would need to drink considerably more to get an appreciable amount of vitamin D. It is important to note that in countries such as New Zealand, most people are able to achieve adequate vitamin D status in the summer months as a result of increased sun exposure, and the long-term health impact of intermittent deficiency is unclear.

### 34.6.3 Iodine

Although severe iodine deficiency is now rare due to the widespread use of iodized salt (UNICEF estimates that 66% of households globally have access to iodized salt), moderate iodine deficiency remains a concern and is associated with reduced cognition in children. Treatment of mild iodine deficiency has been shown to improve aspects of cognition in short-term trials, although it is not known whether mild iodine deficiency impairs cognition. Many families are attempting to reduce their sodium intake, and as a result, are reducing the amount of discretionary salt they use. The unintended impact on iodine intake is being addressed by a number of countries, including Australia and New Zealand, by mandating the use of iodized salt when bread is manufactured. Iodine is discussed in Chapter 10.

## 34.7 Other matters related to diet

### 34.7.1 Dental caries

Dental caries (tooth decay) is the second most common infectious disease after the common cold and is one of the more expensive diet-related health problems. It is also preventable. In the USA, by the age of 5 years, 1 in 5 children have had at least one cavity or filling, and by 19 years of age, more than half have had tooth decay. The consequences include pain and infections.

Dental decay is an interaction between three factors: sugars, oral bacteria, and the tooth. Sugars (glucose and fructose, as well as sucrose) are fermented by oral bacteria (particularly *Streptococcus mutans*) that produce acids, which demineralize the enamel and may eventually break down the tooth. All three factors are required for dental decay, although in treating and preventing dental caries, multiple strategies are likely to be more successful than attempting to eliminate any one single factor.

It is the frequency of exposure to sugars, rather than the amount of sugar, that predicts caries risk. In practice, this means that:

- sugars can be consumed with meals (especially if children are encouraged to brush their teeth afterwards), but should not be consumed between meals
- sticky carbohydrate foods (such as hard sweets, muesli bars, fruit roll-ups, honey, and even dried fruit such as raisins) are particularly bad for teeth because they remain on the tooth surface for longer
- children should only drink plain water between meals, and not sugar-containing beverages (including fruit juice and sports and energy drinks). The practice of using sugar-containing beverages to settle children to sleep at night is particularly risky because the beverage bathes the teeth for a prolonged period—at a time when production of protective saliva is low. Diet drinks are also not encouraged because while they contain no sugar, they are still high in acid, which causes tooth erosion.

The sugars contained in the cellular structure of foods (such as the intrinsic sugars of fresh fruits and vegetables) have little cariogenic potential so these are good food choices for snacks.

Bacterial numbers are controlled by effective twice-daily tooth brushing with a fluoride toothpaste (supervised by an adult until the child is at least 8–9 years of age). Fluoride in the saliva, either from fluoridated water or from fluoride supplements, protects the teeth by making the enamel more resistant to acid attack. In many countries, fluoridation of water supplies and the use of fluoridated toothpaste have resulted in dramatic declines in average levels of dental decay (see section 10.5).

### 34.7.2 Food allergies and intolerances

*Food allergy* in the narrow, technical sense is an abnormal reaction to a food by a mechanism that usually involves a specific immunoglobulin E (IgE) in the plasma.

The most common food allergies in young children are to cows' milk protein, eggs, peanuts, tree nuts, sesame, soy, wheat, fish, and shellfish. Peanuts, tree nuts, and shellfish are the foods most commonly associated with life-threatening anaphylaxis. Most children who are allergic to cows' milk protein, soy, wheat, or eggs have grown out of their food allergy by the time they are 5–9 years of age. Interestingly, this is less likely for those with peanut allergy.

Symptoms of food allergy include hives (urticaria) and/or swelling around the mouth (angioedema), vomiting, cough, wheeze, a runny or blocked nose, stomach pains, or diarrhoea. Food allergy may also result in gastro-oesophageal reflux and eczema.

The three most widely accepted tests used to confirm or exclude potential food triggers are: skin prick allergy tests, blood tests for IgE specific to the allergen (a RAST or EAST test), and a closely supervised temporary elimination diet followed by food challenges with the suspected food. Currently, it is considered that food allergies are best managed by identifying and avoiding the food.

A *food intolerance* is a reaction to a food that does not involve immunoglobulins (i.e. non-IgE mediated)—for example, urticaria in response to consumption of certain food additives such as tartrazine.

Other adverse reactions to food (i.e. reactions other than allergy and food intolerance) include wind and diarrhoea in response to milk in children who have lactase insufficiency; and coeliac disease, an autoimmune enteropathy triggered by sensitivity to wheat gluten.

Diagnosis of food intolerance and other non-immune adverse food reactions is usually based on clinical history and response to removal of the food from the diet, although in some cases a strictly supervised temporary elimination diet followed by food challenges is also used diagnostically.

It is important to avoid unorthodox testing and treatments that can be expensive and harmful if they delay effective treatment or result in a nutritionally inadequate diet (either because major foods such as milk or wheat have been removed from the diet without appropriate replacement or because the diet becomes so unpalatable that the child's energy intake is compromised).

### 34.7.3 Vegetarianism and meat avoidance

There are many different types of vegetarianism, but most vegetarians are either lacto-ovo-vegetarian (avoid meat, poultry, and fish, but consume dairy products and eggs) or, less commonly, vegan (no animal products are consumed). The term 'plant-based' diet is being used increasingly, but the term is non-specific, being used to describe a vegan diet through to a diet with a reduced emphasis on meat and dairy products, depending on the author. Some children are raised as vegetarians within vegetarian families, but many more become vegetarian through personal choice, particularly during adolescence. Reasons for becoming vegetarian include concerns about animal exploitation, the environment, and health. However, adolescents with disordered attitudes to eating may use vegetarianism as a socially acceptable way to restrict their food intake. Many more children and adolescents are meat reducers, and may describe themselves as 'flexitarian,' consuming small amounts of meat (or poultry or fish only) infrequently, and presumably experiencing health benefits and risks that are intermediate between those of vegetarians and non-vegetarians.

In 2014, 4% of 8–18-year-old US children described themselves as vegetarian or vegan, but this masks what are likely to be higher rates in certain sectors of the population, particularly amongst teenage girls.

The nutrient of greatest concern in vegetarian diets is vitamin $B_{12}$, particularly for vegans who have no natural sources of the vitamin in their diets (foods such as spirulina contain inactive analogues). It is essential, therefore, that vegan children and adolescents either use foods fortified with vitamin $B_{12}$ or take vitamin $B_{12}$ supplements. Recommended iron intakes are 80% higher for vegetarians; however, studies in adults suggest that while vegetarians have lower iron stores, they do not have a higher risk of iron-deficiency anaemia than their non-vegetarian counterparts, perhaps because lower iron status is associated with increased iron absorption. Zinc may also be a concern, particularly because of the high phytate content of foods traditionally consumed by vegetarians such as legumes, nuts, and whole grain cereals. In the past there was some concern that protein intakes might be inadequate in diets that avoided meat, but it is now known that as long as a variety of protein-containing foods are consumed during the course of the day, and energy intake is sufficient, protein and amino acid needs are easily met.

Vegetarian diets are associated with a number of potential benefits. In particular, vegetarians tend to have a healthier body mass index, their saturated fat intake can be lower (as long as meat is not replaced with high-fat dairy products such as cheese), they have a higher fibre intake, and fruit and vegetable intake tends to be greater. They have been noted to have lower cholesterol and blood pressure levels than non-vegetarians and, in the long term, vegetarianism is associated with a lower risk of death from ischaemic heart disease. It is uncertain whether these health benefits are due to meat avoidance or the other potentially health-promoting attributes of the diets of most vegetarians. A vegetarian dietary pattern is certainly consistent with good health in children and adolescents provided it is well planned so that it includes a wide variety of nutritious foods. However, some young people remove meat from their diet without ensuring appropriate vegetarian protein sources are consumed, and there is growing concern about the potential impact of the increased availability of ultra-processed foods designed specifically for the vegetarian and vegan consumer, many of which are high in fat and salt, and low in fibre. Of particular concern are plant-based products that use the terms 'milk' or 'cheese' in their title yet do not provide the nutrients these foods are often relied on to provide, in particular protein, calcium, and vitamin $B_{12}$.

# 34.8 Low income countries

## 34.8.1 Introduction

Even in a country that is poor, poverty is not experienced equally across all members of the population. It may not be experienced equally within a family—it is common for children and women to have more limited access to resources, including food, than men. Growing up in a low income country may mean that there is less food and that the nutritional quality of the food is lower. In many cases access to clean water, healthcare, education, and good sanitation is limited or absent. These factors all have an impact on access to food and on nutritional status.

Globally, the WHO has estimated that 690 million people were undernourished in 2019 (8.9% of the world population), the vast majority in low income countries, many of them children. These numbers are expected to increase considerably due to the Covid-19 pandemic. Young children are particularly vulnerable, so that one in five children under five years of age suffer stunted growth. Childhood undernutrition increases mortality such that it has been estimated that more than half of deaths in children under 5 years of age (3.1 million children each year) in low income countries are associated with malnutrition. Malnutrition stunts physical growth, and increases the risk of infections (gastroenteritis, pneumonia, tuberculosis) and of serious obstetric complications. Malnourished children are less able to benefit from education, leading to poorer outcomes at an individual level, and lower economic productivity and slowed socioeconomic development at the local and national level.

In low income countries, and poor communities in other countries, childhood malnutrition is due to a complex interaction of factors including insufficient or poor-quality food and housing (often with overcrowding), and poor sanitation and health services, leading to an increased risk of infection (see Fig. 34.3).

Children may not get sufficient food because of crop failure; conflict that makes it unsafe to plant, tend, or harvest the crops; and, in some cases, poor care and feeding practices. Displacement because of natural disasters such as drought, or conflict and persecution, also affects the lives of millions of children. The classic protein-energy malnutrition conditions marasmus and kwashiorkor are discussed in Chapter 18.

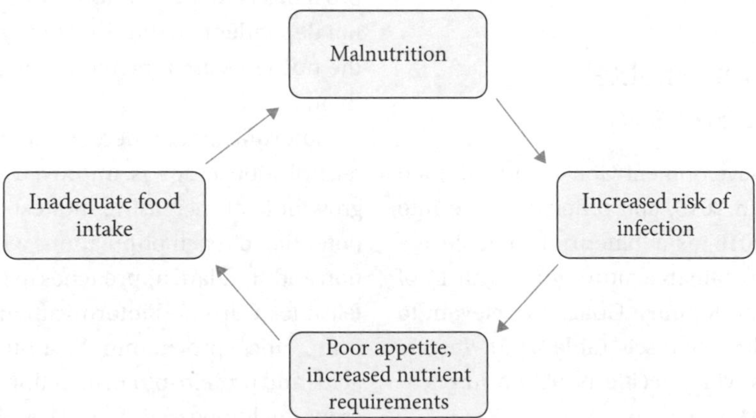


Fig. 34.3 The cycle of malnutrition and infection seen in children living in low income countries.

Consumption of poor-quality food with low nutrient content has been described as the 'hidden hunger' of micronutrient deficiency. In 2021 the World Health Organization described iodine (Chapter 10), vitamin A (Chapter 11), and iron (Chapter 9), as being the most important micronutrients of public health concern because their deficiency poses a major threat to health and development worldwide.

Key anthropometric indices used as measures of nutritional and health status in children in low income countries are:

- *weight-for-age*: measures weight relative to that of a healthy reference population of the same age (used to identify 'underweight')
- *height-for-age*: measures *long-term growth faltering* (used to identify 'stunting')
- *weight-for-height* or weight-for-length: measures acute growth disturbance (used to identify 'wasting').

The terms 'underweight', 'stunting', and 'wasting' are used when a child's measurement is two or more standard deviations below the median reference standards for the index.

There is increasing interest in using *mid upper arm circumference* (MUAC) to identify *severe acute malnutrition* in young children, with evidence that mothers are able to carry out the assessment successfully themselves with minimal training using a colour-banded MUAC tape.

### 34.8.2 The Sustainable Development Goals

The Sustainable Development Goals were adopted by world leaders in 2015, and officially came into force in January 2016, as a 'blueprint to achieve a better and more sustainable future for all'. All 17 of the Sustainable Development Goals are relevant to children and adolescents (see Table 34.5). Targets under these goals with specific nutrition implications are in Table 34.6.

### 34.8.3 Interventions for childhood malnutrition

The main dietary approaches to addressing childhood malnutrition in low income countries include supplementation with specific nutrients (e.g. iron supplementation of children identified with anaemia), fortification of a commonly consumed food vehicle (e.g. the iodization of salt), and dietary change (e.g. increasing the intake of animal foods or encouraging processing methods such as fermentation that decrease the phytate content of unrefined cereals and legumes). Deworming can also markedly decrease iron requirements in children who have hookworm or other gastrointestinal parasites. Less direct approaches, such as vaccination, improved sanitation, and education on good feeding practices, are also important.

Much work has been done in recent years to develop and test home or 'point-of-use' fortification strategies. Two such products being used in programmes for preschool children and school meals are:

- micronutrient sprinkles: these are pre-packaged micronutrients that are sprinkled onto food before it is eaten
- lipid-based nutrient supplements: these are products that contain vitamins and minerals, protein, and essential fatty acids in a palatable form. One example is 'nutributter', which is based on peanut, milk powder, and flavouring such as cocoa.

The advantage of these home fortification approaches is that the amount of nutrient provided is not dependent on the child's energy intake, and that the nutrients are targeted to the person who needs them.

Biofortification, whereby the micronutrient content of food crops is improved during the plant's growth, is of increasing interest because it has the potential to reach populations where supplementation and standard approaches to fortification are not feasible. Current biofortification projects include iron-, zinc-, provitamin A carotenoid-, and amino acid- and protein-biofortification of a range of staple crops, including rice, cassava, and maize.

**Table 34.5 The sustainable development goals for 2016–2030**

| |
|---|
| *Goal 1*: End poverty in all its forms everywhere. |
| *Goal 2*: End hunger, achieve food security and improved nutrition, and promote sustainable agriculture. |
| *Goal 3*: Ensure healthy lives and promote well-being for all at all ages. |
| *Goal 4*: Ensure inclusive and equitable quality education and promote lifelong learning opportunities for all. |
| *Goal 5*: Achieve gender equality and empower all women and girls. |
| *Goal 6*: Ensure availability and sustainable management of water and sanitation for all. |
| *Goal 7*: Ensure access to affordable, reliable, sustainable, and modern energy for all. |
| *Goal 8*: Promote sustained, inclusive and sustainable economic growth, full and productive employment and decent work for all. |
| *Goal 9*: Build resilient infrastructure, promote inclusive and sustainable industrialization and foster innovation. |
| *Goal 10*: Reduce inequality within and among countries. |
| *Goal 11*: Make cities and human settlements inclusive, safe, resilient and sustainable. |
| *Goal 12*: Ensure sustainable consumption and production patterns. |
| *Goal 13*: Take urgent action to combat climate change and its impacts. |
| *Goal 14*: Conserve and sustainably use the oceans, seas and marine resources for sustainable development. |
| *Goal 15*: Protect, restore and promote sustainable use of terrestrial ecosystems, sustainably manage forests, combat desertification, and halt and reverse land degradation and halt biodiversity loss. |
| *Goal 16*: Promote peaceful and inclusive societies for sustainable development, provide access to justice for all and build effective, accountable and inclusive institutions at all levels. |
| *Goal 17*: Strengthen the means of implementation and revitalize the global partnership for sustainable development. |

*Source:* Sustainable Development Knowledge Platform. Transforming Our World: the 2030 Agenda for Sustainable Development. United Nations. https://sustainabledevelopment.un.org/sdgs (accessed 22 July 2022). Courtesy of the United Nations.

**Table 34.6 Sustainable development goals targets with special relevance to nutrition**

| |
|---|
| *Target 2.1*: By 2030, end hunger and ensure access by all people, in particular the poor and people in vulnerable situations, including infants, to safe, nutritious and sufficient food all year round |
| *Target 2.2*: By 2030, end all forms of malnutrition, including achieving, by 2025, the internationally agreed targets on stunting and wasting in children under 5 years of age, and address the nutritional needs of adolescent girls, pregnant and lactating women and older persons |
| *Target 3.2*: By 2030, end preventable deaths of newborns and children under 5 years of age, with all countries aiming to reduce neonatal mortality to at least as low as 12 per 1000 live births and under-5 mortality to at least as low as 25 per 1000 live births |
| *Target 3.3*: By 2030, end the epidemics of AIDS, tuberculosis, malaria and neglected tropical diseases and combat hepatitis, water-borne diseases and other communicable diseases |
| *Target 6.1*: By 2030, achieve universal and equitable access to safe and affordable drinking water for all |

*Source:* Sustainable Development Knowledge Platform. Transforming Our World: the 2030 Agenda for Sustainable Development. United Nations. https://sustainabledevelopment.un.org/sdgs (accessed 20 July 2022). Courtesy of the United Nations.

### 34.8.4 Emerging issues

HIV continues to place many children in low income countries at extreme nutritional risk. The risk to children does not only lie in contracting the condition—it was estimated that in 2008 alone, 17.5 million children and adolescents lost at least one parent to AIDS. Children orphaned by AIDS are more likely to be malnourished, unwell, or subjected to exploitation and discrimination than children orphaned for other reasons.

Tuberculosis (TB) is becoming an increasing concern, and interacts with HIV so that TB is a leading cause of death among people living with HIV. Every year, an estimated 67 million children under the age of 15 years are infected with TB, and at least 1.2 million fall ill. Tuberculosis disease is associated with reduction in appetite, micronutrient malabsorption, and cachexia that can lead to wasting, all of which are likely to impair immunity.

From 2020, the Covid-19 pandemic has placed children at additional nutritional risk by reducing families' ability to produce, distribute, and purchase food. The pandemic is likely to continue to have effects until the large majority of people in low income, as well as high income, countries are vaccinated. This may require ongoing booster shots, and new vaccines as new variants arise.

There is also increasing concern about the 'double burden of disease' in low income countries, the co-existence of undernutrition and over-nutrition in the same country. It is assumed that the substantial changes in traditional diets (to foods that are more energy-dense, but micronutrient-poor) and lifestyles (with less physical activity) that have occurred in recent decades are responsible, particularly in urban settings. The problem may be more widespread than is currently thought because the use of weight-for-age to identify malnutrition can incorrectly identify many overweight stunted children as normal or even underweight.

Finally, climate change is already having substantial impacts on the food supply, with natural disasters including fires and flooding impacting on food production, and the expectation that there will be increasing numbers of 'climate refugees' in future.

## 34.9 Special issues in adolescents

Adolescence is a time of rapid physical and psychological growth, with numerous pressures and influences resulting from the transition from a dependent childhood state to independent living. While many of the nutritional problems that affect children apply in this age group, adolescents also face additional nutrition-related issues.

### 34.9.1 Pregnancy

More than 13 million children are born annually to teenage mothers, the vast majority to women in low income countries. In these countries, young motherhood is both common, and dangerous: complications from pregnancy and childbirth are the leading cause of mortality in girls aged 15–19 years. In high income countries, considerable social stigma can be associated with teenage pregnancy. Teenage mothers are more likely to be unmarried and to live in poverty, and most teenage pregnancies are unplanned. Because of this, nutritional support during adolescent pregnancy is challenging but extremely important. Low pre-pregnancy BMI, poor gestational weight gain, impaired iron status, and poor calcium and vitamin D status can adversely affect pregnancy outcomes for mother and infant. Breastfeeding initiation rates and duration are lower in younger and unmarried mothers, although barriers to effective breastfeeding are thought to be more social and cultural than physical.

### 34.9.2 Dieting and anorexia nervosa

Western culture typically glorifies a thin body, particularly for women, and the media influences how adolescents view their changing body shape. Body dissatisfaction is rife in adolescent girls and it is not uncommon for teenagers to become preoccupied with their physical appearance. However, few adolescents actually go on to develop a diagnosed eating disorder (see Chapter 22), perhaps only 1%. Substantially more adolescents engage in unhealthy dieting behaviours, which could place them at nutritional risk. These include skipping meals, smoking, using diet pills, vomiting, and in more recent times, orthorexia, which refers to an obsession with healthy or 'clean' eating to the point where it interferes with daily life.

### 34.9.3 Alcohol

The increasing independence of adolescents from their family is associated with increased susceptibility to peer pressure, both positive and negative. In many countries, alcohol consumption starts during adolescence and intake is cause for concern. Although alcohol can be consumed safely, if consumed in moderation as part of an adolescent's dietary intake, many adolescents are exposed to a more negative side of alcohol and the dangers of alcohol intoxication. If consumed in excess, alcohol can replace more nutrient-dense foods from the diet. Adolescents who drink regularly tend to engage in other risky health behaviours, and careful campaigns are required in order to reach this target audience. It is unlikely that negative messages are well received, so that it may be better to focus on safe and legal drinking limits, and how to determine the alcohol content of drinks with which adolescents are unfamiliar, rather than graphic images of the consequences of drinking and driving.

### 34.9.4 Long-term conditions

Prevention of long-term conditions such as heart disease and cancer should start early in life. Although the overt symptoms of most of these diseases may not be apparent until mid-adulthood, cancer initiation starts many years before tumours form, post-mortem studies have demonstrated that fatty streaks in the arteries are apparent in adolescents, and some long-term conditions, in particular type 2 diabetes, are now manifesting as early as adolescence. In fact, although the majority of children and adolescents with diabetes have type 1 diabetes, up to 45% of new cases in some population groups now have type 2. This has been attributed to the escalating prevalence of obesity.

There is, therefore, considerable justification for encouraging adolescents to follow widely accepted dietary guidelines. However, adolescence is a time of experimentation, when young people inevitably experience the need to assert themselves and respond to peer pressure. This applies as much to food habits as to other aspects of their lifestyle. Adolescents generally, with their parents' help, manage to consume enough of the essential nutrients in their own individual way. If appropriate eating habits have been established in childhood, there is the expectation that food habits should improve when they have passed through this transitional stage. However, for some adolescents, particularly those who are above a healthy weight, intervention may be required.

## 34.10 Summary

Appropriate nutrition is essential for optimal growth and development in childhood and adolescence. While more than one in five children worldwide suffer from malnutrition, one in three in high income countries are overweight and, in some societies, undernutrition and obesity coexist. In childhood and adolescence, overweight and obesity account for the emergence, in some population groups, of type 2 diabetes as a major health issue amongst young people, in addition to the well-recognized psychosocial consequences, and increased risk of fractures. Overweight also tends to continue into adult life, when the many comorbidities associated with excess body fat become apparent. At the same time, suboptimal micronutrient status, in particular of iron, iodine, and vitamin D, remain an issue for many children and adolescents. Appropriate eating habits established in childhood will encourage healthier food choices in later life, even if in adolescence there is a tendency to adopt less conventional food habits. Thus, implementation of dietary guidelines for young people should be regarded as a priority in families, as well as in the context of public health.

## Further reading


1. **Cole, T.J., Bellizzi, M.C., Flegal, K.M., and Dietz, W.H.** (2000) Establishing a standard definition for child overweight and obesity worldwide: international survey. *BMJ*, **320**, 1240–3.
2. **de Onis, M., Onyango, A., Borghi, E., et al.** (2006) WHO Child Growth Standards. Length/height-for-age, weight-for-age, weight-for-length and body mass index-for-age: methods and development. Geneva: WHO. Available at: https://www.who.int/publications/i/item/924154693X (accessed 14 March 2022).
3. **Han, J.C., Lawlor, D.A., and Kimm, S.Y.S.** (2010) Childhood obesity. *Lancet*, **375**, 1737–49.
4. **Mansoor, Y. and Hale, I.** (2021) Parent perceptions of routine growth monitoring: a scoping review. *Paediatrics & Child Health*, **26**, 154–8.
5. **National Health and Medical Research Council** (2013) *Australian dietary guidelines*. Canberra: National Health and Medical Research Council.
6. **Niinikoski, H., Lagström, H., Jokinen, J., et al.** (2007) Impact of repeated dietary counseling between infancy and 14 years of age on dietary intakes and serum lipids and lipoproteins. *Circulation*, **116**, 1032–40.
7. **Ogata, B.N. and Hayes, D.** (2013) Position of the Academy of Nutrition and Dietetics: nutrition guidance for healthy children ages 2 to 11 years. *J Acad Nutr Diet*, **114**, 1257–76.
8. **United Nations Children's Fund (UNICEF), World Health Organization, International Bank for Reconstruction and Development/The World Bank** (2021) *Levels and trends in child malnutrition: key findings of the 2021 edition of the joint child malnutrition estimates*. Geneva: World Health Organization.


## Useful websites

1. **United States and Canada Dietary Reference Intakes**
https://ods.od.nih.gov/HealthInformation/Dietary_Reference_Intakes.aspx
2. **European Dietary Reference Values**
http://www.efsa.europa.eu/en/topics/topic/drv
3. **Sustainable Development Goals**
https://sdgs.un.org/
4. **Raisingchildren.net.au (the Australian parenting website)**
https://raisingchildren.net.au/
5. **World Health Organization Health Topics: Obesity**
https://www.who.int/health-topics/obesity#tab=tab_1

# Nutrition and Ageing


Sian Robinson and Clare Corish




This chapter provides an overview of why nutrition is important in older age. Focusing on the physiological changes that occur with ageing, it describes the nutritional needs and status of older adults and considers some special issues arising in nutrition-disease relationships in older age.

## 35.1 Population ageing

Across the world, populations are ageing. Improvements in living standards over past decades, alongside advances in medicine and public health, have led to marked increases in life expectancy. Although the rate of change is different across settings, and large disparities persist, life expectancy at birth has risen in all regions, amounting to an increase of more than 10 years in the period since the 1950s in the US and Europe, and almost 30 years in Asia, with further increases projected (see Fig. 35.1) (UN, 2017).

These gains in life expectancy, together with lower fertility rates, are the drivers of population ageing, such that population age structures are changing, with increases both in the number and proportion of older adults. Globally, the number of older adults (aged 60 years and above) is expected to increase from 962 million (in 2017) to 2.1 billion by 2050 and could rise to 3.1 billion by 2100; even greater changes are projected for the number of older adults aged 80 years and above, increasing from 137 million (2017) to 909 million by 2100 (UN, 2017). With variation in the timing of the demographic transition across settings, there are significant differences in the age structures of populations. In Europe, where the transition was early, the population is relatively older: for example, in Germany 28% of the population was aged 60 years and above in 2017, with projected increases to reach 38% by 2050 (see Fig. 35.2). In regions where the demographic transition began later, such as Asia and Latin America, populations are relatively younger: for example, 10% of the population of Mexico was aged 60 years and above in 2017, with increases projected to reach 25% by 2050.

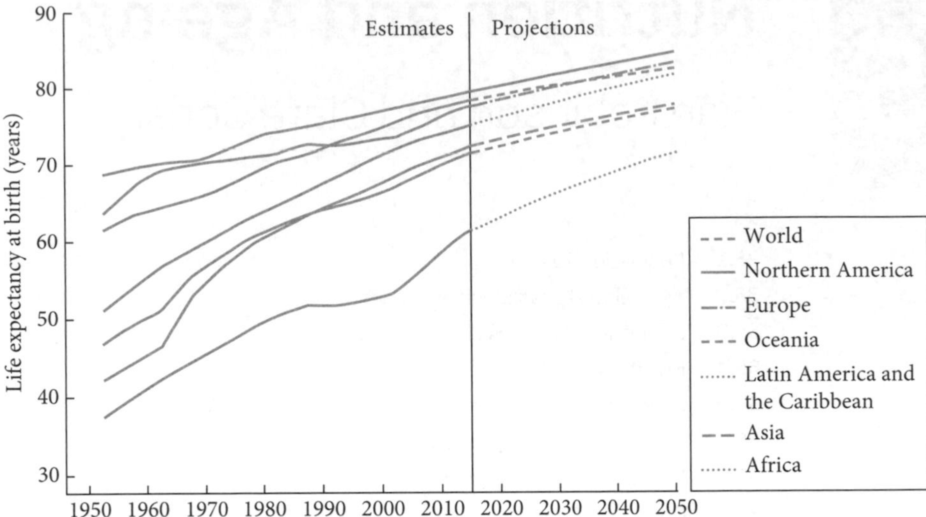


**Fig. 35.1** Life expectancy at birth by region from 1950 to 2050 (both sexes combined) (UN, 2017).

*Source:* United Nations (2017) Department of Economic and Social Affairs, Population Division (2017). World Population Ageing 2017-Highlights (ST/ESA/SER.A/397). https://www.un.org/en/development/desa/population/publications/pdf/ageing/WPA2017_Highlights.pdf

As populations age, they face the challenge that improvements in life expectancy are not matched by gains in healthy life expectancy, so that older adults may face a longer time spent in poor health, with greater likelihood of having multiple chronic and complex health conditions and disability. Importantly, this gap may be more marked in disadvantaged groups. Recognition of changes in the age structure of populations, and the implications for the health of older people, has focused attention on health trajectories in later life—understanding the determinants of differences in the rate and pace of ageing across populations will be key both to predicting the health-care needs of older people in the future, as well as provision of effective support to promote health and to help maintain independence in later life.

The 2015 World Health Organization report on ageing and health emphasizes that healthy ageing is more than a simple absence of disease; it defines healthy ageing as *'the process of developing and maintaining the functional ability that enables well-being in older age'*—with functional ability referring to '*the health-related attributes that enable people to be and to do what they have reason to value*' (WHO, 2015). As populations age, the growing need to promote healthy ageing has renewed interest in the impact of lifestyle on health in older age, including a focus on the role of diet and nutrition.

## 35.2 Age-related physiological changes

Ageing is associated with an accumulation of molecular and cellular damage over time that results in gradual losses of physiological reserves (WHO, 2015); thus, most physiological systems deteriorate with age (Table 35.1). Changes in body composition, fluid regulation, digestive, cognitive, cardiovascular, and immune function, vision, and the microbiome can impact on the risk of nutrition-related conditions such as sarcopenia, undernutrition, frailty, osteoporosis, constipation, and diarrhoea. However, wide variation exists among people in the extent to which functional decline is observed. Successful ageing is dependent on factors including genetics, health status, lifestyle, and social interaction. Sufficient high quality macro- and micronutrient intakes throughout the life course, and specifically in older age, can mitigate the physiological deterioration in body composition and functional status.

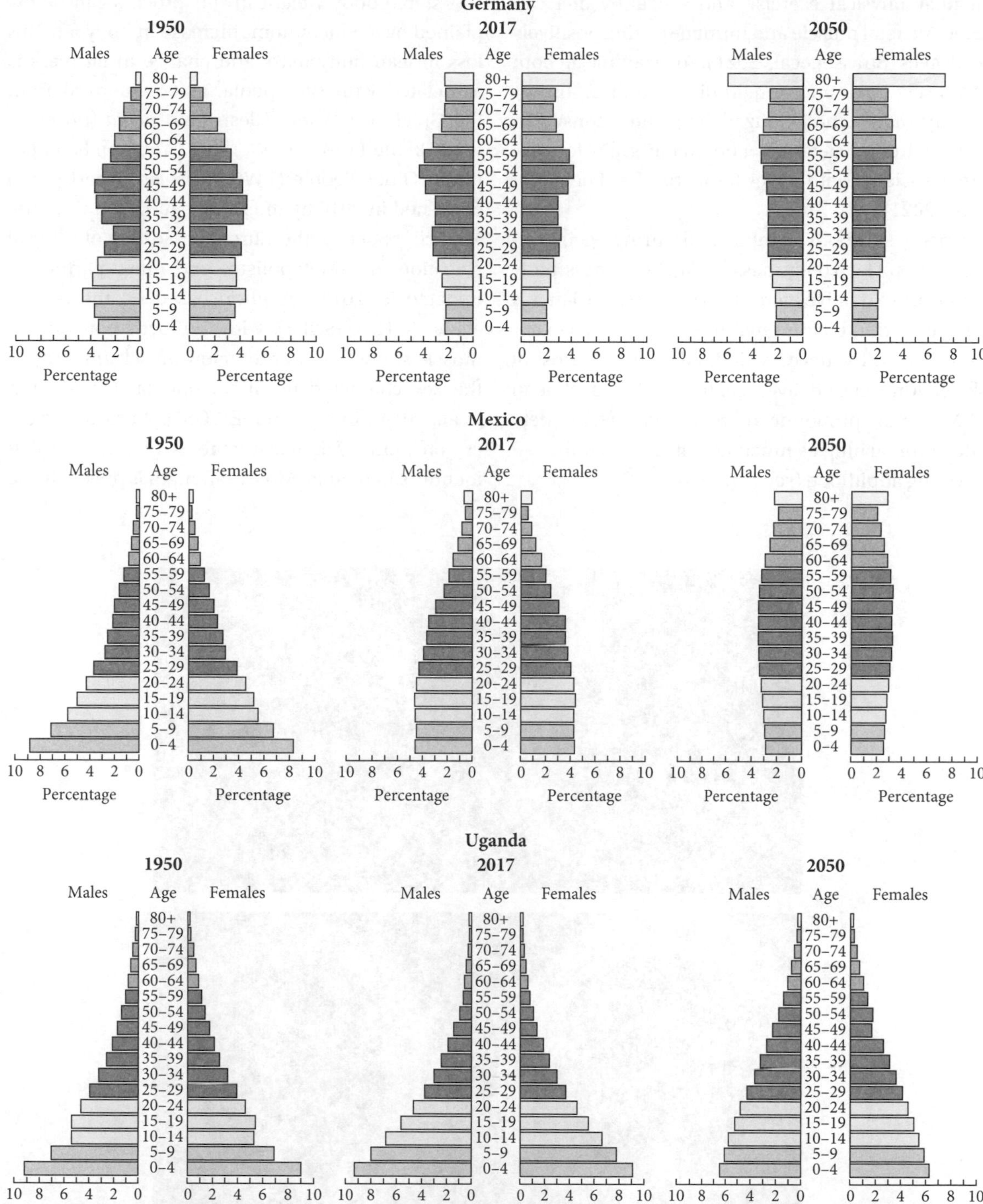


Fig. 35.2 Distribution (%) of population according to sex and age group in 1950, 2017, and 2050: Germany, Mexico, Uganda (from UN, 2017).

*Source:* United Nations (2017) Department of Economic and Social Affairs, Population Division (2017). World Population Ageing 2017-Highlights (ST/ESA/SER.A/397). https://www.un.org/en/development/desa/population/publications/pdf/ageing/WPA2017_Highlights.pdf

Regular physical exercise and a healthy diet can affect skeletal muscle and immune ageing positively at all ages (Strasser et al., 2021). For example, in both cross-sectional and longitudinal studies, higher quality diets—characterized by greater consumption of fruit, vegetables, and whole grain foods—are associated with lower frailty risk (Ni Lochlainn et al., 2021).

Changes in body composition during ageing include loss of lean body mass (LBM), bone mass, body water, and a relative increase of fat mass. The latter is also redistributed from mainly subcutaneous to abdominal fat. The decrease in LBM reflects the loss of skeletal muscle. On average, the marked decline in LBM is most pronounced around the seventh decade, approaching as much as a 40% loss compared to young adulthood (see Fig. 35.3).

A stable body weight in this process can be explained by a concomitant increase in body fat. This loss in lean body mass and change in fat mass is associated with sarcopenia, a term derived from the Greek words *sarx* (flesh) and *penia* (poverty). In 2018, the European Working Group on Sarcopenia in Older People (EWGSOP)—an expert group convened in 2010 by the European Union Geriatric Medicine Society, the European Society of Clinical Nutrition and Metabolism, and other partners—updated its 2010 definition based on the newest evidence (EWGSOP2), with the focus now on low muscle strength rather than low muscle quantity as the key characteristic of sarcopenia (Cruz-Jentoft et al., 2019). In addition, EWGSOP2 provides clear cut-off points for measurements of variables that identify sarcopenia. Muscle strength is presently the

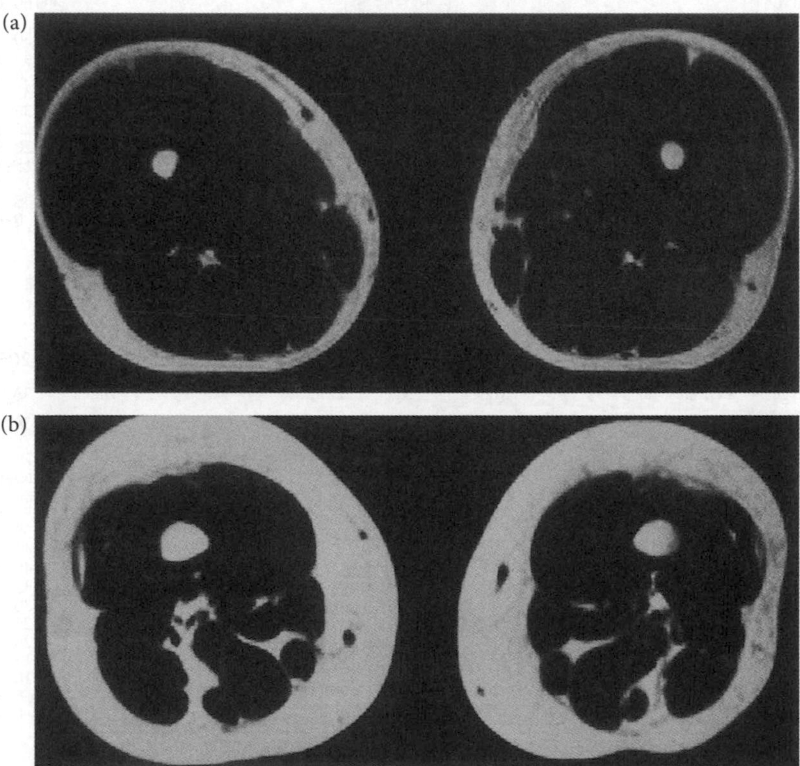


Fig. 35.3 Magnetic resonance images of the thigh showing differences in total muscle, intramuscular fat, subcutaneous fat, and bone between (a) a young woman athlete (age 20 years, BMI 22.6) and (b) an elderly sedentary woman (age 64 years, BMI 30.7).

*Source:* Evans, W.J. and Meredith, E.N. (1989) Exercise and nutrition in the elderly. In Munro, H.N. and Danford, D.E. (eds) *Nutrition, aging and the elderly*. Plenum Press, New York. With permission from Springer.

most used measure of muscle function and sarcopenia is defined as probable when low muscle strength is detected. A sarcopenia diagnosis is confirmed by the presence of low muscle quantity or quality. When low muscle strength, low muscle quantity/quality, and low physical performance are all detected, sarcopenia is considered severe.

Higher levels of physical activity reduce the risk of sarcopenia, and exercise training, particularly resistance exercise, is an established treatment for it. However, physical activity declines in older age. Due to diminished maximum oxygen uptake and muscle fibre atrophy, greater physical effort for the same task is required in older people. Consequently, older people may reduce physical activity further and, therefore, have lower energy expenditure. Fig. 35.4 shows how the physical activity level (PAL), calculated as daily total energy expenditure divided by resting metabolic rate, decreases with age.

Food consumption studies indicate that older people respond to lower physical activity by eating less, which together with other factors such as poor appetite in older age (see section 35.4) can result in a negative energy balance. This phenomenon is called anorexia of ageing and is common in people over 70 years of age, increasing the risk of progressive undernutrition, micronutrient deficiencies, and nutrition-related diseases. Weight loss in later life should be prevented (see Fig. 35.5) because it is associated with poor health outcomes that include frailty and increased mortality (De Groot et al., 2002; Alharbi et al., 2021).

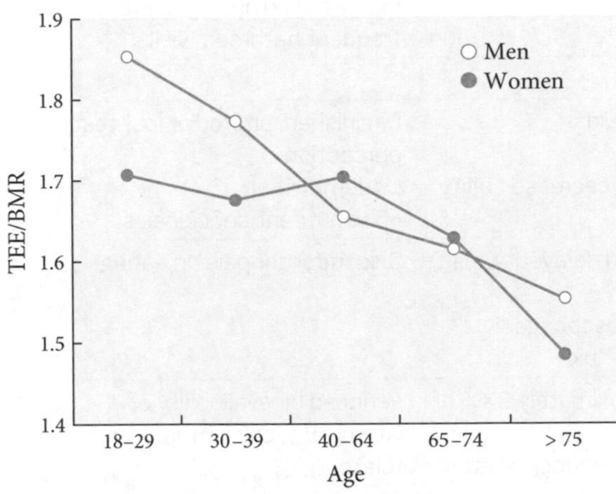


Fig. 35.4 Decrease of physical activity level (PAL) for men and women during ageing. PAL is derived as total energy expenditure (TEE, assessed by doubly labelled water method) divided by basal metabolic rate (BMR, derived from proxy measures).

*Source:* Based on Black, A.E. (1996) Physical activity levels from a meta-analysis of doubly labeled water studies for validating energy intake by dietary assessment. *Nutr Rev*, **54**, 170–4; conclusion confirmed by Manini, T.M. (2010) Energy expenditure and aging. *Ageing Res Rev*, **9**, 1–11.

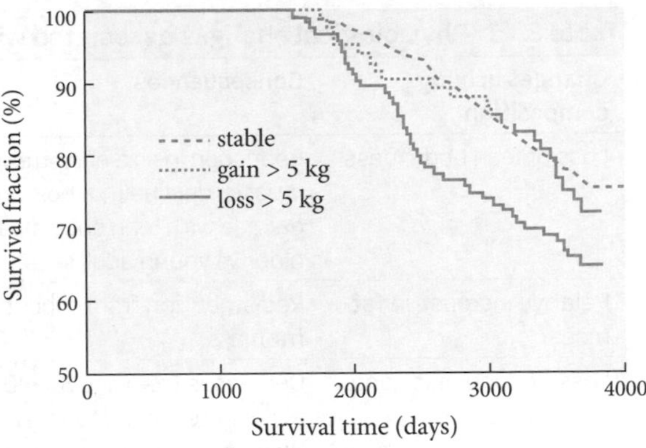


Fig. 35.5 Probability of survival for participants in the SENECA study with and without weight change.

*Source:* De Groot, C.P.G.M. and Van Staveren, W.A. (2002) Undernutrition in the European SENECA studies. *Clin Geriatr Med*, **18**, 699–708.

Bone mass and bone density generally start to decrease from the age of 35–40 years onwards (see Chapter 8). The decrease in bone mass and density is associated with osteoporosis, a multifactorial disease with a higher prevalence in those with a low BMI, inadequate intakes of calcium and vitamin D, and reduced mobility, as well as higher consumption of caffeine and alcohol.

Dehydration is another problem often observed in older adults, which can contribute to greater morbidity and mortality (Volkert et al., 2019a). Due to lower LBM, but also less interstitial fluid, the body of an older adult contains less body water (<50% vs 70% of total body mass in younger adults). Furthermore, impaired kidney function, loss of thirst sensors, fear of incontinence, and increased arthritic pain may inhibit older adults from having an adequate intake of fluid. Dehydration may result in constipation, faecal impaction, cognitive impairment, and even death.

**Table 35.1 Physiological changes associated with ageing**

| Changes in body composition | Consequences | Potential associated morbidities |
|---|---|---|
| Loss of lean body mass | Reduction in size and strength of skeletal muscle; decline most pronounced in seventh decade with lean body mass up to 40% lower in older vs young adults | Sarcopenia, frailty |
| Relative increase in fat mass | Redistribution from subcutaneous to abdominal fat mass | |
| Loss of bone mass and density | Decreases from age 35–40 years onwards; accelerated in menopause<br>Women may lose 50% of their trabecular and 33% of their cortical bone mass<br>Men lose approximately 50–70% bone mass and density vs women | Osteoporosis |
| Loss of body water | Body water is less than 50% of total body mass in older vs 70% in younger adults due to reduction in lean body mass and interstitial fluid | Dehydration resulting in impaired thermoregulation, constipation, faecal impaction, cognitive impairment, and death |
| **Changes in fluid regulation** | | |
| Decreased renal function | Decreased kidney mass and blood flow; 10% decrement in glomerular rate per decade after age 30 years | Dehydration resulting in impaired thermoregulation, constipation, faecal impaction, cognitive impairment, and death |
| Loss of thirst sensors | | |
| Decreased bladder function | Decreased elasticity, muscle tone, and capacity (Amarya et al., 2015) | Incontinence<br>Increased arthritic pain from frequent bathroom visits |
| **Digestive function** | | |
| Changes in oral cavity | Reduction in number of taste buds<br>Reduced saliva production<br>Increased tooth loss/dentures decrease ability to masticate food | Diminished, altered or lost taste perception<br>Undernutrition<br>Micronutrient deficiencies |
| Changes in oesophageal function | Reduced peristaltic activity and delayed transit time<br>Loss of function of the lower oesophageal sphincter can lead to gastric reflux | Discomfort or pain on eating |
| Reduced gastric acid output | Atrophic gastritis affects approximately 33% of adults aged over 60<br>Lower secretion of gastric acid, intrinsic factor, and pepsin | Reduced bioavailability of vitamin $B_{12}$, calcium, iron, and folate |
| Reduced intestinal motility | Slower clearance of intestinal contents<br>Colon becomes hypotonic resulting in increased storage capacity, longer stool transit time, and greater stool dehydration | Constipation, faecal impaction<br>Diarrhoea |

**Table 35.1 Physiological changes associated with ageing (*Continued*)**

| Changes in body composition | Consequences | Potential associated morbidities |
|---|---|---|
| **Changes in cognitive function** | | |
| Reduced sensory perception of food | Changes in neuronal fibres connected to the brain<br>Reduced sense of taste and smell | Undernutrition and frailty<br>Micronutrient deficiencies |
| Reduced appetite | Reduced food intake | |
| Changes in neuromuscular function | Difficulty swallowing<br>Decreased oesophageal and intestinal motility<br>Decreased mobility | |
| **Cardiovascular changes** | | |
| Decreased cardiac output | Cardiac output of an 80-year-old is approximately 50% of a 20-year-old | |
| Increased blood pressure | Greater elevation in systolic than diastolic pressure | Hypertension |
| Increased atherosclerosis | Thickening of arterial walls<br>Increase in fibrous plaques | Cardiovascular disease |
| **Immune system changes** | | |
| Decreased immune function | Reduced antioxidant capacity<br>Reduced lymphoid tissue<br>Impaired activity of immune defence and functional deficits of immune cells<br>Increased inflammation | Increased susceptibility to infection |
| **Changes in vision** | | |
| Vision loss | Preparing and eating meals is more difficult<br>Limits mobility | Undernutrition<br>Micronutrient deficiencies |
| **Changes in microbiome** | | |
| | Reduced microbiotic diversity<br>Decline in beneficial bacteria<br>Less short-chain fatty acid e.g. butyrate | Inflammation<br>Decreased immune function<br>Increased susceptibility to *Clostridium difficile* infection |

Although the efficiency of the digestive and absorptive functions of the gastrointestinal tract declines with age, this can be worsened by the effects of medications or disease states. However, when malabsorption occurs in an older adult, investigation of gastrointestinal disease is required.

Diminished, altered, or lost taste or olfactory system (smell) perception with ageing can impact on food consumption, with lower intakes increasing the risk of undernutrition. Altered taste perception can arise due to changes in the oral cavity (e.g. reduced number of taste buds, saliva production, or dental problems) and/or changes in the neuronal fibres that are connected to the brain. Together with changes in the olfactory system, the overall sensory perception of food declines. It can, however, be difficult to distinguish between taste and smell dysfunctions that are directly related to ageing and those resulting from disease and/or medication. Individuals may not be able to discriminate between the loss of taste and smell and many taste complaints are, in fact, olfactory disorders.

Older adults are susceptible to chronic dental diseases; hence oral health is important, as this can influence dietary intake and nutritional status.

Moreover, other conditions that are prevalent in older adults can also influence dietary intake and nutritional status. Difficulty in swallowing can occur if neurological function is disturbed. Atrophy of the stomach mucosa affects about one-third of those aged over 60 years and results in lowered secretion of acid, intrinsic factor, and pepsin, which reduce the bioavailability of vitamin $B_{12}$, calcium, iron, and folate. This can be magnified in older adults who use proton pump inhibitors. The implications are most profound for vitamin $B_{12}$ due both to the diminished dissociation of the vitamin from food proteins and binding of the small amount of freed vitamin $B_{12}$ by the increased numbers of swallowed bacteria, which can survive in the low-acid environment of the proximal small intestine.

Another change is reduced intestinal motility and additional dystrophy in the large intestine, which may lead to constipation or diarrhoea, both disorders occurring in older people. Constipation occurs more frequently, contributes to morbidity, and detrimentally affects well-being.

## 35.3 Do nutritional needs change in older age?

Older adults are a diverse group, with some being healthy and fit while others have chronic conditions that compromise overall lifestyle and quality of life. Thus, the nutritional needs of older adults are particularly varied because of health, physiological function, and susceptibility to disease. Food intakes fall in older age alongside declining levels of physical activity and lower energy requirements, although the heterogeneity across older populations is such that changes in patterns and amounts of foods consumed are hugely variable between individuals. Overall, the difference in energy intake between younger and older populations is significant—amounting to an estimated 16–20% (when comparing adults aged 26 and 70 years), or around 0.5% per year (Giezenaar et al., 2016). While recent dietary surveys show that energy intakes of many older adults meet or exceed requirements, age-related decline in food intake may be a concern as, compared with younger adults, most older adults need the same or even higher intakes of micronutrients. Unless diets are varied and of sufficient 'quality,' including regular consumption of nutrient-dense foods, nutrient intakes fall in parallel with declining energy intakes, making it more challenging for older adults to meet their nutrient needs. Thus, a nutrient-dense diet is a high priority in older adults.

At the same time, in developed countries, overweight and obesity have the highest prevalence in the older population. Older adults who are overweight, particularly those with abdominal fat distribution, should be encouraged to combine a nutrient-rich diet with physical activity to avoid further weight gain and maintain lean body mass. To avoid loss of muscle mass and function, they should avoid very low energy weight loss diets that can exacerbate the loss of lean body mass (Volkert et al., 2019). In older adults who are obese with weight-related health problems, energy restriction should be moderate, not less than 1000 kcal/day, and combined with physical activity to preserve muscle mass (Volkert et al., 2019; FSAI, 2021). Controlling the amount and type of carbohydrate and fat consumed may help weight control and delay or prevent the onset of type 2 diabetes and its associated macro- and microvascular complications (SACN, 2021). A reduction of saturated fat is associated with a lower risk of developing cardio-metabolic disease and other chronic inflammatory conditions associated with ageing (SACN, 2019; FSAI, 2021). Current guidance is that saturated fatty acids should be replaced with polyunsaturated or monounsaturated fatty acids and long-chain omega-3 polyunsaturated fatty acids, e.g. eicosapentaenoic and docosahexaenoic acids should be included in the diet (FSAI, 2021) to support the immune system. Because of the relationship between overweight, obesity, salt intake, and cardiovascular disease, the recommended salt target for older adults is similar to that for the general adult population, i.e. 6 g salt/day (2.4 g/100 mmol sodium/day).

Both in the US and in Europe, the recommended dietary allowance for protein is generally set at a similar level for all healthy adults: 0.75–0.8 g protein/kg body weight/day. This allowance is based on the minimum amount needed to maintain whole-body protein balance. There is currently debate about whether protein requirements may be greater in older age so that diets of healthier older people should provide at least 1.0–1.2 g protein/kg body weight, with higher levels recommended (1.2–1.5 g/kg body weight) (Deutz et al., 2014) for compromised older adults who are malnourished, sarcopenic, and/or frail or at risk of these conditions. Because of the reduction in overall energy requirements, older adults may need a more protein-dense diet than the general adult population to meet these requirements. Moreover, optimizing the potential for muscle protein anabolism by consuming an adequate amount of high-quality protein at each meal (0.4 g/kg or approximately 25–30 g), together with adequate resistance activity, has been suggested as a promising strategy to counteract or delay the onset of sarcopenia, and sustain muscle mass and function.

The recommended amount of fluid is 2 L/day for women and 2.5 L/day for men from all sources or 1.6 L/d and 2.0 L/d, respectively, from drinks alone (EFSA, 2010). Consideration of types of beverages is important; water is the best fluid to drink but tea, coffee, milk, and unsweetened fruit juice all contribute to meeting fluid requirements (EFSA, 2010; Volkert et al., 2019). Strong tea should be consumed between meals to limit the inhibition of iron and zinc absorption, excess caffeine can disturb sleep, and excess alcohol can increase the risk of comorbidities. Furthermore, alcohol tolerance decreases with age and reduced LBM.

Calcium needs are higher in postmenopausal women (with low oestrogen levels) and this may be reflected in increased recommendations for calcium intakes. An abundant calcium intake throughout life helps protect against osteoporosis, particularly in women (see Chapter 8). As a food-based dietary guideline, consumption of three portions of calcium-rich foods is recommended daily to meet nutritional requirements for calcium (FSAI, 2021). These foods are also rich in other nutrients valuable for older adults, e.g. protein, and can be particularly useful for older adults with poor appetite who are frail and/or malnourished.

The body's requirements for iron are lowest in older age; however, other factors can increase the risk of iron deficiency, for example, chronic blood loss from ulcers or other disease conditions, poor iron absorption due to reduced stomach acid secretion, or medications like aspirin, which can cause blood loss.

Zinc deficiency can occur in older adults, particularly in those who are dependent on residential care. Zinc deficiency is associated with low socioeconomic status, problems with chewing food, and poor gastrointestinal absorption. Even marginal zinc status can impair immune function.

There is also some evidence that older adults may have greater needs for vitamin D, riboflavin, and vitamin $B_6$, and need higher doses of vitamin $B_{12}$ to correct poor status. Vitamin D is essential for bone health; a link between vitamin D deficiency and non-skeletal health, e.g. inflammatory, infectious, and immune disorders, is proposed but causal evidence is lacking. Recommended intakes for older adults for vitamin D are 10–20 μg/day, so foods fortified with vitamin D and supplements are advised. Low status of folate, vitamin $B_{12}$, vitamin $B_6$, and riboflavin commonly occurs in older adults and is associated with a higher risk of cardio-metabolic disease, cognitive dysfunction, and osteoporosis. Inadequate intake, malabsorption, increased requirements, and/or drug-nutrient interactions can all affect B vitamin status. Many older adults have atrophic gastritis and/or use proton pump inhibitors, both of which are associated with low status of $B_{12}$. In recognition of this and the prevalence of atrophic gastritis in the older population, some current dietary recommendations are that older adults meet their vitamin $B_{12}$ needs from supplements or foods fortified with crystalline, free vitamin $B_{12}$. The increase in vitamin $B_6$ requirement with age appears to be unrelated to absorption; however, subclinical deficiency of this vitamin may result in immune dysfunction.

Since vitamin E is the only lipid-soluble, chain-breaking antioxidant found in biological membranes, this vitamin may play an important role in maintaining neuronal integrity and preventing cell loss. Thus, higher intakes may prove useful in neurological disorders where oxidative stress has been implicated, although the evidence for this remains unclear. Vitamin C is a key antioxidant vitamin required for immune function and there is some limited evidence that vitamin C ameliorates the symptoms of viral infections (Colunga Biancatelli et al., 2020). Inadequate intakes are associated with low socioeconomic status and ill health.

Achieving an adequate intake of dietary fibre is important for the maintenance of healthy bowel function and prevention of constipation and associated conditions, e.g. diverticular disease and colon cancer. Dietary fibre also helps protect against cardio-metabolic disease and can enhance the satiety value of food. The most frequently used recommendation for dietary fibre is 25 g/day; however, given the challenges associated with poor appetite or dentition, it may be more realistic to consider a dietary fibre goal of ≥3 g of dietary fibre/MJ/day, as recommended in the recent scientific recommendations for food-based dietary guidelines for older adults in Ireland (FSAI, 2021).

Food palatability is vital in older age to promote intake in those who are underweight or with a small appetite. Offering highly palatable foods in a stimulating environment can increase food intake even in frail older adults. Older adults who are underweight with a small appetite may benefit from fortification of foods with energy and protein. Those with a swallowing disorder, e.g. after stroke, require texture modification and food fortification to meet nutritional requirements.

## 35.4 Nutrition in older age

While age-related declines in the level of food consumption are expected in older age, individual trajectories of change in eating habits in later life are very variable between individuals and poorly understood. This is a subject of current research, as improving understanding will be key to defining effective approaches for future preventive strategies to promote the nutritional health and well-being of older populations. Recent national survey data from the UK show that older adults have fat, free sugars, and salt intakes that exceed recommendations, and low consumption of fruit, vegetables, and oily fish (SACN, 2021). While these findings are consistent with younger adults, there are important differences in nutrient intake, with evidence of low intakes of energy, protein, and some micronutrients in older age groups and declines with age, so that the oldest group studied (75+ years) tended to have lower micronutrient intakes as a percentage of recommended values, when compared with the 65 to 74 years age group (SACN, 2021).

A diverse range of influences on nutrition in older age has been identified (see Fig. 35.6). This highlights the importance of age-related as well as social and psychological influences. Institutionalized older people and those living independently in the community, but restricted in their mobility, may be at greater nutritional risk. Importantly, the role of social factors, such as loneliness and lack of social contact, are increasingly recognized as determinants of poor nutrition. To design effective interventions to support older adults, a clear understanding of the array of influences that affect patterns of food choice and consumption is therefore needed, particularly the importance of social and psychological factors (Robinson, 2018).

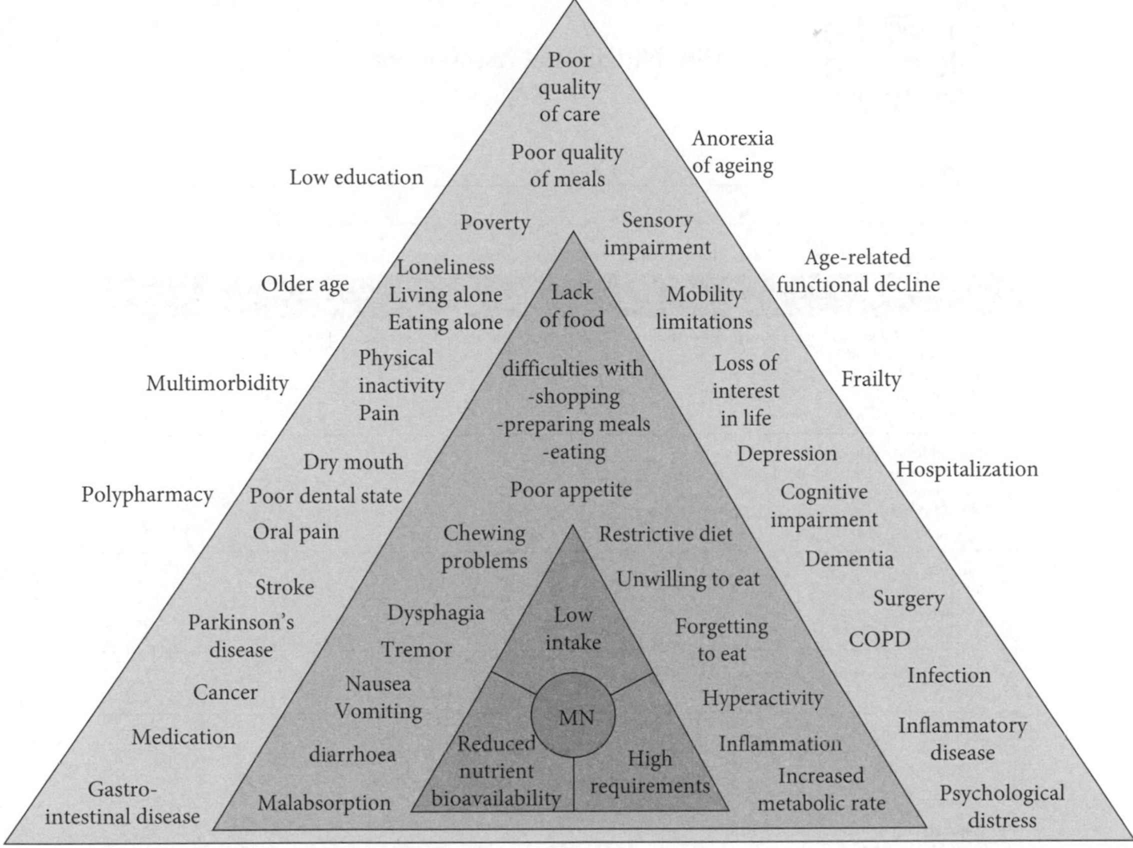


Fig. 35.6 The Malnutrition in the Elderly (MaNuEL) Knowledge Hub Determinants of Protein-energy Malnutrition (PEM) in Aged Persons (DoMAP).

*Source:* Volkert D., Kiesswetter E., Cederholm T., et al. Development of a model on determinants of malnutrition in aged persons: A MaNuEL Project. Gerontol Geriatr Med. 2019;5.

## 35.5 Screening tools for undernutrition in older adults

Inadequate nutritional intake is the predominant cause of undernutrition in older adults. The identification of undernutrition is undertaken using different methods and the population prevalence depends on the methods and cut-off values applied. The literature describes a prevalence of 5–10% in community-dwelling persons above 70 years of age and up to 30–60% in institutionalized people of that age. It is crucial to identify older adults at risk and to make an early diagnosis for effective intervention. Screening tools have been developed from parameters that indicate dietary intake and nutritional status. They are designed for self-assessment or for administration by healthcare professionals. It is important to always use a screening tool for undernutrition that has been validated in older adults and in the healthcare setting (i.e. community, hospital, long-term care, rehabilitation) in which it is used (Power et al., 2018).

Fig. 35.7 presents an example of a screening tool that incorporates several domains including functionality, lifestyle, questions on diet, and subjective health and anthropometric indicators. The sensitivity and specificity of the tool validated against an extensive evaluation by clinicians were 96% and 98%, respectively.

# Mini Nutritional Assessment
# MNA®

Last name: First name:

Sex: Age: Weight, kg: Height, cm: Date:

Complete the screen by filling in the boxes with the appropriate numbers. Total the numbers for the final screening score.

## Screening

**A Has food intake declined over the past 3 months due to loss of appetite, digestive problems, chewing or swallowing difficulties?**
0 = severe decrease in food intake
1 = moderate decrease in food intake
2 = no decrease in food intake ☐

**B Weight loss during the last 3 months**
0 = weight loss greater than 3 kg (6.6 lbs)
1 = does not know
2 = weight loss between 1 and 3 kg (2.2 and 6.6 lbs)
3 = no weight loss ☐

**C Mobility**
0 = bed or chair bound
1 = able to get out of bed / chair but does not go out
2 = goes out ☐

**D Has suffered psychological stress or acute disease in the past 3 months?**
0 = yes 2 = no ☐

**E Neuropsychological problems**
0 = severe dementia or depression
1 = mild dementia
2 = no psychological problems ☐

**F1 Body Mass Index (BMI) (weight in kg) / (height in m²)**
0 = BMI less than 19
1 = BMI 19 to less than 21
2 = BMI 21 to less than 23
3 = BMI 23 or greater ☐

IF BMI IS NOT AVAILABLE, REPLACE QUESTION F1 WITH QUESTION F2.
DO NOT ANSWER QUESTION F2 IF QUESTION F1 IS ALREADY COMPLETED.

**F2 Calf circumference (CC) in cm**
0 = CC less than 31
3 = CC 31 or greater ☐

**Screening score**
(max. 14 points) ☐☐

| | |
|---|---|
| **12-14 points:** | Normal nutritional status |
| **8-11 points:** | At risk of malnutrition |
| **0-7 points:** | Malnourished |

For a more in-depth assessment, complete the full MNA® which is available at **www.mna-elderly.com**

Ref. Vellas B, Villars H, Abellan G, *et al. Overview of the MNA® — Its History and Challenges.* J Nutr Health Aging 2006;**10**:456-65.
Rubenstein LZ, Harker JO, Salva A, Guigoz Y, Vellas B. Screening for Undernutrition in Geriatric Practice: Developing the Short-Form Mini Nutritional Assessment (MNA-SF) . J. Geront 2001;**56A**: M366-377.
Guigoz Y.The Mini-Nutritional Assessment (MNA®) Review of the Literature — What does it tell us? J Nutr Health Aging 2006; 10:466-487.



**For more information: www.mna-elderly.com**

Fig. 35.7 Mini-nutritional assessment.

*Source:* Guigoz, Y. (2006) Mini-nutritional assessment (MNA) Review of the literature—what does it tell us? *J Nutr Health Aging*, **10**, 466–87.

## 35.6 Older adults' perceptions of malnutrition

It is estimated that 10% of the population in developed countries who require care from a general practitioner are at risk of malnutrition. A recent review of studies of older adults' perceptions of their nutritional needs indicates that older community-dwelling adults are unaware of their nutritional requirements, follow similar dietary guidelines to those set for the general adult population, and perceive weight loss as a positive and normal occurrence (Dominguez Castro et al., 2021). The term 'malnutrition' is perceived by older adults as negative and associated with extreme circumstances of severe hunger and famine, with older adults reporting that they feel offended if the term is applied to their own health (Geraghty et al., 2021). Despite the high prevalence of malnutrition risk (Leij-Halfwerk et al., 2019), there is inadequate management of malnutrition within the community, with patients reporting little communication between healthcare providers, a lack of dietetic services available in the community setting, and that they mainly depend on friends, family, and carers for support (Reynolds et al., 2021).

## 35.7 Nutrition-disease relationships in older age

In comparison with younger adults, the influence of diet and lifestyle on non-communicable diseases (NCDs) in older age has been less studied. Despite being most prevalent in older age, lifestyle advice and recommendations to prevent NCDs in older populations are often extrapolated from knowledge of effects in younger adults. However, there is increasing evidence of the importance of dietary patterns, and the opportunity to use dietary change as an effective approach to improve and promote health in older age.

### 35.7.1 Cardio-metabolic disease

Current evidence on the role of dietary patterns and their links to cardio-metabolic disease in older populations is limited, with few intervention trials to test findings from observational studies. Overall, the strongest indication of protective benefits is for the Mediterranean dietary pattern, characterized by greater consumption of fruit, vegetables, and fish and lower consumption of red meat and processed foods. In a recent systematic review of the effects of dietary patterns on cardio-metabolic health in older adults, an intervention to promote compliance with the Mediterranean dietary pattern was shown to lower circulating triglyceride concentrations and systolic blood pressure, with no evidence of harm (Luong et al., 2021). In contrast, the effects of other dietary patterns were more mixed. Importantly, the authors highlight the small body of evidence on cardio-metabolic disease and the need for further research on dietary patterns in older populations (Luong et al., 2021).

### 35.7.2 Cancer

Although more than two-thirds of new cancers are diagnosed among older adults (see Fig. 35.8), cancer prevention efforts have often focused more on younger adult populations, with less consideration of the needs of older adults (White et al., 2019). Diet-cancer links are, therefore, less well-described in older populations, although there is some observational evidence of effects of individual foods in relation to specific cancers. For example, the EPIC Elderly Study, with 100,000 older participants (>65 years) who were followed up over 10 years, suggested benefits of higher consumption of fruit (lung cancer) and lower consumption of red meat (colorectal cancer). Consistent with this evidence, the Healthy Ageing: a Longitudinal Study in Europe (HALE), Europe, that followed up 2339 older participants (70–90 years) over 10 years, reported that greater compliance with a Mediterranean dietary pattern, with similar characteristic foods, was associated with

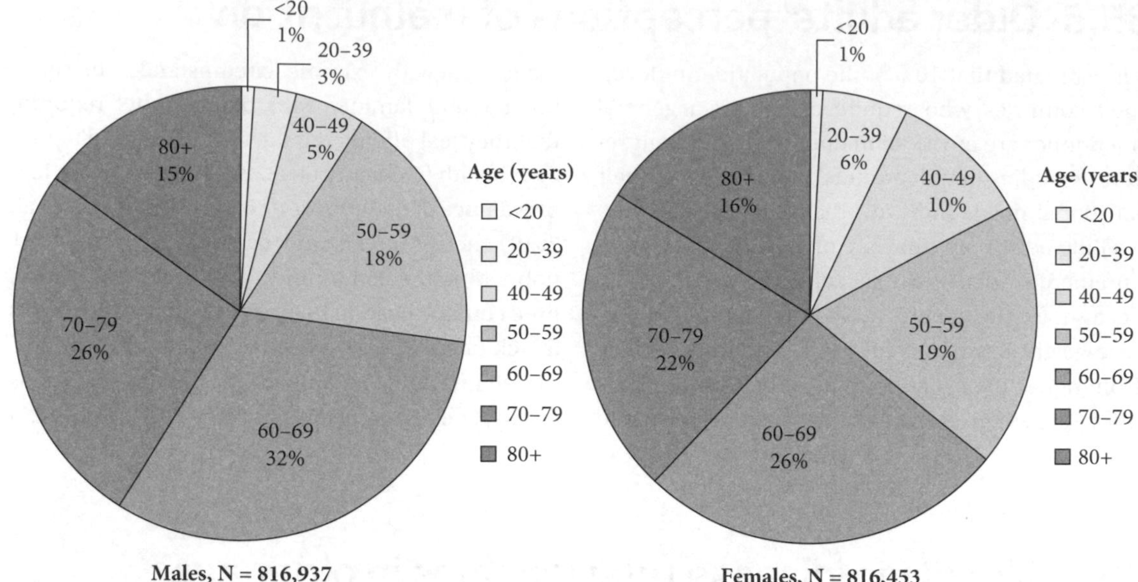


Fig. 35.8 Distribution of all invasive cancers by sex and age at diagnosis, United States, 2015.

*Source:* White, M.C., Holman, D.M., Goodman, R.A., and Richardson L.C. (2019) Cancer risk among older adults: Time for cancer prevention to go silver. Gerontologist, 59, S1–S6.

lower cancer risk (Tyrovolas and Panagiotakos, 2010). However, better evidence is needed to understand the protective effects of diet on cancer in older age.

### 35.7.3 Sarcopenia and impaired physical function

Sarcopenia, the skeletal muscle disorder characterized by the loss of muscle mass and function (see section 35.2), is common in older populations and associated with a range of poor health outcomes that include frailty, hospitalization, and greater mortality. The role of diet in the aetiology of sarcopenia has become a focus of recent research activity and there is now a significant body of evidence that links nutrition to muscle outcomes in older age—for example, there is a recognized overlap between malnutrition and sarcopenia. However, the impact of poorer habitual diets in older populations and/or other differences in dietary patterns on sarcopenia risk is currently less well understood. There is a body of intervention evidence, most frequently based on protein/vitamin D supplementation effects on muscle outcomes, although findings are mixed (Robinson et al., 2018). An important area of current research is on dietary patterns and, in common with cardio-metabolic outcomes and cancer, there is now substantial observational evidence to suggest benefits of 'healthier' dietary patterns for muscle health, characterized by greater consumption of fruit, vegetables, and whole grain foods alongside lower intakes of refined cereals and highly processed foods. The most consistent evidence found is for the 'Mediterranean' pattern, and this is consistent with established mechanistic links that underpin potential beneficial effects on health and ageing (summarized in Fig. 35.9). However, better prospective evidence, together with intervention trials, are needed to inform dietary guidelines to protect muscle health in older populations.

### 35.7.4 Cognitive impairment and dementia

Cognitive impairment and dementia commonly occur in older adults, with prevalence continuing to increase across the older age range, and higher rates found among women. An estimated 5–10% of cases of mild cognitive impairment progress to dementia each year. With increasing severity, and greater decline in cognitive abilities, the losses of cognitive function impact on an individual's ability to carry

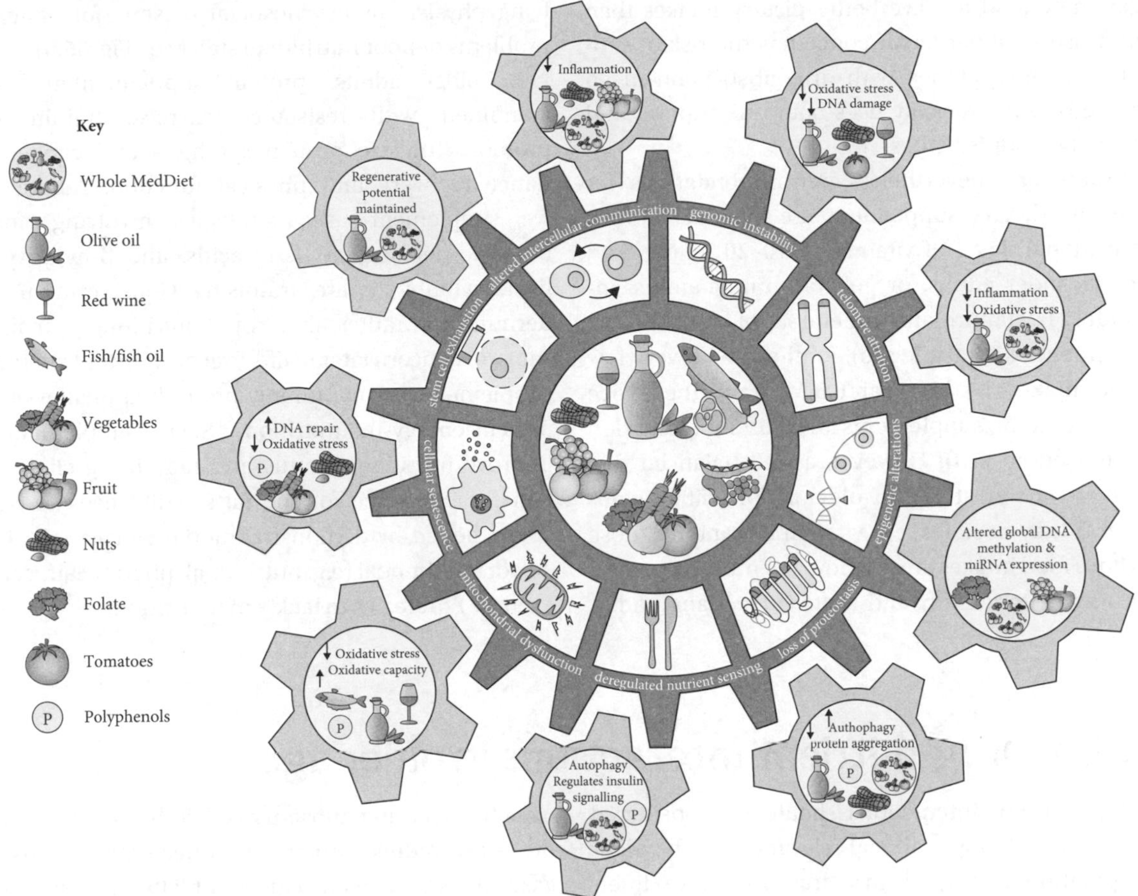


Fig. 35.9 Mechanisms through which the Mediterranean diet may influence the hallmarks of ageing.

*Source:* Shannon, O.M., Ashor, A.W., Scialo, F., et al. (2021) Mediterranean diet and the hallmarks of ageing. Eur J Clin Nutr, 75, 1176–92.

out activities of daily living, with profound consequences for functional independence and quality of life. There is growing evidence that health behaviours, such as being physically active and having a higher consumption of foods that include fish, fruit, and vegetables, have protective effects. Although much of the evidence is observational, there is now a body of evidence that suggests benefits of a Mediterranean dietary pattern, both for risk of mild cognitive impairment as well as dementia (SACN, 2018). The effects of individual dietary components are not known; there is currently insufficient evidence to identify roles of specific nutrients.

## 35.8 Dietary and oral nutritional supplements for older adults

The use of dietary supplements in the USA, Canada, Australia, and New Zealand is widespread amongst older men (35–60%) and women (45–79%). Advertisers often target older people, claiming their products prevent disease or promote longevity. Unfortunately, the nutrient supplements most frequently used are rarely those in shortest supply in the diet; furthermore, dietary supplement users generally tend to have better dietary intakes than non-users. Of particular concern is the risk of supplement interference with drug absorption in an age group that heavily uses both prescription and over-the-counter drugs.

There are, nevertheless, recommendations for specific dietary supplement use by older adults: nutritional doses of vitamin D (10–20 μg per day) for all older adults, or general multivitamins, including B vitamins and mineral supplements (at recommended intake levels), for those with very low food intakes (i.e. less than 6.3 MJ/day). The advice for vitamin $B_{12}$ supplements is still under discussion (see section 35.3). However, a well-balanced diet will provide most healthy older people with the nutrients they need (except vitamin D), and for those whose food intakes are very low, the more important priority is to identify and try to correct any underlying physical or psychosocial reasons for eating problems or poor nutritional state (see Fig. 35.6).

In older adults, protein supplementation—combined with resistance exercise training—promotes muscle protein synthesis and can enhance recovery and physical function. Adjunctive supplement-based strategies involving, for example, leucine, ω3 fatty acids, and β-hydroxy-β-methylbutyrate, are promising. Considering undernutrition rather than over-nutrition, a main cause of concern later in life, oral protein and energy supplements are commonly prescribed to alleviate protein-energy-malnutrition. Such supplements improve nutritional status and may bring clinical benefits. Evidence for the latter still needs to be strengthened by demonstrating the efficacy of targeted, multimodal (e.g. nutritional, pharmaceutical, exercise) strategies to tackle malnutrition.

## 35.9 Drug-nutrient interactions in older age

Drug-nutrient interactions should be considered as an issue of high clinical relevance for the older population. Firstly, because drugs may have different effects and side effects due to changes in body composition and functionality that occur with ageing. Secondly, dietary intake often decreases, and nutrient-nutrient interactions change, as well as drug-nutrient interactions. Thirdly, the use of (multi)medicinal preparations, including over-the-counter products and food or dietary supplements, is, on average, very high in this age group. There are several ways in which drugs can affect nutrition, usually increasing nutrient need (Table 35.2). Many drugs can reduce appetite, produce nausea, cause gastrointestinal disturbances, or alter the senses of taste or smell, and hence affect food intake; long-term laxative use can impair intestinal function; and laxatives and diuretics can lead to severe loss of potassium. As many as 30% of older adults complain that drugs change their sense of taste. It is very likely that undesirable drug-nutrient interactions are underdiagnosed, especially in frail, older people taking several drugs at a time and having low dietary intakes.

## 35.10 Summary

Improvements in diet and maintenance of exercise have been shown to benefit health in older age although, in comparison with younger adults, relatively less is known about the impact of lifestyle on non-communicable health conditions in older age. Ageing is associated with changes in body composition that occur alongside declines in physiological function. However, individual changes are very variable and there is huge heterogeneity across older populations, so that chronological age may be a poor guide to current health status. Age-related decline in food consumption aligns with

Table 35.2 **Commonly used medications and their interactions with various nutrients**

| Medication | Nutrient(s) affected | Effect(s) of interaction |
|---|---|---|
| Opioid painkillers, codeine, calcium channel blockers, cyclic antidepressants, MAOIs, NSAIDS, diuretics | Fibre | Constipation |
| Aspirin | Iron | Causes gastric bleeding, hence loss of iron from the body and iron deficiency |
| Proton pump inhibitors (PPIs), H2-receptor antagonists (H2RAs) | Vitamin $B_{12}$<br>Iron | Gastric acid suppression, leading to food-bound vitamin $B_{12}$ malabsorption and iron malabsorption<br>Vitamin $B_{12}$ deficiency, iron deficiency |
| Metformin | Vitamin $B_{12}$<br>Vitamin $B_6$ | Vitamin $B_{12}$ deficiency<br>Vitamin $B_6$ deficiency |
| Anticonvulsant and sulfasalazine | Folate<br>Vitamin $B_{12}$<br>Vitamin $B_6$ | Negative impact on nutrient status |

Abbreviations:

- MAOIs: monoamine oxidase inhibitors
- NSAIDS: non-steroidal anti-inflammatory drugs

*Source:* Report of the Scientific Committee of the Food Safety Authority of Ireland. Scientific recommendations for food-based dietary guidelines for older adults in Ireland.

Published by: Food Safety Authority of Ireland, The Exchange, George's Dock, IFSC, Dublin 1, D01 P2V6. file:///Users/lesleyday/Downloads/Scientific%20recommendations%20for%20food-based%20dietary%20guidelines%20for%20older%20adults%20in%20Ireland%20(6).pdf (accessed 11August 2022)

lower energy needs but, for some older adults, eating less may make it more challenging to meet nutrient requirements. Ensuring sufficient consumption of nutrient-dense foods and a diet of adequate quality in older age is key to meeting nutrient needs. As undernutrition and malnutrition are common in older populations and associated with poorer health outcomes, including longer duration of hospital stays and greater mortality, routine screening of nutritional status in older populations is essential for early diagnosis. Recognized risk factors for poor nutrition in older age include poorer appetite and physical health, as well as negative effects of social isolation and eating alone. Interventions to support older adults need to consider the complex array of influences on diet. Furthermore, the ageing global population points to the urgent need for better longitudinal data to understand trajectories of nutrition in older age that will inform the design of future preventive measures. Strategies to promote nutritional health in older populations also need to consider opportunities to intervene early in the life course, ahead of changes in physiology and loss of function, as well as the feasibility and acceptability of participatory approaches to working with older adults in their co-design.

## Further reading

1. **Alharbi, T.A., Paudel, S., Gasevic, D., Ryan, J., Freak-Poli, R., and Owen, A.J.** (2021) The association of weight change and all-cause mortality in older adults: a systematic review and meta-analysis. *Age Ageing*, **50**, 697–704.

2. **Amarya, S., Singh, K., and Sabharwal, M.** (2015) Changes during aging and their association with malnutrition. *Journal of Clinical Gerontology and Geriatrics*, **6**(3), 78–84. https://doi.org/10.1016/j.jcgg.2015.05.003

3. **Black, A.E.** (1996) Physical activity levels from a meta-analysis of doubly labelled water studies for validating energy intake by dietary assessment. *Nutr Rev*, **54**, 170–4.

4. **Colunga Biancatelli, R.M.L., Berrill, M., and Marik, P.E.** (2020) The antiviral properties of vitamin C. *Expert Review of Anti-infective Therapy*, **18**, 99–101.

5. **Cruz-Jentoft, A.J., Bahat, G., Bauer, J., et al.** (2019) Writing Group for the European Working Group on Sarcopenia in Older People 2 (EWGSOP2), and the Extended Group for EWGSOP2. Sarcopenia: revised European consensus on definition and diagnosis. *Age Ageing*, **48**, 16–31.

6. **De Groot, C.P.G.M. and Van Staveren, W.A.** (2002) Undernutrition in the European SENECA studies. *Clinics in Geriatric Medicine*, **18**, 699–708.

7. **Deutz, N.E., Bauer, J.M., Barazzoni, R., et al.** (2014) Protein intake and exercise for optimal muscle function with aging: recommendations from the ESPEN Expert Group. *Clin Nutr*, **33**, 929–36. doi: 10.1016/j.clnu.2014.04.007

8. **Dominguez Castro, P., Reynolds, C.M., Kennelly, S., et al.** (2021) An investigation of community-dwelling older adults' opinions about their nutritional needs and risk of malnutrition; a scoping review. *Clin Nutr*, **40**, 2936–45.

9. **European Food Safety Authority Panel on Dietetic Products, Nutrition and Allergies** (2010) Scientific opinion on dietary reference values for water. *EFSA Journal*, **8**, 48.

10. **Evans, W.J. and Meredith, E.N.** (1989) Exercise and nutrition in the elderly. In Munro, H.N. and Danford, D.E. (eds) *Nutrition, aging and the elderly*. New York: Plenum Press.

11. **Food Safety Authority of Ireland** (2021) Report of the Scientific Committee of the Food Safety Authority of Ireland. Scientific recommendations for food-based dietary guidelines for older adults in Ireland. Dublin: Food Safety Authority of Ireland.

12. **Geraghty, A.A., Browne, S., Reynolds, C.M., et al.** (2021) Malnutrition: a misunderstood diagnosis by primary care health care professionals and community-dwelling older adults in Ireland. *Journal of the Academy of Nutrition and Dietetics*, **12**, 2443–53.

13. **Giezenaar, C., Chapman, I., Luscombe-Marsh, N., Feinle-Bisset, C., Horowitz, M., and Soenen, S.** (2016) Ageing is associated with decreases in appetite and energy intake—a meta-analysis in healthy adults. *Nutrients*, **8**(1), 28.

14. **Guigoz, Y.** (2006) Mini-nutritional assessment (MNA): Review of the literature—what does it tell us? *Journal of Nutrition Health and Aging*, **10**, 466–87.

15. **Leij-Halfwerk, S., Verwijs, M.H., van Houdt, S., et al.** (2019) Prevalence of protein-energy malnutrition risk in European older adults in community, residential and hospital settings, according to 22 malnutrition screening tools validated for use in adults ≥65 years: a systematic review and meta-analysis. *Maturitas*, **126**, 80–89.

16. **Luong, R., Ribeiro, R.V., Cunningham, J., Chen, S., and Hirani, V.** (2021) The short- and long-term effects of dietary patterns on cardiometabolic health in adults aged 65 years or older: a systematic review. *Nutr Rev*, doi: 10.1093/nutrit/nuab032

17. **Manini, T.M.** (2010) Energy expenditure and aging. *Ageing Res Rev*, **9**, 1–11.

18. **Ni Lochlainn, M., Cox, N.J., Wilson, T., et al.** (2021) Nutrition and frailty: opportunities for prevention and treatment. *Nutrients*, **13**, 2349.

19. **Power, L., Mullally, D., Gibney, E.R., et al. on behalf of the MaNuEL Consortium** (2018) A review of the validity of malnutrition screening tools used in older adults in community and healthcare settings—A MaNuEL study. *Clin Nutr ESPEN*, **24**, 1–13.

20. **Reynolds, C.M., Dominguez Castro, P., Geraghty, A.A., Bardon, L., and Corish, C.A.** (2021) 'It takes a village': Patient perspectives on the management of malnutrition in the community. *European Journal of Public Health*, **31**, 1284–90.

21. **Robinson, S.M.** (2018) Improving nutrition to support healthy ageing: what are the opportunities for intervention? *Proc Nutr Soc*, **77**, 257–64.

22. **Robinson, S.M., Reginster, J.Y., Rizzoli, R., et al. on behalf of the ESCEO working group** (2018) Does nutrition play a role in the prevention and management of sarcopenia? *Clin Nutr*, **37**(4), 1121–32.

23. **Scientific Advisory Committee on Nutrition** (2018) SACN statement on diet, cognitive impairment and dementias. Available at: https://assets.publishing.service.gov.uk/government/uploads/system/uploads/attachment_data/file/685153/SACN_Statement_on_Diet__Cognitive_Impairment_and_Dementias.pdf (accessed 30 March 2022).

24. **Scientific Advisory Committee on Nutrition** (2019) SACN report on saturated fats and health. Available at: https://www.gov.uk/government/publications/saturated-fats-and-health-sacn-report (accessed 30 March 2022).

25. **Scientific Advisory Committee on Nutrition** (2021) SACN statement on nutrition and older adults living in the community. Available at: https://assets.publishing.service.gov.uk/government/uploads/system/uploads/attachment_data/file/953911/SACN_Nutrition_and_older_adults.pdf (accessed 30 March 2022).

26. **Shannon, O.M., Ashor, A.W., Scialo, F., et al.** (2021) Mediterranean diet and the hallmarks of ageing. *Eur J Clin Nutr*, **75**, 1176–92.

27. **Strasser, B., Wolters, M., Weyh, C., Krüger, K., and Ticinesi, A.** (2021) The effects of lifestyle and diet on gut microbiota composition, inflammation and muscle performance in our aging society. *Nutrients*, **13**(6), 2045. doi: 10.3390/nu13062045

28. **Tyrovolas, S. and Panagiotakos, D.B.** (2010) The role of Mediterranean type of diet on the development of cancer and cardiovascular disease, in the elderly: a systematic review. *Maturitas*, **65**(2), 122–30. doi: 10.1016/j.maturitas.2009.07.003.

29. **United Nations** (2017) *World population ageing 2017—Highlights* (ST/ESA/SER.A/397). Department of Economic and Social Affairs, Population Division.

30. **Volkert, D., Beck, A.M., Cederholm, T., et al.** (2022) ESPEN practical guideline: Clinical nutrition and hydration in geriatrics. *Clin Nutr*, **41**(4), 958–89.

31. **White, M.C., Holman, D.M., Goodman, R.A., and Richardson, L.C.** (2019) Cancer risk among older adults: time for cancer prevention to go silver. *Gerontologist*, **59**, S1–S6.

32. **World Health Organization** (2015) *World report on ageing and health*. Geneva: World Health Organization.

# Part 8

# Nutritional Assessment

# Food Analysis, Food Composition Tables, and Databases

Philippa Lyons-Wall



Food composition tables or databases are designed to describe the composition of the foods in the country of origin. They contain data on foods eaten on a regular basis by the population and generally include some less widely consumed foods that are unique to the culture or eaten on special occasions. The values for nutrient and non-nutrient constituents are based on chemical analyses, sometimes performed by the compiler of the tables (or databases) or in an associated laboratory. Some food composition values may be 'borrowed' from a major overseas food table or represent estimated averages from reports in the literature. Alternatively, they may be imputed from analytical values existing for a similar food or derived from the ingredients of a mixed food. The origins of the nutrient composition values should be specified, although in practice this is not always done. The UK food composition tables (*McCance and Widdowson's: the composition of foods*) and the US data (USDA FoodData Central) are widely used reference sources; both are updated regularly and available as nutrient databases at the UK Government internet site (Composition of foods integrated dataset, CoFID) or the US Department of Agriculture internet site (FoodData Central), respectively.

When compiling food composition data, there are two important considerations. First, food items must be relevant; sampling of an individual food should be representative of the types commonly consumed by the population on a year-round, nationwide basis and should be pertinent to the current food supply. Second, the food composition data must be of high quality; analyses of the foods should be conducted in a rigorous, scientific environment so that values are precise and accurate. Well-established food composition tables have evolved over many years and often combine old and new analytical methods from a variety of different sources. Clear and detailed documentation of sampling and analytical procedures at all stages is as critical as the choice of the analytical procedure itself, so that compilers of tables, faced with the challenge of inevitable changes in the food supply, can continue to evaluate the relevance of the item and quality of the data.

## 36.1 Sampling

How does one sample foods that are truly representative of a particular food item? Does the analyst just go out to the corner shop nearest the laboratory and buy some food or try to include the varieties of that food across the nation? Foods are ultimately based on parts of plants or animals that vary naturally according to many factors. For example, varieties of sweet potato differ widely in β-carotene content according to whether the flesh is orange, yellow, or white in colour. Seasonal variation can markedly influence water and vitamin content, and fruits and vegetables tend to increase their concentration of sugars as they ripen, a process that is highly temperature dependent. Fat depots in animal foods are also extremely variable according to the degree of exercise, type of feed, and age of the animal. Guidelines for sampling protocols that take these variations into account are detailed by Greenfield and Southgate (2003). In general, the greater the natural variation in a particular food, the larger the number of samples required. National food production figures may also indicate the types of foods most widely consumed and therefore most representative of the population.

When the food arrives in the laboratory for analysis, it must first be unambiguously identified with both scientific and local names. Full descriptions are required for the part of the animal or plant used, and its stage of maturity, size, shape, and form (see Table 36.1). Any cooking or processing methods

**Table 36.1 INFOODS guidelines for describing foods**

| **Name and identification** | **Name of food in national language of the country** |
|---|---|
| | Name in local language or dialect |
| | Nearest equivalent name in English, French, or Spanish |
| | Country or area in which sample of food was obtained |
| | Food group and code in database used in the country |
| | Food group and code for food in regional nutrient database |
| | Codex Alimentarius or INFOODS food indexing group |
| **Description** | Food source (common and scientific name) |
| | Variety, breed, strain |
| | Part of plant or animal |
| | Manufacturer's name and address |
| | Other ingredients (including additives) |
| | Food processing and/or preparation |
| | Preservation method |
| | Degree of cooking |
| | Agricultural production conditions |
| | Maturity or ripeness |
| | Storage conditions |
| | Grade |
| | Container and food contact surface |
| | Physical state, shape, or form |
| | Colour |

*Adapted from:* Truswell, A.S., Bateson, D.J., Madafiglio, K.C., Pennington, J.A.T., Rand, W.M., and Klensin, J.C. (1991) INFOODS guidelines for describing foods: a systematic approach to describing foods to facilitate international exchange of food composition data. *J Food Comp Anal*, **4**, 18–38.

used in preparing the item must be documented and the edible portion must be carefully separated from inedible refuse. Analyses may then proceed in one of two directions. Individual samples of the same food from different locations can be analysed separately in order to provide information on the variation between samples, as well as their average nutrient content. This approach, however, may be a luxury that many laboratories cannot afford. Alternatively, a composite sample can be prepared by pooling several individual subsamples of a food from many locations to give a single sample for analyses. Often, a weighting scheme is used to ensure that those varieties and/or locations where the food item is consumed more frequently are proportionately represented in the final composite. Whether derived from an individual or composite sample, the edible portion must be homogenized or ground thoroughly to ensure that the aliquot taken for analysis is representative of the original sample. For trace minerals, it is also important that the sample is not exposed to adventitious sources of contamination during the collection, homogenization, sample preparation, and analytical stages. Similarly, care must be taken that vitamins and other susceptible organic components are not degraded by air or light, for example.

## 36.2 Analysis

The ultimate aim of food tables or databases is to provide nutrient information on food components that are of nutritional importance to the health of the population, over their lifetime and in different disease states (see Table 36.2). Some of the more common methods used to analyse food components are described below. It is important to document the accuracy and precision of all analytical methods and to use reference materials of a similar matrix to the food sample and certified for the nutrient of interest. Reference materials can be obtained from the US National Institute for Standards and Technology, Gaithersburg, Maryland, and the International Atomic Energy Agency, Vienna.

### 36.2.1 Moisture

Moisture (water) is the first component to be analysed in a food, and it is probably the single most important piece of food composition data. Underestimation of water content will lead to an overestimation of other components that are subsequently determined in the dried food. Water can be gained or lost during the cooking process, with changes in the apparent content of an array of nutrients, as well as energy. Water content is an important preliminary consideration when comparing the nutrient content of similar items in tables from different countries, as may happen if a specific food

**Table 36.2 Food components of nutritional importance**

| |
|---|
| **Basic components**[a] |
| Moisture |
| Energy |
| Protein, fat, and carbohydrate |
| Up to 13 vitamins, and 10 or more minerals or trace elements |
| **More detailed profiles**[b] |
| Fatty acid profile (up to 37 fatty acids) |
| Amino acid profile (around 18 amino acids) |
| Carbohydrate components (sugars, starches) |
| Dietary fibre and components (soluble, insoluble) |
| Alcohol |
| **Optional components**[c] |
| Cholesterol |
| Vitamin A-inactive carotenoids (lycopene, lutein, zeaxanthin, and others) |
| Organic acids (malic, citric, lactic, formic, oxalic, and salicylic acids) |
| Biologically active components (e.g. flavonoids and isoflavonoids) |
| Glycaemic index |
| Caffeine |

[a]Essential for growth and maintenance of body tissues and always listed in food tables.
[b]Useful for research into diet and disease risk and usually included in well-established tables.
[c]Not essential nutrients but may influence nutritional status indirectly by exerting physiological effects. Not routinely listed in food tables but may be cited in appendix sections.

is not available in the local database. Furthermore, a high moisture content, typical of many fresh fruits and vegetables, indicates a low energy value, while the reverse is true for items of low moisture content.

The moisture content in foods is determined by simply evaporating off the water and calculating the difference in weight at the beginning and end. Different methods are used. Water may be driven off in an oven at temperatures around its boiling point, provided the sample itself does not decompose or oxidize, or contain other volatiles that would contribute to the weight loss. Alternatively, evaporation can be achieved at lower temperatures by vacuum drying under reduced pressure, or by freeze-drying. In practice, a wide range of methods varying in temperature, time interval, and sample preparation have evolved to optimize this apparently simple process.

### 36.2.2 Protein and amino acids

Protein in foods is determined indirectly by measuring the content of amino nitrogen, an essential constituent of the amino acid units that combine to form proteins. For the amino nitrogen assay, the food sample is digested in hot concentrated sulphuric acid to convert the nitrogen into ammonium ion, which is then quantified by either distillation and titration in the classic *Kjeldahl method* or by *spectrophotometric analysis*. The protein content in the original food is then estimated by multiplying the total nitrogen value by specific conversion factors for different foods, as shown in Table 36.3 (based on their amino acid composition). Alternatively, the general conversion factor of 6.25 is used because nitrogen is assumed to represent about 16% of the protein content.

From a nutritional viewpoint, the importance of a particular dietary protein lies in its ability to sustain growth or replenish tissues, functions that depend more on the quality of the protein (or pattern of content of indispensable amino acids) than the total amount (see Chapter 4). Gelatin, for example, is a food that comprises over 80% protein, one of the highest values listed in food composition tables, yet as the sole protein source it cannot sustain life, no matter how much is eaten, because it is deficient in the indispensable amino acid tryptophan. *Amino acids* are measured by hydrolysing the protein with strong acid or alkali to break down the peptide bonds, followed by separation and measurement of the free amino acids using ion-exchange chromatography. Food tables may list the amino acid composition of major foods in addition to the total amount of protein, either in the main tables or in an appendix. The profile of indispensable amino acids can then be compared with that of a reference protein, such as hen's egg (a protein known to be utilized very efficiently) for adults, or human milk for babies, to develop a measure of the protein quality, referred to as the amino acid score or chemical score. Table 36.4 shows the indispensable amino acids in two common food proteins compared with those in hen's egg.

### 36.2.3 Fat

A characteristic property of fats is their solubility in organic solvents such as $n$-hexane, petroleum ether, or chloroform. Estimation of fat in foods involves extraction with one of these solvents followed by evaporation of the solvent and weighing of the final fat residue.

**Table 36.3 Sample calculations of protein content from nitrogen**

| Item (100 g) | Total nitrogen (g) | | Conversion factor[a] | | Protein (g) |
|---|---|---|---|---|---|
| Rump steak | 3.02 | × | 6.25 | = | 18.9 |
| White rice | 1.23 | × | 5.95 | = | 7.3 |

[a]Conversion factors for converting nitrogen in foods to protein: milk and milk products, 6.38; eggs, meat, fish, 6.25; rice, 5.95; barley, oats, rye, whole wheat, 5.83; soybeans, 5.71; peanuts, brazil nuts, 5.46; almonds, 5.18.

**Table 36.4 Content of indispensable amino acids in food proteins (mg amino acid per g of protein)**

| | Hen's egg | Cheese | Wheat flour |
|---|---|---|---|
| Isoleucine | 54 | 67 | 42 |
| Leucine | 86 | 98 | 71 |
| Lysine | 70 | 74 | 20[a] |
| Methionine + cysteine | 57 | 32 | 31 |
| Phenylalanine + tyrosine | 93 | 102 | 79 |
| Threonine | 47 | 37 | 28 |
| Tryptophan | 17 | 14 | 11 |
| Valine | 66 | 72 | 42 |

The nutritive value of a dietary protein can be estimated by comparing its pattern of indispensable amino acids with those in a reference protein such as hen's egg. The *amino acid score* is the content of the most limiting (inadequate) amino acid, as a percentage of the reference value for this amino acid.

[a]For wheat flour, the most limiting amino acid is lysine and the *amino acid score* is 20/70 or 28%. In real life, however, wheat flour (as bread) is frequently eaten with another protein (e.g. cheese) that has even more lysine than egg, so the *amino acid score* of whole meals is usually much higher than for individual foods. The limiting amino acid in an equal mixture of wheat and cheese protein is methionine + cysteine and the *amino acid score* is 55%.

The accuracy of this estimation depends on the type of fat as well as the mix of other components in the food. The traditional *Soxhlet method* tended to underestimate fat that was bound to other food components such as protein. To overcome this problem, the sample is now pre-digested in concentrated acid or alcoholic ammonia to release the bound portion before extraction. The fat residue may be further analysed in various ways. Thin-layer chromatography can separate the fat into lipid classes, including triglycerides, phospholipids, and sterols. If lipids are hydrolysed to liberate their fatty acids, these can be separated by gas–liquid chromatography. Fatty acid profiles of major foods are provided in several comprehensive food tables such as The New Zealand Food Composition Tables and the US FoodData Central. Total saturated fatty acids (including branched-chain fatty acids), monounsaturated (*cis* and *trans* together), and polyunsaturated fatty acids are given in a supplement to the UK's *McCance and Widdowson's: the composition of foods*.

The presentation of the fat content of foods differs among food tables. In addition to providing the total fat content, compilers may sum the individual fatty acids into classes: polyunsaturated, monounsaturated, and saturated, and list each class separately, or they may present these as a ratio of polyunsaturated and monounsaturated fatty acids to that of saturated fatty acids, termed the PMS ratio. Alternatively, they may list each fatty acid according to its chain length and degree of saturation. The PMS ratio provides a crude estimate of overall atherogenic risk; the higher the proportion of unsaturated fatty acids, the more favourable the ratio. For research purposes, however, the content of individual fatty acids is more useful because each may exert independent effects. (Chapters 3 and 19 discuss further the role of individual fatty acids in health and disease.)

### 36.2.4 Carbohydrates

Food composition tables vary widely in the methods used to measure carbohydrate. One approach is to estimate 'by difference', which defines carbohydrate as the difference between 100 and the sum of the weights (as a percentage of the total food weight) for protein, fat, water, ash, and alcohol (where present). However, inaccuracies arise using this approach because of the summation of the errors in estimating these four constituents. In addition, such values are of limited use because they do not distinguish between different carbohydrates, especially those that are available to the body (i.e. digested, absorbed, and utilized) and those that are unavailable. More accurate methods quantify the available and unavailable carbohydrate by direct measurements.

Available carbohydrate includes sugars (monosaccharides and disaccharides) and starches and dextrins (which are polysaccharides). Individual sugars can be extracted with aqueous alcohol and measured by high-performance liquid chromatography or using specific enzymatic colorimetric tests. Starches and dextrins, which are glucose polymers, are measured in the same way after an initial hydrolysis step to liberate free glucose. Food tables that report carbohydrate by direct analysis may list the monosaccharide equivalents because this is the form in which the carbohydrate is estimated. To obtain the actual values for disaccharides and starch (polysaccharide) in the food, these values should be divided by 1.05 and 1.10, respectively. A published database is also available for the glycaemic index (GI), which ranks carbohydrate-containing foods according to the degree of rise in blood glucose immediately after the food is consumed, compared with that of a standard food—either glucose or white bread (Atkinson et al., 2008). The GI data are derived from studies in which human subjects are fed a range of carbohydrate-containing foods and, therefore, provide a qualitative biological rather than quantitative chemical measure of carbohydrate intake. A searchable international GI database is available at: http://www.glycemicindex.com/ (accessed 28 May 2023).

Unavailable carbohydrate or dietary fibre is the mixture of plant components that are resistant to digestive enzymes in the human small bowel. With advances in food technology, the term has been expanded to include chemically synthesized and also extracted fibre components; these components can be added for functional purposes to non-plant foods, such as dairy products, but are not usually listed as a separate nutrient in food composition databases. The chemical diversity of fibre makes it a very challenging and elusive component to analyse in the laboratory. Accordingly, a number of methods have evolved. All begin with the defatted, dried food sample, but each method measures a different chemical fraction. The *Englyst method* is the most sensitive and perhaps the most useful from a nutritional perspective because it can distinguish between soluble and insoluble fibres, both of which have physiologically distinct effects in the body in relation to chronic diseases such as diabetes, heart disease, and cancer. In the Englyst method, starch is initially removed by digestion with strong amylases and then the constituent sugars of dietary fibre are measured directly after acid hydrolysis to produce the free sugars. This yields estimates of both soluble (pectin, gums, mucilages, and hemicelluloses) and insoluble (cellulose and other hemicelluloses) fibre components, collectively called non-starch polysaccharides (NSPs). Lignin escapes detection because it is not a carbohydrate, but a polymeric phenolic compound. The older *Southgate method* gives a higher value for dietary fibre content as it measures lignin, as well as NSPs.

Other methods are less precise because they measure fibre 'by difference', but they involve less analytical work and so are more economical. In the widely used procedure of the *Association of Official Analytical Chemists* (AOAC), developed by Prosky et al. (1984), starch is first removed by enzymatic hydrolysis and the undigested residue is weighed, analysed for nitrogen, and then ashed. Protein and ash contents are then subtracted from the residue weight. *Van Soest's neutral detergent fibre* method measures only the insoluble cellulose and lignin, and not the soluble fibre; hence, this method underestimates total dietary fibre. The *crude fibre* method is the least accurate, involving rigorous treatment with boiling acid and alkali, which removes much of the dietary fibre itself.

A further complication in the analysis of dietary fibre is the recognition that a variable but small proportion of the dietary starch found in beans, whole grains, potatoes (especially if eaten cold), or unripe bananas is not completely digested and is unavailable to the body for absorption. This starch, termed *'resistant starch'*, escapes digestion because it is physically inaccessible to the enzymes and, instead, it is probably fermented in the colon, thus behaving like soluble dietary fibre. Current values for dietary fibre in food tables do not include separate values for resistant starch, although the AOAC and Southgate methods include some of the resistant starch in the fibre value.

Different analytical methods can result in several-fold variations in the estimate of fibre for the same item, as shown in Table 36.5. Most food tables, however, are internally consistent in their choice of analytical method(s) and these should be stated clearly in the introductory section or in the main tables

Table 36.5 **Dietary fibre content of dried red kidney beans estimated by five different methods**

| Method | Fibre content (g/100 g) | Fibre components |
|---|---|---|
| Englyst[a] | 15.7 | Soluble + insoluble fibre (not lignin or resistant starch) |
| Southgate[a] | 23.4 | Soluble + insoluble fibre, lignin |
| AOAC (Prosky and Asp)[b] | 21.5 | Soluble + insoluble fibre, lignin |
| Neutral detergent fibre (Van Soest)[c] | 10.4 | Insoluble fibre only |
| Crude fibre[c] | 6.2 | Part of the insoluble fibre |

Red kidney beans contain both soluble and insoluble dietary fibre. AOAC, Southgate, and Englyst methods measure total dietary fibre, but Englyst fibre is lower because it does not measure lignin or resistant starch. Neutral detergent fibre is lower because it does not measure soluble fibre. AOAC, Association of Official Analytical Chemists.

*Adapted from:*

[a]Finglas et al. (2015) *McCance and Widdowson's: the composition of foods*, 7th summary edn. Cambridge: Royal Society of Chemistry.

[b]Food Standards Australia New Zealand (2019) *Australian Food Composition Database—Release 1.0.* Available at: https://www.foodstandards.gov.au/science/monitoringnutrients/afcd/pages/default.aspx (accessed 31 July 2021).

[c]USDA (1986) *Composition of foods: legumes and legume products, Handbook No. 8–16.* Washington, DC: US Department of Agriculture. Available at: https://naldc.nal.usda.gov/download/CAT87869981/PDF (accessed 28 May 2023).

alongside the nutrient values. Special care should be taken when comparing carbohydrate values from different food composition tables. For reasons outlined above, values from those tables in which total carbohydrate is analysed by difference (i.e. including dietary fibre) are not compatible with others in which carbohydrate constituents are analysed directly.

## 36.2.5 Energy

The total energy of a food is measured by *bomb calorimetry,* in which a sample of the food is burned with oxygen in a sealed chamber until completely oxidized. The heat released corresponds to the chemical or gross energy of the food. Food energy is reported in kilojoules (kJ) or kilocalories (kcal), where 1 kJ = 0.24 kcal or 1 kcal = 4.187 kJ. When a food is eaten, however, the energy-yielding components—protein, fat, carbohydrate, and (where present) alcohol—are oxidized by enzymatic processes within the body to provide energy, but not with 100% efficiency. Some energy is lost into the faeces, as not all food components are fully absorbed from the digestive tract. Further energy is lost into urine, since dietary protein, unlike carbohydrate and fat, is not completely oxidized by the body and its excretory product, urea, still retains some of the chemical energy from the original protein. The eminent American physiologist Atwater measured the energy losses into faeces and urine by a series of meticulous experiments in humans fed a mixture of foods. In his experiments, 92% of protein, 95% of fat, and 97% of carbohydrate were absorbed by the body, but for every gram of protein ingested, about one-quarter of its gross energy was lost into the urine.

Atwater's experiments, conducted at the turn of the nineteenth century, represent landmark studies from which the energy content of foods in today's food composition tables are derived. Atwater developed a system of four conversion factors, which represent the energy available in: (1) protein (17 kJ/g, 4 kcal/g); (2) fat (37 kJ/g, 9 kcal/g); (3) carbohydrate (16 kJ/g, 4 kcal/g); and (4) alcohol (29 kJ/g, 7 kcal/g), taking into account the estimated energy losses into faeces and urine. The value is slightly higher for starch (due to lower hydration) than that given for all carbohydrates, and slightly lower for sugars. These factors provide values for food energy as it is utilized by the body (i.e. metabolizable energy). The energy content of each food is calculated by first multiplying the weight of each component by its Atwater conversion factor, and second, summing the energy from each component to give a total energy for the food,

**Table 36.6 Calculation of metabolizable energy content in cooked kidney beans using Atwater conversion factors**

| Component | Weight (g/100 g) | | Atwater factor (kcal/g) | | Energy content (kcal/100 g) |
|---|---|---|---|---|---|
| Moisture | 35.3 | – | – | = | – |
| Protein | 7.4 | × | 4 | = | 29.6 |
| Fat | 2.0 | × | 9 | = | 18.0 |
| Carbohydrate total (including fibre) | 54.2 | × | 4 | = | 216.8 |
| Dietary fibre | 2.1 | – | Not *usually* included | = | – |
| Ash | 1.1 | – | | = | – |

The energy content of individual food components is then summed to give a total of 264 kcal/100 g.

Reference: Wu Leung, W.T., Butrum, R.R., Chang, F.H., Rao, M.N., and Polacchi, W. (1972) Food composition table for use in East Asia. Rome: Food and Agricultural Organization.

as shown in Table 36.6. Note that the total weight of each nutrient is obtained by direct analysis of the food, as described previously.

The biggest variable is the carbohydrate value and whether this includes fibre, because Atwater factors are applied to the total carbohydrate content, irrespective of whether this is analysed directly or by difference. In practice, the value for energy is somewhere between the two extremes of 16 kcal/g for available carbohydrate and 0 kJ/g for unavailable carbohydrate. This is because a proportion of dietary fibre is fermented in the colon to short-chain fatty acids, which can be absorbed from the large bowel and oxidized for energy. Livesey (1991) has estimated that about half the dietary fibre can be utilized by the body in this way and proposed an average energy conversion factor of 8 kJ/g (2 kcal/g) for unavailable carbohydrates, about half the value for available carbohydrate. Estimates of the energy from fibre are included in some more recent databases, such as the Australian Food Composition Database (Food Standards Australia New Zealand, 2019).

The energy conversion factors used today vary somewhat from Atwater's original factors. The German, British, Australian, and New Zealand food composition tables, for example, apply the same four factors to all foods, whereas the US and East Asian tables use a range of slightly differing conversion factors, rather lower for components in plant foods than in animal foods, reflecting Atwater's initial observation that the energy from plants was less available. The conversion factors selected to calculate energy should be specified in the introduction or appendix section of all food composition tables and the reader is referred to these for more in-depth information.

## 36.2.6 Inorganic nutrients and vitamins

The range of inorganic nutrients in foods, including calcium, iron, magnesium, zinc, copper, manganese, potassium, and sodium, can be determined by *flame atomic absorption spectrophotometry* (AAS), a method whereby a solution of the ashed or acid-digested food sample is sprayed into the flame of an atomic absorption spectrophotometer and quantified by the degree of absorption at a specified wavelength. For selenium, direct AAS with a Zeeman background correction is required. Graphite-furnace AAS is used for analysing the ultratrace elements, such as chromium, nickel, and manganese. Alternatively, all the minerals can be measured in the one sample using inductively coupled plasma spectrophotometry or X-ray fluorescence. Minerals are a very stable component of foods and these procedures can be highly accurate provided any interfering substances such as plant pigments or organic constituents have first been removed. This is achieved by reducing the food sample to a dry ash by thorough heating in a muffle furnace or by breaking down and oxidizing the organic components by wet ashing with boiling concentrated acids. For trace element analysis, precautions must be used to avoid adventitious contamination, by the use of ultrapure acids, acid-washed glassware, plastic materials for sample preparation and analysis, and high-grade deionized water.

In contrast to the minerals, many of the vitamins in foods are not very stable. Riboflavin and vitamin A are sensitive to light; thiamin, folate, and vitamin C are sensitive to heat; and vitamin E to oxidation. Vitamins are analysed either by the traditional but more time-consuming microbiological methods or by newer, faster chemical techniques. *Microbiological assays* are conducted with a culture of organisms that have a specific growth requirement for the particular vitamin. The assumption is made, however, that the micro-organism reacts in the same way as the human organism. Such methods are available for a wide range of B vitamins, including thiamin, niacin, riboflavin, vitamins $B_6$ and $B_{12}$, folate, biotin, and pantothenic acid, and have the advantage of estimating the total biological potential of the vitamin. Alternative chemical methods, such as *high-performance liquid chromatography* (HPLC), can be used for most of the vitamins. They require an initial extraction step to remove other components in the food, but are useful for separating and quantifying different chemical forms of the vitamin. It should be borne in mind that many of the existing values for vitamins, as well as minerals, in food tables were obtained with older, less specific colorimetric methods that may now be obsolete.

Values in food composition tables represent the total content of each mineral or vitamin in the food and do not address the complex problem of bioavailability, defined as the proportion of a nutrient that is actually absorbed from the food and utilized. When a vitamin exists in two or more forms that are utilized differently in the body, some food composition tables tabulate each form separately. Vitamin A, for example, has two major components obtained from quite different food sources: preformed vitamin A (retinol), which is found in many animal products, and the provitamin A carotenoids derived from plants. The compiler may further attempt to calculate the overall potency of the vitamin in the body by summing the different forms, taking into account their relative biological activities. Some of the assumptions and calculations made in food tables regarding the different forms of niacin, folate, and vitamins A, C, D, and E are shown in Table 36.7.

### 36.2.7 Non-nutrient biologically active constituents

In addition to the nutrient components, such as protein, fat, carbohydrate, vitamins, and minerals, plant foods in particular contain an abundant array of chemical constituents or 'phytochemicals'. Examples of nutritional importance include the flavonoids, isoflavonoids, and non-vitamin A carotenoids. Although not classed as essential nutrients in terms of preventing a specific nutritional deficiency, these constituents are biologically active in the body and are thought to contribute to optimal health and longevity (see Chapter 15).

Phytochemicals occur in relatively small mg or μg quantities per 100 g food. While precise HPLC assays are available for quantification, it is often difficult to estimate levels in foods accurately. For example, the concentration of isoflavonoids in soybeans, a rich natural source, shows up to six-fold variation. This reflects genetic differences and also the fact that isoflavonoids, unlike more stable structural components, such as proteins, are part of the plant's natural response to stress. Insect infestations or climatic considerations, including low temperature and high soil moisture, can trigger dramatic increases in isoflavonoid content. This natural variability means that the isoflavonoid content in the same variety of soybean or soy product available in local retail outlets could vary several-fold between different batches. Despite certain limitations, the identification and quantification of biologically active substances in plants is an area of intense current research. Databases for isoflavonoids and a range of other special-interest constituents have been collated by the Nutrient Data Laboratory at the US Department of Agriculture, and are available at the USDA internet site: https://data.nal.usda.gov/dataset/usda-database-isoflavone-content-selected-foods-release-21-november-2015 (accessed 28 May 2023).

**Table 36.7 Presentation of different vitamin forms in food composition tables**

| Vitamin | Main forms | Unit of total vitamin activity |
|---|---|---|
| Niacin[a] | 1 Preformed in foods (nicotinic acid + nicotamide)<br>2 Derived from tryptophan | Niacin equivalents (NE) |
| Folate[b] | 1 Food folates<br>2 Folic acid enrichment | Dietary folate equivalents (DFE) |
| Vitamin A[c] | 1 Retinol<br>2 Provitamin A carotenoids | Retinol equivalents (RE) or retinol activity equivalents (RAE) |
| Vitamin C[d] | 1 Ascorbic acid<br>2 Dehydroascorbic acid | Vitamin C |
| Vitamin D[e] | 1 Vitamin D (cholecalciferol)<br>2 25-Hydroxyvitamin D | Total vitamin D activity |
| Vitamin E[f] | 1 Tocopherols (α, β, γ, δ)<br>2 Tocotrienols (α, β, γ, δ) | α-Tocopherol equivalents |

[a]Because approximately 1% of protein is tryptophan and 1/60th tryptophan is converted to niacin in the body: NE (mg) = preformed niacin (mg) + dietary protein (g) ÷ 6000 (as mg)

[b]Bioavailability of folate from food is about 50% from foods, 85% from fortified foods or as a supplement (consumed with food), or 100% as a supplement (on an empty stomach) (Institute of Medicine, 2006): 1 μg DFE = 1 μg food folate = 0.6 μg folic acid (taken with meals) = 0.5 μg folic acid (on empty stomach).

[c]Vitamin A can be expressed as RE, where β-carotene has one-sixth the activity of retinol and some other carotenoids have one-twelfth the activity of retinol. To acknowledge lower reported availability from vegetable sources, some databases (e.g. US FoodData Central) now express vitamin A as RAE, where β-carotene has only one-twelfth the activity of retinol (Institute of Medicine, 2006): RAE (μg) = retinol (μg) + β-carotene (μg)/12 + (α-carotene + β-cryptoxanthin (μg))/24.

[d]Both forms have equal activity and are summed to give total vitamin C.

[e]Vitamin D in food is measured as natural cholecalciferol (or ergocalciferol) and 25-hydroxyvitamin D (an active circulating form in animals and hence meats), which has about five times the activity of cholecalciferol (Finglas et al., 2015): total vitamin D activity = sum of cholecalciferol + 5 × 25-hydroxycholecalciferol (in meats).

[f]α-Tocopherol is the most abundant form, with over twice the activity of other tocopherols and tocotrienols. Activities of individual vitamin forms are cited in Finglas et al. (2015).

## 36.3 Compilation, limitations, and uses of food composition data

Compilation of food composition data in the form of either tables or online databases is a very large task. It requires meticulous inspection of a wide range of sources that use a variety of sampling and analytical procedures. Data analysed outside the compiler's laboratory must frequently be traced back to its source and any items without clear documentation discarded because there is no way to evaluate their quality. Values from different sources must be compared and statistical calculations made to provide a meaningful average for the nutrient content of a food. As food patterns within a population are constantly changing and evolving, data must also be scrutinized to determine their relevance in the current food supply.

Not only should data be accurate and relevant, but also the format must be clear so that the user may easily understand the data. Food items are listed alphabetically and usually grouped according to food groups with similar nutritional properties (e.g. vegetables, fruits, grains, meats, and dairy products) or by product use (e.g. snacks, desserts, and breakfast cereals). In cases where foods are collected or prepared with inedible matter, the percentage edible portion, sometimes expressed indirectly as percentage refuse, is also given.

However, irrespective of the proportion of edible matter or the accustomed serving size, nutrient values for items are always presented in terms of 100 g edible portions. Consequently, this does not include the core or stone in fruit, or the bones in meat and chicken, but it does include optional materials such as certain vegetable skins and trimmable meat fat, unless specified otherwise. Most food tables cite both scientific and local names for each item, and some specify the number of food items analysed, and whether a single or composite sample was used for analysis. Rarely do tables include the natural variation around the mean value, but rather provide a single mean representative value. The German (Souci et al., 2008) and the US (FoodData Central) tables are exceptions, citing the range (i.e. highest and lowest values known) in addition to the mean or median value, respectively. Possible limitations of food composition data include changes in the nutrient composition of foods over time, such as natural variation according to location or seasonal change, or availability of new products not included in the published database. Food composition data are also country-specific and designed to be representative of items available at a national level. Therefore, they are not an accurate assessment of individual nutrient intakes and may not be transferrable between countries. When using food composition data, it is important to understand and consider these limitations to ensure appropriate interpretation of the results as required, at the individual, national, or global level.

Ideally, food composition tables or databases should include analyses for all food components of nutritional relevance to the potential user, whether this be a dietitian or nutritionist prescribing advice to a client, a research worker investigating certain nutrients in relation to disease risk, or a food manufacturer seeking accurate nutrient information on their products for the purposes of marketing and food legislation. In practice, however, inclusion is determined more by the analytical resources and public health priorities of the country concerned. A wealth of analytical and descriptive information on food habits and customs already exists within different cultures. Yet, many of these data are not widely accessible outside the country, often because local names are culture-specific, making it challenging to identify the food. In this regard, the INFOODS guidelines (FAO/INFOODS, 2012) have been established to ensure that foods are named and described in a standardized manner with a view to facilitating interchange of food composition data at the international level.

## Further reading


1. **Atkinson, F.S., Foster-Powell, K., and Brand-Miller, J.C.** (2008) International table of glycaemic index and glycaemic load values. *Diabetes Care*, **31**, 2281–3. Available at: http://www.glycemicindex.com/
2. **FAO/INFOODS** (2012) *Guidelines for checking food composition data prior to the publication of a user table/database*, version 1.0. Rome: Food and Agricultural Organization. Available at: http://www.fao.org/infoods/infoods/standards-guidelines/en/
3. **Finglas, P.M., Roe, M.A., Pinchen, H.M., et al.** (2015) *McCance and Widdowson's the composition of foods*, 7th summary edn. Cambridge: Royal Society of Chemistry. Available at: https://www.gov.uk/government/publications/composition-of-foods-integrated-dataset-cofid
4. **Food Standards Australia New Zealand** (2019) *Australian Food Composition Database*, Release 1.0. Available at: https://www.foodstandards.gov.au/science/monitoringnutrients/afcd/pages/default.aspx
5. **Greenfield, H. and Southgate, D.A.T.** (2003) *Food composition data. Production, management and use*, 2nd edn. Rome: Food and Agricultural Organization.
6. **Institute of Medicine** (2006) *Dietary reference intakes: the essential guide to nutrient requirements*. Washington, DC: The National Academies Press. Available at: https://nap.nationalacademies.org/catalog/11537/dietary-reference-intakes-the-essential-guide-to-nutrient-requirements

7. **Livesey, G.** (1991) Calculating the energy values of foods: towards new empirical formulae based on diets with varied intakes of unavailable complex carbohydrates. *Eur J Clin Nutr*, **45**, 1–12.

8. **New Zealand Institute for Plant and Food Research Ltd** (2018) *The concise New Zealand food composition tables*, 13th edn. Sandringham, Auckland: New Zealand Institute for Plant and Food Research Ltd; Ministry of Health. Available at: https://www.foodcomposition.co.nz

9. **Prosky, L., Asp, N.G., Furda, I., De Vries, J.W., Schweizer, T.F., and Harland, B.F.** (1984) Determination of total dietary fibre in foods, food products and total diets: interlaboratory study. *J Assoc Off Anal Chem*, **67**, 1044–52.

10. **Souci, S.W., Fachmann, W., and Kraut, H.** (2008) *Food composition and nutrition tables*, 7th revision and completed edn. Stuttgart: Medpharm Scientific Publishers.

11. **Truswell, A.S., Bateson, D.J., Madafiglio, K.C., Pennington, J.A.T., Rand, W.M., and Klensin, J.C.** (1991) INFOODS guidelines for describing foods: a systematic approach to describing foods to facilitate international exchange of food composition data. *J Food Comp Anal*, **4**, 18–38.

12. **USDA** (1986) *Composition of foods: legumes and legume products*, Agriculture Handbook No. 8–16. Washington, DC: US Department of Agriculture. Available at: https://naldc.nal.usda.gov/download/CAT87869981/PDF

13. **Wu Leung, W.T., Butrum, R.R., Chang, F.H., Rao, M.N., and Polacchi, W.** (1972) *Food composition table for use in East Asia*. Rome: Food and Agricultural Organization. Available at: http://www.fao.org/infoods/infoods/tables-and-databases/asia/en/

## Useful websites

1. **McCance and Widdowson's Composition of Foods Integrated Dataset (CoFID)**
https://www.gov.uk/government/publications/composition-of-foods-integrated-dataset-cofid
2. **US Department of Agriculture, Agricultural Research Service. FoodData Central.**
https://fdc.nal.usda.gov
3. **International Network of Food Data Systems (INFOODS)**
http://www.fao.org/infoods/infoods/en/
4. **GI database**
http://www.glycemicindex.com/
5. **New Zealand Food Composition Database**
https://www.foodcomposition.co.nz
6. **Australian Food Composition Database**
https://www.foodstandards.gov.au/science/monitoringnutrients/afcd/pages/default.aspx
7. **International food composition table/database directory**
http://www.fao.org/infoods/infoods/tables-and-databases/en/
8. **FoodData Central, US Department of Agriculture, Agricultural Research Service**
https://fdc.nal.usda.gov/

# Dietary Assessment

Jim Mann and Silke Morrison



Dietary assessment is one of the specialized interests of nutritionists, used in surveillance of populations, nutritional epidemiology, clinical assessment, and experimental research. There are two basic approaches to estimating food intake: one principally concerned with determining the intake of populations, the other with assessing the diets of individuals. These approaches are not mutually exclusive, since methods for assessing dietary intake of individuals may be used for estimating intake of populations, as happens in most national nutrition surveys, which generally involve individual 24-hour recalls. This chapter describes the various methods used in the different contexts in which knowledge of food and nutrient intakes are required.

Methods used for assessing dietary intake in carefully controlled experimental studies will differ appreciably from those required in large, epidemiological studies. The chapter also considers the reliability of the various methods, considering the conflict between the need for accuracy to establish exactly where an individual lies within the overall distribution of foods and nutrients and the logistics of doing so when very large populations required for epidemiological studies are investigated. True estimates of food consumption can only be obtained by observing the activities of participants, or by developing some other independent way of assessing food intake. This has become possible with the advent of biological markers in biological specimens, such as blood, urine, or hair, that reflect intake sufficiently closely to act as objective indices of true intake. These are discussed in Chapter 39.

## 37.1 Population estimates

Population estimates are needed principally for surveillance, for example, to assess intakes of a particular food or nutrient in relation to reference nutrient intakes and to determine changes over time. Such information would also be used during emergencies when food is in short supply, to make recommendations concerning usual diet, and in considering the case for fortification of foods on a national basis. Population estimates of dietary intake have also been compared with disease rates in different countries or populations or within the same country over time to identify clues as to possible nutritional causes of the disease. Population estimates may be derived from food balance sheets, household food surveys, national nutrition surveys, and, potentially, supermarket records.

*Food balance sheets* are based on national statistics of food produced, imported, and exported with factors for wastage included. The Food and Agricultural Organization (FAO) of the United Nations

### Box 37.1 Food consumption at the national level

Food consumption at the national level is also called food moving into consumption, food disappearance data, and apparent food consumption. The food supply is calculated from estimates of domestic food production plus imports minus exports. Potential food diverted for farm animal feed, non-food industrial use, and wastage at wholesale level are subtracted. The total is divided by the estimated population each year. These statistics are useful for monitoring changes in commodities available for consumption and comparing countries' food habits. FAO food balance sheets are based on these statistics and available each year from some 181 countries. For many countries, these are the only regular measures of food available for consumption. However, these are macro figures. The calorie and nutrient numbers are around 25% (or more) above what individuals actually eat and drink, because there is wastage of food in homes and catering establishments, and food is fed to tourists and pets. They give no idea of distribution of food resources among regions, socioeconomic groups, or within the family.

(https://www.fao.org/home/en/, accessed 29 May 2023) collates these and they have been used extensively for monitoring changes in food available for consumption and for population comparisons, linking, for example, population estimates of fat and cardiovascular disease rates and population estimates of meat and fat, and bowel and breast cancer rates.

Associations identified in this way need to be examined further using experimental and epidemiological approaches to confirm or refute the likelihood of the relationships being causal. They have also been used for estimating and comparing risk of nutrient deficiency in the absence of information based on national food consumption survey data (Box 37.1).

*Household surveys* are records kept by the householder of all food available to the family over a specific period, and the total food entering the household is divided by the number of people living there. This approach has been used in the UK in the National Food Survey, which has been running continuously for over 75 years as a combined food consumption and expenditure survey over 2 weeks, using self-reported diaries, which include food eaten out, and are supported by till receipts of all food purchased (https://www.gov.uk/government/collections/family-food-statistics, accessed 29 May 2023). From this, regional comparisons and secular trends in consumption are available, for example the trend towards a lower saturated fat composition in the diet in the UK.

When using data derived from food balance sheets or household surveys, it is not possible to compare the intake of food or nutrients in different age and sex groups with differential trends in disease incidence or risk factors, nor for individual data to be assessed. This is because the findings relating to, for example, children, the elderly, or males and females cannot be separated out from the overall population average data. Potentially, computerized supermarket records of sales and commercial databases could be used to obtain regional information on nutritional consumption.

Typically, the 24-hour recall (see section 37.2 and Box 37.2) is used in *national nutrition surveys*. Al-

### Box 37.2 National nutrition surveys

National nutrition surveys estimate food intake of individuals and may involve clinical examination and blood tests to further assess nutritional status. Although based on the intakes of individuals, such surveys provide the average and range intakes of foods and nutrients of different age and sex subgroups relating to the population from which the individuals are drawn. Examples are NHANES in the USA and the National Diet and Nutrition Surveys in the UK. National nutrition surveys are all somewhat different. They identify nutrients for which intakes are lower than recommended on the one hand and frequency of overweight/obesity on the other. They also gather information about food usage as a basis for formulating and evaluating health policy and regulatory needs. The food intake method used has to be a 24-hour recall or food record because of the need to know the exact types of food people are eating.

though the information is derived from individuals, the aim is to describe the nutritional intake of the population. This requires the study of large numbers of people who are representative of the population of interest in order to be reasonably confident that the survey yields an unbiased estimate of the usual mean intake of nutrients of interest. National nutrition surveys often involve a repeat 24-hour recall on a subsample of participants.

## 37.2 Individual methods

Several methods are available for measuring the dietary intake of individuals. They generally consist of either the collation of observations from a number of separate days' investigations, as in records, checklists, and 24-hour recalls, or attempts to obtain average intake by asking about the usual frequency of food consumption, as in the diet history and food frequency questionnaires (FFQs). In all methods of dietary assessment, some estimate of the quantity of food consumed is required, and for the determination of nutrient or other food component intake, either an appropriate description that can be matched with an entry in the food tables or an aliquot for chemical analysis (Box 37.3). Each of the methods is described briefly below and further details regarding equipment, protocols, uses, limitations, and best practice guidelines are available in the references (http://www.nutritools.org, accessed 29 May 2023) and in the Further reading. Detailed examples of methods used in particular studies are shown on the websites given.

### 37.2.1 Food records

Food records or food diaries involve subjects being taught to describe and either weigh or estimate the amount of food immediately before eating and to record leftovers. Records are generally completed by the participant on sheets, booklets, or potentially online that, in the case of estimated records, may include photographs to facilitate estimation of portion size. Cups, spoons, rulers, and scales may be provided to aid accurate description. Verbal records, with descriptions of amounts recorded also been used, as have records incorporating bar codes from purchased foods. As this

**Box 37.3 Estimating individual food intake**

Basically, four types of method are used to estimate individual food intakes:

- *Food diary or record*: (Prospective) 'Please write down (and describe) everything you eat and drink (and estimate the amount) for the next 3 (4 or 7) days'. Amounts are usually recorded in household measures, but for more accuracy subjects can be provided with quick-reading scales to weigh food before it goes on the plate and weigh any leftovers.
- *24-hour recall*: (Retrospective) 'Tell me everything you had to eat and drink in the last 24 hours'. This is less subject to wishful thinking about what the subject feels they ought to have eaten. The weakness is that yesterday may have been an unusual day; 24-hour recalls can, however, be repeated (multiple recall) to capture individual variation.
- *Food frequency questionnaire*: (Most commonly retrospective) 'Do you eat meat/fish/bread/milk, etc., on average more than once a day, two or three times a week, once a week, once a month, etc.?' (usually filling in 100 to 150 lines on a questionnaire form) (Fig. 37.1).
- *Dietary history*: (Retrospective) 'What do you eat on a typical day and how does your food intake vary?' This requires a skilled and patient interviewer. Food models, cups, plates, and spoons are used to estimate portion sizes.

method is a record kept at the time of eating and does not involve participants attempting to remember if or how often a food has been eaten. It is generally regarded as providing the most reliable information regarding the usual dietary intake of individuals, provided sufficient days' observations are collected on each individual (see section 37.4.2). In the past, it has been used for the purpose of validating other methods of dietary assessment, but this approach is now recognized to underestimate the extent of measurement error (see section 37.4.6). Although data collection is easier, coding is more time-consuming and respondent burden is higher than with other methods. However, the approach has been used in multicentre cross-sectional comparisons of representative population samples in which instructions to participants have been standardized among different centres and other epidemiological settings such as the large prospective studies of diet and health, for example a study of 25,000 people in the European Prospective Investigation of Cancer in Norfolk (EPIC Norfolk) see https://www.epic-norfolk.org.uk/ accessed 29 May 2023. Weighed records have also been used in surveillance procedures; for example, records from representative population samples have been routinely obtained by the UK Government National Diet and Nutrition Survey (https://www.gov.uk/government/statistics/family-food-201920/family-food-201920, accessed 29 May 2023). See Further reading (Stewart et al., 2022).

### 37.2.2 Twenty-four-hour recalls

This method is also a report of daily habits, but interviewed or written information about the previous day's intake, the 24-hour recall, is obtained. The participant has to remember the actual foods consumed and give information on portion weights from memory. Some information may be forgotten and descriptions of portion size are more difficult to supply, though the interviewers will often use food models or photographs as memory aids and to assist in quantifying portion size. Although the 24-hour recall may consist of a very simple written list completed by the participant, most 24-hour recalls have several stages or multiple passes, in which data are checked and verified by a skilled interviewer, and each recall may take about 40 minutes. The respondent burden for a single 24-hour recall is less than for several days of food records and the method is typically used for determining average usual intakes of a large population or group, for example in national nutrition surveys. For details of methods used in the USA NHANES survey, see https://wwwn.cdc.gov/nchs/nhanes/continuousnhanes/questionnaires.aspx?BeginYear=2019 (accessed 12 February 2022). National nutrition surveys are typically based on a single 24-hour recall, sometimes with a repeat recall being undertaken on a subsample. However, when attempting to assess the diet of individuals using this method, multiple 24-hour recalls may be needed, with a subsequent increase in respondent burden, depending on the level of precision required and nutrient to be studied (see section 37.4). Furthermore, the researcher may apply a statistical method enabling an estimate of usual dietary intake to be made, including episodically consumed foods based on two or more 24-hour recalls (Harttig et al., 2011). The website https://msm.dife.de/ (accessed 31 May 2022) provides free access to a statistical program package that allows calculation of usual intake by combining short- and long-term measurements (multiple sources method).

Increasingly, there are efforts to develop technology-assisted 24-hour recall programmes. Many are self-administered web-based platforms, which can be run on a laptop, but may require participant training if the tool is not intuitive or the participants lack computer skills. Examples of self-administered web-based 24-hour recall technologies are provided in further reading (Bradley et al., 2016; Simpson et al., 2017) and currently available and validated tools are available online at https://www.nutritools.org/tools or https://www.fao.org/infoods/infoods/software-tools/en/ (accessed 29 May 2023). Data for these applications are collected online and most utilize a digital display of graduated food photographs for portion size estimation and are generally linked to an in-country food composition database.

The software used in the EPIC study (EPIC-SOFT) developed by the International Agency for Research on Cancer has been adapted for use in other countries and renamed GloboDiet. It is interviewer-administered, runs on a laptop, and is offline. It has been developed to collect standardized national-level dietary data, for both research and surveillance purposes, in 19 European countries, and successfully adapted for other mid-to-upper income country settings (e.g. Korea, Latin America, and Africa) (Aglago et al., 2017). Like many other technology-assisted 24-hour recall programmes, GloboDiet allows the respondent to report the amount of food consumed using a variety of quantification methods such as household measures, standard units, and direct reporting of known gram or volume amounts consumed, as well as graduated photographs.

### 37.2.3 Food frequency questionnaires

FFQs are designed to assess long-term habits, over weeks, months, or years, and may comprise either a relatively small list of foods that are the major sources of a limited group of nutrients of interest, or a longer list if a full dietary assessment is required. Participants can be supported by interviewers or complete the FFQ themselves, generally after receiving the FFQ in the post or online with detailed instructions regarding completion of the questionnaire. The length of the list of foods can be short ('screeners'), but generally does not exceed 150 items. Various methods to assess portion sizes may be used, for example fitting age and sex-specific average portion weights derived from other data to the respondents' chosen food and frequency selections. To assess the frequency of food consumption, accompanying the food list is a multiple-response grid in which respondents attempt to estimate how often selected foods are eaten. Up to ten categories ranging from never or once a month or less, to six times per day is a usual format. Fig. 37.1 shows an example of an FFQ taken from EPIC Norfolk. Because responses are standardized, FFQs can be analysed in comparatively short periods of time so that large numbers of individuals can be investigated relatively inexpensively. The FFQ has been widely used in large epidemiological cohort studies to classify participants according to quantiles of intake of nutrients and disease incidence in each quantile examined, to identify food patterns associated with excessive or inadequate intakes of nutrients, and to obtain descriptive information on usual intakes of foods. Although their use is increasing, some doubt has been expressed regarding the ability of FFQs to detect associations between diet and disease using this method (Tollosa et al., 2017).

There is an increasing use of online FFQs, such as the online Food4Me FFQ developed to collect dietary intakes across seven countries in Europe (Foster et al., 2014). The food list of Food4Me consists of 157 food items (for the English version) and questions respondents about their consumption of these food items over the past month. Agreement between online FFQs and conventional paper-based FFQs for intakes of food groups and nutrients has been variable.

### 37.2.4 Diet histories

The diet history is a combination of short- and long-term methods, usually conducted by trained interviewers, using a 24-hour recall or food diary, followed by FFQ with more detailed information on cooking methods and meal patterns over the recent past. This method is less commonly used in epidemiological research due to the necessity for face-to-face interviews of up to 90 minutes, and consequent costs, but more often in clinical settings by dieticians with varying protocols (see Gibson, 2005).

### 37.2.5 Food checklists

The checklist is a record to be completed for 7 days or more, is commonly known as a 'screener', and can be completed online. This form of assessment can be suitable for comparing compliance with dietary guidelines, estimating intakes of specific foods or nutrients occurring in high levels in particular foods. The checklist method is a list of representative foods in which participants are asked to check off at the end of each day which foods they have eaten. This

PLEASE PUT A TICK (✓) ON EVERY LINE

| FOODS AND AMOUNTS | AVERAGE USE LAST YEAR | | | | | | | | |
|---|---|---|---|---|---|---|---|---|---|
| DRINKS | Never or less than once/month | 1–3 per month | Once a week | 2–4 per week | 5–6 per week | Once a day | 2–3 per day | 4–5 per day | 6+ per day |
| Tea (cup) | | | | | | | | ✓ | |
| Coffee, instant or ground (cup) | | | | | | ✓ | | | |
| Coffee, decaffeinated (cup) | ✓ | | | | | | | | |
| Coffee whitener, e.g. Coffee-mate (teaspoon) | ✓ | | | | | | | | |
| Cocoa, hot chocolate (cup) | | | | | | ✓ | | | |
| Horlicks, Ovaltine (cup) | ✓ | | | | | | | | |
| Wine (glass) | ✓ | | | | | | | | |
| Beer, lager or cider (half pint) | ✓ | | | | | | | | |
| Port, sherry, vermouth, liqueurs (glass) | ✓ | | | | | | | | |
| Spirits, e.g. gin, brandy, whisky, vodka (single) | ✓ | | | | | | | | |
| Low calorie or diet fizzy soft drinks (glass) | ✓ | | | | | | | | |
| Fizzy soft drinks, e.g. Coca Cola, lemonade (glass) | | | | | | ✓ | | | |
| Pure fruit juice (100%), e.g. orange, apple juice (glass) | ✓ | | | | | | | | |
| Fruit squash or cordial (glass) | | | | | | | ✓ | | |
| **FRUIT** (1 fruit or medium serving)<br>**For very seasonal fruits such as strawberries, please estimate your average use when the fruit is in season** | | | | | | | | | |
| Apples | | | | ✓ | | | | | |
| Pears | | | | ✓ | | | | | |
| Oranges, satsumas, mandarins | | ✓ | | | | | | | |
| Grapefruit | ✓ | | | | | | | | |
| Bananas | | | ✓ | | | | | | |
| Grapes | | | ✓ | | | | | | |
| Melon | ✓ | | | | | | | | |
| Peaches, plums, apricots | | | | ✓ | | | | | |
| Strawberries, raspberries, kiwi fruit | | | | | | ✓ | | | |
| Tinned fruit | | ✓ | | | | | | | |
| Dried fruit, e.g. raisins, prunes | ✓ | | | | | | | | |
| | Never or less than once/month | 1–3 per month | Once a week | 2–4 per week | 5–6 per week | Once a day | 2–3 per day | 4–5 per day | 6+ per day |

**Please check that you have a tick (✓) on EVERY line**

Fig. 37.1 Example of a food frequency questionnaire. This is one of several different pages that have to be filled in.

*Source:* Reproduced with permission from EPIC Norfolk.

means that participants do not have to estimate how often the food is eaten, thus avoiding problems in the estimation of the frequency of food consumption that occur in the FFQs. Like the FFQs, however, the foods can be pre-coded for rapid data entry and computerized linkage to food tables. In one published version, the checklist took the form of a booklet, which comprised one page of instructions, one of an example, and seven pages (one for each day over 1 week) of the checklist. When selecting foods, participants were asked to count half for a small portion and two for a large portion. A space was left to record foods not present on the printed list, but otherwise the list was pre-coded for nutrient analysis. The list of 160 foods was then used for an FFQ and, where possible, 'units' (slices, cups, etc.) were specified. This method has been further developed and validated in older adults for 14 nutrients. Food checklist tools have been developed to assess diet in different age groups, such as emphasizing fruit and vegetable intake using CADET—Child and Diet Evaluation Tool in children (Christian et al., 2014).

### 37.2.6 Duplicate diets

If there is inadequate food table information, 'precise weighing' of duplicate diets may be necessary, for example if food composition tables with values for cooked foods are not available, in carefully controlled experiments when precise knowledge of nutrients and energy intake is required or if exposure to phytochemicals and contaminants are being investigated. Raw ingredients, the cooked food, meal, or snack, plus the individual portions are generally prepared in duplicate, one for consumption, the other for weighing and chemical analysis if required. Amounts not consumed are taken into account. This method is very labour-intensive compared with the records outlined above and it is usual for skilled fieldworkers to carry out this survey, rather than the subjects themselves. This method requires sophisticated chemical analysis based in a specialist laboratory.

### 37.2.7 Retrospective assessment

Methods that are designed to assess recent past diet (the 24-hour recall, FFQs, and diet history) can, in theory, also be used to assess distant past diet, for example in case-control investigations where dietary habits before the onset of symptoms (and possible change in diet) are required. However, there is evidence to suggest that individuals cannot remember past diet and instead report present diet. This may introduce bias into case-control studies if dietary habits have changed as a result of the symptoms of the disease in question. For this reason, more weight is placed on results of prospective studies than retrospective case-control studies in nutritional epidemiology.

## 37.3 Calculation of nutrient intake

Once the primary data concerning foods consumed are obtained, the information is converted to nutrient intake using tables of food composition. In the past, this was generally done manually, perhaps with the assistance of a calculator, and the information obtained was generally restricted to a narrow range of nutrients, for example energy and macronutrient consumption. Computerized databases of food composition revolutionized the amount of information that could be obtained, but in some data-entry systems the matching of the description of the food consumed to the correct computer code must be done manually, which leads to errors. In present-day surveys, necessitated partly by the growth in the variety of foods consumed, this procedure is now usually entirely computerized, usually by the investigators themselves. Programmes are more expensive to run and develop for record or 24-hour recall methods than for FFQs, since at least 150,000 different food items are available in Westernized food supplies, all of which require estimation of portion size and individual computer coding (see https://www.epic-norfolk.org.uk/about-epic-norfolk/nutritional-methods/, accessed 12 February 2022). Furthermore, most investigators will incorporate some means of calculating nutrient intake from individual recipes used in home cooking and specific brands of food, since these can have a marked effect on some nutrients, for example on specific fatty acids or sugar content. The checklist and FFQ methods require much simpler methods and considerably less coding time by the investigator, although much information on actual foods consumed is lost.

## 37.4 Measurement error in dietary assessments

Methods of measuring diet are associated with both random and systematic error. Both types of error can arise in the assessment of portion size, daily variation, frequency of food consumption, and failure to report usual diet, because of either changes in habits while taking part in an investigation or misreporting of food choice or amount. Error may also result from the use of food tables.

### 37.4.1 Assessment of portion size

Information about the weight of food consumed may be obtained by either asking subjects to weigh out individual items of food onto the plate as it is being served (weighed records), or requiring that portions of food be described in terms of household measures, volume models, photographs, average portions, units, or pack sizes (estimated records, diet histories, and 24-hour recalls). Errors are reduced when weights are obtained, but participants need to be given a set of scales accurate to 1–2 g with a capacity of 2 kg. Participants need instruction on the use of the scales and on the detail of information required, including description of recipes used (see Further reading). Estimated records are much easier for participants to complete, but conversion of descriptions of food into weights requires considerable investment by the investigator and may necessitate the determination of density of separate foods, as well as a detailed database of weights of foods equivalent to the photographs, models, package sizes, and household measures used. On balance, there appears to be little or no systematic bias in group averages of nutrients obtained by records with estimates of food, compared with group averages obtained by weighed dietary records. Nevertheless, despite the absence of overall bias in a population, the estimation of portion size rather than direct weighing is associated with imprecision at the individual level. In general, this is in the order of 50% (coefficient of variation) for foods, but less, about 20%, for nutrients, probably due to cancellation of error from the use of food tables. Food models and photographs may incur less error in the estimation of portion weights, at least when compared with estimations from household measures and dimensions. However, studies have found that some children find the estimation of portion sizes using photographs difficult and have a tendency to over-report, in particular for foods such as cheese or muesli (see Further reading).

Sometimes when conducting 24-hour recalls on children, age-appropriate graduated portion size photographs depict foods 'as served', as well as photographs of possible 'leftovers portions' (Foster et al., 2014). Inclusion of a 'size cue' such as a household utensil (e.g. spoon) helps respondents to conceptualize the size depicted in the digital photograph. All photographs used should be region- and population-specific; they can be displayed on the computer or carried to the interview as an atlas. In some recent studies, respondents have taken digital photos of their foods consumed for use later during the 24-hour recall interviews as prompts to help recalling food brands and quantities consumed or left over, although this method has not been extensively tested. Digital imaging or passive camera technologies using devices such as sensors or wearable cameras for manual or automated dietary analysis is also being developed and its suitability is under investigation. These innovations may have potential in terms of obtaining a 'real time' dietary analysis with reduced participant burden (see Further reading).

### 37.4.2 Daily variation

Individuals do not consume the same food from day to day and substantial error is introduced when diet is assessed from a single day's dietary investigation in records or 24-hour recalls. Thus, daily variation is one of the main factors in reducing precision of individual estimates in either of these methods of assessing diet. The variability from day to day is closely related to the nutrient under study. The early descriptions of record techniques specified that subjects should be observed for 7 days, and this practice has been followed for over 60 years. When the average intake of a group is required, then only one day per person is required provided all days of the week are equally represented in the final sample of days. Increasing the number of individuals will increase the precision of the estimate for the average intake for a group, whereas to enhance the precision at the individual level, increasing the number of days per individual is necessary.

Seven days is generally accepted as the minimum length of time required to gain precision in observations on each individual, although shorter periods of time with correction for the within-subject component error are under investigation. The actual number of records required to classify individuals in any specific population according to quantiles of nutrient intake will depend on the ratio of the average within-person daily variation and the between-person variation (see section 37.2.2). Thus, whereas a 7-day record is probably sufficient to classify into thirds of the distribution for energy and energy-yielding nutrients, longer periods are necessary for items such as some vitamins and minerals and cholesterol.

### 37.4.3 Frequency of food consumption

Overestimation, compared with records of food consumption, particularly of vegetables, but also of energy and energy-yielding nutrients, is a usual finding with FFQs. The cause of this is uncertain but may result from the use of lists. Restriction of the choice of food into a comparatively short list of around 150 foods or fewer means that error associated with estimation of amounts of single items is more likely to be biased than when the full variety of foods is analysed, as occurs, for example, in a 24-hour recall or record of food consumption. Participants using the FFQ may also have difficulty in choosing the correct category of how often food is consumed, so that overestimation occurs of the number of times foods are eaten over a defined period. FFQs routinely overestimate intake of fruits and vegetables.

### 37.4.4 Under-reporting

The term 'under-reporting' particularly applies to methods that attempt to assess total energy intake. This problem has been demonstrated by comparing the group average intake of energy from diet assessment methods with group energy expenditure estimated from body weight, or, more accurately, the doubly labelled water technique. Under-reporting has been documented with all methods of dietary assessment, including 24-hour recalls, weighed records, diet histories, and FFQs designed to assess total diet. Studies found that individuals who are overweight, in particular, are likely to under-report the amount they eat. Table 37.1 is an example showing that some, but not all, nutrients and foods are under-reported: protein, sugars, and fat, and foods such as cakes and sweets tend to be under-reported, but nutrients such as carotene, non-starch polysaccharides and vitamin C, and vegetables are not.

**Table 37.1 Intakes of energy and macronutrients expressed as reported and after energy adjustment in individuals who under-reported dietary intake and those who did not, as judged by the urine to dietary nitrogen ratio**

| Nutrient | Valid records n=126 (mean ± SE) | Under-reporters n=33 (mean ± SE) | P-value |
|---|---|---|---|
| **Reported intakes** | | | |
| MJ | 8.14 ± 0.04 | 6.65 ± 0.23 | $<0.001$ |
| Protein (g) | 71 ± 0.2 | 60 ± 1.7 | $<0.001$ |
| Fat (g) | 80 ± 0.3 | 62 ± 2.5 | $<0.001$ |
| Carbohydrate (g) | 231 ± 4.3 | 191 ± 8.2 | $<0.001$ |
| **Energy-adjusted intakes** | | | |
| Protein (g) | 69 ± 0.8 | 67 ± 1.4 | $>0.05$ |
| Fat (g) | 77 ± 0.9 | 75 ± 0.9 | $>0.05$ |
| Carbohydrate (g) | 223 ± 2.7 | 225 ± 4.0 | $>0.05$ |

The problem of under-reporting is particularly difficult when mean intakes are to be compared with reference nutrient intakes for surveillance and clinical work or amounts of nutrients eaten by a different population or group (such as people who are obese compared those who are lean). Cut-offs based on estimated energy expenditure calculated from body weight have been devised, but they are imprecise when used in the absence of information on energy expenditure. Ideally, all dietary studies should include independent measures of validity (see Chapter 39).

## 37.4.5 Energy adjustment

Energy adjustment can be carried out by a variety of methods, including expressing results for nutrients as a percentage of the total energy, or using regression techniques. One reason for attempting to correct for energy intake in dietary assessments is to reduce extraneous variation from the general correlation of nutrients with total energy intake, brought about by differences in body size and hence (in sedentary populations) energy expenditure. In addition, the correlation between results from one method and another is sometimes improved by energy adjustment. Furthermore, although there are significant differences in absolute macronutrient intake between individuals who give valid records and those who do not, these differences are substantially reduced after energy adjustment, although the overall mean within a population is not altered. Table 37.1 shows reported and energy-adjusted intakes of fat, carbohydrate, and protein in a group of women. Differences between under-reporters and those who gave valid records were no longer significant after energy adjustment.

The effect of energy adjustment depends on the correlation between the nutrient concerned and energy intake, and also on the correlation between the errors of measurement for these two quantities. The latter is heavily dependent on the dietary method used. Hence, the relation between nutrient intakes derived from FFQs and weighed records can be much improved by energy adjustment, but to a lesser extent between nutrient intakes derived from weighed records and 24-hour recalls. Energy adjustment is inappropriate (and without effect) if there are zero correlations between energy intake and the nutrient concerned, for example in the case of some vitamins. More details on different techniques for energy adjustment used are given in Tomova et al. (2022).

## 37.4.6 Effects of measurement error

Measurement error is a serious problem in dietary assessment. The effect of measurement error may be to introduce bias, so that, for example, group mean intakes may be over- or underestimated when population intakes are investigated for comparison against recommended levels (see section 37.4.4). In epidemiological research, individuals may be mis-classified in the distribution of nutritional intakes so that a null or attenuated relationship may be obtained and the true effect between diet and disease missed. For example, in a prospective study relating diet to breast cancer risk, diet was assessed using both an FFQ and a detailed 7-day diary of food and drink in 13,070 women in 1993–97. By 2002, there were 168 incident breast cancer cases for analysis. When their baseline dietary intake was compared with matched controls (four for each of the breast cancer cases), the hazard ratio for breast cancer for each quintile increase of energy-adjusted fat was strongly associated with saturated fat intake measured using the food diary (1.219 (95% CI: 1.061–1.401), P = 0.005). However, with saturated fat measured using the FFQ, the comparable ratio was 1.100 (0.941–1.285, P = 0.229) (Bingham et al., 2003).

Different methods of dietary assessment have different types of error structure, so that the magnitude of the error varies according to the method and may not always be predictable in different populations. In large prospective epidemiological studies, it is now common practice to correct for measurement error in the assessment of relative risk by regression calibration when the correction factors are derived by comparison of the method in use, such as a FFQ, with a 'reference' method, such as a record. This 'relative validation' relies on the assumptions that errors in the reference instrument are not correlated with both 'true' intake and errors in the method in use. However, errors associated with the method under investigation may be correlated with those of the reference method, so that correction for regression dilution is substantially underestimated. For example, an individual who under-reports using one dietary assessment method such as a food record will also do so with another, such as an FFQ. The validity of dietary assessment methods for determining intake of 15 priority micronutrients and ω3 fatty acids has been reviewed by the EURopean micronutrient RECommendations Aligned Network of Excellence (EURRECA). The conclusion of this substantial report generally supports the view that in epidemiological studies, FFQs provide reasonable estimates of intake when their preference is compared with a presumed superior method or biomarker. Most correlations were regarded as being in the acceptable to good range ($r = 0.30$–$0.70$). Whether such correlations are acceptable and whether it is appropriate to compare one dietary intake method with another will continue to be debated. It is generally accepted that wherever possible and appropriate, biomarkers rather than 'relative validation studies' should be used to validate methods for assessing dietary intake. Further discussion on this topic is given in the commentary by Subar et al. (2017).

### 37.4.7 Innovative dietary assessment technology

There are numerous new technology-based dietary assessment methods available or under development. All are designed to objectively measure diet without relying on user-reported food intake using methods that are cheaper with a lower burden for both the respondents and the researchers compared to the conventional methods.

Increasingly, smart phones are being used by individuals for dietary self-management (e.g. weight loss). Currently, they are not used for research purposes or for the population-level assessment. The user tracks daily food consumption in real time, usually in connection with food composition databases of varying quality. Dietary information can be entered into these devices by a variety of methods, including bar code scanners, voice recognition software, photographs, text entry, or text-entry by browsing a list of food items. Methods for portion size estimation also differ, and may include digital images of graduated portion sizes, single photographs, or a list of items in household measures. The information provided by these applications may be limited to energy intakes alone or include intakes of energy, macronutrients, and micronutrients.

Several image-based methods are also available. These are defined as any device that uses images for either identification of foods or portion size estimation. The images can be captured passively or actively and aim to enhance the accuracy of self-reported intakes. For the methods relying on passive image capture, wearable devices, often digital cameras, take pictures unobtrusively and automatically, usually at timed intervals. These methods have the potential to measure intake over multiple days without any effort on the part of the user. For those methods based on active image capture, a range of devices are used, including smart phones, digital cameras, and cell phones. For most of the image-based methods, processing is not yet fully automated, and often dietitians identify and quantify the foods and calculate the nutrient values. Several other wearable devices employing sensor technologies, worn around the neck, lapel, or as wristbands, are under development. Some rely on chewing and bite-related wearable sensors, whereas others track vibrations in the neck to detect muscle contraction and motion of the skin during ingestion. Technical problems in safe data transfer, the issue of addressing privacy concerns, frequency of images required to capture all foods consumed, camera placement and size, battery life, and other concerns must be overcome. However, these new technologies have potential, but may still result in altered food intake as experienced with other dietary assessment methods.

## Further reading

1. **Aglago, E.K., Landais, E., Nicolas, G., et al.** (2017) Evaluation of the international standardized 24-h dietary recall methodology (GloboDiet) for potential application in research and surveillance within African settings. *Global Health*, **13**, 35. doi.org/10.1186/s12992-017-0260-6

2. **Bingham, S.A., Luben, R., Welch, A., et al.** (2003) Fat and breast cancer: are imprecise methods obscuring a relationship? Report from the EPIC Norfolk prospective cohort study. *Lancet*, **362**, 212–14.

3. **Bradley, J., Simpson, E., Poliakov, I., et al.** (2016) Comparison of INTAKE24 (an online 24-h dietary recall tool) with interviewer-led 24-h recall in 11–24-year-olds. *Nutrients*, **8**, 358. doi:10.3390/nu8060358

4. **Christian, M.S., Evans, C.E., Nykjaer, C., et al**. (2014) Measuring diet in primary school children aged 8-11 years: validation of the Child and Diet Evaluation Tool (CADET) with an emphasis on fruit and vegetable intake. *Eur J Clin Nutr*, **69**(2), 234–41.

5. **Foster, E., Delve, J., Simpson, E., and Breininger, S.** (2014) *Comparison study: INTAKE24 vs interviewer led recall.* Newcastle: Food Standards Agency/Newcastle University.

6. Gibson, R.S. (2005) *Principles of nutrition assessment*, 2nd edn. Oxford: Oxford University Press.

7. **Harttig, U., Haubrock, J., Knüppel, S., et al.** (2011) The MSM program: web-based statistics package for estimating usual dietary intake using the Multiple Source Method. *Eur J Clin Nutr*, **65**, S87–S91. doi.org/10.1038/ejcn.2011.92

8. **Simpson, E., Bradley, J., Poliakov, I., et al.** (2017) Iterative development of an online dietary recall tool: INTAKE24. *Nutrients*, **9**, 118. doi:10.3390/nu9020118

9. **Stewart, C., Bianchi, F., Frie, K., et al.** (2022) Comparison of three dietary assessment methods to estimate meat intake as part of a meat reduction intervention among adults in the UK. *Nutrients*, **14**, 411. doi:10.3390/nu14030411

10. **Subar, A.F., Kushi, L.H., Lerman, J.L., et al.** (2017) Invited commentary: the contribution to the field of nutritional epidemiology of the landmark 1985 publication by Willett et al. *Am J Epidemiol*, **185**(11), 1124–9. https://doi.org/10.1093/aje/kwx072

11. **Tollosa, D.N., Van Camp, J., Huybrechts, I., et al.** (2017) Validity and reproducibility of a food frequency questionnaire for dietary factors related to colorectal cancer. *Nutrients*, **9**(11), 1257. https://doi.org/10.3390/nu9111257

12. **Tomova, G.D., Arnold, K.F., Gilthorpe, M.S., et al.** (2022) Adjustment for energy intake in nutritional research: a causal inference perspective. *Am J Clin Nutr*, **115**(1), 189–98. https://doi.org/10.1093/ajcn/nqab266

# Assessment of Nutritional Status

A. Stewart Truswell



## 38.1 Nutritional status versus dietary intake

Dietary intake estimation, described in the preceding chapter, cannot always prove that an individual or community is well nourished or poorly or over-nourished. This has to be confirmed or established by one or more methods of examination.

Food intake can be distorted by intrusion of investigators; intake over a day or a few days may not represent intake over time. It is difficult to relate mixed dishes to lines in the food tables. Nutrients in food tables are only averages, often from another country and an earlier time. Not all nutrients are in the food tables. Nutrient reference intakes (recommended dietary intakes) may not be enough for everyone, especially if they have an illness. Then is an intake below the recommended dietary intake serious or covered by safety factors? Health professionals cannot rely on the history of quantity and type of food when assessing a patient's state of nourishment. You cannot diagnose obesity, over-nutrition, from someone's dietary history, and undernourished people may not be able to tell you what they have and have not eaten, for a variety of reasons (Box 38.1).

Food intake measurement—really estimation—is ultimately *subjective*. It depends on the memory, cooperation, and honesty of individuals. Assessment of nutritional status is, by contrast, ultimately *objective*. A person's weight, height, and chemical concentration in blood or urine is measured by an outside observer and if a second and third observer repeats the measurement, they should obtain about the same result.

**Box 38.1 Nutritional status**

Nutritional status is a multidimensional concept, a jargon term used by nutrition professionals. It ultimately means whether a person has functioning in their body enough, but not too much, of all the nutrients, from calories down to the micronutrients. Nutritional status can be obvious, with wasting or obesity, or only discovered with high-technology chemical methods on samples of body fluids or a biopsy. Different aspects of nutritional status are the focus for hospital dietitians and clinicians, for public health workers and epidemiologists, and for nutrition researchers.

## 38.2 Uses of nutritional assessment

### 38.2.1 Evolution of assessment methods: looking for malnutrition

The scientific methods of assessing nutritional status were put together after World War II when there was widespread malnutrition across Europe. They were used to detect people who were poorly nourished. Nutrition surveys were done in communities considered at risk by nutrition specialists from Britain and North America.

In the 1950s, these methods were applied in the rest of the world, especially in less-developed countries (where kwashiorkor (Chapter 18) was rediscovered in 1952). The US Interdepartmental Committee for Nutrition in National Defense carried out surveys of military personnel and civilians between 1956 and 1967 in 26 countries that had alliances with the USA, each as a separate operation with its own report. The standardized methods for the nutritional surveys (national food supply, sampling, clinical examination, biochemical studies, and dietary data) were published in a manual in 1963. This enabled investigators to compare results between communities, to plan applied nutrition programmes, and to advise on national food and nutrition policy.

In the same year, the World Health Organization (WHO) Expert Committee on Medical Assessment of Nutritional Status commissioned Derek Jelliffe to prepare a standard guide for nutrition surveys everywhere in the world. He consulted 25 top nutrition experts in various countries and wrote a WHO monograph, *The assessment of the nutritional status of the community (with special reference to field surveys in developing regions of the world)*, in 1966. This classic of nutrition literature is the foundation of examining people systematically to see whether they are malnourished.

Since the 1960s, the biggest change is that most of the 50 clinical signs of malnutrition in Jelliffe's book are little used today. They are rare in industrial countries, require experienced medical personnel to diagnose, and many have other causes as well as malnutrition. Pallor (anaemia), oedema, sore lips, inflamed tongue, and enlarged liver can all be due to poor nutrition, but other causes are more common in most countries. A small number of clinical signs are important in nutrition work, but it depends where you are. Hair changes of kwashiorkor in a toddler suggest protein deficiency in deprived parts of Africa, but not in the developed world. Bitôt's spots or the skin changes of pellagra are reliable signs of vitamin A deficiency or niacin deficiency, respectively, in places where these deficiencies are known to occur and when observed by an experienced clinician. Thyroid enlargement in teenagers (likely to be endemic goitre from iodine deficiency) and mottled teeth (likely to indicate mild excess of fluoride in early life) are reliable signs of nutritional status and so are some of the signs of rickets (enlarged radial epiphysis, beading of costochondral junction).

The two main types of methods used today for nutritional assessment are anthropometry, measuring weights and heights and other body measurements, and biochemical tests, usually on blood, sometimes on urine. These are described in detail in section 38.3 and Chapter 39.

### 38.2.2 Parenteral and enteral nutrition

Modern formulae for total parenteral nutrition, including balanced amino acids and safe intravenous lipid preparation, have been available and approved since 1977. Special nutrition support teams have been set up in major hospitals and there are international societies for the nutritional speciality of enteral and parenteral nutrition (ASPEN in America, ESPEN in Europe). For assessing and monitoring hospital patients' nutritional status, clinical teams use selections from the general methods for nutritional assessment. Rapid biochemical tests are available for monitoring, but critically ill patients lying in bed with lines and tubes attached cannot be weighed, so other anthropometric measurements have to be used. Chapter 43 discusses nutritional support for the hospitalized patient in more detail.

### 38.2.3 Over-nutrition

With the increase of overweight and obesity in the last 40 years (Truswell, 2013) there has naturally been a focus on reliable and accessible indicators of the amount and effects of a person's excess accumulated calories. The two established simple anthropometric measures are (see Chapter 17):

- body mass index (BMI), where:

  BMI = weight/height$^2$ (see section 38.3.3)

  where weight is in kg and height is in metres
- waist circumference (either alone or expressed against hip circumference).

Reference numbers have been derived for different ages and nations for BMI and for men and women (for the waist measurement). Along with these physical measurements, biochemical tests can indicate if overweight is accompanied by metabolic abnormalities. The most usual are plasma low-density and high-density lipoprotein cholesterol, fasting triglycerides, and fasting or postprandial glucose.

### 38.2.4 Biomarkers to check, support, or replace some food intake estimates

Some biochemical tests are increasingly being used for this purpose in human nutrition experiments and epidemiological studies. Not many are suitable, but four examples are:

- for checking protein intake (and hence roughly energy intake)—24-hour urinary nitrogen
- to support change of type of fat—plasma fatty acid pattern
- more reliable than estimating salt intake—24-hour urinary sodium
- the only way to gauge iodine intake (because it varies greatly among foods)—urinary iodine.

Biomarkers are further discussed in Chapter 39.

## 38.3 Anthropometric assessment

The basic anthropometric measurements are simple, straightforward, inexpensive, and safe and anyone can do them. For research purposes, weight and height are measured more precisely, and for clinical work some other measurements are made.

### 38.3.1 Body mass (body weight)

In affluent communities, most people know their body weight and many weigh themselves regularly on an electronic bathroom scale. These are not as accurate as health professionals require in a clinic or consulting room. The best weighing machines are beam balances with non-detachable weights (Fig. 38.1), but these are bulky and difficult to move. They should stand on a level, hard surface and be checked with a known weight regularly. People should be weighed to the nearest 0.1 kg, wearing minimal clothing. If changes in body weight are being followed, the measurement should, if possible, be made at the same time of day because meals, drink, a full bladder, and bowel action can all affect the reading. When people in a steady state are weighed repeatedly, the day-to-day fluctuation can be ±1.0 kg.

Nutritionists are very interested in the measurement of body weight and its interpretation, whether someone is overweight or underweight, and whether their weight is increasing or going down. They want as well to estimate what components inside the body make up the weight and the change in weight (see Fig. 38.2).

Chemically, the largest component is water, then there are about equal amounts of protein and fat (each around 20%), and the rest is bone mineral. In adults, weight gain is nearly all fat; weight loss from negative energy balance is of fat, but some fat-free mass, i.e. protein, is lost as well.

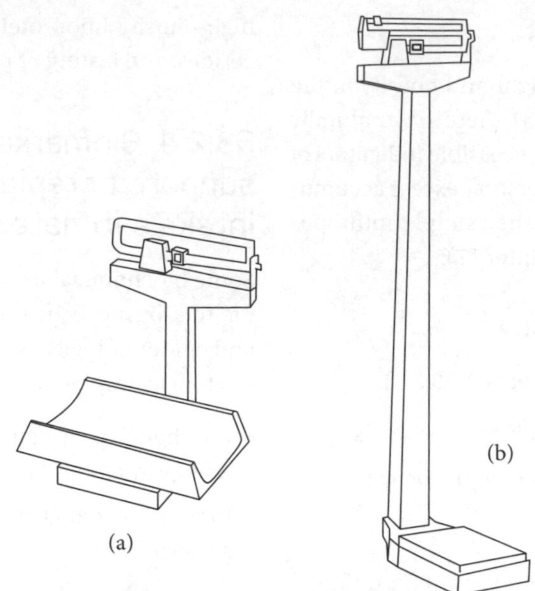


Fig. 38.1 Measurement of weight. (a) Paediatric scale for infants, and (b) beam balance for a child or adult.

*Source:* Gibson, R.S. (2005) *Principles of nutritional assessment*, 2nd edn. Oxford: Oxford University Press. By permission of Oxford University Press, USA.

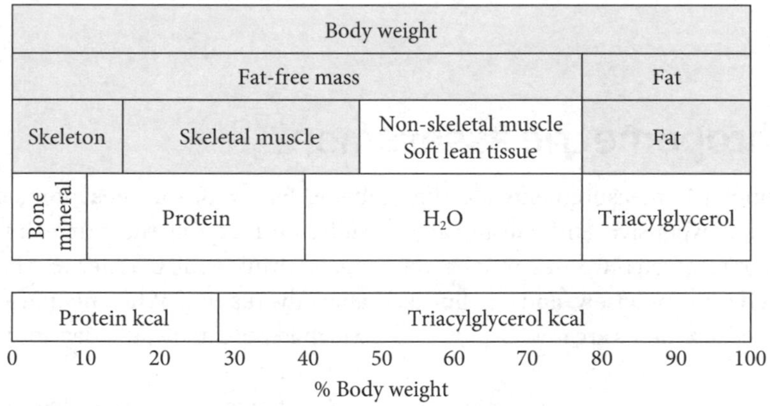

| Body weight | | | | |
|---|---|---|---|---|
| Fat-free mass | | | | Fat |
| Skeleton | Skeletal muscle | | Non-skeletal muscle Soft lean tissue | Fat |
| Bone mineral | Protein | $H_2O$ | | Triacylglycerol |

| Protein kcal | Triacylglycerol kcal |
|---|---|

0 10 20 30 40 50 60 70 80 90 100

% Body weight

Fig. 38.2 Compartments of the body. Relationship between anthropometry (shaded), body composition, and energy reserves. Different compartments in an average, healthy weight adult. About 20% of this average person is fat and the rest is fat-free mass. Women have proportionally more fat and less muscle than men. Of the fat-free mass, part is muscle (about 40% of total body weight), part is bones, and the rest is all the other organs, in descending order of weight, skin, blood, gut, liver, brain, lungs, heart, and so on.

*Source:* Heymsfield, S.B. (1984) Anthropometric assessment of adult protein-energy malnutrition. In: Wright, R.A., Heymsfield, S.B., and McManus, C.B. (eds) *Nutritional assessment*. Boston, MA: Blackwell Scientific.

When total body mass has been measured, we need to judge whether it is in the healthy range or whether the person is too thin or too fat. First, an adjustment has to be made for size. A lean giant must weigh more than a dwarf. Weight has to be considered in terms of height (or stature). Different indices have been proposed and tested. Of all the possibilities, weight/height$^2$ has been found the most generally useful (Cole, 1991) as a direct measure of over- or underweight. It only requires two measurements, weight and height, and a simple calculation. This BMI is expressed in kg (for weight) and metres (for height), i.e. in SI units. Quite different numbers would be obtained if height were in centimetres. The Belgian mathematician Quetelet first recommended this index.

## 38.3.2 Measurement of height (stature)

Height is more difficult to measure than weight. Consequently, people's heights are seldom measured and often not accurately remembered. For adults and children, a level floor and straight wall are needed (Fig. 38.3). The subject has to stand straight with buttocks, shoulders, and back of the head touching the wall, with heels flat and together, shoulders relaxed, and arms hanging down. The head should be erect and look straight forward, the lower border of the orbit in line with the external auditory meatus (the Frankfurt plane). The headpiece (a metal bar or wooden block) is lowered gently, pressing down the hair. Ideally, this headpiece should be on a sliding scale, but if a loose object is used, such as a firm book, this must be kept horizontal. In practice, two people can better manage an accurate reading. For research work, a specially designed stadiometer can be obtained. There is a circadian variation in height. People are taller in the morning by 1–2 cm. Then, during the day, the intervertebral discs get somewhat compressed.

For very young children that cannot yet stand properly, their length is best measured lying supine on a specially designed measuring board. Two examiners are needed to position the infant correctly and comfortably.

For adults who are deformed (e.g. with scoliosis) or bedridden, estimates of what their height would be if they could stand straight can be made from *knee height* or *arm span* (right fingertips outstretched to tips of left fingers) or *demispan* (from sternal notch to webspace between middle and ring fingers). These measurements are used in geriatrics. Equations to give estimated height differ for gender and ethnic group.

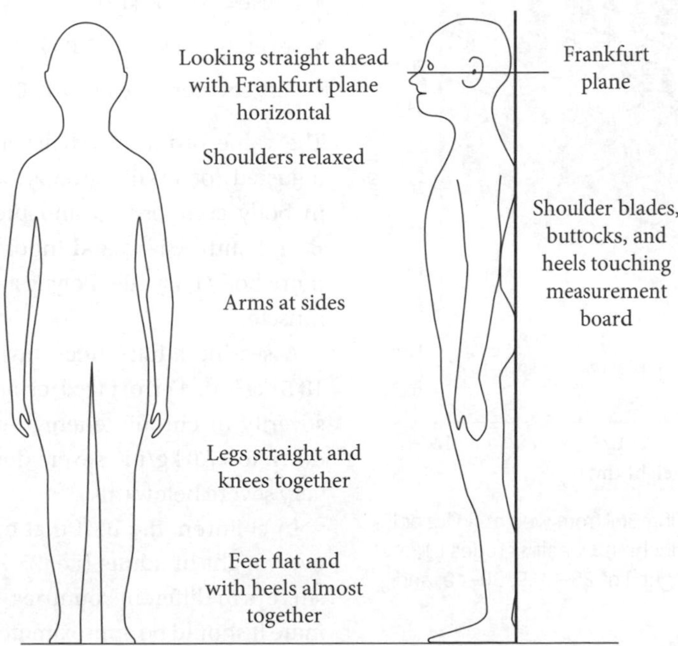


Fig. 38.3 Positioning of person for height measurement.

*Source:* Gibson, R.S. (2005) *Principles of nutritional assessment*, 2nd edn. Oxford: Oxford University Press. By permission of Oxford University Press, USA.

### 38.3.3 Interpretation of body mass index

The WHO and government health departments of the major countries have all adopted BMI as the standard way of diagnosing overweight and obesity. With the same cut-offs for men and women, this is much simpler than the earlier tables of desirable weights for height, with different numbers for frame size and for gender. For those who are put off by, say,

$$65\ (\text{kg}) \div (1.73 \times 1.73(\text{m})) \rightarrow 21.67\ \text{kg/m}^2$$

graphs like Fig. 38.4 are available.

Increased BMI indicates increased adiposity, but the correlation is not, of course, 100%. People with broad frame and weightlifters (with big muscles) can have a high BMI without excess body fat. In older people (see Chapter 35), muscle bulk declines and percentage of body fat increases. As

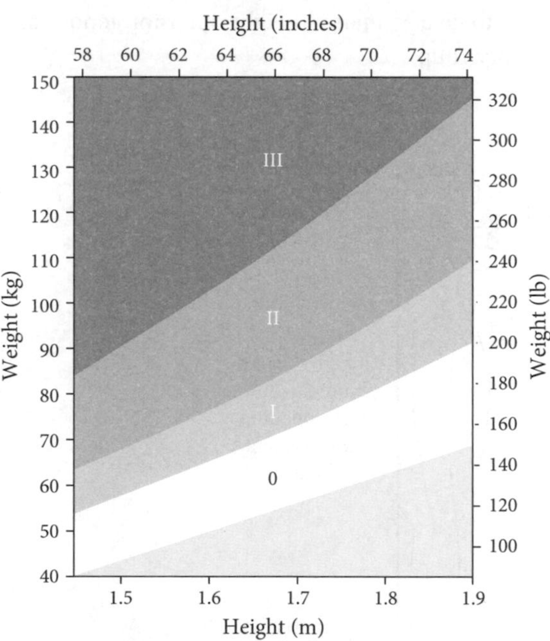


Fig. 38.4 Graph for reading BMI from weight (in kg or lb) and height (in metres or inches). Obesity grades I, II, and III have a BMI (weight/height$^2$) of 25–29.9, 30–40, and over 40, respectively.

*Source:* Garrow, J.S. (1981) *Treat obesity seriously*. Edinburgh, Churchill Livingstone.

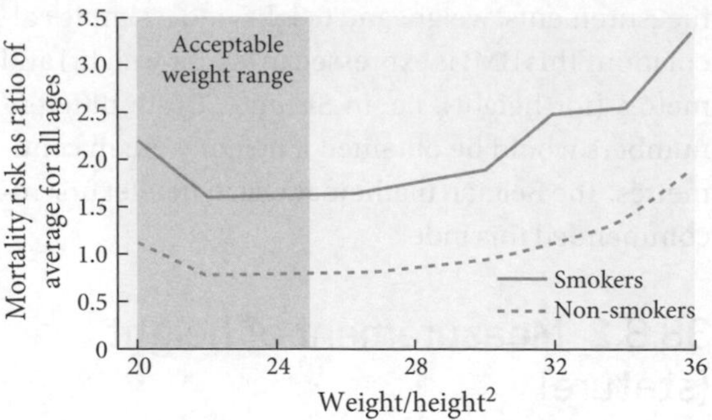


Fig. 38.5 Variations in mortality by body mass index among 750,000 US men and women.

*Source:* Lew, E.A. and Garfinkel, L. (1979) Variations in mortality by weight among 750,000 men and women. *J Chron Dis*, **32**, 563–7.

BMI increases above 25 kg/m$^2$, mortality increases gently at first and then (above BMI 30 kg/m$^2$) more steeply (Fig. 38.5).

The WHO classifies adults according to BMI:

- underweight = <18.50 kg/m$^2$
- normal range = 18.5–24.99 kg/m$^2$
- overweight = 25.00–29.99 kg/m$^2$
- obese = >30.00 kg/m$^2$
- severely obese = >35.00 kg/m$^2$
- very severely obese = >40.00 kg/m$^2$.

The value of these BMI levels probably has to be adjusted for ethnic groups because of differences in body composition and proportions. At a given BMI, South Asians and Indonesians have relatively more body fat, while Polynesians tend to have more muscle.

Assessing adult undernutrition, below a BMI of 18.5 kg/m$^2$, Ferro Luzzi et al. (1992) suggest that severity of chronic energy deficiency is moderate down to 17.0 kg/m$^2$, severe down to 16.0 kg/m$^2$, and very severe below this.

In children, the BMI that corresponds to start of overweight in adults (i.e. 25 kg/m$^2$) is lower. From surveys in different countries, Cole et al. (2000) estimate it should be approximately 17.3 kg/m$^2$ at age 5, 19.9 kg/m$^2$ at age 10, and 23.6 kg/m$^2$ at age 15.

### 38.3.4 Waist circumference

The metabolic complications of overweight and obesity (see Chapter 17) have been found to be more likely if some of the adipose tissue is specially concentrated inside the abdomen (in the omentum, mesentery, etc.). This abdominal visceral obesity can be estimated with a simple tape measure around the waist. The subject should be standing upright (with shirt off), arms at the sides. An inelastic measuring tape is applied horizontally midway between the lowest (10th) rib margin and the iliac crest. The subject breathes quietly and the measurement is taken after breathing out. Some authorities prefer measurement of the waist/hip ratio to allow for people with a heavy frame, but it is more complicated to do and does not seem to have better predictive value.

Table 38.1 gives the WHO recommendations for Caucasians. The identification of risk using waist circumference is population-specific and depends on the levels of obesity and other risk factors for cardiovascular diseases and diabetes in the ethnic group.

### 38.3.5 Measuring children's growth

The primary index of growth, used universally, is *weight for age*. At any particular age, the infant's or the child's weight (unclothed) is compared against a reference, a sort of standard, to see if their weight is at, below, or above the average. If it is far off the average, percentile lines on the reference graph will show how it compares with the reference population.

**Table 38.1 WHO recommendations for Caucasians**

| Risk of metabolic complications | Waist circumference (cm) | |
|---|---|---|
| | **Men** | **Women** |
| Increased | ≥94 | ≥80 |
| Severely increased | ≥102 | ≥88 |

The most used weight-for-age set for older children is probably the US Centers for Disease Control and Prevention growth reference. Graphs for boys and girls aged 2–20 years are reproduced in Chapter 34.

For boys and girls from birth to 5 years there is a WHO child growth standard (WHO, 2006). Infants are weighed more frequently, so a broader horizontal scale is needed; infants' stature is measured lying down and length is 1.0–2.0 cm longer than height. The reference subjects were healthy and included all ethnic groups, excluding those who had very low birth weights.

In Britain, the current weight-for-age reference (Freeman et al., 1995) is based on measurements of 25,000 white children between 1978 and 1990. These references replace earlier standards because more infants are now breastfed and children have fewer infections and grow taller.

In weight-for-age graphs, the average is the 50th percentile, i.e. median of the reference sample. An individual's difference up or down from this median can be read, firstly, from the *percentile lines*. A second way used for distance from the median is the *standard deviation* or *z-score* above or below the median. One SD below the median is a *z*-score of minus one (−1). Below 80% of the median, a child is 'underweight'; this weight is near the third percentile line and near a *z*-score of −2. A child under 60% of the median is seriously underweight and has marasmus. A third indicator used for wasting is weight for height.

In developing countries, a simplified version of the weight-for-age graph, a 'Road to Health' card, can be kept by the mother for her child and brought back to the clinic on each visit (see Chapter 18).

There is little difference in weight for age of modern children of the privileged class between different countries and ethnic groups. The reference data can thus be used internationally. If a child is somewhat heavy for age (say 80th percentile) or light (say 20th percentile), this does not mean over-nutrition or undernutrition. The child may be larger (taller) or smaller (shorter) than average. The weight has to be judged against the height (or length).

*Length-for-age* (for infants) and *stature-for-age* references will show if a child's longitudinal growth is taller or average, or shorter than the reference population. In developing countries, *stunting* is an important measure of poor nutrition and/or other adverse environment. It is usually defined as 2 standard deviations below the international median reference height for age, i.e. a *z*-score of −2.

Excessive thinness or *wasting* is recognized anthropometrically from *weight-for-height* of 2 standard deviations below the median for age, i.e. a *z*-score of −2.

Prevalence of wasting, stunting, and underweight (as defined by the WHO) in preschool children in different countries are given in Table 38.2. This shows the value of simple anthropometry in monitoring the nutrition situation in the world's young children.

In adolescence, the growth references have to be used cautiously because there is a growth spurt around the time of puberty followed by slowing of growth, and some girls mature early and some late, with about 5 years between their peaks (see Fig. 34.1). There is the same sort of range in peak height velocity in boys.

Table 38.2 **Wasting, stunting, and underweight in 0–5-year-old children, 2009–2013 (%)**

| | Wasting | Stunting | Underweight |
|---|---|---|---|
| Afghanistan | 9 | 59 | 33 |
| Bangladesh | 16 | 41 | 37 |
| Brazil | 2 | 7 | 2 |
| China | 2 | 9 | 3 |
| Cuba | 2 | 7 | 3 |
| Ethiopia | 10 | 44 | 29 |
| Guatemala | 1 | 48 | 13 |
| India | 20 | 48 | 44 |
| Kenya | 7 | 35 | 16 |
| South Africa | 5 | 24 | 9 |
| USA | 1 | 2 | 1 |

Wasting, more than 2 SD below international reference median (WHO/NCHS) weight for height.
Stunting, more than 2 SD below reference median height-for-age.
Underweight, more than 2 SD below reference median weight-for-age.

*Adapted from:* UNICEF (2015) *The state of the world's children*. New York, United Nations Children's Fund.

## 38.4 Estimating body composition: simple methods

The methods available that anyone can do measure body fat or muscle at one or more sites and from these, total body fat and/or muscle can be estimated approximately. The following methods are used by clinical nutritionists.

### 38.4.1 Skinfold thickness

With special precision callipers (Fig. 38.6), a pinch of subcutaneous fat is gently taken up and the width measured. Caught between the jaws of the callipers is a double layer of fat and skin. This fold is measured in mm. Skinfolds could be measured at many sites, but the best established are:

- *triceps skinfold*: over the triceps muscle midway down the back of the upper arm
- *subscapular skinfold*: a vertical fold taken just below and lateral to the inferior angle of the scapula, with shoulder and arm relaxed.

The skinfold is first picked up between finger and thumb, clean away from the underlying muscle, before closing the callipers on the fold.

Mid triceps and subscapular skinfolds are easily accessible and have been the most published sites. Special precision callipers must be used, which give a constant pressure on the skinfold: Harpenden or Holtain are recommended. Exact positioning of the skinfold is important; training is desirable. The mean of three repeat measurements is used. Once the callipers have been bought, skinfolds can be used without further expense to demonstrate undernutrition, or excess fat, and to follow changes clinically with

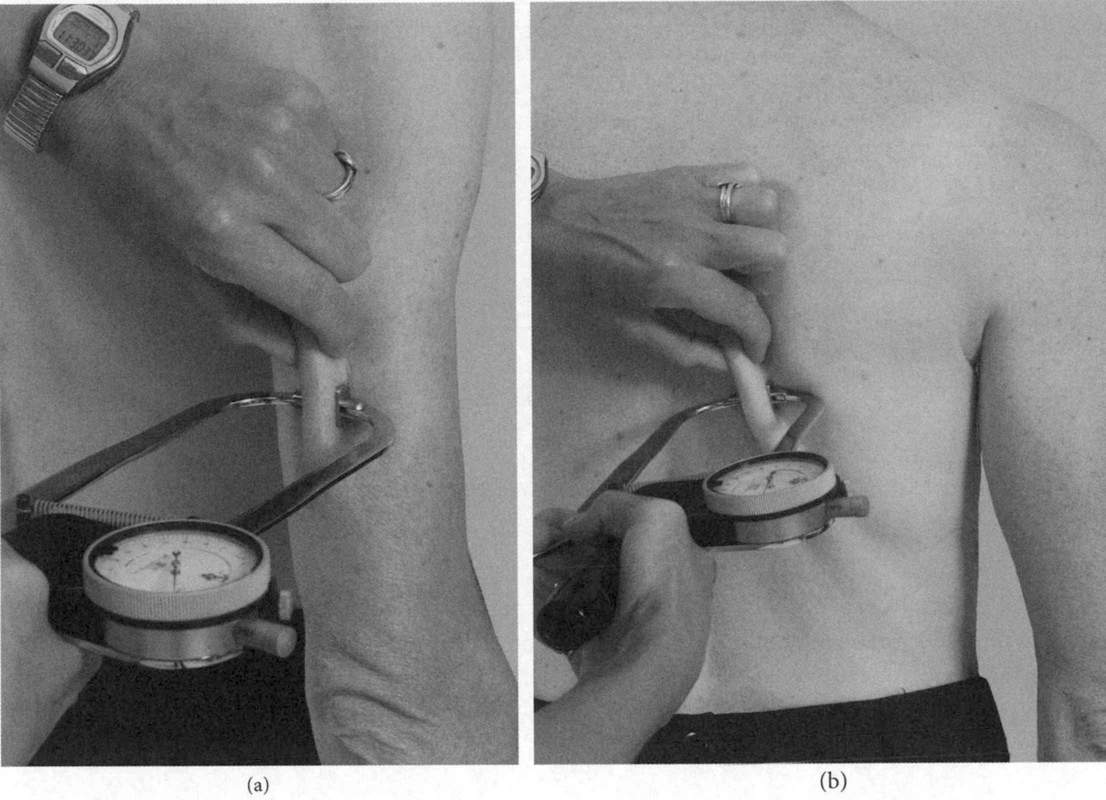

(a) (b)

Fig. 38.6 Measuring skinfold thickness—(a) triceps (left) and (b) subscapular (right) sites with Harpenden callipers.

*Source:* Truswell, A.S. (2003) *ABC of nutrition*, London: BMJ Books.

refeeding or weight reduction. In classic research, Durnin and Womersley (1974) measured triceps, subscapular, biceps, and supra-iliac skinfolds in 480 white men and women and compared the sum of these with body fat assessed from body density by underwater weighing. They then derived equations to predict total body fat from these skinfolds in men and women at different ages. For perspective, Table 38.3 shows median (50th percentile) skinfolds from NHANES in white Americans.

In very fat people, skinfolds at some trunk sites are not practicable as it is impossible to take up and measure a fold or the fold is too large for the callipers.

## 38.4.2 Limb circumferences

By simply measuring the circumference of an arm or leg and skinfolds at the same level, it is possible to calculate the circumference or the cross-sectional area of the limb's muscle. An assumption for the area of bone inside the muscle section improves the muscle area for estimating total muscle in the body.

Mid-upper arm is the usual site. The other two sites are mid-thigh and mid-calf. The arm is not always accessible. Heymsfield and Baumgartner (2006) found that the sum of limb muscle area for arm, thigh, and calf predicted quite well the skeletal muscle measured by whole-body magnetic resonance imaging:

$$\text{Limb muscle circumference} = C - \pi \times SF$$

$$\text{Limb muscle area} = ([C - \pi \times SF]^2)/4\pi$$

where $C$ is a limb circumference (upper arm, thigh, or calf) and $SF$ is the skinfold taken at the same level.

The uncorrected mid-upper arm circumference reflects muscle and fat in the limb and is used for screening preschool children in the field where no weighing scales are available. A tape is used, marked

Table 38.3 **50th percentile values for triceps and subscapular skinfolds in white Americans**

| | Triceps | | Subscapular | |
|---|---|---|---|---|
| | Men | Women | Men | Women |
| 20–29 | 10.7 | 18.8 | 13.8 | 14.2 |
| 30–39 | 11.8 | 24.5 (21.5) | 18.3 | 18.1 |
| 40–49 | 12.5 | 25.1 | 19.5 | 21.1 |
| 50–59 | 12.9 | 26.7 | 19.5 | 23.4 |
| 60–69 | 13.1 | 24.3 | 20.7 | 21.4 |
| 70–79 | 12.4 | 21.6 | 18.7 | 18.9 |
| ≥80 | 11.3 | 18.4 | 16.7 | 14.5 |

*Adapted from:* Gibson RS (2005) *Principles of nutritional assessment*, Appendices A 12.27 to A 12.31. Oxford: Oxford University Press.

at 12.5 and 13.5 cm. From 12 to 60 months of age, the arm circumference of healthy children, boys and girls, stays the same. A circumference over 13.5 cm is normal, between 12.5 and 13.5 cm suggests mild undernutrition, and under 12.5 cm indicates definite malnutrition.

## 38.5 High-technology methods for body composition

Total body water by isotope dilution This method uses a tracer dose of water, labelled with deuterium $^{2}H$, tritium $^{3}H$, or $^{18}O$. After time for equilibration, the concentration of the label is measured in serum or urine.

Bone mineral density by DEXA Dual-energy X-ray absorptiometry (DEXA) scanning is usually of the lumbar spine and hip with X-rays at two energies. There is relatively more attenuation of the lower voltage rays by more dense tissue (i.e. bone mineral). Results are related to average bone mineral density for age. All major hospitals have this equipment.

Body fat by DEXA Dual-energy X-ray absorptiometry is also used with different settings to estimate total body fat, i.e. subcutaneous plus inside the body. The radiation dose is much smaller than for an ordinary chest X-ray. Results are affected by body size, i.e. thickness, and hydration of the fat-free mass. Although errors can occur, and there is cost, the method is convenient and used now in well-funded research.

Underwater weighing using Archimedes principle The subject is weighed in air (i.e. in the usual way) and again when completely submerged in water in a large tank. Body volume is given by the apparent loss of body weight in water (i.e. the difference between weight in air and in water), which corresponds to the displaced water. Weight (in air)/volume gives body density. Knowing that the density of fat is 0.90, the percentage body fat can be calculated. Adjustments have to be made for residual air in the lungs.

Bioelectrical impedance Relatively inexpensive equipment is used to measure the impedance of an electrical current (typically 800 mA) passed between an electrode on the right foot and another on the right wrist, where there is a voltage sensor. The fat-free mass is a good conductor of electricity, while fat is not. The voltage drop between foot and wrist is greater in subjects with more fat. A computer calculates fat-free mass and fat when data on weight, height, age, gender, and level of physical activity are entered. A number of factors, including hydration state, meals, and length of limbs affect the reliability of the results.

Total body potassium, by counting $^{40}K$ This depends on the fact that 0.012% of potassium everywhere (and in our bodies) is the γ-emitting isotope $^{40}K$. The amount of this can be counted in a whole-body counter from which background radiation has been screened with thick steel or lead. Potassium occurs in the body almost exclusively inside the cells so that from total body potassium the body's cell mass can be calculated.

Total body nitrogen by *in vivo* neutral activation analysis This is an ingenious method for determining total body nitrogen and hence total protein in the body (N × 6.25). The patient is 'bombarded,' while lying on a special table, with a low neutron flux from a neutron source (such as californium-252). This converts a proportion of the nitrogen, $^{14}N$, to a very short-lived state of $^{15}N$, which emits a γ-ray at 10.83 MeV. This is counted with a gamma counter as the neutron source is moved over the subject's body on a motorized bed. The method is very expensive; only a few institutions can use it, and it gives a significant radiation dose.

Ultrasound This is harmless, not expensive, and has some uses, e.g. in quantifying the size of the thyroid gland.

Magnetic resonance imaging and computed tomography These have also been used for some research studies on body composition. Both use very expensive, bulky equipment.

## Further reading


1. **Cole, T.J.** (1991) Weight-stature indices to measure underweight, overweight and obesity. In: Himes, J.H. (ed) *Anthropometric assessment of nutritional status*, pp. 83–111. New York, NY: Wiley-Liss.
2. **Cole, T.J., Bellizzi, M.C., Flegal, K.M., and Dietz, W.H.** (2000) Establishing a standard definition for child overweight and obesity worldwide: international survey. *BMJ*, **320**, 1240–3.
3. **Durnin, J.V.G.A. and Womersley, J.** (1974) Body fat assessed from total body density and its estimation from skinfold thickness: measurements on 481 men and women aged from 16 to 72 years. *Br J Nutr*, **32**, 77–97.
4. **Eston, R.G., Rowlands, A.V., Charlesworth, S., Davies, A., and Hoppitt, T.** (2005) Prediction of DXA-determined whole body fat from skinfolds: importance of including skinfolds from the thigh and calf in young, healthy men and women. *Eur J Clin Nutr*, **59**, 695–702.
5. **Ferro-Luzzi, A., Sette, S., Franklin, M., and James, W.P.T.** (1992) A simplified approach to assessing adult chronic energy deficiency. *Eur J Clin Nutr*, **46**, 17–86.
6. **Freeman, J.V., Cole, T.J., Chinn, S., Jones, P.R.M., White, E.M., and Preece, M.A.** (1995) Cross sectional stature and weight reference curve for the UK. *Arch Dis Child*, **73**, 17–24.
7. **Gibson, R.S.** (2005) *Principles of nutritional assessment*, 2nd edn. Oxford: Oxford University Press.
8. **Heymsfield, S.B. and Baumgartner, R.N.** (2006) Body composition and anthropometry. In: Shils, M.E. (ed) *Modern nutrition in health and disease*, pp. 751–70. Philadelphia, PA: Lippincott Williams and Wilkins.
9. **Jelliffe, D.B.** (1966) *The assessment of the nutritional status of the community (with special reference to field surveys in developing regions of the world)*, WHO Monograph Series No. 53. Geneva: World Health Organization.
10. **Kukzmarski, R.J., Ogden, C.L., Grummer-Strawn, L.M., et al.** (2000) CDC growth charts: United States. *Adv Data*, **314**, 1–27.

11. **Lew, E.A. and Garfinkel, L.** (1979) Variations in mortality by weight among 750,000 men and women. *J Chron Dis*, **32**, 563–7.
12. **Sauberlich, H.E.** (1999) *Laboratory tests for the assessment of nutritional status*, 2nd edn. Boca Raton, FL: CRC Press.
13. **Steindel, S.J. and Howanitz, P.J.** (2001) The uncertainty of hair analysis for trace metals. *JAMA*, **285**, 83–5.
14. **Truswell, A.S.** (2013) Medical history of obesity. *Nutrition and Medicine*, **1**(1), 2–25.
15. **UNICEF** (2008) *The state of the world's children*. New York, NY: United Nations Children's Fund.
16. **WHO** (2006) *WHO child growth standards. Length/height-for-age, weight-for-age, weight-for-length, weight-for-height and body mass index-for-age. Methods and development*. Geneva: World Health Organization.

# Biomarkers

Nita G. Forouhi and Albert Koulman



The definition of a biomarker is broad, namely a characteristic that is objectively measured, as an indicator of normal biological processes, pathogenic processes, or responses to an exposure or intervention. Within that definition, all objective measurements in nutrition research that are indicators of the diet or nutrition-related processes, such as the assessment of body mass index as an indicator of body size, can therefore be called biomarkers. Anthropometric assessment reflecting undernutrition or over-nutrition is covered in Chapters 17 and 38. In this chapter, we focus on a nutritional biomarker specifically to indicate any biological specimen that is an indicator of nutritional status with respect to intake or metabolism of dietary constituents. An operational definition is that nutritional biomarkers include biochemical methods that assess dietary intake or nutritional status based on exogenous intake, endogenous processes, or a combination of both. Though there are multiple uses for nutritional biomarkers, a chief reason for their utility is to serve as complementary objective information rather than as a replacement for subjective dietary information from self-report methods.

## 39.1 Types of biomarkers

Biomarkers of dietary exposure assess dietary intake of different nutrients, non-nutritive components, foods, food groups, or dietary patterns. These biomarkers can provide useful complementary information alongside that from subjective reporting of dietary intakes, which is prone to measurement error (see Chapter 37).

Biomarkers of nutritional status assess not only intake but also physiological processes and metabolism, thus these biomarkers can indicate the nutrient status to which body tissues are exposed. The overall nutrient status can be more strongly associated with health outcomes than the association with nutrient intake. For example, vitamin D status measured by serum 25-hydroxy vitamin D concentration reflects both vitamin D intake as well as endogenous production of vitamin D through a chain of processes stimulated by sunshine exposure of the skin.

A classification scheme of nutritional biomarkers distinguishes recovery, concentration, predictive, and replacement biomarkers, of which recovery and concentration markers are most frequently used in nutritional epidemiology. The classification is not mutually exclusive.

*Recovery biomarkers* are based on the concept of metabolic balance between intake and excretion over a fixed period of time. Recovery biomarkers are directly associated with dietary intake and can be used to assess absolute intake and to correct for measurement error as reference instruments, currently most often used to measure under-reporting of self-reported energy intake. For example, doubly labelled water is a biomarker that measures average energy expenditure and in weight stable individuals this can be used as a marker of energy intake and compared against estimates derived from subjective reports. There are relatively few other validated recovery biomarkers and these include urinary nitrogen, urinary potassium, and urinary sodium, which are used as recovery biomarkers for each of protein intake and potassium and sodium intakes, respectively.

*Concentration biomarkers* correlate with dietary intake and are used for ranking of individuals with respect to their intakes. They are not used to determine absolute intake because they are related to metabolism, personal characteristics (e.g. age, sex), and health behavioural factors such as smoking and physical activity. Most nutritional biomarkers are concentration biomarkers. Some examples of concentration biomarkers are plasma vitamin C or plasma carotenoids (such as beta carotene) as indicators of fruit and vegetable intake or plasma or adipose tissue concentration of *trans*-fatty acids or polyunsaturated fatty acids (PUFA) as markers of dietary intakes of *trans*- and polyunsaturated fat.

*Predictive biomarkers* do not completely reflect dietary intake but can predict intakes to some extent, and similarly to recovery biomarkers, these are time dependent and demonstrate a dose-response with intake, but with overall lower recovery. Examples include urinary sucrose and fructose as markers of dietary sugars consumption.

*Replacement biomarkers* serve as a proxy for dietary intake when it is not possible to capture this reliably from subjective assessment due to lack of or insufficient information in nutrient (food composition) databases. For example, assessing 24-hour urinary sodium is superior to trying to estimate dietary salt intake, which is highly prevalent in multiple foods consumed daily and hard to assess subjectively. Other examples include polyphenols and phyto-oestrogens.

## 39.2 Uses of nutritional biomarkers

Historically, biomarkers were used to confirm the clinical diagnosis of nutritional deficiency diseases such as scurvy, beriberi, rickets, xerophthalmia, or Wernicke's encephalopathy, and to diagnose acute malnutrition for which the clinical signs are non-specific, e.g. potassium deficiency. Biomarkers are used extensively in haematological diagnosis and management, e.g. iron, folate, and vitamin $B_{12}$ measurements. They are also used for monitoring nutritional management in intensive care, with parenteral nutrition or tube feeding, or to diagnose nutritional supplement overdosing (e.g. with vitamin A, pyridoxine).

Over time, the use of nutritional biomarkers has become an important part of population nutrition surveys, such as in the UK National Diet and Nutrition Survey, to detect subclinical micronutrient deficiency, e.g. with respect to the status of iodine deficiency, iron deficiency, folate deficiency, or vitamin D deficiency. Biomarkers can also be used to objectively demonstrate the response to a nutrition education programme, e.g. reduction of plasma cholesterol or of urinary sodium for a salt reduction campaign.

We next highlight three specific uses of nutritional biomarkers in nutritional epidemiology research.

## 39.3 Validation of dietary methods with biomarkers

Biomarkers are useful for validating data from dietary assessment based on self-reporting. The objective nature of biomarkers provides an advantage over comparing one subjective method against another subjective method, which are prone to correlated errors, such as when comparing intakes from a food diary and food frequency questionnaire (FFQ) or 24-hour recall. For example, the recovery biomarker doubly labelled water is recognized as the 'gold standard' reference method to estimate energy intake. This biomarker consists of water that is labelled with non-radioactive isotopes heavy oxygen ($^{18}O$) and heavy hydrogen ($^{2}H$) and after consumption of this doubly labelled water, at least two samples of body water are collected, typically from urine. The difference in the rate of loss of labelled oxygen, lost in both water and carbon dioxide, and hydrogen, lost only in water, is calculated from the urine samples and allows the calculation of energy expenditure over the time period between the two (or more) urine collections (e.g. over one week or two weeks). The underlying assumption is that in weight stable individuals, total energy intake is equivalent to total energy expenditure over a period of time. A comparison of the estimates of energy intake from this biomarker with estimates of energy intake derived from self-reported methods allows the calculation of degree of energy under-reporting (see Chapters 5 and 17). The use of doubly labelled water has been limited by its high cost, but it has been used to assess energy misreporting in smaller sub-samples in national surveys, e.g. in the UK National Diet and Nutrition Survey for data from the food diary (food records) and in epidemiological studies, e.g. the Fenland Study in the UK and the OPEN Study in the USA using 24-hour recalls or food frequency questionnaires.

Beyond energy intake, biomarkers of individual nutrients are useful for estimating validity of a dietary instrument. To assess the validity of several different methods of dietary assessment, in the EPIC-Norfolk Study, 160 women were asked to complete 16 days of weighed-food records over 1 year, as four repeated weighed 4-day records. They also provided an estimated 7-day food diary. The volunteers were also asked to provide eight 24-hour urine collections, as four repeated 2-day collections. The volunteers were provided with para-aminobenzoic acid (PABA) tablets with their three main meals during the day to assess the completeness of the urine collections, since PABA is known to have near 100% excretion in urine. Correlations were greater between the biomarker 24-hour urine nitrogen and estimates of nitrogen intake from food records (r=0.7) than from estimates of intake from other methods, including FFQs (r=0.4). A similar pattern was evident with the urinary potassium biomarker. This study showed that using recovery biomarkers (nitrogen, sodium, potassium), the 7-day food diary was associated with less measurement error compared with the FFQ. However, the degree of validity depends on the nutrient. For example, in the EPIC-Norfolk Study the 7-day food diary and the FFQ performed similarly for the assessment of PUFA intakes against plasma PUFA biomarkers. Similarly, a US-based study reported that the correlation of adipose tissue PUFA was similar with both the food diary (r=0.5) and the FFQ (r=0.49), suggesting that both methods (food diary and FFQ) have equivalent validity as a measure of PUFA intake.

## 39.4 Assessing diet-disease associations

The use of biomarkers has more recently expanded as objective markers of dietary intakes and metabolism in relation to health outcomes. Evidence from prospective studies linking dietary factors and disease endpoints has often been inconsistent or weak due in part to measurement error and bias in subjective dietary assessment, while evidence from randomized controlled trials is sparse due to the challenges of conducting nutritional intervention studies. The assumption is that biomarkers can help with better ranking of individuals for their exposure to a particular food or nutrient compared with subjective dietary assessment tools and therefore avoids misreporting biases. Among prospective cohort studies assessing incident non-communicable diseases, the EPIC Study provides an example of the difference in observed associations of dietary intakes measured by habitual intake with subjective instruments and with nutritional biomarkers. In a proof-of-principle study, a striking inverse association between plasma vitamin C concentration, as a biomarker of fruit and vegetable intake, and the risk of incident diabetes was demonstrated, while, by contrast, the association with total fruit/vegetable intake measured by the food frequency questionnaire was modest and lacked a dose-response effect.

This research was further extended in the EPIC-InterAct Study to generate a composite biomarker score as the average of standardized values of plasma vitamin C and six individual carotenoids (α-carotene, β-carotene, lycopene, lutein, zeaxanthin, and β-cryptoxanthin). A total of 340,234 people from eight European countries were followed up, among whom approximately 10,000 individuals developed type 2 diabetes over approximately 10 years and a representative sample of 13,500 individuals remained free of diabetes. By measuring the nutritional biomarkers in approximately 23,000 people with and without type 2 diabetes, this study is the largest of its kind in the world. The study found that the higher the composite biomarker score level, the lower the risk of developing diabetes in the future. Compared with those people who were in the lowest fifth of the distribution of the composite biomarker score, those in increasing fifths of the biomarker score (2nd, 3rd, 4th, and highest categories) had a reduced relative risk of diabetes by 23%, 34%, 41% and 50%, respectively. Looked at another way, a one standard deviation difference in the composite biomarker score, equivalent to a 66 g/day difference in fruit and vegetable intake, could potentially reduce the risk of developing new-onset type 2 diabetes by a quarter. With the use of nutritional biomarkers it was possible to infer that diets rich in even a modestly higher consumption of fruit and vegetables, close to one 80 g portion of fruit or vegetable per day, could help to prevent the onset of type 2 diabetes, therefore below the threshold of 'five-a-day' or five portions of fruit and vegetables per day.

This approach has also been extended to other nutritional biomarkers, such as but not limited to examining associations with health outcomes of dietary vitamin D versus circulating 25-hydroxy vitamin D, or dietary saturated and polyunsaturated fat intake versus circulating fatty acids. Notwithstanding the challenges in interpretation, as discussed later in this chapter, taken together the research to date demonstrates the potential for using individual and combinations of biomarkers for investigating associations of dietary factors with disease risk.

## 39.5 Calibration

Dietary data obtained from a subjective, less accurate instrument can be re-scaled using information from another, more accurate, method and used to reduce measurement error and estimate population means more accurately. Methods for calibration using two subjective dietary instruments have been well described and used extensively, for instance, in the pan-European EPIC Study. In EPIC, dietary intakes were assessed by country-specific questionnaires (mostly FFQ) in the full sample of approximately 500,000 adults and additionally by a 24-hour diet recall in an 8% random sample of the cohort. Biomarkers can further enhance calibration and provide important information on subjective assessment methods at lower cost than doing a full validation study. Linear regression of biomarker values on corresponding self-report values, together with other relevant participant characteristics such as age, sex, and body mass index, can be used to derive calibration equations for dietary consumption from which calibrated consumption estimates can be calculated. A recent example was provided by the development of calibration equations using mean daily self-reported citrus fruit intake from semi-weighed food diaries and biomarker-derived citrus intake using urinary proline betaine. A simulation study from Ireland suggested that in large population studies, obtaining biomarker data on 20–30% of the participants can enable reliable estimation of calibration equations. An additional approach to calibration is the 'method of triads'. This is an application of a factor analysis model using data from subjective dietary intake such as from the FFQ, a reference method (e.g. 24-hour recall), and a biomarker.

### 39.5.1 Integrating information from subjective and objective nutritional assessments in diet-disease association studies

To maximize the advantages of each type of dietary assessment and to reduce their limitations, it is ideal to combine the information. Combining data from biomarkers (objective) and subjective methods can account partly for the disadvantages of the other method, such that biomarkers account for dietary misreporting while subjective methods account for variations related to absorption or metabolism of foods and nutrients, and may help to eliminate some of the errors associated with each of the methods of nutrient estimation. Combining information from two types of exposure variables (subjective and objective) may also lead to less misclassification than using either variable in isolation and therefore increase statistical power to detect diet-disease associations.

Statistical approaches have been developed and some are still in development to integrate such data. One approach is to create new dietary exposure variables based on a combination of information from both reported intakes and biomarkers, to create a prediction score for specific food intakes. Regression approaches such as 'reduced rank regression' can be used to derive food-based scores that predict the status for multiple nutrients rather than predicting solely single nutrient status.

## 39.6 Methodological issues for biomarkers

**Specimen types:** Nutritional biomarkers can be measured in a variety of biological specimens, including:

- Blood serum or plasma are the most common form of specimen type, reflecting short-term intake from a few days to a month; or erythrocytes with longer-term intake reflecting the average 120-day half-life of erythrocytes.
- Urine samples provide information on a wide variety of biomarkers including each of recovery, concentration, and prediction biomarkers. For most biomarkers it is necessary to collect all voided urine over a 24-hour period, but it may be impractical for many people to carry around a collection bottle with them. This reduces participation rates in studies and often leads to higher attrition, while incomplete urine collections can be a major source of measurement error. An alternative, easier approach is an early-morning sample or spot urine samples. Limitations of spot urine samples include differing degrees of dilution owing to the amount of fluid intake, but statistical adjustments can be made using urine creatinine concentrations, because creatinine excretion is assumed to be constant from day to day. However, this does not account for within- and between-person variability dependent on muscle mass and meat consumption.
- Adipose tissue samples are invasive, but will reflect long-term intake, which is useful to assess exposure to fatty acids and fat-soluble vitamins.
- Hair and nail specimens reflect long-term intakes, particularly of trace elements such as selenium or toxicants such as mercury. In practice, though these specimens are easy to collect and store,

limitations include environmental contaminants that adhere, e.g. to hair.

- Stool samples have traditionally been difficult to collect but more latterly with the increased interest in the gut microbiome, easier ways to collect such samples have emerged. However, stool samples are still uncommonly used for nutritional biomarkers as not many such biomarkers are available or validated.
- Rarer forms of specimens include non-invasive skin sample concentrations such as for carotenoid measurements, or cheek cells or saliva, which are simple to collect and are gaining interest but are not routinely used for nutritional biomarker assessment.

### 39.6.1 Sample collection and timing

The collection of samples should be standardized following 'standard operating procedures' (SOP). The time of day, fasting or non-fasting status, and season of measurement can all potentially affect within- and between-person variability. For example, for some biomarkers diurnal variation may be important and for some the concentrations will vary by season, such as higher 25-hydroxy vitamin D levels in the summer than winter (due to greater sun exposure in the summer), or blood vitamin C or carotenoid levels due to variation in availability of fruit and vegetable intake by season.

### 39.6.2 Sample storage

Most nutritional biomarker measurements make use of stored samples and stability is best achieved at low temperatures such as at –80°C or even lower (liquid nitrogen, –196°C) to minimize sample degradation. Some assays require specific storage conditions, such as for some vitamins that are photo-sensitive (e.g. vitamin K) light should be avoided, or for plasma vitamin C, where the addition of metaphosphoric acid stabilizes the sample against oxidation. Traceability and retrievability are paramount when dealing with human samples and in modern laboratories dedicated computerized and robotics-assisted sample retrieval systems are increasingly available. Many countries have strict regulations on the collection and storage of human samples.

### 39.6.3 Sample analysis

Laboratory analysis should be standardized according to SOPs and follow quality control and quality assurance schemes. When possible, laboratories should aim to sign up to external quality assurance and assessment schemes. Samples should be anonymized and laboratory staff should be blinded to participant information where that can introduce biases. Samples should be measured in random order or consecutively to reduce between-assay variation. Reliability should be assessable using the quality control samples and can be tested by conducting blinded duplicate analysis of specimens. Batch effects need to be monitored and any corrections applied as appropriate.

## 39.7 Specific biomarkers

### 39.7.1 Micronutrients

As aforementioned, in the past there was a focus on assessing micronutrient deficiencies. A large range of biomarkers has become available for this purpose, including but not limited to various vitamins such as vitamin A, B vitamins, vitamins C, D, E, and K, and minerals such as iodine, iron, magnesium, zinc, selenium, sodium, and potassium (see Table 39.1). Many of these are included in national nutrition surveys to monitor population level nutritional adequacy.

Table 39.1 **Biochemical methods for diagnosing nutritional deficiencies**

| Nutrient | Indicating reduced intake | Indicating impaired function (IF) or cell depletion (CD) | Supplementary method |
|---|---|---|---|
| Protein | Urinary nitrogen | Plasma albumin (IF) | Fasting plasma amino acid pattern |
| Vitamin A | Plasma β-carotene | Plasma retinol | Relative dose-response |
| Thiamin | Urinary thiamin | Red cell transketolase and TPP effect (IF) | |
| Riboflavin | Urinary riboflavin | Red cell glutathione reductase and FAD effect (IF) | |
| Niacin | Urinary *N*′-methyl nicotinamide or 2-pyridone, or both | Red cell NAD/NADP ratio | Fasting plasma tryptophan |
| Vitamin $B_6$ | Urinary 4-pyridoxic acid | Plasma pyridoxal 5′-phosphate | Urinary xanthurenic acid after tryptophan load |
| Folate | Plasma folate | Red cell folate (CD) | |
| Vitamin $B_{12}$ | Plasma holotranscobalamin II | Plasma vitamin $B_{12}$<br>Plasma methylmalonate | Schilling test |
| Vitamin C | Plasma ascorbate | Leukocyte ascorbate (CD) | Urinary ascorbate |
| Vitamin D | Plasma 25-hydroxy vitamin D | Raised plasma alkaline phosphatase (bone isoenzyme) (IF) | Plasma 1,25-dihydroxy vitamin D |
| Vitamin E | Ratio of plasma tocopherol to cholesterol + triglyceride | Red cell haemolysis with $H_2O_2$ *in vitro* (IF) | |
| Vitamin K | Plasma phylloquinone | Plasma prothrombin (IF) | Plasma des-γ-carboxyprothrombin |
| Sodium | Urinary sodium | Plasma sodium | |
| Potassium | Urinary potassium | Plasma potassium | Total body potassium by counting $^{40}K$ |
| Iron | Plasma iron and transferrin | Plasma ferritin (CD) | Free erythrocyte protoporphyrin |
| Magnesium | Plasma magnesium | Red cell magnesium (CD) | |
| Iodine | Urinary (stable) iodine | Plasma thyroxine (IF) | Plasma TSH |
| Zinc | Plasma zinc | Red cell zinc | |
| Selenium | Plasma selenium | Red cell glutathione peroxidase | Toenail selenium |
| Fluoride | Urinary fluoride | Plasma ionic fluoride | (Bone fluoride) |

$^{40}K$, natural radioactive potassium; FAD, flavin adenine dinucleotide; NAD, nicotinamide adenine dinucleotide; NADP, NAD phosphate; TPP, thiamine pyrophosphate; TSH, thyroid-stimulating hormone.

## 39.8 Biomarkers of macronutrient status and their food sources

Over time there has been a shift in interest for the role of nutritional biomarkers from detecting and monitoring micronutrient deficiencies to the ability to assess the long-term effects of macronutrients and their food sources. All three classes of macronutrients—carbohydrates, fat, and protein—are hugely complex mixtures of molecules that partly overlap with the endogenous human metabolism, making the identification or development of biomarkers of macronutrients very challenging.

### 39.8.1 Biomarkers for dietary carbohydrates and related food groups

Chapters 2 and 27 of this book outline detailed information on carbohydrates and the current dietary recommendations on carbohydrate intake. Finding appropriate biomarkers for carbohydrate intakes has been challenging. Refined carbohydrates and starch are predominant components of many processed foods, as well as the main source of carbohydrates and often of energy intake in many dietary patterns. However, their circulating concentrations are not directly related to intake and mostly driven by metabolic processes.

Some monosaccharides (e.g. glucose and fructose) and disaccharides (e.g. sucrose, lactose) will pass through into the urine. When the disaccharide sucrose is consumed, most is broken down to glucose and fructose and absorbed in the small intestine but small amounts (in mg quantities) of intact sucrose are also absorbed in the small intestine and, along with fructose, excreted in the urine. In the circulation, sucrose and fructose, unlike glucose, are not hormonally regulated by insulin, and hence, non-metabolized sucrose and fructose are excreted in the urine as predictive biomarkers. Although the amount is small (i.e. approximately ~0.05% of consumed sucrose and fructose is excreted in the urine), this small amount correlates well with sugar intake under controlled dietary intake and urination conditions and a dose-response relation has been demonstrated between urinary sugar excretion and actual sugar intake. These sugars can be measured either by liquid chromatography coupled with mass spectrometry or after derivatization by gas chromatography coupled with mass spectrometry in urine samples. To enable absolute quantitation, 24-hour urine samples are necessary, but spot urine samples have been used for relative differences. This approach has been applied in a few research studies but has not yet been widely used.

Changes in dietary fibre, both qualitative and quantitative, can have a range of effects on metabolism, many of which will be driven by changes in the microbial composition and activity in the gut. Research evidence has accumulated studying the effect of dietary fibre on the gut microbiome composition. So far, this work has not developed specific biomarkers of fibre consumption and with current high costs of gut microbiome genotyping, it is unlikely that this will be a practical tool to assess fibre consumption. However, genotyping technology is still developing and becoming more cost effective, which might make this approach achievable in the future.

There are some biomarkers for certain wholemeal flours and cereals. In some cases, it is possible to measure specific compounds that accumulate within cereals, like lignans (such as secoisolariciresinol and lariciresinol) in rye. These lignans can also be bioactive and were originally described as phyto-oestrogens. This high specificity for rye makes these biomarkers less relevant in most populations, as rye is only commonly consumed in a few countries (e.g. Finland). Plasma enterolactone may be useful as a biomarker of lignin-containing foods, but it is of limited value as a specific biomarker of carbohydrate-rich plant foods because coffee, tea, and alcoholic beverages also influence it. More recently, studies have suggested that proteins from wheat can be used as a biomarker for wheat consumption, such as wheatgerm agglutinin, a lectin specific to wholemeal wheat. Some studies have examined plasma alkylresorcinol and its metabolite concentrations as a possible wholegrain wheat and rye intake biomarker but this approach is still at a developmental stage.

We already referred to concentration biomarkers of fruit and vegetable intake, including plasma vitamin C and carotenoids, and these have been used in nutritional epidemiology research as valid biomarkers. Moreover, polyphenol concentrations can be used as biomarkers of fruit and vegetable intake. Polyphenols include phenolic acids, lignans, stilbenes, and flavonoids which include flavonols, and isoflavones. The very large number of polyphenol compounds offer the potential to distinguish between different types of fruits and vegetables but they are also found in beverages like coffee, tea, and wine, as well as in chocolate and whole grains.

Biomarkers of a habitual diet very low in carbohydrates have been considered. Such a diet, consisting largely of protein and fat, will lead to a reduction of oxaloacetate in cells. This metabolite is necessary to fully oxidize fats and have a functioning Krebs cycle. The lack of oxaloacetate leads to poor fat oxidation and the production of ketone bodies (acetone, beta-hydroxybutyrate, and acetoacetate). As this is an inefficient use of energy, weight loss is one of the consequences and therefore this diet has become popular. Thus, the measurement of ketone bodies can be used to assess reduced carbohydrate intake in individuals that are healthy. However, people that are hyperglycaemic, or exposed to prolonged fasting or exercise, can all have increased ketone body production. Ketone body measurements for the assessment of carbohydrate intake should be used when there is sufficient clinical understanding.

### 39.8.2 Biomarkers for dietary fats and related food sources

Most fat is consumed in the form of oil and butter (mainly triglycerides) and from dairy products, eggs, meat, fish, and nuts, while certain fruit and vegetables can also contain a range of triglycerides, lipids, and cholesterol that all contribute to fat intake.

Common to all fats and lipids are fatty acids (see Chapter 3). The number of carbon atoms in the fatty acid determines the chain length. This can be short chain (2–6), medium chain (8–12), long chain (14–18), and very long chain (+20). Within the carbon chain, all bonds can be saturated (saturated fatty acids) or have one (monounsaturated) or multiple (polyunsaturated, or PUFA) double bonds. These double bonds can be in different locations within the chain. In most mammalian systems, these double bonds will be in the *cis* position, but microbial enzymes and organic synthesis can lead to double bonds in the *trans* position, so called *trans*-fatty acids. Different fatty acids have varying biological properties and can be used as biomarkers of types of fat intake from various foods.

For instance, the consumption of food products that contain considerable amounts (>2% w/w) of *trans* fats (such as hydrogenated fats or often reused deep-frying oils) will directly result in increased concentrations of the resulting *trans*-fatty acids in the circulation. To a similar extent this is also true for the essential fatty acids, especially n-3 PUFAs (docosahexaenoic acid, DHA, C22:6n3 and eicosapentaenoic acid, EPA C20:5n3), which are fatty acids commonly found in marine organisms (e.g. fatty fish). The intake of the foods with a high n-3 PUFA content will directly affect circulating levels. However, for many other fatty acids the circulating concentrations are less dependent on dietary intakes. In particular, saturated fatty acids like palmitic acid and stearic acid do not change substantially with dietary intake and are strongly regulated. Reducing saturated fat intake, and accompanying isocaloric rise in carbohydrate intakes, results in an increase in hepatic *de novo* lipogenesis, which helps to maintain circulating levels of these fatty acids.

Odd chain fatty acids (especially C15:0) and specific *trans*-fatty acids (C16:1t) are good indicators of dairy fat intake. The circulating concentration of odd chain fatty acids is not only dependent on their intake, but some microbial species in the gut can produce propionic acid that could be incorporated in fatty acid synthesis, and alpha-oxidation can result in the conversion of stearic acid (C18:0) into heptadecanoic acid (C17:0). Odd chain fatty acids are also present in other ruminant-derived fat, like beef dripping, but the consumption of this type of fat is usually small in comparison to dairy fat consumption.

### 39.8.3 Biomarkers for dietary proteins and related foods

Total urinary nitrogen concentration in 24-hour urine samples is a long established recovery biomarker for protein consumption and has been used to validate subjective dietary instruments. Under stable conditions the body is in nitrogen balance when, on average, dietary nitrogen intake will be equal to nitrogen loss and therefore correlation values are high between multiple assessments of urinary nitrogen and protein intake over the same period. However, there are different types of protein and protein sources, and urinary nitrogen cannot distinguish between these. The components of protein are amino acids, which include essential amino acids (obtained from the diet) as well as non-essential amino acids. The consumption of amino acids is required to maintain the constant natural turnover of muscle, with muscle catabolism during the fasting state and muscle anabolism in the postprandial state. This turnover enables the maintenance of amino acid homeostasis and adequate amino acid concentrations in the circulation. Amino acids can serve as biomarkers in relation to disease risk. For instance, it has been reported that higher concentrations of circulating branched chain amino acids (valine, leucine, and isoleucine) and aromatic amino acids (phenyl alanine and tyrosine) are associated with an increased risk for type 2 diabetes. How or whether this is linked to muscle mass (lean body mass) remains unclear. The physiological role of circulating amino acid concentrations complicates their use as status biomarkers.

The consumed proteins can also enter the circulation as peptides, and especially tripeptides and dipeptides can be markers of dietary protein, such as anserine (a dipeptide of alanine and 3-methylhistidine), which is a marker of chicken meat, and pro-hyp (proline-hydroxyproline) for fish. The application of these short peptides as markers of protein was only recently validated, mainly in urine, as dietary biomarkers. The kinetics of these short peptides in plasma or serum has only had limited attention. Together with other components of the protein source (such as choline metabolites for meat and fish, or phyto-oestrogens for plant proteins), these peptides hold the potential to help to determine an individual's protein status and pinpoint the protein source.

Meats may contain compounds such as 3-methyl histidine or creatine that may serve as potential predictive biomarkers of total meat consumption. Red and processed meat can increase the formation of N-nitroso compounds (NOCs) in the gastrointestinal tract. The measurement of NOCs requires faecal samples, which are often not readily available in epidemiological studies.

### 39.8.4 Biomarkers of dietary patterns

Research to date has focused on identifying and using dietary biomarkers of single nutrients, individual foods, or food groups. There are no biologically plausible single biomarkers for dietary patterns that reflect overall diets. A recent analysis was able to distinguish between vegetarian and non-vegetarian diet groups using biomarkers in plasma, urine, and adipose tissue. Specifically, biomarker profiles differed between five diet groups: vegan, ovo-vegetarian, pesco-vegetarian, semi-vegetarian, and non-vegetarian. In this study, vegans had significantly higher plasma carotenoids and excretion of urinary isoflavones and enterolactone, lower abundance of saturated fatty acids, and higher linoleic acid relative to non-vegetarians. Methylhistidine was 92% lower in vegans and lower in lacto-ovo- and pesco-vegetarians by 90% and 80%, respectively, relative to non-vegetarians. This work shows how different biomarkers can be used to assess dietary patterns, but this approach has not yet been replicated and will require further development and independent validation.

Another approach combines multiple biomarkers of individual dietary characteristics into composite biomarker scores and offers a potential framework to derive objective indicators of dietary patterns. For instance, an American study combined data from individualized habitual diets and data-driven variable selection of nutritional biomarkers to derive a biomarker score, including blood carotenoids, fatty acids, urinary potassium, and doubly labelled water. The validity and external generalizability of known biomarkers for overall dietary patterns, such as Mediterranean diet, the DASH (Dietary Approaches to Stop Hypertension) diet, or others, remains largely unestablished thus far, though nutritional metabolomics is promising in this regard (see below).

## 39.9 Limitations

As objective measures, nutritional biomarkers are assumed to be independent of the biases and errors associated with study participants and dietary assessment methods. It is indeed unlikely that participants will be able to consciously alter biomarker results. However, it is important to recognize that all biomarkers have limitations, though the degree and type of limitations vary by biomarker. Some limitations could be reduced by focusing on the quality of the sampling, assay, and statistical adjustment for errors and confounding. However, some limitations are currently unresolved. For instance, sample collection can be invasive and biomarker measurements are relatively expensive; though repeat measures would be useful, both the cost and logistics of sample collection often limit this possibility. Only a limited range of validated biomarkers is currently available. Biomarkers are not specific solely to individual foods or food groups. Intra- and inter-individual differences in absorption and metabolism may affect the actual or measured concentration, which may depend not only on diet composition but also on genetic variability. Non-nutritional influences can raise or lower concentrations of nutrients, for instance in the presence of inflammatory disease. Some compounds may undergo extensive metabolism while others may be endogenously synthesized; some biomarkers may not reflect the preferred reference period of intake, determined by a biomarker's half-life in blood or urine.

## 39.10 Future directions

The recent advances in 'omics' technologies, such as metabolomics, have opened up the study of small molecules or metabolites. Using hypothesis-driven, targeted, and hypothesis-free, untargeted discovery approaches have paved the way for the identification, development, and validation of new nutritional biomarkers. Metabolomic signatures of food groups and foods have been identified, such as for coffee intake, meat intake, and sugar-sweetened beverages consumption. In addition, metabolomics markers have been identified as potential biomarkers for healthy dietary patterns. In one study with three prospective US cohorts, metabolomics signatures of plant-based diets were identified to distinguish between an overall plant-based dietary index, and a healthy and an unhealthy plant-based dietary index. In another study, 32 serum metabolites were identified for four different healthy dietary patterns. Further work is needed for the validation and application of these nutritional metabolomics markers, but this holds great promise for objective assessment of dietary factors, including for evaluating adherence to dietary recommendations issued by public health agencies.

The availability of a broad range of nutritional biomarkers will open up possibilities for more robust testing of causality of proposed diet–disease associations using genetic Mendelian randomization experiments that can overcome issues of confounding and reverse causality that are known problems in observational epidemiology.

Conventional ways of sample collection have limitations outlined earlier in this chapter. The current logistics of blood collection, for example, for nutritional biomarkers pose challenges for researchers and for participants, including lower response rates, high costs, and issues of preparation and long-term storage of samples. A range of new technologies have emerged, such as near-person or point-of-care testing, with the advantage of rapid or immediate reporting, or a dry blood spot that is easier, faster, and cheaper to collect and store. Both methods require only a finger prick sample and would not need a trained phlebotomist, and collect only a tiny volume compared with traditional large volume blood collections. Though clearly attractive, very few nutritional biomarkers have currently been measured and validated using these methods. This is an exciting area of development that opens up the possibility of collecting repeated samples efficiently, which has the potential to revolutionize the field in several ways, including overcoming the current substantial limitation of single-time biomarker measurements in medical research and national surveys, and it could also enhance the use of nutritional biomarkers for personalized nutrition.

## Further reading


1. **Bingham, S.A.** (2003) Urine nitrogen as a biomarker for the validation of dietary protein intake. *J Nutr*, **133**, 921S–924S.
2. **Bingham, S.A., Cassidy, A., Cole, T.J., et al.** (1995) Validation of weighed records and other methods of dietary assessment using the 24-hour urine nitrogen technique and other biological markers. *Br J Nutr*, **73**, 531–50.
3. **Costabile, A., Klinder, A., Fava, F., et al.** (2008) Whole-grain wheat breakfast cereal has a prebiotic effect on the human gut microbiota: a double-blind, placebo-controlled, crossover study. *Br J Nutr*, **99**(1), 110–20.
4. **D'Angelo, S., Gormley, I.C., McNulty, B.A., et al.** (2019) Combining biomarker and food intake data: calibration equations for citrus intake. *Am J Clin Nutr*, **110**(4), 977–83.
5. **Day, N., McKeown, N., Wong, M., Welch, A., and Bingham, S.** (2001) Epidemiological assessment of diet: a comparison of a 7-day diary with a food frequency questionnaire using urinary markers of nitrogen, potassium and sodium. *Int J Epidemiol*, **30**(2), 309–17.
6. **Foster, E., Lee, C., Imamura, F., et al.** (2019) Validity and reliability of an online self-report 24-h dietary recall method (Intake24): a doubly labelled water study and repeated-measures analysis. *J Nutr Sci*, **8**, e29.
7. **Garcia-Aloy, M., Rabassa, M., Casas-Agustench, P., et al.** (2017) Novel strategies for improving dietary exposure assessment: multiple-data fusion is a more accurate measure than the traditional single-biomarker approach. *Trends Food Sci Technol*, **69**, 220–9.
8. **Gormley, I.C., Bai, Y., Brennan, L.** (2020) Combining biomarker and self-reported dietary intake data: a review of the state of the art and an exposition of concepts. *Stat Methods Med Res*, **29**(2), 617–35.
9. **Harding, A.H., Wareham, N.J., Bingham, S.A., et al.** (2008) Plasma vitamin C level, fruit and vegetable consumption, and the risk of new-onset type 2 diabetes mellitus: the European prospective investigation of cancer—Norfolk prospective study. *Arch Intern Med*, **168**(14), 1493–9.
10. **Huang, Y., Zheng, C., Tinker, L.F., Neuhouser, M.L., and Prentice, R.L.** (2022) Biomarker-based methods and study designs to calibrate dietary intake for assessing diet-disease associations. *J Nutr*, **52**(3), 899–906.
11. **Hunter, D.J., Rimm, E.B., Sacks, F.M., et al** (1992). Comparison of measures of fatty acid intake by subcutaneous fat aspirate, food frequency questionnaire, and diet records in a free-living population of US men. *Am J Epidemiol*, **135**(4), 418–27.

12. **Jenkins, B.J., Seyssel, K., Chiu, S., et al.** (2017) Odd chain fatty acids; new insights of the relationship between the gut microbiota, dietary intake, biosynthesis and glucose intolerance. *Sci Rep*, **7**, 44845.
13. **Kabagambe, E.K., Baylin, A., Allan, D.A., Siles, X., Spiegelman, D., and Campos, H.** (2001) Application of the method of triads to evaluate the performance of food frequency questionnaires and biomarkers as indicators of long-term dietary intake. *Am J Epidemiol*, **154**(12), 1126–35.
14. **Killilea, D.W., McQueen, R., and Abegania, J.R.** (2020) Wheat germ agglutinin is a biomarker of whole grain content in wheat flour and pasta. *J Food Sci*, **85**(3), 808–15.
15. **Linko, A.M., Juntunen, K.S., Mykkänen, H.M., and Adlercreutz, H.** (2005) Whole-grain rye bread consumption by women correlates with plasma alkylresorcinols and increases their concentration compared with low-fiber wheat bread. *J Nutr*, **135**(3), 580–3.
16. **Livingstone, M.B. and Black, A.E.** (2003) Markers of the validity of reported energy intake. *J Nutr*, **133**(Suppl 3), 895s–920s.
17. **McCullough, M.L., Maliniak, M.L., Stevens, V.L., et al.** (2019) Metabolomic markers of healthy dietary patterns in US postmenopausal women. *Am J Clin Nutr*, **109**(5), 1439–51.
18. **Miles, F.L., Lloren, J.I., Haddad, E., et al.** (2019) Plasma, urine, and adipose tissue biomarkers of dietary intake differ between vegetarian and non-vegetarian diet groups in the Adventist Health Study-2. *J Nutr*, **149**(4), 667–75.
19. **Mitry, P., Wawro, N., Rohrmann, S., Giesbertz, P., Daniel, H., Linseisen, J.,** (2019) Plasma concentrations of anserine, carnosine and pi-methylhistidine as biomarkers of habitual meat consumption. *Eur J Clin Nutr*. **73**(5):692–702.
20. **Mozaffarian, D., de Oliveira, O.M.C., Lemaitre, R.N., et al.** (2013) Trans-palmitoleic acid, other dairy fat biomarkers, and incident diabetes: the Multi-Ethnic Study of Atherosclerosis (MESA). *Am J Clin Nutr*, **97**(4), 854–61.
21. **Neuhouser, M.L., Pettinger, M., Lampe, J.W., et al.** (2021) Novel application of nutritional biomarkers from a controlled feeding study and observational study toward dietary pattern characterization in postmenopausal women. *Am J Epidemiol*, **190**(11), 2461–73.
22. **Prentice, R.L., Sugar, E., Wang, C.Y., Neuhouser, M., and Patterson, R.** (2002) Research strategies and the use of nutrient biomarkers in studies of diet and chronic disease. *Public Health Nutr*, **5**(6A), 977–84.
23. **Prentice, R.L., Tinker, L.F., Huang, Y., and Neuhouser, M.L.** (2013) Calibration of self-reported dietary measures using biomarkers: an approach to enhancing nutritional epidemiology reliability. *Curr Atheroscler Rep*, **15**(9), 353.
24. **Schatzkin, A., Kipnis, V., Carroll, R.J., et al.** (2003) A comparison of a food frequency questionnaire with a 24-hour recall for use in an epidemiological cohort study: results from the biomarker-based OPEN study. *Int J Epidemiol*, **32**, 1054–62.
25. **Slimani, N., Kaaks, R., Ferrari, P., et al.** (2002) European Prospective Investigation into Cancer and Nutrition (EPIC) calibration study: rationale, design and population characteristics. *Public Health Nutr*, **5**(6B), 1125–45.
26. **Tasevska, N., Runswick, S.A., McTaggart, A., and Bingham, S.A.** (2005) Urinary sucrose and fructose as biomarkers for sugar consumption. *Cancer Epidemiol Biomarkers Prev*, **14**(5), 1287–94.
27. **Wang, F., Baden, M.Y., Guasch-Ferré, M., et al.** (2022) Plasma metabolite profiles related to plant-based diets and the risk of type 2 diabetes. *Diabetologia*, **65**(7), 1119–32.
28. **Welch, A.A., Bingham, S.A., Ive, J.E., et al.** (2006) Dietary fish intake and plasma phospholipid n-3 polyunsaturated fatty acid concentrations in men and women in the European Prospective Investigation into Cancer-Norfolk United Kingdom cohort. *Am J Clin Nutr*, **84**(6), 1330–9.
29. **Zheng, J.S., Sharp, S.J., Imamura, F., et al.** (2020) Association of plasma biomarkers of fruit and vegetable intake with incident type 2 diabetes: EPIC-InterAct case-cohort study in eight European countries. *BMJ*, **370**, m2194.

# Part 9

# Applications

# 40 Sports Nutrition

Louise M. Burke



This chapter focuses on dietary strategies to enhance performance of sport for competitors at various levels of sporting ability. The strategies also offer the opportunity to enhance the benefits of exercise for the health of those not involved in professional or competitive sports.

Although the desired outcomes of those involved in sport at various levels are widely divergent, there is a shared interest in goals such as gaining lean mass, reducing body fat stores, building strong bones, and improving metabolic capacity. Knowledge relating to sports nutrition has evolved rapidly since the first edition of this textbook and recommendations are based on a firmer evidence base. Social media have enabled rapid dissemination of evidence-based knowledge but have also provided opportunities for promoting untested and potentially deleterious approaches to enhancing performance.

## 40.1 New themes in sports nutrition

The jointly prepared 2016 Position Stand on Nutrition for Athletic Performance by the American College of Sports Medicine, the Academy of Nutrition, and Dietetics and Dietitians of Canada identified ten key themes, which underpin the important strategies in sports nutrition. These themes provide a way to unify the advice offered across a range of different sports and sporting goals.

**Nutrition goals and requirements for each athlete are changing rather than static** Athletes organize their training and competition activities into a periodized programme, which integrates different types of workouts into various cycles of the training calendar (e.g. weekly cycles, monthly cycles) with the overall aim of achieving optimal performance at targeted events. The athlete's diet also needs to be periodized and is constructed by assembling nutritional support for each of the different training sessions into a bigger plan that takes into account the overall nutritional goals for any period.

**The athlete's nutrition plan needs to be individualized** Nutrition plans need to be personalized for each athlete to take into account the uniqueness of their event or goals, their individual experiences and responses to the same strategies, and specific practical challenges and eating preferences. One size does not fit all, even for athletes in the same sport or for the same athlete in different scenarios. High performance athletes and recreational athletes may share some similar challenges and nutritional strategies, as well as differences. For example, an elite runner trying to break the 2-hour marathon time barrier will glycogen load pre-race and consume carbohydrate-containing drinks during the event (see section 40.3), with rapid handling of personal bottles containing a bespoke solution that are made available at special aid stations over the race-course. He is likely to become substantially dehydrated over the course of the race, because his rate of sweat loss is high (high heat production) and it is difficult to grab and swallow significant amounts of fluid while running at 21 km/hour. He may complete 120–160 km per week in training to prepare for the race, involving a range of specific sessions and specialized training techniques (speed sessions, low glycogen training, double sessions in a day, gut training). Meanwhile, a 4-hour marathon runner in the same race should also glycogen load and consume extra carbohydrate over the race duration. However, he will run at a substantially lower absolute intensity (speed) and at a lower percentage of his maximal aerobic capacity, so his absolute rates of burning carbohydrate fuel are lower but for a longer duration. He will probably undertake 70–100 km/week of training, and include some specialized techniques, but his major focus is likely to be on completing the distance. He might carry some of his own carbohydrate sources (gels) during the race or make use of the commercial sports drink and water supplied at general aid stations. A lower speed will mean lower sweat rates, and because he is likely to be able to drink larger amounts of fluid while running more slowly, he will likely be less dehydrated by the end of the race. His friend, a 6-hour marathon runner, might actually overhydrate (gain weight) during the race, because of his very slow speed (low sweat rates) and his choice to walk through the aid stations and to follow (mistaken) advice to 'drink as much as possible'.

**Training and competition nutrition strategies may have different goals and principles** While competition nutrition strategies provide support to maximize performance, a key goal of training is to adapt the body to develop metabolic and functional adaptations that may be best developed by withholding nutrition support (e.g. undertaking a lengthy session without consuming carbohydrate for additional fuel). Optimal performance requires optimal fuel use and adequate fuel stores. Training helps the body to develop optimal use of body fuels, while competition strategies should ensure that fuel stores are adequate for the needs of the event.

**Energy availability sets an important foundation for health and performance goals** To maximize health and performance outcomes, an athlete's energy intake must cover the cost of all of the body's systems and activities. Energy availability is a concept that focuses on the difference between the athlete's energy intake and the energy they commit to exercise, which provides the fuel for a range of body systems needed to keep the body in optimal function. When this gap becomes too small, the body adapts to become more energy efficient by reducing the expenditure (and optimal function) of these systems. This is the basis of the syndromes now known as Relative Energy Deficiency in Sport (REDs) and Female and Male Athlete Triad (Triad).

**The achievement of the body composition associated with optimal performance is important, but challenging** The athlete should manipulate training and nutrition to achieve a level of body mass, body fat, and muscle mass that is consistent with good health and good performance. However, manipulation of body composition should be individualized and periodized to avoid practices that create unacceptably low energy availability and psychological stress. Many sporting environments struggle with a high risk of disordered eating or poor mental health due to an excessive and misunderstood focus on achieving and maintaining low body fat/weight levels. High performance systems are now trying to implement programmes that provide a more evidence-based and supportive approach to physique management.

**The way we express nutritional recommendations is important** Guidelines for some nutrients (e.g. energy, carbohydrate, and protein) should be expressed using guidelines per kg body mass to allow

11. **Schultz, W.M., Kelli, H.M., Lisko, J.C., et al.** (2018) Socioeconomic status and cardiovascular outcomes: challenges and interventions. *Circulation*, **137**(20), 2166–78.
12. **Seligman, H.K. and Berkowitz, S.A.** (2019) Aligning programs and policies to support food security and public health goals in the United States. *Annu Rev Public Health*, **40**, 319–37.
13. **Tarasuk, V.S. and Beaton, G.H.** (1999) Women's dietary intakes in the context of household food insecurity. *J Nutr*, **129**, 672–9.
14. **Williams, C. and Dowler, E.A.** (1994) *A working paper for the Nutrition Task Force Low Income Project Team*. London: Department of Health.

## Useful website

**PROOF Food Insecurity Policy Research**
https://proof.utoronto.ca/

# Food in Hospitals

Suzie Ferrie



## 42.1 The challenge of food in hospital: 'harder than running a restaurant'

Like any other institutional catering or food service, a hospital has to provide meals for a large number of people with differing needs and preferences. Allowing for customers' diverse cultural and social food habits, and making mass-produced food that is palatable and safe, while paying attention to environmental sustainability, is a challenge for any large-scale catering endeavour. However, hospitals have an additional problem to solve: they are serving customers who are ill. During illness, it is normal to experience a loss of appetite, and hospital patients may have other issues preventing them from achieving an adequate nutritional intake (see Table 42.1).

Variation between individuals' food preferences also increases during illness, as different people may have very different, and quite personal, ideas about what they should eat when they are sick. An institutional food service cannot provide every patient with their own idiosyncratic selection of 'comfort foods.' Illness can also increase the requirements for particular nutrients, and large-scale catering systems usually involve increased nutrient losses from foods, due to extended storage of chilled food and/or extended holding at hot temperatures for service. For example, most of the content of heat-labile vitamins such as folate or vitamin C may be lost due to these delays after cooking. Hospital patients are also at higher risk of food poisoning, making safety a greater concern than in other large-scale food systems, and within a public health system, taxpayer-funded hospitals may also be working within a very limited food service budget, which reduces the scope for varying serving hours and food choices to suit patients' preferences.

It is important to try to solve this problem because hospital is the one place in an affluent environment that malnourished people can always be found—in different studies across the world, researchers repeatedly find that around 40% of hospital patients are malnourished to some degree. There are several reasons for this. People with chronic or debilitating disease processes are often already malnourished on admission to hospital. However, this can become worse in hospital, and well-nourished people can deteriorate too, not just because they are ill, but because they do not receive adequate nutrition to meet their needs, or are unable to consume enough of it. Repeated or prolonged periods of fasting, required for some of the tests and treatments in hospital, may also lead to periods without any nutrition at all. Malnutrition can slow or prevent recovery from illness or surgery. Malnourished people have

**Table 42.1 Barriers to adequate intake in hospital patients**

| Contributing factors | | Effect |
|---|---|---|
| Physical factors | • Reduced mobility (e.g. unable to get up or sit up for meals)<br>• Difficulty in managing cutlery | Reduced access to food |
| Institutional/ environmental factors | • Enforced fasting for hospital tests and procedures<br>• Interruptions to meal times due to tests and procedures (e.g. absence from ward when meal is served, or meal is interrupted by clinical visit)<br>• Limitations of the food service system (budget, timing, flexibility, infection control, patient choice) | Reduced access to food |
| Patient subjective factors | • Lack of appetite<br>• Dislike of the limited food choices<br>• Altered taste sensation<br>• Nausea (e.g. due to reduced gastric emptying, or some medications) | Reduced intake of food |
| Patient physical factors | • Reduced level of consciousness (e.g. drowsy or confused)<br>• Shortness of breath (interferes with chewing, or may need mask ventilation)<br>• Chewing problems (e.g. due to poor dentition, dry mouth, or sore mouth)<br>• Mechanical obstruction in mouth or gastrointestinal tract (e.g. cancers of the head and neck)<br>• Neuromuscular swallowing problems | Reduced intake of food |
| External factors | • Limitations to allowed intake (e.g. clear fluids diet)<br>• Dietary restrictions due to illness (e.g. fluid restriction, potassium restriction) | Reduced intake of food |
| Increased nutritional needs | • Increased nutritional requirements due to illness (e.g. higher protein) | Increased requirements |
| Increased nutritional losses | • Altered digestive functioning or nutrient losses (e.g. malabsorption) | Increased requirements |

poorer wound healing, a higher risk of contracting new infections, and ongoing weakness, which interferes with rehabilitation (see Fig. 42.1). All of this can mean a longer and more costly hospital stay and longer-term problems after discharge. Costs are further increased if food and resources are wasted when patients do not, or cannot, eat their food. Making sure that patients receive optimal nutrition is therefore an essential part of achieving the hospital's overall aim of getting people better and getting them home.

Unfortunately, malnutrition in hospitals is still common despite being recognized long ago and discussed in detail for decades, and being investigated in numerous research studies. Across the world media reports recur regularly about unappetizing

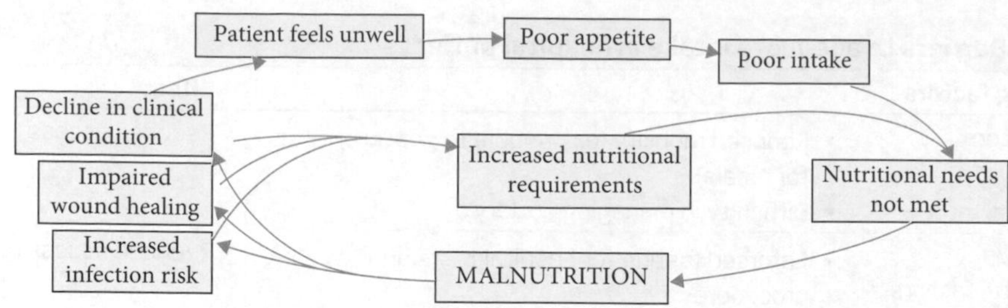


Fig. 42.1 Cycle of malnutrition in hospital.

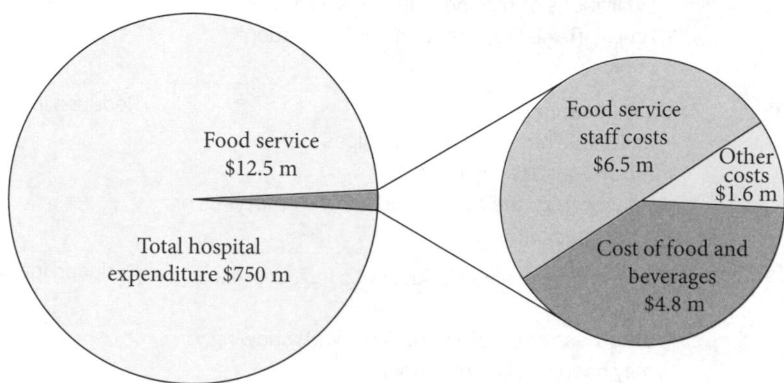


Fig. 42.2 Typical yearly budget for a large teaching hospital of over 750 beds (2021–2022).

hospital meals going uneaten, and malnourished hospital patients unable to reach their meal tray or open the food packaging, reluctant to complain for fear of being labelled a 'bad patient'. Usually such emotive stories are followed by a call to improve food service funding and hospital staffing so that the food can be improved and patients can receive the assistance they need at meal times. However, such calls mostly seem to go unheeded. While funding arrangements vary enormously between countries and health systems, healthcare budgets are generally limited. The food service is actually quite a small proportion of the hospital's budget, and doubling the expenditure on food might add only 1–2% to the total costs of running the hospital (see Fig. 42.2), but often it appears that patients' food is not considered to be a part of the medical treatment, unlike expensive drugs and equipment. Rather, it is seen as a 'hotel service', like bedlinen and cleaning, that can and should be economized.

## 42.2 Policy approaches to the problem

There have been many suggestions about ways to address the problem of hospital food that do not require an overhaul of the food service budget, and in countries where there is a strong public health system, numerous guidelines, policies, and regulations apply to institutional food services with the goal of ensuring adequate nutrition for all patients. Such policies may set standards for the nutritional contents of hospital diets, direct the provision of assistance at meal times, regulate food safety procedures, and dictate suitable meal timings. There may also be hospital policies governing environmental

**Box 42.1** Example of nutrition policy standards (Ministry of Health, NSW, Australia, 2017)

- All clinical staff are responsible for ensuring hospital patients eat their meal.
- Patients should have the opportunity to choose foods no more than one day before the meal, and have assistance with menu choice where needed.
- Meals should be served promptly to optimize temperature, nutrient content, and quality; food should also be available when needed outside the set meal times.
- Wards should be adequately staffed at meal times and the importance of providing timely and individualized assistance with eating and drinking should be recognized in work allocations.
- Meal time environment should be conducive to eating.
- Feeding assistance should be provided where required—carers, relatives, or volunteers can be involved if deemed safe by clinical staff and if training has been provided. Clinical staff still have the primary responsibility for this.
- Ward/medication rounds, teaching, and diagnostic procedures should be scheduled to avoid interrupting patients' meal times.
- Patients should be assisted in preparing for eating prior to the meal delivery (e.g. positioning, toileting, dentures, handwashing, and making room for the meal tray).
- Mobility assistance should be provided to patients who are able to sit out of bed for meals.
- Staff should ensure that patients are able to reach and manage their tray and open food packaging.
- Appropriate modifications should be made, where needed, to enable eating (e.g. adaptive aids, modified cutlery, drinking devices).
- Assistance with feeding should be provided as needed.
- Independence should be supported and dignity maintained throughout meals and preparation for the meal.
- Patients should be given adequate time (at least 30 minutes) to consume their meal before the tray is collected.
- A system should be in place for assessing food, packaging, and dinnerware for ease of use by hospital patients.

sustainability that affect the food service. In many countries there are national healthcare standards that include aspects of patients' nutrition and hydration (see Box 42.1), and hospitals may undergo regular accreditation checks to make sure they are meeting those standards.

## 42.3 Structural influences on hospital food

The hospital's mode of food service can make a difference to patients' nutrition too. Patients may be more inclined to eat if they are served meals at their usual times, but this possibility may be limited by the food service system. It can take longer than an hour to serve one meal to all the patients in a large hospital, leading to staggering of meal times across the hospital. When appetite is poor, a common strategy to increase intake is to offer smaller, more frequent meals spread over the day, but this increases staffing costs, particularly in a cook-fresh system, as cooking fresh food for immediate service requires a larger number of food service staff to be rostered over the three main meals, and costs can be reduced by shortening the food service day (with the breakfast served late and the evening meal early), which leads to an undesirably long overnight period without food. Some healthcare councils mandate a limit to the overnight fast of no more than 15 hours, but even that is too long for many patients. This has become less of a problem with the recent trend away from individual hospital kitchens, towards centralized food production facilities using cook/chill methods. These allow lower staffing levels at meal service times. Such systems also enable the nutritional content and food safety of hospital meals to be more standardized, but also mean that there is less scope to cater to the preferences of an individual patient and to support local food producers and suppliers. The food service system also determines which types of foods can be provided, as in any system there will be some foods that end up unpalatable, or not microbiologically safe, when produced that way.

Different hospitals may structure the meal service differently, which can also indirectly affect patients' food intake. Some use an individual meal tray for each patient, taken by trolley to the patient's bed, which enables the meal to be tailored in advance, but does not allow for on-the-spot changes. With such a system, variety is ensured by using a menu cycle, often 14 days long, in which different meal choices are offered each day of the cycle. The patient may have to order their desired options more than 24 hours before the meal, which can be difficult during illness as the patient may not be able to predict how they will be feeling at the meal time. Other facilities may have bulk food delivery to a ward kitchen area, which allows different options to be offered at the point of meal service. This enables the patient to choose exactly what they feel like eating at the time, but makes it more difficult to plan meals in advance or to structure and monitor the nutrient intakes over the day. Lack of variety can contribute to poor intake, as longer-term patients tire of repetitive menu choices. In recent years, for medical and budget reasons, the average length of hospital stay has decreased markedly so that patients tend to be in acute-care hospitals for days rather than weeks. This means most patients will be discharged before the same food items reappear on the menu, but it does become monotonous for the one-quarter of patients whose admission lasts weeks or even months.

The choice of food service system can also have more direct effects on intake. For example, all of the food for the meal served together on one tray may be off-putting for patients whose appetites are poor, and the small plates that fit on most hospitals' food trays can also make normal serves of food look intimidatingly large. Flexibility to alter serve size may be minimal if the reheating method delivers the same 'dose' of heat to every plate—inadequately warming a serve that is too large, and drying out or even scorching a serve that is too small. To eat sitting in bed can be awkward even for patients with no loss of function, as this position makes it difficult to cut up meat, manage soup, or put spreads onto bread.

## 42.4 Strategies to improve food intake in hospital

Improving nutrition for patients in hospital is a multidimensional challenge, and many different strategies have been tried or are currently being used to try to address this issue. See Box 42.2 for change management strategies that have been shown to deliver improvement in hospital trials. Some of these are practical only in particular settings. For example, in an acute-care hospital, most patients may be too sick to sit up in a dining area, and not have enough energy to be sociable, or infection control policies may require patients to remain isolated. Coloured meal trays can be used only in settings with a tray system; and it may be hard to schedule meal assistance when meal times are flexible. Attempts to improve food intake are rarely successful when implemented singly. For example, if there is no extra meal time assistance available, it makes little difference to have coloured trays identifying patients requiring it. Eliminating interruptions at meal times may result in fewer clinical staff present in the ward during meals, so that eating problems are then less likely to be noticed and addressed. Also, some interruptions can have a positive effect on intake, for example if a doctor concludes an examination of the patient by helping to set up the meal tray, opening packets, and verbally emphasizing the importance of eating well to get better. The benefit of this sort of 'interruption' can be lost if all interruptions are discouraged at meal times.

Ultimately, the success of any approach seems to depend on a cultural shift in the hospital, so that

**Box 42.2 Examples of successful strategies to improve meal intake in hospital**

- 'Protected meal times', where interruptions are avoided for a defined period at each main meal, and staff are mobilized from other tasks and their own meal breaks delayed, to assist patients with eating.
- Coloured meal trays to indicate that a particular patient needs help with eating.
- Rostered volunteers attending at meal times to help with meal set up and/or feeding.
- Specific allocated staff to 'host' each meal and supervise intake.
- Encouraging family and friends to visit at meal times and help with feeding.
- Setting up a group dining area for any patients able to sit up for meals, with a pleasant social environment and peer encouragement to help motivate and normalize eating.
- Consecutive scheduling of tests or procedures that require fasting, to avoid repeated periods without food.
- Minimized fasting times before and after tests and procedures.
- Tests and procedures prioritized or 'fast-tracked' for malnourished or critically ill patients or those who have already been fasting.
- Flexible hospital menus that allow nutritious between-meal snacks, or delayed meals (e.g. after a procedure).
- Meal ordering close to each meal time so that patients do not have to predict how they will feel and what they will want to eat too far in advance.

all staff learn to place high priority on patients' nutrition. Part of the problem here is that, in many hospitals, it remains unclear which staff should be responsible for making sure that patients can and do consume the food. Traditionally, nurses provided holistic care of the patient, including their feeding. Florence Nightingale noted that 'the proper selection and administration of diet' was as important as fresh air, and nurses were responsible for ensuring that the patient had plenty of both. Hospital dietetics arose from this nursing involvement in the patients' food, but now is a separate discipline in its own right.

As nursing has become more clinically focused, and less about general care of the patient, many aspects of the 'care' have become medicalized, including food. Several studies have indicated that nursing staff now see their role in hospital meals as ensuring that the correct therapeutic diet is ordered, and restrictions adhered to, rather than responding to the patients' cues (food preferences, or difficulties in eating). Nurse staffing levels may also not reflect the need to assist many patients at the bedside during meals, or this task may be delegated to non-nursing, unskilled helpers.

Although nurses are the main staff at the patients' bedside who are able to assist and monitor eating, and obtain extra items when needed outside mealtimes, they usually have little or no influence on the food service system or the selection of foods. Hospital dietitians may have significant power to alter food selections, and have some input into aspects of the food service system, but the scope for the latter is often limited, particularly where food production is centralized and the budget controlled externally. Assembling the meal tray and delivering it to the patient is usually the responsibility of the food service staff, who may have little or no nutritional training. Typically such staff are not considered to be an influence on food intake, yet their role may make a nutritional difference if, for example, the meal tray is simply dumped on a dirty bedside table out of the patient's reach, or if there are errors in the composition of the meal.

As early as 1963, Platt et al. (in the book *Food in hospitals*) referred to this 'cumbersome tripartite structure' of hospital food provision, noting potential for conflict, evasion of responsibility, and poor communication between nurses, dietitians, and catering staff. Since then, several studies have investigated the values underlying the work of these three groups, and these identify some differences that may contribute to conflict. Dietitians may have goals determined by clinical principles, while the catering staff are constrained by the practical demands of running a food service for three meals a day, seven days a week. Nursing staff may view the patients' food as just one more of the many things that have to be managed over the day, perhaps less important than dressing a wound or administering medication. Rather than arguing about whose responsibility it is, or viewing it as 'cumbersome', it should be possible to build a sense of three-way partnership. Ensuring good communication between these three groups of staff is essential, to enable shared values to be developed that prioritize patients' nutrition, and to emphasize that responsibility for this issue is shared between all three groups and the rest of the multidisciplinary team.

## 42.5 The patient's experience of food in hospital

Patients' impressions of the food service may vary considerably. In many developing countries, the hospital provides only a minimal meal and there is a strong expectation that families will provide food and often also help with the physical care of the patient. People in industrialized countries usually assume that the hospital will provide all necessary care during the patient's admission, including a balanced and adequate diet that meets the patient's full nutritional needs more accurately than their diet at home. However, patients themselves may not expect the food to be enjoyable, or to be able to eat when they want to—they may accept the hospital's system, as part of accepting their 'patient role'. Meals, even unpalatable ones, may be appreciated as a distraction from the monotony of a day in hospital. Several studies have found that patients are often pleasantly surprised when their meals do not fit the common stereotype of horrible hospital food. It may be that food preparation has improved since these stereotypes were formed. Another possible reason is hospitals' increasing reliance on portion-controlled packs for items such as stewed fruit, juice, cheeses, desserts, biscuits, and cakes, which means that the foods come with familiar brand names, in packages that are available in supermarkets, more recognizable and less institutional-looking. Patients (and, more often, their families) may, however, judge their overall hospital care partly in terms of what the food was like. Relatives may be outraged if the food does not look palatable or if a meal is incorrect or omitted. As Briony Thomas's *Manual of dietetic practice* points out, 'Patients often cannot tell the difference between good treatment and bad treatment, but they can always identify poor food' (Thomas and Bishop, 2013). See Table 42.2 for a list of the common complaints about hospital food.

## 42.6 Special diets and nutritional supplementation

The food that patients receive may be subject to various dietetic alterations. Traditionally, a 'diet kitchen' was staffed by a dietetic nurse (later a dietitian) who would prepare the foods for special therapeutic diets. These might include reduced-salt items, sugar-free items, special foods high or low in protein, etc. Decreased length of stay in hospital now means that patients are quite unwell for most of their stay, and may also be subjected to many tests and procedures in that time. There is therefore less opportunity to have a proper meal, often only on the last day or so in hospital. Dietary restrictions consequently have become less important in patient care, and the difference between the general hospital meal and a special diet meal has decreased. Another factor contributing to this change has been the emergence of new evidence supporting 'normal' foods for many conditions. For example, people with diabetes are no longer advised to avoid all sugar (or to eat any fat they want!), and instead are encouraged to have

Table 42.2 **Common complaints about hospital food**

| Complaint | Why it happens |
|---|---|
| Hospital food is bland | Many sick people find the smell and taste of spicy food unappealing, so spicing is often reduced or avoided in hospital meals. Fresh ingredients also lose flavour during prolonged holding at hot temperatures for service and extended periods of chilled storage |
| Hospital food is packaged like an aeroplane meal, with lots of individual packs | Labour costs are a high proportion of the food service cost, so significant savings can be made with individual packs, since staff are not required to prepare or portion the food. Individual packs may also be subject to less spillage |
| Hospital food is overcooked<br>Hospital food is undercooked | Hospital food service systems often reheat food by giving the same amount of heat to each meal. Variation in food consistency or serve size may result in overcooked or undercooked items. Overcooking can also result from holding food at safe hot service temperature for the length of time required to serve a large number of meals |
| Hospital meals seem to contain little fresh fruit or salads | Raw fruits and vegetables pose a microbiological risk that is too high for many patients or food service systems |
| Packaging is difficult for weak or elderly patients to open | While many health departments do consult with manufacturers about packaging, seals need to be robust to survive transport and handling |

a healthy balanced diet with 'good' fats and carbohydrate choices. Additionally, many conditions, such as gout, gastric ulcer, or pancreatitis, are now largely managed with medication instead of dietary restriction. Special diets in hospital now have more emphasis on supporting intake, rather than restricting it. This is a good thing because intake is so often impaired, and poor intake may also mean restrictions are unnecessary: if a patient is restricted to half of a normal daily sodium intake, for example, they do not need a special 'low-sodium' diet if they are managing only half of their meals! Maintaining such restrictions may improve outcomes less effectively than making sure that the patient is eating enough food generally for good nutrition. Responding to this, the American Dietetic Association and the ESPEN society recently recommended avoiding any dietary restrictions in older people, as they are already at higher risk of malnutrition than the general population.

When usual meals are inadequate to meet the patient's needs, there are some options for supplementing the diet. Firstly, extra snack items (such as yoghurt, a small sandwich, nuts, or a milk drink) may be added to an individual patient's regimen for consuming between meals. These may be chosen to be high in energy and/or high in protein, depending on the patient's needs.

Secondly, fortifying existing foods can boost the energy, protein, or micronutrient content of particular items without the patient having to consume greater bulk. This may be done with normal food ingredients (such as by adding egg or cheese to a dish for protein, or margarine for fat) or using modular products such as protein powder, oil, or glucose polymer syrup that can be cooked in the food or mixed in before serving.

Thirdly, formulated nutrition supplement drinks and puddings may be used to replace snack items. These are usually high in energy and/or protein, and many are also nutritionally complete, providing all the essential micronutrients in some achievable volume of product. These make it very easy to boost intake. For example, a 200 mL drink might contain 1200–1600 kJ (300–400 kcal) and 15–20 g protein with one-seventh of an adult's daily micronutrient requirements; that is, more than twice the energy and protein found in the same volume of flavoured milk, which is not nutritionally complete. The increasing use of these products in hospitals has been challenged, with critics suggesting that it would be more cost-effective to improve the hospital food so that patients ate more of their meal, rather than relying on supplement drinks to make up the deficit. Supplements have less flexibility than food to meet patients' preferences, and many patients find them too sweet, particularly if they have experienced taste changes as part of their illness. Supplement products are also more expensive than ordinary food, but it is doubtful whether increased intake of improved hospital food could match the concentrated complete nutrition that a supplement can provide.

Some hospitals are now conducting research to investigate better ways to improve patients' nutrition. Success may require a combination of changes to hospital policies, hospital food service systems, dietary strategies, and various multidisciplinary ward-based interventions, along with an institution-wide cultural shift, because hospitals are complex systems and changing them is a complex process. It is important to do this because good hospital food has the potential to achieve significant benefits for healthcare costs, and for patients' health and well-being.

## Further reading

1. **American Dietetic Association** (2005) Position of the American Dietetic Association: liberalization of the diet prescription improves quality of life for older adults in long-term care. *J Am Diet Assoc*, **105**, 1955–65.
2. **Platt, B.S., Eddy, T.P., and Pellett, P.L.** (1963) *Food in hospitals: a study of feeding arrangements and the nutritional value of meals in hospitals.* London: Oxford University Press for the Nuffield Provincial Hospitals Trust.
3. **Thomas, B. and Bishop, J. (eds)** (2013) *Manual of dietetic practice*, 4th edn. Oxford: Wiley-Blackwell.

# 43 Nutritional Support for Hospital Patients

Ross C. Smith



## 43.1 Enteral nutrition

Surgical staff in the 1960s and 1970s would provide puréed food directly through a large-bore gastrostomy tube into the stomach of patients who were unable to eat. The concept of 'enteral nutrition' is much more refined than this and has developed as a specialized form of nutrition therapy out of the space programme, where a balanced nutrition that resulted in minimal excretion was a distinct advantage (see Box 43.1). It is interesting that in the late 1960s the concept of parenteral nutrition was also at a phase of rapid development in the era of early space exploration, and this probably helped to evolve the use of 'space diets' as a type of complete nutrition in a liquid form that could be delivered by a tube directly into the gastrointestinal tract for the treatment of hospitalized patients. The first such product was an elemental formula, but polymeric products are more easily metabolized and cheaper. Following on from the success of enteral nutrition for complex hospitalized patients, the concept has been successfully evolved for nutritional treatment for nursing-home patients and outpatients who are otherwise unable to eat. The most recent research has demonstrated the most effective means of feeding the critically ill patient.

**Box 43.1** Enteral and parenteral nutrition

- *Enteral nutrition*: feeding by tube directly into the stomach or upper small intestine by using a formula that is a mixture of nutrient sources that can be passed in water emulsion through a fine tube.
- *Parenteral feeding*: provides the patient's nutrition into a vein because the gastrointestinal tract is not functioning, or has to be rested. Parenteral feeding can be total or partial (supplying part of the total nutrition requirements).

Currently, patients entering hospital are screened for nutritional status by admitting doctors, nurses, and dietitians. The at-risk patient is referred to the ward dietitian, who undertakes a more detailed nutritional assessment and considers the best method of treatment: specialized oral diet, enteral nutrition, or parenteral nutrition. Nutritional support becomes indicated for all patients who are unable to nourish themselves with oral intake for more than 5 days, or sooner if they are malnourished on admission and when it is predicted that the time to oral intake will be more than 5 days. Enteral nutrition should always be used in preference to parenteral nutrition when it can be administered safely because it is cheaper, more physiological, and less complicated than parenteral nutrition. It is important to understand the safety issues when deciding on a patient's suitability for treatment.

### 43.1.1 Indications for enteral nutrition

Enteral nutrition is indicated as a means of nutritional support for patients who are unable to sustain themselves with an oral diet and who have a sufficient normal intestine available for absorption of enteral formula. The frequently quoted statement 'If the gut works, use it!' should be remembered in all patients who are referred for nutritional therapy. There are many different clinical situations where enteral nutrition becomes the preferred method of feeding (see Table 43.1).

Patients with a stroke or other neurological deficit who are unable to swallow satisfactorily are a good example of a situation where enteral nutrition has been found to be invaluable. Critically ill patients who have a functioning gut are also suitable, but there may be an initial period when the gut is not functioning and parenteral nutrition should be considered. A delay in nourishing a critically ill patient is associated with increased complications, length of stay, and mortality.

**Table 43.1 Indications for enteral nutrition**

| Indication | Examples |
|---|---|
| Anorexia due to illness | Eating disorders |
| | Weakness due to illness or surgery |
| | Cancer |
| Swallowing disorders | Cerebrovascular disease |
| | Motor neurone disease |
| | Oesophageal stricture |
| Gastric stasis (gastroparesis) | Post-operative, intensive-care patients |
| Inability to take sufficient oral nutrition | Burns pancreatitis |
| | Open, infected wounds |
| | Trauma |
| | Inflammatory bowel disease |
| | Some patients with sepsis |
| | Critical illness |

### 43.1.2 Contraindications to enteral nutrition

The main contraindication to the use of enteral nutrition is poor functioning of the gastrointestinal tract. In general, in this circumstance the enteral nutrition causes vomiting, gastrointestinal distension, or severe diarrhoea. Patients who are frail and have loss of sensation in the pharynx are at great risk of aspiration of enteral nutrition into the lung; an example of such a patient in this circumstance would be following a stroke involving the muscles of the pharynx. Aspiration should be particularly considered when patients are vomiting or have a respiratory disorder. This is a most important issue and needs to be emphasized.

Another important contraindication to the use of enteral nutrition is the presence of inadequate circulation to the gut, which occurs during conditions such as septic shock and the use of high doses of vasopressor agents. Other extreme conditions, such as mesenteric artery thrombosis and abdominal compartment syndrome, cause gross impairment of absorption.

Ethical issues are also important when considering the need for enteral nutrition. Because enteral nutrition allows complete nutritional therapy, withholding such treatment can be seen as denying a patient their normal and natural requirements. Withholding nutritional support from a patient who is unable to eat is an emotionally charged ethical problem. It is inappropriate to withhold treatment from a patient who has reasonable potential for extended quality of life. Alternatively, it is unreasonable to extend a patient's life when there is no hope of relief from pain and suffering. Unfortunately, these decisions are not always easily resolved and may lead to conflict between the expectations of some relatives and of medical and nursing staff. Mostly these issues can be settled by discussion and by helping all parties understand the potential outcomes of the patient's condition. Funding issues are also an important consideration. In some communities, enteral nutrition is funded by government bodies, while in other communities, the patients' families have to pay.

### 43.1.3 Access to the gastrointestinal tract

Enteral nutrition can be given by mouth as a bolus but, because it is unpalatable and uninteresting when repeated over a number of days, this is not frequently successful. It is therefore mostly given through a nasogastric/jejunal tube directly into the stomach or intestine (see Fig. 43.1). The tube is usually thin and soft and positioned with a guide wire. Because it is small and made of soft material, it is quite acceptable for many patients. Although these tubes are fine and can be used for bolus feeding to allow a patient to be disconnected for mobility, enteral nutrition is tolerated better when given as a continuous infusion over a 24-hour cycle. In some situations, nutritional requirements are achieved as an overnight infusion to provide supplementation. Such a programme would require daily insertion of the tube, which would present difficulties.

A gastrostomy tube can be safely inserted through the abdominal skin into the stomach to allow for direct delivery of the nutrition. Gastrostomy tubes are most frequently placed with the aid of the endoscope. Under vision with an endoscope in the stomach, a needle is passed through the skin into the stomach and a guide is passed through the needle. This guide is pulled back into the mouth and attached to the tube. The tube is then pulled in through the mouth and pulled out of the stomach along the needle, leaving a bulbar portion of the tube in the stomach to anchor it in position. These tubes are designed to lie comfortably in the stomach and to be fixed to the skin. Such tubes are more easily used for intermittent feeding and can be disconnected from the nutrition infusion to allow mobility.

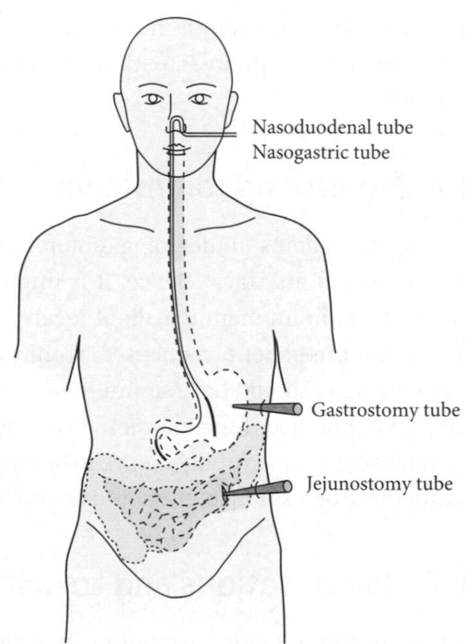


Fig. 43.1 Routes of enteral (tube) feeding.

Jejunostomy tubes can be placed to lie in the small bowel for conditions that result in poor gastric emptying. Again, they can be inserted via the nose, via a gastrostomy, or directly into the small intestine. When the tube is placed into the jejunum, it is best used for continuous infusion.

### 43.1.4 Enteral formulations

In each class of enteral nutrition formulation, there are numerous commercial products. The reason for choosing one over the other often depends on the presentation or method of delivery. Some formulas are presented as dry powder, while others are presented already prepared in a sterile container with easily connected delivery tubing. Most enteral nutrition products are well balanced and can be given as a complete diet, but some formulas are developed for specialized indications. In early formulations, milk was a predominant protein source and with this, lactose was a significant carbohydrate; intolerance to lactose leads to diarrhoea in many patients. In general, lactose has been replaced by corn starch.

The majority of popular formulae use casein hydrolysates for protein and a mixture of different oils for the provision of lipid. They provide 1 kcal/mL of a solution containing about 110 kcal/g N (nitrogen) and have a balance of vitamins, trace elements, and electrolytes with an osmolality of 300–700 mOsm/kg.

**Polymeric diets** These are most commonly used because they are well utilized and are cheap to manufacture. They contain maltodextrins, milk protein, and vegetable oils. They provide recommended dietary intakes (RDIs) of vitamins and minerals and are presented in various manners for ease of clinical use.

Elemental diets Elemental diets were developed for space travel to have low residue. They mainly use amino acids as nitrogen source and are enriched with glutamine. They have a low fat content, but provide essential fatty acids along with the RDI of vitamins, trace elements, and electrolytes. Although the amino acids are free, they are not necessarily absorbed more easily because the peptidases in the brush border improve uptake of peptides, hydrolysing them to amino acids on absorption.

Added liquid fibre This is added to some enteral nutrition solutions, which may be important for reducing diarrhoea. The fibre is an important nutrient for the colonic mucosa and may be useful in patients with colitis.

Prebiotics, probiotics, and symbiotics These have been used for ill patients who are on long-term antibiotics and have an alteration of the bowel flora that causes diarrhoea and may lead to endotoxaemia. Strains of lactobacilli, which are normal commensal organisms in the small bowel and may help with digestion of fibre, are able to compete with and reduce the effect of the pathogenic organisms. Different strains of lactobacilli have been recommended. A recent study indicated that these products are dangerous in acutely ill and immune-compromised patients, where they increased the risk of bowel perforation.

#### 43.1.4.1 Special formulations

- *Immunonutrition*: these solutions contain added amounts of different nutrients considered to be useful in promoting the immune system. They include: ω3 fatty acids and medium-chain triglycerides, L-arginine and L-glutamine, and dietary nucleotides. These solutions have been shown to reduce post-operative wound infections.
- *Hepatic failure with encephalopathy*: these patients require solutions containing added branched-chain amino acids (leucine, isoleucine, and valine).
- *Renal failure*: it is important in these patients to lower solute load to reduce the need for dialysis. The past concept of limiting protein is considered to lead to malnutrition, but there may be improvement in outcome when a greater proportion of essential amino acids are given.
- *Acutely stressed patients*: these patients may need high-protein enteral nutrition and it has been shown that commencing enteral nutrition early in their course improves outcome. These patients need to be monitored for insulin resistance and each intensive care unit will have a protocol for these complicated patients.

### 43.1.5 Initiating enteral nutrition

Once the tube is in place, the enteral nutrition is commenced slowly to ensure that there is good tolerance. It is recommended to reduce the rate rather than the concentration of the solution. This author's preference is for continuous feeding because bolus feeding can induce vomiting and dumping with diarrhoea. Regular monitoring of blood glucose, urea and electrolytes, and liver function is important. After 12–24 hours, the rate can generally be increased to provide daily calorie and protein requirements. Frequent assessment of the patients for respiratory distress or abdominal distension should be part of routine care in these patients. Continuous enteral nutrition was not found to reduce the appetite in patients recovering from surgery.

### 43.1.6 Monitoring for efficacy

Monitoring of patients undergoing enteral nutrition therapy is not an exact science. It is important to monitor the amount that the patient receives and whether or not the patient has nausea, vomiting, or diarrhoea. Patients should have serum urea, electrolytes, albumin, and liver function monitored. By following weight each week, a judgement can be made as to whether more or less nutrition should be provided.

### 43.1.7 Complications and toxicity

Major complications include aspiration pneumonia, which is a particular threat in patients with weakness and reduced pharyngeal reflexes. It is particularly risky in patients with poor gastric emptying, who may aspirate when vomiting. Other possible major complications are:

- acute intestinal distension and ischaemic damage to the intestine causing intestinal necrosis and disruption
- dislodgement of the catheter into the peritoneal cavity with spillage of nutrient solution into the peritoneal cavity
- severe diarrhoea causing electrolyte disturbances.

Moderate to minor complications include:

- diarrhoea
- metabolic complications
- glucose intolerance
- low serum sodium
- low serum potassium
- low serum phosphate
- low serum magnesium
- essential fatty acid deficiency
- low serum zinc.

Mechanical issues include:

- blocked feeding tubes
- dislodged feeding tubes
- bacterial contamination.

## 43.2 Parenteral nutrition

Parenteral feeding is indicated when patients cannot be nourished with oral nutrition or enteral feeding for more than 5 days. Although the desire to feed directly into the vein had tempted doctors over the centuries, it was not until the development of pyrogen-free fluids in the 1920s and the development of protein hydrolysates and lipid emulsions by Arvid Wretland in the 1940s that this became possible. The technique of complete feeding was further advanced in 1968 by Stanley Dudrick and colleagues with central vein (line) placement and care, so that hypertonic glucose and amino acid solutions could be delivered. They neatly demonstrated that beagle puppies could grow with parenteral nutrition, and subsequently that human infants could develop at a similar rate to breastfed infants. Their system was termed 'hyperalimentation' because it allowed the delivery of large amounts of nutrition that sometimes caused problems of overfeeding. There were fears about the use of lipid emulsions for many years, but their safety with admixtures of amino acids and glucose in a three-in-one solution was demonstrated by French workers, and this has gradually become a most common presentation. Three-in-one solutions allow the provision of the daily nutrient requirements in a 3 L bag with an osmolality of about 1000 mosm/L. These solutions can be delivered into peripheral veins, particularly if a fine catheter is used in a larger vein such as the basilic vein. Parenteral nutrition also needs to provide all necessary electrolytes, trace elements, and vitamins in balanced amounts.

### 43.2.1 Constituents of parenteral nutrition

The glucose concentration ranges from 10% to 50%, which is much higher than the usual 5% of replacement intravenous fluids. One gram of glucose provides approximately 4 kcal (16.8 kJ), so 1 L of 25% glucose will provide about 1000 kcal (4200 kJ).

Lipid emulsions are presented as 10%, 20%, and 30% solutions and provide about 8 kcal/g lipid, so 500 mL of 20% solution provides about 1000 kcal (4200 kJ). They have traditionally been composed of soybean oil with lecithin as emulsifying agent. Some also contain medium-chain triglyceride or safflower oil. They provide the requirement for essential fatty acids (see section 3.4.2). It is generally considered that 'lipid burns in a carbohydrate fire' and therefore it is usual that no more than 50% of non-protein calories are given as lipid.

Amino acids are provided as crystalline *laevo*-form in solutions providing from 5.5 to 11.4 g amino acid

Table 43.2 **Parenteral nutrition, per day**

| | |
|---|---|
| Nitrogen | 1 litre 18 g/L (114 g protein) |
| Glucose | 1 litre 25% (1000 kcal) |
| Lipid emulsion | 500 mL 20% (1000 kcal) |
| Sodium | 70–100 mmol |
| Potassium | 60–80 mmol |
| Magnesium | 10–15 mmol |
| Phosphate | 10–20 mmol |
| *Balanced water-soluble and lipid-soluble vitamins* | |
| **Balanced trace element solution (zinc, selenium, copper)** | |
| Water | 2500 mL |

per 100 mL. This value needs to be divided by 6.25 to derive grams of nitrogen. The amino acid pattern corresponds to that of high-quality dietary protein, although it has been difficult to give insoluble amino acids like tyrosine and unstable amino acids like cysteine and glutamine.

The glucose, lipid, and amino acid solutions are combined to provide between 110 and 150 non-protein calories to 1 g N in the final solution. Electrolytes, vitamins, and trace elements are also necessary. It is particularly important to provide a balance of the intracellular electrolytes potassium, magnesium, and phosphate because with the provision of nutrition the intracellular compartment expands. If one nutrient is deficient, the response to nutrition is impaired.

An example of an adult formula is given in Table 43.2. This solution is run to provide about 0.25–0.3 g N/kg/day to the acute patient. Long-term patients may require less than this when they are not stressed. The maximum nutritional input that stressed hypermetabolic patients can metabolize is up to twice their basal metabolic rate, but there has been a tendency to be conservative to prevent the complications of overfeeding in these patients.

## 43.2.2 Methods of venous access

Concentrated glucose and amino acids solutions have an osmolality of about 2000 mOsm and a pH approaching 5, which is very irritating to the endothelium in veins and leads to the rapid development of *thrombophlebitis* if such solutions are delivered into small peripheral veins. However, when delivered into a large pool of blood it can be buffered and diluted. Therefore, this solution has to be delivered through a catheter, placed in a large central vein like the superior vena cava (see Fig. 43.2). Central lines require special care to prevent sepsis, and the catheter used for parenteral nutrition should be dedicated to this purpose. Antibiotics and other drugs should be delivered through a separate line. Apart from the risks of insertion of central lines, there is a constant risk of major sepsis. Sepsis is acquired either around or through a central line, which must be dressed to support the catheter from moving in and out and the patient must be monitored carefully to prevent overwhelming sepsis. One always needs to be alert to the possibility of sepsis.

The combination of lipid emulsion with the glucose and amino acids results in a complex three-in-one solution. Three-in-one solution allows the buffering of the solution and a marked reduction of osmolarity. When this is at or below 1000 mOsm/kg, it can be delivered into peripheral veins using normal cannulae, but these need to be rotated regularly to prevent thrombophlebitis. An improved thrombophlebitis rate can be achieved with a midline placed about 15 cm into the basilic vein or a peripherally inserted central line (see Fig. 43.2). There are a number of advantages to midlines because they can be placed without radiological guidance and can be monitored clinically for any tenderness over the vein. These lines need to be 2–3 Fr in size and can only be used for continuous infusion for rates up to 120 mL/hour. They cannot be used for resuscitation. Studies have demonstrated a reduced risk of sepsis with the use of midlines. Another advantage of these lines is that they can be safely managed by nursing staff without the need for referral to the radiology department.

A peripherally inserted central catheter (PICC) line is a further choice because it is safer to insert. Again, it has to be managed very carefully to ensure that sepsis is not introduced around or through the catheter.

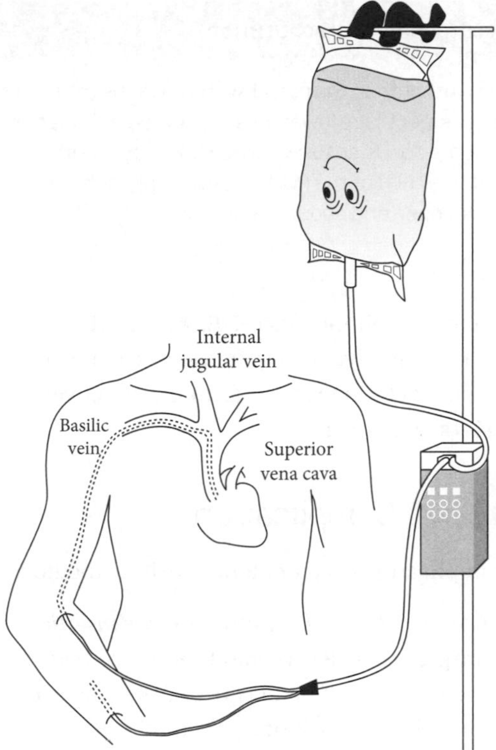


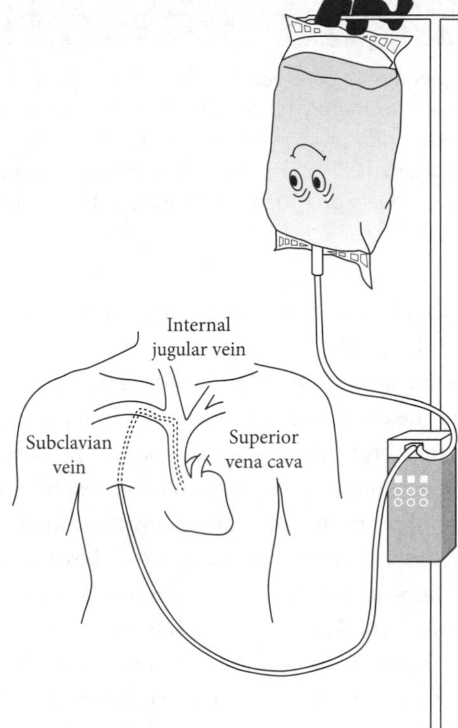


Fig. 43.2 Routes of parenteral nutrition.

## 43.2.3 Monitoring when on parenteral nutrition

Monitoring of electrolytes in patients treated with parenteral nutrition is required more frequently in the acute setting because the introduction of nutrition can induce severe acute depression of blood values of many nutrients that are marginally deficient. The *refeeding syndrome* (see Box 43.2) is prevented by the slow introduction of the parenteral nutrition and the provision of increased amounts of phosphate, potassium, and magnesium. Patients at risk of refeeding syndrome may need to have their electrolytes checked twice daily until they are stable.

Blood glucose is also important to monitor to prevent hyperglycaemia. In intensive-care patients, this requires a 4-hourly finger-prick for blood glucose. In the more stable ward patient, a urinalysis is undertaken twice daily and followed by blood glucose measures if urine glucose is elevated.

**Box 43.2 The refeeding syndrome**

This occurs in very malnourished patients who are refed too fast. The starved body adapts to use less carbohydrate and more fat. The introduction of artificial nutrition stimulates insulin secretion. There is enhanced uptake into the cells of glucose, water, phosphate, potassium, and magnesium, with consequent subnormal extracellular fluid and plasma levels of these electrolytes. This can lead to cardiac arrhythmia. Thiamin deficiency is also a risk.

Acid–base balance should be measured on arterial blood samples if the patient's condition suggests a metabolic disturbance. After the patient has reached goal amounts of parenteral nutrition infusion, the frequency of blood electrolyte measures can be reduced to every other day if the patient is stable. Home parenteral nutrition patients are frequently sufficiently

**Box 43.3 Daily requirements of micronutrients with total parenteral nutrition**

Requirements of several *vitamins* given parenterally are approximately twice the RDI (i.e. food by mouth): thiamin, riboflavin, niacin, pantothenate, vitamin $B_{12}$, and vitamin C. This is because of oxidation in the bag or faster urinary excretion when they go into a peripheral vein rather than the portal system. On the other hand, requirements are lower for several *inorganic* elements that are poorly absorbed when taken by mouth: about half or a third of the RDI for calcium, phosphorus, zinc, and copper, and about a tenth for iron.

stable that electrolytes only need to be measured at monthly intervals.

Monitoring of vitamins and trace elements needs to be undertaken at weekly intervals in the acute setting and 3-monthly in the long-term patients. The vitamins and trace elements most frequently depressed during parenteral nutrition are folate, vitamin C and D, vitamin $B_6$, zinc, selenium, and copper (see Box 43.3). Essential fatty acid deficiency occasionally occurred in patients who had glucose as the only non-protein calorie source and was prevented by twice-weekly lipid infusions. However, one needs to be aware that there may be depression of any nutrient that is not included in the parenteral nutrition formula.

Protein-calorie nutritional status is monitored with blood values that reflect protein metabolism, such as plasma proteins, albumin, transferrin, and C-reactive protein. Serum creatinine and urea are depressed in patients with malnutrition and it is important to observe their return to the normal range. Measures of body composition can be simple, by anthropometric measures of skinfolds and arm muscle circumference. There are a number of techniques for estimating muscle and fat mass, including electrical bioimpedence, DEXA, and CT assessment. Experimentally, the requirements can be defined by special techniques such as neutron activation analysis (see Chapter 38).

### 43.2.4 Complications

Complications of parenteral nutrition include:

- *Complications of venous catheter insertion*: these are greatest when a central venous catheter is used; there is a risk of significant injury to the subclavian and related vessels or a pneumothorax can occur.
- *Sepsis is a constant risk for patients with parenteral nutrition*: again, this is greater with central venous access than with peripheral venous access, and any spike in fever or persistent fever has to alert the medical team to the possibility of line sepsis. Such events should be treated with removal of the line and blood culture.
- Metabolic complications are mostly related to hyperglycaemia. The many possible nutrient deficiencies are discussed in the monitoring section (see section 43.2.3).

## 43.3 Nutritional support teams

Many studies have demonstrated the benefit of nutritional support teams. With complex medical treatment for patients in intensive care, gastroenterology, surgical, oncology, and renal failure wards, many patients have poor nutritional status, and the prevention of significant nutritional deficiency has been shown to improve outcome. Multidisciplinary teams help make this process happen in a complex tertiary hospital, but all medical and paramedical staff need to be aware of the importance of good nutritional care.

### 43.3.1 Decision path for nutritional therapy

When a patient is at risk of developing malnutrition, Figure 43.3 may help in considering nutritional therapy. Clearly, it is unacceptable to allow patients to

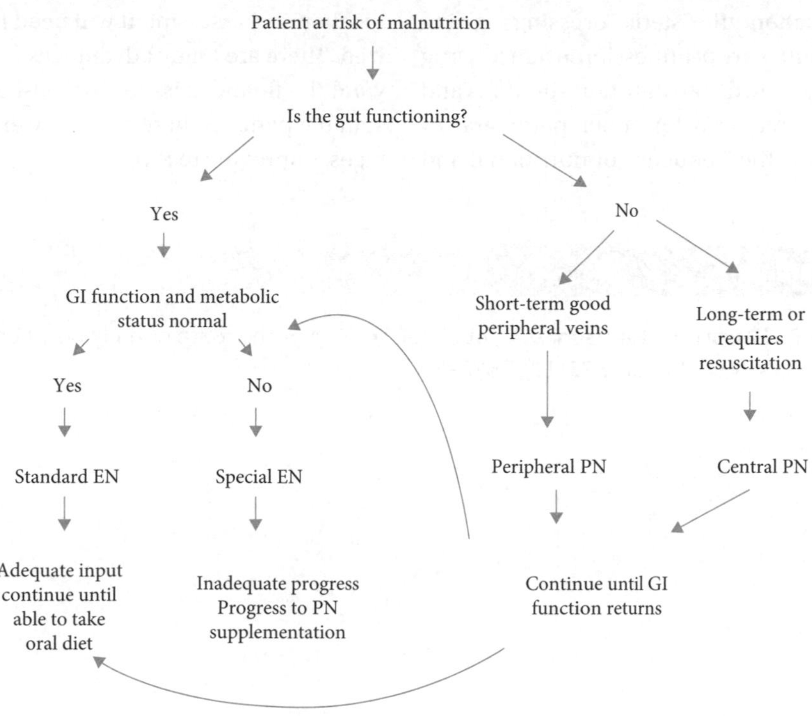


Fig. 43.3 Some factors in the decision process of providing nutritional support. EN, enteral; GI, glycaemic index; PN, parenteral.

become malnourished in hospital when there are well-designed therapies that have proven efficacy for preventing this. There is a large body of evidence that malnutrition increases the risk of complications and death. Nutritional assessment of patients entering hospital is important, and there should be a greater stress on recording body weight and weight loss, plasma proteins, lymphocyte count, and haemoglobin values on a nutritional assessment form.

Enteral nutrition should be used if the gut is available and functioning. Enteral nutrition is most commonly delivered through a fine-bore nasogastric tube. In complex patients in intensive care, it may be important to pass the tube into the jejunum. Long-term access may require the placement of a feeding gastrostomy or jejunostomy tube. Enteral nutrition is frequently prescribed to provide 20–25 kcal/kg/day, while the infusion rate for parenteral nutrition is frequently 25–35 kcal/kg/day. The decision to use either peripheral or central parenteral nutrition depends on the presence of suitable peripheral veins. Peripheral parenteral nutrition may be preferred for short-term treatment to limit complications of sepsis and venous access. Infusion rates need to be individualized, depending on tolerance and complication, and whether there is a positive response to treatment. A recent multi-centre study demonstrated no difference in complications in intensive-care patients whether enteral nutrition or parenteral nutrition was commenced within 36 hours after admission to intensive care.

## 43.4 Conclusion

Although nutritional support is usually established during inpatient care, when tolerance and metabolic consequences can be monitored and stabilized, home nutritional support is certainly a reasonable goal when a patient's medical team is confident that they can manage at home after a period of education. The sterile dressings, giving sets, and techniques are of utmost importance. Drug companies may provide the nutrition solutions and infusion sets as required for either parenteral or enteral nutrition. The frequency of nutritional and biochemical assessment will need to be individualized. There are many advantages for home care beyond the financial issues to well-being, and for the suitable patients there are many emotional advantages, improving recovery.

## Further reading

1. **Harvey, S.E., Parrott, F., Harrison, D.A., et al.** (2014) Trial of the route of early nutritional support in critically ill adults. *N Engl J Med*, **371**(18), 1673–84.

# Index

Tables, figures, and boxes are indicated by an italic *t*, *f*, and *b* following the page number.